Plant Phenolics in Biotic Stress Management

Rafiq Lone • Salim Khan •
Abdullah Mohammed Al-Sadi
Editors

Plant Phenolics in Biotic Stress Management

Springer

Editors
Rafiq Lone
Department of Botany
Central University of Kashmir
Srinagar, India

Salim Khan
Department of Botany and Microbiology
King Saud University
Riyadh, Saudi Arabia

Abdullah Mohammed Al-Sadi
Department of Crop Sciences, College of
Agricultural & Marine Science
Sultan Qaboos University
AlKhoud, Oman

ISBN 978-981-99-3333-4 ISBN 978-981-99-3334-1 (eBook)
https://doi.org/10.1007/978-981-99-3334-1

This Springer imprint is published by the registered company Springer Nature Singapore Pte Ltd.
The registered company address is: 152 Beach Road, #21-01/04 Gateway East, Singapore 189721, Singapore

Paper in this product is recyclable.

Dedicated To Great Teachers and Botanists

Acknowledgments

Prof. K. K. Koul

I am trying to place on record my sincere and humble thanks to my mentor and supervisor Prof. (Rtd.) **K. K. Koul**, School of Studies in Botany, Jiwaji University, Gwalior. I have been fortunate to have him as supervisor during my PhD. He is one who gave me the freedom to explore on my own and at the same time guided me to recover when my steps faltered. His unflinching courage, conviction, and steadfastness will always inspire me. It will be really hard to describe him in words and would feel successful if I emulate even a fraction of him.

Prof. Azra N. Kamili

My special gratitude to my mentor and teacher Prof. Azra N. Kamili, Department of Botany, Central University of Kashmir Ganderbal Jammu and Kashmir-India for providing me a wonderful opportunity to work on a project during my N-PDF. Madam! You are a wonderful teacher, mentor, and able leader who always believes in unity. Your direction and insightful comments have made me a better professional. Thank you for showing me how to stand my ground and have patience in difficult situations. I count myself lucky for having had your mentorship. I have learned from you the value of tolerance, patience, and trust, which are the pillars of success.

From

Rafiq Lone, PhD, N-PDF
Assistant Professor
Department of Botany
Central University of Kashmir, Ganderbal, Jammu and Kashmir, India

Contents

Editors and Contributors

About the Editors

Rafiq Lone is presently working as Assistant Professor of Botany at the Central University of Kashmir, Jammu and Kashmir, India. He has completed his MSc and PhD in Botany from Jiwaji University, Gwalior, Madhya Pradesh, India. Dr. Lone have been awarded National Post-Doctoral Fellowship (N-PDF) by SERB, DST, Government of India. Previously, he was working as Assistant Professor in Botany at SBBS University, Khiala Jalandhar, Punjab, Lecturer in higher education, Jammu and Kashmir, Teaching Assistant at the School of Studies in Botany, Jiwaji University, Gwalior-India and Project Fellow in a project funded by MPCST-Bhopal (MP), India. Dr. Lone has 8 years of teaching and 12 years of research experience. He has published more than 30 research papers in National and International journals besides published more than 20 book chapters and four edited books are also to his credit. Dr. Lone has attended 25 national and international conferences across the country and has presented papers on plant–microbe interaction and plant phenolics. He has many awards to his credit. His main area of research interest focuses on plant–microbe interaction, plant nutrition, plant phenolics, and invasion biology.

Salim Khan has been working as a Research Associate since October 2008 in the Department of Botany and Microbiology, King Saud University, Saudi Arabia. He has done his PhD in Biotechnology from Jamia Hamdard University, New Delhi, India. He has done MSc in Molecular Biology and Biotechnology from G.B. Pant University of Agriculture and Technology, Uttarakhand, India. Dr. Khan has also worked as a Guest Lecturer in Jamia Millia Islamia, New Delhi. He has published many national and international research articles. Dr. Khan has completed many research projects funded by the Research Centre of Science College, Centre of Excellence in Biotechnology, Deanship of Scientific Research and National Plan for Science and Technology, Saudi Arabia. He has many patents nationally and internationally. His main research interest concentrates on molecular markers, seed cryo-banking, next-generation sequencing, and plant DNA banking.

Abdullah M. Al-Sadi is a Professor of Plant Pathology and the Dean of College of Agricultural and Marine Sciences at Sultan Qaboos University, Oman. He received his PhD in Plant Pathology from the University of Queensland, Australia in 2007. His research focuses on the etiology, plant–pathogen interactions, and management of plant diseases using chemical and biological control. Al-Sadi has 378 publications, of which 256 are refereed papers in journals indexed in Scopus/WoS with a cumulative impact factor of 727. He has supervised 64 PhD and MSc students. Al-Sadi is in the editorial board of two international journals and received 38 national and international awards.

Contributors

Ambreen Ahmed Institute of Botany, University of the Punjab, Quaid-e-Azam Campus, Lahore, Pakistan, India

Abdullah Mohammed Al-Sadi Department of Crop Sciences, College of Agricultural & Marine Science, Sultan Qaboos University, AlKhoud, Oman

Abdul Azeez Institute of Biological Chemistry, Washington State University, Pullman, WA, USA

Irfan Ashraf Badroo Department of Zoology, Sant Gadge Baba Amravati University, Amravati, Maharashtra, India

Iqra Bashir Department of Botany, University of Kashmir, Srinagar, Jammu and Kashmir, India

Renu Bhardwaj Department of Botanical and Environmental Sciences, Guru Nanak Dev University, Amritsar, Punjab, India

Shruti Bhasin Department of Biotechnology, Banasthali Vidyapith, Aliyabad, Rajasthan, India

Abid Bhat Department of Botany, Central University of Kashmir, Ganderbal, Jammu and Kashmir, India

Shivani Bhutani Department of Biotechnology, Invertis Village, Bareilly, Uttar Pradesh, India

Luis A. Cabanillas-Bojórquez Post-Doct CONAHCyT-Centro de Investigación en Alimentación y Desarrollo, Culiacán, Sinaloa, México

Priyanka Chakraborty Department of Pharmaceutical Chemistry, BCDA College of Pharmacy and Technology, Hridaypur, West Bengal, India

Parul Chaudhary Department of Animal Biotechnology, Animal Genomics Lab, NDRI, Karnal, India

Sailee Chowdhury Department of Pharmaceutical Chemistry, BCDA College of Pharmacy and Technology, Hridaypur, West Bengal, India

Azharuddin B. Daphedar Department of Botany, Anjuman Arts, Science and Commerce College, Vijayapura, Karnataka, India

Diksha Division of Zoology, Department of Biosciences, Career Point University, Hamirpur, Himachal Pradesh, India

Cristina A. Elizalde-Romero Post-Doct CONAHCyT-Centro de Investigación en Alimentación y Desarrollo, Culiacán, Sinaloa, México

Naveed Gulzar Center of Research for Development, University of Kashmir, Srinagar, Jammu and Kashmir, India

Erick P. Gutiérrez-Grijalva Cátedras CONAHCyT-Centro de Investigación en Alimentación y Desarrollo, Culiacán, Sinaloa, México

Younis Ahmad Hajam Department of Life Sciences and Allied Health Sciences, Sant Baba Bhag Singh University, Jalandhar, Punjab, India

Nowsheen Hassan Department of Environmental Sciences, SKUAST-Kashmir, Srinagar, Jammu and Kashmir, India

J. Basilio Heredia Post-Doct CONAHCyT-Centro de Investigación en Alimentación y Desarrollo, Culiacán, Sinaloa, México

Arsalan Hussain Department of Life Science, Presidency University, Calcutta University, Kolkata, West Bengal, India

Ebru Kafkas Faculty of Agriculture, Department of Horticulture, University of Çukurova, Adana, Turkey

Siddappa Kakkalamel Department of Botany, Davangere University, Davangere, Karnataka, India

Koyel Kar Department of Pharmaceutical Chemistry, BCDA College of Pharmacy and Technology, Hridaypur, West Bengal, India

Amanpreet Kaur Department of Biochemistry, Punjab Agricultural University, Ludhiana, Punjab, India

Manpreet Kaur Department of Biochemistry, Punjab Agricultural University, Ludhiana, Punjab, India

Anjali Khajuria Department of Zoology, Guru Nanak Dev University, Amritsar, India

Ashiq Hussain Khanday Department of Botany, Sant Gadge Baba Amravati University, Amravati, Maharashtra, India

Prattusha Khan Department of Microbiology, St. Xavier's College, Kolkata, West Bengal, India

Salim Khan Department of Botany and Microbiology, King Saud University, Riyadh, Saudi Arabia

Sandeep Kour Department of Zoology, Guru Nanak Dev University, Amritsar, India

Kuldeep Department of Agricultural & Food Engineering, IIT Kharagpur, Kharagpur, West Bengal, India

Deepak Kumar Department of Zoology, Guru Nanak Dev University, Amritsar, India

Rajesh Kumar Department of Biosciences, Himachal Pradesh University, Shimla, Himachal Pradesh, India

Yogesh Kumar Department of Botany, Central University of Jammu, Jammu, Jammu and Kashmir, India

Rafiq Lone Department of Botany, Central University of Kashmir, Ganderbal, Jammu and Kashmir, India

Junaid Ahmad Malik Department of Zoology, Government Degree College, Kulgam, Jammu and Kashmir, India

Nazir Ahmad Malla Department of Botany, Government Degree College of Women, Anantnag, Jammu and Kashmir, India

Javid Manzoor Department of Environmental Sciences, Shri JJT University, Jhunjhunu, Rajasthan, India

Kamalika Mazumder Department of Pharmaceutical Chemistry, BCDA College of Pharmacy and Technology, Hridaypur, West Bengal, India

Anshi Mehra Department of Biotechnology, Invertis Village, Bareilly, Uttar Pradesh, India

Truong Ngoc Minh Center for Research and Technology Transfer, Vietnam Academy of Science and Technology (VAST), Hanoi, Vietnam

Heba I. Mohamed Biological and Geological Sciences Department, Faculty of Education, Ain Shams University, Cairo, Egypt

Gulshan Mushtaq Department of Botany, Central University of Kashmir, Ganderbal, Jammu and Kashmir, India

Sagnik Nag Department of Bio-Sciences, School of Bio-Sciences & Technology (SBST), Vellore Institute of Technology (VIT), Vellore, Tamil Nadu, India

Nibedita Naha Biochemistry Department, Biological Sciences Division, ICMR-National Institute of Occupational Health (NIOH), Ahmedabad, Gujarat, India

Subzar Ahmad Nanda Department of Botany, University of Kashmir, Srinagar, Jammu and Kashmir, India

Naveena Nazim College of Temperate Sericulture, Mirgund, SKUAST-Kashmir, Srinagar, Jammu and Kashmir, India

Puja Ohri Department of Zoology, Guru Nanak Dev University, Amritsar, India

Semran Parvaiz Department of Zoology, Central University of Kashmir, Ganderbal, Jammu and Kashmir, India

Mahima Praharaju Department of Environmental Sciences, IGNOU University, Hyderabad, India

Shah Rafiq Plant Tissue Culture Laboratory, Department of Botany, University of Kashmir, Srinagar, Jammu and Kashmir, India

Irfan Rashid Department of Botany, University of Kashmir, Srinagar, Jammu and Kashmir, India

Nazima Rasool Department of Botany, North Campus, University of Kashmir, Delina, Baramulla, Jammu and Kashmir, India

Younis Ahmad Rather Department of Zoology, Government Degree College, Ramban, Jammu and Kashmir, India

Deepsi Rathore Biochemistry Department, Biological Sciences Division, ICMR-National Institute of Occupational Health (NIOH), Ahmedabad, Gujarat, India

Shahid Razzak Department of Zoology, Central University of Kashmir, Ganderbal, Jammu and Kashmir, India

Sumaiya Rehmaan Department of Botany, University of Kashmir, Srinagar, Jammu and Kashmir, India

Ishfaq Ul Rehman Department of Botany, University of Kashmir, Srinagar, Jammu and Kashmir, India

Zafar Ahmad Reshi Department of Botany, Main Campus, University of Kashmir, Srinagar, Jammu and Kashmir, India

Gulab Khan Rohella Biotechnology Section, Moriculture Division, Central Sericultural Research and Training Institute, Central Silk Board, Ministry of Textiles, Government of India, Pampore, Jammu and Kashmir, India

Aditi Sharma College of Horticulture and Forestry, Thunag, Mandi, Dr. Y. S. Parmar University of Horticulture and Forestry, Solan, Himachal Pradesh, India

Ajay Sharma Department of Chemistry, Career Point University, Hamirpur, Himachal Pradesh, India

Ashutosh Sharma Faculty of Agricultural Sciences, DAV University, Jalandhar, Punjab, India

Indu Sharma Department of Life Sciences, Sant Baba Bhag Singh University, Jalandhar, Punjab, India

Nandni Sharma Department of Zoology, Guru Nanak Dev University, Amritsar, India

Pooja Sharma Department of Botanical and Environmental Sciences, Guru Nanak Dev University, Amritsar, Punjab, India
Department of Microbiology, DAV University, Jalandhar, Punjab, India

Ishfaq Ahmad Sheergojri Department of Botany, University of Kashmir, Srinagar, Jammu and Kashmir, India

Manali Singh Department of Life Sciences, Parul Institute of Applied Sciences, Parul University, Vadodara, Gujarat, India

Shraddha Singh Biochemistry Department, Biological Sciences Division, ICMR-National Institute of Occupational Health (NIOH), Ahmedabad, Gujarat, India

Yamini Tak Agricultural Research Station, Ummedganj, Agriculture University, Kota, Rajasthan, India

Tarikere C. Taranath Environmental Biology Laboratory, P. G. Department of Studies in Botany, Karnataka University, Dharwad, Karnataka, India

Aqsa Tariq Institute of Botany, University of the Punjab, Quaid-e-Azam Campus, Lahore, Pakistan, India

Nasir Aziz Wagay Department of Botany, Government Degree College Baramulla (Boys), Baramulla, Jammu and Kashmir, India

Khursheed Ahmad Wani Department of Environmental Science, Government Degree College, Thindim Kreeri, Jammu and Kashmir, India

Aadil Farooq War Department of Botany, University of Kashmir, Srinagar, Jammu and Kashmir, India

Tran Dang Xuan Transdisciplinary Science and Engineering Program, Graduate School of Advanced Science and Engineering, Hiroshima University, Hiroshima, Japan

Parvaiz Yousuf Department of Zoology, Central University of Kashmir, Ganderbal, Jammu and Kashmir, India

Role of Phenolics in Plant–Microbe Interaction: A Review

1

Rafiq Lone, Abid Bhat, Naveena Nazim, Nazir Ahmad Malla, Gulab Khan Rohella, and Heba I. Mohamed

Abstract

Plants, being immobile, face many types of stresses in the environment. To combat these stresses, they have phenolics as secondary metabolites. There are many classes of phenolics such as simple hydroxybenzoic acid, free and hydroxycinnamic acids, coumarins, flavonoids and stilbenes, etc. These are classified on the basis of number of carbon atoms and basic arrangement of carbon skeleton in their structures. The types of phenol that are implicated in defense differ greatly and depend on plant species. Plants form phenolics mainly through shikimic acid pathway. The first step in the synthesis of phenolic compounds from phenylalanine in plants is deamination of phenylalanine by the enzyme phenylalanine ammonia lyase (PAL). These compounds have diverse functions and are immensely important in plant–microbe interaction/symbiosis. Phenolic compounds act as signaling molecules in the initiation of legume

R. Lone (✉) · A. Bhat
Department of Botany, Central University of Kashmir, Ganderbal, Jammu and Kashmir, India

N. Nazim
College of Temperate Sericulture, Mirgund, SKUAST, Kashmir, Jammu and Kashmir, India

N. A. Malla
Department of Botany, Government Degree College of Women's, Anantnag, Jammu and Kashmir, India

G. K. Rohella
Biotechnology Section, Moriculture Division, Central Sericultural Research & Training Institute, Central Silk Board, Ministry of Textiles, Government of India, Pampore, Jammu and Kashmir, India

H. I. Mohamed
Biological and Geological Sciences Department, Faculty of Education, Ain Shams University, Cairo, Egypt

R. Lone et al. (eds.), *Plant Phenolics in Biotic Stress Management*,
https://doi.org/10.1007/978-981-99-3334-1_1

rhizobia symbiosis and establishment of arbuscular mycorrhizal symbiosis and can act as agents in plant defense. In plants, phenolics play many important roles in plant–microbe symbiosis, rhizobium–legume symbiosis, and arbuscular mycorrhiza symbiosis and provide defense system to the plant. They also help in quorum sensing as well as act as signaling molecules.

Keywords

Phenolics · Symbiosis · AMF · Quorum sensing · Signaling molecules

1.1 Introduction

Plants growth and development depend on several factors, among which major influencing factors are availability of water, sunlight, and plant–microbial interactions and coping up with abiotic and biotic stresses (Stout et al. 2006). Plants' ability to counter a stress (biotic or abiotic) is mostly influenced by their genetical nature of synthesizing specific compounds which can help survive the plant under stressful conditions (Paul et al. 2000). Plants synthesize diverse compounds to escape from different types of stresses in order to maintain phenotypic plasticity (Kaplan et al. 2008). Majorly, a class of compounds named phenolics with phenolic acids, polyphenols, flavonoids, and stilbenoids helps plants to defend themselves from these stresses (Wallis and Galarneau 2020). Aromatic compounds possessing at least one hydroxyl groups are termed as phenolics. Among the various plant secondary metabolites, phenolics are possibly the superior ones, located in vacuolar cells of the epidermis and subepidermis besides the cell wall (Lattanzio et al. 2006). Plants in the early growing phase have phenolics mainly in the nucleus or in chloroplasts because they do not have well-developed vacuoles (Khlestkina 2013). The accretion of phenolic compounds in chloroplasts is directly related to the chloroplast's ultrastructural organization and functional activities (Zaprometov and Nikolaeva 2003). The various types of stress-relieving secondary metabolites found in cereals, oilseeds, fruits, vegetables, legumes, and many other plants are mostly phenolics (Tsao 2010). Nearly more than 8000 phenolics and polyphenolics metabolites have structural properties similar to common phenolics (Bravo 1998). In plants, these phenolics may be in free or in conjugated forms (Babenko et al. 2019).

Phenolics range from simple to complex; simple phenolics like phenolic acids are low-molecular-weight compounds with one aromatic ring, while complex ones like tannins possess more than one aromatic ring (Veberic 2016). Flavonoids having two aromatic rings are an important class of phenolics (Rispail et al. 2005). In the plant kingdom, phenolics are the essential and universal class of compounds (Naczk and Shahidi 2004). The structure of phenolics varies from simple to complex regarding their molecular masses, and they contain an aromatic ring with not less than one hydroxyl group (Balasundram et al. 2006). These phenolic compounds are called secondary metabolites as they do not have any role in energy processes and growth of a plant directly (Harnly et al. 2007). Foods from plants like cereals, fruits, nuts, and vegetables contain phenolics in considerable amounts (Aura 2008). To counter the biotic and abiotic stress, most of the tree species form phenolics-based secondary metabolites (Chomel et al. 2016; Zwetsloot et al. 2018).

Depending on the structure of chemical, concentration, and environmental conditions, phenolics can be a source of food or sometimes may be toxic (Shaw et al. 2006). The respiration rate and microbial community composition show divergence because of phenolics (Badri et al. 2013; Zwetsloot et al. 2018). In higher plants, phenolics shows many important properties like antiallergic, anticancerous, antihypertensive, anti-inflammatory, antimicrobial, and antioxidant (Cesco et al. 2012; Daglia 2012). The various important processes in plants, like growth, development, reproduction, and defense, are aided by phenolics (Dixon 2001).

1.2 Microbes

In an ecosystem, detrimental microbes are everywhere, ensuing a deleterious effect on the environment (Gupta et al. 2017). Being omnipresent, microbes are important for supporting life on earth, but we are not much aware of most of the microbes (Gilbert et al. 2010). Their role in ecological processes was examined to become aware of various microbes present in the biosphere (Gilbert et al. 2010). The growth, development, and productivity of a plant are supported by the plant–microbe interaction (Berendsen et al. 2012). Moreover, in recent times, this topic has received much more attention (Lebeis et al. 2012). Based on their interaction, microbes may have positive or negative effect on the plants (Turner et al. 2013). Many vital interactions in plants, like symbiosis between Rhizobium and legumes, are mediated by microbes (Oldroyd et al. 2011).

The essential functions of an ecosystem are maintained by the soil microbes like archaea, bacteria, and fungi (Aislabie et al. 2013). These microbes, such as the maintenance of nutrient cycling of various important elements, support the essential services of our environment. The diversity of these microbes also affects the functioning of the biosphere (Aislabie et al. 2013). These microbes not only manage essential soil processes but also have a role in protecting the plants from various types of stresses (Aislabie et al. 2013). Consistently changing the behavior of these microbes leads to changes in soil phosphorous (Hermans et al. 2017), carbon pool (Ramírez et al. 2020), soil pH (Delgado-Baquerizo et al. 2018), and the moisture content of the soil (Isobe et al. 2020). The intimacy of plants with the microbes helps them in growth, development, and defense at all tropical levels (Yan et al. 2019).

1.3 Classification of Phenolics

Based on the total number of carbons present in a molecule, Harborne and Simmonds categorized phenolics into various groups (Harborne and Simmonds 1964). As this system of classification contains additionally ten classes, it is challenging to order them into higher groups (Tsimogiannis et al. 2006).

On the basis of a biological rank system of taxonomy, phenolics were classified into three main classes:

1. Compounds containing only one benzene ring, i.e., (C_6 class).
2. Compounds in which the benzene ring has more than one carbon atoms attached, i.e., (C_6–C_n class).
3. Compounds with two benzene rings attached together (C_6–C_n–C_6 class).

Every class further splits into subclasses, with flavonoids falling within the C_6–C_3–C_6 subclass (Tsimogiannis et al. 2006).

1.3.1 C_6 Phenolic Compounds

Medicinal plants contain simple phenolics like phloroglucinol, catechol, and hydroquinone (Tsimogiannis et al. 2006). These simple phenolics are found in different parts of various plants, like catechol in Gaultheria species, arbutin in Vaccinium species, and in pear trees (*Pyrus communis* L., Rosaceae) (Lattanzio 2013). Likewise, arbutin has also been reported in *Bergenia crassifolia* (L.), covering plants dry weight up to 23%, according to Pop et al. (2009).

1.3.2 C_6–C_n Phenolic Compounds

This group of phenolic compounds are present everywhere, with ordinary or sparse families. The C_6–C_1 class is a restricted one that incorporates benzaldehyde derivatives and carboxybenzene (Tsimogiannis et al. 2006). Salicylic acid, a member of this class, is present in small concentrations ($\leq$1 ppm) in essential species of vegetables and fruits (Herrmann 1989; Tomás-Barberán and Clifford 2000). Vallin is likely the major notable phenolic aldehyde commonly used as an element of flavor and aroma (Tsimogiannis et al. 2006). The commonly named phenylethanoids, such as acetophenone, phenethyl alcohol, hydroxylated, and methoxylated derivatives of phenylacetic acid, are under the C_6–C_2 group (Tsimogiannis et al. 2006). The C_6–C_3 class contains lignols, 3-phenylpropionic acid, and aldehydes of cinnamic, propenylphenol, coumarins, isochromen-1-one, and chromones (Tsimogiannis et al. 2006). Hydrocinnamic acid is the most notable one in this class because it is distributed in almost all the plant species, such as *p*-coumaric, 3,4-dihydroxy, and ferulic acid (Rabe et al. 1994). The initial material for the anabolism of monolignols is included in the C_6–C_3 group, such as *p*-coumaryl alcohol (Chiang 2006). Those coumarins, which were made up of single molecules, were further divided into simple coumarins, furocoumarins, and pyranocoumarins (Borges et al. 2005; Bourgaud et al. 2006).

1.3.3 C_6–C_N–C_6 (1 ≤ N ≤ 3, or N57) Phenolic Compounds

The phenolic compounds under this class are boundless such as xanthonoids under the C6–C_1–C_6 group; stilbenoids, anthraquinones, and acridone under the C_6–C_2–C_6 group; flavonoids under C_6–C_3–C_6 group; and the diarylheptanoids under C_6–C_7–C_6. In the plant kingdom, due to their wide distribution nature, flavonoids are mostly studied (Tsimogiannis et al. 2006).

1.3.4 C_6–C_3–C_6 Phenolics: Flavonoids

Flavonoids are placed into a separate group based on a bridge present in them which may be open or make a heterocyclic ring (third ring) (Tsimogiannis et al. 2006). These constitute the main group of phenolics present in plants (Erlund 2004). There are two benzene rings in flavonoids attached together by a pyrene ring containing an oxygen atom (Brodowska 2017). The Flavan system is present in all the chemical structures of flavonoids (Symonowicz and Kolanek 2012). Chalcone structure is formed after the condensation of two benzene rings, which upon condensation makes flavanone, an initiator for other flavonoid groups (Brodowska 2017). Further, they are divided into different types like flavones, anthocyanins, and flavanols, based on the dissimilarity in their structure (Brodowska 2017).

1.4 Synthesis of Phenolic Compounds

Because they are immobile, plants must contend with biotic and abiotic environmental challenges. To combat these pressures, plants produce low-molecular-weight secondary metabolites. These types of metabolites are obtained from different pathways such as fatty acid pathway, alkaloid pathway, isopropanol, and phenylpropanoid pathway (Dixon 2001). Plants form phenolics mainly through three separate pathways:

- The shikimic acid pathway which produces phenyl propanoid as a by-product.
- The mevalonate pathway, which forms aromatic terpenoids.
- The malonate pathway which forms phenyl propanoids (Bhattacharya et al. 2010).

Phenolics, which every plant produces as a secondary metabolite, possess an aromatic ring structure, and these phenolics are mostly studied in plant research (Boudet 2007; Whiting 2001). In plants, bacteria and fungi phenolics are majorly synthesized by shikimate pathway, which changes simple carbohydrate forerunners obtained from hexose monophosphate shunt and glycolysis into phenylalanine and tryptophan (Mandal et al. 2010). In bacteria and fungi, propanedioic acid pathway is the main pathway for the biosynthesis of phenolics in comparison with plants (Mandal et al. 2010). *p*-coumaroyl CoA obtained from shikimate and malonyl

CoA from obtained acetate give rise to flavonoids having two phenolic rings attached together by a unit containing three carbon atoms (Mann 1978). The biosynthesis of cinnamate from phenylalanine through deamination reaction catalyzed by phenylalanine ammonia-lyase (PAL) is the main switch in the synthesis of a large number of phenolics (Koukol and Conn 1961; Macheix et al. 2005). Oxidation with the help of phenylalanine leads to tyrosine biosynthesis along with its role in the phenolic formation (Kulma and Szopa 2007). The various types of mechanisms, such as feedback regulation changes after transcription and channeling of metabolites (Stafford 1974; Blount et al. 2000; Cheng et al. 2001), initiate protein activity and expression of PAL genes (Hahlbrock and Scheel 1989) that direct the formation of phenolics. The shikimic acid pathway forms not only essential amino acids like tyrosine, tryptophan, and phenylalanine but also cofactors and important vitamins (Paiva 2000). Carbohydrates obtained from hexose monophosphate shunt and glycolysis are changed into chorismate, the forerunner of synthesis of aromatic amino acids through the shikimic acid pathway (Zabalza et al. 2017). Nearly 20% of the carbon fixed by plants under ordinary conditions passes through the shikimic acid pathway, whereas plants facing any stress or showing fast growth fix more carbon (Tohge et al. 2013). The main site for biosynthesis of most of aromatic natured amino acids is plastids, but three of those amino acids and a few intermediates pathways are transported in to cytosol for protein synthesis and extra compounds obtained through the pathway of shikimic acid (Marchiosi et al. 2020). Along with phenolic compounds, the shikimic acid route gives carbon as a basic structure for many important compounds like auxin, salicylic acid, folic acid, pigments of plants, and simple quinones, which are important in energy transport along with coupling of electrons in thylakoids (Galili and Tzin 2010; Maeda and Dudareva 2012). Oxidation with the help of phenylalanine leads to tyrosine biosynthesis along with its role in phenolic formation (Kulma and Szopa 2007) (Fig. 1.1).

1.5 Role of Phenolics in Plant–Microbe Symbiosis

Phenolics are the main polyphenols produced by plants which not only protect the plants from various types of stresses but also has a major role in the interaction of plants with microbes (Mandal et al. 2010). The region which surrounds the plant root system is known as rhizosphere, and this region helps the plant to develop interaction with the micro-organisms present in its vicinity as this region is rich in nutrients (Tyagi et al. 2018). Phenolics not only provide a defense mechanism to the plants but also play a vital role for the interaction of rhizobia and legumes by acting as a signaling molecule (Khare et al. 2020). Phenolic compounds released by injured plants, like phytoalexins, are toxic for some microbes, while other microbes having unique mechanisms use these phenols for beneficial purposes (Khare et al. 2020). A redox reaction mediated by phenolic compounds affects the growth of the soil microbes through enzyme activity, hormone production, access to plant-based nutrients, and competition between microbes and nearby plants (Hättenschwiler and Vitousek 2000; Bhattacharya et al. 2010).

Fig. 1.1 A schematic representation of the biosynthesis of phenolic compounds in plants. (Source: Redrawn from Lttanzio 2013)

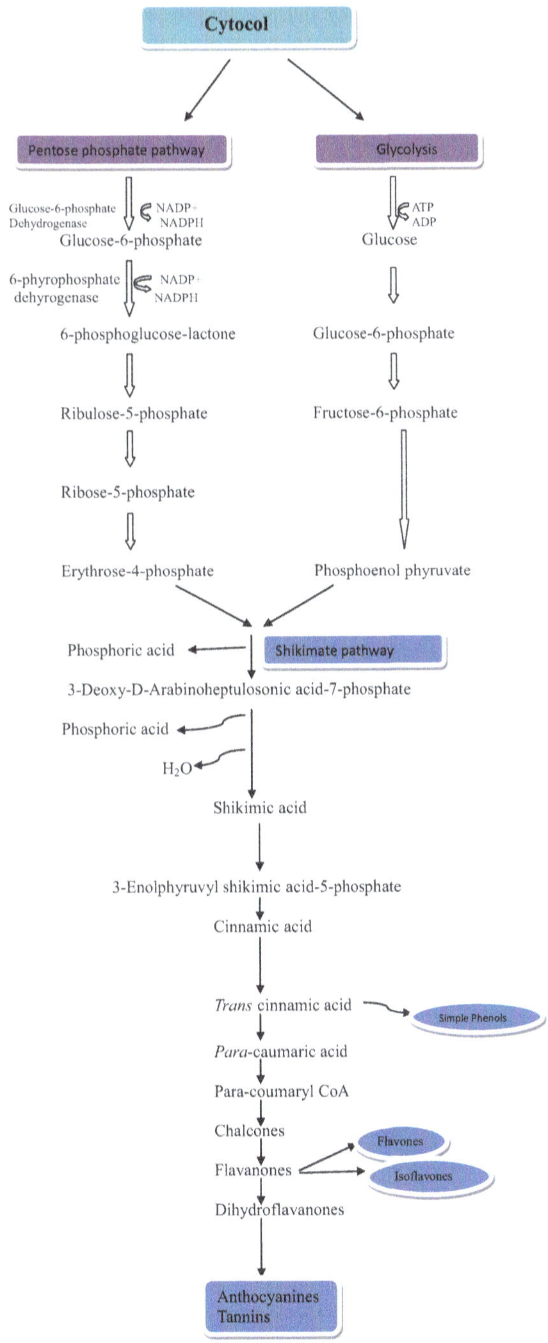

Flavonoids are a separate class of polyphenols that are gaining immense attention because of their role in mediating the relationship between microbes and plants (Khare et al. 2020). The nodulation genes which are essential for establishing rhizobia symbiosis are induced by flavonoids that act as chemical attractants (Perret et al. 2000). Ruan et al. observed that the flavonoids secreted by the legumes through exudates of roots have a prominent role in the germination of pathogens present in pea and beans (Ruan et al. 1995).

The plant–microbe-based association is gaining a lot of attention in the field of research because these types of interactions are very helpful in maintaining the equilibrium of the soil's ecosystem through different ways like availability of nutrients for plants, enhancing the soil's fertility, the health of soil, and defense toward various types of stresses (Tiwari and Singh 2017; Singh and Gupta 2018). For the maintenance of the soil's ecosystem, the relation between microbes and plants is important, which is beneficial for the growth of the plants through signaling, but we do not have enough information about this signaling and the processes by which this interaction helps in defense and symbiosis (Shastri and Kumar 2019). Plant roots secrete various functional compounds in the rhizosphere; these compounds have an essential part in the physicochemical associations and biological interaction of plants with surrounding environment. These compounds, through signaling, provide benefits to the microbes and to the plants involved with these microbes. Thus, these interactions in between the plants and microbes may be useful, dangerous, and neutral depending on the molecular signals exchanged (Shastri and Kumar 2019). To attract these beneficial microbes, plants prepare various signaling compounds, which may include both primary as well as secondary metabolites (Singh et al. 2016). Microbes present in the surrounding of the plants secrete various small and volatile molecules like plant hormones, hydrogen cyanide, enzymes, siderophores, homoserine, etc. These molecules are identified by the plant, and plants start to prepare strategies for their normal growth and development (Ortíz-Castro et al. 2009). Plants and soil microbes frequently engage in long-lasting close interactions with other species on this planet, like pathogenesis and symbiosis. The plant and microbe interactions present in the soil may be for a short period of time or long-lasting; interactions like symbiosis as well as pathogenesis are the commonest ones of long-lasting interactions, and the role of these interactions in nitrogen fixation and mycorrhizae formation is well studied. With the help of these types of interactions, the plant attains a healthy growth as well as defense against diseases (Shastri and Kumar 2019) (Fig. 1.2).

1.6 Role of Phenolics in Rhizobium–Legume Symbiosis

The roots of the leguminous plants have a specialized structure, i.e., nodule, which originates because of the symbiosis with the rhizobia bacteria. The bacteria present in the nodules of the roots discriminate into bacteroids fixing nitrogen in reduced form for the plant in return for glucose and shield (Becker et al. 1998; Schauser et al. 1999). The physiology behind the rhizobium and legume interaction and the

Fig. 1.2 Effect of plant microbe interaction on the growth and development of the plant

mechanism involved in the fixation of nitrogen has mainly been explored by many researchers (Broughton et al. 2000; Jones et al. 2007).

The microbe–plant-based interaction involves various signal molecules, and this interaction is vital for the root nodules formation. Consequently, a number of characteristics involved in plant–bacteria identification, nodule emergence, and the fixation of nitrogen have been well-researched (Kondorsi and Schultze 1998; Geurts et al. 2005). In recent times, in *Arachis hypogaea*, it has been found that the expression and assemblage of phenolics occur during nodulation (Chakraborty and Mandal 2008). The leguminous plants secrete various signaling molecules such as betaines, aldonic acids, and flavonoids inside seeds as well as in roots which help in plant–rhizobium symbiosis (Phillips and Torrey 1972). In the interaction between the bacteria and rhizobia inside the rhizosphere, the phenolics not only help in their growth but also act as chemoattractants for legume–rhizobia symbiosis (Hartwig et al. 1991; Caetano-Anolles et al. 1988). Phenolic compounds having free –OH groups can hinder in flavonoid signaling involved in symbiosis of rhizobia–legume as well as in the fixation of nitrogen (Fox et al. 2001). The rhizobia being occupied in symbiosis also helps in the production of flavonoids in roots as well as in their nodules, sets Nod factor formation during infection, and also earlier to infection (Zuanazzi et al. 1998). Different phenolic compounds are secreted from various parts of leguminous plants, like flavonols from Vicia faba, vanillin from peanuts, and isoflavonoids from soybeans (Bekkara et al. 1998; Zawoznik et al. 2000; D'Arcy-Lameta and Jay 1987). These phenolics control the gene expression of nodulation and also show a prominent effect on legume–rhizobial interaction (Mandal et al. 2010). The various phenolics involved in rhizobia–legume symbiosis differ in their structure, but they all have a nonreducing side chain of long fatty-acid having an *N*-acyl group, biologically active in the leguminous plant, bring about the prevention of root hairs formation as well as cell division in cortical cells inside the roots which causes the formation of nodules (Mandal et al. 2010). In a mixture containing both the legumes and rhizobia, it has been found that both of these are important for this interaction (Zawoznik et al. 2000; Novikova 1994). In *Rhizobium trifolii*, it has been found that some phenolics help in the activation of the nod genes while others in its suppression (Djordjevic and Rolfe 1988).

1.7 Role of Phenolics in Arbuscular Mycorrhiza Symbiosis

Glomeromycota-related fungi produce distinctive vesicles and arbuscules, which define arbuscular mycorrhizas (AM). A special phylum of fungi has been found to have a vital role in the formation of vesicles as well as arbuscules in Arbuscular mycorrhizas (AM) (Mandal et al. 2010). In terms of growth, mycorrhizal plants grow faster than nonmycorrhizal ones because this association helps them in accumulation of nutrients and water uptake from the soil (Gerdemann 1968; Bago et al. 2003; Mosse 1988). It has been found that for the successful symbiotic interaction among fungi especially AM with their host plant, following three steps are required, i.e., formation of spores, growth of hyphae, and the emergence of appressoria (Nagahashi and Douds Jr. 2000). Hyphal branching occurs only in some roots of the host plant suitable for this interaction (Mosse and Hepper 1975; Giovannetti et al. 1996). External application of flavonoids positively impacts hyphal emergence during symbiosis (Gianinazzi-Pearson et al. 1989; Tsai and Phillips 1991; Poulin et al. 1997) (Fig. 1.3).

1.8 Phenolic Acid-Induced Molecular Response of Microsymbionts

Phenolics not only regulate the expression of nod genes responsible for symbiosis but also bring changes in legume–rhizobium interaction (Kondorsi and Schultze 1998). During legume–rhizobial symbiosis, rhizobium secretes chemical molecules with the help of nod genes, which regulate the nodule formation on roots. The activation of these nod genes is controlled by the activator protein, i.e., nod D protein (Horvath et al. 1987). The signaling molecules secreted by the leguminous plants are identified by the rhizobia, which in return secrete lipo-chitooligosaccharide nod factor which brings the changes in the morphology of root hairs of legumes, causes the development of infection thread inside the root cortical, nodule formation and nitrification (Mandal et al. 2010). Phenolics are essential for nodule formation, e.g., flavonoids released from various regions of the roots changes the gene expression of nod genes and nodule organogenesis (Redmond et al. 1986). The Rhizobial nod genes (NodD1/NodD2/NodD3) are activated by the secreted flavonoids from the roots, and the genes' products trigger the transcriptional activation of all other nod genes. Root-based flavonoids trigger the expression of other nod genes by binding to their transcriptional activation site. The inner membrane of the cell contains flavonoids that the NodD proteins are targeted to and interact with (van Rhijn and Vanderleyden 1995). These flavonoids produced from the roots fasten with the

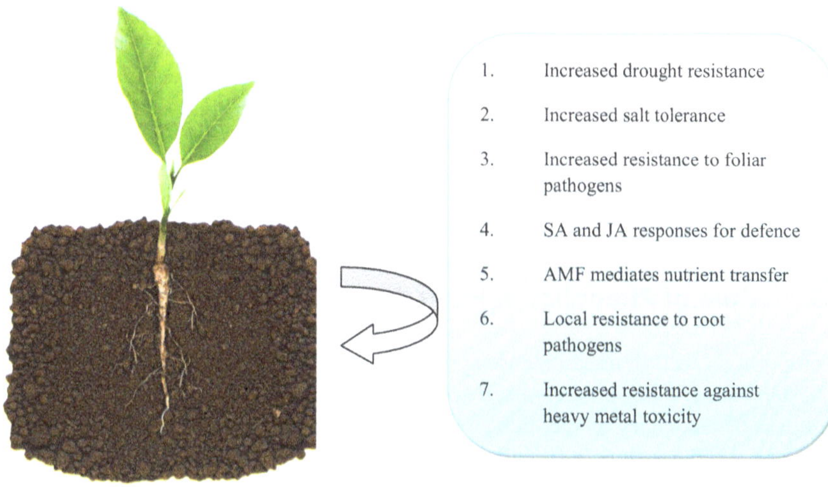

Fig. 1.3 Beneficial effects of arbuscular mycorrhiza (AM) colonization

transcription active regions of Nod genes of rhizobia, and as a result of this, their occurs the transcriptional activation of further essential nod genes. These proteins of NodD are present inside the membrane of cytoplasm, and it interacts with flavonoids in the interior of the same membrane (van Rhijn and Vanderleyden 1995). It has been found that flavonoids also control cell division through the transport of auxin or through its direct impact on the adjustment of the cell cycle (Ferguson and Mathesius 2003). Flavonoids show a positive impact on the synthesis of indole-3-acetic acid (IAA) during symbiosis, probably through the formation of molecules responsible for nod signaling (Prinsen et al. 1991). Host plant-based flavonoids help in nodule organogenesis through the activation of the nod gene y4wEFG present on the downstream region of the nod box, which contains protein information necessary for the formation of IAA by Rhizobia (Theunis et al. 2004). It has been reported that for the formation of nodules, there must be high levels of auxin, which helps in the cell division and also in the formation of nodule primordia (deBilly et al. 2001). The stimulation of nodulating components of nod genes causes the first stage in the production of root nodules. The main member of the Rhizobiaceae family, i.e., *Agrobacterium tumefaciens* is causing tumors in plants, mostly in dicotyledonous at their injury sites (Lippincott et al. 1981). The plasmid of this bacterium has a specific segment, i.e., T-DNA, which helps in its transport as well as its attachment to the host plant's genome. A gene present on this plasmid, i.e., virulence (vir) gene, has a main role in the processing and transport of T-DNA (Mandal et al. 2010). The transcriptional regulation of vir genes are regulated by virA and virG (Heath C. The control of the vir gene is maintained by two members, i.e., virA and virG (Heath et al. 1995)).The protein virA is the main one because it understands the signals coming from the plant and sends these signals to VirG protein which then activates other vir genes through binding to their upstream sites. The locus of virA protein regulates which phenolic compound may act as the stimulator of vir genes, and this protein also identifies these phenolics (Lee et al. 1992). Requena et al. recognize a new gene (GmGin1) having a domain with two proteins that gets downregulated during symbiosis, and this gene may act as a detector for signaling in plants. This protein present in the cell membrane has amino and carboxy terminus. After sensing the signals of a plant, its carboxy end experiences splicing, while its amino side is still fused with the plant signals serving as a nucleophile.

1.9 Role of Phenolics in Plant Defense

Plants contain well-defined defense systems to protect themselves against pathogens (Chester 1993). Plant protects itself from different types of pathogens through the synthesis of secondary metabolites, which provides systemic and local defense (Aist and Gold 1987; Metraux 2001). Phenolics that are secreted from roots, seeds as well as from the remaining parts of the plants helps to fight against the pathogens present inside the soil as well as against insects feeding on the roots (Ndakidemi and Dakora 2003). To fight against soil-based pathogens and to protect the plants from the competition, their roots have particular natural products (Inderjit and Duke 2003; Bais et al. 2006). Plants deal with the various types of soil microbes, such as pathogens, nematodes as well as plant-feeding insects, with the help of phenolics (Dakora 1995; Dakora and Phillips 1996). Based on the variation in the structure, phenolics have different roles to play like isoflavonoid, medicarpin, cajanin, rotenone, phaseolin, glyceollin, coumestrol, flavonoids, and phaseolin and work as phytoanticipins, phytoalexins, and nematicides against pathogens present in the soil as well as against plant-feeding insects (Ndakidemi and Dakora 2003; Dakora and Phillips 1996). Phenolics also show antifungal properties (Ndakidemi and Dakora 2003; Dakora and Phillips 1996). In agricultural crops, to counter the pathogens phenolics can be used in place of chemicals (Dakora and Phillips 1996). The growth of phenolics at the site where the pathogen has attacked strengthens the cell wall through reactive oxygen species, leading to cross-linking of the cell wall and thus providing antimicrobial properties as well as helping in the signaling of defense (Field et al. 2006). The occurrence of micro-organisms at the rhizosphere affects the quantity as well as quality of flavonoids via alteration and catabolism of root exudates. Microbial modification and weakening of phenolic signaling allowed to have environmental effects on the plant–microbe interactions (Shaw et al. 2006).

Phenolics released amid pathogen–host interlinkage helps in the plants defense system (Dixon 2001; Walker et al. 2003; Pieterse et al. 2001). Rhizobia provide many defense systems in plants, thus aiding in resistance against plant diseases (Pieterse et al. 2001; Castro-Sowinski et al. 2007). The plants growth is also being promoted by the rhizobacteria known as plant growth promoting rhizobacteria (PGPR) which causes abnormal conditions for plant pathogens (Castro-Sowinski et al. 2007).

1.10 Plant–Environment Interactions and Functions of Phenolics

As phenolics are everywhere, they have a crucial role in interacting with plants and the environment (Iakimova et al. 2005). Plants are frequently invaded by different organisms, such as insects, herbivores, and pathogens (War et al. 2012). During the invasion by the pathogen, some parts of the plant may get infected only in certain circumstances (Iakimova et al. 2005; Lam 2004). The diseased plant uses a quick and potent hypersensitive response (HR) defense mechanism to respond with the infection (Lam 2004). In plants, this hypersensitive response is considered as a quick defense mechanism to fight against biotic and abiotic stress, and this defense system is known as hypersensitive response (Coll et al. 2011). This process may be damaging for the cell, e.g., it may cause cell death in many cases (War et al. 2012; Coll et al. 2011). Like animals, plants, too, require programmed cell death as a strategy for growth, expansion, and defense in opposition to diseases (Coll et al. 2011; Danon et al. 2000). UV radiation, chemical toxins, and oxygen starvation cause damage to a plant's cells, and the affected tissue actively reacts to these environmental stresses (Lois and Buchanan 1994; Iriti and Faoro 2009; Ellis et al. 1999; Van Doorn and Woltering 2005). During hypersensitive response (HR), plant cells may undergo death, and their cell membranes deposit phenolics to fortify themselves (Iakimova et al. 2005). As the hypersensitive response HR is turned on inside the plant, stressed plant cells undergo structural changes (such as biotic stress) (Coll et al. 2011; Lam et al. 2001). Infected tissues in plants receive phenolics during hypersensitive response (HR), which changes the chemical structure of those tissues. An example of this is the accumulation of phenolics in damaged tissues of tomato and the accumulation of lignin polymers (Zhu and Yao 2004; Mandal and Mitra 2007; Panina et al. 2007). Cells close to wounded tissues are activated, boosting phenolic compound production to limit local infections (Ferreira et al. 2007). Some plants produce phytoalexins in response to pathogen infections, such as hydroxycoumarins and conjugates of hydroxycinnamate (Bhattacharya et al. 2007; Zipfel 2008). Abiotic stresses can also encourage plants to produce phenolics (Treutter 2010). The relationship between temperature and the buildup of phenolics in tracheophytes may be either positive or negative (Treutter 2010). Compared to summer berries, phenolics are elevated in winter berries, according to research by Xu and colleagues (2011). According to several studies, when hypersensitive response (HR) happens in plants, the damaged cells turn brown, indicating Phenolic production in that area (Iakimova et al. 2005; Xu et al. 2011). Numerous plant defensive mechanisms against microbial invaders depend on the production, release, and storage of phenolics (Bhattacharya et al. 2010). Plant pattern recognition receptors produce phenolics when they identify potential pathogens through specific pathogen-associated molecular patterns (PAMPs), which results in immunity (Bhattacharya et al. 2010; Zipfel 2008). As a consequence, the infection's progression is limited before the virus completely takes control of the plant (Zipfel 2008; Bittel and Robatzek 2007). Lignins, their polymers, and further phenolic compounds connected to the cell wall are deposited in response to pathogenic infection, abiotic stress, or wounds (Hématy et al. 2009).

1.10.1 Antioxidant Properties of PCs in Plants

According to Mierziak et al. (2014), antioxidants quench free radicals and lower oxidative stress in order to stop the oxidation of oxidizable compounds. Because of its cellular metabolism and due to oxidative stress, living organisms generate reactive oxygen compounds (Tamang and Fukao 2015). Reactive oxygen species are highly reactive in nature that can alter how cells function as well as destroy biological components like carbohydrates, proteins, nucleic acids, and lipids (Møller et al. 2007). Oxidative stress is thought to have a major part in the rise of chronic degenerative diseases like coronary heart disease, cancer, and aging (Umeno et al. 2016; Wahle et al. 2010). They prevent the oxidation of lipids and other compounds by rapidly donating an atom of hydrogen to radicals (R) (Dai and Mumper 2010; Cotelle 2001).

1.11 Role of Phenolics in Hypersensitive Response (HR)

According to detailed research, some enzymes, including peroxidase (POX), polyphenol oxidase (PPO), and phenyl ammonia-lyase (PAL), are increased during HR activity (Dicko et al. 2005; Sandoval-Chávez et al. 2015). The major enzyme responsible for producing phenolics is phenyl ammonia-lyase (PAL), as was previously mentioned (Ghasemzadeh and Ghasemzadeh 2011). During plant–pathogen interactions, the peroxidase (POX) enzyme changes phenol to lignins (Kavitha and Umesha 2008). As a result, increased peroxidase (POX) activity and polyphenol oxidase (PPO) the accumulation of Phenolics through their oxidation may be related to plant protection (Poiatti et al. 2009).

1.12 Phenolics in Plant Defense

Phenolics have a dual purpose in the plant's environment, acting as both an attractant and a repellent to certain species (Table 1.1).

In mammals, phenols are known to act as plant-derived estrogens (Adams 1989), and for plants and weeds, they also act as allelochemicals while competing with other plants (Xuan et al. 2005; Weir et al. 2004). Numerous straightforward and complicated phenolics build up in tissues of plants and function as phytoanticipins, phytoalexins as well as nematicides against plant-feeding insects and soil-borne diseases (Lattanzio et al. 2006; Akhtar and Malik 2000). As a result, chemicals obtained from phenolics have long been suggested as effective alternatives for the control of agricultural crop diseases in place of chemical pesticides (Langcake et al. 1981). Hydroxycoumarins and conjugates of hydroxycinnamate type of

Table 1.1 Phenolic and their purpose in plant's environment

	Examples	Functions
Phenolics	Salicylic acid	Quorum quencher in *Agrobacterium*
	Hydroquinones	Allochemical for plant competition
	Cinnamic acid, coniferyl alcohol, sinapinic acid	*Vir* gene inducer, symbiotic microbes in plants, determinants of scents and attractants of pollinators
	Hydroxybenzoate, hydroxycinnamates, 5-hydroxyanthraquinones	Allochemical for plant competition
	Umbelliferone, *p*-hydroxybenzoic acid, vanillyl alcohol, isoflavones	Chemoattractants in *Rhizobium*
	p-Hydroxybenzoate, ferulic acid, vanillyl alcohol, bromo acetosyringone	Inhibitors of *vir* gene induction in *Agrobacterium*
	Acetosyringone, α-hydroxyacetosyringone, *p*-hydroxybenzoate	Chemoattractants in *Agrobacterium* and *Rhizobium*
	Coumarins, xanthones, anthocyanidins	Determinants of color and attractants of pollinators in plants
	Caffeic acid	*vir* gene inducer in *Agrobacterium*
	3,4-Dihydroxybenzoic acid	Chemoattractants in *Rhizobium*
	Chlorogenic acid	Precursor for lignin and suberin synthesis in plants
	Lignin, suberin, and tannins	Structural components of plant cells
	Flavonoids, flavonols, flavones, genistein, daidzein, *o*-acetyldaidzein	*nod* gene inducer in *Rhizobium*
	Apigenin, naringenin, luteolin	Chemoattractants in *Agrobacterium* and *Rhizobium*
	Isoflavonoids	Chemoattractants and *nod* gene inducer in *Rhizobium*
	Gallate, gallic acid, pyrogallic acid, syringic acid, kaempferol	*Vir* gene inducers in *Agrobacterium*
	Catechins	Plant defense
	Flavanones, quercetin	*nod* gene inducer in *Rhizobium*
	Cajanin, medicarpin, glyceollin, rotenone, coumestrol, phaseolin, phaseolin, limonoids, tannins, flavonoids	Phytoalexins, phytoanticipins, nematicides in plant defense

phytoalexins are produced by plants in response to pathogen attack and are accumulated by plants (Karou et al. 2005; Mert-Türk 2002). Plants' defense tactics to fend off microbial invaders depend on the biosynthesis, secretion, and accumulation of phenolic compounds, especially salicylic acid (Koornneef and Pieterse 2008; Tsuda et al. 2008; Boller and He 2009; Lu 2009). When plant pattern identification receptors identify prospective pathogens through conserved PAMPs (Schuhegger et al. 2006; Newman et al. 2007; Ongena et al. 2007; Tran et al. 2007), phenolics are produced, resulting in PAMP-triggered immunity (Zipfel 2008). Because of this, the infection's progression is limited before the virus takes the complete control of plant (Bittel and Robatzek 2007; Nicaise et al. 2009).

1.13 Role of Phenolics in Protection Against Extreme Temperatures

Controversial findings have been found in studies on how temperature affects phenolic content. In *Solanum lycopersicum* plants after a thermal stress (35°C), whose ideal temperature for growth is between 22 and 26°C, and then cold stress in *Citrullus lanatus* plants, whose ideal temperature for growth is between 33 and 35°C, aggregation of phenolic compounds has been occurred (Rivero et al. 2001). The levels of phenolics in the cold-sensitive leaves of grapevine were lesser than that of resistant types (Król et al. 2015). The leaves, as well as the roots of grapes, have more phenolics after exposure to cold stress (Amarowicz et al. 2010; Weidner et al. 2009). Wheat's level of flavonoids significantly increased after its acclimatization to cold temperatures, notably in frost-resistant types (Olenichenko et al. 2008). Numerous plants are reported to induce anthocyanin production in response to cold temperatures (Chalker-Scott 1999). In comparison to stoichiometric ratio preparations of vitamin C and vitamin E, anthocyanin solutions are four times more effective at neutralizing a large number of reactive oxygen (ROS) as well as nitrogen species (Gould et al. 2000). The mutants of *Arabidopsis thaliana,* which are faulty in the manufacturing of anthocyanins, experienced substantial oxidative losses under the impact of excessive light as well as low temperatures (Harvaux and Kloppstech 2001).

1.14 Phenolics in Protection Against UV Radiation

Prevention of genetic alterations, photosynthesis, and cellular systems is greatly aided by tolerance to UV light (Hectors et al. 2012). The UV-resistant cells of *Rosa damascena* in vitro revealed that they possess 15 times higher flavonoids (Murphy and Hamilton 1979). The biosynthesis and deposition of various phenolics in the tissues of numerous plant species have been demonstrated to be promoted by UV radiation, particularly in short wavelengths (Hectors et al. 2012; Kondo and Kawashima 2000; Ou et al. 2018). It has been demonstrated that flavonoids can block UV light with a wavelength between 280 and 320 nm. They filter out 95% of the UV rays while permitting up to 80% of visible light (Ou et al. 2018). Anthocyanins shield the photosynthetic machinery by absorbing extra photons that chlorophyll b could otherwise absorb. Red leaves generally receive higher light than green ones, but their photosynthetic regions get fewer photons because the energy collected by the vacuole where the anthocyanins are concentrated cannot be transferred to the chloroplasts (Gould and Lister 2006). The shielding effect of Phenolics against UV radiation is also dependent on their interactions with peroxides as well as free radicals, which stop free radical events from occurring in UV-damaged cells (Kondo and Kawashima 2000).

1.15 Adaptive Protective Functions of Phenolics in the Pathogenesis

One important role of phenolic compounds is to protect the plants from the pathogens. Plant organs and their tissues contain endogenous phenolics that have bactericidal, fungistatic, and insecticidal properties (Punyasiri et al. 2005; Yao et al. 1995). Lignification is among the most crucial defense mechanisms in sick and wounded plant tissues, which creates a physicochemical obstacle to pathogens (Vanholme et al. 2010). Biosynthesis of a particular type of phenolic compound is initiated whenever a plant is attacked by pathogens, and this is associated with a rise in the enzyme production of 4-hydroxylase, phenylalanine ammonia-lyase, as well as *trans*-cinnamic acid chalcone synthase (Amthor 2003; Beckman 2000). The toxic effects of phytoalexins, which become poisonous to pathogens and are missing from or present in minute amounts in normal tissue, are also significantly rising in the overlying tissues at the same time. Phenolic compounds make up the vast bulk (more than 80%) of phytoalexins (Zaprometov and Nikolaeva 2003). Numerous isoflavonoids and phytoalexins from legumes are among the most researched phytoalexins (Lozovaya et al. 2004). The pathogen causes the formation of phytoalexins of the stilbene, flavan, coumarin, and auron classes (Dmitriev 1999).

1.16 Role of Phenolics in Mediating Quorum Sensing

Unicellular prokaryotic species have complicated processes that allow them to manage their activity in changing environmental conditions, as was discovered with quorum sensing (QS) which is a density-dependent type of bacterial intracellular signaling (Fuqua et al. 1994; Yu et al. 2019). According to findings on soil microbes, quorum sensing (QS) systems are primarily a trait of rhizosphere microorganisms (Elasri et al. 2001). Numerous genes involved in the development of biofilms, nitrification, the creation of proteolytic enzymes, exopolysaccharides, and toxicant, cell movement, and conjugation, as well as many others are activated by QS and occupies a significant role in the relationships between plants and microorganisms (Von Bodman et al. 2003; Gonzalez and Marketon 2003). Plant metabolites that engage target gene expression through *N*-acyl-homoserine lactones (AHL) receptor interaction are particularly relevant. *N*-acyl-homoserine lactones (AHL) imitation refers to their ability to function as a substitute for *N*-acyl-homoserine lactones (AHLs) in terms of how they connect with specific receptors (Teplitski et al. 2000). The connection among higher plants, as well as the accompanying microbiota, can be significantly regulated by plant metabolites that imitate bacterial communication molecules (Yu et al. 2019).

The Rh1lR transcriptional regulator is activated in *Pseudomonas aeruginosa* by the phenylpropanoid derivative rosmarinic acid (Corral-Lugo et al. 2016). Plant roots can produce the lignin progenitor *p*-coumaric acid to create *p*-coumaroyl-homoserine lactone, a unique quorum sensing (QS) message that certain bacteria can detect (Schaefer et al. 2008). Additionally, several plant species include the flavonoids catechin as well as naringenin, which can imitate the QS function (Vikram et al. 2010; Vandeputte et al. 2010). It is fascinating to know that plants' *N*-acyl-homoserine lactones (AHL) imitation manufacturing is triggered by rhizobacteria's quorum sensing (QS) signals (Yu et al. 2019).

Roots secreted rosmarinic acid as a result of *Pseudomonas aeruginosa* PAO1 variant plant colonization (Corral-Lugo et al. 2016). These findings strongly suggest that plants may understand bacterial cues and produce AHL copies in response to these cues (Yu et al. 2019).

Quorum sensing cues, which are sent between cells, help to control broad range of behaviors in the rhizosphere where bacteria reside (Fuqua et al. 2001). Most bacteria produce quorum sensing cues, among which the family of acyl-homoserine lactones (AHLs) is used as one of their preferred signaling molecules by gram negative bacteria (Hassan and Mathesius 2012). When quorum sensing signals reach a specific level, they spread between and within the bacteria and, therefore, can attach to receptors there (Fuqua et al. 1994). Numerous bacterial genes, namely those linked to biofilm creation, nitrification, the production of exopolysaccharides, hydrolytic enzymes and toxins, as well as movement and coupling, are activated as a result (Von Bodman et al. 2003; Gonzalez and Marketon 2003). Halogenated furanones generated by *Rhodophyceae* are among the chemicals that were discovered in recent years to obstruct quorum sensing (Manefield et al. 1999). While the majority of these substances are yet unknown, it has been demonstrated that a variety of terrestrial plants produce quorum sensing duplicates that are able to suppress and activate AHL-dependent genetic traits in different reporter lines (Gao et al. 2003; Teplitski et al. 2000). Lumichrome, a riboflavin-derived compound, became the first duplicate signal discovered in plants (Rajamani et al. 2008). The *p*-coumaric acid is a phenolic substance, and a lignin analogue that roots could secrete into the soil seems to be another possible AHL imitator (Bodini et al. 2009). Another way to create *p*-coumaric acid is to degrade flavonoids in root exudates (Rao and Cooper 1995). Another important note is that sensitivity to quorum-sensing cues from rhizobia activates the flavonoid cascade in legumes. Additionally, bacterial AHLs (at 50 M concentrations) were demonstrated to encourage *M. truncatula* to produce AHL analogues (Mathesius et al. 1998). These findings clearly imply a connection between plants' sensing of AHL, initiation of the flavonoid system, and potential bacterial feedback through the creation of AHL imitators (Hassan and Mathesius 2012). Within a wide range of ecological niches, quorum sensing facilitates bacterial cell–cell interaction and encourages behavior that is both favorable for the existence and the establishment of pathogenic or beneficial partnerships (Joint et al. 2002). For a good association, rhizobia must behave as complex organisms and synchronize itself to phenolic cues on a community scale (Amita et al. 2010). Quorum sensing in Rhizobium improves nodulation effectiveness, symbiosome formation, exopolysaccharide production, nitrification, and stress tolerance (Danino et al. 2003; Gonzalez and Marketon 2003).

1.17 Phenolics as Signaling Molecules

Various types of phenolic chemicals are synthesized by plants, which includes phenolic acids along with its derivatives. In rhizospheric plant–microbe association, phenolics are well-known to carry out diverse functions (Martens 2002). Shikimic acid, a result of the cinnamic acid pathway, the monolignol pathway, and the degradation output of lignin along with its cell wall polymers in tracheophytes are the three pathways by which phenolics are formed in plants (Harkin 1973; Croteau et al. 2000). Furthermore, certain phenolic acids have microbiological origins (Moorman et al. 1992). In contrast to those of microbes, which have the chemical makeup C_6–C_1 (phenylmethyl type), phenolics found in plant cell walls as well as in lignin have a distinctive C_6–C_3 (phenyl-propanoid form) structure (Sarakanen and Ludwig 1971). Plants engage defensive mechanisms in reply to microbial infection, which induces a wide variety of antimicrobial chemicals, a few of which are species-specific (van Loon 2000). These triggered defenses defend the plant against infection progress and additional attacks by being expressed both at the attack region (hypersensitive response) and distantly (methyl salicylate signaling) (Mandal et al. 2010). A complex and interconnected network of signal transmission pathways controls induced resistance, and essential signaling molecules in these pathways include phenolics (Pieterse and Van Loon 1999; Feys and Parker 2000). Phenolics are reasonably hardy biochemical entities that change in the soil since some bacteria can utilize these as carbohydrates (Kefeli et al. 2003). The capacity of some soil bacteria to oxidize aromatic chemicals is well known (Chen et al. 1984; Carmona et al. 2009). Decomposing plant remains frequently create simple phenolics like methoxy and hydroxybenzoic acid, along with cinnamic acids (Whitehead et al. 1983). In constrained conditions, certain diazotrophs may be able to obtain alternate byproducts from phenolics. These phenolics may also act as contributors in the manufacture of phenolic lipids (Funa et al. 2006).

1.18 Phenolic Signaling in Rhizobium Legume Symbiosis

Leguminous plants plus bacteria of the genera Rhizobium work together in mutual ways to produce the distinctive and coordinated structure known as the root nodule (Mandal et al. 2010). The invasive bacteria (*Rhizobium* sp.) in the root nodule evolve into nitrogen-fixing bacteroids, which thus give the plant less nitrogen in return for carbohydrates and protection (Becker et al. 1998; Schauser et al. 1999). Numerous researchers have thoroughly studied various facets of the biology of said legume–rhizobium relationship and nitrification process (Broughton et al. 2000; Jones et al. 2007). Both partners must signal and recognize each other for the partnership to form the nodules; different signaling molecules must be transferred between the plant and bacterium to control formation of nodule, division, and activity (Mandal et al. 2010). As a result, many aspects of plant identification, nodule creation, and nitrification have been extensively researched (Kondorsi and Schultze 1998; Geurts et al. 2005), but there is scanty information about the systems through which root nodules' phenylpropanoids control rhizobia's ability to infect plants and cause nodule growth (Mandal et al. 2010). In order to communicate with the microbial symbiont, the plant produces flavonoids, chromones, and adonic acids in its seed, as well as root secretions during the creation of nodules in legumes (Phillips and Torrey 1972). Although the principal polyphenols involved are phenolic acids, flavonoids have garnered more focus on their control. According to Hartwig et al. (1991), some phenolic from symbiotic legumes works as chemical attractants to draw rhizobia bacteria toward root hairs of legumes and to encourage the proliferation of rhizobia in the rhizosphere (Caetano-Anolles et al. 1988). According to research by Fox et al. (2001), planar phenolics with free hydroxyl molecules can affect the symbiosis between rhizobia and legumes that fixes nitrogen. Rhizobia in roots, as well as nodules, which control the production of Nod (nodulation) factor before and after infection (Cooper and Rao 1992), also promote the generation of flavonoids (Zuanazzi et al. 1998). Because flavonoids, such as phenolic acids, operate as antioxidants, it is likely that they shield cells that divide from oxidative harm (Rice-Evans 2001). Leguminous plants release a range of phenolics from their roots, including vanillin in peanuts (Zawoznik et al. 2000), isoflavonoids in soybean (D'Arcy-Lameta and Jay 1987), and flavonoids and flavone from Vicia faba (Bekkara et al. 1998). These phenolic chemicals alter the legume–rhizobial association by controlling the symbiont (Rhizobiumnod)'s expression of genes (Mandal et al. 2010).

During the activation of several *nod* genes, the host root emits phenolic chemicals that serve as molecules of signaling (Mandal et al. 2010). Chito-lipooligosaccharides are a unique family of glycolipids that are produced by certain of these nod genes' encoded enzymes (Mandal et al. 2010). Although these signal molecules have a variety of structural characteristics, their *N*-acyl long-chain fatty acid with nonreducing terminus, causes the root hair deflections, cortical cell divisions in root leading to the formation of nodules (Mandal et al. 2010). Potential options for signaling during the establishment of mycorrhiza were plant phenolic chemicals (Mandal et al. 2010). This has been demonstrated that changes in the gene expression, which participates in flavonoid, phenylpropanoid, as well as metabolism of isoflavonoid caused by the mycorrhizal development, change the composition of flavonoid in root extracts (Harrison and Dixon 1993, 1994). An individual flavonoid might have a favorable, neutral or detrimental effect, on certain fungi, and they are only found in mycorrhizal symbioses (Poulin et al. 1997; Siqueira et al. 1991). According to current research, each flavonoid has distinct properties that could help to explain this (Scervino et al. 2005a, b). Several researchers have found that AM injection causes the host's levels of phenolics to significantly increase (Ling-Lee et al. 1977; Selvaraj and Subramanian 1990). Giovannetti et al. (1996) hypothesized that the invasion of root cell by the fungus is dependent on the genetic material and is only facilitated by recognition of the proper signal molecules arriving from the plant roots. As a result, phenolic chemicals may help with the early phases of mycorrhizal establishment, but root invasion and the subsequent interactions between the fungal partner and host plant are probably what control how AM develops (Mandal et al. 2010). Phenolic substances change the symbiosis between legumes and rhizobia by strongly regulating the nod expression of genes of the symbiont (Rhizobium) (Kondorsi and Schultze 1998). Rhizobium produces effectors by activating the root nodule organogenesis-controlling nodulation (nod) genes (Mandal et al. 2010). The nodD enhancer proteins control the expression of these genes (Horvath et al. 1987). In response to the phenolic cue, rhizobia release lipo-chitooligosaccharide Nod elements that alter the morphology of legume root hairs, resulting in the production of infection threads, nodules, and symbiotic Nitrogen metabolism (Mandal et al. 2010). The nod genes (NodD1/NodD2/NodD3) of rhizobia are activated by plant-based flavonoids, and the genes' outputs in turn trigger other nod genes expression. The flavonoids inside the inner membrane connect with the NodD proteins, which are located in the cytoplasmic membrane (van Rhijn and Vanderleyden 1995). The susceptibility of the receiving strain to the flavonoid is altered via the transfer of a nodD gene across Rhizobium strains that vary in their reactivity to various flavonoids (Horvath et al. 1987; Spaink 2000). The production of various nodulating components, which act to trigger the initial stages of root nodule formation, is catalyzed by the Nod gene protein (Mandal et al. 2010). Recently, it has been discovered that several phenolics present in root nodules enhance Rhizobia's ability to produce IAA (Mandal et al. 2009). According to Lippincott et al. (1981), *Agrobacterium tumefaciens* infects the injury sites of primarily dicotyledonous plus also a small number of monocotyledonous plants to produce crown gall tumors. A particular section of the tumor-inducing (Ti) plasmid known as T-DNA is transported to interacting plant cells and then integrated into genome of the plant. Particularly, virulence (vir) genes involved in the handling and transmission of the T-DNA are found on Ti plasmid. Two members, virG and virA, control the transcription of other vir genes (Heath et al. 1995). Plant signal chemicals are reacted to by the VirA protein, which subsequently transmits the message to the response regulator VirG protein. After that, each of the vir genes' located at the upstream site bind to this protein, which stimulates the transcription of the vir genes. The virA gene regulates which phenolic chemicals can operate as vir gene inducers, and the VirA protein immediately senses the phenolics for vir gene activation, according to Lee et al. (1992). The transcriptional alterations in fungi caused by plant cues have recently been discovered in a few experiments (Martin-Laurent et al. 1997; Harrier et al. 1998). A small proportion of fungal genes which are regulated during growth were discovered by Requena et al. (2002) using suppressive subtractive hybridization (SSH) to produce a subtractive cDNA library from *Glomus mosseae*. They found a brand-new gene (GmGin1) that expresses a two-domain protein whose transcription is suppressed during symbiosis entry. These researchers hypothesized that GmGin1 might function as a plant signal detector. The carboxy end of such a protein, which is found at the cell surface, is subject to splicing in reaction to cues from the plant. The amino end of the plant signal is kept covalently linked after splicing and functions as a nucleophile. According to Requena et al. (2007), a changed Gin1 may be able to influence other downstream signaling proteins and perform a signaling role through its ATPase metabolism. In order to promote the growth of effective symbioses, it was also demonstrated that chemical interaction with the plant symbiont altered fungal gene transcription and caused posttranscriptional changes of fungal proteins.

1.19 Conclusion

This chapter focuses on the importance of plant phenolic as a whole while as their importance in plant microbe symbiosis in particular. Phenolic compounds are widely found in plants with many essential functions while some act as defense elements against herbivores and pathogens. Phenolic compounds are mainly classified into phenolic acids, flavonoids, tannins, phenolic lignans, and phenolic stilbenes. Plants respond to beneficial microbes by increasing phenolic compound levels. Many studies were conducted to show the importance and effectiveness of phenolic compounds, which could influence the plant–microbe relationship. Plant phenolics are considered to have a key role as defense compound under environment stresses such as heavy metal stress, salinity stress, high light, low temperature, pathogen infection, herbivores and nutrient deficiency, etc. Phenolics play many important roles in plants like in plant–microbe symbiosis, rhizobium–legume symbiosis, and arbuscular mycorrhiza symbiosis, and provide defense system to the plant. Higher plants phenolic patterns have been developed by a dialog between the environment and the plants for the advantage of the plants and for improved environmental adaption.

References

Adams NR (1989) Phytoestrogens. In: Cheeke PR (ed) Toxicants of plant origin, pp 23–51

Aislabie J, Deslippe JR, Dymond J (2013) Soil microbes and their contribution to soil services. In: Ecosystem services in New Zealand–conditions and trends, vol 1(12). Manaaki Whenua Press, Lincoln, New Zealand, pp 143–161

Aist JR, Gold RE (1987) Prevention of fungal ingress: the role of papillae and calcium. Japan Sci Soc Press, Tokyo, pp 47–58

Akhtar M, Malik A (2000) Roles of organic soil amendments and soil organisms in the biological control of plant-parasitic nematodes: a review. Bioresour Technol 74:35–47

Amarowicz R, Weidner S, Wójtowicz I, Karmać M, Kosińska A, Rybarczyk A (2010) Influence of low-temperature stress on changes in the composition of grapevine leaf phenolic compounds and their antioxidant properties. Funct Plant Sci Biotechnol 4:90–96

Amita M, Takahashi T, Tsutsumi S, Ohta T, Takata K, Henmi N, Hara S, Igarashi H, Takahashi K, Kurachi H (2010) Molecular mechanism of the inhibition of estradiol-induced endometrial epithelial cell proliferation by clomiphene citrate. Endocrinology 151(1):394–405

Amthor JS (2003) Efficiency of lignin biosynthesis: a quantitative analysis. Ann Bot 91(6):673–695

Aura AM (2008) Microbial metabolism of dietary phenolic compounds in the colon. Phytochem Rev 7:407–429

Babenko LM, Smirnov OE, Romanenko KO, Trunova OK, Kosakivska IV (2019) Phenolic compounds in plants: biogenesis and functions. Ukr Biochem J 91(3):5–18

Badri DV, Chaparro JM, Zhang R, Shen Q, Vivanco JM (2013) Application of natural blends of phytochemicals derived from the root exudates of Arabidopsis to the soil reveal that phenolic-related compounds predominantly modulate the soil microbiome. J Biol Chem 288:4502–4512

Bago B, Pfeffer PE, Abubaker J, Jun J, Allen JW, Brouillette J et al (2003) Carbon export from arbuscular mycorrhizal roots involves the translocation of carbohydrate as well as lipid. Plant Physiol 131:1496–1507

Bais HP, Weir TL, Perry LG, Gilroy S, Vivanco JM (2006) The role of root exudates in rhizosphere interactions with plants and other organisms. Annu Rev Plant Biol 57:233–266

Balasundram N, Sundram K, Samman S (2006) Phenolic compounds in plants and agri-industrial by-products: antioxidant activity, occurrence, and potential uses. Food Chem 99:191–203

Becker BU, Kosch K, Parniske M, Müller P (1998) Exopolysaccharide (EPS) synthesis in Bradyrhizobium japonicum: sequence, operon structure and mutational analysis of an exo gene cluster. Mol Gen Genet 259:161–171

Beckman CH (2000) Phenolic-storing cells: keys to programmed cell death and periderm formation in wilt disease resistance and in general defence responses in plants? Physiol Mol Plant Pathol 57(3):101–110. ISSN 0885-5765

Bekkara F, Jay M, Viricel MR, Rome S (1998) Distribution of phenolic compounds within seed and seedlings of two Vicia faba cvs differing in their seed tannin content, and study of their seed and root phenolic exudations. Plant Soil 203:27–36

Berendsen RL, Pieterse CMJ, Bakker P (2012) The rhizosphere microbiome and plant health. Trends Plant Sci 17:478–486

Bhattacharya P, Welch AH, Stollenwerk KG, McLaughlin MJ, Bundschuh J, Panaullah G (2007) Arsenic in the environment: biology and chemistry. Sci Total Environ 379:109–120. ISSN 0048-9697

Bhattacharya A, Sood P, Citovsky V (2010) The roles of plant phenolics in defence and communication during Agrobacterium and Rhizobium infection. Mol Plant Pathol 11:705–719. [PMC free article] [PubMed]

Bittel P, Robatzek S (2007) Microbe-associated molecular patterns (MAMPs) probe plant immunity. Curr Opin Plant Biol 10:335–341

Blount JW, Korth KL, Masoud SA, Rasmussen S, Lamb C, Dixon RA (2000) Altering expression of cinnamic acid 4-hydroxylase in transgenic plants provides evidence for a feedback loop at the entry point into the phenylpropanoid pathway. Plant Physiol 122:107–116

Bodini SF, Manfredini S, Epp M, Valentini S, Santori F (2009) Quorum sensing inhibition activity of garlic extract and p-coumaric acid. Lett Appl Microbiol 49:551–555

Boller T, He SY (2009) Innate immunity in plants: an arms race between pattern recognition receptors in plants and effectors in microbial pathogens. Science 324:742–744

Borges F, Roleira F, Milhazes N, Santana L, Uriarte E (2005) Simple coumarins and analogues in medicinal chemistry: occurrence, synthesis and biological activity. Curr Med Chem 12:887–916

Boudet AM (2007) Evolution and current status of research in phenolic compounds. Phytochemistry 68:2722–2735

Bourgaud F, Hehn A, Larbat R, Doerper S, Gontier E, Kellner S, Matern U (2006) Biosynthesis of coumarins in plants: a major pathway still to be unravelled for cytochrome P450 enzymes. Phytochem Rev 5:293–308

Bravo L (1998) Polyphenols: chemistry, dietary sources, metabolism, and nutritional significance. Nutr Rev 56(11):317–333

Brodowska KM (2017) Natural flavonoids: classification, potential role, and application of flavonoid analogues. Eur J Biol Res 7(2):108–123

Broughton WJ, Jabbouri S, Perret X (2000) Keys to symbiotic harmony. J Bacteriol 182:5641–5645

Caetano-Anolles G, Wall LG, De Micheli AT, Macchi EM, Bauer WD, Favelukes G (1988) Role of motility and chemotaxis in efficiency of nodulation by Rhizobium meliloti. Plant Physiol 86: 1228–1235

Carmona M, Zamarro MT, Blázquez B, Durante Rodríguez G, Juárez JF, Valderrama AJ et al (2009) Anaerobic catabolism of aromatic compounds: a genetic and genomic view. Microbiol Mol Biol Rev 73:71–133

Castro-Sowinski S, Herschkovitz Y, Okon Y, Jurkevitch E (2007) Effects of inoculation with plant growth-promoting rhizobacteria on resident rhizosphere microorganisms. FEMS Microbiol Lett 276:1–11

Cesco S, Mimmo T, Tonon G, Tomasi N, Pinton R, Terzano R, Neumann G, Weisskopf L, Renella G, Landi L, Nannipieri P (2012) Plant-borne flavonoids released into the rhizosphere: impact on soil bio-activities related to plant nutrition. A review. Biol Fertil Soils 48:123–149

Chakraborty D, Mandal SM (2008) Fractional changes in phenolic acids composition in root nodules of Arachis hypogaea L. Plant Growth Regul 55:159–163

Chalker-Scott L (1999) Environmental significance of anthocyanin in plant stress responses. Photochem Photobiol 70(1):1–9

Chen YP, Dilworth MJ, Glenn AR (1984) Uptake and oxidation of aromatic substances by Rhizobium leguminosarum MNF 8841 and Rhizobium trifolii TA1. FEMS Microbiol Lett 21: 201–205

Chester KS (1993) The problem of acquired physiological immunity in plants. Q Rev Biol 8:275–324

Cheng SH, Sheen J, Gerrish C, Bolwell GP (2001) Molecular identification of phenylalanineammonia-lyase as a substrate of aspecific constitutively active Arabidopsis CDPK expressed in maize protoplasts. FEBS Lett 503:185–188

Chiang VL (2006) Monolignol biosynthesis and genetic engineering of lignin in trees, a review. Environ Chem Lett 4:143 146

Chomel M, Guittonny-Larchevêque M, Fernandez C, Gallet C, DesRochers A, Paré D, Jackson BG, Baldy V (2016) Plant secondary metabolites: a key driver of litter decomposition and soil nutrient cycling. J Ecol 104:1527–1541. https://doi.org/10.1111/1365-2745.12644

Coll NS, Epple P, Dangl JL (2011) Programmed cell death in the plant immune system. Cell Death Differ 18:1–10

Cooper JE, Rao R (1992) Localised changes in flavonoid biosynthesis in roots of Lotus pendiculatus after infection by Rhizobium loti. Plant Physiol 100:444–450

Corral-Lugo A, Daddaoua A, Ortega A, Espinosa-Urgel M, Krell T (2016) Rosmarinic acid is a homoserine lactone mimic produced by plants that activates a bacterial quorum-sensing regulator. Sci Signal 9(409):ra1

Cotelle N (2001) Role of flavonoids in oxidative stress. Curr Trends Med Chem 1:569–590

Croteau R, Kutchan TM, Lewis NG (2000) Natural products (secondary metabolites). In: Buchanan et al (eds) Biochemistry and molecular biology of plants. American Society of Plant Physiologists, Rockville, MD, pp 1250–1318

D'Arcy-Lameta A, Jay M (1987) Study of soybean and lentil root exudates III. Influence of soybean isoflavonoids on the growth of rhizobia and some rhizospheric microorganisms. Plant Soil 101: 267–272

Daglia M (2012) Polyphenols as antimicrobial agents. Curr Opin Biotechnol 23:174–181

Dai J, Mumper RJ (2010) Plant phenolics: extraction, analysis and their antioxidant and anticancer properties. Molecules 15:7313–7352

Dakora FD (1995) Plant flavonoids: biological molecules for useful exploitation. Aust J Plant Physiol 22:87–99

Dakora FD, Phillips DA (1996) Diverse functions of isoflavonoids in legumes transcend ant-microbial definitions of phytoalexins. Physiol Mol Plant Pathol 49:1–20

Danino VE, Wilkinson A, Edwards A, Downie JA (2003) Recipient-induced transfer of the symbiotic plasmid pRL1JI in Rhizobium leguminosarum bv. viciae is regulated by a quorum-sensing relay. Mol Microbiol 50:511–525

Danon A, Delorme D, Mailhac N, Gallois PG (2000) Plant programmed cell death: a common way to die. Plant Physiol Biochem 38:647–655

deBilly F, Grosjean C, May S, Bennett M, Cullimore JV (2001) Expression studies on AUX1-like genes in Medicago truncatula suggest that auxin is required at two steps in early nodule development. Mol Plant Microbe Interact 14:267–277

Delgado-Baquerizo M, Oliverio AM, Brewer TE, Benavent-González A, Eldridge DJ, Bardgett RD, Maestre FT, Singh BK, Fierer N (2018) A global atlas of the dominant bacteria found in soil. Science 359:320–325

Dicko MH, Gruppen H, Barro C, Traore AS, van Berkel WJH, Voragen AGJ (2005) Impact of phenolic compounds and related enzymes in sorghum varieties for resistance and susceptibility to biotic and abiotic stresses. J Chem Ecol 31:2671

Dixon RA (2001) Natural products and plant disease resistance. Nature 411:843–847

Djordjevic M, Rolfe B (1988) Recognition process in the Rhizobium trifolii-white clover symbiosis. In: Bothe H, de Bruijn F, Newton F (eds) Nitrogen fixation: hundred years after. Gustav Fischer, Stuttgart, Germany, pp 431–436

Dmitriev AP (1999) Phytoalexins and their role in plant resistance. Institute of Cell Biology and Genetic Engineering of the National Academy of Sciences, Kyiv. 207 p

Elasri M, Delorme S, Lemanceau P, Stewart G, Laue B, Glickmann E, Oger PM, Dessaux Y (2001) Acyl-homoserine lactone production is more common among plant-associated Pseudomonas spp. than among soilborne Pseudomonas spp. Appl Environ Microbiol 67:1198–1209

Ellis MH, Dennis ES, Peacock JW (1999) Arabidopsis roots and shoots have different mechanisms for hypoxic stress tolerance. Plant Physiol 119:57–64

Erlund I (2004) Review of the flavonoids quercetin, hesperetin, and naringenin. Dietary sources, bioactivities, bioavailability, and epidemiology. Nutr Res 24(10):851–874

Ferguson BJ, Mathesius U (2003) Signaling interactions during nodule development. Plant Growth Regul 22:47–72

Ferreira RB, Monteiro S, Freitas R, Santos CN, Chen Z, Batista LM, Duarte J, Borges A, Teixeira AR (2007) The role of plant defence proteins in fungal pathogenesis. Mol Plant Pathol 8:677–700

Feys B, Parker JE (2000) Interplay of signaling pathways in plant disease resistance. Trends Genet 16:449–455

Field B, Jordon F, Osbourn A (2006) First encounters-deployment of defence related natural products by plants. New Phytol 172:193–207

Fox JE, Starcevic M, Kow KY, Burow ME, McLachlan JA (2001) Endocrine disrupters and flavonoid signaling. Nature 413:128–129

Funa N, Ozawa H, Hirata A, Horinouchi S (2006) Phenolic lipid synthesis by type III polyketide synthases is essential for cyst formation in Azotobacter vinelandii. Biol Sci 103(16):6356–6361

Fuqua WC, Winans SC, Greenberg EP (1994) Quorum sensing in bacteria—the luxR–LuxI family of cell density-responsive transcriptional regulators. J Bacteriol 176:269–275

Fuqua C, Parsek MR, Greenberg EP (2001) Regulation of gene expression by cell-to-cell communication: acyl-homoserine lactone quorum sensing. Annu Rev Genet 35:439–468

Galili G, Tzin V (2010) New insights into the shikimate and aromatic amino acids biosynthesis pathways in plants. Mol Plant 3:956–972. https://doi.org/10.1093/mp/ssq048

Gao MS, Teplitski M, Robinson JB, Bauer WD (2003) Production of substances by Medicago truncatula that affect bacterial quorum sensing. Mol Plant Microbe Interact 16:827–834

Gerdemann JW (1968) Vesicular-arbuscular mycorrhiza and plant growth. Annu Rev Phytopathol 6:397–418

Geurts R, Fedorova E, Bisseling T (2005) Nod factor signaling genes and their function in the early stages of Rhizobium infection. Curr Opin Plant Biol 8:346–352

Ghasemzadeh A, Ghasemzadeh N (2011) Flavonoids and phenolic acids: role and biochemical activity in plants and human. J Med Plants Res 5:6697–6703

Gianinazzi-Pearson V, Branzanti B, Gianinazzi S (1989) In vitro enhancement of spore germination and early hyphal growth of a vesicular-arbuscular mycorrhizal fungus by host root exudates and plant flavonoids. Symbiosis 7:243–255

Gilbert JA, Meyer F, Jansson J, Gordon J, Pace N, Tiedje J, Ley R, Fierer N, Field D, Kyrpides N, Glöckner FO, Klenk HP, Wommack KE, Glass E, Docherty K, Gallery R, Stevens R, Knight R (2010) The Earth Microbiome Project: meeting report of the "1st EMP meeting on sample

selection and acquisition" at Argonne National Laboratory October 6 2010. Stand Genomic Sci 3:249–253
Giovannetti M, Sbrana C, Citernesi AS, Avio L (1996) Analysis of factors involved in fungal recognition responses to host derived signals by arbuscular mycorrhizal fungi. New Phytol 133: 65–71
Gonzalez JE, Marketon MM (2003) Quorum sensing in nitrogen-fixing rhizobia. Microbiol Mol Biol Rev 67:574–592
Gould KS, Lister C (2006) Flavonoid functions in plants. In: Andersen OM, Markham KR (eds) Flavonoids: chemistry, biochemistry and applications. Taylor & Francis Group, LLC, pp 397–442
Gould KS, Markham KR, Smith RH, Goris JJ (2000) Functional role of anthocyanins in the leaves of Quintinia serrata A. Cunn. J Exp Bot 51(347):1107–1115
Gupta A, Gupta R, Singh RL (2017) Microbes and environment. In: Principles and applications of environmental biotechnology for a sustainable future. Springer, Singapore, pp 43–84
Hahlbrock K, Scheel D (1989) Physiology and molecular biology of phenylpropanoid metabolism. Plant Mol Biol 40:347–369
Harborne JB, Simmonds NW (1964) Biochemistry of phenolic compounds. Academic, London, p 77
Harkin JM (1973) Lignin (forage plants). In: Butler GW, Bailey RW (eds) Chemistry and biochemistry of herbage. Academic, New York, pp 323–373
Harnly JM, Bhagwat S, Lin LZ (2007) Profiling methods for the determination of phenolic compounds in foods and dietary supplements. Anal Bioanal Chem 389:47–61
Harrier LA, Wright F, Hooker JE (1998) Isolation of the 3-phosphoglycerate kinase gene of the arbuscular mycorrhizal fungus Glomus mosseae (Nicol. & Gerd.) Gerdemann & Trappe. Curr Genet 34:386–392
Harrison MJ, Dixon RA (1993) Isoflavonoid accumulation and expression of defense gene transcripts during the establishment of vesicular arbuscular mycorrhizal associations in roots of Medicago truncatula. Mol Plant-Microbe Interact 6:643–654
Harrison MJ, Dixon RA (1994) Spatial patterns of expression of flavonoid/isoflavonoid pathway genes during interactions between roots of Medicago truncatula and the mycorrhizal fungus Glomus versiforme. Plant J 6:9–20
Hartwig U, Joseph C, Phillips DA (1991) Flavonoids released naturally from alfalfa seeds enhance growth rate of Rhizobium meliloti. Plant Physiol 95:797–803
Harvaux M, Kloppstech K (2001) The protective functions of carotenoid and flavonoid pigments against excess visible radiation at chilling temperature investigated in Arabidopsis npq and tt mutants. Planta 213(6):953–966
Hassan S, Mathesius U (2012) The role of flavonoids in root-rhizosphere signalling: opportunities and challenges for improving plant-microbe interactions. J Exp Bot 63(9):3429–3444
Hättenschwiler S, Vitousek PM (2000) The role of polyphenols in terrestrial ecosystem nutrient cycling. Trends Ecol Evol 15(6):238–243. https://doi.org/10.1016/s0169-5347(00)01861-9. PMID: 10802549
Heath JD, Charles TC, Nester EW (1995) Two-component signal transduction. In: Hoch JA, Silhavy TJ (eds) Curr Opin Microbiol. Am Soc Microbiol, Washington, DC, pp 367–385
Hectors K, van Oevelen S, Guisez Y, Prinsen E, Jansen MA (2012) The phytohormone auxin is a component of the regulatory system that controls UV-mediated accumulation of flavonoids and UV-induced morphogenesis. Physiol Plant 145(4):594–603
Hématy K, Cherk C, Somerville S (2009) Host-pathogen warfare at the plant cell wall. Curr Opin Plant Biol 12:406–413
Hermans SM, Buckley HL, Case BS, Curran-Cournane F, Taylor M, Lear G (2017) Bacteria as emerging indicators of soil condition. Appl Environ Microbiol 83:e02826–e02816
Herrmann K (1989) Occurrence and content of hydroxycinnamic and hydroxybenzoic acid compounds in foods. Crit Rev Food Sci Nutr 28:315–347

Horvath B, Bachem CWB, Schell J, Kondorsi A (1987) Host specific regulation of nodulation genes in Rhizobium is mediated by a plant signal, interacting with the nodD gene product. EMBO J 6: 841–848

Iakimova ET, Lech M, Woltering JE (2005) Hypersensitive cell death in plants—its mechanism and role in plant defence against pathogens. J Fruit Ornam Plant Res 13:135–158

Inderjit, Duke SO (2003) Ecophysiological aspects of allelopathy. Planta 217:529–539

Iriti M, Faoro F (2009) Chemical diversity and defence metabolism: how plants cope with pathogens and ozone pollution. Int J Mol Sci 10(8):3371–3399. https://doi.org/10.3390/ijms1008337

Isobe K, Bouskill NJ, Brodie EL, Sudderth EA, Martiny JBH (2020) Phylogenetic conservation of soil bacterial responses to simulated global changes. Philos Trans R Soc Lond Ser B Biol Sci 375:20190242

Joint I, Trait K, Callow ME, Callow JA, Milton D, Williams P, Cámara M (2002) Cell-to-cell communication across the prokaryote–eukaryote boundary. Science 298:1207

Jones KM, Kobayashi H, Davies BW, Taga ME, Walker GC (2007) How rhizobial symbionts invade plants: the Sinorhizobium-Medicago model. Nat Rev Microbiol 5:619–633

Kaplan I, Halitschke R, Kessler A, Sardanelli S, Denno RF (2008) Constitutive and induced defenses to herbivory in above- and belowground plant tissues. Ecology 89:392–406. https://doi.org/10.1890/07-0471.1

Karou D, Dicko MH, Simpore J, Traore AS (2005) Antioxidant and antibacterial activities of polyphenols from ethnomedicinal plants of Burkina Faso. Afr J Biotechnol 4:823–828

Kavitha R, Umesha S (2008) Regulation of defense-related enzymes associated with bacterial spot resistance in tomato. Phytoparasitica 36:144–159

Kefeli VI, Kalevitch MV, Borsari B (2003) Phenolic cycle in plants and environment. J Cell Mol Biol 2:13–18

Khare E, Tyagi S, Patil KS (2020) Language of plant-microbe-microbe interactions in rhizospheric ecosystems. In: Molecular aspects of plant beneficial microbes in agriculture, pp 59–76. https://doi.org/10.1016/b978-0-12-818469-1.00005-5

Khlestkina E (2013) The adaptive role of flavonoids: emphasis on cereals. Cereal Res Commun 41(2):185–198

Kondo N, Kawashima M (2000) Enhancement of the tolerance to oxidative stress in cucumber (Cucumis sativus L.) seedlings by UV-B irradiation: possible involvement of phenolic compounds and antioxidative enzymes. J Plant Res 113(3):311–317

Kondorsi A, Schultze M (1998) Regulation of symbiotic root nodule development. Annu Rev Genet 32:33–57

Koornneef A, Pieterse CMJ (2008) Cross talk in defense signaling. Plant Physiol 146:839–884

Koukol J, Conn EE (1961) The metabolism of aromatic compounds in higher plants. IV Purification and properties of the phenylalaninedea minase of Hordeum vulgare. J Biol Chem 236:2692–2698

Król A, Amarowicz R, Weidner S (2015) The effects of cold stress on the phenolic compounds and antioxidant capacity of grapevine (Vitis vinifera L.) leaves. J Plant Physiol 189:97–104

Kulma A, Szopa J (2007) Catecholamines are active compounds in plants. Plant Sci 172:433–440

Lam EL (2004) Controlled cell death, plant survival and development. Nat Rev Mol Cell Biol 5: 305–315

Lam E, Kato N, Lawton M (2001) Programmed cell death, mitochondria and the plant hypersensitive response. Nature 411:848–853

Langcake P, Irvine JA, Jeger MJ (1981) Alternative chemical agents for controlling plant disease. Philos Trans R Soc Lond Ser B Biol Sci 295:83–101

Lattanzio V (2013) Phenolic compounds: introduction. In: Natural products. Springer, pp 1543–1580

Lattanzio V, Lattanzio VM, Cardinali A (2006) Role of phenolics in the resistance mechanisms of plants against fungal pathogens and insects. In: Imperato F (ed) Phytochemistry advances in research. Research Signpost, India, pp 23–67

Lebeis SL, Rott M, Dangl JL, Schulze-Lefert P (2012) Culturing a plant microbiome community at the cross-Rhodes. New Phytol 196:341–344
Lee K, Dudley MW, Hess KM, Lynn DG, Joerger RD, Binns AN (1992) Mechanism of activation of Agrobacterium virulence genes: identification of phenol-binding proteins. Proc Natl Acad Sci U S A 89:8666–8667
Ling-Lee M, Chilvers GA, Ashford AE (1977) A histochemical study of phenolic materials in mycorrhizal and uninfected roots of Eucalyptus fastigata Dean & Maid. New Phytol 78:313–328
Lippincott JA, Lippincott BB, Starr MP (1981) In: Starr MP, Stoip H, Trupper HG, Balows A, Schiegel HG (eds) The prokaryotes, vol 1. Springer, New York, pp 842–855
Lois R, Buchanan BB (1994) Severe sensitivity to ultraviolet radiation in an Arabidopsis mutant deficient in flavonoid accumulation. Planta 194:504–509
Lozovaya VV, Lygin AV, Zernova OV, Li S, Hartman GL, Widholm JM (2004) Isoflavonoid accumulation in soybean hairy roots upon treatment with Fusarium solani. Plant Physiol Biochem 42(7–8):671–679
Lu H (2009) Dissection of salicylic acid-mediated defense signaling networks. Plant Signal Behav 4:713–771
Macheix J-J, Fleuriet A, Jay-Allemand C (2005) Les composés phénoliques des végétaux: un exemple de métabolites secondaires d'importance économique. Presses Polytechniques et Universitaires Romandes, Lausanne, Suisse
Maeda H, Dudareva N (2012) The shikimate pathway and aromatic Amino acid biosynthesis in plants. Annu Rev Plant Biol 63:73–105
Mandal S, Mitra A (2007) Reinforcement of cell wall in roots of lycopersicon esculentum through induction of phenolic compounds and lignin by elicitors. Physiol Mol Plant Pathol 71:201–209
Mandal SM, Mandal M, Das AK, Pati BR, Ghosh AK (2009) Stimulation of indoleacetic acid production in a Rhizobium isolate of Vigna mungo by root nodule phenolic acids. Arch Microbiol 191:389–393
Mandal SM, Chakraborty D, Dey S (2010) Phenolic acids act as signaling molecules in plant-microbe symbioses. Plant Signal Behav 5(4):359–368
Manefield M, de Nys R, Kumar N, Read R, Givskov M, Steinberg P, Kjelleberg SA (1999) Evidence that halogenated furanones from Delisea pulchra inhibit acylated homoserine lactone (AHL)-mediated gene expression by displacing the AHL signal from its receptor protein. Microbiology 145:283–291
Mann J (1978) Secondary metabolism. Clarendon Press, Oxford, p 316
Marchiosi R, dos Santos WD, Constantin RP, de Lima RB, Soares AR, Finger-Teixeira A et al (2020) Biosynthesis and metabolic actions of simple phenolic acids in plants. Phytochem Rev 19(4):865–906
Martens DA (2002) Relationship between plant phenolic acids released during soil mineralization and aggregate stabilization. Soil Sci Soc Am J 66:1857–1867
Martin-Laurent F, van Tuinen D, Dumas-Gaudot E, Gianinazzi-Pearson V, Gianinazzi S, Franken P (1997) Differential display analysis of RNA accumulation in arbuscular mycorrhiza of pea and isolation of a novel symbiosis-regulated plant gene. Mol Gen Genomics 256:37–44
Mathesius U, Schlaman HRM, Spaink HP, Sautter C, Rolfe BG, Djordjevic MA (1998) Auxin transport inhibition precedes root nodule formation in white clover roots and is regulated by flavonoids and derivatives of chitin oligosaccharides. Plant J 14:23–34
Mert-Türk F (2002) Phytoalexins: defense or just a response to stress? J Cell Mol Biol 1:1–6
Metraux JP (2001) Systemic acquired resistance and salicylic acid: current state of knowledge. Eur J Plant Pathol 107:13–18
Mierziak J, Kostyn K, Kulma A (2014) Flavonoids as important molecules of plant interactions with the environment. Molecules 19:16240–16265
Møller IM, Jensen PE, Hansson A (2007) Oxidative modifications to cellular components in plants. Annu Rev Plant Biol 58:459–481

Moorman TB, Becerril JM, Lydon J, Duke SO (1992) Production of hydroxybenzoic acids by Bradyrhizobium japonicum strains after treatment with glyphosphate. J Agric Food Chem 40: 289–293
Mosse B (1988) Some studies relating to "independent" growth of vesicular-arbuscular endophytes. Can J Bot 66:2533–2540
Mosse B, Hepper CE (1975) Vesicular-arbuscular mycorrhizal infections in root organ cultures. Physiol Plant Pathol 5:2015–2223
Murphy TM, Hamilton CM (1979) A itrain of Rosa damascena cultured cells resistant to ultraviolet light. Plant Physiol 64(6):936–941
Naczk M, Shahidi F (2004) Extraction and analysis of phenolics in food. J Chromatogr A 1054:95–111
Nagahashi G, Douds DD Jr (2000) Partial separation of root exudates components and their effects upon the growth of germinated spores of AM fungi. Mycol Res 104:1453–1464
Ndakidemi PA, Dakora FD (2003) Legume seed flavonoids and nitrogenous metabolites as signals and protectants in early seedling development. Rev Funct Plant Biol 30:729–745
Newman MA, Dow JM, Molinaro A, Parrilli M (2007) Priming, induction and modulation of plant defense responses by bacteria lipopolysaccharides. J Endotoxin Res 13:69–84
Nicaise V, Roux M, Zipfel C (2009) Recent advances in PAMP triggered immunity against bacteria: pattern recognition receptors watch over and raise the alarm. Plant Physiol 150: 1638–1647
Novikova TL (1994) Influence of natural phenols on Trifolium pratense-Rhizobium trifoli symbiosis. Acta Hort 381:421–424
Oldroyd GED, Murray JD, Poole PS, Downie JA (2011) The rules of engagement in the legume-rhizobial symbiosis. Annu Rev Genet 45:119–144
Olenichenko NA, Zagoskina NV, Astakhova NV, Trunova TI, Kuznetsov Yu V (2008) Primary and secondary metabolism of winter wheat under cold hardening and treatment with antioxidants. Appl Biochem Microbiol 44(5):535–540
Ongena M, Jourdan E, Adam A, Paquot M, Brans A, Joris B, Arpigny JL, Thonart P (2007) Surfactin and fengycin lipopeptides of Bacillus subtilis as elicitors of induced systemic resistance in plants. Environ Microbiol 9:1084–1090
Ortíz-Castro R, Contreras-Cornejo HA, Macías-Rodríguez L, López-Bucio J (2009) The role of microbial signals in plant growth and development. Plant Signal Behav 4(8):701–712
Ou S, Lu S, Yang S (2018) Effects of enhanced UV-B radiation on the content of flavonoids in mesophyll cells of wheat. Imaging Radiat Res 1(1):1–9
Paiva NL (2000) An introduction to the biosynthesis of chemicals used in plant-microbe communication. J Plant Growth Regul 19(2):131–143
Panina Y, Fravel DR, Baker CJ, Shcherbakova LA (2007) Biocontrol and plant pathogenic fusarium oxysporum-induced changes in phenolic compounds in tomato leaves and roots. J Phytopathol 155:475–481
Paul ND, Hatcher PE, Taylor JE (2000) Coping with multiple enemies: an integration of molecular and ecological perspectives. Trends Plant Sci 5:220–225. https://doi.org/10.1016/S1360-1385(00)01603-4
Perret X, Staehelin C, Broughton WJ (2000) Molecular basis of symbiotic promiscuity. Microbiol Mol Biol Rev 64(1):180–201. https://doi.org/10.1128/MMBR.64.1.180-201.2000. PMID: 10704479; PMCID: PMC98991
Phillips DA, Torrey JG (1972) Studies on cytokinin production by Rhizobium. Plant Physiol 49:11
Pieterse C, Van Loon LC (1999) Salicylic acid-independent plant defence pathways. Trends Plant Sci 4:52–58
Pieterse CMJ, Van Pelt JA, Van Wees SCM, Ton J, L'eon-Kloosterziel KM, Keurentjes JJB et al (2001) Rhizobacteria-mediated induced systemic resistance: triggering, signalling and expression. Eur J Plant Pathol 107:51–61
Poiatti VAD, Dalmas FR, Astarita LV (2009) Defense mechanisms of Solanum tuberosum L. in response to attack by plant-pathogenic bacteria. Biol Res 42:205–215

Pop C, Vlase L, Tamas M (2009) Natural resources containing arbutin. Determination of arbutin in the leaves of Bergenia crassifolia (L.) Fritsch. acclimated in Romania. Not Bot Horti Agrobot Cluj-Napoca 37:129–132

Poulin MJ, Simard J, Catford JG, Labrie F, Piche Y (1997) Response of symbiotic endomycorrhizal fungi to estrogen and antiestrogens. Mol Plant Microbe Interact 10:481–487

Prinsen E, Chauvaux N, Schmidt J, John M, Wieneke U, Greef JD et al (1991) Stimulation of indole-3-acetic acid production in Rhizobium by flavonoids. FEBS Lett 282:53–55

Punyasiri PA, Abeysinghe SB, Kumar V (2005) Preformed and induced chemical resistance of tea leaf against Exobasidium vexans infection. J Chem Ecol 31(6):1315–1324

Rabe C, Steenkamp JA, Joubert E, Burger JFW, Ferreira D (1994) Phenolic metabolites from rooibos tea (Aspalathus linearis). Phytochemistry 35:1559–1565

Rajamani S, Bauer WD, Robinson JB, Farrow JM, Pesci EC, Teplitski M, Gao MS, Sayre RT, Phillips DA (2008) The vitamin riboflavin and its derivative lumichrome activate the LasR bacterial quorum-sensing receptor. Mol Plant Microbe Interact 21:1184–1192

Ramírez PB, Fuentes-Alburquenque S, Díez B, Vargas I, Bonilla CA (2020) Soil microbial community responses to labile organic carbon fractions in relation to soil type and land use along a climate gradient. Soil Biol Biochem 141:107692

Rao JR, Cooper JE (1995) Soybean nodulating rhizobia modify nod gene inducers daidzein and genistein to yield aromatic products that can influence gene-inducing activity. Mol Plant Microbe Interact 8:855–862

Redmond JW, Batley M, Djordjevic MA, Innes RW, Kuempel PL, Rolfe BG (1986) Flavones induce expression of nodulation genes in Rhizobium. Nature 323:632–635

Requena N, Mann P, Hampp R, Franken P (2002) Early developmentally regulated genes in the arbuscular mycorrhizal fungus Glomus mosseae: identification of GmGin1 a novel gene with homology to the C-terminus of metazoan hedgehog proteins. Plant Soil 244:129–139

Requena N, Serrano E, Ocon A, Breuninger M (2007) Plant signals and fungal perception during arbuscular mycorrhiza establishment. Phytochemistry 68:33–40

Rice-Evans C (2001) Flavonoid antioxidants. Curr Med Chem 8:797–807

Rispail N, Morris P, Webb K (2005) Phenolic compounds: extraction and analysis. In: Márquez A (ed) Lotus Japonicus handbook. Springer, Berlin, pp 349–354

Rivero RM, Ruiz JM, García PC, López Lefebre LR, Sánchez E, Romero L (2001) Resistance to cold and heat stress: accumulation of phenolic compounds in tomato and watermelon plants. Plant Sci 160(2):315–321

Ruan Y, Kotraiah V, Straney DC (1995) Flavonoids stimulate spore germination in Fusarium solani pathogenic on legumes in a manner sensitive to inhibitors of cAMP-dependent protein kinase. Mol Plant Microbe Interact 8(6):929–938

Sandoval-Chávez RA, Martínez-Peniche RÁ, Hernández-Iturriaga M, Teixidó-Espasa N, UsallRodié J, Viñas-Almenar I, Torres-Sanchis R (2015) Mechanisms of resistance in postharvest fruit-pathogen interaction. Rev Chapingo Ser Aortic 21:185–198

Sarakanen KV, Ludwig CH (1971) Lignins: occurance, formation, structure and reactions. Wiley-Interscience, New York

Scervino JM, Ponce MA, Erra-Bassells R, Vierheilig H, Ocampo JA, Godeas A (2005a) Flavonoids exhibit fungal species and genus specific effects on the presymbiotic growth of Gigaspora and Glomus. Mycol Res 109:789–794

Scervino JM, Ponce MA, Erra-Bassells R, Vierheilig H, Ocampo JA, Godeas A (2005b) Arbuscular mycorrhizal colonization of tomato by Gigaspora and Glomus species in the presence of root flavonoids. J Plant Physiol 162:625–633

Schaefer AL, Greenberg EP, Oliver CM, Oda Y, Huang JJ, Bittan-Banin G, Peres CM, Schmidt S, Juhaszova K, Sufrin JR, Harwood CS (2008) A new class of homoserine lactone quorum-sensing signals. Nature 454:595–599

Schauser L, Roussis A, Stiller J, Stougaard J (1999) A plant regulator controlling development of symbiotic root nodules. Nature 402:191–195

Schuhegger R, Ihring A, Gantner S, Bahnweg G, Knappe C, Vogg G, Hutzler P, Schmid M, van Breusegem F, Eberl L, Hartmann A, Langebartels C (2006) Induction of systemic resistance in tomato by N-acyl-L-homoserine lactone-producing rhizosphere bacteria. Plant Cell Environ 29: 909–918

Selvaraj JT, Subramanian G (1990) Phenols and lipids in mycorrhizal and non-mycorrhizal roots of Sesamum indicum. Curr Sci 59:471–473

Shastri B, Kumar R (2019) Microbial secondary metabolites and plant–microbe communications in the rhizosphere. In: New and future developments in microbial biotechnology and bioengineering. Elsevier, pp 93–111

Shaw L, Phil Morris J, Hooker JE (2006) Perception and modification of plant flavonoid signals by rhizosphere microorganisms. Environ Microbiol 8:1867–1880. https://doi.org/10.1111/j.1462-2920.2006.01141.x

Singh JS, Gupta VK (2018) Soil microbial biomass: a key soil driver in management of ecosystem functioning. Sci Total Environ 634:497–500

Singh JS, Abhilash PC, Gupta VK (2016) Agriculturally important microbes in sustainable food production. Trends Biotechnol 34(10):773–775

Siqueira J, Safir G, Nair M (1991) Stimulation of vesicular arbuscular mycorrhiza formation and growth of white clover by flavonoid compounds. New Phytol 118:87–93

Spaink HP (2000) Root nodulation and infection factors produced by rhizobial bacteria. Annu Rev Microbiol 54:257–288

Stafford HA (1974) Possible multienzyme complexes regulating the formation of C6–C3 phenolic compounds and lignins in higher plants. Recent Adv Phytochem 8:53–79

Stout MJ, Thaler JS, Thomma BPHJ (2006) Plant-mediated interactions between pathogenic microorganisms and herbivorous arthropods. Annu Rev Entomol 2006(51):663–689

Symonowicz M, Kolanek M (2012) Flavonoids and their properties to form chelate complexes. Biotechnol Food Sci 76(1):35–41

Tamang BG, Fukao T (2015) Plant adaptation to multiple stresses during submergence and following desubmergence. Int J Mol Sci 16:30164–30180

Teplitski M, Robinson JB, Bauer WD (2000) Plants secrete substances that mimic bacterial N-acyl homoserine lactone signal activities and affect population density-dependent behaviors in associated bacteria. Mol Plant Microbe Interact 13:637–648

Theunis M, Kobayashi H, Broughton WJ, Prinsen E (2004) Flavonoids, NodD1, NodD2 and nod-box NB15 modulate expression of the y4wEFG locus that is required for indole-3-acetic acid synthesis in Rhizobium sp. strain NGR234. Mol Plant Microbe Interact 10:1153–1161

Tiwari P, Singh JS (2017) A plant growth promoting rhizospheric Pseudomonas aeruginosa strain inhibits seed germination in Triticum aestivum (L) and Zea mays (L). Microbiol Res 8(2):7233

Tohge T, Watanabe M, Hoefgen R, Fernie AR (2013) Shikimate and phenylalanine biosynthesis in the green lineage. Front Plant Sci 4:1–13

Tomás-Barberán FA, Clifford MN (2000) Dietary hydroxybenzoic acid derivatives—nature, occurrence and dietary burden. J Sci Food Agric 80:1024–1032

Tran H, Ficke A, Aslimwe T, Hofte M, Raaijmakers JM (2007) Role of the cyclic lipopolypeptide massetolide A in biological control of Phytophthora infestans and in colonization of tomato plants by Pseudomonas fluorescens. New Phytol 175:731–742

Treutter D (2010) Managing phenol contents in crop plants by phytochemical farming and breeding—visions and constraints. Int J Mol Sci 11:807–857

Tsai SM, Phillips DA (1991) Flavonoids released naturally from alfalfa promote development of symbiotic Glomus spores in vitro. Appl Environ Microbiol 57:85–88

Tsao R (2010) Chemistry and biochemistry of dietary polyphenols. Nutrients 2(12):1231–1246

Tsimogiannis D, Stavrakaki M, Oreopoulou V (2006) Isolation and characterisation of antioxidant components from oregano (Origanum heracleoticum). Int J Food Sci Technol 41:39–48

Tsuda K, Glazebrook J, Katagiri F (2008) The interplay between MAMP and SA signaling. Plant Signal Behav 3:359–361

Turner TR, James EK, Poole PS (2013) The plant microbiome. Genome Biol 14:209

Tyagi S, Mulla SI, Lee K-J, Chae J-C, Shukla P (2018) VOCs-mediated hormonal signaling and crosstalk with plant growth promoting microbes. Crit Rev Biotechnol 38(8):1277–1296
Umeno A, Masanori H, Murotomi K, Nakajima Y, Yoshida Y (2016) Antioxidative and antidiabetic effects of natural polyphenols and isoflavones. Molecules 21:708
Van Doorn WG, Woltering JE (2005) Many ways to exit? Cell death categories in plants. Trends Plant Sci 10:117–122
van Loon LC (2000) Systemic induced resistance. In: Slusarenko AJ, Fraser RSS, Van Loon LC (eds) Mechanisms of resistance to plant diseases. Kluwer Academic, Dordrecht, The Netherlands, pp 521–574
van Rhijn P, Vanderleyden J (1995) The Rhizobium-plant symbiosis. Microbiol Rev 59:124–142. [Crossref], [PubMed]
Vandeputte OM, Kiendrebeogo M, Rajaonson S, Diallo B, Mol A, El Jaziri M, Baucher M (2010) Identification of catechin as one of the flavonoids from Combretum albiflorum bark extract that reduces the production of quorum-sensing-controlled virulence factors in Pseudomonas aeruginosa PAO1. Appl Environ Microbiol 76:243–253
Vanholme R, Demedts B, Morreel K, Ralph J, Boerjan W (2010) Lignin biosynthesis and structure. Plant Physiol 153(3):895–905
Veberic R (2016) The impact of production technology on plant phenolics. Horticulturae 2:8. https://doi.org/10.3390/horticulturae2030008
Vikram A, Jayaprakasha GK, Jesudhasan PR, Pillai SD, Patil BS (2010) Suppression of bacterial cell-cell signalling, biofilm formation and type III secretion system by citrus flavonoids. J Appl Microbiol 109:51
Von Bodman SB, Bauer WD, Coplin DL (2003) Quorum sensing in plant-pathogenic bacteria. Annu Rev Phytopathol 41:4
Wahle KW, Brown I, Rotondo D, Heys SD (2010) Plant phenolics in the prevention and treatment of cancer. Adv Exp Med Biol 698:36–35
Walker T, Bias H, Grotewold E, Vivanco J (2003) Root exudation and rhizosphere biology. Plant Physiol 132:44–51
Wallis CM, Galarneau ERA (2020) Phenolic compound induction in plant-microbe and plant-insect interactions: meta-analysis. Front Plant Sci 11
War AR, Paulraj MG, Ahmad T, Buhroo AA, Hussain B, Ignacimuthu SI, Sharma HC (2012) Mechanisms of plant defense against insect herbivores. Plant Signal Behav 7:1306–1320
Weidner S, Karolak M, Karamać M, Amarowicz R (2009) Phenolic compounds and properties of antioxidants in grapevine roots (Vitis vinifera) under drought stress followed by recovery. Acta Soc Bot Pol 78(2):97–103
Weir TL, Park SW, Vivanco JM (2004) Biochemical and physiological mechanisms mediated by allelochemicals. Curr Opin Plant Biol 7:472–479
Whitehead DC, Dibb H, Hartley RD (1983) Bound phenolic compounds in water extract of soils, plant roots and leaf litter. Soil Biol Biochem 15:133–136
Whiting DA (2001) Natural phenolic compounds 1900–2000: a bird's eye view of a centuries chemistry. Nat Prod Rep 18:583–606
Xu C, Zhang Y, Zhu L, Huang Y, Lu J (2011) Influence of growing season on phenolic compounds and antioxidant properties of grape berries from vines grown in subtropical climate. J Agric Food Chem 59:1078–1086
Xuan TD, Shinkichi T, Khanh TD, Chung IM (2005) Biological control of weeds and plant pathogens in paddy rice by exploiting plant allelopathy: an overview. Crop Protect 24(3): 197–206
Yan L, Zhu J, Zhao X, Shi J, Jiang C, Shao D (2019) Beneficial effects of endophytic fungi colonization on plants. Appl Microbiol Biotechnol 103(8):3327–3340
Yao K, De Luca V, Brisson N (1995) Creation of a metabolic sink for tryptophan alters the phenylpropanoid pathway and the susceptibility of potato to phytophthora infestans. Plant Cell 7(11):1787–1799

Yu K, Pieterse CMJ, Bakker P, Berendsen RL (2019) Beneficial microbes going underground of root immunity. Plant Cell Environ 42:2860–2870

Zabalza A, Orcaray L, Fernández-Escalada M et al (2017) The pattern of shikimate pathway and phenylpropanoids after inhibition by glyphosate or quinate feeding in pea roots. Pestic Biochem Physiol 141:96–102

Zaprometov MN, Nikolaeva TN (2003) Chloroplasts isolated from kidney bean leaves are vapable of phenolic compound biosynthesis. Russ J Plant Physiol 50(5):623–626

Zawoznik MS, Garrido LM, del Pero Martinez MA, Tomaro ML (2000) Effect of vanillin on growth and symbiotic ability of Bradyrhizobium sp. (Arachis) strain. Proc Int plant growth promoting rhizobacteria. http://www.ag.auburn.edu/argentina

Zhu HH, Yao Q (2004) Localized and systemic increase of phenols in tomato roots induced by glomus versiforme inhibits ralstonia solanacearum. J Phytopathol 152:537–542

Zipfel C (2008) Pattern-recognition receptors in plant innate immunity. Curr Opin Immunol 20:10–16

Zuanazzi JAS, Clergeot PH, Quirion J-C, Husson HP, Kondorosi P, Ratet P (1998) Production of Sinorhizobium meliloti nod gene activation and repressor flavonoids from Medicago sativa roots. Mol Plant Microbe Interact 11:784–794

Zwetsloot MJ, Kessler A, Bauerle TL (2018) Phenolic root exudate and tissue compounds vary widely among temperate forest tree species and have contrasting effects on soil microbial respiration. New Phytol 218:530–541

Role of Phenolics in Establishing Mycorrhizal Association in Plants for Management of Biotic Stress

2

Rafiq Lone, Gulshan Mushtaq, Nowsheen Hassan, Nazir Ahmad Malla, Gulab Khan Rohella, and Salim Khan

Abstract

Phenolic compounds are the most significant secondary metabolites produced by plants for the defense. Arbuscular mycorrhiza fungi (AMF), obligate symbionts, are the prominent one with an expanded host range and have an important role in designing ecosystems and associated productivity. Nearly up to 70% of the vascular plants are capable to form symbiotic association with AMF. AMF are primarily dependent on the host plant for photosynthates but offer much more benefit in return for the well-being of the host. Notably, they are able to modulate the tolerance of the host plant against various types of biotic stresses like fungi, bacterial, viral, phytopathogens, herbivores and nematodes. To protect themselves from the stress, plants have modified themselves with different sensory systems which can detect biotic invasion and combat the harm it causes to growth, productivity and survival. The establishment of AMF with the plants starts with the recognition of signal molecules or mostly phenolics. Among

R. Lone (✉) · G. Mushtaq
Department of Botany, Central University of Kashmir, Ganderbal, Jammu and Kashmir, India

N. Hassan
Department of Environmental Sciences, SKUAST, Kashmir, Jammu and Kashmir, India

N. A. Malla
Department of Botany, Government Degree College of Women's, Anantnag, Jammu and Kashmir, India

G. K. Rohella
Biotechnology Section, Moriculture Division, Central Sericultural Research & Training Institute, Central Silk Board, Ministry of Textiles, Government of India, Pampore, Jammu and Kashmir, India

S. Khan
Department of Botany and Microbiology, King Saud University, Riyadh, Saudi Arabia

R. Lone et al. (eds.), *Plant Phenolics in Biotic Stress Management*,
https://doi.org/10.1007/978-981-99-3334-1_2

phenolics, flavonoids are the abundant compounds which are able to accelerate the development of AMF at micromolar concentrations. In addition, strigolactones molecules are also responsible for the germination of spore and growth of hyphae in fungi. The increase in phenolics compound concentrations is effective in inducing enhanced resistance against these biotic stress agents.

Keywords

Phenolics · Mycorrhizae · Resistance · Signaling molecules · Biotic stress

2.1 Introduction

Plants are constantly exposed to different types of abiotic and biotic stresses during their growth and developmental phases. Among the various stresses, to some extent, plants can able to handle the abiotic stresses by making several changes at physiological, molecular and biochemical levels (Lamers et al. 2020). But, most of the plant species are sensitive toward biotic stress, and due to which, the net yield of commercial crops is affected by the attack of pathogens, pests, herbivorous animals and competition in nutrient uptake from the competitive weeds (Cheng et al. 2012). When a plant experiences stress, a number of molecular and cellular processes are activated (Rejeb et al. 2014; Lamers et al. 2020), which in turn triggers a cross-wired mesh of physiological, morphological and biochemical variations (Cheng et al. 2012; Nejat and Mantri 2017; Saijo and Loo 2020).

To protect themselves from the stress, plants have modified themselves with different sensory systems which can detect biotic invasion and combat the harm it causes to growth, productivity and survival (Rizhsky et al. 2004; Lamers et al. 2020). As a result, plants have developed an abundance of defense mechanisms to protect themselves from an array of pathogens and pests such as viruses, nematodes, bacteria, fungus and herbivorous insects (Hammond-Kosack and Jones 2000). To counteract the detrimental effect on their survival, plants often achieve a balance between their reaction and biotic stress (Peck and Mittler 2020). The molecular processes underlying plant defense responses have been extensively studied (Wang et al. 2019).

Mycorrhizal symbiosis begins with the two partners recognizing each other's signal molecules. The establishment of the relationship of arbuscular mycorrhizal fungi (AMF) with plants starts with the recognition of signal molecules released by the plant and uniquely recognized by the fungus. These substances, namely flavonoids, are secondary metabolites. They can accelerate AMF development in vitro at micromolar concentrations and are found in elevated amounts in the exudates and roots of plants that are phosphate-deficient. Furthermore, mycorrhization can be induced by flavonoids, and mycorrhized roots have altered flavonoid contents. Flavonoids are one class of signal molecules, but they are not the only ones. Strigolactones can also cause the spores to germinate and the hyphae of fungi to proliferate (Akiyama et al. 2005). The fungus significantly alters the metabolism of the plant once the mycorrhizal symbiosis is established. Plants and AMF interact in a way that alters more than only the production of carotenoid derivatives. The presence of the fungus also affects other chemicals, especially phenolics. By examining the differences in secondary metabolite biosynthesis seen in mycorrhized plants, authors have attempted to comprehend why the plant alters the synthesis of particular metabolites. Phenolics were given particular attention among the several classes affected by these alterations. Actually, the presence of AMF causes tannins to rise. They are well known for being involved in a variety of defense mechanisms against microbes. One characteristic that can support plant resistance is the tannins' ability to precipitate the enzymes released by necrotrophic phytopathogenic fungi. The sense of stress can sometimes trigger the manufacture of tannins, which is then regulated by signaling systems involving ethylene or jasmonate. Cinnamic acids may play a role in the production of lignin, among other possibilities. The thickening of the plant wall as a result of the increased cinnamic acid production may prevent infection by several cellulolytic pathogens. A biotic interaction can lead to significant changes in produced metabolites because of the reactivity of secondary metabolites and their rapid turnover. These are beginning to interact thanks to systems that allow diverse partners to recognize these signaling molecules, enabling a chemical conversation.

2.2 Biotic Stress

Biotic stressors are brought on by different type of living things, including insects, bacteria, weeds and fungi (Lattanzio et al. 2006). They seriously hamper the development of plants. Fungi pose a more serious harm than other biotic organisms among those present (Waller et al. 2005). About 8100 fungal species, which is a staggering number, have been found to be plant diseases (Tarkowski and Vereecke 2014). Similar to how fungi harm crops more than viruses do worldwide, viruses are likewise regarded as hazardous pathogens as reported by Bai et al. (2002). In 2004, Anderson et al. reported that these microbes can create a variety of signs in plants, comprising seed damage, stains on leaves, wilting and root rot. They can harm every part of the plant, including the stem, leaves, bark and blossoms (Walling 2008), and they are also a source of bacterial and viral transmission from infected to noninfected plants as reported by Mann et al. (2012). Wild, self-replicating plant species known as weeds either directly harm other plants or indirectly prevent them from growing by increasing competition for space and nutrients (Dass et al. 2016; Dorn et al. 2013). Weeds grow quickly, and because of their great production, they can occasionally take over the area before the necessary crop plants can (Dass et al. 2016). Numerous variables, including causing species, environmental circumstances and corresponding levels of crops and causal creatures, affect the intensity of these effects and the subsequent loss of crops (Waller et al. 2005). Plant diseases are mostly brought by pathogens such as nematodes, viruses, bacteria and fungi. Fungi and bacteria can affect many plant organs and create symptoms like vascular wilts, leaf spots and cankers. Nematodes consume plant cell material and attack every part

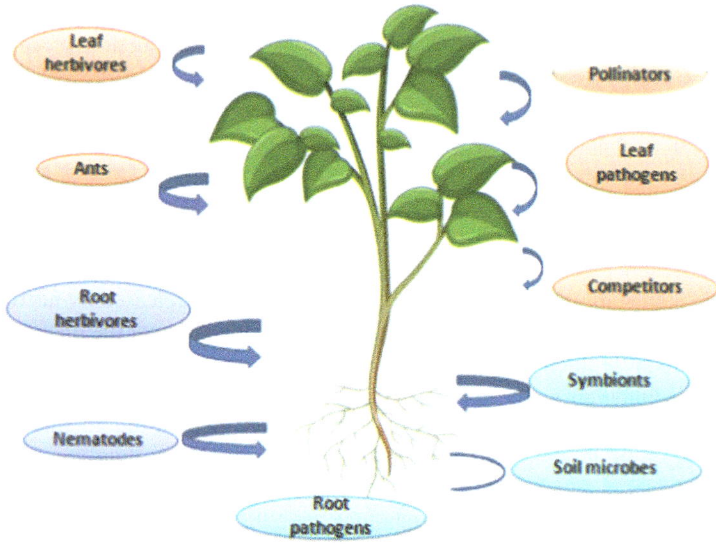

Fig. 2.1 Almost each plant part is evolved to coexist with its environment and surrounding organisms. (After Van-Dam 2009)

of the plant. They can also make it easier for soil-borne pathogens to enter the roots, which results in symptoms of nutrient deficiency like stunting or wilting. Even though their hosts are frequently not killed, viruses often result in chlorosis, deformities and stunting in different plant parts in addition to local lesions (Agrios 2005). On the other side, emphasis needs to be placed on insects and mites. Plants are harmed by them when they lay eggs or feed. Through their own stylets, piercing-sucking insects can act as virus vectors and transfer viruses to plants (Schumann and D'Arcy 2010). The control of environmental stresses is crucial for increased plant productivity since both abiotic and biotic factors have an adverse influence on crop development and growth (Fig. 2.1).

2.2.1 Bacterial Infection

Saprotrophic bacteria, which make up the majority of plant pathogenic bacteria, rarely harm the plant itself. The ability to harm plants is shared by roughly 100 identified bacterial species (Purcell and Hopkins 1996). The prevalence of bacterial illness worldwide is higher in tropical as well as subtropical regions (Ashbolt 2004). In 2012, Mansfield et al. reported that most plants are contaminated by rod-shaped bacilli. These bacteria are able to spread disease across plants thanks to a few distinct pathogenicity factors. In this context, there are five widely recognized major types of factors. Effector proteins, enzymes that break down cell walls, exopolysaccharides, phytohormones and toxins, are some of these components (Gagne-Bourgue et al. 2013). Changes in auxin levels caused by some *Agrobacterium* species result in the development of tumors (Lee et al. 2009). Similar to this, bacteria produce exopolysaccharides that often lead to plant death by blocking xylem arteries (Thorne et al. 2006)

2.2.2 Viral Infection

Viruses pose one of the greatest risks to agricultural productivity among the different biotic stressors. Plant viruses can be divided into *Reoviridae* and *Geminiviridae*. According to Reddy et al. (2009), they are tiny genome viruses, and Fereres and Moreno (2009) reported that they enter fields through contaminated seeds or pests that carry the virus and are observed on infected plants. Because they offer a direct route for the viral genetic material to enter plant cells, damaged or injured plant parts are typically where viral infection spreads in plants. Here, genome duplication and subsequent propagation through the vascular tissues and symplastic pathway may pose a greater threat to plant development (Taliansky et al. 2008). By adopting transgenic crops, regulated plant material through crop trapping and expanded agricultural practices, we can limit viral infections (Prins et al. 2008).

2.2.3 Fungal Infection

Biotrophs and necrotrophs are two groups of fungi that are parasitic on plants and are based on how they live. Unlike biotrophs which severely harm plant tissues before turning them into food source for themselves, biotrophs obtain their nutrition straight from the host's live tissue (Campe et al. 2014). Plant disease causing fungi, also referred as hemibiotrophs, that can adopt either mechanism, depends on the environmental circumstances (Mang et al. 2009). Fungi initially serve as biotrophs throughout the infection process, but according to earlier findings, they were thought of as necrotrophs.

2.2.4 Herbivory: Insects and Weeds

Numerous nematodes severely harm a variety of plants, including citrus. The most well-known nematode is *Tylenchulus semipenetrans*, sometimes known as the citrus nematode, and it may be found practically everywhere (Duncan 2005). *T. semipenetrans* is the root cause of the citrus slow decline disease (Safdar et al. 2013). The worm *Radopholus similis* also causes "vast dispersion decline disease," which is a citrus disease (Kaplan and Opperman 2000). Root-knot, lesion, sting and stubby root are a few additional nematode-caused illnesses that contribute to the devastation of citrus (Etebu and Nwauzoma 2014).

The root-knot nematode is the culprit behind the sickness that it causes (Ramamoorthy et al. 2001). Lesion, reniform, lance and dagger nematodes are some more harmful nematodes that attack cotton (Noe 2004). Potato-rot, root-knot and lesion nematodes are further nematodes that contribute to the devastation of potatoes (Balasubramanian et al. 2009). Lesions are produced on tubers and roots by root-lesion nematodes (George et al. 2016). Numerous nematodes harm rice plants' leaves, panicles and stems, among other components of the plant (Bridge 1996). Cyst nematode, a soybean pest, is among the most destructive nematodes (Matthews et al. 2013). Soybeans are known to be harmed by about 50 different types of nematodes. Most well-known among them are root-knot nematodes and cyst (Vuong et al. 2015). Vegetable output in tropical and subtropical regions has been significantly diminished as a result of root-knot nematodes, which are extremely harmful to vegetables (Lamovsek et al. 2013). Others are well-known vegetable pests including sting, reniform, stubby root and cyst (Grabau et al. 2017). Wheat and barley are harmed by lesion, root-knot, seed gall and cereal cyst nematodes (Nicol et al. 2011).

The "ear-cockle" illness caused by seed gall worm is well known among cereal worms (McDonald and Linde 2002). An economically significant parasite known as wheat gall nematode spreads to many areas of the world where wheat is produced through infected seeds (Nicol and Rivoal 2008). Reduced food availability and metabolic activity result in stressed growth and yield, which is one of the most detrimental effects of nematodes (Kerry 2000).

2.2.5 Weeds

Weeds are able to thrive in a variety of ecological niches where other plants cannot because of their multiple unique growth habits and adaptations. The competition between weeds and agricultural plants is not just between them (Ramesh et al. 2017). Among the major significant adaptations associated with competitive advantage are precisely rapid establishment, coordinated germination and seedling growth, resistance to shadowing outcomes through beneficial plants or by another weeds for quick progression, quick response to earlier existing moisture of soil and nutrient, adjustment to the severe climatic circumstances of the natural environment, modifications toward edaphic mechanisms, comparative resistance to postseeding soil condition, and agriculture strategy (Sardana et al. 2017).

Weeds as well as crops competed fiercely for the limited resources available during the early phases of weed invasion in ecological niches. Even when the connection between crops and weeds has been established, resource competition is more pronounced (Burger et al. 2015). Crop along with weed plants contribute to attain the highest linked development and yield due to the natural phenomena known as plant competition (Benaragama et al. 2016).

2.3 Plants' Resistance to Biotic Stress

Plants frequently have a relatively sophisticated immune system to fight biotic stresses. First, physical obstacles on plants, such as unique trichomes, thick cuticles and waxes, avoid insects or pathogens from attaching to them. In order to defend themselves against pathogens and herbivores, plants also produce chemical compounds (Zaynab et al. 2018). Two mechanisms that trigger defense reactions are also used by plants to sense pathogens. The first, motif recognition receptor, that picks up on pathogen-associated molecular patterns (PAMPs) like fungal chitin, peptidoglycans, quorum sensing, flagellin and bacterial lipopolysaccharides. PAMP-triggered immunity (PTI) (Monaghan and Zipfel 2012) is the term for this fundamental sort of defense. Plant resistance R proteins, the secondary immune system, are able to identify the specific effectors (Avr proteins, i.e., Avirulence proteins) of pathogens or pests and activate the plant susceptible response, which is termed as Effector-Triggered Immunity (ETI). This type of mechanism can cause hypersensitivity reactions (HR), that can trigger necrosis in the affected cells as well as their surroundings (Spoel and Dong 2012). Several plant hormones stand out in the signaling pathways activated by ETI (Effector-Triggered Immunity) and PTI (Pattern-Triggered Immunity): Salicylic acid (SA), Ethylene (ET) and Jasmonic acid (JA). *Arabidopsis* frequently induces the JA and ET pathways in reaction to necrotrophic infections and chewing insects, and however, the SA mechanism increases constant protection against biotrophic and hemi-biotrophic pathogens (De Vleesschauwer et al. 2014). Three hormones, namely ET, SA and JA, typically serve as signaling molecules for two powerful defense mechanisms against the diseases of plant.

The first, referred as systemic acquired resistance (SAR), is stimulated following primary infection along with a malignant pathogen as well as associated by coupled pathogenesis proteins increasing amounts of SA (Grant and Lamb 2006). Another class of innate plant-based resistance is ISR-induced systemic resistance, and it is a type of plant resistance that is produced by particular strains of nonpathogenic root-colonizing bacteria and whose signaling calls for ET (Ethylene) and SA (Salicylic Acid) (Loon et al. 1998). By recognizing preserved herbivore-linked elicitors of the conquering insect, phytophagic insects compel plants to expire volatiles in order to draw their enemies and alert their neighbor plants of hovering threats (Santamaria et al. 2013).

Numerous microorganisms, including bacteria, fungus and insect herbivores, pose a risk to plants and can cause damage (Ramegowda and Senthil-Kumar 2015). In 1997, Durner et al. highlighted that many of the defense mechanisms that plants have developed include those that are produced in response to pathogen attack. Plants can detect microbial attacks and can identify pathogen-linked patterns have through their extracellular sensors present on their membranes (Chisholm et al. 2006) and sources of contamination in them before the detrimental bacteria infect the entire plant (Boller and He 2009). However, pathogens triggered by pathogen-associated molecular pattern (PAMP) have created a way to suppress immunity by secreting certain proteins into the plant cell's cytoplasm that alter signals for plant resistance and the development of defensive mechanisms (Chisholm et al. 2006).

Plants acquired a more definite system to recognize germs, known as effector triggered immunity (ETI), once diseases were able to overcome the primary defenses (Bilgin et al. 2010). Resistance genes (R) expression in the ETI situation typically identifies the receptor molecules initiated by the microorganism that conquers the host cell. Plants underwent alteration as a result of the effector and receptor's contact, which could lead to resistance (Miedes et al. 2015). The second type of technique of defense, a type of cell death intended to prevent the growth of pathogens from nutrients and water, is dependent on the oxidative burst brought on by the resistance by expression of R genes (Sato et al. 2010).

2.4 Plant's Reaction to These Organisms

2.4.1 Viral Infections and Plant Responses

Viruses harm agriculture plants and reduce yields as diseases (Anderson et al. 2004). Because viruses have a small genome and the ability to spread disease, they must come into contact with plants. Plant proteins and viral proteins work together to accomplish the life cycle of virus. The resistance and molecular processes in viral illnesses are still mostly unknown (Kang et al. 2005). As a host to numerous viruses, Arabidopsis as well as Cherry leaf roll virus of cherry, it is confirmed that the virus is spreading by through infected seed (Goodin et al. 2008; Rumbou et al. 2009), similarly *Nicotiana benthamiana* is recommended as a representative host to review numerous phytophages. The primary host used to investigate how plant viruses interact with eukaryotic nuclear systems is yeast (Nagy 2008). Proteomics studies the relationship between plants and viruses via subcellular separation (Haynes and Roberts 2007).

Vector control, the utilization of plant-made substances that are always virus free and a reduction in developing strategies are all examples of genetic resistance (Wu et al. 2010). A majority of resistance toward viruses are monogenic, while the remaining are inherited quantitatively (Kang et al. 2005). In several cases a different pedigree identified on host protection, discovered that several plants with a wide range of pathogens experience necrotic type of programmed cell death, which reduces susceptibility to the subsequent attack of pathogen (Van-Doorn et al. 2011; Durrant and Dong 2004).

2.5 Responses of Plants to Fungal Infection

Typically, plants communicate resistance to biotrophs utilizing salicylic acid (SA) to demonstrate tolerance against pathogenic assault. On the other hand, ethylene (ET) and jasmonic acid (JA) signaling pathways play a pivotal part in resistance to necrotrophs (Caarls et al. 2015). Resistance of gene-for-gene to the *Blumeria* fungus is visible in barley (Schulze-Lefert and Vogel 2000). Hence, during biotrophic fungus attack, gene resistance and SA-dependent protective strategies are both in use (Thomma et al. 2001). It has been discovered that the activation of the R gene, that is employed in the manufacture of plant defense polypeptides, can help plants defend themselves against fungi (Collins et al. 2003). The indicated gene activation plays a pivotal part in the natural defense mechanism of plant actively contributing to the defense system's enhancement to create resistance (Lee et al. 2009). Fungal infection may result in the regulation of numerous transcriptional factors, according to research on synthesis of RNA of various host–pest systems of fungi (Pande et al. 2005). The above-mentioned genes belonged to several classes, including those that control stress, signaling, cell metabolic activities and a number of unidentified processes (Jaiswal et al. 2014).

Chitin serves as a membrane identifier in the cell walls of fungi, and as such, it is essential for producing defense mechanisms against many plant pests (Petutschnig et al. 2010). Because of the responses displayed during the defense mechanism, stress signaling of plant leaves revealed such chitin possesses a challenging impact on various hereditary variables (Povero et al. 2011). Additionally, it was discovered that a few genes function as an essential mediator in the stimulation of the expressing genes related to SA and JA signaling (Birkenbihl and Somssich 2011). Similar to JA and SA signaling, WRKY70 (Zinc-finger type), a transcription factor in plants, induces SA and JA signaling (Li et al. 2004). Following the discovery of fungus PAMPs, plants initiate the mitogen-activated protein kinase (MAPK) signaling cascade, which ultimately results in tryptophan production and camalexin collection (Pitzschke et al. 2009). Such enzymes are involved in the manufacture of free radicals of oxygen-unbound nature, signaling governed by its formation and cue transduction by abundant types in view of auxin and are examples of proteins that are typically generated (Sharma et al. 2007). Hence, research on the function of proteins is useful for understanding how hazardous fungus reproduces and invades new areas (Fernandes et al. 2014).

2.6 Responses of Plants to Herbivory

The aligned evolutionary histories of plants and insects demonstrated numerous strategies for mutual development of tolerance mechanisms (Verhage et al. 2010). Due to this link, plants were able to evolve strong defense mechanisms that were effective enough to detect impulses from tissues that had been harmed by insect attacks (Hare 2011). Plants create specialized body parts, metabolic processes and proteins to ward off infection by creating effects that are repellent, dangerous and toxic for insects (War et al. 2011). By altering their reproductive activity, plants defend themselves against pests directly (direct defense) and indirectly (indirect defense) by luring other creatures that are pest foes or predators (Dudareva et al. 2006). Plant attributes that irritate insects, like plant surface defense through the development of outgrowths (such as thorns, hairs, spines and trichomes), or the production of noxious substances like phenols, terpenoids and anthocyanins, which hamper the development of pests, are examples of direct defenses (Sharma and Dietz 2009). Secondary protection includes luring the predators of pests with lodging and food (Arimura et al. 2009).

However, plants have created resistant genotypes in the race against herbivores (Sharma et al. 2009). The complex and active defense mechanism used by plants in response to herbivore wounds includes physical barriers and harmful substances (Karban 2011). Infected for the management of pests, plant response is an important method for the limitation of insects in crops (Sharma et al. 2009). The above mentioned are crucial if they are created to respond to the pests infection and are meant to reduce common pressures (Miranda et al. 2007). Dissimilarities in a plant's defense mechanism in opposition to pest attack enable the plant to manufacture some poisons against insects; however, they also interfere with the strength and function of arthropod pests (War et al. 2011). If a produced reaction occurs immediately, it is extremely beneficial to the crops and reduces subsequent herbivore attacks while also improving the overall appropriateness of valued crops (Agrawal 2011). The emergence of insect plant communication has rectified our understanding of how plants evolved defense mechanism against herbivory (Karban 2010).

2.7 AMF (Arbuscular Mycorrhizal Fungi)

2.7.1 General Preface

Arbuscular mycorrhizal fungi and soil microorganisms collaborate with 80% to 90% of tracheophytes along with 90% of cultivable plants, including most cultivable crops, especially horticultural plants, vegetables and cereals (Smith and Read 2010). They are widely distributed in world ecosystems, which are essentially determined by the distribution of recognized plant hosts worldwide (Kivlin et al. 2011; Wang and Qiu 2006). AMF (arbuscular mycorrhiza fungi) are categorized as pertaining to the *Glomeromycota* phylum, including three classes (*Paraglomeromycetes*, *Archaeosporomycetes* and *Glomeromycetes*), as well as the subkingdom *Mucoromyceta* (Tedersoo et al. 2018). AMF comprise about 250 species, 11 families and 25 genera (Schüβler et al. 2001; Spatafora et al. 2016). *Glomeromycota* are obligatory symbionts that are likely 20% dependent on plant-established carbon from their hosts to thrive (Siddiqui and Pichtel 2008; Johns 2014). This symbiosis predates the earliest emergence of terrestrial plants by 400–450 million years, according to fossil records and molecular evidence (Smith and Read 2010). The symbiosis of AM fungi is arguably the extremely extensive favorable connection in between microbes and plants (Parniske 2008). A number of considerations have indicated that they perform an analytical part in plant development and nutrition under challenged settings and improve a tally of vital ecosystem processes (Siddiqui and Pichtel 2008; Nakmee et al. 2016).

Arbuscular mycorrhizal fungi assist host plants to have a normal growth phase under stressful conditions. Numerous studies discuss how fungal symbiosis increases plant resilience to an array of stressors, including as herbivory, salinity, drought, diseases, metals and temperature (Salam et al. 2017; Ahanger et al. 2014; Rodriguez et al. 2008). In 2010, Zhu et al. and, in 2014, Ahanger et al. highlighted that about 90% of plant division, comprising ferns, flowering plants and bryophytes, can form harmonious association with AMF. Moreover, hyphae and spores hyphae within rhizosphere, AMF give rise to hyphae, vesicles and arbuscules in roots. The ability of roots to access a vast area of soil surface is considerably improved by the development of a hyphal mesh through the AMF with the roots of plant, leading to improved plant development (Bowles et al. 2016). By increasing the transport and accessibility of distinct nutrients, AMF strengthen plant nutrition (Rouphael et al. 2015). Arbuscular mycorrhiza fungi (AMF) enhance soil quality by affecting the texture and structure of the soil, which benefits plant health (Zou et al. 2016; Thirkell et al. 2017). Organic materials in soil can decompose more quickly thanks to fungus hyphae (Paterson et al. 2016). Additionally, by intensifying the "sink effect" and working photo-assimilates from the ethereal sections to the roots, mycorrhizal fungi may affect how well host plants fix atmospheric carbon dioxide.

Given the significance of AMF and developments in research pertaining to implementation in agriculture sector, the current review emphasizes the role of AMF in reducing different types of biotic stresses caused by various living organisms for the synchronization of plant for nutrient uptake and maximum yield during unfavorable circumstances, as well as the degree to what end AMF could amplify the development of plant.

2.7.2 AMF's Part in Reducing Different Types of Biotic Stress

2.7.2.1 AMF's Effect on Soil-Borne Pathogens

The harm caused by soil-borne diseases is generally considered to be lessened by AM symbioses. Numerous research studies have shown the decrease in the frequency of diseases like wilting or root rot caused by a variety of bacteria, fungi and oomycetes, including *Macrophomina*, *Fusarium*, *Verticillium*, *Rhizoctonia*, *Aphanomyces*, *Phytophthora* and *Pythium*. A comprehensive analysis of the investigations was published by Whipps (2004). Similar to this, parasitic nematodes like *Pratylenchus* and *Meloidogyne* have been demonstrated to mitigate the detrimental effects on inoculated plants (de la Peña et al. 2006; Pinochet et al. 1996; Li et al. 2006). By proving defense against additional soil pathogens like *Armillaria melea* in grapevine, later investigations broadened the pathosystems for which AM symbiotic associations could have beneficial impact (Nogales et al. 2009), and the effectiveness against such a wide range of enemies lends credence by connecting AM symbiosis to the induced resistance broad-spectrum nature. Comparative investigations among many fungal strains or cultures have demonstrated that the AMF in question has a significant influence on the degree of protection (Kobra et al. 2009). Contrary to common perception, many researches show that *Glomus mosseae* has a better protective effect compared to the other two AM fungi (Utkhede 2006; Pozo et al. 2002; Ozgonen and Erkilic 2007). The evolved resistance's systemic nature suggested that phytoresponses were involved (Elsen et al. 2008).

2.7.2.2 Effects of Symbiosis of AM Fungi on Root Parasitic Plants

Striga and *Orobanche* plant species are among the most destructive agricultural pests because they parasitize a variety of hosts worldwide. These parasitic obligate organisms connect to such root system of numerous plants and support in uptake of nutrients and water via their donor (Bouwmeester et al. 2003). AMF inoculation considerably decreased the number of pathogenic plants such as hemiparasite *Striga hermonthica* for maize as well as sorghum cultivars. Mycorrhizae have thus been suggested for the effective control of pathogenic weed species (López-Ráez et al. 2009; Lendzemo et al. 2005). Strigolactones induce the germination of spores of root pathogenic plants (Bouwmeester et al. 2003). Strigolactones identification as host recognition markers to AM fungi of rhizosphere allowed for the establishment of a contributory relationship between AM fungi and its effects on parasitic plants (Akiyama et al. 2005). In fact, another research conducted under controlled circumstances has shown that AMF inoculation reduces the impact of *Striga*, which is reportedly associated to a decrease in strigolactone biosynthesis (Lendzemo et al. 2007). The germination of *Orobanche ramosa* seeds is also inhibited by isolates in tomatoes inhabited with *G. mosseae* when compared to extracts of nonmycorrhizal species; additionally, a tomato mutant's decreased synthesis of strigolactones was associated with a decreased hypersensitivity against *Orobanche* (López-Ráez et al. 2008). By the above literature, it is certain that such a decline in synthesis of strigolactone is associated with the decline in the prevalence of root parasite plants on mycorrhizal plants.

2.7.2.3 Impact of AM Symbioses on Plant-Eating Insects

Apparently, mycorrhizal state of such host species could also influence the activity of insect herbivores; however, the strength and direction of either the effect rely on the insect's manner of feeding and lifestyle (Koricheva et al. 2009; Hartley and Gange 2009). Various researches on mycorrhizal plant–insect interactions in lab and outdoor settings include a broad spectrum of topics. Hartley and Gange (2009) came to the general conclusion that mycorrhizae generally exhibit strong negative consequences on rhizophagous pests, whereas impacts upon shoot-feeding pests seem milder and thus more varied. Despite this diversity, several broad trends become apparent: whereas specialist insects frequently benefit from mycorrhizae, generalist insects are typically negatively impacted. Furthermore, leaf-chewing insects typically suffer as a result of the symbiosis, but aphids typically perform better on AM plants.

The detrimental impact against leaf-chewers may be probably due to their vulnerability toward jasmonate-dependent responses that are potentiated in mycorrhizal plants (Peña-Cortés et al. 2004). The act of predators, herbivores and parasitoids can also be impacted by AM. For example, in tomato, aromatic blend generated by either and plants tends to be more seductive toward aphid parasitoids in comparison to nonmycorrhizal species (Guerrieri et al. 2004).

2.7.3 Plant Diseases and AMF

Plants that have received mycorrhizal inoculation are more resistant to disease (Bagyaraj 1984; Graham 2001), particularly diseases brought on by soil-borne microbes (Reddy et al. 2006). Plants are assumed to be able to tolerate illnesses because of physiologic changes brought on by better nutritional condition, the generation of phenolic chemicals and anatomical alterations in the host plant (Morandi and Gianinazzi-Pearson 1986; Morandi et al. 1984; Zhang et al. 2013). Isoflavonoid and phytoalexins chemicals, which offer defense against diseases, were shown to be produced by AMF-associated plants, according to Morandi et al. (1984). The interaction between AMF and plants increases the production of plant growth regulators such as gibberellin-related compounds, indole-3-acetic acid and cytokinins (Smith and Read 1997). By serving as chemical signals as in beginning of the legume–rhizobia association, assisting in the establishment of symbiosis with arbuscular mycorrhizae but also possibly acting as mediators in defense response, phenolics have an important role in the growth and defense of plants (Santi et al. 2010). The mycorrhizal fungi race with diseases for colonization sites, enhance the plant's nutrition pool and raise the host's resistance to pathogen attack. Plants with mycorrhizal associations may have had increased resilience to the development of disease signs (Smith and Read 1997). AM fungi are widespread throughout most temperate biomes and are obligatory mutualistic symbionts to most of the plants (Smith and Read 2008).

2.7.4 AMF as Biocontrol Agents in Parasitic Nematodes

Parasitic nematodes of plants, which include endoparasites AMF, frequently coexist in the same region of host plant roots in the rhizosphere. AMF and nematodes have been linked in the past (Elsen et al. 2003; Hussey and Roncadori 1982; de la Peña et al. 2006); in earlier times, there has also been evidence of roundworm decrease, no influence or an increase in nematodes (Atilano et al. 1981; Siddiqui and Akhtar 2007; Hasan and Jain 1987; Sankaranarayanan and Sundarababu 1994; Kantharaju et al. 2005; John and Bai 2004). In 2008, Elsen et al. reported that in many plant/nematode systems, studies have shown that AM fungi increase host resistance or tolerance in various nematode/plant systems, and in the plant roots, it induces the systemic resistance against parasitic nematodes (Elsen et al. 2008).

The *Nacobbus aberrans*, a plant parasitic nematode, generates gall formation in the root system of plants and severely damages a variety of crops (Marro et al. 2014). It is difficult to handle nematodes in this group since few of them have preferences for particular hosts or exhibit other behaviors. The use of arbuscular mycorrhizal fungus was suggested by the authors as a potential biological control method to lessen the harm created by this species (AMF). AMF application encouraged biomass of tomato and reduced the amount of galls in plants transplanted with nematode inoculums. In 2007, Shreenivasa et al. investigated the association between the mycorrhizal fungus *Glomus fasciculatum* and the root-knot nematode

Meloidogyne incognita and how this affected the tomato plant. Reduced root-knot index, nematode population and number of galls and increased plant biomass, phosphorous uptake and productivity were observed in tomato plants inoculated with *G. fasciculatum* alone in comparison to nematode inoculation. Arbuscular mycorrhizal (AM) fungus has been used to biocontrol nematodes, and Hajra et al. (2013) found substantial differences in a number of parameters. The mycorrhized plants outgrew nonmycorrhizal plants with regard to leaf area and plant's height, while mycorrhizal plants showed a dramatic decline in nematode-contaminated plants caused by xylem vessel damage. Nematodes are broad class of parasites that include both plant and animal parasites and free-living nematodes that can live in a variety of settings worldwide (Ferraz and Brown 2002). Several agricultural crops of economically significance can become infested by numerous species of parasitic nematodes.

Biological control is a more sustainable method of managing plant parasitic nematodes (PPN), particularly AMF (Bajaj et al. 2015, 2017). Plant stress brought on by biotic and abiotic factors, such as PPN, may be lessened by AMF (Vos et al. 2012a; Singh et al. 2011; Gianinazzi et al. 2010). Necrotrophic and biotrophic pathogens are inhibited either directly or indirectly by AMF (Veresoglou and Rillig 2012). *Strigam hermonthica*, an obligate root parasitic plant species, showed significantly reduced seed germination in response to AMF, suggesting that it may be used as a biocontrol agent for obligate root parasitic weeds (Gworgwor and Weber 2003; Lendzemo et al. 2005, 2006; Hearne 2009; Louarn et al. 2012). Plants inoculated with AM fungus prevented the development of haustoria in *Pedicularis tricolor*, a facultative root hemi-parasitic plant (Li et al. 2012). After AM inoculation, hemiparasite growth was suppressed, indicating that AMF may be useful in treatment of *Pedicularis* (Li et al. 2013).

In order to combat *Striga hermonthica*, the effects of soil-borne biocontrol agent (BCA) *Fusarium oxysporum* on AMF fungi were examined in rhizospheric sandy and clayey soil of maize. Together, arbuscular mycorrhizal fungus and endoparasitic nematodes during their interaction occupy the similar region of the host plant's rhizosphere. Members of microbial communities, such as AM fungus and root knots of plants, compete with one another for the similar place in the rhizosphere. Thus, to control root-knot nematode infection, AMF can be employed as biocontrol agents. Higher plants' roots developed systemic resistance to plant-parasitic nematodes, which enhanced host nutrition. It has been shown that antagonistic behavior results from or may be caused by enhanced host nutrition (Youssef and El-Nagdi 2015). The most prevalent and widespread nonmycorrhizal plants that stimulate plant development by improving nutrient uptake and increasing growth hormones are those that are arbuscular and ectomycorrhizal.

AMF can be employed as both a biofertilizer and a biocontrol agent since it increases plants' resilience to plant diseases and the extent of their root systems, which allows them to better absorb nutriments from the soil (Goicoechea et al. 2010). Upon root colonization by AMF, the host plant undergoes significant physiological changes that have an influence on interactions with a diversity of species. Different plant species, including kinds of crucial crops for agriculture, have been shown to benefit from the symbiosis' defenses against infections, pests and parasitic plants (Jung et al. 2012).

2.7.5 Mechanism of AMF-Induced Relief from Biotic Stress

Arbuscular mycorrhizal fungi can reduce biotic stresses through a variety of methods, including direct connections such as opposition with the attacker and indirect, plant-moderated impacts. Challenging for niches either particular infection sites, as well as competition for carbon, nitrogen and other growth nutrients are direct impacts. In 2003, Filion et al. reported that the immediate competition has been proposed as a way through which arbuscular mycorrhizal fungi can lessen the number of harmful fungus in root system. It's likely that pathogenic and AM fungi use the same root resources, such as infection/colonization sites, photosynthates and space (Whipps 2004).

The relevance of microsymbionts mostly in the control of biotic challenging is emerging. AMF minimizes the harm caused by several plant diseases, according to numerous studies (Nguvo and Gao 2019; Kumari and Prabina 2019; Vos et al. 2012b). It has been revealed that AM treatment can minimize the severity of charcoal root-rot diseases in soybean (Spagnoletti et al. 2020). Arbuscular colonization enhanced shoot dry weight in the presence of *Fusarium* (Vos et al. 2012b). Mycorrhiza-induced resistance (MIR) is a protective effect of AM colonization, which offers systemic defense against a variety of invaders and shares traits with ISR-induced systemic resistance after root colonization by nonpathogenic rhizobacteria and systemic acquired resistance (SAR) upon pathogen infection (Cameron et al. 2013). In roots and soil, pathogenic bacteria have been linked negatively to the number of AM fungal structures (St Arnaud and Elsen 2005; Filion et al. 2003). Additionally, evidence of complete exclusion of the infectious *Phytophthora*, an oomycete fungi from the cells of arbuscules, was found (Cordier et al. 1998). The architecture, morphology and root exudates of the root system are all altered as a result of AM fungus colonizing the roots (Pivato et al. 2008; Norman et al. 1996; Schellenbaum et al. 1991). Such modifications could affect the advances of pathogen infection or the mycorrhizosphere's microbial population, favoring microbiota elements that can combat root infections (Badri and Vivanco 2009; Barea et al. 2005). Root exudation changes have a direct effect on nematodes and microbial pathogens (Vos et al. 2011; Norman and Hooker 2000). According to the latest research (López-Ráez et al. 2010a, Campos-Soriano et al. 2012; Liu et al. 2007), the fundamental system of controlling of pathogen is the potential of AMF to change the induction of plant genes.

Investigators determined that the existence of AMF improves plants' manufacturing of enzymes of antioxidant nature enzymes and can act as a defense against other stresses and infections (Cameron et al. 2013).Various aspects, including altered root development and morphology, improved host nutrient status, altered root growth and morphology, competition for host photosynthates and colonization sites, as well as microbial alterations in the mycorrhizosphere, have been suggested as additional explanations for the reduced pathogen damage caused by AMF (Siddiqui and Mahmood 1998; Akhtar and Siddiqui 2007; Vos et al. 2012b). Because AM can assist in the restoring of tissues after pathogen outbreaks, the improvement in plant growth can have a favorable impact. Although when plant nutrition enhances, it also becomes more beneficial or captivating to herbivores, which could have adverse impacts. The colonization by *F. Mosseae* in tomato roots caused systemic resistance amid the *Pratylenchus penetrans*—a migratory nematode as well as the *Meloidogyne incognita*—a sedentary nematode (Cobb 1917; Kofoid and White 1919; Filipjev and Schuurmans Stekhoven 1941; Vos et al. 2012b). The existence of AM fungi has reduced nematode infection by about 45% and 87% for *M. incognita* and *P. penetrans*, respectively, in plants colonized by AMF in comparison to controls. In contrast to either nonmycorrhizal root exudates or water, the administration of root exudates from mycorrhized plants additionally impeded nematode perforation in mycorrhizal plants as well as briefly paralyzed nematodes. In tomato plants, inoculation of *Meloidogyne javanica* (Treub 1885) and *F. mosseae* has resulted in a decrease in nematode reproduction, galling and female morphometric characteristics (Siddiqui and Mahmood 1998).

In terms of infectious fungi, inoculated with mycorrhiza, *F. mosseae* remarkably reduced, respectively, the severity of *Alternaria solani* and *Fusarium oxysporum* (Schltdl., 1824; Jones & Grout 1896; Ellis & G. Martin) and caused tomato diseases (Song et al. 2015; El-Khallal 2007; Akhtar et al. 2010). This advantageous effect was amplified when plants were sprinkled with hormone inducers (Salicylic and jasmonic acids) and cultured with AMF, characterizing a cooperative and synergistic effect among them that increased the evocation and management of resistance to diseases (El-Khallal 2007). Similar advantages of AMF were also reported in pathogen-infected chickpea (Akhtar and Siddiqui 2007) and potato. There aren't many studies on how AMF affects herbivorous insects, in comparison with its well-known influence on pathogenic roundworms and fungus (Koricheva et al. 2009; Gange et al. 2003). In contrast to other fungal species investigated, certain AM fungi, like *R. intraradices*, shows a negative effect on performance of chewers (Koricheva et al. 2009).

In 2003, Gange et al. had demonstrated that mycorrhizal plants have decreased *Diglyphus isaea* (Walker, 1838) parasitism of *Chromatomyia syngenesiae*. For instance, mycorrhizal infection in *Plantago lanceolate* L. increased the leaf's resistance in *Arctia caja* (Linnaeus, 1758) against chewing insect, whereas *Mysus persicae* (Sulzer) performed better on mycorrhizal plants (Gange and West 1994). According to Gange et al. (2003), *Diglyphus isaea* (Walker, 1838) was less likely to parasitize *Chromatomyia syngenesiae* in mycorrhized plants, although in the lab, the influence of three AM fungal species on the rate of parasitism is diverse based on species type.

As a result, changes to the plant's main and secondary metabolism occur, with most of the modifications being connected to the plants defense (Schliemann et al. 2008; López-Ráez et al. 2010b; Hause et al. 2007). In fact, like other biotrophs, AM fungus can initiate early plant defensive responses (Paszkowski 2006). The fungus must therefore adapt to these responses and actively control plant defense responses in order for colonization to succeed. This modification could prime the tissues for effective stimulation of plants defense in response to attack by a challenger (Pozo and Azcón-Aguilar 2007). When a plant is "primed," it enters a state of alertness in which its defenses are not overtly expressed but where it responds to an attack more quickly and/or forcefully than other plants, effectively boosting plant resistance. Because priming increases plant fitness, defensive priming by AM has significant ecological significance (Walters and Heil 2007; Conrath et al. 2006; Jung et al. 2012).

2.8 The Signaling Involved

Sensitization of the tissue in response to appropriate stimulation can result in the expression of basal defense mechanisms more effectively in response to subsequent pathogen attack, which can lead to the initiation of resistance following root colonization through AMF (Pozo and Azcón-Aguilar 2007; Jung et al. 2012). When interacting with helpful microbes, plants frequently experience priming of their innate immune systems, which has significant fitness advantages over direct mobilization of defense of plant (Van Wees et al. 2008; Van Hulten et al. 2006; Conrath 2009). A number of defense-coupled regulatory agents, including chromatin changes, transcription factors and MAP kinases, have been suggested as mediating mechanisms for the establishment of the primed state (Pozo et al. 2008; Van der Ent et al. 2009; Beckers et al. 2009; Pastor et al. 2012). The development of MYC2, a transcription factor essential to the control of JA responses, is linked to rhizobacteria-activated systemic resistance in Arabidopsis (Pozo et al. 2008). The first root tissue samples showing primed defense responses in AM plants were noted. Carrot roots with Mycorrhizal transformation showed more robust defense reactions in *Fusarium*-affected areas (Benhamou et al. 1994). Systematic protection against *Phytophthora parasitica* infection was provided by AM colonization in tomato roots. By depositing nonesterified pectins and callose in around positions of pathogen infection, mycorrhizal plants were the only ones to form the papilla-like frameworks that prevented the pathogen from progressing further. Cordier et al. (1998) and Pozo et al. (1999, 2002) highlighted that they also deposited remarkably high PR-1a and basic -1,3-glucanases than nonmycorrhizal plants did in response to *Phytophthora* attack. Similar to this, when *Rhizoctonia* infection occurred, mycorrhizal potatoes had increased deposition of the phytoalexins *solavetivine* and *rishitin*, although AM fungi alone had no effect on these chemicals' concentrations (Yao et al. 2003). According to Jaiti et al. (2007), primed deposition of phenolics in AMF inoculated date palm trees confers resistance to *F. oxysporum* and aids in the induction of nematode resistance (Li et al. 2006; Hao et al. 2012). The initiated reaction, however, is not only limited to the root system, but also been demonstrated in the shoots of AM colonized plants (Pozo et al. 2010). As a result of AM symbiosis, the tomato plant has developed systemic resistance via the *Botrytis cinerea*, a necrotrophic foliar disease. Mycorrhizal plants had considerably less pathogen in their leaves, but those plants also had increased expression of several jasmonate-modulated, defense-coupled genes (Pozo et al. 2010).

Transcript profiling of leaves following exogenous jasmonic acid application indicated that JA-dependent defenses were primed since they were triggered earlier as well as to a greater extent in AM colonized plants (Pozo et al. 2009). The application of mutants of tomato deficient in jasmonic acid signaling demonstrated such JA is necessary for AMF-produced *Botrytis* resistance. This finding supports the notion that MIR is comparable to the most investigated rhizobacteria-based ISR-induced systemic resistance within *Arabidopsis* as well as necessitates an effective jasmonic acid signaling route for effective stimulation of resistance (Pieterse et al. 1998).

2.8.1 Phenolics

Environmental changes that are detrimental to plant development and growth expose plants to an array of abiotic stresses (Zhu 2016). Such stress of abiotic type encompass heavy metals, salinity, water (flooding and drought), mechanical factors, extremely low and high temperatures (freezing and chilling), extreme amounts of light (low and high), ozone, extreme or insufficiency of nutrients, radiation (UV-B and UV-A) sulfur dioxide and other less often happening stressors (Pereira 2016). The build-up of phenolics within plant tissues is thought to represent a plant's robust reaction to these unfavorable environmental conditions because plants are grounded in the surroundings they develop in and must adjust to the dynamic conditions carried on by abiotic stresses (Pereira 2016; Lattanzio 2013).

The vast majority of the compounds classified as secondary and primary metabolites are manufactured by plants. All plants include the principal metabolites—nucleic acids, sugars, amino acids and fatty acids—which are necessary for plant development and growth (Fiehn 2002; Wu and Chappell 2008). Specialized molecules, secondary metabolites which are not instantly necessary for the fundamental metabolism of plant but are necessary for plants to survive in the habitat due to their structural and chemical diversity in comparison to primary metabolites. Secondary metabolites, the most prevalent class with significant morphological and physiological significance within plants, are known as polyphenols or plant phenolics.

Randhir et al. (2004) reported that phenolics are hydroxylated aromatic compounds that come from the polyketide acetate/malonate pathway or the shikimate/phenylpropanoid pathway to produce polyphenols and polymeric and monomeric phenols. Plant phenolics execute a crucial role in reproduction, development and plant growth as well as defense countering abiotic stresses like nutrient deficiency, UV-B radiation, high light, heavy metals and low temperatures, (Lattanzio 2013), defense against predators and pathogens (Bravo 1998), production of color and receptive features of vegetables and fruits, and other vital effects like antiallergic (Alasalvar et al. 2001; Balasundram et al. 2006).

Plant phenolic acids, also known as polyphenols, the most prevalent secondary metabolites, are found in abundance throughout the plant world. Swain (1975) and Harborne (1980) reported that bryophytes regularly create polyphenols like flavonoids, whereas fungi, algae and bacteria make distinct phenolic substances, but vascular plants contain the whole spectrum of polyphenols or phenolic compounds. According to estimates, phenolic chemicals make up around 2% of all the carbon that plants use for photosynthetic energy (Robards and Antolovich 1997). Higher plants are investigated to generate heaps of phenolic substances, and more and more of these chemicals are being characterized. Esters, hydroxycinnamic acid (HCA), amides and glycosides, and glycosylated flavonoids, notably flavonols, proanthocyanidins and their derivatives, are all found in plant leaves. There are a few more polyphenolic polymers, including lignin, suberin and pollen sporopollenin. One-of-a-kind biomarkers for taxonomic investigations are soluble phenolics since some, like chlorogenic acid, are generally present while others are confined to particular genera or families. Through the shikimic acid pathway, phenylalanine and shikimic acid are used as a metabolic intermediary in the production of plant phenolics. As intermediates of the hexose monophosphate shunt and glycolysis consequently, Phospho-Enol Pathway (PEP) and erythose-4-phosphate are the pathway's beginning metabolites.

2.8.2 Phytochemicals as Signaling Molecules Beneath the Ground

The assembly of secondary, frequently diffusible substances allows plant to interact actively with its biotic surroundings. Thus, the physical and chemical characteristics of roots' background soil, or rhizosphere, are altered by root exudation (Hiltner 1904). Because the exudates of the plants frequently contain enzymes, ions and molecules abundant in biological carbon (secondary and primary metabolites), which are a hub of nutrients required to terrestrial organisms, this habitat is favorable for the viability of microbes throughout the plants (Bertin et al. 2003). Secondary metabolites that are expelled from plants are frequently a source of chemotaxis for a particular group of organisms (commensals or pathogens) near roots. The creation of diverse relationships, particularly those involving symbiotic bacteria, is encouraged by this chemical environment.

2.8.2.1 Rhizobium–Legume Symbiosis

The rhizobium and legume symbiosis is specific coordinated symbioses that are most frequently described. This contact, which is rooted primarily at the roots, enables the development of particular structures known as nodules in which bacteroides fix atmospheric nitrogen. The indicated symbiosis, which is comparably new in evolutionary terms, evolved from an earlier mycorrhizal symbiosis, which was rather abundant in the kingdom of plants and enabled the plant to get vital phosphate minerals (Harrison 2005). Rhizobium symbiosis-related genes are frequently found in chromosomal blocks or plasmids, even as the case with *Sinorhizobium meliloti*, in which the genes for nitrogen fixation (*nif* and *nix*) and the display of nodulation-related factors (*nod*, *dol*, *Noah*) are found on the *pSymA* plasmid (Galibert et al. 2001). Additionally, it has been demonstrated that the particular chromosomal genes or plasmids can be transferred horizontally between *Rhizobia* species (Biondi et al. 2006). Through their roots, legumes release several secondary metabolites into the soil, including flavonoids, particularly isoflavonoids and flavones in the rhizospheric zone. These flavonoids control the activation of further *nod genes* namely A, B and C as well as the creation of Nod factors (constituting lipo-chitin-oligosaccharide), although they are identified (oxidation, location of the side chain groups) by the bacteria's oligomerization domain D (*NodD*) and nucleotide-binding receptors. The ability of compatible bacteria to travel to the flavonoid flux via chemotactic signals suggests that the specificity of the host is established prior to the bacterium's actual physical contact with its host plant. The host plant later uniquely recognizes *Nod* factors produced at the rhizobial based on their shape (unsaturation and extent of the acyl chain, substitutions, monomeric number of *N*-acetyl-glucosamine) (Gibson et al. 2008). The flavonoids in roots appear to be necessary to promote the production of *Nod* factors within the infection peg (Zhang et al. 2009).

Therefore, it is possible that *Nod* factors might cause the PHP pathway's enzymes to be overexpressed, which would then lead to an increase in the synthesis of phenolic chemicals, particularly flavonoids. As a result, it's likely that the existence of *Nod* components in the plant's environment causes the plant to respond favorably to Rhizobial infection. The plant boosts the synthesis of phenolic acids, like *cis* and *trans* caffeic acid, chlorogenic acid, vanillic acid, *cis* and *trans* ferulic acid and *para* coumaric acid, specifically in the infected condition. Rhizobium does not impact or reduce the synthesis of other substances like proto-catechuic acid and *p*-hydroxybenzoic acid. Rhizobium does not impact or reduce the synthesis of other substances like protocatechuic acid and *p*-hydroxybenzoic acid. Plants may in fact interact with soil microbes by diffusing (active or passive) phenolic chemicals (Nicholson and Hammerschmidt 1992). It has been demonstrated that *Bradyrhizobium loti* nodulating *Lotus subbiflorus* Lag. was capable to utilize coumaric and ferulic acid as a source of carbon. These phenolic acids, which are among the most prevalent simple phenols in plants, include 4-hydroxybenzoic acid, vanillic acid, caffeic acid and coumaric acid (Van Rossum et al. 1995). Rhizobium exhibits a favorable chemotaxis as a result of the host plant's synthesis of these substances, and flavonoids also facilitate the induction of *Nod* factors (Prinsen et al. 1991).

2.8.3 Flavonoids in the Arbuscular Mycorrhizal Symbiosis

Mycorrhizae, in contrast to rhizobial symbiosis, are more common than not in nature. In this mutualistic connection, hyphae of mycorrhiza, that may be likely 100 times lengthy in addition to root hairs, intervene as an expansion of the plants root, penetrating deeper soil external to the P (phosphate) consumption zone (Mohammadi et al. 2011; Balemi and Negisho 2012). Denison and Kiers (2011) proposed that mycorrhization thus benefits roots by giving plants a wider range of nutrient feeding systems, and this feeding gives fungi a way to discover new hosts. Therefore, a deeper comprehension of this microbial connection may hold the key to figuring out if agricultural systems will be profitable and sustainable (Mohammadi et al. 2011). Therefore, in this article, we will examine the methods through which plant perceives, particularly flavonoids, mutualistic symbiosis and fungal contamination.

2.8.3.1 Flavonoids' Developing Signaling Role Assigned to Arbuscular Mycorrhiza (AM) Development

When the soil's temperature, CO_2, and matric potential are favorable, round, thick-walled multinucleate fungal spores (Hijri and Sanders 2005) germinate to create haploid coenocytic hyphae (Mandal et al. 2010). Requena et al. (2007) reported that asymbiotic stage refers to this stage of growth that occurs devoid of plant signal. The fungal colonies during this period, which mostly depend on triacylglyceride stores, extend a few centimeters intermittent hyphal branching and display a distinctive developmental pattern with clear apical dominance. Giovannetti et al. (2010) and Bonfante and Genre (2010) reported that spore of AMF, without the host root, development stops at the point in which septation of hyphae originating from the apex pauses growth and turns protoplasm reversing in spore, remaining dormant once more before using up the spore pool. Gachomo et al. (2009) and Bonfante and Genre (2010) highlighted that, in light of this, even while spores have ability to germinate even in absence of host, the transition starting with developmental stage of asymbiotic phase up to dynamic phase of presymbiotic development only occurs in return to initial detection by complex, carefully calibrated signaling events. Host signal-mediated increases in branching of hyphae and metabolic actions, which ensure directed growth of hyphal branches facing host root, are a critical developmental phase in the life cycle of mycorrhizal fungi prior to colonization (Pinior et al. 1999). Limited in their ability to expand, germ tubes react in presence of "branching factor" root exudates of host nearby by changing the fungal morphology to one that favors improved growth of hyphae and widespread branches of hyphae (Buee et al. 2000). All mycotrophic plants have the branch-inducing component in their root exudates, whereas nonhost plants do not. Host secretions that promote branches of hyphae in soil, spore germination, arbuscule production within the root and fungal invasion, frequently in a mutualistic specific manner, recognized as flavonoids as contrary to the legume-Rhizobium-legume symbiosis (Scervino et al. 2007; Steinkellner et al. 2007). They may be identified by effector proteins connected to the plasma membrane of fungi, along with the rhizodermis signals of thigmotropic and the chemical signals released by the plant (Requena et al. 2007). Fascinatingly, it was hypothesized that flavonoids might either abolish AM fungal self-inhibition or directly boost fungal growth (Siqueira et al. 1991b). Akiyama et al. (2010) reported that the hyphal branching of mycorrhizal fungus was inhibited by pyrano isoflavones generated from white lupin, a plant not being a host to mycorrhizal fungi, indicating that flavonoids may either have stimulating or inhibiting effects on symbiotic fungi present in soil. By changing the profile of flavonoid exudates, host and nonhost

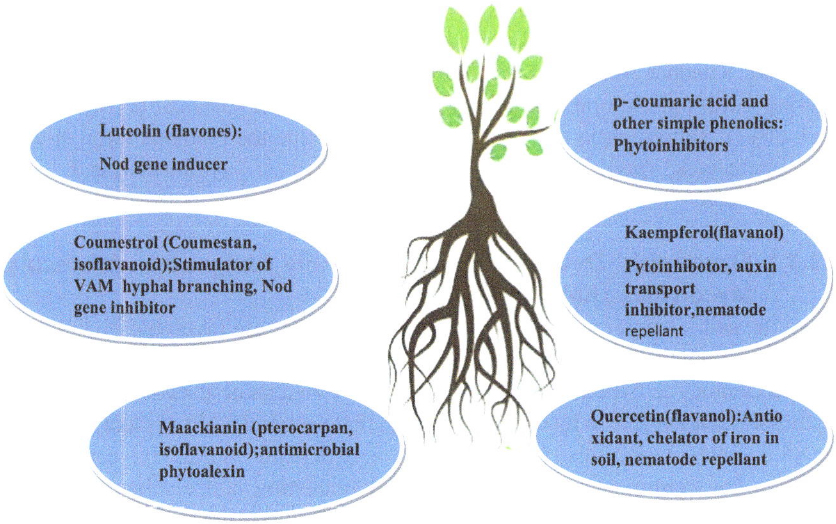

Fig. 2.2 Role of flavonoids in the rhizosphere

plants may be able to influence the development of symbiosis (Hassan and Mathesius 2012). Type of flavonoid and concentration, host specificity and fungal species of AM and genera all affect how these substances affect AM fungi (investigated by Vierheilig et al. 1998). Different flavonoids can promote hyphal development, hyphal branching and AM fungal spore germination at low concentrations, and even at high concentrations, the very flavonoid becomes inhibitor (Nair et al. 1991; Baptista and Siqueira 1994). According to Siqueira et al. (1991b), 5 mg/L concentrations of isoflavonoids (*Biochanin* A and *Formononetin*) in *T. repens* L. activated AM colonization, although flavones *chrysin* enhanced root colonization at 40–60 mg/L concentrations (Fig. 2.2).

2.8.4 The Root Epidermis Contains a *Myc* Factor Signaling Pathway

In order to coordinate the AM infection process, branching fungal hyphae secrete substance known as "*Myc* factor" that causes different physiological and morphological changes in the hosts. This substance blocks the defense system of plant and increases the stimulation of genes related to symbiosis (Kloppholz et al. 2011). Initially in the process of development of mycorrhiza, chitinases that constitute the plant may partly break the elicitors, resulting in the response of temporary defenses of plant (Maillet et al. 2011; Antunes and Goss 2005). Contrary to pathogens of fungi, *Myc* factors diffused via AMF increase the formation of lateral roots, cause a brief elevation in calcium level in cytosol within one or two minutes and cause the activation of particular genes only in those cells of roots that come into direct proximity with the pervasive fungus (Navazio et al. 2007). Conversely, suppression is pronounced in nearby cells that are not colonized by the fungus (Kosuta et al. 2003; Genre et al. 2005). Stracke et al. (2002) reported that the intracellular kinase domain of the MtDMI2/MtSYM2, which is a receptor-like leucine-rich kinase *Sym*RK, which is intricated either directly or indirectly in the transmission of rhizobial or fungal signals to the cytoplasm, is one common signaling component. *Nod* factors, which are more recent evolutionary targets, also share structural and functional similarities with *Myc* symbiotic signals (Limpens et al. 2015; Laparre et al. 2014). The calcium oscillations caused by these channels in both symbioses activate the MtSYMI3/DMI3/CCAMK, a calcium-calmodulin-dependent protein kinase in the perinuclear cytoplasm and nucleus (Riely et al. 2007; Mitra et al. 2004). The last SYM gene to be discovered, CYCLOPS, is known to be phosphorylated by CCAMK. Messinese et al. (2007) reported gene codes for the IPD3/CYCLOPS protein that combines with DMI3. Legumes need to distinguish between the two symbiotic signal types concerning to turn on the proper symbiotic program. Due to the fact that the growth of arbuscular structures and the emergence of infection threads are both CYCLOPS dependent, whereas organogenesis of nodule is independent on CYCLOPS, CYCLOPS serves as a bifurcation point in the usual SYM pathway (Yano et al. 2008). In eudicots and monocots of nonnodulating type, a parallel *Sym* pathway also exists and facilitates AM signaling (Mukherjee and Ané 2011). In the mycorrhizal condition, the symbiotic fungi might be depleting these phytoalexin precursors which act as precursors of carbon or combination of medicarpin might eliminate a possibly harmful product. The debate over the role of flavonoids as regulating composites during mycorrhizal infection signaling was revived by studies on flavonoids and AM by Akiyama et al. (2002) and Guenoune et al. (2001). For the function of flavonoids during the AM connection, Vierheilig and Piche's (2002) model was complex. They suggested that a basic molecular conversation takes place that controls later colonization in addition to the initial growth of symbiotic AM, which needs to be stabilized for the formation of true symbiosis of symbiosis.

2.9 Flavonoid Role in Mycorrhizal Autoregulation

Once more, flavonoids are involved in the autoregulation of mycorrhization since they can have either stimulatory effect or an inhibitory on fungi, depending on the flavonoids and their concentration (Subramanian et al. 2007; Vierheilig et al. 1998; Lira Jr et al. 2015). The fact that medicarpin mostly accumulates at the phase of appressoria development, as soon as fungus has developed its initial root structures, suggests that after the fungal companion has developed its appressoria. Later, the plant's perception of fungus altered, and during a period of intensive AM root colonization, medicarpin accumulation was unaltered, although coumestrol accumulation continued to rise. According to Vierheilig and Piche's research from 2002, the build-up of *cholesterol*, *medicarpin* and/or *formononetin* at later stages was associated favorably with the autoregulation of mycorrhization. *Genistein* was the only substance that concentrated to comparable amounts in all roots colonized by

AMF. In 2002, Larose et al. proposed that this consistent motif of *genistein* accumulation in mycorrhized roots independent of the colonized AMF root suggested that this substance had a general regulatory role during mycorrhization that was neither species- nor genus-specific. Subsequently, in 2005a, Scervino et al. observed that *rhamnetin* and *acacetin* hindered the development of AM piercing structures and colonization and hinted at a potential role in the autoregulation process of the symbiotic AM by specific two flavonoids. For the first time, in 2005b, Scervino et al. connected variations in mycorrhizal roots in flavonoids shortly to the governing mechanisms of the symbiotic AM. Interestingly, RR4-2 and RR4 induced the majority stages of the presymbiosis in *Gigaspora* (excepting those of two *Glomus* species), confirming the hypothesis of Akiyama et al. (2002) that flavonoids promoted root colonization via AMF fungi arise mostly in mycorrhized roots and absent from non-mycorrhizal roots. They also hypothesized that the substance may be intricated in one phase of regulation of the fungus except that in another because RR-4 expanded the proportion of *Gi. Margarita* spore formation, but it prohibited the hyphal development and had no consequence on hyphal extension or the development of axillary cell clumps. All AM fungi studied were blocked from developing presymbiotic relationships by NM7, which was only found in white clover roots that were not mycorrhizal. The hyphal development and/or spore formation of *Gigaspora* species were stimulated by freshly produced *quercetin*. In 2005b, Scervino et al. investigated that although almost all fungal variables of the two AM species (*Glomus* and *Gigaspora*) evaluated showed an inhibitory effect, *rhamnetin* and *acacetin* showed a hindering effect, suggesting that they may play a role in the self-regulation of mycorrhiza process. In 2004b, Vierheilig and, in 2005, Meixner et al. found that root colonization on one part of split-root networks of barley, soybean and alfalfa severely inhibited the mycorrhiza process of "autoregulated roots" on the alternative side. The mutated receptor kinase gene GmNARK has created a supernodulating mutant of soybean (*nts*1007) which does not autoregulate AMF root colonization because it lacked the controlled nodulation through autoregulatory mechanism (Searle et al. 2003; Meixner et al. 2005). These studies thus suggested that both symbioses shared a similar autoregulation mechanism. Flavonoids also play a role in the synchronization of mycorrhization, similar to the rhizobium–legume symbiosis, since roots treated with certain flavonoids showed an increase in colonized root of AM fungi (Vierheilig 2004a; Vierheilig and Piché 2002; Scervino et al. 2005a, b). Exogenous administration of isoflavonoids inactivated the roots or counteracted the mutualistic-suppressive effects brought on by extended self-regulation signals as well as enhanced the development of nodules and the symbiosis of AM fungi. *Formononetin* induced nodulation in regions of the root system that are autoregulated, whereas *ononin* (except neither *formononetin*) encouraged colonization of AMF root. Accordingly, despite the fact that mycorrhization and autoregulation of nodulation appear to share certain identical signaling processes as reported by Vierheilig (2004a), Vierheilig et al. (2008) and Meixner et al. (2005), the outcomes of organized modulated isoflavonoids on the development of symbiosis existed contrasting. The three-way symbiosis between rhizobia and arbuscular mycorrhizal fungus is advantageous to many legume plants.

According to Siqueira et al. (1991b), rhizobium nodulation was increased when AM fungi were present in plants that had been treated with the "anti-nod inducers" isoflavonoid *formononetin* and *biochanin* A. This might have happened because the plants had better mycorrhizal colonization and growth. For AM fungi inoculation, soil treatments of similar chemicals had also been validated to increase mycorrhizal formation and development of nonnodulated plants (Siqueira et al. 1991a), demonstrating that the stimulation for development of isoflavonoid was no more moderated by nodulation. According to Antunes et al. (2006), coumestrol, that was missing from the soybean seed, was recently produced in the root of the plant and had an amount of *daidzein* partially four times higher than that of the seed.

Daidzein and *coumestrol* levels significantly increased, and this appeared to be primarily due to the emergence of AM fungi, that may have facilitated symbiosis of brady-rhizobial (*coumestrol* and *daidzein* are intricated in symbiosis of *bradyrhizobium*) and as a consequence played a significant part in the initial phases of the trilateral symbiosis. A new mutant B9 of *M. truncatula* was described by Morandi et al. (2009); it was hypermycorrhizal but defective for nodulation. It offered a brand-new method for researching plant metabolites that differentially control nodulation and mycorrhizal symbioses, especially those involving autoregulation processes.

Over time, the germination of the living roots and roots of leguminous crops releases a large number of flavonoids such as *luteolin*, *quercetin* and other alternated flavanones and flavones. In 2012, Hassan and Mathesius highlighted that these communications among fungal pathogens indicate that, in addition to exhilarating mycorrhizal fungi, these flavonoids might also be crucial for the defense of plants from seed-borne pathogens. From a field trial, Cordeiro et al. (2015) have shown that a well-known formononetin can increase colonization of mycorrhiza and nodule count, and lessen the detrimental results of fungicides Thiram + Carbendazim within soybean productivity. In contemplation of increasing crop yield in an economical way as well as with less agricultural synthetic inputs, using flavonoids is likely a great chance to use and exploit AM fungi and Rhizobium. The promotion of rhizobium–legume symbiosis and N_2 fixation in agricultural operations is currently done commercially using seed secretions, which contain a combination of flavonoids (Skorupska et al. 2010). According to this study, several flavonoids play an important role in the development tripartite symbiosis in addition to synthesized and delivered within the rhizosphere as being a part of the colonization activity (Skorupska et al. 2010).

2.10 Conclusion

This chapter highlighted the significance of phenolics particularly flavonoids in the establishment of two agronomically important symbiotic association of roots with the nodule formation and mycorrhizal interactions. By deciphering the complexity of flavonoids induced responses and their combined role with other modulators of symbioses like hormones make flavonoids a key player in plant microbe interactions.

As the plant comes in contact with the different agents of biotic stress, phenolics are produced for the defense strategy to make the plant tolerant against these agents. AMF has been mainly used as beneficial entities for better nutrient uptake from soil; however, it is clearly assigned that plants inoculated with AMF effectively combat various environmental stresses, thereby increasing the productivity of various vegetables as well as crops. The preliminary focus of future research should be the recognition and identification of genes and host as well as AMF-specific protein factors controlling the AMF-modulated growth and symbiotic association cellular and metabolic pathways under stressful environmental conditions.

References

Agrawal AA (2011) Current trends in the evolutionary ecology of plant defence. Funct Ecol 25(2): 420–432

Agrios GN (2005) Preface. In: Agrios GN (ed) Plant pathology, 5th edn. Academic, San Diego, CA

Ahanger MA, Tyagi SR, Wani MR, Ahmad P (2014) Drought tolerance: role of organic osmolytes, growth regulators, and mineral nutrients. In: Ahmad P, Wani MR (eds) Physiological mechanisms and adaptation strategies in plants under changing environment, vol 1. Springer, New York, pp 25–55. https://doi.org/10.1007/978-1-4614-8591-9_2

Akhtar MS, Siddiqui ZA (2007) Glomus intraradices, Pseudomonas alcaligenes, and Bacillus pumilus: effective agents for the control of root-rot disease complex of chickpea (Cicer arietinum L.). J Gen Plant Pathol 74:53–60

Akhtar MS, Siddiqui ZA, Wiemken A (2010) Arbuscular mycorrhizal fungi and Rhizobium to control plant fungal diseases. In: Lichtfouse E (ed) Alternative farming systems, biotechnology, drought stress and ecological fertilisation, Sustainable agriculture reviews, vol 6. Springer, Dordrecht., pp 263–292

Akiyama K, Matsuoka H, Hayashi H (2002) Isolation and identification of a phosphate deficiency induced C-glycosyl-flavonoid that stimulates arbuscular mycorrhiza formation in melon roots. Mol Plant Microbe Interact 15:334–340

Akiyama K, Matsuzaki K, Hayashi H (2005) Plant sesquiterpenes induce hyphal branching in arbuscular mycorrhizal fungi. Nature 435:824–827

Akiyama K, Tanigawa F, Kashihara T, Hayashi H (2010) Lupin pyranoisoflavones inhibiting hyphal development in arbuscular mycorrhizal fungi. Phytochemistry 71:1865–1871

Alasalvar C, Grigor JM, Zhang D, Quantick PC, Shahidi F (2001) Comparison of volatiles, phenolics, sugars, antioxidant vitamins, and sensory quality of different colored carrot varieties. J Agric Food Chem 49:1410–1416

Anderson PK, Cunningham AA, Patel NG, Morales FJ, Epstein PR, Daszak P (2004) Emerging infectious diseases of plants: pathogen pollution, climate change and agrotechnology drivers. Trends Ecol Evol 19(10):535–544

Antunes PM, Goss MJ (2005) Communication in the tripartite symbiosis formed by arbuscular mycorrhizal fungi, Rhizobia and legume plants: a review. Am Soc Agron 48:199–122

Antunes PM, de Variennes A, Rajcan I, Goss MJ (2006) Accumulation of specific flavonoids in soybean as a function of the early tripartite symbiosis with arbuscular mycorrhizal fungi and Bradyrhizobium japonicum. Soil Biol Biochem 38:1234–1242

Arimura GI, Matsui K, Takabayashi J (2009) Chemical and molecular ecology of herbivore-induced plant volatiles: proximate factors and their ultimate functions. Plant Cell Physiol 50(5):911–923

Ashbolt NJ (2004) Microbial contamination of drinking water and disease outcomes in developing regions. Toxicology 198(1):229–238

Atilano RA, Menge JA, van Gundy SD (1981) Interactions between Meloidogyne arenaria and Glomus fassiculatus in grape. J Nematol 13:52–57
Badri DV, Vivanco JM (2009) Regulation and function of root exudates. Plant Cell Environ 32: 666–681
Bagyaraj DJ (1984) Biological interactions with mycorrhizal fungi. In: Powell CL, Bagyaraj DJ (eds) VA mycorrhiza. CRC Press, New York, pp 131–153
Bai FW, Yan J, Qu ZC, Zhang HW, Xu J, Ye MM, Shen DL (2002) Phylogenetic analysis reveals that a dwarfing disease on different cereal crops in China is due to rice black streaked dwarf virus (RBSDV). Virus Gene 25(2):201–206
Bajaj R, Hu W, Huang Y, Chen S, Prasad R, Varma A, Bushley K (2015) The beneficial root endophyte Piriformospora indica reduces egg density of the soybean cyst nematode. Biol Control 90:193–199
Bajaj R, Prasad R, Varma A, Bushley KE (2017) The role of arbuscular mycorrhizal fungi and the mycorrhizal-like fungus Piriformospora indica in biocontrol of plant parasitic nematodes. In: Varma A, Prasad R, Tuteja N (eds) Mycorrhiza. Springer International Publishing AG, Cham, pp 43–56
Balasubramanian N, Hao YJ, Toubarro D, Nascimento G, Simões N (2009) Purification, biochemical and molecular analysis of a chymotrypsin protease with prophenoloxidase suppression activity from the entomopathogenic nematode Steinernema carpocapsae. Int J Parasitol 39(9): 975–984
Balasundram N, Sundram K, Samman S (2006) Phenolic compounds in plants and agri-industrial by-products: antioxidant activity, occurrence, and potential uses. Food Chem 99:191–203
Balemi T, Negisho K (2012) Management of soil phosphorus and plant adaptation mechanisms to phosphorus stress for sustainable crop production: a review. J Soil Sci Plant Nutr 12:547–562
Baptista MJ, Siqueira JO (1994) Efeito de flavonoides na germinação de esporos e no crescimento assimbiótico do fungo micorrízico arbuscular Gigaspora gigantea. Rev Bras Fisiol Veg 6:127–134
Barea JM, Pozo MJ, Azcon R, Azcon-Aguilar C (2005) Microbial co-operation in the rhizosphere. J Exp Bot 56:1761–1778
Beckers GJM, Jaskiewicz M, Liu Y, Underwood WR, He SY, Zhang S, Conrath U (2009) Mitogen-activated protein kinases 3 and 6 are required for full priming of stress responses in Arabidopsis thaliana. Plant Cell 21:944–953
Benaragama D, Shirtliffe SJ, Johnson EN, Duddu HSN, Syrovy LD (2016) Does yield loss due to weed competition differ between organic and conventional cropping systems? Weed Res 56(4): 274–283
Benhamou N, Fortin JA, Hamel C, St Arnaud M, Shatilla A (1994) Resistance responses of mycorrhizal Ri T-DNA-transformed carrot roots to infection by Fusarium oxysporum f. sp. chrysanthemi. Phytopathology 84:958–968
Bertin C, Yang X, Weston A (2003) The role of root exudates and allelochemicals in the rhizosphere. Plant Soil 256:67–83
Bilgin DD, Zavala JA, Zhu JIN, Clough SJ, Ort DR, Delucia E (2010) Biotic stress globally downregulates photosynthesis genes. Plant Cell Environ 33(10):1597–1613
Biondi EG, Reisinger SJ, Skerker JM, Arif M, Perchuk BS, Ryan KR, Laub MT (2006) Regulation of the bacterial cell cycle by an integrated genetic circuit. Nature 444:899–904
Birkenbihl RP, Somssich IE (2011) Transcriptional plant responses critical for resistance towards necrotrophic pathogens. Front Plant Sci. https://doi.org/10.3389/fpls.2011.00076
Boller T, He SY (2009) Innate immunity in plants: an arms race between pattern recognition receptors in plants and effectors in microbial pathogens. Science 324(5928):742–744
Bonfante P, Genre A (2010) Mechanisms underlying beneficial plant-fungus interactions in mycorrhizal symbiosis. Nat Commun 1:48. https://doi.org/10.1038/ncomms1046
Bouwmeester HJ, Matusova R, Zhongkui S, Beale MH (2003) Secondary metabolite signalling in host-parasitic plant interactions. Curr Opin Plant Biol 6:358–364

Bowles TM, Barrios-Masias FH, Carlisle EA, Cavagnaro TR, Jackson LE (2016) Effects of arbuscular mycorrhizae on tomato yield, nutrient uptake, water relations, and soil carbon dynamics under deficit irrigation in field conditions. Sci Total Environ 566:1223–1234. https://doi.org/10.1016/j.scitotenv.2016.05.178

Bravo L (1998) Polyphenols: chemistry ,dietary sources, metabolism and nutritional significance. Nutr Rev 56:317–333

Bridge J (1996) Nematode management in sustainable and subsistence agriculture. Annu Rev Phytopathol 34(1):201–225

Buee M, Rossignol M, Jauneau A, Ranjeva R, Becard G (2000) The pre-symbiotic growth of arbuscular mycorrhizal fungi is induced by a branching factor partially purified from plant root exudates. Mol Plant Microbe Interact 13:693–698

Burger J, Darmency H, Granger S, Guyot SH, Messéan A, Colbach N (2015) Simulation study of the impact of changed cropping practices in conventional and GM maize on weeds and associated biodiversity. Agric Syst 137:51–63

Caarls L, Pieterse CM, Van Wees S (2015) How salicylic acid takes transcriptional control over jasmonic acid signaling. Front Plant Sci. https://doi.org/10.3389/fpls.2015.00170

Cameron DD, Neal AL, van Wees SC, Ton J (2013) Mycorrhiza induced resistance: more than the sum of its parts? Trends Plant Sci 18:539–545. https://doi.org/10.1016/j.tplants.2013.06.004

Campe R, Loehrer M, Conrath U, Goellner K (2014) Phakopsora pachyrhizi induces defense marker genes to necrotrophs in Arabidopsis thaliana. Physiol Mol Plant Pathol 87:1–8

Campos-Soriano L, García-Martínez J, Segundo BS (2012) The arbuscular mycorrhizal symbiosis promotes the systemic induction of regulatory defence-related genes in rice leaves and confers resistance to pathogen infection. Mol Plant Pathol 13:579–592

Cheng X, Tian C, Li A, Qiu J (2012) Advances on molecular mechanisms of plant-pathogen interactions. Yi Chuan Hereditas 34:134–144. https://doi.org/10.3724/SP.J.1005.2012.00134

Chisholm ST, Coaker G, Day B, Staskawicz BJ (2006) Host-microbe interactions: shaping the evolution of the plant immune response. Cell 124(4):803–814

Cobb NA (1917) The mononchs (Mononchus Bastian 1866), a genus of free-living predatory nematodes. Soil Sci 3:431–486

Collins NC, Thordal-Christensen H, Lipka V, Bau S (2003) SNARE-protein-mediated disease resistance at the plant cell wall. Nature (London) 425(6961):973–977

Conrath U (2009) Priming of induced plant defense responses. In: Loon LCV (ed) Advances in botanical research. Academic, Burlington, MA, pp 361–395

Conrath U, Beckers GJM, Flors V, García-Agustín P, Jakab G, Mauch F, Newman MA, Pieterse CMJ, Poinssot B, Pozo MJ, Pugin A, Schaffrath U, Ton J, Wendehenne D, Zimmerli L, MauchMani B (2006) Priming: getting ready for battle. Mol Plant Microbe Interact 11:1017–1028

Cordeiro MAS, Ferreira DA, Paulino HB, Souza CRF, Siqueira JO, Carneiro MAC (2015) Mycorrhization stimulant based in formononetin associated to fungicide and doses of phosphorus in soybean in the Cerrado. Biosci J 31:1062–1070

Cordier C, Pozo MJ, Barea JM, Gianinazzi S, Gianinazzi-Pearson V (1998) Cell defense responses associated with localized and systemic resistance to Phytophthora induced in tomato by an arbuscular mycorrhizal fungus. Mol Plant Microbe Interact 11:1017–1028

Dass A, Shekhawat K, Choudhary AK, Sepat S, Rathore SS, Mahajan G, Chauhan BS (2016) Weed management in rice using crop competitiona review. Crop Prot 95:45–52

de la Peña E, Rodriguez-Echevarria S, van der Putten WH, Freitas H, Moens M (2006) Mechanism of control of root-feeding nematodes by mycorrhizal fungi in the dune grass, Ammophila arenaria. New Phytol 169:829–840

De Vleesschauwer D, Xu J, Hãfte M, Höfte M (2014) Making sense of hormone-mediated defense networking: from rice to Arabidopsis. Front Plant Sci 5:1–15

Denison RF, Kiers ET (2011) Life histories of symbiotic rhizobia and mycorrhizal fungi. Curr Biol 21:775–785

Dorn KM, Fankhauser JD, Wyse DL, Marks MD (2013) De novo assembly of the pennycress (Thlaspi arvense) transcriptome provides tools for the development of a winter cover crop and biodiesel feedstock. Plant J 75(6):1028–1038

Dudareva N, Negre F, Nagegowda DA, Orlova I (2006) Plant volatiles: recent advances and future perspectives. Crit Rev Plant Sci 25(5):417–440

Duncan LW (2005) Nematode parasites of citrus. In: Luc M, Sikora RA (eds) Plant parasitic nematodes in subtropical and tropical agriculture. CABI Bioscience, Egham, UK. (Chapter 11)

Durner J, Shah J, Klessig DF (1997) Salicylic acid and disease resistance in plants. Trends Plant Sci 2(7):266–274

Durrant WE, Dong X (2004) Systemic acquired resistance. Annu Rev Phytopathol 42:185–209

El-Khallal SM (2007) Induction and modulation of resistance in tomato plants against Fusarium wilt disease by bioagent fungi (arbuscular mycorrhiza) and/or hormonal elicitors (Jasmonic acid & Salicylic acid): 2-Changes in the antioxidant enzymes, phenolic compounds and pathogen related-proteins. Aust J Basic Appl Sci 1:717–732

Elsen A, Beeterens R, Swennen R, De Waele D (2003) Effects of an arbuscular mycorrhizal fungus and two plant-parasitic nematodes on Musa genotypes differing in root morphology. Biol Fertil Soils 38:367–376

Elsen A, Gervacio D, Swennen R, De Waele D (2008) AMF-induced biocontrol against plant parasitic nematodes in Musa sp.: a systemic effect. Mycorrhiza 18:251–256

Etebu E, Nwauzoma AB (2014) A review on sweet orange (Citrus sinensis L Osbeck): health, diseases and management. Am J Res Commun 2(2):33–70

Fereres A, Moreno A (2009) Behavioural aspects influencing plant virus transmission by homopteran insects. Virus Res 141(2):158–168

Fernandes I, Alves A, Correia A, Devreese B, Esteves AC (2014) Secretome analysis identifies potential virulence factors of Diplodia corticola, a fungal pathogen involved in cork oak (Quercus suber) decline. Fung Biol 118(5):516–523

Ferraz L, Brown D (2002) An introduction to nematodes-plant nematology. Pensoft, Sofia

Fiehn O (2002) Metabolomics the link between genotypes and phenotypes. Plant Mol Biol 48:155–171

Filion M, St-Arnaud M, Jabaji-Hare SH (2003) Quantification of Fusarium solani f. sp phaseoli in mycorrhizal bean plants and surrounding mycorrhizosphere soil using real-time polymerase chain reaction and direct isolations on selective media. Phytopathology 93:229–235

Filipjev IN, Schuurmans Stekhoven JH (1941) A manual of agricultural helminthology. Brill, Leiden, p 878

Gachomo E, Allen JW, Pfeffer PE, Govindarajulu M, Douds DD, Jin HR, Nagahashi G, Lammers PJ, Shachar-Hill Y, Bücking H (2009) Germinating spores of Glomus intraradices can use internal and exogenous nitrogen sources for de novo biosynthesis of amino acids. New Phytol 184:399–411

Gagne-Bourgue F, Aliferis KA, Seguin P, Rani M, Samson R, Jabaji S (2013) Isolation and characterization of indigenous endophytic bacteria associated with leaves of switchgrass (Panicum virgatum L.) cultivars. J Appl Microbiol 114(3):836–853

Galibert F, Finan TM, Long SR, Puhler A, Abola P, Ampe F, BarloyHubler F, Barnett MJ, Becker A, Boistard P, Bothe G, Boutry M, Bowser L, Buhrmester J, Cadieu E, Capela D, Chain P, Cowie A, Davis RW, Dreano S, Federspiel NA, Fisher RF, Gloux S, Godrie T, Goffeau A, Golding B, Gouzy J, Gurjal M, Hernandez-Lucas I, Hong A, Huizar L, Hyman RW, Jones T, Kahn D, Kahn ML, Kalman S, Keating DH, Kiss E, Komp C, Lelaure V, Masuy D, Palm C, Peck MC, Pohl TM, Portetelle D, Purnelle B, Ramsperger U, Surzycki R, Thebault P, Vandenbol M, Vorholter FJ, Weidner S, Wells DH, Wong K, Yeh KC, Batut J (2001) The composite genome of the legume symbiont Sinorhizobium meliloti. Science 293: 668–672

Gange AC, West HM (1994) Interactions between arbuscular mycorrhizal fungi and foliar-feeding insects in Plantago lanceolata L. New Phytol 128:79–87

Gange AC, Brown VK, Aplin DM (2003) Multitrophic links between arbuscular mycorrhizal fungi and insect parasitoids. Ecol Lett 6:1051–1055
Genre A, Chabaud M, Timmers T, Bonfante P, Barker DG (2005) Arbuscular mycorrhizal fungi elicit a novel intracellular apparatus in Medicago truncatula root epidermal cells before infection. Plant Cell 17:3489–3499
George C, Kohler J, Rillig MC (2016) Biochars reduce infection rates of the root-lesion nematode Pratylenchus penetrans and associated biomass loss in carrot. Soil Biol Biochem 95:11–18
Gianinazzi S, Gollotte A, Binet MN, van Tuinen D, Redecker D, Wipf D (2010) Agroecology: the key role of arbuscular mycorrhizas in ecosystem services. Mycorrhiza 20:519–530
Gibson KE, Kobayashi H, Walker GC (2008) Molecular determinants of a symbiotic chronic infection. Annu Rev Genet 42:413–441
Giovannetti M, Avio L, Sbrana C (2010) Fungal spore germination and presymbiotic mycelial growth—physiological and genetic aspects. In: Koltai H, Kapulnik Y (eds) Arbuscular mycorrhizas: physiology and function. Springer, New York, pp 3–32
Goicoechea N, Garmendia I, Sánchez-Díaz M, Aguirreolea J (2010) Review. Arbuscular mycorrhizal fungi (AMF) as bioprotector agents against wilt induced by Verticillium spp. in pepper. Span J Agric Res 8(S1):S25–S42
Goodin MM, Zaitlin D, Naidu RA, Lommel SA (2008) Nicotiana benthamiana: its history and future as a model for plant–pathogen interactions. Mol Plant Microbe Interact 21(8):1015–1026
Grabau ZJ, Maung ZTZ, Noyes DC, Baas DG, Werling BP, Brainard DC, Melakeberhan H (2017) Effects of cover crops on Pratylenchus penetrans and the nematode community in carrot production. J Nematol 49(1):114
Graham JH (2001) What do root pathogens see in mycorrhizas. New Phytol 148:357–359
Grant M, Lamb C (2006) Systemic immunity. Curr Opin Plant Biol 9:414–420
Guenoune D, Galili S, Phillips DA, Volpin H, Chet I, Okon Y, Kapulnik Y (2001) The defense response elicited by the pathogen Rhizoctonia solani is suppressed by colonization of the AM-fungus Glomus intraradices. Plant Sci 160:925–932
Guerrieri E, Lingua G, Digilio MC, Massa N, Berta G (2004) Do interactions between plant roots and the rhizosphere affect parasitoid behaviour? Ecol Entomol 29:753–756
Gworgwor NA, Weber HC (2003) Arbuscular mycorrhizal fungi–parasite–host interaction for the control of Striga hermonthica (Del.) Benth. in sorghum Sorghum bicolor (L.) Moench. Mycorrhiza 13:277–281
Hajra N, Shahina F, Firoza K (2013) Biocontrol of root-knot nematode by arbuscular mycorrhizal fungi in Luffa cylindrica. Pak J Nematol 31:77–84
Hammond-Kosack K, Jones JDG (2000) Responses to plant pathogens. Biochem Mol Biol Plants 1: 1102–1156
Hao Z, Fayolle L, van Tuinen D, Chatagnier O, Li X, Gianinazzi S, Gianinazzi-Pearson V (2012) Local and systemic mycorrhiza-induced protection against the ectoparasitic nematode Xiphinema index involves priming of defence gene responses in grapevine. J Exp Bot 63: 3657–3672
Harborne JB (1980) Plant phenolics. In: Bell EA, Charlwood BV (eds) Encyclopedia of plant physiology, Secondary plant products, vol 8. Springer, Berlin, pp 329–402
Hare P (2011) Antimalarial fungal pesticide. Nat Biotechnol 29(4):330–331
Harrison MJ (2005) Signaling in the arbuscular mycorrhizal symbiosis. Annu Rev Microbiol 59: 19–42
Hartley SE, Gange AC (2009) Impacts of plant symbiotic fungi on insect herbivores: mutualism in a multitrophic context. Annu Rev Entomol 54:323–342
Hasan N, Jain RK (1987) Parasitic nematodes and vesicular-arbuscular mycorrhizal (VAM) fungi associated with berseem (Trifolium alexandrinum L.) in the Bundelkhan region. Indian J Nematol 17:184–188
Hassan S, Mathesius U (2012) The role of flavonoids in root-rhizosphere signalling: opportunities and challenges for improving plant-microbe interactions. J Exp Bot 63:3429–3444

Hause B, Mrosk C, Isayenkov S, Strack D (2007) Jasmonates in arbuscular mycorrhizal interactions. Phytochemistry 68:101–110

Haynes PA, Roberts TH (2007) Subcellular shotgun proteomics in plants: looking beyond the usual suspects. Proteomics 7(16):2963–2975

Hearne SJ (2009) Control-the Striga conundrum. Pest Manag Sci 65:603–614

Hijri M, Sanders IR (2005) Low gene copy number shows that arbuscular mycorrhizal fungi inherit genetically different nuclei. Nature 433:160–163

Hiltner L (1904) Uber neure Erfahrungen und probleme auf dem gebeit der bodenbackteriologie und unter besonderer berucksichtigung der grundungung und brache. Arb Deut Landwirsch Ges 98:59–78

Hussey RS, Roncadori RW (1982) Vesicular arbuscular mycorrhizal fungi may limit nematode activity and improve plant growth. Plant Dis 66:9–14

Jaiswal S, Fontanillas P, Flannick J, Manning A, Grauman PV, Mar BG, Lindsley RC, Mermel CH (2014) Age-related clonal hematopoiesis associated with adverse outcomes. N Engl J Med 371: 2488–2498

Jaiti F, Meddich A, El Hadrami I (2007) Effectiveness of arbuscular mycorrhizal fungi in the protection of date palm (Phoenix dactylifera L.) against bayoud disease. Physiol Mol Plant Pathol 71:166–173

John A, Bai H (2004) Evaluation of VAM for management of root knot nematodes in Brinjal. Indian J Nematol 34:22–25

Johns CD (2014) Agricultural application of mycorrhizal fungi to increase crop yields, promote soil health and combat climate change. Future Directions International. https://www.futuredirections.org.au/publication/agricultural-application-of-mycorrhizal-fungi-toincrease-crop-yields-promote-soil-health-and-combat-climate-change/. Accessed 11 Aug 2020

Jung SC, Martinez-Medina A, Lopez-Raez JA, Pozo MJ (2012) Mycorrhiza-induced resistance and priming of plant defenses. J Chem Ecol 38:651–664

Kang KW, Lee SJ, Kim SG (2005) Molecular mechanism of nrf2 activation by oxidative stress. Antioxid Redox Signal 7(11–12):1664–1673

Kantharaju V, Krishnappa K, Ravichandra NG, Karuna K (2005) Management of root-knot fungus, Glomus fasciculatum. Indian J Nematol 35:32–36

Kaplan DT, Opperman CH (2000) Reproductive strategies and karyotype of the burrowing nematode, Radopholus similis. J Nematol 32(2):126

Karban R (2010) Neighbors affect resistance to herbivory–a new mechanism. New Phytol 186(3): 564–566

Karban R (2011) The ecology and evolution of induced resistance against herbivores. Funct Ecol 25(2):339–347

Kerry BR (2000) Rhizosphere interactions and the exploitation of microbial agents for the biological control of plant-parasitic nematodes. Annu Rev Phytopathol 38(1):423–441

Kivlin SN, Hawkes CV, Treseder KK (2011) Global diversity and distribution of arbuscular mycorrhizal fungi. Soil Biol Biochem 43:2294–2303

Kloppholz S, Kuhn H, Requena N (2011) A secreted fungal effector of Glomus intraradices promotes symbiotic biotrophy. Curr Biol 21:1204–1209

Kobra N, Jalil K, Youbert G (2009) Effects of three Glomus species as biocontrol agents against verticillium-induced wilt in cotton. J Plant Protect Res 49:185–189

Kofoid CA, White WA (1919) A new nematode infection of man. J Am Med Assoc 72:567–569

Koricheva J, Gange AC, Jones T (2009) Effects of mycorrhizal fungi on insect herbivores: a meta-analysis. Ecology 90:2088–2097

Kosuta S, Chabaud M, Lougnon G, Gough C, Dénarie J, Barker DG, Bécard G (2003) A diffusible factor from arbuscular mycorrhizal fungi induces symbiosis-specific MtENOD11 expression in roots of Medicago truncatula. Plant Physiol 131:1–11

Kumari SMP, Prabina BJ (2019) Protection of Tomato, Lycopersicon esculentum from Wilt Pathogen, Fusarium oxysporum f.sp. lycopersici by Arbuscular Mycorrhizal Fungi, Glomus sp. Int J Curr Microbiol Appl Sci 8:1368–1378

Lamers J, Van Der Meer T, Testerink C (2020) How plants sense and respond to stressful environments. Plant Physiol 182:1624–1635. https://doi.org/10.1104/pp.19.01464

Lamovsek J, Urek G, Trdan S (2013) Biological control of root-knot nematodes (Meloidogyne spp.): microbes against the pests/BIOTIcNO ZATIRANJE OGORcIC KORENINSKIH SISK (Meloidogyne spp.): MIKROORGANIZMI PROTI SKODLJIVCEM. Acta Agric Slov 101(2): 263

Laparre J, Malbreila M, Letissec F, Portaisc JC, Rouxa C, Bécarda G, Puech-Pagésa V (2014) Combining metabolomics and gene expression analysis reveals that propionyl- and butyryl carnitines are involved in late stages of arbuscular mycorrhizal symbiosis. Mol Plant 7:554–566

Larose G, Chênevert R, Moutoglis P, Gagné S, Piché Y, Vierheilig H (2002) Flavonoid levels in roots of Medicago sativa are modulated by the developmental stage of the symbiosis and the root colonizing arbuscular mycorrhizal fungus. J Plant Physiol 159:1329–1339

Lattanzio V (2013) Phenolic compounds: introduction. In: Ramawat KG, Merillon JM (eds) Natural products. Springer, Berlin. https://doi.org/10.1007/978-3-642-22144-6_57

Lattanzio V, Lattanzio VM, Cardinali A (2006) Role of phenolics in the resistance mechanisms of plants against fungal pathogens and insects. Phytochem Adv Res 661:23–67

Lee CW, Efetova M, Engelmann JC, Kramell R, Wasternack C, Ludwig-Müller J, Deeken R (2009) Agrobacterium tumefaciens promotes tumor induction by modulating pathogen defense in Arabidopsis thaliana. Plant Cell 21(9):2948–2962

Lendzemo VW, Kuyper TW, Kropff MJ, van Ast A (2005) Field inoculation with arbuscular mycorrhizal fungi reduces Striga hermonthica performance on cereal crops and has the potential to contribute to integrated Striga management. Field Crop Res 91:51–61

Lendzemo VW, Van Ast A, Kuyper TW (2006) Can arbuscular mycorrhizal fungi contribute to Striga management on cereals in Africa? Outlook Agric 35:307–311

Lendzemo VW, Kuyper TW, Matusova R, Bouwmeester HJ, van Ast A (2007) Colonization by arbuscular mycorrhizal fungi of sorghum leads to reduced germination and subsequent attachment and emergence of Striga hermonthica. Plant Behav 2:58–62

Li J, Brader G, Palva ET (2004) The WRKY70 transcription factor: a node of convergence for jasmonate-mediated and salicylate-mediated signals in plant defense. Plant Cell 16(2):319–331

Li HY, Yang GD, Shu HR, Yang YT, Ye BX, Nishida I, Zheng CC (2006) Colonization by the arbuscular mycorrhizal fungus Glomus versiforme induces a defense response against the rootknot nematode Meloidogyne incognita in the grapevine (Vitis amurensis Rupr.), which includes transcriptional activation of the class III chitinase gene VCH3. Plant Cell Physiol 47: 154–163

Li AR, Smith SE, Smith FA, Guan KY (2012) Inoculation with arbuscular mycorrhizal fungi suppresses initiation of haustoria in the root hemiparasite Pedicularis tricolor. Ann Bot 109: 1075–1080

Li AR, Guan KY, Stonor R, Smith SE, Smith FA (2013) Direct and indirect influences of arbuscular mycorrhizal fungi on phosphorus uptake by two root hemiparasitic Pedicularis species: do the fungal partners matter at low colonization levels? Ann Bot 112:1089–1098

Limpens E, van Zeijl A, Geurts R (2015) Lipochitooligosaccharides modulate plant host immunity to enable endosymbioses. Annu Rev Phytopathol 53:311–334

Lira MA Jr, Nascimento LRS, Fracetto GGM (2015) Legume-rhizobia signal exchange: promiscuity and environmental effects. Front Microbiol 6:945. https://doi.org/10.3389/fmicb.2015.00945

Liu J, Maldonado-Mendoza I, Lopez-Meyer M, Cheung F, Town CD, Harrison MJ (2007) Arbuscular mycorrhizal symbiosis is accompanied by local and systemic alterations in gene expression and an increase in disease resistance in the shoots. Plant J 50:529–544

Loon LCV, Bakker PAHM, Pieterse CMJ (1998) Systemic resistance induced by rhizosphere bacteria. Annu Rev Phytopathol 36:453–483

López-Ráez JA, Charnikhova T, Mulder P, Kohlen W, Bino R, Levin I, Bouwmeester H (2008) Susceptibility of the tomato mutant high pigment-2dg (hp-2dg) to Orobanche spp infection. J Agric Food Chem 56:6326–6332

López-Ráez JA, Matusova R, Cardoso C, Jamil M, Charnikhova T, Kohlen W, Verstappen F, Ruyter-Spira C, Bouwmeester HJ (2009) Strigolactones: ecological significance and use as a target for parasitic plant control. Pest Manag Sci 64:471–477

López-Ráez JA, Verhage A, Fernández I, García JM, Azcón-Aguilar C, Flors V, Pozo MJ (2010a) Hormonal and transcriptional profiles highlight common and differential host responses to arbuscular mycorrhizal fungi and the regulation of the oxylipin pathway. J Exp Bot 61:2589–2601

López-Ráez JA, Flors V, García JM, Pozo MJ (2010b) AM symbiosis alters phenolic acid content in tomato roots. Plant Signal Behav 5:1138–1140

Louarn J, Carbonne F, Delavault P, Becard G, Rochange S (2012) Reduced germination of Orobanche cumana seeds in the presence of arbuscular mycorrhizal fungi or their exudates. PLoS One 7(11):e49273. https://doi.org/10.1371/journal.pone.0049273

Maillet F, Poinsot V, Andre O, Puech-Pages V, Haouy A, Gueunier M, Cromer L, Giraudet D, Formey D, Niebel A, Martinez EA, Driguez H, Bécard G, Dénarié J (2011) Fungal lipochitooligosaccharide symbiotic signals in arbuscular mycorrhiza. Nature 469:58–64

Mandal SM, Chakraborty D, Dey S (2010) Phenolic acids act as signaling molecules in plantmicrobe symbioses. Plant Signal Behav 5:359–368

Mang HG, Laluk KA, Parsons EP, Kosma DK, Cooper BR, Park HC, Chilosi G (2009) The Arabidopsis RESURRECTION1 gene regulates a novel antagonistic interaction in plant defense to biotrophs and necrotrophs. Plant Physiol 151(1):290–305

Mann RS, Ali JG, Hermann SL, Tiwari S, Pelz-Stelinski KS, Alborn HT, Stelinski LL (2012) Induced release of a plant-defense volatile 'deceptively' attracts insect vectors to plants infected with a bacterial pathogen. PLoS Pathog 8(3):111–124

Mansfield J, Genin S, Magori S, Citovsky V, Sriariyanum M, Ronald P, Toth IAN (2012) Top 10 plant pathogenic bacteria in molecular plant pathology. Mol Plant Pathol 13(6):614–629

Marro N, Paola L, Cabello M, Doucet ME, Becerra AG (2014) Use of the arbuscular mycorrhizal fungus Glomus intraradices as biological control agent of the nematode Nacobbus aberrans parasitizing tomato. Braz Arch Biol Technol 57:668–674

Matthews BF, Beard H, MacDonald MH, Kabir S, Youssef RM, Hosseini P, Brewer E (2013) Engineered resistance and hypersusceptibility through functional metabolic studies of 100 genes in soybean to its major pathogen, the soybean cyst nematode. Planta 237(5):1337

McDonald BA, Linde C (2002) The population genetics of plant pathogens and breeding strategies for durable resistance. Euphytica 124(2):163–180

Meixner C, Ludwig-Müller J, Miersch O, Gresshoff P, Staehelin C, Vierheilig H (2005) Lack of mycorrhizal autoregulation and phytohormonal changes in the supernodulating soybean mutant nts1007. Planta 222:709–715

Messinese E, Mun JH, Yeun LH, Jayaraman D, Rougé P, Barre A, Lougnon G, Schornack S, Bono JJ, Cook DR, Ané JM (2007) A novel nuclear protein interacts with the symbiotic DMI3 calcium- and calmodulin-dependent protein kinase of Medicago truncatula. Mol Plant Microbe Interact 20:912–921

Miedes E, Vanholme R, Boerjan W, Molina A (2015) The role of the secondary cell wall in plant resistance to pathogens. Plant Cell Wall Pathog Parasit Symb 78:213–221

Miranda M, Ralph SG, Mellway R, White R, Heath MC, Bohlmann J, Constabel CP (2007) The transcriptional response of hybrid poplar (Populus trichocarpa × P. deltoids) to infection by Melampsora medusae leaf rust involves induction of flavonoid pathway genes leading to the accumulation of proanthocyanidins. Mol Plant Microbe Interact 20(7):816–831

Mitra RM, Gleason CA, Edwards A, Hadfield J, Downie JA, Oldroyd GED, Long SR (2004) A Ca^{2+}/calmodulin-dependent protein kinase required for symbiotic nodule development: gene identification by transcript-based cloning. Proc Natl Acad Sci U S A 101:4701–4705

Mohammadi K, Khalesro S, Sohrabi Y, Heidari G (2011) A review: Beneficial effects of the mycorrhizal fungi for plant growth. J Appl Environ Biol Sci 1:310–319

Monaghan J, Zipfel C (2012) Plant pattern recognition receptor complexes at the plasma membrane. Curr Opin Plant Biol 15:349–357

Morandi D, Gianinazzi-Pearson V (1986) Influence of mycorrhiza and phosphate nutrition on secondary metabolite contents of soybean roots. In: Gianinazzi-Pearson V, Gianinazzi S (eds) Physiological and genetical aspects of mycorrhizae. INRA, Paris, pp 787–791

Morandi D, Bailey JA, Gianinazzi-Person V (1984) Isoflavonoid accumulation in soybean roots infected with vesicular-Arbuscular mycorrhizal fungi. Physiol Plant Pathol 24:357–364

Morandi D, Le Signor C, Gianinazzi-Pearson V, Duc G (2009) A Medicago truncatula mutant hyper-responsive to mycorrhiza and defective for nodulation. Mycorrhiza 19:435–441

Mukherjee A, Ané JM (2011) Germinating spore exudates from arbuscular mycorrhizal fungi: molecular and developmental responses in plants and their regulation by ethylene. Mol Plant Microbe Interact 24:260–270

Nagy PD (2008) Recombination in plant RNA viruses. In: Plant virus evolution, vol 26. Springer, Berlin, pp 133–156

Nair MG, Safir GR, Siqueira JO (1991) Isolation and identification of vesicular-arbuscular mycorrhiza-stimulatory compounds from clover (Trifolium repens) roots. Appl Environ Microbiol 57: 434–439

Nakmee PS, Techapinyawat S, Ngamprasit S (2016) Comparative potentials of native arbuscular mycorrhizal fungi to improve nutrient uptake and biomass of Sorghum bicolor Linn. Agric Nat Resour 50:173–178

Navazio L, Moscatiello R, Genre A, Novero M, Baldan B, Bonfante P, Mariani P (2007) A diffusible signal from arbuscular mycorrhizal fungi elicits a transient cytosolic calcium elevation in host plant cells. Plant Physiol 144:673–681

Nejat N, Mantri N (2017) Plant immune system: crosstalk between responses to biotic and abiotic stresses the missing link in understanding plant defence. Curr Issues Mol Biol 23:1–16. https://doi.org/10.21775/cimb.023.001

Nguvo KJ, Gao X (2019) Weapons hidden underneath: bio-control agents and their potentials to activate plant induced systemic resistance in controlling crop Fusarium diseases. J Plant Dis Prot 126:177–190

Nicholson RL, Hammerschmidt R (1992) Phenolic compounds and their role in disease resistance. Annu Rev Phytopathol 30:369–389

Nicol JM, Rivoal R (2008) Global knowledge and its application for the integrated control and management of nematodes on wheat. Integr Manag Biocontrol Veg Grain Crops Nematodes 2: 251–294

Nicol JM, Turner SJ, Coyne DL, Den-Nijs L, Hockland S, Maafi ZT (2011) Current nematode threats to world agriculture. In: Genomics and molecular genetics of plant-nematode interactions, vol 45. Springer, Netherlands, pp 21–43

Noe JP (2004) Plant-parasitic nematodes. In: Trigiano RN (ed) Plant pathology: concepts and laboratory exercises. CRC Press

Nogales A, Aguirreolea J, Santa María E, Camprubí A, Calvet C (2009) Response of mycorrhizal grapevine to Armillaria mellea inoculation: disease development and polyamines. Plant Soil 31: 177–187

Norman JR, Hooker JE (2000) Sporulation of Phytophthora fragariae shows greater stimulation by exudates of non-mycorrhizal than by mycorrhizal strawberry roots. Mycol Res 104:1069–1073

Norman J, Atkinson D, Hooker J (1996) Arbuscular mycorrhizal fungal-induced alteration to root architecture in strawberry and induced resistance to the root pathogen Phytophthora fragariae. Plant Soil 185:191–198

Ozgonen H, Erkilic A (2007) Growth enhancement and Phytophthora blight (Phytophthora capsici Leonian) control by arbuscular mycorrhizal fungal inoculation in pepper. Crop Prot 26:1682–1688

Pande S, Siddique KHM, Kishore GK, Bayaa B, Gaur PM, Gowda CLL, Crouch JH (2005) Ascochyta blight of chickpea (Cicer arietinum L.): a review of biology, pathogenicity, and disease management. Crop Past Sci 56(4):317–332

Parniske M (2008) Arbuscular mycorrhiza: the mother of plant root endosymbioses. Nat Rev Genet 6:763–775

Pastor V, Luna E, Mauch-Mani B, Ton J, Flors V (2012) Primed plants do not forget. Environ Exp Bot. https://doi.org/10.1016/j.envexpbot.2012.02.013

Paszkowski U (2006) Mutualism and parasitism: the yin and yang of plant symbioses. Curr Opin Plant Biol 9:364–370

Paterson E, Sim A, Davidson J, Daniell TJ (2016) Arbuscular mycorrhizal hyphae promote priming of native soil organic matter mineralization. Plant Soil 408:243–C254. https://doi.org/10.1007/s11104-016-2928-8

Peck S, Mittler R (2020) Plant signaling in biotic and abiotic stress. J Exp Bot 71:1649–1651. https://doi.org/10.1093/jxb/eraa051

Peña-Cortés H, Barrios P, Dorta F, Polanco V, Sánchez C, Sánchez E, Ramírez I (2004) Involvement of jasmonic acid and derivatives in plant response to pathogen and insects and in fruit ripening. J Plant Growth Regul 23:246–260

Pereira A (2016) Plant abiotic stress challenges from the changing environment. Front Plant Sci 7: 1123

Petutschnig EK, Jones AM, Serazetdinova L, Lipka U, Lipka V (2010) The lysin motif receptor-like kinase (LysM-RLK) CERK1 is a major chitinbinding protein in Arabidopsis thaliana and subject to chitin induced phosphorylation. J Biol Chem 285(37):28902–28911

Pieterse CMJ, Van Wees SCM, Van Pelt JA, Knoester M, Laan R, Gerrits H, Weisbeek PJ, Van Loon LC (1998) A novel signaling pathway controlling induced systemic resistance in Arabidopsis. Plant Cell 10:1571–1580

Pinior A, Wyss U, Piché Y, Vierheilig H (1999) Plants colonized by AM fungi regulate further root colonization by AM fungi through altered root exudation. Can J Bot 77:891–897

Pinochet J, Calvet C, Camprubi A, Fernandez C (1996) Interactions between migratory endoparasitic nematodes and arbuscular mycorrhizal fungi in perennial crops – a review. Plant Soil 185: 183–190

Pitzschke A, Schikora A, Hirt H (2009) MAPK cascade signalling networks in plant defence. Curr Opin Plant Biol 12(4):421–426

Pivato B, Gamalero E, Lemanceau P, Berta G (2008) Colonization of adventitious roots of Medicago truncatula by Pseudomonas fluorescens C7R12 as affected by arbuscular mycorrhiza. FEMS Microbiol Lett 289:173–180

Povero G, Loreti E, Pucciariello C, Santaniello A, Di Tommaso D, Di Tommaso G, Perata P (2011) Transcript profiling of chitosan-treated Arabidopsis seedlings. J Plant Res 124(5):619–629

Pozo MJ, Azcón-Aguilar C (2007) Unraveling mycorrhiza-induced resistance. Curr Opin Plant Biol 10:393–398

Pozo MJ, Azcón-Aguilar C, Dumas-Gaudot E, Barea JM (1999) β-1,3-Glucanase activities in tomato roots inoculated with arbuscular mycorrhizal fungi and/or Phytophthora parasitica and their possible involvement in bioprotection. Plant Sci 141:149–157

Pozo MJ, Cordier C, Dumas-Gaudot E, Gianinazzi S, Barea JM, Azcón-Aguilar C (2002) Localized versus systemic effect of arbuscular mycorrhizal fungi on defence responses to Phytophthora infection in tomato plants. J Exp Bot 53:525–534

Pozo MJ, Van Der Ent S, Van Loon LC, Pieterse CMJ (2008) Transcription factor MYC2 is involved in priming for enhanced defense during rhizobacteria-induced systemic resistance in Arabidopsis thaliana. New Phytol 180:511–523

Pozo MJ, Verhage A, García-Andrade J, García JM, Azcón-Aguilar C (2009) Priming plant defence against pathogens by arbuscular mycorrhizal fungi. In: Azcón-Aguilar C, Barea JM, Gianinazzi S, Gianinazzi-Pearson V (eds) Mycorrhizas—functional processes and ecological impact. Springer, Berlin, pp 123–135

Pozo MJ, Jung SC, López-Ráez JA, Azcón-Aguilar C (2010) Impact of arbuscular mycorrhizal symbiosis on plant response to biotic stress: the role of plant defence mechanisms. In: Kapulnick Y, Douds DD (eds) Arbuscular mycorrhizas: physiology and function. Springer, Dordrecht, pp 193–207

Prins M, Laimer M, Noris E, Schubert J, Wassenegger M, Tepfer M (2008) Strategies for antiviral resistance in transgenic plants. Mol Plant Pathol 9(1):73–83

Prinsen E, Chauvaux N, Schmidt J, John M, Wieneke U, De Greef J, Schell J, Van Onckelen H (1991) Stimulation of indole-3-acetic acid production in rhizobium by flavonoids. FEBS Lett 282:53–55

Purcell AH, Hopkins DL (1996) Fastidious xylem-limited bacterial plant pathogens. Annu Rev Phytopathol 34(1):131–151

Ramamoorthy V, Viswanathan R, Raguchander T, Prakasam V, Samiyappan R (2001) Induction of systemic resistance by plant growth promoting rhizobacteria in crop plants against pests and diseases. Crop Prot 20(1):1–11

Ramegowda V, Senthil-Kumar M (2015) The interactive effects of simultaneous biotic and abiotic stresses on plants: mechanistic understanding from drought and pathogen combination. J Plant Physiol 176:47–54

Ramesh K, Rao AN, Chauhan BS (2017) Role of crop competition in managing weeds in rice, wheat, and maize in India: a review. Crop Prot 95:14–21

Randhir R, Lin Y-T, Shetty K (2004) Phenolics, their antioxidant and antimicrobial activity in dark germinated fenugreek sprouts in response to peptide and phytochemical elicitors. Asia Pac J Clin Nutr 13:295–307

Reddy N, Raghavender CR, Sreevani A (2006) Approach for enhancing Mycorrhiza mediated disease resistance of tomato damping-off. Indian Phytopathol 59(3):299–304

Reddy DVR, Sudarshana MR, Fuchs M, Rao NC, Thottappilly G (2009) Genetically engineered virus-resistant plants in developing countries: status and future prospects. Adv Virus Res 75: 185–220

Rejeb IB, Pastor V, Mauch-Mani B (2014) Plant responses to simultaneous biotic and abiotic stress: molecular mechanisms. Plants 3:458–475. https://doi.org/10.3390/plants3040458

Requena N, Serrano E, Ocón A, Breuninger M (2007) Plant signals and fungal perception during arbuscular mycorrhizal establishment. Phytochemistry 68:33–40

Riely BK, Lougnon G, Ané JM, Cook DR (2007) The symbiotic ion channel homolog DMI1 is localized in the nuclear membrane of Medicago truncatula roots. Plant J 49:208–216

Rizhsky L, Liang H, Shuman J, Shulaev V, Davletova S, Mittler R (2004) When defense pathways collide. The response of Arabidopsis to a combination of drought and heat stress. Plant Physiol 134:1683–1696. https://doi.org/10.1104/pp.103.033431

Robards R, Antolovich M (1997) Analytical chemistry of fruit bioflavonoids. A review. Analyst 122:11R–34R

Rodriguez RJ, Henson J, Van Volkenburgh E, Hoy M, Wright L, Beckwith F et al (2008) Stress tolerance in plants via habitat-adapted symbiosis. Int Soc Microb Ecol 2:404–416. https://doi.org/10.1038/ismej.2007.106

Rouphael Y, Franken P, Schneider C, Schwarz D, Giovannetti M, Agnolucci M (2015) Arbuscular mycorrhizal fungi act as bio-stimulants in horticultural crops. Sci Hort 196:91–108. https://doi.org/10.1016/j.scienta.2015.09.002

Rumbou A, von Bargen S, Büttner C (2009) A model system for plant-virus interaction—infectivity and seed transmission of Cherry leaf roll virus (CLRV) in Arabidopsis thaliana. Eur J Plant Pathol 124(3):527–532

Safdar A, Javed N, Khan SA, Safdar H, Haq IU, Abbas H, Ullah Z (2013) Synergistic effect of a fungus, Fusarium semitectum, and a nematode, Tylenchulus semipenetrans, on citrus decline. Pak J Zool 45(3):643–651

Saijo Y, Loo EPI (2020) Plant immunity in signal integration between biotic and abiotic stress responses. New Phytol 225:87–104. https://doi.org/10.1111/nph.15989

Salam EA, Alatar A, El-Sheikh MA (2017) Inoculation with arbuscular mycorrhizal fungi alleviates harmful effects of drought stress on damask rose. Saudi J Biol Sci 25(8):1772–1780. https://doi.org/10.1016/j.sjbs.2017.10.015

Sankaranarayanan C, Sundarababu R (1994) Interaction of Glomus fasciculatum with Meloidogyne incognita inoculated at different timings on blackgram (Vigna mungo). Nematol Medit 22:35–36

Santamaria ME, Martínez M, Cambra I, Grbíc V, Diaz I (2013) Understanding plant defence responses against herbivore attacks: an essential first step towards the development of sustainable resistance against pests. Transgenic Res 22:697–708
Santi MM, Dipjyoti C, Satyahari D (2010) Phenolic acids act as signaling molecules in plantmicrobe symbioses. Plant Signal Behav 5(4):359–368
Sardana V, Mahajan G, Jabran K, Chauhan BS (2017) Role of competition in managing weeds: an introduction to the special issue. Crop Prot 95:1–7
Sato K, Naito M, Yukitake H, Hirakawa H, Shoji M, McBride MJ, Nakayama K (2010) A protein secretion system linked to bacteroidete gliding motility and pathogenesis. Proc Natl Acad Sci 107(1):276–281
Scervino JM, Ponce MA, Erra-Bassells R, Vierheilig H, Ocampo JA, Godeas A (2005a) Arbuscular mycorrhizal colonization of tomato by Gigaspora and Glomus species in the presence of root flavonoids. J Plant Physiol 162:625–633
Scervino JM, Ponce MA, Erra-Bassells R, Vierheilig H, Ocampo JA, Godeas A (2005b) Flavonoids exhibit fungal species and genus specific effects on the presymbiotic growth of Gigaspora and Glomus. Mycol Res 109:789–794
Scervino JM, Ponce MA, Erra-Bassells R, Bornpadre J, Vierheilig H, Ocampo JA, Godeas A (2007) The effect of flavones and flavonols on colonization of tomato plants by arbuscular mycorrhizal fungi of the genera Gigaspora and Glomus. Can J Microbiol 53:702–709
Schellenbaum L, Berta G, Ravolanirina F, Tisserant B, Gianinazzi S, Fitter AH (1991) Influence of endomycorrhizal infection on root morphology in a micropropagated woody plant species (Vitis vinifera L.). Ann Bot 68:135–141
Schliemann W, Ammer C, Strack D (2008) Metabolite profiling of mycorrhizal roots of Medicago truncatula. Phytochemistry 69:112–146
Schulze-Lefert P, Vogel J (2000) Closing the ranks to attack by powdery mildew. Trends Plant Sci 5(8):343–348
Schumann GL, D'Arcy CJ (2010) Essential plant pathology. American Phytopathological Society (APS Press), St. Paul, MI
Schüβler A, Schwarzott D, Walker C (2001) A new fungal phylum, the Glomeromycota: phylogeny and evolution. Mycol Res 105:1413–1421
Searle IR, Men AE, Laniya TS, Buzas DM, Iturbe-Ormaetxe I, Carroll BJ, Gresshoff PM (2003) Long-distance signaling in nodulation directed by a CLAVATA1-like receptor kinase. Science 299:109–112
Sharma SS, Dietz KJ (2009) The relationship between metal toxicity and cellular redox imbalance. Trends Plant Sci 14(1):43–50
Sharma SK, Bryan GJ, Winfield MO, Millam S (2007) Stability of potato (Solanum tuberosum L.) plants regenerated via somatic embryos, axillary bud proliferated shoots, microtubers and true potato seeds: a comparative phenotypic, cytogenetic and molecular assessment. Planta 226(6): 1449–1458
Sharma HC, Sujana G, Rao DM (2009) Morphological and chemical components of resistance to pod borer, Helicoverpa armigera in wild relatives of pigeonpea. Arthropod Plant Interact 3(3): 151–161
Shreenivasa R, Krishnappa K, Ravichandra NG (2007) Interaction effects of arbuscular mycorrhizal fungus Glomus fasciculatum and root-knot nematode, Meloidogyne incognita on growth and phosphorous uptake of tomato. Karnataka J Agric Sci 20:57–61
Siddiqui ZA, Akhtar MS (2007) Effects of AM fungi and organic fertilizers on the reproduction of the nematode Meloidogyne incognita and on the growth and water loss of tomato. Biol Fertil Soils 43:603–609
Siddiqui ZA, Mahmood I (1998) Effect of a plant growth promoting bacterium, an AM fungus and soil types on the morphometrics and reproduction of Meloidogyne javanica on tomato. Appl Soil Ecol 8:77–84
Siddiqui ZA, Pichtel J (2008) Mycorrhizae: an overview. In: Mycorrhizae: sustainable agriculture and forestry. Springer Science and Business Media LLC, Berlin, Germany, pp 1–35

Singh LP, Gill SS, Tuteja N (2011) Unraveling the role of fungal symbionts in plant abiotic stress tolerance. Plant Signal Behav 6:175–191
Siqueira JO, Safir GR, Nair MG (1991a) Significance of phenolic compounds in plant-soil microbial systems. Crit Rev Plant Sci 10:63
Siqueira JO, Safir GR, Nair MG (1991b) Stimulation of vesicular arbuscular mycorrhiza formation and growth of white clover by flavonoid compounds. New Phytol 118:87–93
Skorupska A, Wielbo J, Kidaj D, Marek-Kozaczuk M (2010) Enhancing Rhizobium-legume symbiosis using signaling factors. In: Khan MS (ed) Microbes for legume improvement. Springer, Wien, pp 27–54
Smith SE, Read DJ (1997) Mycorrhizal symbiosis. Academic, San Diego
Smith SE, Read DJ (2008) Mycorrhizal symbiosis, 3rd edn. Elsevier Academic Publishers, New York
Smith SE, Read DJ (2010) Mycorrhizal symbiosis, Academic, Cambridge, MA. ISBN 978-0-08-055934-6
Song Y, Chen D, Lu K, Sun Z, Zeng R (2015) Enhanced tomato disease resistance primed by arbuscular mycorrhizal fungus. Front Plant Sci 6:6
Spagnoletti FN, Cornero M, Chiocchio V, Lavado RS, Roberts IN (2020) Arbuscular mycorrhiza protects soybean plants against Macrophomina phaseolina even under nitrogen fertilization. Eur J Plant Pathol 156:839–849
Spatafora JW, Chang Y, Benny GL, Lazarus K, Smith ME, Berbee ML, Bonito G, Corradi N, Grigoriev IV, Gryganskyi A et al (2016) A phylum-level phylogenetic classification of zygomycete fungi based on genome-scale data. Mycologia 108:1028–1046
Spoel SH, Dong X (2012) How do plants achieve immunity? Defence without specialized immune cells. Nat Rev Immunol 12:89–100
St Arnaud M, Elsen A (2005) Interaction of arbuscular mycorrhizal fungi with soil-borne pathogens and non-pathogenic rhizosphere micro-organisms. In: Declerck S, Fortin JA, Strullu D-G (eds) In vitro culture of mycorrhizas. Springer, Berlin, pp 217–231
Steinkellner S, Lendzemo V, Langer I, Schweiger I, Khaosaad T, Toussaint J-P, Vierheilig H (2007) Flavonoids and strigolactones in root exudates as signals in symbiotic and pathogenic plant-fungus interactions. Molecules 12:1290–1306
Stracke S, Kistner C, Yoshida S, Mulder L, Sato S, Kaneko T, Tabata S, Sandal N, Stougaard J, Szczyglowski K, Parniske M (2002) A plant receptor-like kinase required for both bacterial and fungal symbiosis. Nature 417:959–962
Subramanian S, Stacey G, Yu O (2007) Distinct, crucial roles of flavonoids during legume nodulation. Trends Plant Sci 12:282–285
Swain T (1975) Evolution of flavonoid compounds. In: Harborne JB, Mabry TJ, Mabry H (eds) The flavonoids. Chapman & Hall, London, pp 1096–1138
Taliansky M, Torrance L, Kalinina NO (2008) Role of plant virus movement proteins. Methods Mol Biol 451:33–54
Tarkowski P, Vereecke D (2014) Threats and opportunities of plant pathogenic bacteria. Biotechnol Adv 32(1):215–229
Tedersoo L, Sánchez-Ramírez S, Kõljalg U, Bahram M, Döring M, Schigel DS, May T, Ryberg M, Abarenkov K (2018) High-level classification of the Fungi and a tool for evolutionary ecological analyses. Fungal Divers 90:135–159
Thirkell TJ, Charters MD, Elliott AJ, Sait SM, Field KJ (2017) Are mycorrhizal fungi our sustainable saviours considerations for achieving food security. J Ecol 105:921–929. https://doi.org/10.1111/1365-2745.12788
Thomma BP, Tierens KF, Penninckx IA, Mauch-Mani B, Broekaert WF, Cammue BP (2001) Different micro-organisms differentially induce Arabidopsis disease response pathways. Plant Physiol Biochem 39(7):673–680
Thorne ET, Young BM, Young GM, Stevenson JF, Labavitch JM, Matthews MA, Rost TL (2006) The structure of xylem vessels in grapevine (Vitaceae) and a possible passive mechanism for the systemic spread of bacterial disease. Am J Bot 93(4):497–504

Utkhede R (2006) Increased growth and yield of hydroponically grown greenhouse tomato plants inoculated with arbuscular mycorrhizal fungi and Fusarium oxysporum f. sp radicislycopersici. BioControl 51:393–400

Van-Dam NM (2009) How plants cope with biotic interactions. Plant Biol 11:1–5

Van der Ent S, Van Wees SCM, Pieterse CMJ (2009) Jasmonate signaling in plant interactions with resistance-inducing beneficial microbes. Phytochemistry 70:1581–1588

Van Hulten M, Pelser M, Van Loon LC, Pieterse CMJ, Ton J (2006) Costs and benefits of priming for defense in Arabidopsis. Proc Natl Acad Sci U S A 103:5602–5607

Van Rossum D, Schuurmans FP, Gillis M, Muyotcha A, Van Verseveld HW, Stouthamer AH, Boogerd FC (1995) Genetic and phenetic analyses of Bradyrhizobium strains nodulating peanut (Arachis hypogaea L.) roots. Appl Environ Microbiol 61:1599–1609

Van Wees SCM, Van der Ent S, Pieterse CMJ (2008) Plant immune responses triggered by beneficial microbes. Curr Opin Plant Biol 11:443–448

Van-Doorn WG, Beers EP, Dangl JL, Franklin-Tong VE, Gallois P, Hara-Nishimura I, Mur LAJ (2011) Morphological classification of plant cell deaths. Cell Death Differ 18(8):1241–1246

Veresoglou SD, Rillig MC (2012) Suppression of fungal and nematode plant pathogens through arbuscular mycorrhizal fungi. Biol Lett 8:214–217

Verhage A, van Wees SC, Pieterse CM (2010) Plant immunity: it's the hormones talking, but what do they say? Plant Physiol 154(2):536–540

Vierheilig H (2004a) Regulatory mechanisms during the plant-arbuscular mycorrhizal fungus interaction. Can J Bot 82:1166–1176

Vierheilig H (2004b) Further root colonization by arbuscular mycorrhizal fungi in already mycorrhizal plants is suppressed after a critical level of root colonization. J Plant Physiol 161:339–341

Vierheilig H, Piché Y (2002) Signalling in arbuscular mycorrhiza: facts and hypotheses. In: Buslig B, Manthey J (eds) Flavonoids in cell functions. Kluwer/Plenum, New York, pp 23–39

Vierheilig H, Bago B, Albrecht C, Poulin MJ, Piché Y (1998) Flavonoids and arbuscular mycorrhizal fungi. In: Manthey J, Buslig B (eds) Flavonoids in the living system. Plenum Press, New York, pp 9–33

Vierheilig H, Steinkellner S, Khaosaad T, Garcia-Garrido JM (2008) The biocontrol effect of mycorrhization on soilborne fungal pathogens and the autoregulation of the AM symbiosis: one mechanism, two effects? In: Varma A (ed) Mycorrhiza: state of the art, genetics and molecular biology, eco-function, biotechnology, eco-physiology, structure and systematics. Springer, Berlin, pp 307–320

Vos C, Claerhout S, Mkandawire R, Panis B, De Waele D, Elsen A (2011) Arbuscular mycorrhizal fungi reduce root-knot nematode penetration through altered root exudation of their host. Plant Soil 354:335–345

Vos C, Claerhout S, Mkandawire R, Panis B, de Waele D, Elsen A (2012a) Arbuscular mycorrhizal fungi reduce root-knot nematode penetration through altered root exudation of their host. Plant Soil 354:335–345

Vos C, Tesfahun A, Panis B, De Waele D, Elsen A (2012b) Arbuscular mycorrhizal fungi induce systemic resistance in tomato against the sedentary nematode Meloidogyne incognita and the migratory nematode Pratylenchus penetrans. Appl Soil Ecol 61:1–6

Vuong TD, Sonah H, Meinhardt CG, Deshmukh R, Kadam S, Nelson RL, Nguyen HT (2015) Genetic architecture of cyst nematode resistance revealed by genome-wide association study in soybean. BMC Genomics 16(1):593

Waller F, Achatz B, Baltruschat H, Fodor J, Becker K, Fischer M, Franken P (2005) The endophytic fungus Piriformospora indica reprograms barley to salt-stress tolerance, disease resistance, and higher yield. Proc Natl Acad Sci U S A 102(38):13386–13391

Walling LL (2008) Avoiding effective defenses: strategies employed by phloem-feeding insects. Plant Physiol 146(3):859–866

Walters D, Heil M (2007) Costs and trade-offs associated with induced resistance. Physiol Mol Plant Pathol 71:3–17

Wang B, Qiu Y-L (2006) Phylogenetic distribution and evolution of mycorrhizas in land plants. Mycorrhiza 16:299–363

Wang Z, Ma L-Y, Cao J, Li Y-L, Ding L-N, Zhu K-M et al (2019) Recent advances in mechanisms of plant defense to Sclerotinia sclerotiorum. Front Plant Sci 10:1314. https://doi.org/10.3389/fpls.2019.0131

War AR, Sharma HC, Paulraj MG, War MY, Ignacimuthu S (2011) Herbivore induced plant volatiles: their role in plant defense for pest management. Plant Signal Behav 6(12):1973–1978

Whipps JM (2004) Prospects and limitations for mycorrhizas in biocontrol of root pathogens. Can J Bot 82:1198–1227

Wu S, Chappell J (2008) Metabolic engineering of natural products in plants; tools of the trade and challenges for the future. Curr Opin Biotechnol 19:145–152

Wu Q, Luo Y, Lu R, Lau N, Lai EC, Li WX, Ding SW (2010) Virus discovery by deep sequencing and assembly of virus-derived small silencing RNAs. Proc Natl Acad Sci 107(4):1606–1611

Yano K, Yoshida S, Muller J, Singh S, Banba M, Vickers K, Markmann K, White C, Schuller B, Sato S, Asamizu E, Tabata S, Murooka Y, Perry J, Wang TL, Kawaguchi M, Imaizumi-Anraku H, Hayashi M, Parniske M (2008) CYCLOPS, a mediator of symbiotic intracellular accommodation. Proc Natl Acad Sci U S A 105:20540–20545

Yao MK, Désilets H, Charles MT, Boulanger R, Tweddell RJ (2003) Effect of mycorrhization on the accumulation of rishitin and solavetivone in potato plantlets challenged with Rhizoctonia solani. Mycorrhiza 13:333–336

Youssef MMA, El-Nagdi WMA (2015) Vesicular arbuscular mycorrhizae: a promising trend for biocontrolling plant parasitic nematodes. A review. Sci Agric 11:76–80

Zaynab M, Fatima M, Abbas S, Sharif Y, Umair M, Zafar MH, Bahadar K (2018) Role of secondary metabolites in plant defense against pathogens. Microb Pathog 124:198–202

Zhang J, Subramanian S, Stacey G, Yu O (2009) Flavones and flavonols play distinct critical roles during nodulation of Medicago truncatula by Sinorhizobium meliloti. Plant J 57:171–183

Zhang R, Zhub H, Zhao H, Yao Q (2013) Arbuscular mycorrhizal fungal inoculation increases phenolic synthesis in clover roots via hydrogen peroxide, salicylic acid and nitric oxide signaling pathways. J Plant Physiol 170(1):74–79

Zhu JK (2016) Abiotic stress signalling and responses in plants. Cell 167(2):313–324

Zhu XC, Song FB, Xu HW (2010) Arbuscular mycorrhizae improve low temperature stress in maize via alterations in host water status and photosynthesis. Plant Soil 331:129–137. https://doi.org/10.1007/s11104-009-0239-z

Zou YN, Srivastava AK, Wu QS (2016) Glomalin: a potential soil conditioner for perennial fruits. Int J Agric Biol 18:293–297. https://doi.org/10.17957/IJAB/15.0085

Insights into Biotic Stress Management by Plants Using Phenolic Compounds

3

Amanpreet Kaur, Manpreet Kaur, and Yamini Tak

Abstract

Plants are sessile in nature and are not able to escape from various pathogens including bacteria, fungi, viruses, nematodes, insects, weeds, and herbivores which cause biotic stress. These agents can right away deprive host by absorbing their nutrients and lead to plants' death. In this way, it could be the major reason of pre/postharvest crop loss. Plants counter biotic stress through different strategies or defense mechanisms including resistant genes and antioxidant defense system. Plants sense external stimuli by sensors present on cell membrane or cytoplasm, and it transfers to nucleus and ultimately activates signal transduction machinery. When the plant is under any kind of stress, there is the excessive reactive oxygen species (ROS) or reactive nitrogen species (RNS) production, and their homeostasis is very necessary to prevent from oxidative damage. Phenolics are a group of plant secondary metabolites that contain one or more hydroxy derivatives of benzene rings and have an ideal structure to scavenge ROS. Phenolics protect the plant by binding to the cell wall, deposition of lignin to interrupt the entry of pathogens, and conversion of simplex phenolics to complex under stress. This chapter will highlight the role, functions, targets, and mode of action of various phenolics in amelioration of biotic stress.

Keywords

Amelioration · Antioxidant · Biotic · Phenolics · Reactive oxygen species

A. Kaur (✉) · M. Kaur
Department of Biochemistry, Punjab Agricultural University, Ludhiana, Punjab, India

Y. Tak
Agricultural Research Station, Ummedganj, Agriculture University, Kota, Rajasthan, India

R. Lone et al. (eds.), *Plant Phenolics in Biotic Stress Management*,
https://doi.org/10.1007/978-981-99-3334-1_3

3.1 Introduction

There is a need to reduce the crop loss caused by diseases in order to meet the rising food demand with population growth (Pawlak and Kołodziejczak 2020). Plants' growth and development depend upon the quality of seed, soil composition and fertility (Schjoerring et al. 2019), and environmental conditions (Heins et al. 1998). From past many decades, plants in their wild progeny state are much more resistant to stress conditions (Kovalchuk 2021). But modern cultivars formed by plant breeders for economy benefits by increasing its yield (da Silva Dias 2020) by using pesticides make them more susceptible to environmental conditions. Various biotic and abiotic stresses affect the crop productivity (Gull et al. 2019). Biotic stress includes the interaction of crops with harmful organisms like bacteria, viruses, fungi, nematodes, weeds, herbivore, etc. and abiotic factors like UV, salinity, heavy metal, and water stress (Gull et al. 2019). All these factors alone or in combination have an effect on the growth, development, yield, and efficiency of crop (Hassan et al. 2021; Rivero et al. 2022). Plants overcome or suppress biotic stress up to some extend by using its own machinery including antioxidant defense system (Das et al. 2016). Primary defense systems in plants are waxes on leaves, cuticle, thorns, and release of volatile compounds (Belete 2018). R-genes are present in plants against biotic stress (Maiti et al. 2014). Many signaling compounds synthesized in higher amount to act as defense compounds and activate defense genes (Rojo et al. 2003). When plants are attacked by bacteria or any other biotic agent, unnecessary accumulation of reactive oxygen species (ROS) and reactive nitrogen species (RNS) occurs, leading to oxidative stress which is ultimately harmful for plants (Czarnocka and Karpiński 2018). As membrane fluidity increases due to lipid peroxidation caused by oxidative stress, organelles flow out of cell and degraded. In order to protect the plant from oxidative stress, scavenging of ROS takes place by antioxidant defense system which includes antioxidative enzymes like superoxide dismutase (SOD), peroxidase (POD), catalase (CAT), ascorbate peroxidase (APX), and glutathione peroxidase (GPX); nonenzymatic compounds including dehydroascorbate, glutathione, and oxidized glutathione; and numerous phenolic compounds acting as antioxidants (Das and Roychoudhury 2014).

Biotic stress is caused by several biotic agents like bacteria, fungi, and viruses which affect the plant in different ways (Gull et al. 2019). Bacteria enter the plant by wound or small opening of stomata, and then reproduce intercellularly, and when produced in sufficient density, it affects the plant and forms lesions in leaf and then necrosis of the leaf, which ultimately causes defoliation, patchy fruits, and plant death. Bacteria absorb the xylem sap material and make it unavailable to the shoot (Graham et al. 2004; Gulya et al. 1997). Symptoms of bacterial infection in plants are leaf yellowing, leaf spot, blits, wilts, sacks, cankers, soft rot of roots, storage organs and fruits, and overgrowth. Fungi can attack plants in three ways (1) biotrophic way which acts on living tissue, overcomes host machinery and nutrients (Koeck et al. 2011); (2) necrotrophic way which acts on dead tissue and gain nutrients from them and attacks on plant by secreting cell wall-degrading enzymes like oxidases, cutinases, and lipases, then necrotic lesions and finally plant death (Williamson et al. 2007); and (3) hemi-biotrophic way which has half-life as biotrophs and then necrotrophs (Koeck et al. 2011). Generally, fungi attach to cell and then form hyphae and arbuscular in the host. Biotrophic fungi affect the crops in adverse manner because they take all carbohydrates and nutrients from the host. Starvation occurs in host and plant dies. Fungi reproduce in the form of spores. Spores move from one host to other by air, water, insects, etc. In this way, fungal disease easily transmits from unhealthy plant to healthy plant, e.g., powdery mildew and rust fungi. Viruses cause disease in many plants and are composed of outer cover of proteins and nucleic acids inside it (Lodish et al. 2000). When they enter the plant cell, outer membrane is degraded and the nucleic acids enter the hosts' nuclei and integrate there to form many more viruses. In this way, viruses overcome the host machinery to their use. Viruses are transmitted from one plant to another by insect carriers. Insects, nematodes, and pest move from one plant species to other and some are specific to attack on one species. When insects attack plant, they absorb all their nutrients from leaves (Hansen and Moran 2014) and leave them hungry. Some insects act only on fruits, other on leaves. Herbivores eat the plant by chewing the plant tissues. Weeds compete with plants by absorbing nutrients, and some weeds also excrete enzymes that retard the growth of plants or even seeds do not germinate in that condition.

In the counter action, to protect from biotic stress, plants activate ROS scavenging system through antioxidant defense mechanism. Phenolics are secondary metabolites which got enhanced during plethora of biotic stress conditions as well as play a fundamental role in first line defense to protect from several diseases caused by them (Tak and Kumar 2020). Phenolics can act as superior antioxidant due to various structural and other properties like presence of one or more hydroxyl group on benzene ring, high ROS scavenging power, metal chelation, and high antioxidant capacity. Phenolics scavenge ROS with the help of OH group present in their structure which acts as a hydrogen donor to free radical, and phenoxy radical intermediates are formed by donation of hydrogen to free radical. In terms to counteract biotic stress, phenolics increase the permeability of pathogens' membranes by lipid peroxidation and denaturation of their membranes deform the structure of membrane-bound proteins which in turn changes the pH gradient, electron transport cycle, and ATP production. All these changes in metabolism of pathogens lead to their death. In this way, phenolics act as safeguarding options to fight against various biotic pathogens like viruses, nematodes, insects, bacteria, or fungi. In this chapter, we intended to furnish the current understanding on the different types and biosynthesis of phenolics, functions, and how they mitigate the biotic stress in plants.

3.2 Types, Structure, and Biosynthesis of Phenolics

The primary structure of phenolics consists of one or more OH groups attached on the benzene ring which form phenols and polyphenols, respectively (Bhattacharya et al. 2010). Further, phenolics are divided into different categories like phenolics, flavonoids, cinnamic acid, tannins, anthocyanins, lignins, and phytoalexins (Wallis

Table 3.1 Different types of phenolic compounds and their structures

Type of phenolic	Name of compound	Carbon skeleton	Structure
Simple phenolics	Phenolic acid	C6–C1	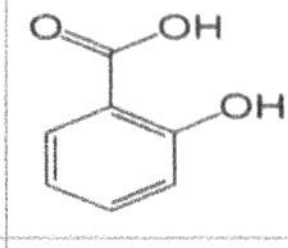
	Benzoic acid derivatives	C6–C1	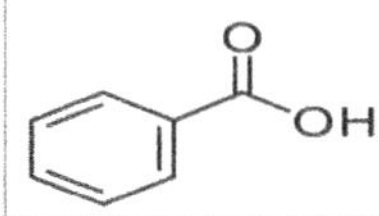
	Cinnamic acid derivatives	C6–C3	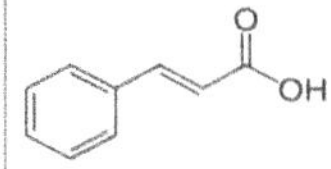
Flavonoids	Flavones Flavanols	C6–C3–C6	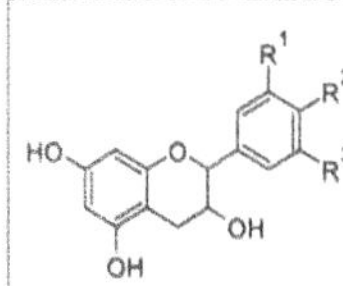
	Anthocyanins		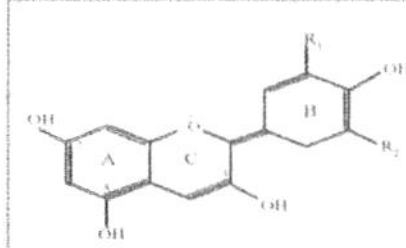
Complex phenolics	Tannins	$(C6–C3–C6)_n$	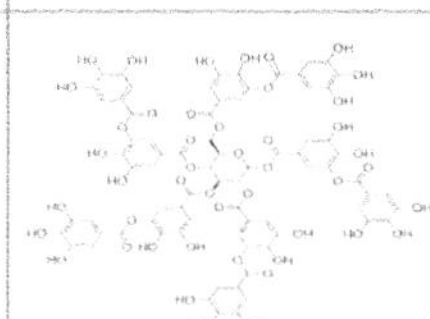
	Lignans	$(C6–C3)_2$	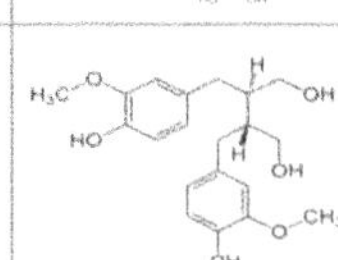
	Stilbenes	C6–C2–C6	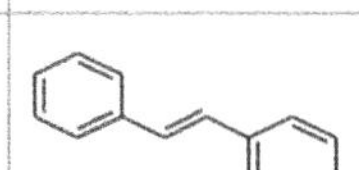

and Galarneau 2020) (Table 3.1). Phenolics are present in all plants in different tissues at different growth stages, and their concentration varies from tissue to tissue, growth stage, and during stress (Thomas and Ravindra 1999; Ozyigit et al. 2007; Bhattacharya et al. 2010). Accumulation of phenolics was observed in the stress conditions like wounding, nutrient stress, and abiotic and biotic stresses under the subepidermal layer of plants. Phenolics also protect plants from photoinhibition (Kefeli et al. 2003). Plant secondary metabolites are synthesized from the compounds of primary metabolism. Phenolics are synthesized by two pathways

(1) the phenylalanine/shikimic acid pathway and (2) the acetate/malonate pathway. The phenylalanine pathway is the most common pathway of phenol synthesis in higher plants (Tak and Kumar 2020). Two precursors of a shikimic acid pathway are erythrose-4 phosphate and phosphoenolpyruvic acid, which form shikimate by several intermediate steps. Then shikimate converts into chorismate by shikimate kinase, 5-enolpyruvylshikimate, and chorismate synthase. Chorismate is the branch point for forming anthranilate and prephenate. Prephenate dehydratase and aromatic aminotransferase convert prephenate to phenylalanine. Phenylalanine is the precursor of cinnamic acid which further forms simple phenolics, flavonoids, lignin, and tannins. Phenylalanine ammonia-lyase (PAL) is an important enzyme for the synthesis of cinnamic acid. From the cinnamic acid, a number of secondary metabolites are formed like stilbenes. They act as osmoregulants, deposit at the site of infection, are insect repellent, prevent photoinhibition by UV light, act as sunscreen, and signal molecule (Tak and Kumar 2020).

3.3 Role of Enzymes in Phenolic Biosynthesis Under Biotic Stress

It has been reported in the literature that to mitigate the effect of biotic stress, activities of various enzymes like peroxidases (Higuchi 2004), phenylalanine ammonia lyase (Kubalt 2016), tyrosine ammonia lyase (Poppe and Rétey 2005), and polyphenoloxidase (Constabel and Barbehenn 2008) enhance, which ultimately contributes in the biosynthesis of phenolics. Peroxidases forms the lignin or complex phenols by polymerization of phenols, phenylalanine ammonia-lyase is the key enzyme for the synthesis of phenols, phytoalexins, and other defense-related chemicals, tyrosine ammonia lyase uses tyrosine to form coumaric acid which is the intermediate of phenolic biosynthesis, and polyphenol oxidases convert phenols to quinines which act as bactericidal, fungicidal, and detoxify the pathogen phytotoxins by the oxidative process. The biosynthesis and accumulation of phenolic acid, flavonoids, and proanthocyanidins were elevated during both infections, but the phenolic defense was stronger and faster in endophyte as compared to pathogenic fungi by increasing the activity of phenolic biosynthesis enzymes (PAL) (Schulz et al. 1999; Laukkanen et al. 2000). Patel et al. (2015) studied infestation of *Fusarium udum* Butler (Wilt) on two resistant and two susceptible pigeon pea genotypes and identified higher activity of enzymes PAL and PPO in leaves and roots of resistant genotypes as compared to susceptible genotypes. Another study concluded that lower accumulation of phenolics in susceptible cultivars of chickpea and pigeon pea makes plants unable to restrict the progression of fungal diseases. On the other hand, resistant cultivars were able to restrict the fungal infection by increasing the accumulation of phenolics in different issues (Datta and Lal 2012). Mehta et al. (1992) observed that the content of total phenols and orthohydroxyphenols in the root exudates of moderately susceptible and highly susceptible cowpea against Rhizoctonia infection and concluded that phenolics were higher in moderately susceptible cultivars as compared to highly susceptible cultivar.

3.4 Extraction and Estimation of Phenolics in Plants Under Stress

Both organic and inorganic solvents are used in the extraction of phenolics (Dai and Mumper 2010). Flavonoids are extracted by using water, ethanol, and methanol, while extraction of anthocyanin requires a low concentration of HCl along with an organic solvent. Proanthocyanidins are complex tannins that are extracted by using ethanol. New extraction techniques for phenolics are Soxhlet, i.e., heated to reflux and maceration; soxetec which is a modified soxlet technique; ultrasound-assisted extraction (UAE); microwave-assisted extraction (MAE); ultrasound microwave-associated extraction (UMAE); supercritical fluid extraction (SFE); subcritical water extraction (SCWE); and high hydrostatic pressure processing (HHPP) (Dai and Mumper 2010). Estimation of phenolics is done by using different instruments including spectrophotometer, gas chromatography, high-performance liquid chromatography, paper chromatography, thin layer chromatography, and capillary electrophoresis. It has been reported that in comparison to 100% ethanol and 80% ethanol/methanol, 50% concentration solvent is much more effective in the extraction of phenolics. Free phenolics are extracted by using 80% methanol, and supernatant is used in the estimation of free phenolics, while residue was dried in air. Then, this residue is used to extract bound phenolics as illustrated by Qiu et al. (2010). A spectrophotometric method is quick, simple, inexpensive, and can rapidly be used for the estimation of secondary metabolites. Estimation of both free and bound phenolics can be done by using the spectrophotometric method of Swain and Hillis (1959). Folin Ciocalteu reagent binds to phenols in the sample and gives blue color, which is read at 760 nm (Singleton et al. 1999). Methanolic preparation of aluminum chloride binds to flavonoids in a sample and gives yellow color which is read at 410 to 423 nm (Bao et al. 2005). In this way, other phenolics can be estimated by using standard reading.

3.5 Mode of Action of Phenolic Compounds

Phenolic compounds contain hydroxyl groups, which are likely to dissociate into phenolate ions. The hydrogen group released from the phenolic compound binds to ROO-, to make it less harmful to plant cells (Dai and Mumper 2010). Phenolics are often called antifeedants, as they reduce digestibility, and act as toxins to insects and herbivores (Pratyusha 2022). When insects eat plants, phenolics present in plant tissue enter the digestive tract of insects and form reactive oxygen species, particularly at low pH present in the mid-gut. These ROS directly damage the mid-gut tissue and forms pore in the insects, leading to the death of the insect. Phytoalexins also hinder the pathogen metabolism (Barz et al. 2007). Complex phenolics like tannins

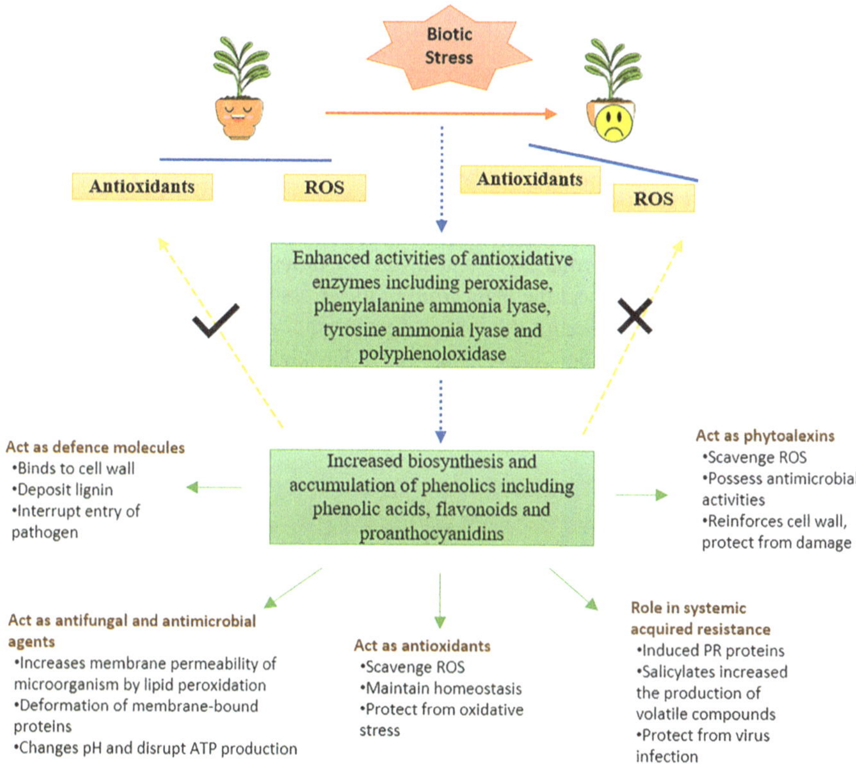

Fig. 3.1 Mode of action of phenolics under biotic stress

lead to protein inactivation in insects and inactivate digestive enzymes including trypsin and chymotrypsin. Insects/herbivores that ingest a high amount of tannins are unable to grow and eventually die (War et al. 2012). Lignin provides a physical barrier to the plants by depositing in the cell wall (Miedes et al. 2014). Figure 3.1 depicts the diagrammatic representation of mode of action of phenolics under biotic stress. It has been studied that furanocoumarins formed in response to herbivores are activated by UV light and denature the DNA of insects, eventually leading to death (War et al. 2012). The action of phenolic defense was depends on the type of infection and the host species. Like in annual grasses, the content of flavonoids was less important in defense (Logemann and Hahlbrock 2002). Flavan-3-ols, that are catechin, epicatechin, and proanthocyanidins, had antifungal properties toward pathogenic fungi (*B. cinerea*) (Hébert et al. 2002). Proanthocyanidins that were in oligomeric state were act as storage compounds and become activated during biotic stress (Koskimäki et al. 2009). The antioxidant action mechanism of phenolics can either through the transfer of hydrogen atom, transition metal ion chelation, sequential proton loss and transfer of a single electron.

3.6 Role and Targets of Phenolic Compounds in Amelioration of Biotic Stress

3.6.1 Phenolics as a Defense Agent

Phenolics are a class of plant secondary metabolites that contain one or more hydroxy derivative of benzene rings. Phenolics are widely distributed in plants and are used for defensive functions in many plant species. Mostly phenolics are under the subepidermal or epidermal layer and in the vacuoles of the plant cell (Hutzler et al. 1998). Phenolics protect the plant from biotic stress by binding to the cell wall, deposition of lignin to interrupt the entry of pathogens and conversion of simplex phenolics to complex under stress (Fig. 3.2). Koskimäki et al. (2009) studied phenolic defense response of bilberry against *Paraphaeosphaeria* sp. (endophyte)

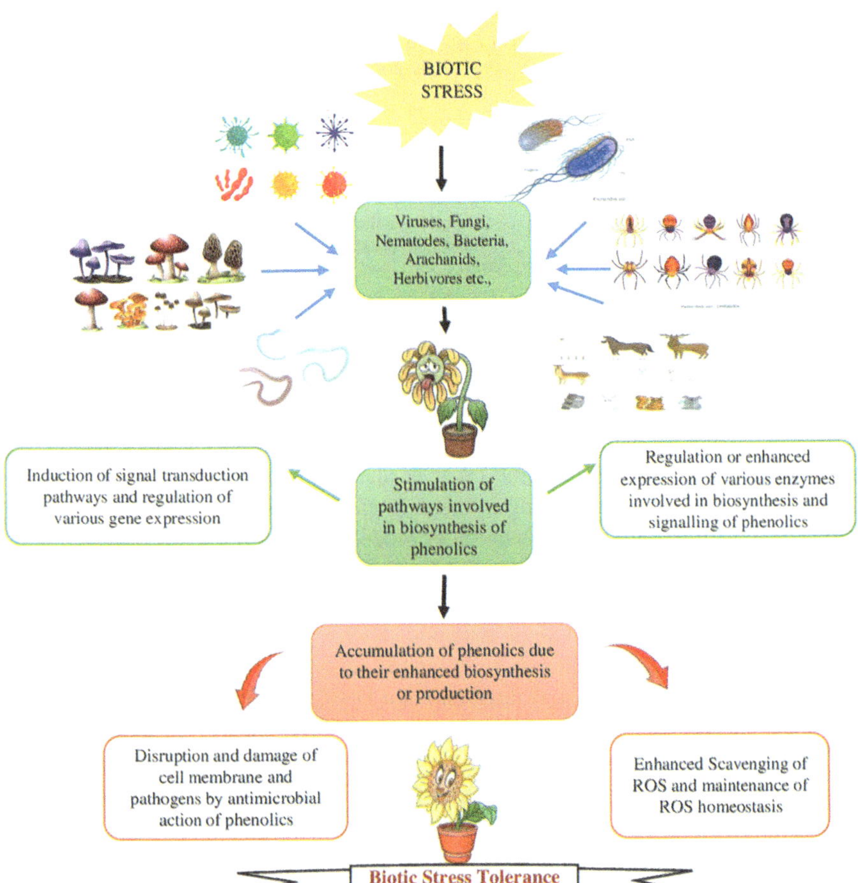

Fig. 3.2 Role of phenolics in biotic stress amelioration

and *Botrytis cinerea* (pathogenic fungi). Phenolics are recognized as phytoestrogen in animal cells and allelochemicals in plant cells (Bhattacharya et al. 2010). Allelochemicals are secondary metabolites like phenolics, tannins, saponins, and terpenoids that may be volatile or water-soluble compounds. Some secondary metabolites deposit in plants against phytopathogens (Pusztahelyi et al. 2015). Chlorosis is the major symptom of insect feeding in wheat plants which is indicative of chlorophyll loss (Heng-Moss et al. 2004). Chakraborty et al. (2002) showed that response to tea plant to blister infection increases the accumulation of phenolics in infested leaves which ultimately affect the quality of tea. Therefore, phenolics can be proposed as a useful alternative to the toxic chemicals that are used to control the phytopathogens in crops (Jamiołkowska 2020). Khattab and Khattab (2005) studied alterations of different biochemical constituents of Eucalyptus leaves infected by Gall-Forming Psyllid insect. Insect feeding disturbs the leaf tissue by forming cavities in the xylem tissue. They took leaves from diseased (galled) and healthy (un-galled) eucalyptus and studied total phenols and phenolic acid fractions. The content of total phenols and fractions like salicylic, ferulic, coumaric, caffeic acids, and vanillin is higher in infested leaves as compared to healthy leaves.

3.6.2 Phenolics as Antimicrobial and Antifungal Agent

Several studies suggested that phenolics and flavonoids have antifungal and antibacterial constituents (Carvalho et al. 2018; Bouarab-Chibane et al. 2019) (Table 3.2). Phenols are usually present in fruit skin and leaves in high concentrations which take part in the defense mechanism against phytopathogens, UV resistance, and pigmentation. In terms to overcome biotic stress, phenolics increase the membrane permeability of microorganisms by lipid peroxidation, denaturation of membranes leads to deforming the structure of membrane-bound proteins which in turn changes the pH gradient, electron transport cycle, and ATP production. All these changes in microorganisms' metabolism led to the death of microbes. Some phenolics are also antimicrobial compounds that disrupt/hinder the plant growth in the apoplast. For example, tomato saponin (α-tomatine) has strong antifungal activity (Arneson and Durbin 1968). It leads to ROS burst and cell component leakage in the fungus *Fusarium oxysporum* (Ruiz-Rubio et al. 2001). Coumarins are an active group of phenolic compounds that has antipathogenic action against bacteria and fungi (Ahmed et al. 2017). Some coumarin-derived halogenated compounds like 7-hydroxylated simple coumarins prevent the *Orabanche cernua* growth in vitro. Furano-coumarins also protect plants against fungal toxicity, e.g., Psaoralin (Ahmed et al. 2017). Maddox et al. (2010) found 12 phenolic compounds which disrupt the attack of *Xylella fastidiosa* in almonds and grapes. Xylem sap of grapes has proteins, amino acids, and low-molecular-weight phytoalexins. Some phytoalexins target particular pathogens, while others have broader-spectrum activities against a variety of pathogens. Nicholson and Hammerschmidt (1992) reported that besides the reason for cell death, plants accumulate phenolics at the infection site and isolate the pathogen to that place

Table 3.2 Role of several phenolic compounds in different crops in biotic stress tolerance

S. No.	Crop	Disease-causing agent	Name of phenolic compound formed	References
1.	Rice	Fungus (*Magnaporthe oryzae*)	Momilactone A and B	Schmelz et al. (2014)
		Bacteria (*Xanthomonas oryzae* pv. *oryzicola*)	Sakuranetin	Park et al. (2014)
2.	Wheat	Fungus (Fusarium)	Phenolic acids like Gallic acid, Vanillic acid, Syringic acid, Benzoic acid	Stuper-Szablewska et al. (2019)
3.	Blackgram	Urdbean Leaf Crinkle Virus	Total phenols	Karthikeyan et al. (2009)
4.	Lentil	Oomycete (*Aphanomycea euteiches*)	Coumaric acid, Naringenin, Vanillic acid	Bazghaleh et al. (2018)
5.	Maize	Fungus (*Fusarium verticillioides*)	Anthocyanin	Bernardi et al. (2018)
6.	Tomato	Fungus (*Alternaria solani*)	Gallic acid	El-Nagar et al. (2020)
		Bacteria (*Pseudomonas syringae*)	Anthocyanin Ferulic acid, Benzoic acid	Dadáková et al. (2020)
7.	Potato	Oomycete (*Phytophthora infestans*)	Hydroxycinnamic acid	Pushpa et al. (2014)
8.	Capsicum	Fungus (*Phytophthora capsici*)	Lignin	Hu-zhe et al. (2005)
9.	Barley	Fungus (*Ustilago hordei*)	Total phenols	Singh et al. (2021)
10.	Grape berries	Fungus (*Rhizopus stolonifer)*	Resveratrol	Sarig et al. (1997)
11.	Oat	Fungus (*Puccinia coronate*)	Avenanthramides	Dimberg and Peterson (2009)
12.	Cotton	Bacteria (Bacillus *coagulans, cereus, subtilis*)	Total phenols	Aly et al. (2022)
13.	Fenugreek	Fungus (*Erysiphe polygoni*)	Total phenols	Avtar et al. (2003)
14.	Flax	Fungus (*Oidium lini*)	Total phenols	Mohamed et al. (2012)
15.	Sugarcane	Fungus (*Sporisorium scitamineum*)	Total phenols and lignin	Deng et al. (2020)

(continued)

Table 3.2 (continued)

S. No.	Crop	Disease-causing agent	Name of phenolic compound formed	References
16.	Onion	*Botrytis allii*	Total phenol content	Hussein et al. (2018)
17.	Strawberry	*Tetranych usurticae* Koch	Catechol-based phenolics	Steinite and Levinsh (2002, 2003)
18.	Cucumber and chayote	Melon fly (*Bacterocera cucurbitae*)	Phenolics and total flavonoids	Pramod (2017)
19.	Coconut	*Ganoderma lucidum*	Phenolics	Karthikeyan et al. (2006)
20.	Cabbage	Fungus (*Alternaria brassicicola*)	Free phenols and lignin	Nowakowska et al. (2019)
21.	Chickpea	*Sclerotium rolfsii*	Gallic acid, vanillic acid, and salicylic acid	Maurya et al. (2005)

only. When fungi attack, *Sorghum bicolor* exhibits a change in metabolism and produces a variety of protective phenolic compounds, such as apigeninidin, luteolin, 3-deoxyanthocyanidin phytoalexins, etc. (Tugizimana et al. 2019; Gautam et al. 2020). Gautam et al. (2020) reported that plants store the antifungal phenolic compounds in the vacuoles or organelles. In the organelles or vacuoles, plants store the antifungal phenolic compounds. According to the extent of the fungus's tissue damage, the concentration of phenols that inhibit the fungus will increase. In this response, plants lead to the formation of lignin. Lignin stiffs the cell wall and prevents the entry and movement of pathogens.

3.6.3 Phenolics as Antioxidant

Antioxidants are defined as the compounds that protect the plant by converting harmful ROS to less harmful ones or by delaying the formation of ROS/RNS (Dai and Mumper 2010). In this way, oxidative stress is delayed by antioxidants. Oxidative stress damages the plant cells by lipid peroxidation, damaging DNA, and changing conformation of enzymes and proteins. Various properties of phenolics make them superior antioxidants as compared to vitamins C and E (Rice-Evans et al. 1995, 1996) like more free radical scavenging power, metal chelators, total antioxidant capacity, etc. Phenolics have an ideal chemical structure to scavenge free radicals as OH group in phenolics act as a hydrogen donor to free radical, and phenoxy radical intermediate formed by donation of hydrogen to free radical is also available to bind other free radicals. Flavonoids like quercetin have the number of hydroxyl groups and double bonds which has electron-donating and delocalization properties (Cotelle 2001; Dziedzic and Hudson 1983). Active metal ions like Cu^{+} and Fe^{2+} react with H_2O_2 (Hydrogen peroxide) to produce hydroxyl radical which is among the most reactive oxygen species. Phenolics inhibit metal ion-induced oxygen radicals and lead to autooxidation of metal ions. Gallic acid and catecholate also bind to metal ions and form inactive complexes (Shahidi and Wanasundara 1992). However, at high pH conditions, some small phenolic compounds like gallic acid act as prooxidants by autooxidation, while high-molecular-weight phenolics like tannins have very little or no activity of prooxidants (Hagerman et al. 1998; Dai and Mumper 2010).

3.6.4 Role of Phenolics in Systemic Acquired Resistance

Systemic acquired resistance (SAR) is developed in plants in response to biotic stresses. It does not harm insects/pests/microorganisms but strengthens the plants to overcome further infection by physical or chemical barriers. Salicylic acid (SA) is a phenolic compound that acts as a signal against biotic stress. Salicylic acid is found in the leaves of both infected and uninfected tobacco plants (Sticher and Mauch-Mani 1997). Sticher and Mauch-Mani (1997) reported that salicylic acid treatment protects the plants against TMV. Kang et al. (2003) found that increase in salicylic acid involved in protection of eucalypt against reactive oxygen species and also activate antioxidant defense enzymes. Viruses are one of the microorganism pathogens that affect crop plants the most. Plant viral diseases can easily spread in the field and are very hazardous to crop plants (Gautam et al. 2020). The viral pathogen is deflected by a number of plant phenolic compounds, and salicylic acid is the most important compound that has antiviral property. Salicylates also induce PR proteins, giving bean (Pennazio et al. 1987) and cowpea (Hooft van Huijsduijnen et al. 1986) resistance to alfalfa mosaic virus, and they diminished the symptoms of tobacco necrosis virus (TNV) disease in asparagus beans. SA was discovered by White (1979) as an acetylsalicylic acid that acts as endogenous signal in tobacco against *tobacco mosaic virus* (TMV). Mohamed et al. explained the role of salicylic acid to mitigate biotic stress. Thaler (1999) reported the role of salicylates in protecting plants against herbivores by increasing the production of volatile compounds. Low concentrations of SA also encourage "priming," a process that aids in the induction of defense mechanisms, by promoting the quicker and stronger activation of callose accumulation and gene expression in response to pathogen or microbial elicitors (Kohler et al. 2002; Mohamed et al. 2020).

3.6.5 Phenolics as Phytoalexins

Phytoalexin are low molecular weight compound that are formed by the phenylpropanoid pathway under biotic stress conditions (Cho and Lee 2015). In plants, most genes for this pathway belong to multigene families, of which a set of genes are induced and implicated in biotic and abiotic stress triggered by the synthesis of phenolic phytoalexins. These phytoalexins act by reinforcing cell walls (Hamann 2012; Underwood 2012) and scavenge ROS. Sakuranetin is an example of phenolic phytoalexin (Kodama et al. 1992) which is produced by rice and has health benefits as it has antibiotic activity against *Helicobacter pylori*. Phenylamine phytoalexins, such as CinTrp and CouSer, are reported to have antimicrobial properties against pathogenic bacteria, as well as anti-inflammatory and antiatherogenic effects (Takimoto et al. 2011). Oat produces nitrogen-containing phenolic phytoalexins, the avenalumins (Mayama et al. 1982), and these compounds accumulated only in incompatible host–pathogen interactions. The Vitaceae family of plants, which includes grapes, exhibits increased metabolism, accumulation of phytoalexins, and production of PR proteins as a protection against pathogenic fungi (Jeandet et al. 2002). Datta and Lal (2012) identified two major isoflavonoids (Medicarpin and maackiain) as phytoalexins in pigeon pea which has antimicrobial activity and are produced via phenyl propanoid pathway. One response involves the plant-producing compounds that are off-target and effective a long way from the infection site. In another reaction, the plant produces substances that specifically target the infection site (Kumar et al. 2020). In general, phytoalexin refers to stress metabolites or the substances created during infections and stress. Pathogens exhibit a specific toxicity to phytoalexin. A majority of phytoalexins are members of the flavonoid and isoflavonoid group and have antioxidant and antimicrobial properties. An example of phytoalexin is danielone that had antifungal activity (Echeverri et al. 1997). Phenolic allochemicals has power to inhibit or improve the growth of other plants nearby. The presence of phenolic allelochemicals in both naturally occurring and managed ecosystems has been linked to a number of ecological and economic issues, such as decreased crop yield due to soil disease, difficulties with orchard replanting, and failure of natural forest regeneration. They effect the weeds by changing the permeability of cells; inhibition of nutrient absorption; impairs photosynthesis, respiration and protein synthesis (Palanisamy et al. 2020).

3.7 Conclusion

Phenolics are extensively distributed in plants and are used for defensive functions in many plant species. Mostly phenolics are under the subepidermal or epidermal layer and in the vacuoles of the plant cell. Under various biotic stress conditions, the production of phenolics or polyphenols enhanced due to overexpression of various enzymes involved in their biosynthesis. Due to the specific structural and other properties, phenolics can act as strong antioxidants by scavenging ROS and have high antioxidant capacity, disturb cell membrane, increase membrane permeability, deform membrane-bound proteins and structure of pathogens, disrupt their life cycle, and have strong antimicrobial activity which ultimately can disrupt or hinder the growth of various pathogens or microbes which grow in apoplast, take the xylem sap, and block the transfer of nutrients from roots to other parts of the plants.

References

Ahmed E, Arshad M, Khan MZ, Amjad MS, Sadaf HM, Riaz I, Sabir S, Ahmad N (2017) Secondary metabolites and their multidimensional prospective in plant life. J Pharmacogn Phytochem 6(2):205–214

Aly AA, El-Mahdy OM, Habeb MM, Elhakem A, Asran AA, Youssef MM, Mohamed HI, Hanafy RS (2022) Pathogenicity of Bacillus strains to cotton seedlings and their effects on some biochemical components of the infected seedlings. Plant Pathol J 38(2):90–101

Arneson PA, Durbin RD (1968) Studies on the mode of action of tomatine as a fungitoxic agent. Plant Physiol 43(5):683–686

Avtar R, Rathi AS, Jatasra DS, Joshi UN (2003) Changes in phenolics and some oxidative enzymes in fenugreek leaves due to powdery mildew infection. Acta Phytopathol Entomol Hung 38(3–4): 237–244

Bao J, Cai Y, Sun M, Wang G, Corke H (2005) Anthocyanins, flavonols, and free radical scavenging activity of Chinese bayberry (*Myrica rubra*) extracts and their color properties and stability. J Agric Food Chem 53(6):2327–2332

Barz W, Bless W, Börger-Papendorf G, Gunia W, Mackenbrock U, Meier D, Otto CH, Süper E (2007) Phytoalexins as part of induced defence reactions in plants: their elicitation, function and metabolism. In: Bioactive compounds from plants: bioactive compounds from plants: Ciba Foundation Symposium 154. Wiley, Chichester, UK, pp 140–156

Bazghaleh N, Prashar P, Purves RW, Vandenberg A (2018) Polyphenolic composition of lentil roots in response to infection by *Aphanomyces euteiches*. Front Plant Sci 9:1131

Belete T (2018) Defense mechanisms of plants to insect pests: from morphological to biochemical approach. Trends Tech Sci Res 2:30–38

Bernardi J, Stagnati L, Lucini L, Rocchetti G, Lanubile A, Cortellini C, De Poli G, Busconi M, Marocco A (2018) Phenolic profile and susceptibility to Fusarium infection of pigmented maize cultivars. Front Plant Sci 9:1189

Bhattacharya A, Sood P, Citovsky V (2010) The roles of plant phenolics in defence and communication during Agrobacterium and Rhizobium infection. Mol Plant Pathol 11(5):705–719

Bouarab-Chibane L, Forquet V, Lantéri P, Clément Y, Léonard-Akkari L, Oulahal N, Degraeve P, Bordes C (2019) Antibacterial properties of polyphenols: characterization and QSAR (quantitative structure–activity relationship) models. Front Microbiol 10:829

Carvalho RS, Carollo CA, De Magalhães JC, Palumbo JMC, Boaretto AG, e Sá IN, Ferraz AC, Lima WG, De Siqueira JM, Ferreira JMS (2018) Antibacterial and antifungal activities of phenolic compound-enriched ethyl acetate fraction from Cochlospermum regium (mart. Et. Schr.) Pilger roots: mechanisms of action and synergism with tannin and gallic acid. S Afr J Bot 114:181–187

Chakraborty BN, Dutta S, Chakraborty U (2002) Biochemical responses of tea plants induced by foliar infection with Exobasidium vexans. Indian Phytopathol 55(1):8–13

Cho MH, Lee SW (2015) Phenolic phytoalexins in rice: biological functions and biosynthesis. Int J Mol Sci 16(12):29120–29133

Constabel CP, Barbehenn R (2008) Defensive roles of polyphenol oxidase in plants. In: Induced plant resistance to herbivory. Springer, Dordrecht, pp 253–270

Cotelle N (2001) Role of flavonoids in oxidative stress. Curr Top Med Chem 1:569–590

Czarnocka W, Karpiński S (2018) Friend or foe? Reactive oxygen species production, scavenging and signaling in plant response to environmental stresses. Free Radic Biol Med 122:4–20

da Silva Dias JC (2020) Integrated plant breeding strategies for harmony between modern agriculture production and the environment. In: Cutting-edge research in agricultural sciences, vol 3, pp 1–37

Dadáková K, Heinrichová T, Lochman J, Kašparovský T (2020) Production of defense phenolics in tomato leaves of different age. Molecules 25(21):4952

Dai J, Mumper RJ (2010) Plant phenolics: extraction, analysis and their antioxidant and anticancer properties. Molecules 15(10):7313–7352

Das K, Roychoudhury A (2014) Reactive oxygen species (ROS) and response of antioxidants as ROSscavengers during environmental stress in plants. Front Environ Sci 2:53

Das SK, Patra JK, Thatoi H (2016) Antioxidative response to abiotic and biotic stresses in mangrove plants: a review. Int Rev Hydrobiol 101(1–2):3–19

Datta J, Lal N (2012) Temporal and spatial changes in phenolic compounds in response to Fusarium wilt in chickpea and pigeonpea. Cell Mol Biol 58(1):96–102

Deng Q, Wu J, Chen J, Shen W (2020) Physiological mechanisms of improved smut resistance in sugarcane through application of silicon. Front Plant Sci 11:568130

Dimberg LH, Peterson DM (2009) Phenols in spikelets and leaves of field-grown oats (*Avena sativa*) with different inherent resistance to crown rust (*Puccinia coronata* f. sp. avenae). J Sci Food Agric 89(11):1815–1824

Dziedzic SZ, Hudson BJF (1983) Polyhydroxy chalcones and flavanones as antioxidants for edible oils. Food Chem 12:205–212

Echeverri F, Torres F, Quiñones W, Cardona G, Archbold R, Roldan J, Brito I, Luis JG, Lahlou EH (1997) Danielone, a phytoalexin from papaya fruit. Phytochemistry 44:255–256

El-Nagar A, Elzaawely AA, Taha NA, Nehela Y (2020) The antifungal activity of gallic acid and its derivatives against *Alternaria solani*, the causal agent of tomato early blight. Agronomy 10(9): 1402

Gautam AK, Singh PK, Aravind M (2020) Defensive role of plant phenolics against pathogenic microbes for sustainable agriculture. In: Plant phenolics in sustainable agriculture. Springer, Singapore, pp 579–594

Graham JH, Gottwald TR, Cubero J, Achor DS (2004) *Xanthomonas axonopodis*pv. citri: factors affecting successful eradication of citrus canker. Mol Plant Pathol 5(1):1–15

Gull A, Lone AA, Wani NUI (2019) Biotic and abiotic stresses in plants. In: Abiotic and biotic stress in plants, pp 1–19

Gulya T, Rashid KY, Masirevic SM (1997) Sunflower diseases. Sunflower technology and production 35:263–379

Hagerman AE, Riedl KM, Jones GA, Sovik KN, Ritchard NT, Hartzfeld PW, Riechel TL (1998) High molecular weight plant polyphenolics (tannins) as biological antioxidants. J Agric Food Chem 46:1887–1892

Hamann T (2012) Plant cell wall integrity maintenance as an essential component of biotic stress response mechanisms. Front Plant Sci 3:77

Hassan MU, Chattha MU, Khan I, Chattha MB, Barbanti L, Aamer M, Iqbal MM, Nawaz M, Mahmood A, Ali A, Aslam MT (2021) Heat stress in cultivated plants: nature, impact, mechanisms, and mitigation strategies—a review. Plant Biosyst 155(2):211–234

Hansen AK, Moran NA (2014) The impact of microbial symbionts on host plant utilization by herbivorous insects. Mol Ecol 23(6):1473–1496

Hébert C, Charles MT, Gauthier L, Willemot C, Khanizadeh S, Cousineau J (2002) Strawberry proanthocyanidins: biochemical markers for Botrytis cinerea resistance and shelf-life predictability. Acta Hortic 567:659–662

Heins RD, Liu B, Runkle ES (1998) Regulation of crop growth and development based on environmental factors. In: XXV International Horticultural Congress, Part 3: Culture techniques with special emphasis on environmental implications, vol 513, pp 17–28

Heng-Moss T, Sabath G, Baxendale F, Novak D, Bose S, Ni X, Quisenberry S (2004) Characterization of oxidative enzyme changes in buffalograsses challenged by *Blissus occiduus*. J Econ Entomol 97:1086–1095

Higuchi T (2004) Microbial degradation of lignin: role of lignin peroxidase, manganese peroxidase, and laccase. Proc Jpn Acad B 80(5):204–214

Hooft van Huijsduijnen RAM, Alblas SW, de Rijk RH, Bol JF (1986) Induction by salicylic acid of pathogenesis-related proteins and resistance to alfalfa mosaic virus infection in various plant species. J Gen Virol 67(2):135–143

Hussein M, Abo-Elyousr KA, Hassan MA, Hashem M, Hassan EA, Alamri SA (2018) Induction of defense mechanisms involved in disease resistance of onion blight disease caused by *Botrytis allii*. Egypt J Biol Pest Control 28(1):1–11

Hutzler P, Fischbach R, Heller W, Jungblut TP, Reuber S, Schmitz R, Veit M, Weissenböck G, Schnitzler JP (1998) Tissue localization of phenolic compounds in plants by confocal laser scanning microscopy. J Exp Bot 49(323):953–965

Hu-zhe Z, Chun-lan C, Yu-ting Z, Dan W, Yu J, Yong KK (2005) Active changes of lignification-related enzymes in pepper response to Glomus intraradices and/or Phytophthora capsici. J Zhejiang Univ Sci B 6(8):778–786

Jamiołkowska A (2020) Natural compounds as elicitors of plant resistance against diseases and new biocontrol strategies. Agronomy 10(2):173

Jeandet P, Douillet-Breuil AC, Bessis R, Debord S, Sbaghi M, Adrian M (2002) Phytoalexins from the Vitaceae: biosynthesis, phytoalexin gene expression in transgenic plants, antifungal activity, and metabolism. J Agric Food Chem 50(10):2731–2741

Kang G, Changhu W, Guchou S, Zhengxun W (2003) Salicylic acid changes activities of H_2O_2-metabolizing enzymes and increases the chilling tolerance of banana seedlings. Environ Exp Bot 50:9–15

Karthikeyan M, Radhika K, Mathiyazhagan S, Bhaskaran R, Samiyappan R, Velazhahan R (2006) Induction of phenolics and defense-related enzymes in coconut (*Cocos nucifera* L.) roots treated with biocontrol agents. Braz J Plant Physiol 18:367–377

Karthikeyan G, Doraisamy S, Rabindran R, Ganapathy T (2009) Evaluation of antiviral principles for the induction of systemic resistance in blackgram (*Vigna mungo*) against Urdbean Leaf Crinkle Virus. Arch Phytopathol Pflanzenschutz 42(12):1172–1186

Kefeli VI, Kalevitch MV, Borsari B (2003) Phenolic cycle in plants and environment. J Cell Mol Biol 2:13–18

Khattab H, Khattab I (2005) Responses of Eucalypt trees to the insect feeding (Gall forming Psyllid). Int J Agric Biol 7(6):979–984

Kodama O, Miyakawa J, Akatsuka T, Kiyosawa S (1992) Sakuranetin, a flavanone phytoalexin from ultraviolet-irradiated rice leaves. Phytochemistry 31:3807–3809

Koeck M, Hardham AR, Dodds PN (2011) The role of effectors of biotrophic and hemibiotrophic fungi in infection. Cell Microbiol 13(12):1849–1857

Kohler A, Schwindling S, Conrath U (2002) Benzothiadiazole-induced priming for potentiated responses to pathogen infection, wounding, and infiltration of water into leaves require the NPR1/NIM1 gene in Arabidopsis. Plant Physiol 128:1046–1056

Koskimäki JJ, Hokkanen J, Jaakola L, Suorsa M, Tolonen A, Mattila S, Pirttilä AM, Hohtola A (2009) Flavonoid biosynthesis and degradation play a role in early defence responses of bilberry (*Vaccinium myrtillus*) against biotic stress. Eur J Plant Pathol 125(4):629–640

Kovalchuk I (2021) Transgenerational genome instability in Plants. In: Genome stability. Academic, pp 659–678

Kubalt K (2016) The role of phenolic compounds in plant resistance. Biotechnol Food Sci 80(2): 97–108

Kumar S, Abedin M, Singh AK, Das S (2020) Role of phenolic compounds in plant-defensive mechanisms. In: Plant phenolics in sustainable agriculture. Springer, Singapore, pp 517–532

Laukkanen H, Soini H, Kontunen-Soppela S, Hohtola A, Viljanen M (2000) A mycobacterium isolated from tissue cultures of mature *Pinus sylvestris* interferes with growth of Scots pine seedlings. Tree Physiol 20:915–920

Lodish H, Berk A, Zipursky SL, Matsudaira P, Baltimore D, Darnell J (2000) Viruses: structure, function, and uses, Molecular cell biology, 4th edn. WH Freeman

Logemann E, Hahlbrock K (2002) Crosstalk among stress responses in plants: pathogen defence overrides UV protection through an inversely regulated ACE/ACE type of light-responsive gene promoter unit. Proc Natl Acad Sci U S A 99:2428–2432

Maddox CE, Laur LM, Tian L (2010) Antibacterial activity of phenolic compounds against the phytopathogen Xylella fastidiosa. Curr Microbiol 60(1):53–58

Maiti S, Basak J, Pal A (2014) Current understanding on plant r-genes/proteins and mechanisms of defense responses against biotic stresses. Rev Plant Pathol 6:93–126

Maurya S, Singh UP, Singh DP, Singh KP, Srivastava JS (2005) Secondary metabolites of chickpea (*Cicer arietinum*) and their role in pathogenesis after infection by *Sclerotium rolfsii*. J Plant Dis Protect 112(2):118–123

Mayama S, Matsuura Y, Iida H, Tani T (1982) The role of avenalumin in the resistance of oat to crown rust, *Puccinia coronata* f. sp. avenae. Physiol Plant Pathol 20(2):189–199

Mehta S, Sharma S, Sindhan GS (1992) Analysis of root exudates of cowpea and their influence on the growth of *Rhizoctonia solani*. Indian J Mycol Plant Pathol 22:227–231

Miedes E, Vanholme R, Boerjan W, Molina A (2014) The role of the secondary cell wall in plant resistance to pathogens. Front Plant Sci 5:358

Mohamed H, El-Hady AA, Mansour M, El-Samawaty AER (2012) Association of oxidative stress components with resistance to flax powdery mildew. Trop Plant Pathol 37:386–392

Mohamed HI, El-Shazly HH, Badr A (2020) Role of salicylic acid in biotic and abiotic stress tolerance in plants. In: Plant phenolics in sustainable agriculture. Springer, Singapore, pp 533–554

Nicholson RL, Hammerschmidt R (1992) Phenolic compounds and their role in disease resistance. Annu Rev Phytopathol 30(1):369–389

Nowakowska M, Wrzesińska M, Kamiński P, Szczechura W, Lichocka M, Tartanus M, Kozik EU, Nowicki M (2019) Alternaria brassicicola–Brassicaceae pathosystem: insights into the infection process and resistance mechanisms under optimized artificial bio-assay. Eur J Plant Pathol 153(1):131–151

Ozyigit II, Kahraman MV, Ercan O (2007) Relation between explant age, total phenols and regeneration response of tissue cultured cotton (*Gossypium hirsutum* L). Afr J Biotechnol 6: 3–8

Palanisamy CP, Gunasekaran VP, Dominic S, Xuan TD (2020) Phenolic allelochemicals from crops and weed management. In: Plant phenolics in sustainable agriculture. Springer, Singapore, pp 183–199

Park HL, Yoo Y, Hahn TR, Bhoo SH, Lee SW, Cho MH (2014) Antimicrobial activity of UV-induced phenylamides from rice leaves. Molecules 19:18193–18151

Patel NJ, Kandoliya UK, Talati JG (2015) Induction of phenol and defence-related enzymes during wilt (*Fusarium udum* Butler) infestation in pigeon pea. Int J Curr Microbiol Appl Sci 4(2): 291–299

Pawlak K, Kołodziejczak M (2020) The role of agriculture in ensuring food security in developing countries: considerations in the context of the problem of sustainable food production. Sustainability 12(13):5488

Pennazio S, Colariccio D, Roggero P, Lenzi R (1987) Effect of salicylate stress on the hypersensitive reaction of asparagus bean to tobacco necrosis virus. Physiol Mol Plant Pathol 30:347–357

Poppe L, Rétey J (2005) Friedel–crafts-type mechanism for the enzymatic elimination of ammonia from histidine and phenylalanine. Angew Chem Int Ed 44(24):3668–3688

Pramod SN (2017) Biotic stress induced by *Bacterocera cucurbitae* (Melon Fly) triggers defense related phenylpropanoid pathway (PPP) and ROS detoxifying enzymes in cucurbits as adaptation. Asian J Plant Sci Res 7(4):18–29

Pratyusha S (2022) Phenolic compounds in the plant development and defense: an overview. In: Plant stress physiology-perspectives in agriculture. https://doi.org/10.5772/intechopen.102873

Pushpa D, Yogendra KN, Gunnaiah R, Kushalappa AC, Murphy A (2014) Identification of late blight resistance-related metabolites and genes in potato through nontargeted metabolomics. Plant Mol Biol Rep 32(2):584–595

Pusztahelyi T, Holb IJ, Pócsi I (2015) Secondary metabolites in fungus-plant interactions. Front Plant Sci 6:573

Qiu Y, Liu Q, Bet T (2010) Antioxidant properties of commercial wild rice and analysis of soluble and insoluble phenolic acids. Food Chem 121:140–147

Rice-Evans CA, Miller NJ, Bolwell PG, Bramley PM, Pridham JB (1995) The relative antioxidant activities of plant-derived polyphenolic flavonoids. Free Radic Res 22:375–383

Rice-Evans CA, Miller NJ, Paganga G (1996) Structure-antioxidant activity relationships of flavonoids and phenolic acids. Free Radic Biol Med 20:933–956

Rivero RM, Mittler R, Blumwald E, Zandalinas SI (2022) Developing climate-resilient crops: improving plant tolerance to stress combination. Plant J 109(2):373–389

Rojo E, Solano R, Sánchez-Serrano JJ (2003) Interactions between signaling compounds involved in plant defense. J Plant Growth Regul 22(1):82–98

Ruiz-Rubio M, Pérez-Espinosa A, Lairini K, Roldán-Arjona T, Dipietro A, Anaya N (2001) Metabolism of the tomato saponin α-tomatine by phytopathogenic fungi. Stud Nat Prod Chem 25:293–326

Sarig P, Zutkhi Y, Monjauze A, Lisker N, Ben-Arie R (1997) Phytoalexin elicitation in grape berries and their susceptibility to Rhizopus stolonifer. Physiol Mol Plant Pathol 50(5):337–347

Schjoerring JK, Cakmak I, White PJ (2019) Plant nutrition and soil fertility: synergies for acquiring global green growth and sustainable development. Plant Soil 434(1):1–6

Schmelz EA, Huffaker A, Sims JW, Christensen SA, Lu X, Okada K, Peters RJ (2014) Biosynthesis, elicitation and roles of monocot terpenoid phytoalexins. Plant J 79:659–678

Schulz B, Römmert AK, Dammann U, Aust HJ, Strack D (1999) The endophyte-host interaction: a balanced antagonism? Mycol Res 103:1275–1283

Shahidi F, Wanasundara PK (1992) Phenolic antioxidants. Crit Rev Food Sci Nutr 32:67–103

Singh J, Sharma VK, Kaur A, Sharma S (2021) Differential biochemical response in resistant and susceptible barley (*Hordeum vulgare*) cultivars upon covered smut (*Ustilago hordei*) infection. Indian Phytopathol 74(1):173–179

Singleton VL, Orthofer R, Lamuela-Raventós RM (1999) Analysis of total phenols and other oxidation substrates and antioxidants by means of Folin-Ciocalteu reagent. In: Methods in enzymology, vol 299. Academic, pp 152–178

Steinite I, Ievinsh G (2002) Wound-induced responses in leaves of strawberry cultivars differing in susceptibility to spider mite. J Plant Physiol 159:491–497

Steinite I, Ievinsh G (2003) Possible role of trichomes in resistance of strawberry cultivars against spider mite. Acta Univ Latviensis 662:59–65

Sticher L, Mauch-Mani B, Métraux JP (1997) Systemic acquired resistance. Annu Rev Phytopathol 35(1):235–270

Stuper-Szablewska K, Kurasiak-Popowska D, Nawracała J, Perkowski J (2019) Quantitative profile of phenolic acids and antioxidant activity of wheat grain exposed to stress. Eur Food Res Technol 245(8):1595–1603

Swain T, Hillis WE (1959) The phenolic constituents of Prunus domestica. I.—The quantitative analysis of phenolic constituents. J Sci Food Agric 10(1):63–68

Tak Y, Kumar M (2020) Phenolics: a key defence secondary metabolite to counter biotic stress. In: Plant phenolics in sustainable agriculture. Springer, Singapore, pp 309–329

Takimoto T, Suzuki K, Arisaka H, Murata T, Ozaki H, Koyama N (2011) Effect of N-(p-coumaroyl) serotonin and N-feruloylserotonin, major anti-atherogenic polyphenols in safflower seed, on vasodilation, proliferation and migration of vascular smooth muscle cells. Mol Nutr Food Res 55:1561–1571

Thaler JS (1999) Jasmonate-inducible plant defenses cause increased parasitism of herbivores. Nature 399:686–688

Thomas P, Ravindra MB (1999) Shoot tip culture in mango: influence of medium, genotype, explant factors, season and decontamination treatments on phenolic exudation, explant survival and axenic culture establishment. J Hortic Sci 72:713–722

Tugizimana F, Djami-Tchatchou AT, Steenkamp PA, Piater LA, Dubery IA (2019) Metabolomic analysis of defense-related reprogramming in Sorghum bicolor in response to Colletotrichum sublineolum infection reveals a functional metabolic web of phenylpropanoid and flavanoid pathways. Front Plant Sci 4(9):1840

Underwood W (2012) The plant cell wall: a dynamic barrier against pathogen invasion. Front Plant Sci 3:85

Wallis CM, Galarneau ERA (2020) Phenolic compound induction in plant-microbe and plant-insect interactions: a meta-analysis. Front Plant Sci 11:580753

War AR, Paulraj MG, Ahmad T, Buhroo AA, Hussain B, Ignacimuthu S, Sharma HC (2012) Mechanisms of plant defense against insect herbivores. Plant Signal Behav 7(10):1306–1320

White RF (1979) Acetylsalicylic acid (aspirin) induces resistance to tobacco mosaic virus in tobacco. Virology 99:410–412

Williamson B, Tudzynski B, Tudzynski P, Van Kan JA (2007) Botrytis cinerea: the cause of grey mould disease. Mol Plant Pathol 8(5):561–580

Plant Phenolics: Role in Biotic Stress Alleviation and Plant Microbe Interactions

4

Nazima Rasool and Zafar A. Reshi

Abstract

Phenolic compounds are structurally and functionally diverse organic compounds synthesized by plants, which apart from being involved in various plant life processes, significantly contribute to plant defense. As plant defense molecules, their production may be induced after the pathogen has attacked the plant or the synthesis may be constitutive. The phenolics act by different modes of action depending on their chemical structure and, nature, and position of the substituent groups. They may disrupt membrane integrity, inhibit enzyme activity, or block energy metabolism in the target pathogenic species. Phenolics provide color and aroma to the fruits. These are the universal signaling molecules at low concentrations, but at higher concentrations, they serve as allelopathic chemicals and antimicrobial phytoalexins. Phenolics are important in the plant–microbe interactions; interaction between leguminous plants and *Rhizobium* sp. and symbioses between mycorrhiza and the host plant roots require two-way exchange of phenolic signals between the symbiotic partners. Owing to their high antibacterial and antifungal activities, plant phenolics are seen as an eco-friendly alternative to the chemical pesticides.

Keywords

Phenolics · Plant defense · Plant–microbe interaction · Biopesticides

N. Rasool (✉)
Department of Botany, North Campus, University of Kashmir, Delina, Baramulla, Jammu and Kashmir, India

Z. A. Reshi
Department of Botany, Main Campus, University of Kashmir, Srinagar, Jammu and Kashmir, India

R. Lone et al. (eds.), *Plant Phenolics in Biotic Stress Management*,
https://doi.org/10.1007/978-981-99-3334-1_4

4.1 Introduction

Biochemically phenols are six C aromatic compounds that contain hydroxyl groups, one or more than one. Phenols (phenolic acids or phenolics) are structurally diverse and ubiquitous in the plant kingdom; about 8000 different types have been reported till date (Cosme et al. 2020). Harborne and Simmonds (1964) classified plant phenolics on the basis of number of carbon atoms. A general outline of the classification of phenols is given in Table 4.1. Phenolic compounds are synthesized in plant cells via three biosynthetic pathways which include, the shikimate/chorismate or succinylbenzoate pathway, the acetate/malonate polyketide pathway, and the acetate/mevalonate pathway (Kumar et al. 2020). These three pathways give rise to phenyl-propanoid derivative (C6–C3), the phenyl propanoids with elongated side chains that include the large group of flavonoids (C6–C3–C6) apart from some quinones and aromatic terpenoids (mostly monoterpenes which are produced by dehydrogenation reactions), in that order (Bhattacharya et al. 2010; Naikoo et al. 2019). The usual definition of phenols as the compounds containing a benzene ring with an attached hydroxyl group and that of polyphenols as compounds containing more than one such structures also includes some compounds that are basically of the terpenoid origin, e.g., the female sex hormone estrone (II) and the phenolic carotenoid 3-hydroxyisorenieratene (I) known as gossypol (Harborne 1989; Lattanzio 2013). Therefore, it has been argued that the term "plant phenolics" be restricted to natural secondary metabolites that arise either from the shikimate/phenylpropanoid pathway, which synthesizes the phenylpropanoids directly, or the "polyketide" acetate/malonate pathway (Harborne 1989; Quideau et al. 2011; Lattanzio 2013; Cheynier et al. 2013). Phenolics play critical role in a variety of functions in the plant body including cell wall thickening, hormone production, fruit flavoring and defense against various kinds of stresses (UV, herbivory, pathogenic organisms, etc.). They are central to the defense against biotic stress (Tak and Kumar 2020). Plants depend on phenolic compounds for eliminating free radicals, chelating metals and preventing lipid peroxidation (Takó et al. 2020). Phenolics offer protection from ROS by scavenging the free radicals (Leon et al. 1995; Matkowski 2006; Torres et al. 2006; Mathew et al. 2015). Due to their antimicrobial nature plant phenolics are also used as natural food preservatives (Takó et al. 2020). Besides, these are essential in the plant–microbe interactions (Dixon 2001; Wallis and Galarneau 2020). A summary of their roles is presented in Fig. 4.1.

Phenolics are ubiquitous in higher plants (the types and content may vary among different plant groups) but rare in bacteria, fungi and algae (Cheynier et al. 2013). Bryophytes produce some phenols including polyphenols (flavonoids) (Cheynier et al. 2013). Leaves of higher plants contain amides, esters and glycosides of hydroxycinnamic acids, glycosylated flavonoids (flavonols, flavones), etc. Sporopollenin present in pollen walls, suberin and lignin in secondary walls of plant cells are phenolic containing polymers (Swain 1975; Harborne 1980; Lattanzio et al. 2008; Cheynier et al. 2013). The nature and content of various phenolics in plants vary with season, the stages of growth and the genotype of species (Lynn and Chang 1990; Thomas and Ravindra 1999; Scalzo et al. 2005; Ozyigit et al. 2007;

Table 4.1 General classification of phenols (based on their basic carbon skeleton) and functions in plants

Serial no.	Carbon skeleton	Group (example)	Functions in plants	References
1	C6	Simple phenols (phloroglucinol)	Phloroglucinol is a precursor in lignin biosynthesis and acts like a hormone and improves plant growth. It has antimicrobial activity	da Silva et al. (2013), Biessy and Filion (2021)
2	C6–C1	Phenolic acids and aldehydes (salicylic acid)	It is a plant hormone and signal molecule. It is critical in numerous physiological activities besides its role in abiotic stress response	Vicente and Plasencia (2011), Yan and Dong (2014)
3	C6–C2	Acetophenones and phenylacetic acids (2-hydroxyacetophenone)	Antimicrobial properties	Hettiarachchi et al. (2011)
4	C6–C3	Hydroxycinnamic acids, coumarins, phenylpropanes, chromones (*p*-coumaric acid)	Antioxidant, antimicrobial, important in plant–microbe interactions	Kannan et al. (2013), Gyawali and Ibrahim (2014)
5	C6–C4	Naphthoquinones (naphthazarin)	Free radical scavenging, antibacterial properties	Kourounakis et al. (2002)
6	C6–C1–C6	Xanthones	Antimicrobial, cytotoxic, and enzyme inhibitory effects	Mazimba et al. (2013)
7	C6–C2–C6	Stilbenes, anthraquinones (resveratrol)	Phytoalexin inhibits multidrug efflux pumps	Lechner et al. (2008), Ferreira et al. (2014)
8	C6–C3–C6	Flavonoids, isoflavonoids, neoflavonoids (quercetin)	Antioxidant, important in stress response (biotic and abiotic), involved in seed germination and plant growth and development	Singh et al. (2021)
9	(C6–C3–C6) 2/3	Di- or trimers of flavonoids and proanthocyanidin (pelargonidin)	Pigment molecule, protective against bleaching effects of UV, quenches O_2, precursor in the synthesis of other pigments, e.g., nudicaulins	Monici et al. (1993), Warskulat et al. (2016)
10	(C6–C3)2	Lignans, neolignans [dimmers or oligomers of lignols] (podophyllotoxin)	Potent antimitotic agent	Oliva et al. (2002), Bégué and Bonnet-Delpon (2008)
11	(C6–C3)*n*	Lignin	Structural polymer and an important first line of defense against biotic and abiotic stress	Campbell and Sederoff (1996)

(continued)

Table 4.1 (continued)

Serial no.	Carbon skeleton	Group (example)	Functions in plants	References
12	(C6)*n*	Catecholmelanins, phlorotannins (eckol)	Bactericidal, antioxidant	Nagayama et al. (2002), Shibata et al. (2007)
13	(C6–C3–C6)*n*	Condensed tannins	Plant defense response	Cooper and Owen-Smith (1985), Gourlay and Constabel (2019)

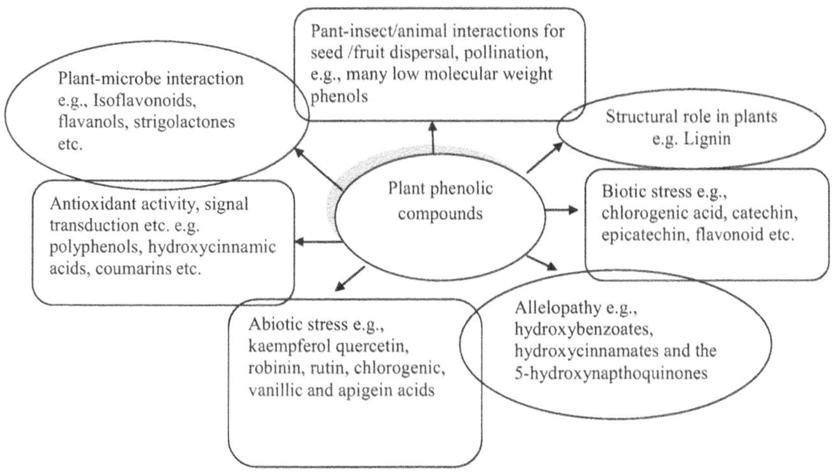

Fig. 4.1 Role of phenolic compounds in the life of plants

Bhattacharya et al. 2010; Vagiri et al. 2017). The content of phenols in plants is also affected by several other factors including, wounding, trauma, pathogen infections, etc. (Zapprometov 1989; Ke and Saltveit Jr 1989; Kefeli et al. 2003; Bhattacharya et al. 2010). Light stimulates biosynthesis of phenolics in chloroplast and their accumulation in the cellular vacuoles (Kefeli et al. 2003; Bhattacharya et al. 2010). Deficiency of mineral nutrients including N, P, B, S, Mg, and Fe induce phenylpropanoid biosynthesis in some plants (Dixon and Paiva 1995; Jin et al. 2007; Bhattacharya et al. 2010). On an average, plants allocate 2% of their photosynthates toward formation of flavonoids and allied compounds (Robards and Antolovich 1997). Phenolic compounds account for 40% of the organic carbon circulating in the biosphere; they decompose very slowly and form a slow carbon pool in the soil (Chapman and Regan 1980; Min et al. 2015; Croteau et al. 2000). Despite there being thousands of phenolics, the research has been done on only a few of them (Wallis and Galarneau 2020).

4.2 Phenolics in the Root Exudates

The root exudate is a term for the suite of substance secreted into the rhizosphere by the roots of living plants (Singh and Singla 2020). Root exudates represent 20% to 40% of the carbon assimilated by the plants (Prescott et al. 2020). Root exudates include a large variety of phenolics. On an average, monocots root exudates contain 2.1% to 4.4% phenolics in their root exudates whiles the dicots contain relatively lesser amounts (0.1–0.6%) by volume (Hartley and Harris 1981; Bhattacharya et al. 2010). Root exudates affect soil porosity, and availability of mineral nutrient including boron, calcium, copper, iron, magnesium, manganese, molybdenum, and zinc because of their metal chelating capacity (Seneviratne and Jayasinghearachchi 2003; Bhattacharya et al. 2010; Baumert et al. 2018). Phenolics in the root exudates significantly influence the soil biochemistry and shape the community composition of the soil microorganisms around the plant roots. Phenolics affect availability of phytonutrients, plant hormonal balance, competition between neighboring plants and soil enzymatic activity (Northup et al. 1998; Hättenschwiler and Vitousek 2000; Kraus et al. 2003; Zhang et al. 2019). Root exudates affect growth and colonization of soil microflora considerably. The soil microflora in turn metabolize the excreted phenolics affecting the soil nutrient cycling and buildup of the soil humus (Halvorson et al. 2009; Bhattacharya et al. 2010). Notwithstanding their capacity to improve mineral nutrition of plants phenolics in these extracts may act to the contrary also (Lodhi et al. 1987; Bhattacharya et al. 2010). Some phenolics or their metabolic products may have adverse effects on seed germination and may be phytotoxic; examples include coumarins, *para*hydroxybenzoic acid, syringic acid, *trans*-cinnamic acid, salicylic acid, and benzoic acid (Bhattacharya et al. 2010). These may inhibit seed germination by inhibiting prolyl aminopeptidase and phosphatase which are involved in seed germination (Shankar et al. 2009; Bhattacharya et al. 2010). Phenolics in the root exudates also serve as a chemotactic stimulus for the soil microbial organisms that tend to colonize the rhizosphere under their influence (Perret et al. 2000; Gray and Smith 2005; Taylor and Grotewold 2005; Bhattacharya et al. 2010). These compounds are used by plants to engage in positive or negative interactions depending on whether the organism is the potential mutualist or antagonist. In comparison to the bulk soil, phenolics in the root exudates make rhizosphere metabolically an active zone which favorably affects root colonization by a variety of plant growth-promoting rhizobacteria including nitrogen-fixing proteobacteria and a host of others that improve biotic and abiotic stress tolerance in plants (Moulin et al. 2001; Bais et al. 2004; Bardgett and Walker 2004; Schardl et al. 2004; Nannipieri et al. 2008; Bhattacharya et al. 2010). Some of these bacteria form a protective sheath of biofilm or symplasmata and secrete antiobiotics in the microhabitat to deter pathogenic bacteria (Bhattacharya et al. 2010). Growth of germinative hyphae of pathogenic fungi, spore germination, and sporulation is inhibited by plant phenolic compounds (Walters et al. 2007; Dakora 1995; Jamiolkowska 2020). Many of the phenolic acids are metabolized by soil microorganism and these serve as their carbon source (Whitehead et al. 1983; Chen et al. 1984; Revillas et al. 2000; Chan 2006; Carmona et al. 2009; Mandal et al. 2010).

4.3 Role of Phenolics in Defense Against Herbivores and Plant Pathogens

Plants synthesize a variety of chemical species on being attacked by pathogenic microorganisms; these substances may act locally or away from the site of infection (Feys and Parker 2000; Compant et al. 2005; Mandal et al. 2010). Some of these antimicrobial products are species specific and others more general (van Loon 2000; Mandal et al. 2010). A large content of the molecules acting in the plant defense are phenolic compounds; phenolics may be synthesized postinfection (induced) or they may be preformed (constitutive). Phenolics are an important defense against pathogenic organisms, herbivores, nematodes, phytophagous insects, and even avian herbivores (Ravin et al. 1989; Lee 1991; Dakora and Phillips 1996; Bhattacharya et al. 2010); with their antifungal, antibacterial, and antiviral properties (Martini et al. 2009; Slatnar et al. 2016), their role in plant defense becomes undeniable (Mandal et al. 2010). In the induced resistance, phenolics primarily are the signal transduction molecules (Pieterse and Van Loon 1999; Mandal et al. 2010). Not only do they protect plant from the attacks of pathogens, they are equally important in safeguarding human health against various diseases (Aviles-Gaxiola et al. 2020). A recent meta-analysis indicates that phenolics are increased in plants as a generic response to colonization by pathogenic bacteria, symbiotic bacteria and symbiotic fungi but not in case of pathogenic fungi and sucking and wood-boring insects (Wallis and Galarneau 2020). Phenolic acids, tannins and other complex phenols on the plant surfaces make the plant material unpleasant to the invaders; these also interact with the gut microflora of herbivores and decrease the digestibility of the plant material. The role of phenolics as nematicides, phytoanticipins, and phytoalexins against phytophagous insects and soil-borne pathogens is well documented (Akhtar and Malik 2000; Wuyts et al. 2006; Lattanzio et al. 2006; Bhattacharya et al. 2010). Plants produce hydroxycoumarins and hydroxycinnamin conjugates in response to pathogen attack. Salicylic acid is at the center stage of many plant defense reactions (Koornneef and Pieterse 2008; Boller and He 2009; Lu 2009; Tsuda et al. 2008; Bhattacharya et al. 2010). Derivatives of phenolics like quinines, chlorogenic acid, etc. resulting from their oxidation and other transformations are important in their action against plant pathogens. Resistance of potato plants to *Streptomyces scabies* was ascribed to chlorogenic acid; in the roots of carrot infected with *Thielaviopsis basicola* the content of chlorogenic acid increased two times as compared with the non-infected plants (Garcion et al. 2007; Jamiolkowska 2020). Low incidence of fungal disease in black currants has been linked with high phenolic content. The low disease incidence of mildews (about 5%)

has been reported with increased phenolic content (Vagiri et al. 2017). Better resistance of grapevine to *Erysiphae necator* has been linked with its phenolic content (Keller et al. 2003; Vagiri et al. 2017). Resistance to *Botrytis cinerea* in strawberries is related with their catechin and epicatechin contents (Terry et al. 2004). Phenolic compounds inhibit *Xylella fastidiosa* which causes many diseases in economically important plants such as Pierce's disease in grapes, variegated chlorosis in citrus and leaf scorch in almond under in vitro conditions (Maddox et al. 2010). There are many reports of increase in phenolic content post pathogen attack, such as significant increase in flavonoid content in bean plants after infection by *Collelotrichum lindemuthianum* (Slatnar et al. 2016). Flavonol content increases significantly post-infection in apple, bean, and strawberry (Slatnar et al. 2016).

4.4 Mechanism of Action

Phenolics employ various strategies for countering plant pathogens which include, cellular membrane damaging, inhibition of enzymes such as, NADH-cytochrome c reductase topoisomerase, ATP synthase, etc. (Chinnam et al. 2010; Cushnie and Lamb 2011; Górniak et al. 2019). Some phenolics reduce membrane fluidity and inhibit energy metabolism, synthesis of nucleic acids and cell wall components (Górniak et al. 2019; Quideau et al. 2011; Takó et al. 2020). Phenolic acids not only counter the pathogens when inside the plant body but these, released into the environment from root, seed and the decomposing plant residue act as antibacterial and antifungal agents and deter pests (Ndakidemi and Dakora 2003; Mandal et al. 2010). Phenolics are deposited in the cell walls as a reliable first line of defense against penetration by pathogens (Underwood 2012; Slatnar et al. 2016). Some phenolics have also been listed as "pre-infection inhibitors" referring to their ability to provide first line of defense against the pathogenic organisms (Wink 2003; Singh et al. 2005; Treutter 2006; Vagiri et al. 2017). Quick accumulation of phenolic acids at the sites of microbial attack has been reported to prevent further spread of the infection (Treutter and Feucht 1990; Modafar et al. 1996; Slatnar et al. 2016). Phenolics slow down the growth of pathogens till the plant gears up its next line of defense that involves phytoalexin synthesis (Bailey and Mansfield 1982; Bell 1980). Phenolics that are engaged in the defense response as phytoanticipins, phytoalexins, and nematicides include glyceolin, coumestrol, cajanin, rotenone, medicarpin, isoflavonoids, flavonoids, phaseolin, phaseolinin, etc.; these act against various phytophagous insects and soil-borne pathogens (Ndakidemi and Dakora 2003; Dakora and Phillips 1996; Mandal et al. 2010). The derivatives of hydroxycinnamic acid are fungitoxic and impair fungal growth and sporulation (Sammi and Masud 2009; Chardonnet et al. 2003; Slatnar et al. 2016).

Phenolics attacking cell membranes have stirred much scientific interest (Tsuchiya 2015; Verstraeten et al. 2015). Numerous bioassays have been devised to understand the sequence of events during attack of phenolics on the cell membranes (Rempe et al. 2017). Membrane disruption is the most common mechanism to neutralize both Gram-negative and Gram-positive bacteria and is assayed by monitoring influx of hydrophobic dyes into the cytoplasm or efflux of the cellular constituents and by microscopic examination. Phenolics like ferulic and gallic acids alter membrane hydrophobicity in Gram-negative and Gram-positive bacteria (Borges et al. 2013; Rempe et al. 2017). On the whole, however, Gram-positive bacteria have been found to be more sensitive to phenolic extracts from fruits than the Gram-negative bacteria (Coman et al. 2018; Wafa et al. 2017; Lima et al. 2019). Enhanced lipophillic character of some plant phenolics facilitate their interaction with cell membrane components of the attacking microorganisms leading to membrane disruption, coagulation of the cytoplasm and finally inhibition of the intracellular enzymes of the pathogen (Sikkema et al. 1995; Bouarab-Chibane et al. 2019). Carvacrol causes membrane destabilization in *Bacillus cereus* cells, it increases membrane fluidity which leads to loss of ions from the cells leading to cell death (Ultee et al. 2002). Phenolics like catechin penetrate the cellular membranes and cause membrane leakage and aggregation of liposome (Ikigai et al. 1993; Taylor et al. 2005; Borges et al. 2013; Bouarab-Chibane et al. 2019; Takó et al. 2020). Nonmembrane disrupting modes of action are employed by many phenolics. Phenolics, e.g., ellagic acid, quercetin, kaempferol, apigenin, nobiletin, etc., inhibit DNA gyrase activity (Ohemeng et al. 1993; Wu et al. 2013a; Rempe et al. 2017), compounds, e.g., baicalein, quercetin, and myricanol have type III secretion inhibition activity which is very important for the virulence of the bacteria invading human intestines (Xu et al. 2011; Khokhani et al. 2013; Tsou et al. 2016; Takó et al. 2020). Compounds like luteolin, morin, myricetin, etc. inhibit helicase activity (Xu et al. 2001), some like catechol, gingerol, resveratrol biochanin A, inhibit multidrug efflux pumps (Smith et al. 2007; Lechner et al. 2008; Bag and Chattopadhyay 2014; Prasch and Bucar 2015; Shriram et al. 2018). Protein kinase inhibition has been reported for ellagic acid quercetin and kaempferol (Mistry et al. 1997; Shakya et al. 2011), some phenolic compounds inhibit dehydratase activity (Zhang et al. 2008); inhibition of urase activity (Lin et al. 2015), iron binding (Khokhar and Apenten 2003), malate and succinate dehydrogenase inhibition (Yao et al. 2012), intercalation into the DNA (Lou et al. 2012), inhibition of FtsZ assembly after binding FtsZ protein (Rai et al. 2008) has been reported for phenolics. Some compounds suppress pathogens by more than one pathways, e.g., quercetin. It causes inhibition of DNA gyrase, inhibition of type III secretion, protein kinase inhibition, membrane disruption, DNA intercalation and inhibition of dehydratase activity (Rempe et al. 2017). Electron transport chain enzymes are also specifically inhibited by phenolics (Hirsch et al. 2003; Bhattacharya et al. 2010). Phenolics form H-bonds with enzyme active site using their hydroxyl group to inhibit the enzyme's activity (Jeong et al. 2009; Takó et al. 2020). Tannins may bind with cell membrane and inactivate the bacterial enzymes (Ya et al. 1988; Tamokou et al. 2017; Bouarab-Chibane et al. 2019). Phenolics suppress microbial virulence by preventing formation of the microbial biofilm and decreasing host ligand adhesion. These also neutralize toxins secreted by microbes. The growth of *Bacillus subtilis*, *Staphylococcus aureus,* and *E. coli* is inhibited by gallotannins isolated from the mango kernels due to their iron-complexing and enzyme-inhibiting properties. Gallotannins also interact with bacterial proteins (Engels et al. 2009; Takó et al. 2020). Inhibition of nucleic acid biosynthesis (Wu et al. 2013a; Takó et al. 2020), cytoplasmic membrane damage by inhibition of energy metabolism (Tsuchiya 2015; Sanver et al. 2016; Eumkeb and Chukrathok 2013; Takó et al. 2020), and inhibition of biofilm formation (Zambrano et al. 2019; Takó et al. 2020) have been reported for plant phenolics. Phenolics can inhibit quorum sensing in bacteria (Kumar et al. 2014; Slobodníková et al. 2016; Oliveira et al. 2017; Asfour 2018; Takó et al. 2020). This happens by their interfering with different regulatory mechanisms without affecting growth, e.g., reducing the bacterial mobility and decreasing the superficial adhesion (Eydelnant and Tufenkji 2008; Takó et al. 2020).

The ease with which the phenolics undergo oxidation affects their biological activity (Garcion et al. 2007; Jamiolkowska 2020). Many of the phenolic based defense responses depend upon peroxidases which hydrolyze cell wall components in pathogens (Koide and Schreiner 1992); presence of hydroxyl groups determines their antimicrobial activity besides their antioxidant properties. Interaction of phenolics with the cell membrane is significantly affected by the position of the hydroxyl groups (Oteiza et al. 2005; Figueiredo et al. 2008; Takó et al. 2020). Position and number of hydroxyl groups, double bonds, and length of the saturated side chain affect electron distribution which in turn affects antimicrobial activity of phenols (Shapiro and Guggenheim 1998; Ultee et al. 2002; Hyldgaard et al. 2012; Wu et al. 2013a, b; Rempe et al. 2017). The types of possible membrane interactions are influenced by location of electron and other features related to the structure of these molecules. Owing to its higher hydroxyl group substitution, antimicrobial activity of caffeic acid is stronger than *p*-coumaric acid (Cueva et al. 2010; Maddox et al. 2010; Stojković et al. 2013; Gyawali and Ibrahim 2014). Presence of hydroxyl group and the localized electron system has been stated to be critical for antimicrobial properties of carvacrol (Ultee et al. 2002; Takó et al. 2020).

Phenolics may polymerize to present a physical barrier to the pathogen (Spanu and Bonfante-Fasolo 1988; Brett and Waldron 1990; Koide and Schreiner 1992). Flavonoids may form complexes after linking with soluble proteins outside the cell walls; these may affect RNA and protein synthesis by affecting DNA synthesis and energy metabolism (Tsuchiya et al. 1996; Haraguchi et al. 1998; Bouarab-Chibane et al. 2019). Modification of the cellular pH with attendant effects on energy metabolism have been reported for some phenolics (Djilani and Dicko 2012; Bouarab-Chibane et al. 2019).

4.5 Role of Phenolics in Plant–Microbe Interactions

4.5.1 Mycorrhizae

Phenolics are at the central stage of the plant–microbe interaction (Martens 2002; Mandal et al. 2010); however, very little research is available regarding the effect of plant phenolics on the fungal symbionts of plants (Harrier et al. 1998; Martin-Laurent et al. 1997; Mandal et al. 2010). Flavonoids have been stated to be important in the communication between mycorrhizae and their plant hosts; different flavonoids may range in their effect on different fungal species from neutral through positive to negative (Poulin et al. 1997; Scervino et al. 2005; Mandal et al. 2010; Kaur and Suseela 2020). The flavonoids increase the likelihood of encounter of the mycorrhizal species with the plant roots (Silva-Junior and Siqueira 1998). Flavonoids both initiate as well as restrict mycorrhizal symbiosis. These serve as potent signals in mediating relay of signals between fungi and their potential hosts to forge mycorrhizal symbiosis (Tsai and Phillips 1991; Mandal et al. 2010; Hassan and Mathesius 2012). Formononetin and ononin regulate mycorrhizal symbiosis (Catford et al. 2006; Kaur and Suseela 2020). *p*-Hydroxybenzoic acid, *p*-coumaric acid and quercetin have a stimulatory effect on colonization of *Trifolium repens* and *Sorghum bicolor* by the *Glomus* sp. at lower concentrations but are inhibitory at higher concentrations (Fries et al. 1997; Sanchez-Lizarraga et al. 2017). Phenolic signals from the host plant lead to improved spore germination, formation of secondary spores, enhanced hyphal growth and branching (Glenn et al. 1988; Koide and Schreiner 1992; Mandal et al. 2010; Sanchez-Lizarraga et al. 2017). The growth response of hyphae varies with the type of flavonoids (Kaur and Suseela 2020). External applications of phenolics also induce hyphal branching in the compatible fungal species. Strigolactones synthesized by plants under the conditions of phosphate deficiency induce branching of hyphae in mycorrhizal fungi (Akiyama 2007; Akiyama et al. 2005; Steinkellner et al. 2007; Bhattacharya et al. 2010). The signals from the host plant may affect rate and direction of growth of the hyphal branches of the fungi apart from germination of spores and extension of germ tube (Koide and Schreiner 1992). Adhesion to the surface of host root tissue and penetration into it may be influenced by chemical cues from the plant at the immediate surface of root (Koide and Schreiner 1992). Only the root exudates from the host plant and not the nonhost plant induce hyphal branching in the mycorrhizal species (Siqueira et al. 1991; Giovannetti et al. 1996; Mandal et al. 2010). Root exudates encourage numerous rhizospheric bacteria which in turn synthesize molecules that influence the interaction of the mycorrhizal fungi with the plant roots (Azcon-Aguilar et al. 1986; Daniels Hetrick 1984; Koide and Schreiner 1992). The inability of mycorrhizas to grow on a nutrient medium may be partly attributed to the nonavailability of the signals from the plant roots.

AM inoculation increases the phenolic complement of the host (Selvaraj and Subramanian 1990; Ling-Lee et al. 1977; Mandal et al. 2010). Plants initiate defense response against the mycorrhizal fungi at initial stage and many phenolics are synthesized; however, this response is later depressed (Koide and Schreiner 1992; Volpin et al. 1994; Mandal et al. 2010). After the mycorrhizal association has developed, the flavonoid profile of the root exudates is considerably modified (Zuanazzi et al. 1998; Silva-Junior and Siqueira 1998; Mandal et al. 2010). Isoflavonoids like diadzein, ononin, and malonylonin were upregulated in *Medicago truncatula* after forming mycorrhizal association (Schliemann et al. 2008; Kaur and Suseela 2020). Quercetin increases in grapes (Mandal et al. 2010; Kaur and Suseela 2020). Increased levels of medicarpin malonyl glycoside, formononetin malonyl glucoside, diadzein, and coumestrol have been found to be correlated with increased stress resistance in plants (Harrison 1993; Jaiti et al. 2007; Kaur and Suseela 2020).

Furthermore, increased disease resistance in mycorrhizal plants has been found to be associated with increased phenolic content in their cell walls (Jaiti et al. 2007; Kaur and Suseela 2020). Phenolics increase in peanuts after AMF inoculation (Devi and Reddy 2002; Kaur and Suseela 2020), hydroxycinnamate amides increase in barley roots (Devi and Reddy 2002), coumarins and their hydroxylated forms increase in *Salix* (Aliferis et al. 2015), and the effect of AM inoculation on different kinds of phenolics may however be variable; ferulic acid increases, while caffeic and chlorogenic acids decrease in tomato after the formation of mycorrhizal association (López-Ráez et al. 2010; Kaur and Suseela 2020). Soybean secretes phenolics compounds to develop mutualistic association with the AM fungi which considerably improves plant's nutritional status and leads to improved plant growth (Siqueira et al. 1991; Bhattacharya et al. 2010; Tidke et al. 2018).

4.5.2 Legume–Rhizobia Interaction

The rhizobium–leguminous plant symbiotic relationship is established in a stepwise manner in which phenolics play an important role (Bhattacharya et al. 2010). The first step is nonspecific and reversible while the second step is more specific, stronger and irreversible (Wheatley and Poole 2018). This biphasic mode of binding is common to a host of bacteria apart from rhizobia, including pathogenic and plant growth promoting bacteria. Flavonoids, betaines, and aldonic acids form important constituents of the root exudates of the leguminous plants that serve as signals for the rhizobia (Phillips and Torrey 1972; Mandal et al. 2010). As the legume roots grow in the soil, they secrete a number of phenolic compounds into the soil (Zaat et al. 1988; Staman et al. 2001; Mandal et al. 2010), causing compatible *Rhizobium* strains to undergo certain changes enabling them to enter into symbiotic relationship (Blum et al. 2000; Hassan and Mathesius 2012; Mandal et al. 2010). Phenolics provide chemotactic stimulus for the bacteria and guide them toward the plant roots. *Sinorhizobium meliloti* and two species of Rhizobium, *Rhizobium leguminosarum* bv. phaseoli and *R. leguminosarum* bv. Trifolii, are strongly attracted by umbelliferone, vanillyl alcohol, *p*-hydroxybenzoic acid and 3,4-dihydroxybenzoic acid (Bhattacharya et al. 2010; Compton and Scharf 2021). Flavones, flavonols, isoflavonoids, vanillin, etc. produced by different leguminous plants regulate expression of nodulation genes in bacteria (Zawoznik et al. 2000; Bekkara et al. 1998; Mandal et al. 2010). Phenolics are potent inducers of nodABC genes that control organogenesis of nodules in presence of *nodD* gene (Subramanian et al. 2006; Bhattacharya et al. 2010; Hassan and Mathesius 2012). *nodD* gene of *Rhizobium* sp. which regulates nodABC genes, is in turn regulated by flavones and flavonones from the host (Bekkara et al. 1998; Lynn and Chang 1990; Koide and Schreiner 1992; Stougaard 2000; Cohen et al. 2001; Bhattacharya et al. 2010). Phenolics secreted by the different zones of roots variously affect the expression of nod genes altering nodule organogenesis (Redmond et al. 1986; Mandal et al. 2010). The sites of transcription activation of the nod genes including *NodD1*, *NodD2*, and *NodD3* bind the flavonoids released by the plant roots activating their expression and the products of these genes in turn activate expression of other genes involved in nodule formation (Mandal et al. 2010). *nodD* gene product catalyzes initial steps in the establishing symbiosis (Mandal et al. 2010). Phenolics involved in nodule morphogenesis have been reported from the *Arachis hypogea* (Chakraborty and Mandal 2008; Mandal et al. 2010). Phenolic acids present in the root nodules of *Vigna mungo* stimulate auxin production in the rhizobial strains and control nodule morphogenesis (Mandal et al. 2009, 2010). nod genes of different rhizobial strains are responsive to specific phenolics; *nodD* determines specificity of different rhizobial strains toward different kinds of plant flavonoids, exchange of nod genes among different rhizobial strains changes their specificity to the different flavonoids (Hassan and Mathesius 2012; Spaink 2000; Mandal et al. 2010). Not all the phenolics affect all the strains of bacteria in the same way; some may act as week attractants or even deterrents for bacteria. Naringenin evokes only a weak to no chemotactic response in *R. leguminosarum* bv. viciae and *R. leguminosarum* bv. trifolii (Bhattacharya et al. 2010; Hassan and Mathesius 2012). It even weakens the chemotactic response of the *Sinorhizobium meliloti* induced by luteolin (Caetano-Anolles et al. 1988; Bhattacharya et al. 2010). Some bacterial strains may require different biochemical species for chemotactic stimuli and inducers of nod genes, e.g., isoliquiritigenin (2′,4,4′-trihydroxychalcone) has been found to induce nod genes; it, however, does not act as a chemoattractant (Kape et al. 1992; Bhattacharya et al. 2010), while for some others, the same chemical species may perform the dual function (of chemoattractant and nod gene inducer). Moreover, the same compounds may be inhibitory for one bacterial species but stimulatory for the other, e.g., genistein, diadzein, and some soybean isoflavonoids induce nod genes in some strains of *Bradyrhizobium japonicum* while as these inhibit luteolin induced nod genes of *S. meliloti* (Kosslak et al. 1987; Subramanian et al. 2006; Bhattacharya et al. 2010).

Bacteria respond to the signals from plants by secreting chitooligosaccharide *Nod*-factors that modify the root hair architecture leading to the formation of infection thread and establishment of a symbiotic union (Mandal et al. 2010). Flavonoids have been reported to inhibit auxin transport toward the site after the rhizobial cells enter the host plant roots (Subramanian et al. 2007; Bhattacharya et al. 2010). Auxin transport in Arabidopsis is regulated by flavonoid binding protein complexes AtAPM and AtMDR. Auxins are very important in root nodule formation and studies indicate that flavonoids influence auxin turnover or its transport (Ferguson and Mathesius 2003; deBilly et al. 2001; Mandal et al. 2010). Quorum sensing in rhizobia is strongly dependent upon homoserine lactones or their acylated forms this enables these bacteria to function as a multicellular organism, synchronize and respond to the phenolic signals on a population level scale. This enables development of successful symbiosis, in turn improving nodulation efficiency, development of symbiosome, production of exopolysaccharide, nitrogen fixation, and response to stress (Gonzalez and Marketon 2003; Sanchez-Contreras et al. 2007; Bhattacharya et al. 2010).

4.6 Phenolics as Organic Pesticides

Because of their high antifungal and antibacterial properties (Friend 1979), phenolic compounds are seen as potential candidates for natural control of plant pathogens (Langcake et al. 1981; Nicholson and Hammerschmidt 1992; Treutter 2006; Lattanzio et al. 2006; Mandal et al. 2010; Gautam et al. 2020). Preventive sprays of plants with plant-based antipathogenic extracts like grapefruit leads to production of phenolics in *Capsicum annuum* (Jamiołkowska 2011; Jamiolkowska 2020). These extracts trigger many defense related reactions in plants apart from eliciting phenol synthesis and lead to plant immunization and thus can be highly useful for the agricultural systems (Villaverde et al. 2016; Jamiolkowska 2020). Naringin, found in grapefruit pulp and seeds, has been reported to be active against many plant pathogens including *Phytophthora cryptogea*, *P. cinnamomi*, *Fusarium oxysporum*, *Phomopsis phaseoli*, *Sclarotinia sclerotiorum,* and *Phoma exigua* (Hassan and Mathesius 2012; Jamiolkowska 2020). Availability of active compounds with pesticidal properties, e.g., phenolics, in the era of biotechnology and genetic engineering creates an opportunity for developing plant-based pesticides. Phenolics are already known to stimulate growth, help overcome stress-induced limitations on growth, regulate and modify plant physiological processes and increase the productivity of plants. Thus, the possibility of using them as alternative option for the pest control needs to be explored. Many bioactive compounds with pesticidal properties have already been isolated from plants and successfully commercialized. These formulations are nontoxic, nonhazardous, and nonpolluting giving them an edge over the chemical formulations. Resistance to chemical pesticides can be effectively minimized by the use of organic formulations. Preventive use of bio-based products can be an effective and eco-friendly alternative to the chemical pesticides. However, the production of organic formulations is not without limitations; the raw material is required in large quantities, the product may be unstable and may breakdown during the production processes, and the formulations may be inconsistent in terms of their efficacy (Koziara et al. 2006; Yu et al. 2015; Jamiolkowska 2020). Notwithstanding these limitations, organic pesticides and compounds with phytosanitary activity are still captivating for the scientists as their molecular structure is amenable to modification for producing new molecules structurally more stable and with improved efficacy. These formulations can be improved by combining them with other bio-based products (Sultana et al. 2011; Jamiolkowska 2020). For example, combination of an efflux pump blocker with a toxin can lead to rapid killing of the pathogen by preventing efflux of the toxic compound (Oh and Jeon 2015; Prasch and Bucar 2015; Rempe et al. 2017; Lima et al. 2019); membrane-disrupting compounds can assist in the rapid access of the toxic compounds into the cell cytoplasm (Hemaiswarya and Doble 2010; Aiyegoro and Okoh 2009; Amin et al. 2015; Oh and Jeon 2015; Rempe et al. 2017; Walsh et al. 2019), and these can be combined with other toxic compounds. This synergistic action between two or more compounds can highly increase the efficacy of the bio-based products and overcome the efficacy-related problems.

In all, phenolics, given their enormous capacity of acting as deterrents against a wide variety of plant pests and pathogens, may provide environment-friendly method of pest control and help secure the agricultural productivity better against the losses suffered due to pest attacks. The role of phenolics in plant disease resistance and their mechanism of action need to be fully investigated. Improved understanding of the how the phenolics work in providing immunity to plants will help counter various pathogens. The field of plant pathology is expected to gain immensely from the research in plant phenolics.

References[1]

Aiyegoro OA, Okoh AI (2009) Use of bioactive plant products in combination with standard antibiotics: implications in antimicrobial chemotherapy. J Med Plants Res 3(13):1147–1152

Akhtar M, Malik A (2000) Roles of organic soil amendments and soil organisms in the biological control of plant-parasitic nematodes: a review. Bioresour Technol 74:35–47

Akiyama K (2007) Chemical identification and functional analysis of apocarotenoids involved in the development of arbuscular mycorrhizal symbiosis. Biosci Biotechnol Biochem 71:1405–1414

Akiyama K, Matsuzaki K, Hayashi H (2005) Plant sesquiterpenes induce hyphal branching in arbuscular mycorrhizal fungi. Nature 435:824–827

Aliferis KA, Chamoun R, Jabaji SH (2015) Metabolic responses of willow (*Salix purpurea* L.) leaves to mycorrhization as revealed by mass spectrometry and 1H NMR spectroscopy metabolite profiling. Front Plant Sci 6:1–15

Amin MU, Khurram M, Khattak B, Khan J (2015) Antibiotic additive and synergistic action of rutin, morin and quercetin against methicillin resistant Staphylococcus aureus. BMC Complement Altern Med 15:59. https://doi.org/10.1186/s12906-015-0580-0

Asfour HZ (2018) Anti-quorum sensing natural compounds. J Microsc Ultrastruct 6:1–10

Aviles-Gaxiola S, Olivo-Vázquez G, Cabanillas-Bojórquez LA, Gutiérrez-Grijalva EP, Heredia JB (2020) Plants as biofactories for phenolic compounds. In: Lone R, Shuab R, Kamili N (eds) Plant phenolics in sustainable agriculture. Springer, Singapore, pp 467–500

Azcon-Aguilar C, Diaz-Rodriguez RM, Barea M (1986) Effect of soil micro-organisms on spore germination and growth of the vesicular-arbuscular mycorrhizal fungus *Glomus mosseae*. Trans Br Mycol Soc 86:337–340

Bag A, Chattopadhyay RR (2014) Efflux-pump inhibitory activity of a gallotannin from Terminalia chebula fruit against multi-drug resistant uropathogenic *Escherichia coli*. Nat Prod Res 28:37–41. https://doi.org/10.1080/14786419.2014.895729

Bailey JA, Mansfield JW (eds) (1982) Phytoalexins. Blackie, Glasgow

Bais HP, Park SW, Weir TL, Callaway RM, Vivanco JM (2004) How plants communicate using the underground information superhighway. Trends Plant Sci 9:1360–1385

Bardgett RD, Walker LR (2004) Impact of coloniser plant species on the development of decomposer microbial communities following deglaciation. Soil Biol Biochem 36:555–559

Baumert VL, Vasilyeva NA, Vladimirov AA, Meier IC, Kögel-Knabner I, Mueller CW (2018) Root exudates induce soil macroaggregation facilitated by fungi in subsoil. Front Environ Sci 6:140

Bégué JP, Bonnet-Delpon D (2008) Biological impacts of fluorination: pharmaceuticals based on natural products. In: Fluorine and health. Elsevier, pp 553–622

[1] *Reference not seen in original

Bekkara F, Jay M, Viricel MR, Rome S (1998) Distribution of phenolic compounds within seed and seedlings of two *Vicia faba* cvs differing in their seed tannin content, and study of their seed and root phenolic exudations. Plant Soil 203:27–36

Bell AA (1980) The time sequence of defense. In: Horsfall JG, Cowling EB (eds) Plant disease. An advanced treatise, How plants defend themselves, vol 5. Academic, New York, pp 53–73

Bhattacharya A, Sood P, Citovsky V (2010) The roles of plant phenolics in defense and communication during Agrobacterium and Rhizobium infection. Mol Plant Pathol 11(5):705–719

Biessy A, Filion M (2021) Phloroglucinol derivatives in plant-beneficial Pseudomonas spp.: biosynthesis, regulation, and functions. Metabolites 11(3):182

Blum U, Stamen KL, Flint LJ, Shafer SR (2000) Induction and/or selection of phenolic acid-utilizing bulk-soil and rhizosphere bacteria and their influence on phenolic acid phytotoxicity. J Chem Ecol 26:2059–2078

Boller T, He SY (2009) Innate immunity in plants: an arms race between pattern recognition receptors in plants and effectors in microbial pathogens. Science 324:742–744

Borges A, Ferreira C, Saavedra MJ, Simões M (2013) Antibacterial activity and mode of action of ferulic and gallic acids against pathogenic bacteria. Microb Drug Resist 19:256–265. https://doi.org/10.1089/mdr.2012.0244

Bouarab-Chibane LA, Pierre FVL, Lucie CY, Nadia O, Pascal D, Claire C (2019) Antibacterial properties of polyphenols: characterization and QSAR (quantitative structure-activity relationship) models. Front Microbiol 10:829

Brett C, Waldron K (1990) Physiology and biochemistry of plant cell walls. Springer. ISBN 978-94-010-9641-6

Caetano-Anolles GD, Crist-Estes K, Bauer WD (1988) Chemotaxis of Rhizobium meliloti to the plant flavone luteolin requires functional nodulation genes. J Bacteriol 170:3164–3169

Campbell MM, Sederoff RR (1996) Variation in lignin content and composition (mechanisms of control and implications for the genetic improvement of plants). Plant Physiol 110(1):3

Carmona M, Zamarro MT, Blázquez B, Durante-Rodríguez G, Juárez JF, Valderrama AJ et al (2009) Anaerobic catabolism of aromatic compounds: a genetic and genomic view. Microbiol Mol Biol Rev 73:71–133

Catford JG, Staehelin C, LaRose G, Piché Y, Vierheilig H (2006) Systemically suppressed isoflavonoids and their stimulating effects on nodulation and mycorrhization in alfalfa split-root systems. Plant Soil 285:257–266

Chakraborty D, Mandal SM (2008) Fractional changes in phenolic acids composition in root nodules of *Arachis hypogaea* L. Plant Growth Regul 55:159–163

Chan YK (2006) Utilization of simple phenolics for dinitrogen fixation by soil diazotrophic bacteria. Plant Soil 90:141–150

Chapman DJ, Regan MA (1980) Evolution of a biochemical pathway: evidence from comparative biochemistry. Annu Rev Plant Physiol 31:639–645

Chardonnet CO, Charron CS, Sams CE, Conway WS (2003) Chemical changes in the cortical tissue and cell walls of calcium in filtrated 'Golden Delicious' apples during storage. Postharvest Biol Technol 28:97–111

Chen YP, Dilworth MJ, Glenn AR (1984) Uptake and oxidation of aromatic substances by Rhizobium leguminosarum MNF 8841 and *Rhizobium trifolii* TA1. FEMS Microbiol Lett 21: 201–205

Cheynier VA, Comte G, Davies KM, Lattanzio L, Martens S (2013) Plant phenolics: recent advances on their biosynthesis, genetics, and ecophysiology. Plant Physiol Biochem 72:1–20

Chinnam N, Dadi PK, Sabri SA, Ahmad M, Kabir MA, Ahmad Z (2010) Dietary bioflavonoids inhibit Escherichia coli ATP synthase in a differential manner. Int J Biol Macromol 46:478–486

Cohen MF, Sakihama Y, Yamasaki H (2001) Roles of plant flavonoids in interactions with microbes: from protection against pathogens to the mediation of mutualism. Recent Res Dev Plant Physiol 2:157–173

Coman MM, Oancea AM, Verdenelli MC, Cecchini C, Bahrim GE, Orpianesi C, Cresci A, Silvi S (2018) Polyphenol content and in vitro evaluation of antioxidant, antimicrobial and prebiotic properties of red fruit extracts. Eur Food Res Technol 244:735–745

Compant S, Duffy B, Nowak J, Clément C, Barka EA (2005) Use of plant growth-promoting bacteria for biocontrol of plant diseases: principles, mechanisms of action, and future prospects. Am Soc Microbiol Appl Environ Microbiol 71(9):4951–4959

Compton KK, Scharf BE (2021) Rhizobial chemoattractants, the taste and preferences of legume symbionts. Front Plant Sci 12:818

Cooper SM, Owen-Smith N (1985) Condensed tannins deter feeding by browsing ruminants in a South African savanna. Oecologia 67(1):142–146

Cosme P, Rodríguez AB, Espino J, Garrido M (2020) Plant phenolics: bioavailability as a key determinant of their potential health-promoting applications. Antioxidants 9:1263

Croteau R, Kutchan TM, Lewis NG (2000) Natural products (secondary metabolites). In: Buchanan B, Gruissem W, Jones R (eds) Biochemistry and molecular biology of plants. American Society of Plant Physiologist, Rockville, pp 1250–1318

Cueva C, Moreno-Arribas MV, Martín-Álvarez PJ, Bills G, Vicente MF, Basilio A, Rivas CL, Requena T, Rodríguez JM, Bartolomé B (2010) Antimicrobial activity of phenolic acids against commensal, probiotic and pathogenic bacteria. Res Microbiol 161:372–382

Cushnie TPT, Lamb AJ (2011) Recent advances in understanding the antibacterial properties of flavonoids. Int J Antimicrob Agents 38:99–107

da Silva TJA, Dobránszki J, Ross S (2013) Phloroglucinol in plant tissue culture. In Vitro Cell Dev Biol Plant 49:1–16. https://doi.org/10.1007/s11627-013-9491-2

Dakora FD (1995) Plant flavonoids: biological molecules for useful exploitation. Aust J Plant Physiol 22:87–99

Dakora FD, Phillips DA (1996) Diverse functions of isoflavonoids in legumes transcend ant-microbial definitions of phytoalexins. Physiol Mol Plant Pathol 49:1–20

Daniels Hetrick BA (1984) Ecology of VA mycorrhizal fungi. In: Powell CII, Bagyaraj DJ (eds) VA Mycorrhiza. CRC Press, Boca Raton, FL, pp 35–55

deBilly F, Grosjean C, May S, Bennett M, Cullimore JV (2001) Expression studies on AUX1-like genes in Medicago truncatula suggest that auxin is required at two steps in early nodule development. Mol Plant Microbe Interact 14:267–277

Devi MC, Reddy M (2002) Phenolic acid metabolism of groundnut (*Arachis hypogaea* L.) plants inoculated with VAM fungus and Rhizobium. Plant Growth Regul 37:151–156

Dixon RA (2001) Natural products and plant disease resistance. Nature 411:843–847

Dixon RA, Paiva NL (1995) Stress-induced phenylpropanoid metabolism. Plant Cell 7:1085–1097

Djilani A, Dicko A (2012) The therapeutic benefits of essential oils. In: Bouayed J (ed) Nutrition, well-being and health. InTech, pp 155–178

Engels C, Knödler M, Zhao Y-Y, Carle R, Gänzle MG, Schieber A (2009) Antimicrobial activity of gallotannins isolated from mango (*Mangifera indica* L.) kernels. J Agric Food Chem 57:7712–7718

Eumkeb G, Chukrathok S (2013) Synergistic activity and mechanism of action of ceftazidime and apigenin combination against ceftazidime-resistant Enterobacter cloacae. Phytomedicine 20: 262–269

Eydelnant IA, Tufenkji N (2008) Cranberry derived proanthocyanidins reduce bacterial adhesion to selected biomaterials. Langmuir 24:10273–10281

Ferguson BJ, Mathesius U (2003) Signaling interactions during nodule development. Plant Growth Regul 22:47–72

Ferreira S, Silva F, Queiroz JA, Oleastro M, Domingues FC (2014) Resveratrol against Arcobacter butzleri and Arcobacter cryaerophilus: activity and effect on cellular functions. Int J Food Microbiol 180:62–68

Feys B, Parker JE (2000) Interplay of signaling pathways in plant disease resistance. Trends Genet 16:449–455

Figueiredo AR, Campos F, De Freitas V, Hogg T, Couto JA (2008) Effect of phenolic aldehydes and flavonoids on growth and inactivation of Oenococcus oeni and *Lactobacillus hilgardii*. Food Microbiol 25:105–112

Friend J (1979) Phenolic substances and plant disease. In: Swain T, Harbone JB, Van Sumere CF (eds) Biochemistry of plant phenolics. Recent advances in phytochemistry, vol 12. Springer, Boston, MA. https://doi.org/10.1007/978-1-4684-3372-2_17

Fries LLM, Pacovsky RS, Safir GR, Siqueira JO (1997) Plant growth and arbuscular mycorrhizal fungal colonization affected by exogenously applied phenolic compounds. J Chem Ecol 23: 755–1767

Garcion C, Lamotte O, Métraux JP (2007) Mechanism of defence to pathogens: biochemistry and physiology. In: Walters D, Newton A, Lyon G (eds) Induced resistance for plant defense: a sustainable approach to crop protection, 1st edn. Blackwell Publishing Ltd., Oxford, UK, pp 109–132

Gautam AK, Singh PK, Aravind M (2020) Defensive role of plant phenolics against pathogenic microbes for sustainable agriculture. In: Lone R, Shuab R, Kamili N (eds) Plant phenolics in sustainable agriculture. Springer, Singapore, pp 579–594

Giovannetti M, Sbrana C, Citernesi AS, Avio L (1996) Analysis of factors involved in fungal recognition responses to host derived signals by arbuscular mycorrhizal fungi. New Phytol 133: 65–71

Glenn MG, Chew FS, Williams PH (1988) Influence of glucosinolate content of Brassica (Cruciferae) roots on growth of vesicular-arbuscular mycorrhizal fungi. New Phytol 11(2): 17–25

Gonzalez JE, Marketon MM (2003) Quorum sensing in nitrogen fixing rhizobia. Microbiol Mol Biol Rev 67:574–592

Górniak I, Bartoszewski R, Króliczewski J (2019) Comprehensive review of antimicrobial activities of plant flavonoids. Phytochem Rev 18:241–272

Gourlay G, Constabel CP (2019) Condensed tannins are inducible antioxidants and protect hybrid poplar against oxidative stress. Tree Physiol 39(3):345–355

Gray EJ, Smith DL (2005) Intracellular and extracellular PGPR: commonalities and distinctions in the plant–bacterium signaling processes. Soil Biol Biochem 37(3):395–412

Gyawali R, Ibrahim SA (2014) Natural products as antimicrobial agents. Food Control 46:412–429. ISSN 0956-7135

Halvorson JJ, Gonzalez JM, Hagerman AE, Smith JL (2009) Sorption of tannin and related phenolic compounds and effects on soluble-N in soil. Soil Biol Biochem 41:2002–2010

Haraguchi H, Tanimoto K, Tamura Y, Mizutani K, Kinoshita T (1998) Mode of antibacterial action of retrochalcones from Glycyrrhiza inflata. Phytochemistry 48:125–129

Harborne JB (1980) Plant phenolics. In: Bell EA, Charlwood BV (eds) Encyclopedia of plant physiology, new series, secondary plant products, vol 8. Springer, Berlin, pp 329–402

Harborne JB (1989) General procedures and measurement of total phenolics. In: Harborne JB (ed) Methods in plant biochemistry, Plant phenolics, vol 1. Academic, London, pp 1–28

Harborne JB, Simmonds NW (1964) Natural distribution of the phenolic aglycones. In: Harborne JB (ed) Biochemistry of phenolic compounds. Academic, London, pp 77–128

Harrier LA, Wright F, Hooker JE (1998) Isolation of the 3-phosphoglycerate kinase gene of the arbuscular mycorrhizal fungus Glomus mosseae (Nicol. & Gerd.) Gerdemann & Trappe. Curr Genet 34:386–392

Harrison MJ (1993) Isoflavonoid accumulation and expression of defense gene transcripts during the establishment of vesicular-arbuscular mycorrhizal associations in roots of Medicago truncatula. Mol Plant Microbe Interact 6:643

Hartley RD, Harris PJ (1981) Phenolic constituents of the cell walls of dicotyledons. Biochem Syst Ecol 9:189–203

Hassan S, Mathesius U (2012) The role of flavonoids in root–rhizosphere signalling: opportunities and challenges for improving plant–microbe interactions. J Exp Bot 63(9):3429–3444. https://doi.org/10.1093/jxb/err430

Hättenschwiler S, Vitousek PM (2000) The role of polyphenols in terrestrial ecosystem nutrient cycling. Trends Ecol Evol 15:238–242

Hemaiswarya S, Doble M (2010) Synergistic interaction of phenylpropanoids with antibiotics against bacteria. J Med Microbiol 59:1469–1476. https://doi.org/10.1099/jmm.0.022426-0
Hettiarachchi DS, Locher C, Longmore RB (2011) Antibacterial compounds from the root of the indigenous Australian medicinal plant Carissa lanceolata R. Br Nat Prod Res 25(15):1388–1395
Hirsch AM, Bauer WD, Bird DM, Cullimore J, Tyler B, Yoder JI (2003) Molecular signals and receptors: controlling rhizosphere interactions between plants and other organisms. Ecology 84: 858–868
Hyldgaard M, Tina M, Rikke R (2012) Essential oils in food preservation: mode of action, synergies, and interactions with food matrix components. Front Microbiol 3:12
Ikigai H, Nakae T, Hara Y, Shimamura T (1993) Bactericidal catechins damage the lipid bilayer. Biochimica et Biophysica Acta (BBA)-Biomembranes 1147(1):132–136
Jaiti F, Meddich A, El Hadrami I (2007) Effectiveness of arbuscular mycorrhizal fungi in the protection of date palm (Phoenix dactylifera L.) against bayoud disease. Physiol Mol Plant Pathol 71:166–173
Jamiołkowska A (2011) Laboratory effect of azoxystrobin (Amistar 250 SC) and grapefruit extract (Biosept 33SL) on growth of fungi colonizing zucchini plants. Acta Sci Pol Hortoru 10:245–257
Jamiolkowska A (2020) Natural compounds as elicitors of plant resistance against diseases and new biocontrol strategies. Agronomy 10(2):73
Jeong K-W, Lee J-Y, Kang D-I, Lee J-U, Shin SY, Kim Y (2009) Screening of flavonoids as candidate antibiotics against Enterococcus faecalis. J Nat Prod 72:719–724
Jin WC, You GY, He YF, Tang C, Wu P, Zheng SJ (2007) Iron deficiency-induced secretion of phenolics facilitates the reutilization of root apoplastic iron in red clover. Plant Physiol 144(1): 278–285. https://doi.org/10.1104/pp.107.095794
Kannan RRR, Arumugam R, Thangaradjou T, Anantharaman P (2013) Phytochemical constituents, antioxidant properties and p-coumaric acid analysis in some seagrasses. Food Res Int 54(1): 1229–1236
Kape R, Parniske M, Brandt S, Werner D (1992) Isoliquiritigenin, a strong nod gene and glyceollin resistance-inducing flavonoid from soybean root exudate. Appl Environ Microbiol 58:1705–1710
Kaur S, Suseela V (2020) Unraveling arbuscular mycorrhiza-induced changes in plant primary and secondary metabolome. Metabolites 10(8):335. https://doi.org/10.3390/metabo10080335
Ke D, Saltveit Jr ME (1989) Wound-induced ethylene production, phenolic metabolism and susceptibility to russet spotting in iceberg lettuce. Physiol Plant 76(3):412–418. https://doi.org/10.1111/j.1399-3054.1989.tb06212.x
Kefeli VI, Kalevitch MV, Borsari B (2003) Phenolic cycle in plants and environment. J Cell Mol Biol 2:13–18
Keller M, Viret O, Cole FM (2003) Botrytis cinerea infection in grape flowers: defense reaction, latency, and disease expression. Biochem Cell Biol 93:316–322
Khokhani D, Zhang C, Li Y, Wang Q, Zeng Q, Yamazaki A, Hutchins W, Zhou S-S, Chen X, Yang C-H (2013) Discovery of plant phenolic compounds that act as type III secretion system inhibitors or inducers of the fire blight pathogen, Erwinia amylovora. Appl Environ Microbiol 79:5424–5436
Khokhar S, Apenten RKO (2003) Iron binding characteristics of phenolic compounds: some tentative structure–activity relations. Food Chem 81(1):133–140. ISSN 0308-8146
Koide RT, Schreiner RP (1992) Regulation of the vesicular-arbuscular mycorrhizal symbiosis. Annu Rev Plant Physiol Plant Mol Biol 43:557–581
Koornneef A, Pieterse CMJ (2008) Cross talk in defense signaling. Plant Physiol 146:839–844
Kosslak RM, Bookland R, Barkei J, Paaren HE, Edward R, Appelbaum ER (1987) Induction of Bradyrhizobium japonicum common nod genes by isoflavones isolated from Glycine max. Proc Natl Acad Sci U S A 84:7428–7423
Kourounakis AP, Assimopoulou AN, Papageorgiou VP, Gavalas A, Kourounakis PN (2002) Alkannin and Shikonin: effect on free radical processes and on inflammation-a preliminary pharmacochemical investigation. Arch Pharm 335(6):262–266

Koziara W, Sulewska H, Panasiewicz K (2006) Effect of resistance stimulator application to some agricultural crops. J Res Appl Agric Eng 51:82–87
Kraus TEC, Dahlgren RA, Zasoski RJ (2003) Tannins in nutrient dynamics of forest ecosystems – a review. Plant Soil 256:41–66
Kumar VN, Murthy PS, Manjunatha JR, Bettadaiah BK (2014) Synthesis and quorum sensing inhibitory activity of key phenolic compounds of ginger and their derivatives. Food Chem 159: 451–457
Kumar S, Abedin MM, Singh AK, Das S (2020) Role of phenolic compounds in plant-defensive mechanisms. In: Plant phenolics in sustainable agriculture. Springer, Singapore, pp 517–532
Langcake P, Irvine JA, Jeger MJ (1981) Alternative chemical agents for controlling plant disease. Philos Trans R Soc Lond Ser B Biol Sci 295:83–101
Lattanzio V (2013) Phenolic compounds: introduction. Nat Prod 50:1543–1580
Lattanzio V, Lattanzio VMT, Cardinali A (2006) Role of phenolics in the resistance mechanisms of plants against fungal pathogens and insects. In: Imperato F (ed) Phytochemistry advances in research. Research Signpost, India, pp 23–67
Lattanzio V, Kroon PA, Quideau S, Treutter D (2008) Plant phenolics e secondary metabolites with diverse functions. In: Daayf F, Lattanzio V (eds) Recent advances in polyphenol research, vol 1. Wiley-Blackwell, Oxford, pp 1–35
Lechner D, Gibbons S, Bucar S (2008) Plant phenolic compounds as ethidium bromide efflux inhibitors in *Mycobacterium smegmatis*. J Antimicrob Chemother 62(2):345–348. https://doi.org/10.1093/jac/dkn178
Lee K (1991) Glutathione S-transferase activities in phytophagous insects: induction and inhibition by plant phototoxins and phenols. Insect Biochem 21(4):353–361
Leon J, Lawton MA, Raskin I (1995) Hydrogen peroxide stimulates salicylic acid biosynthesis in tobacco. Plant Physiol 108:1673–1678
Lima MDC, de Sousa CP, Fernandez-Prada C, Harel J, Dubreuil JD, De Souza EL (2019) A review of the current evidence of fruit phenolic compounds as potential antimicrobials against pathogenic bacteria. Microb Pathog 130:259–270
Lin YT, Kwon YI, Labbe RG, Shetty K (2015) Inhibition of Helicobacter pylori and associated urease by oregano and cranberry phytochemical synergies American Society for Microbiology. Appl Environ Microbiol 71(12):8558–8564. https://doi.org/10.1128/AEM.71.12.8558-8564.2005
Ling-Lee M, Chilvers GA, Ashford AE (1977) A histochemical study of phenolic materials in mycorrhizal and uninfected roots of Eucalyptus fastigata Dean & Maid. New Phytol 78:313–328
Lodhi MAK, Bilal R, Malik KA (1987) Allelopathy in agroecosystems: wheat phytotoxicity and its possible roles in crop rotation. J Chem Ecol 13:1881–1891
López-Ráez JA, Flors V, García JM, Pozo MJ (2010) AM symbiosis alters phenolic acid content in tomato roots. Plant Signal Behav 5:1138–1140
Lou Z, Wang H, Rao S, Sun J, Ma C, Li J (2012) p-Coumaric acid kills bacteria through dual damage mechanisms. Food Control 25:550–554
Lu H (2009) Dissection of salicylic acid-mediated defense signaling networks. Plant Signal Behav 4:713–717
Lynn DG, Chang M (1990) Phenolic signals in cohabitation: implications for plant development. Annu Rev Plant Physiol Plant Mol Biol 41:497–526
Maddox CE, Laur LM, Tian L (2010) Antibacterial activity of phenolic compounds against the phytopathogen Xylella fastidiosa. Curr Microbiol 60:53–58
Mandal SM, Mandal M, Das AK, Pati BR, Ghosh AK (2009) Stimulation of indoleacetic acid production in a Rhizobium isolate of *Vigna mungo* by root nodule phenolic acids. Arch Microbiol 191:389–393
Mandal SM, Chakraborty D, Dey S (2010) Phenolic acids act as signaling molecules in plant-microbe symbioses. Plant Signal Behav 5(4):359–368. https://doi.org/10.4161/psb.5.4.10871

Martens DA (2002) Relationship between plant phenolic acids released during soil mineralization and aggregate stabilization. Soil Sci Soc Am J 66:1857–1867

Martini S, D'Addario C, Colacevich A, Focardi S, Borghini F, Santucci A, Figura N, Rossi C (2009) Antimicrobial activity against Helicobacter pylori strains and antioxidant properties of blackberry leaves (*Rubus ulmifolius*) and isolated compounds. Int J Antimicrob Agents 34(1): 50–59. https://doi.org/10.1016/j.ijantimicag.2009.01.010

Martin-Laurent F, van Tuinen D, Dumas-Gaudot E, Gianinazzi-Pearson V, Gianinazzi S, Franken P (1997) Differential display analysis of RNA accumulation in arbuscular mycorrhiza of pea and isolation of a novel symbiosis-regulated plant gene. Mol Gen Genomics 256:37–44

Mathew S, Abraham TE, Zakaria ZA (2015) Reactivity of phenolic compounds towards free radicals under in vitro conditions. J Food Sci Technol 52(9):5790–5798. https://doi.org/10.1007/s13197-014-1704-0

Matkowski A (2006) Plant phenolic metabolites as antioxidants and antimutagens. In: Blume Y, Durzan DJ, Petro Smertenko Mayr U, Michalek S (eds) Cell biology and instrumentation: UV radiation, nitric oxide and cell death in plants. IOS Press

Mazimba O, Nana F, Kuete V, Singh GS (2013) Xanthones and anthranoids from the medicinal plants of Africa. In: Medicinal plant research in Africa. Elsevier, pp 393–434

Min K, Freeman C, Kang H, Choi S-U (2015) The regulation by phenolic compounds of soil organic matter dynamics under a changing environment. Biomed Res Int. https://doi.org/10.1155/2015/825098

Mistry KJ, Krishna M, Bhattacharya RK (1997) Modulation of aflatoxin B1 activated protein kinase C by phenolic compounds. Cancer Lett 121(1):99–104. ISSN 0304-3835

Modafar CE, Clerivet A, Flavan JJM (1996) Accumulation in stems of Platanus xacerifolia seedlings inoculated with Ceratocystis firnbriata f.sp. platani, the canker stain disease agent. Can J Bot 74:1982–1987

Monici M, Baglioni P, Mulinacci N, Baldi A, Vincieri FF (1993) A research model on flavonoids as photoprotectors: studies on the photochemistry of kaempferol and pelargonidin. Int Symp Nat Phenols Plant Resist 381:340–347

Moulin L, Munive A, Dreyfus B, Boivin-Masson C (2001) Nodulation of legumes by members of the b-subclass of proteobacteria. Nature 411:948–950

Nagayama K, Iwamura Y, Shibata T, Hirayama I, Nakamura T (2002) Bactericidal activity of phlorotannins from the brown alga Ecklonia kurome. J Antimicrob Chemother 50(6):889–893

Naikoo MI, Dar MI, Raghib F, Jaleel H, Ahmad B, Raina A, Khan FA, Naushin F (2019) Role and regulation of plants phenolics in abiotic stress tolerance: an overview. In: Khan IR, Reddy PS, Ferrante A, Khan NA (eds) Plant signaling molecules. Woodhead

Nannipieri P, Ascher J, Ceccherini MT, Landi L, Pietramellara G, Renella G, Valori F (2008) Effects of root exudates in microbial diversity and activity in rhizosphere soils. In: Nautiyal CS, Dion P (eds) Molecular mechanisms of plant, Microbe coexistence. Soil biology, vol 15

Ndakidemi PA, Dakora FD (2003) Legume seed flavonoids and nitrogenous metabolites as signals and protectants in early seedling development. Rev Funct Plant Biol 30:729–745

Nicholson RL, Hammerschmidt R (1992) Phenolic compounds and their role in disease resistance. Annu Rev Phytopathol 30:369–389

Northup RR, Dahlgren RA, McColl JG (1998) Polyphenols as regulators of plant–litter–soil interactions in northern California's pygmy forest: a positive feedback? Biogeochemistry 42: 189–220

Oh E, Jeon B (2015) Synergistic anti-Campylobacter jejuni activity of fluoroquinolone and macrolide antibiotics with phenolic compounds. Front Microbiol 6:1129

Ohemeng KA, Schwender CF, Fu KP, Barrett JF (1993) DNA gyrase inhibitory and antibacterial activity of some flavones. Bioorg Med Chem Lett 3:225–230

Oliva A, Moraes RM, Watson SB, Duke SO, Dayan FE (2002) Aryltetralin lignans inhibit plant growth by affecting the formation of mitotic microtubular organizing centers. Pestic Biochem Physiol 72(1):45–54

Oliveira BDA, Rodrigues AC, Bertoldi MC, Taylor JG, Pinto UM (2017) Microbial control and quorum sensing inhibition by phenolic compounds of acerola (Malpighia emarginata). Int Food Res J 24(5):2228–2237
Oteiza PI, Erlejman AG, Verstraeten SV, Keen CL, Fraga CG (2005) Flavonoid-membrane interactions: a protective role of flavonoids at the membrane surface? Clin Dev Immunol 12: 19–25
Ozyigit II, Kahraman MV, Ercan O (2007) Relation between explant age, total phenols and regeneration response of tissue cultured cotton (Gossypium hirsutum L). Afr J Biotechnol 6: 3–8
Perret X, Staehelin C, Broughton WJ (2000) Molecular basis of symbiotic promiscuity. Microbiol Mol Biol Rev 64:180–201
Phillips DA, Torrey JG (1972) Studies on cytokinin production by Rhizobium. Plant Physiol 49: 11–15
Pieterse C, Van Loon LC (1999) Salicylic acid-independent plant defence pathways. Trends Plant Sci 4:52–58
Poulin MJ, Simard J, Catford JG, Labrie F, Piche Y (1997) Response of symbiotic endomycorrhizal fungi to estrogen and antiestrogens. Mol Plant Microbe Interact 10:481–487
Prasch S, Bucar F (2015) Plant derived inhibitors of bacterial efflux pumps: an update. Phytochem Rev 14:961. https://doi.org/10.1007/s11101-015-9436-y
Prescott CE, Grayston SJ, Helmisaari H-S, Kaštovská E, Körner C, Lambers H, Meier IC, Millard P, Ostonen I (2020) Surplus carbon drives allocation and plant–soil interactions. Trends Ecol Evol 35(12):1110–1118. ISSN 0169-5347
Quideau S, Deffieux D, Douat-Casassus C, Pouységu L (2011) Plant polyphenols: chemical properties, biological activities, and synthesis. Angew Chem Int Ed 50:586–621
Rai D, Singh JK, Roy N, Panda D (2008) Curcumin inhibits FtsZ assembly: an attractive mechanism for its antibacterial activity. Biochem J 410:147–155
Ravin H, Andary C, Kovacs G, Molgaard P (1989) Caffeic acid esters as in vitro inhibitors of plant pathogenic bacteria and fungi. Biochem Syst Ecol 17:175–184
Redmond JW, Batley M, Djordjevic MA, Innes RW, Kuempel PL, Rolfe BG (1986) Flavones induce expression of nodulation genes in Rhizobium. Nature 323:632–635
Rempe CS, Burris KP, Lenaghan SC, Neal SC (2017) The potential of systems biology to discover antibacterial mechanisms of plant phenolics. Front Microbiol 8:422
Revillas JJ, Rodelas B, Pozo C, Martínez-Toledo MV, González-López J (2000) Production of B-group vitamins by two Azotobacter strains with phenolic compounds as sole carbon source under diazotrophic and adiazotrophic conditions. J Appl Microbiol 89:486–493
Robards R, Antolovich M (1997) Analytical chemistry of fruit bioflavonoids. A review. Analyst 122:11R–34R
Sammi S, Masud T (2009) Effect of different packaging systems on the quality of tomato (Lycopersicon esculentum var. Rio Grande) fruits during storage. Int J Food Sci Technol 44: 918–926. https://doi.org/10.1111/j.1365-2621.2007.01649.x
Sanchez-Contreras M, Bauer WD, Gao M, Robinson JB, Allan Downie J (2007) Quorum-sensing regulation in rhizobia and its role in symbiotic interactions with legumes. Philos Trans R Soc Lond Ser B Biol Sci 362(1483):1149–1163. https://doi.org/10.1098/rstb.2007.2041
Sanchez-Lizarraga AL, Dendooven L, Marino-Marmolejo EN, Davila-Vazquez G, Hernández-Cuevas L, Arenas-Montaño V, Contreras-Ramos SM (2017) Presence and diversity of arbuscular mycorrhizal fungi in soil regularly irrigated with vinasses. J Soil Sci Plant Nutr 17(4):1116–1129
Sanver D, Murray BS, Sadeghpour A, Rappolt M, Nelson AL (2016) Experimental modeling of flavonoid–biomembrane interactions. Langmuir 32:13234–13243
Scalzo J, Politi A, Pellegrini N, Mezzetti B, Battino M (2005) Plant genotype affects total antioxidant capacity and phenolic contents in fruit. Nutrition 21(2):207–213. ISSN 0899-9007
Scervino JM, Ponce MA, Erra-Bassells R, Vierheilig H, Ocampo JA, Godeas A (2005) Flavonoids exhibit fungal species and genus specific effects on the presymbiotic growth of *Gigaspora* and *Glomus*. Mycol Res 109:789–794

Schardl CL, Leuchtmann A, Spiering MJ (2004) Symbiosis of grasses with seedborne fungal endophytes. Annu Rev Plant Biol 55:315–340

Schliemann W, Ammer C, Strack D (2008) Metabolite profiling of mycorrhizal roots of Medicago truncatula. Phytochemistry 69:112–146

Selvaraj JT, Subramanian G (1990) Phenols and lipids in mycorrhizal and non-mycorrhizal roots of *Sesamum indicum*. Curr Sci 59:471–473

Seneviratne G, Jayasinghearachchi HS (2003) Mycelial colonization by bradyrhizobia and azorhizobia. J Biosci 28:243–247

Shakya T, Stogios PJ, Waglechner N, Evdokimova E, Ejim L, Blanchard JE et al (2011) A small molecule discrimination map of the antibiotic resistance kinome. Chem Biol 18:1591–1601

Shankar S, Madhan R, Girish R, Karthik N, Rajendran R, Mahendran VS (2009) Allelopathic effects of phenolics and terpenoids extracted from *Gmelina arborea* on germination of Black gram (*Vigna mungo*) and Green gram (*Vigna radiata*). Allelopathy J 23:323–332

Shapiro S, Guggenheim B (1998) Inhibition of oral bacteria by phenolic compounds. Part 1. QSAR analysis using molecular connectivity. Quant Struct Relat 17:327–337

Shibata T, Ishimaru K, Kawaguchi S, Yoshikawa H, Hama Y (2007) Antioxidant activities of phlorotannins isolated from Japanese Laminariaceae. In: Nineteenth international seaweed symposium. Springer, Dordrecht, pp 255–261

Shriram V, Khare T, Bhagwat R, Shukla R, Kumar V (2018) Inhibiting bacterial drug efflux pumps via phyto-therapeutics to combat threatening antimicrobial resistance. Front Microbiol 9:2990

Sikkema J, de Bont JA, Poolman B (1995) Mechanisms of membrane toxicity of hydrocarbons. Microbiol Rev 59:201–222

Silva-Junior JP, Siqueira JO (1998) Colonização micorrízica e crescimento da soja com diferentes fungos e aplicação do isoflavonóide formononetina*. Pesq Agrop Brasileira 33:953–959

Singh AK, Singla P (2020) Root phenolics profile modulates microbial ecology of rhizosphere. In: Lone R, Shuab R, Kamili N (eds) Plant phenolics in sustainable agriculture. Springer, Singapore, pp 555–578

Singh UP, Maurya S, Singh DP (2005) Phenolic acids in neem (Azadirachta indica). J Herb Pharmacother 5(1):35–43. https://doi.org/10.1080/J157v05n01_05

Singh P, Arif Y, Bajguz A, Hayat S (2021) The role of quercetin in plants. Plant Physiol Biochem 166:10–19

Siqueira JO, Safir GR, Nair MG (1991) Stimulation of vesicular arbuscular mycorrhiza formation and growth of white clover by flavonoid compounds. New Phytol 118:87–93

Slatnar A, Mikulič-Petkovšek M, Veberič R, Štampar F (2016) Research on the involvement of phenoloics in the defence of horticultural plants. Acta Agric Slovenica 107(1):183–189. https://doi.org/10.14720/aas.2016.107.1.18

Slobodníková L, Fialová S, Rendeková K, Kováč J, Mučaji P (2016) Antibiofilm activity of plant polyphenols. Molecules 21:1717

Smith ECJ, Kaatz GW, Seo SM, Wareham N, Williamson EM, Gibbons S (2007) The phenolic diterpene totarol inhibits multidrug efflux pump activity in Staphylococcus aureus. Antimicrob Agents Chemother 51:4480–4483

Spaink HP (2000) Root nodulation and infection factors produced by rhizobial bacteria. Annu Rev Microbiol 54:257–288

Spanu P, Bonfante-Fasolo P (1988) Cell-wall-bound peroxidase activity in roots of mycorrhizal *Allium porrum*. New Phytol 109(1):19–12

Staman K, Blum U, Louws F, Robertson D (2001) Can simultaneous inhibition of seedling growth and stimulation of rhizosphere bacterial populations provide evidence for phytotoxin transfer from plant residue in the bulk soil to the rhizosphere of sensitive species? J Chem Ecol 27:807–829

Steinkellner S, Lendzemo V, Langer I, Schweiger P, Khaosaad T, Toussaint JP, Vierheilig H (2007) Flavonoids and strigolactones in root exudates as signals in symbiotic and pathogenic plant–fungus interactions. Molecules 12:1290–1306

Stojković D, Petrović J, Soković M, Glamočlija J, Kukić-Marković J, Petrović S (2013) In situ antioxidant and antimicrobial activities of naturally occurring caffeic acid, p-coumaric acid and rutin, using food systems. J Sci Food Agric 93:3205–3208

Stougaard J (2000) Regulators and regulation of legume root nodule development. Plant Physiol 124:531–540

Subramanian S, Stacey G, Yu O (2006) Endogenous isoflavones are essential for the establishment of symbiosis between soybean and Bradyrhizobium japonicum. J Biotechnol 126:69–77

Subramanian S, Stacey G, Yu O (2007) Distinct, crucial roles of flavonoids during legume nodulation. Trends Plant Sci 12:282–285

Sultana V, Baloch GN, Ara J, Esteshamul-Haque S, Tariq RM, Athar M (2011) Seaweeds as alternative to chemical pesticides for the management of root diseases of sunflower and tomato. J Appl Bot Food Qual 84:162–168

Swain T (1975) Evolution of flavonoid compounds. In: Harborne JB, Mabry TJ, Mabry H (eds) The flavonoids. Chapman & Hall, London, UK, pp 1096–1138

Tak Y, Kumar M (2020) Phenolics: a key defence secondary metabolite to counter biotic stress. In: Lone R, Shuab R, Kamili N (eds) Plant phenolics in sustainable agriculture. Springer, Singapore, pp 309–329

Takó M, Kerekes EB, Zambrano C, Kotogán A, Papp T, Krisch J, Vágvölgyi C (2020) Plant phenolics and phenolic-enriched extracts as antimicrobial agents against food-contaminating microorganisms. Antioxidants 9:165. https://doi.org/10.3390/antiox9020165

Tamokou JD, Mbaveng AT, Kuete V (2017) Antimicrobial activities of African medicinal spices and vegetables. In: Medicinal spices and vegetables from Africa. Academic Press, pp 207–237

Taylor LP, Grotewold E (2005) Flavonoids as developmental regulators. Curr Opin Plant Biol 8 (3):317–323

Taylor PW, Hamilton-Miller JM, Stapleton PD (2005) Antimicrobial properties of green tea catechins. Food Sci Technol Bull 2:71–81

Terry LA, Joyce DC, Adikaram NK, Khambay BP (2004) Preformed antifungal compounds in strawberry fruit and flower tissues. Postharvest Biol Technol 31:201–212

Thomas P, Ravindra MB (1999) Shoot tip culture in mango: influence of medium, genotype, explant factors, season and decontamination treatments on phenolic exudation, explant survival and axenic culture establishment. J Hortic Sci 72:713–722

Tidke SA, Devappa R, Vasist SK, Kosturkova P, Gokare RA (2018) Soybean plants treated with vesicular arbuscular mycorrhiza fungi exhibit enhanced plant growth and nutraceutically important metabolites. J Plant Sci 13:1–11

Torres MA, Jones JDG, Dangl JL (2006) Reactive oxygen species signaling in response to pathogens. Plant Physiol 141:373–378

Treutter D (2006) Significance of flavonoids in plant resistance: a review. Environ Chem Lett 4: 147–157

Treutter D, Feucht W (1990) Accumulation of flavan-3-ols in fungus-infected leaves of Rosaceae. Zeitschrift für Pflanzenkrankheiten und Pflanzenschutz 97:634–641

Tsai SM, Phillips DA (1991) Flavonoids released naturally from alfalfa promote development of symbiotic Glomus spores in vitro. Appl Environ Microbiol 57:85–88

Tsou LK, Lara-Tejero M, Rosefigura J, Zhang ZJ, Wang YC, Yount JS et al (2016) Antibacterial flavonoids from medicinal plants covalently inactivate type III protein secretion substrates. J Am Chem Soc 138:2209–2218

Tsuchiya H (2015) Membrane interactions of phytochemicals as their molecular mechanism applicable to the discovery of drug leads from plants. Molecules 20:18923–18966

Tsuchiya H, Sato M, Miyazaki T, Fujiwara S, Tanigaki S, Ohyama M (1996) Comparative study on the antibacterial activity of phytochemical flavanones against methicillin-resistant Staphylococcus aureus. J Ethnopharmacol 50:27–34

Tsuda K, Glazebrook J, Katagiri F (2008) The interplay between MAMP and SA signaling. Plant Signal Behav 3:359–361

Ultee A, Bennik MHJ, Moezelaar R (2002) The phenolic hydroxyl group of carvacrol is essential for action against the food-borne pathogen Bacillus cereus. Appl Environ Microbiol 68:1561–1568

Underwood W (2012) The plant cell wall: a dynamic barrier against pathogen invasion. Front Plant Sci 3:85. https://doi.org/10.3389/fpls.2012.00085

Vagiri M, Johansson E, Rumpunen K (2017) Phenolic compounds in black currant leaves – an interaction between the plant and foliar diseases? J Plant Interact 12(1):193–199

van Loon LC (2000) Systemic induced resistance. In: Slusarenko AJ, Fraser RSS, Van Loon LC (eds) Mechanisms of resistance to plant diseases. Kluwer Academic, Dordrecht, The Netherlands, pp 521–574

Verstraeten SV, Fraga CG, Oteiza PI (2015) Interactions of flavan-3-ols and procyanidins with membranes: mechanisms and the physiological relevance. Food Funct 6:32–40

Vicente R-SM, Plasencia J (2011) Salicylic acid beyond defence: its role in plant growth and development. J Exp Bot 62(10):3321–3338

Villaverde JJ, Sandin-Espana P, Sevilla-Morañ B, López-Goti C, Alonso-Prados JL (2016) Biopesticides from natural products: current development, legislative framework and future trends. Bioresources 11:5618–5640

Volpin H, Elland Y, Okon Y, Kapulnik Y (1994) A vesicular arbuscular mycorrhizal fungus (*Glomus intraradix*) induces a defence response in alfalfa roots. Plant Physiol 104:683–689

Wafa BAM, Makni S, Ammar L, Khannous AB, Hassana M, Bouaziz NE, Es-Safi RG (2017) Antimicrobial effect of the Tunisian Nana variety *Punica granatum* L. extracts against *Salmonella enterica* (serovars Kentucky and Enteritidis) isolated from chicken meat and phenolic composition of its peel extract. Int J Food Microbiol 241:123–131

Wallis CM, Galarneau ER-A (2020) Phenolic compound induction in plant-microbe and plant-insect interactions: a meta-analysis. Front Plant Sci 11:2034

Walsh DJ, Livinghouse T, Goeres DM, Mettler M, Stewart PS (2019) Antimicrobial activity of naturally occurring phenols and derivatives against biofilm and planktonic bacteria. Front Chem 7:653

Walters D, Newton A, Lyon G (2007) Induced resistance for plant defense: a sustainable approach to crop protection. Blackwell Publishing Ltd., Oxford, UK, p 258

Warskulat AC, Tatsis EC, Dudek B, Kai M, Lorenz S, Schneider B (2016) Unprecedented utilization of pelargonidin and indole for the biosynthesis of plant indole alkaloids. ChemBioChem 17(4):318

Wheatley RM, Poole PS (2018) Mechanisms of bacterial attachment to roots. FEMS Microbiol Rev 014(42):448–461. https://doi.org/10.1093/femsre/fuy014

Whitehead DC, Dibb H, Hartley RD (1983) Bound phenolic compounds in water extract of soils, plant roots and leaf litter. Soil Biol Biochem 15:133–136

Wink M (2003) Evolution of secondary metabolites from an ecological and molecular phylogenetic perspective. Phytochemistry 64(1):3–19. ISSN 0031-9422

Wu T, Zang X, He M, Pan S, Xu X (2013a) Structure–activity relationship of flavonoids on their anti-*Escherichia coli* activity and inhibition of DNA gyrase. J Agric Food Chem 61:8185–8190

Wu T, He M, Zang X, Zhou Y, Qiu T, Pan S et al (2013b) A structure-activity relationship study of flavonoids as inhibitors of E. coli by membrane interaction effect. Biochim Biophys Acta 1828: 2751–2756. https://doi.org/10.1016/j.bbamem.2013.07.029

Wuyts N, Swennen R, De Waele D (2006) Effects of plant phenylpropanoid pathway products and selected terpenoids and alkaloids on the behaviour of the plant-parasitic nematodes Radopholus similis, Pratylenchus penetrans and Meloidogyne incognita. Nematology 8(1):89–101. https://doi.org/10.1163/156854106776179953

Xu H, Ziegelin G, Schröder W, Frank J, Ayora S, Alonso JC, Lanka E, Saenger W (2001) Flavones inhibit the hexameric replicative helicase RepA. Nucleic Acids Res 29(24):5058–5066

Xu X, Zhou XD, Wu CD (2011) The tea catechin epigallocatechin gallate suppresses cariogenic virulence factors of Streptococcus mutans. Antimicrob Agents Chemother 55:1229–1236

Ya C, Gaffney SH, Lilley TH, Haslam E (1988) Carbohydrate-polyphenol complexation. In: Hemingway RW, Karchesy JJ (eds) Chemistry and significance of condensed tannins. Plenum Press, New York, NY, p 553

Yan S, Dong X (2014) Perception of the plant immune signal salicylic acid. Curr Opin Plant Biol 20:64–68

Yao X, Zhu X, Pan S, Fang Y, Jiang F, Phillips GO et al (2012) Antimicrobial activity of nobiletin and tangeretin against Pseudomonas. Food Chem 132:1883–1890

Yu D, Wang J, Shao X, Xu F, Wang H (2015) Antifungal modes of action of tea tree oil and its two characteristic components against Botrytis cinerea. J Appl Microbiol 119:1253–1262

Zaat SAJ, Wijffelman CA, Mulders IHM, van Brussel AAN, Lugtenberg BJJ (1988) Root exudates of various host plants of *Rhizobium leguminosarum* contain different sets of inducers of *Rhizobium* nodulation genes. Plant Physiol 86(4):1298–1303. https://doi.org/10.1104/pp.86.4.1298

Zambrano C, Kerekes EB, Kotogán A, Papp T, Vágvölgyi C, Krisch J, Takó M (2019) Antimicrobial activity of grape, apple and pitahaya residue extracts after carbohydrase treatment against food-related bacteria. LWT Food Sci Technol 100:416–425

Zapprometov M (1989) The formation of phenolic compounds in plant cell and tissue cultures and possibility of its regulation. Adv Cell Cult 7:240–245

Zawoznik MS, Garrido LM, del Pero Martinez MA, Tomaro ML (2000) Effect of vanillin on growth and symbiotic ability of Bradyrhizobium sp. (Arachis) strain. Proc Int Plant Growth-promoting Rhizobacteria. http://www.ag.auburn.edu/argentina

Zhang L, Kong Y, Wu D, Zhang H, Wu J, Chen J et al (2008) Three flavonoids targeting the b-hydroxyacyl-acyl carrier protein dehydratase from Helicobacter pylori: crystal structure characterization with enzymatic inhibition assay. Protein Sci 17:1971–1978

Zhang X, Dippold M, Kuzyakov Y, Razavi B (2019) Spatial pattern of enzyme activities depends on root exudate composition. Soil Biol Biochem. https://doi.org/10.1016/j.soilbio.2019.02.010

Zuanazzi JAS, Clergeot PH, Quirion J-C, Husson HP, Kondorosi P, Ratet P (1998) Production of *Sinorhizobium meliloti* nod gene activation and repressor flavonoids from *Medicago sativa* roots. Mol Plant Microbe Interact 11:784–794

Elucidating the Role of Flavonoids in Countering the Effect of Biotic Stress in Plants

5

Sandeep Kour, Nandni Sharma, Anjali Khajuria, Deepak Kumar, and Puja Ohri

Abstract

In constantly changing environmental conditions, plants are exposed to a lot of biotic as well as abiotic stresses that have an adverse effect on their growth and production. Biotic stressors that include pests and pathogens cause several diseases in plants, deteriorate their health, and lower their produce. In order to have a grip on these stresses, some secondary metabolites are produced by plants that confer resistance against various pathogenic microorganisms. Among those secondary metabolites are flavonoids that represent a class of plant phenolic compounds. Flavonoids are spectacular, low-molecular-weight, structurally diverse group of phenylpropanoids that are found in flowers, stems, roots, vegetables, fruits, grains, etc. These naturally occurring phenolic compounds are further categorized into various subgroups like flavone, isoflavone, flavanone, flavanonol, flavonol, anthocyanidin, flavanol, etc. They play defensive roles in plants against a wide variety of disease-causing microorganisms like bacteria, fungi, nematodes, viruses, and insect pests. Plants are endowed with biosynthetic machinery to produce these secondary metabolic compounds and consist of various transporters for translocating them to different parts where they provide protection against pathogens and pests. The present chapter personifies a comprehensive elucidation of the biogenesis, transport, and protective role of flavonoids against biotic stressors in plants.

Keywords

Phenolics · Flavonoids · Biotic stress: defense molecules · Environment

S. Kour · N. Sharma · A. Khajuria · D. Kumar · P. Ohri (✉)
Department of Zoology, Guru Nanak Dev University, Amritsar, India

R. Lone et al. (eds.), *Plant Phenolics in Biotic Stress Management*,
https://doi.org/10.1007/978-981-99-3334-1_5

5.1 Introduction

Plants fabricate a myriad of low-molecular-weight compounds, whose number is reported to exceed 200,000 (Afendi et al. 2012). These compounds, known as plant-specialized metabolites, perform various biological activities in plants like growth promotion, stress alleviation, and regulation of plant–microbe relations in the rhizosphere (Sugiyama 2021). It is said that the genetic machinery to biosynthesize these metabolic compounds has been acquired by plants during evolution in order to adapt to the environment (Kim and Buell 2015; Lichman et al. 2020; Gani et al. 2021). One such kind of plant-specialized metabolites is flavonoids. Flavonoids, having polyphenolic structure, are an accomplished set of organic compounds produced by plants (Shah and Smith 2020). They are naturally present in different parts of plants like fruits, vegetables, stem, bark, grains, roots, flowers, and certain beverages and are thus called as dietary flavonoids (Panche et al. 2016). They are known to impart characteristic aroma, flavor, and color to the plants (Dias et al. 2021).

Flavonoids are structurally diverse compounds having different subgroups that include chalcones, flavone, isoflavone, flavanone, flavanonol, flavonol, anthocyanidins, flavanol, isoflavonoids, etc. (Palanisamy et al. 2020). Structurally, these aromatic compounds comprise of 15-carbon flavone skeleton, having two (A and B) benzene rings, and a third 3-carbon pyran ring (C) interlinking the two benzene rings (Dias et al. 2021). Flavonoids are fabricated naturally by the plants. Plants are endowed with biosynthetic machinery to produce these phenolic compounds through phenylpropanoid and acetate-malonate pathways (Yonekura-Sakakibara et al. 2019). After their synthesis in the cytosolic compartments of the plants, they are supposed to be amassed into the vacuoles as glucosides (Sugiyama 2021) and display long-distance translocation by employing transport proteins or vesicles (Ku et al. 2020).

Flavonoids perform a diverse array of physiological functions in plants, animals, and microorganisms. In plants, they are in charge of the fragrance as well as the color of flowers, attraction of pollinators, dispersal of fruits, seed germination, and sprouting and maturation of seedlings (Griesbach 2005). Presently, around 6000 flavonoids are known to be the reason behind the colorful pigments of several fruits, herbs, vegetables, medicinal, and aromatic plants (Panche et al. 2016; Gautam et al. 2020). In addition, these phenylpropanoid compounds perform protective functions in plants. They counter the ill effects of different biotic as well as abiotic stresses and provide protection against them (Gutiérrez-Grijalva et al. 2020; Kumar et al. 2020; Tak and Kumar 2020). Various biotic stressors like bacteria, nematodes, fungi, herbivores, etc. cause several diseases and hamper plant growth. Stressed conditions lead to generation and accretion of reactive oxygen species (ROS) that can damage the components of plant cells like DNA, carbohydrates, proteins, and lipids. So, to scavenge ROS and lower the oxidative damage, plants produce flavonoids that are among the crucial nonenzymatic antioxidants (Baskar et al. 2018). Flavonoids guard plants from these pathogenic organisms by acting as phytoalexins, allopathic compounds, antimicrobial compounds, and detoxifying agents (Panche et al. 2016). This book chapter summarizes the plant flavonoids' biosynthesis, classification, translocation, and stress-alleviating potential against various disease-causing pathogens and pests.

5.2 Classification of Flavonoids

Flavonoids are plant secondary metabolites having a fundamental structure that consists of three phenolic rings, viz., A, B, and C with six, six and three carbons, respectively, and C ring acting as a linker or connector ring (Aherne and O'Brien 2002). Depending upon the level of unsaturation and oxidation of ring C, and also the position of carbon that links B and C rings together, flavonoids can be categorized into many subgroups. For instance, in isoflavones (3-benzopyrans), ring B is linked to the third position of ring C; in neoflavonoids (4-benzopyrans) B ring is attached at fourth position. The flavonoids where ring B is joined to second position ring C are further divided into different subgroups that include flavanonols, anthocyanins, chalcones, flavanones, flavonols, and flavones (Samanta et al. 2011; Panche et al. 2016).

5.2.1 Chalcones

Chalcones are the only subtype of flavonoids lacking "ring C" in their fundamental structure, and because of this feature, chalcones are called as "open-chain flavonoids" as well. Some prominent instances of chalcones are chalconaringenin, phloretin, arbutin, and phloridzin. Their dietary sources are tomatoes, pears, apples, berries, and certain wheat products (Panche et al. 2016; Tsao 2010).

5.2.2 Anthocyanins

Anthocyanins provide colors to the plants, flowers, and fruits. For instance, red and blue color in apples and berries is attributed to the presence of anthocyanin or anthocyanin glycosides (Erlund 2004). Peonidin, malvidin, delphinidin, pelargonidin, and cyanidin are the most frequently researched anthocyanins. Different fruits like red grapes, blueberries, blackcurrants, bilberries, cranberries, blackberries, merlot grapes, strawberries, and raspberries contain these pigments primarily in their outer cell layers. The stability and health advantages of anthocyanin make it one of the most prominent components that can be utilized in food sector for various purposes (Giusti and Wrolstad 2003). Moreover, the anthocyanin color is influenced by acylation or methylation at -OH group present on rings, A and B (Iwashina 2013).

5.2.3 Flavones

Similar to anthocyanins, flavones are profusely allocated in flowers, fruits, and leaves of plants. Their main dietary sources are mint, pepper, parsley, chamomile, peppers, celery, ginkgo, and olive oil (López-Lázaro 2009). Flavones include apigenin, tangeritin, and luteolin. In citrus fruit peels, there is abundance of polymethoxylated flavones, tangeretin, sinensetin, and nobiletin. In these flavonoids, at the fourth position, a ketone is present, and between second and third positions of ring C, there is the presence of a double bond. Based on the taxonomic classification, the flavones of most fruits and vegetables have -OH group at fifth position of ring A, although hydroxylation may vary in other places, typically at seventh position ring A or third and fourth position of ring B (Manach et al. 2004).

5.2.4 Flavonols

Ketone-group-containing flavonoids, i.e., flavonols, are considered as the most prevalent and largest category of flavonoids in fruits and vegetables. Proanthocyanidins can be generated by using them. A large number of fruits and vegetables contain flavonols. Among flavonols, fisetin, quercetin, kaempferol, and myricetin are the ones that have been studied the most. Flavonols are present abundantly in foods including berries, leafy vegetables, onions, lettuce, tomatoes, strawberries, medicinal plants, and apples. Flavonols are also present in red wine and tea (Sultana and Anwar 2008). Flavonols, being a rich source of antioxidants, impose various health benefits, and its consumption may decrease the chances of vascular diseases (Panche et al. 2016).

5.2.5 Flavanones

Flavanones also known as "dihydroflavones" is another essential subcategory of flavonoids with reactive oxygen species quenching potential. Flavanones are mostly present in grapes, lemons, and oranges. The only difference between the structures of flavanones and flavones is that a single bond is present between second and third carbon of the C ring. The major examples of flavanones are eriodictyol, hesperitin, and naringenin (Heim et al. 2002; Panche et al. 2016; Rashad et al. 2020).

5.2.6 Isoflavonoids

Isoflavonoids differ structurally from other flavonoid groups because in them ring B connected to ring C at its third position. In legume–rhizobia symbioses, isoflavonoids are discovered to be particularly beneficial for microbial signaling and nodule induction. Aglycone, the glycosides of genistein and daidzein, and other isoflavonoids are typical examples. Legumes like soybean are the principal natural sources of isoflavonoids (Del Valle et al. 2020).

5.2.7 Flavanols

Similar to anthocyanins, in flavanols also ketone group is absent at fourth position of ring C. The main sources of flavanols are grapes, berries, apple, peach, pear, wine, and tea (Anderson and Jordheim 2006; Arts et al. 2000). The two most prevalent types of flavanols include (−)-epicatechin and (+)-catechin.

5.3 Biosynthesis of Flavonoids

The process of biosynthesis of flavonoids is carried out in the plant cells with the utilization of chalcone synthase (CS, EC 2.3.1.74) that catalyzes the consolidation of malonyl-CoA (3 molecules) with 4-coumaryl-CoA (1 molecule) to form chalcone, (*E*)-1-(2-hydroxyphenyl)-3-phenylprop-2-en-1-one (Fig. 5.1). Both the precursors involved in the first step are basically derived from two separate metabolic routes occurring within the cell: the acetate pathway and the shikimate pathway, which respectively produce rings A and B, and also chain linkages that form ring C (heterocycle pyrene ring) of flavonoids. Coumaroyl-CoA gives rise to B and C rings via shikimate pathway, whereas malonyl-CoA produces ring A, through acetate pathway, that involves carboxylation of acetyl-CoA. Coumaroyl-CoA, however, is generated through phenylpropanoid pathway with the involvement of

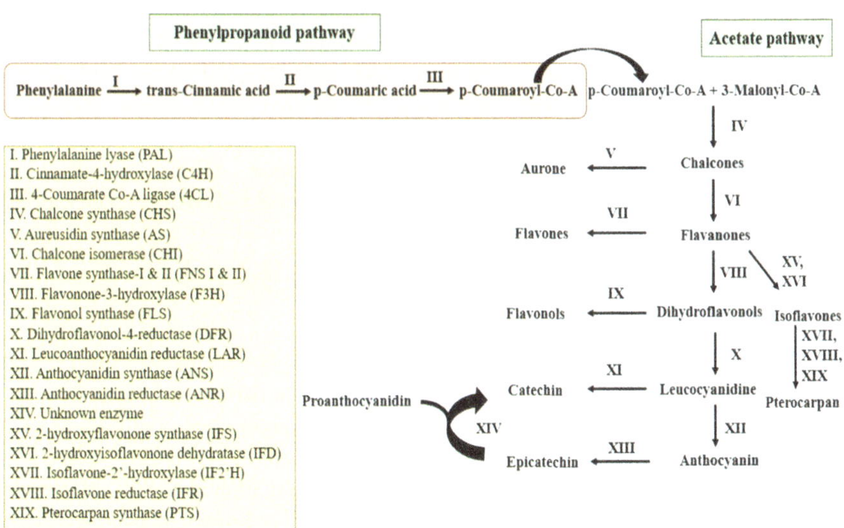

Fig. 5.1 Graphical representation of biosynthesis of flavonoids

phenylalanine as the main substrate for enzymatic reaction (Nabavi et al. 2020). In the first step of coumaroyl-CoA generation, phenylalanine is converted into cinnamate (cinnamic acid) by the elimination of amine by phenylalanine lyase (PAL). After that, cinnamate-4-hydroxylase (C4H) catalyzes hydroxylation at C4 to create *p*-coumaric acid (*p*-coumarate), which is then acylated by 4-coumarate-CoA ligase (4CL) with acetyl-CoA to create *p*-coumaroyl-CoA. In order to regulate subsequent events in the downstream phases of biosynthesis of flavonoids, all the enzymes and intermediates products engaged in the upstream of entire pathway are significant (Fowler and Koffas 2009; Khan et al. 2014; McIntosh and Owens 2016).

The chalcone thus formed undergoes chalcone isomerase (CHI)-based cyclization, thus resulting in the formation of chiral flavanone structure, also known as (2*S*)-flavanones, a central intermediate that is further involved in the formation of various types of flavonoids by undergoing various modifications like methylation, hydroxylation, or glycosylation. For instance, (2*S*)-flavanones are converted to flavones by flavone synthase (FNS). Similarly, the flavanone 3-hydroxylase transforms (2*S*)-flavanones into the corresponding (2*R*, 3*R*)-dihydroflavonols, that act as major intermediates in the biogenesis of flavonols, anthocyanins, and catechins. Both dihydroflavonol 4-reductase (DFR, EC 1.1.1.219) and flavonol synthase (FLS) catalyze the conversion of dihydroflavonols to leucocyanidins (Miranda et al. 2012). In grape berries, the 2,3-*cis*-flavan-3-ols ((+)-catechin) and 2,3-*cis*-flavan-3-ols ((−)-epicatechin) are converted by the enzymes leucoanthocyanidins reductase (LAR, EC 1.17.1.3) and anthocyanidin reductase (ANR), which are present in the inner and outer layers of the seed coat of grape berries, respectively. In grapevine, the leaves and stem contain anthocyanin synthase (ANS, EC 1.14.11.19) that produces both anthocyanins and proanthocyanidins. The route leading to the synthesis of isoflavones and pterocarpans include the unidentified enzyme that consolidates catechin and epicatechin and synthesizes proanthocyanidins. Similarly, 2-hydroxyisoflavanone is produced when isoflavone synthase (IFS) converts (2*S*)-naringenin to the isoflavone genistein. An enzyme 2-hydroxyisoflavanone dehydratase (IFD) catalyzes the conversion of 2-hydroxyisoflavanone into isoflavone. The 5-deoxy-2′-hydroxyisoflavones are converted to 3*R*-isoflavonone descendants by the enzyme isoflavone reductase (IFR). Finally, 2′--hydroxyisoflavanones are converted to the equivalent 3,9-dihydroxypterocarpans by pterocarpan synthase (PTS) (Miranda et al. 2012).

5.4 Accumulation and Transport

Flavonoids are multifaceted class of plant phenolics that performs diverse functions in various physiological processes of plants. They are synthesized ubiquitously in all plant cells and are subjected to vacuolar transportation to various subcellular components for accumulation and biological activity (Agati et al. 2007). In general, they are present in actively functioning organs like fruits, flowers, leaves, seedlings, and ground tissues with bright color appearance as well as in subcellular compartments, viz., chloroplast, vacuole, endoplasmic reticulum (ER), nucleus, and extracellular spaces. Vacuole is the primary site where most of the conjugated flavonoids like flavonol, flavone glycosides, and anthocyanins reside (Dixon et al. 2005; Zhao and Dixon 2009). Accumulation and transportation of flavonoids are dependent on several factors, viz., class of flavonoid, type of cell, geographical factors like latitude and altitude, light intensity, and various environmental conditions (Idris et al. 2018; Pucker and Selmar 2022). Accumulation of flavonoids in cellular and subcellular compartments is restricted to their multiplicity role against various biotic and abiotic stresses. Various reports are available that reveal stress-dependent accumulation of flavonoids and their derivatives in different parts of plants such as epidermal cells, chloroplasts, mesophyll cells, nucleus, tonoplasts, trichomes, etc. The accumulation of dihydroxy-B-ring substituted flavonoids like quercetin as potential antioxidant is reported in severely stressed plants (Kuhn et al. 2011). These derivatives inhibit the production of reactive hydroxyl radical, thus preventing nuclear oxidative damage. During cellular dehydration in *P. latifolia* leaves, flavonoids, such as kaempferol, accumulate epidermal cells, preserving the integrity of chloroplast membrane, and hence prevent oxidative damage (Moellering et al. 2010; Inoue 2011). In response to UV-screening and its absorption, intense amassing of flavonoids occurs in outer parts, like trichomes, and surface of leaves as antiherbivory agents (Agati et al. 2002).

Though reports are available for flavonoid biosynthesis, yet flavonoid location and trafficking require detailed exploration. Flavonoid biosynthesis, transport, and cellular deposition are tightly interlinked with one another (Zhao and Dixon 2009). They are transported probably from the cytosolic face of ER to various plant parts, which implies that the plants must possess some competent transportation system that delivers flavonoid across several membrane-bound compartments. However, various transport mechanisms that are involved in the translocation of flavonoids are still under research. Transportation of flavonoids may occur broadly in two manners: membrane vesicle-mediated transport and membrane transporter-mediated transport.

5.4.1 Membrane Vesicle-Mediated Transport

The concept of vesicle-mediated flavonoid transport is associated with anthocyanoplasts which were first assumed to be the site of biosynthesis or transport vesicles of anthocyanin (Braidot et al. 2008). These cytoplasmic anthocyanin bodies synthesized at cytosolic site of ER are membrane enclosed and are formed from the fusion of tony vesicle like structures resembling with anthocyanic vacuolar inclusions (AVIs), in their movements, present in the vacuoles (Irani and Grotewold 2005). AVIs are regarded as storage complexes and cytoplasmic vesicles containing anthocyanins associated with them following their fusion in the vacuole (Fig. 5.2). These vesicles are moved to the central vacuole by the merger of prevacuolar compartments (PVCs). On the other hand, vesicle-like structures that contain anthocyanins coinhabit with protein storage vacuoles (PSVs). They carry anthocyanins in a *trans*-Golgi network (TGN)-independent ER-to-PVC vesicle trafficking pathway (Zhao and Dixon 2009). Like anthocyanoplasts, the directional

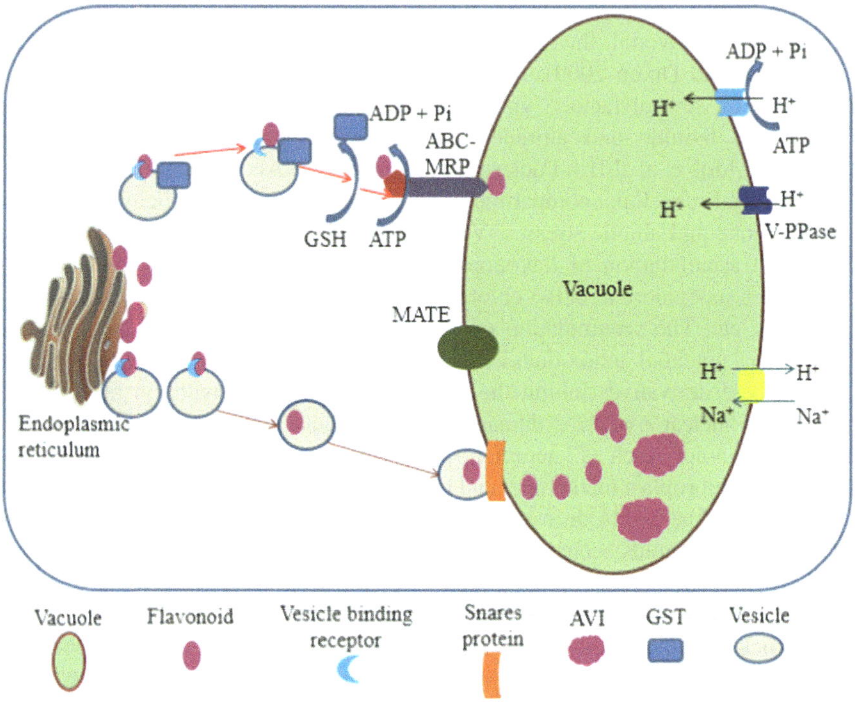

Fig. 5.2 The collective flavonoids transport from endoplasmic reticulum to vacuole mediated through vesicle trafficking and membrane transporters on vacuole. Transport through vesicle trafficking involves vesicles or PVC (prevacuolar compartments) which fuses with central vacuole and release the flavonoids. Membrane transporters such as ABC-MRP transporter (multidrug resistance-associated protein type ABC transporter) and MATE (multidrug and toxic compound extrusion) transporter mediate transportation by using ATP and electrochemical proton gradient respectively on vacuolar membrane

conduct of flavonoids is mediated by vesicles; however, direct biochemical or molecular evidence that supports vesicle-mediated translocation of flavonoid is still missing.

5.4.2 Membrane Transporter-Mediated Transport

Recent advances ascribed the transport of flavonoids into vacuole/cell wall by membrane transporter-mediated transport (MTT) initially observed in carnation petals. The main mainspring for transport of flavonoids (anthocyanins) into vacuole is the presence of H^+-ATPases and H^+-PPases in the vacuolar membrane (tonoplast) necessary for establishing a proton gradient amid the cytosol and the vacuole or cell wall (Fig. 5.2) (Gomez et al. 2009). Acylation of flavonoids and acidic pH in the vacuole are necessary for conformational modification of flavonoids, responsible for the confinement and sequestration of metabolites (Kitamura 2006).

Studies revealed two main membrane transporters which mediate the flavonoid and its derivative transport along the vacuole and different membranes. First, the ABC (ATP binding cassette) transporter which is the primary transporter. Second, the MATE (multidrug and toxic compound extrusion) transporter which is a secondary transporter for flavonoids and their derivatives.

5.4.3 ABC Transporters

ABC transporter consists of a large and pervasive family of proteins that transports wide range of flavonoids across membranes into vacuole. These proteins use the energy from ATP hydrolysis to sequester flavonoids following their coupling with glutathione (GSH), the reaction being accelerated by the enzyme glutathione *S*-transferases (GST) (Fig. 5.2) (Rea 2007). ABC transporters transport flavonoid glycosides and glucuronides into the vacuole by a directly roused (primary) mechanism (Czemmel et al. 2002). Vacuolar confiscation of glutathione-conjugated anthocyanins is carried out by multidrug resistance-associated protein (MRP) type ABC transporter (ABC-MRP) (Fig. 5.2). Similarly, isoflavonoid transport is mediated by phytoalexin medicarpin. Studies based on genetic and biochemical cross-complementation revealed that anthocyanin sequestration requires the enzyme GST itself. By binding to anthocyanin or flavonol, GST forms a GST-anthocyanin or GST-flavonol complex that guides flavonoids to the central vacuole and/or protects them from oxidative damage (Mueller et al. 2000; Zhao and Dixon 2010).

ABC-MRP transporters are associated with unidirectional long-distance cell–cell translocation of flavonoids like dihydrokaempferol, dihydroquercetin, and naringenin between roots and shoot (Buer et al. 2007). Further exploration is required for experimental validation to reveal the underlined mechanism behind the distribution and transportation of flavonoids at the cell, tissue, or organ level.

5.4.4 MATE Transporter

MATE transporters consist of a large family of transporters that are abundantly found in prokaryotes as well as eukaryotes (Moriyama et al. 2008). MATE transporters require an established vacuolar electrochemical proton gradient or sodium ions, created by V-type ATPases, or V-pyrophosphatase of the vacuole or P-type ATPases of the plasma membrane (Gaxiola et al. 2002). Several MATE transporters have been recognized in various plant species, e.g., AtTT12 that serves as a flavonoid/H^+ antiporter for the accumulation of proanthocyanidin in the seed coat of *Arabidopsis thaliana* (Marinova et al. 2007). In case of *M. truncatula*, MtMATE1 carries epicatechin 3′-*O*-glucoside, the acting precursor for proanthocyanidin, while MtMATE2 is involved in the transport of anthocyanins and flavone glycosides (Zhao and Dixon 2009; Zhao et al. 2011). In *Vitis vinifera*, two MATE transporters, namely, anthoMATE1 and anthoMATE3, are contained in the vacuolar membrane that helps in transport of acetylated anthocyanins into the vacuole for their storage (Gomez et al. 2009).

MATEs are distinguished from the other solute transporters by the presence of 12 transmembrane α-helical domains. They are substrate/proton or substrate/Na^+ antiporters involved in the accumulation and transport of secondary metabolites or removal of toxins through secondary active transport (Liu et al. 2006; Takanashi et al. 2014). The function and activity of these transporters rely on various types of H^+-ATPases such as V-type H^+-ATPase or vacuolar pyrophosphatase (V-PPase) proton pumps that provide and maintain electrochemical gradients across the vacuolar membrane and P-type H^+-ATPase that provide and maintain electrochemical gradients across the plasma membrane of cell.

5.5 Role in Biotic Stress Mitigation

In plants, the inducers of biotic stress are various pests and pathogens like fungi, bacteria, insects, nematodes, etc. These biotic stressors infect the crops and cause a huge amount of loss to the agricultural sector. Owing to the immobile nature of the plants, they are exposed to a diversity of these disease-causing pathogens, but they have no means to physically detach themselves from theses biotic stressors. So, to obstruct these unsolicited tenants, plants have developed certain defense strategies such as nonhost resistance, incompatible interactions, and many more (Shah and Smith 2020). However, susceptible plant species could be intensely infected and might not be able to escape the damage caused by these pathogens. Thus, to ward off the disease-causing infectious agents, plants produce flavonoids that serve as antioxidants and help in ameliorating the damage caused to the plants. Flavonoids play a considerable role in the morphology, physiology, and defense mechanisms of many plant species. They possess antipathogenic properties and can restrain several root pathogens in the vicinity of the plants (Hassan and Mathesius 2012; Mierziak et al. 2014). Flavonoids can impede the functioning of DNA gyrase and have the proficiency to disintegrate cell membrane (Weinstein and Albersheim 1983; Wu et al. 2013, 2019). Some of them act as herbivore repellents and are also described as powerful phytoalexins against nematodes, bacterial, and fungal pathogens (Shah and Smith 2020).

5.5.1 Bacteria

Plants being sessile are prone to various bacterial infections in the environment. Studies revealed that plants synthesize complex flavonoids in response to bacterial infection and hence prove very effective against bacteria (Karak 2019) (Table 5.1). Plants flavonoids disrupt bacterial membrane by interfering with lipid bilayers and hinder various physiological growth processes, viz., the synthesis of cell envelope, nucleic acid, ATP, electron transport chain, and biofilm formation (Gorniak et al.

Table 5.1 Antibacterial activity of plant flavonoids against different species of bacteria

Flavonoid	Source	Biotic stress	Effect	Reference
Chrysoeriol	*Artemisia rupestris*	*Staphylococcus aureus* 1199B	NorA pump inhibition, enhanced membrane permeability	Lan et al. (2021)
Sophoraflavanone G	*Sophora alopecuroides*	*S. aureus* 1199B	EtBr efflux inhibition	Sun et al. (2020)
Sarothrin	*Alkanna orientalis*	*Mycobacterium smegmatis* and *S. aureus*	Inhibit growth by blocking NorA efflux pump activity	Gorniak et al. (2019)
Apiofuranosyl Xylopyranoside	*Graptophyllum grandulosum*	*S. aureus*	Disruption of bacterial cell membrane and cell lysis	Tagousop et al. (2018)
Jaceosidin	*Artemisia californica*	*Escherichia coli*	Enoyl reductase enzyme inhibition	Allison et al. (2017)
Baicalein	*Scutellaria baicalensis*	*S. aureus*	ATP inhibition	Qiu et al. (2016)
Techtochrysin and negletein	*Scutellaria oblonga*	*E. coli*, *Bacillus subtilis* and *Enterococcus faecalis*	Growth inhibition	Rajendran et al. (2016)
5-Carbomethoxymethyl-4′,7-dihydroxyflavone	*Selaginella moellendorffii*	*E. coli*	Growth inhibition	Zou et al. (2016)
Corylifol C	*Psoralea corylifolia*	*S. aureus*	Growth and colony inhibition	Cui et al. (2015)
Luteolin, kaempferol, kaempferol glucoside, apigenin, and apigenin glucoside	*Tripleurospermum disciforme*	*S. aureus*, *S. epidermidis*	Growth and colony inhibition	Tofighi et al. (2015)
Free and bound	*Terminalia arjuna* Bark	*Enterobector aerogens*, *B. subtilis*, *Pseudomonas aruginosa*, *Agrobacterium tumefaciens*	Growth suppression	Jaiswal and Kumar (2014)

(continued)

Table 5.1 (continued)

Flavonoid	Source	Biotic stress	Effect	Reference
Licochalcones A and C	*Glycyrrhiza inflate* roots	*S. aureus* and *Micrococcus luteus*	Inhibit electron transfer between CoQ and cytochrome *c*	Kumar and Pandey (2013)

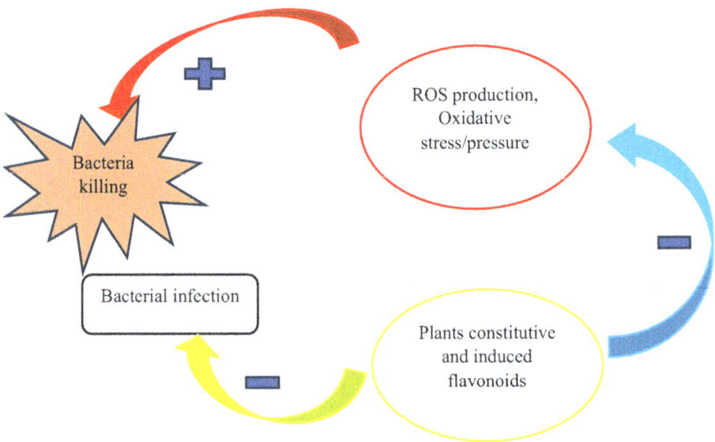

Fig. 5.3 The hypothetical events during flavonoids action against bacterial infection

2019; Jucá et al. 2020). For instance, gallate, catechin, epicatechin, flavonol quercetin, and epigallocatechin increase ROS production, thereby inducing an oxidative burst that increases the permeability of bacterial cell membrane (Fathima and Rao 2016) (Fig. 5.3). They may cause disorientation of lipids and interfere with integrity of membrane, inducing leakage (Gorniak et al. 2019). Some flavonoids inhibit bacterial DNA replication and ATP synthesis such as quercetin, naringenin, genistein, kaempferol, epigallocatechin gallate, baicalein, α-mangostin (AMG), and isobavachalcone (IBC) (Xu et al. 2012; Song et al. 2021).

5.5.2 Fungi

As the primary cause of plant disease, fungi have a negative effect on the crop yield. Crop output is drastically reduced by the extensive range of fungi-caused illnesses (Liu et al. 2022). Nevertheless, plants have evolved to build defenses against biotic stressors. In order to counteract biotic stress, plants produce phytoalexins in response to pathogenic invasion, primarily caused by fungi. For instance, the effect of mango peel extract on *Colletotrichum gloeosporioide* infecting mango fruits was studied, and subsequent transcriptome analysis in response to fungal pathogen was done. Results demonstrated that mango peel extract repressed the conidial germination and reduced the growth of fungi. Further, fungal infection elevated the expression of glucosidase genes that were involved in host defense, and also in fungal pathogenicity leading to enhanced antifungal activity of mango peel (Sudheeran et al. 2020). Another study reported that mutant cultivar of cotton exhibited significantly enhanced resistance against *Verticillium dahliae* compared to control cultivars, and this increased resistance was attributed to upregulated expression of flavonoid biosynthetic genes and increased amounts of flavonoids in the leaves of mutant cultivar (Long et al. 2019). Similarly, woods of *Melia azedarach* treated with various concentrations of acetone extract of *Withania somnifera* fruit exhibited resistance against fungal pathogen like *Fusarium culmorum* and *Rhizoctonia solani*. Treatment with extract also reduced the mycelial growth, and this antifungal activity of extract was attributed to the flavonoids (rutin and myricetin) present in themes revealed by HPLC (El-Hefny et al. 2020). Similar results were obtained while using methanolic extract of peels of *Musa paradisiaca* against different disease-causing fungi (Behiry et al. 2019). Further, Mayo-Prieto et al. (2019) investigated the effect of *Rhizoctonia solani* and biochemical changes in *Phaseolus vulgaris* upon fungal invasion. Results revealed that fungal infection upregulated the accumulation of certain flavonoids which was further enhanced upon inoculation of antagonist fungus *Trichoderma velutinum* providing resistance to common bean fungal infection. In another experiment, it was observed that endophytic bacteria *Pseudomonas aeruginosa* and *P. pseudoalcaligenes* increased flavonoid production in paddy plants in the absence of pathogens and upon infection with fungal pathogen such as *Pyricularia grisea,* bacterial isolates further increased the flavonoid production that led to increased resistance against fungal pathogens (Jha 2019). Table 5.2 summarizes the role of flavonoids against fungal pathogens.

5.5.3 Nematodes

Plant parasitic nematodes (PPNs) develop root galls or cysts and significantly reduce crop production (Ahmad et al. 2021). Plants release a variety of compounds against nematodes either to boost resistance or decrease nematode stress (Jan et al. 2021). Nematode invasion frequently promotes the production and amassing of flavonoids inside the plants. This induced synthesis of flavonoids helps the plants fight off nematode infections (Chin et al. 2018). First encounter of a PPN with flavonoids is in the soil while it is searching for its host. Flavonoids can prevent PPN eggs from hatching when they are in the egg stage. They have the ability to influence the migration of juvenile PPNs toward the roots by repelling, inactivating or killing them. At micromolar concentrations, the flavonols, myricetin, kaempferol, and quercetin repelled and retarded *Meloidogyne incognita* juveniles (Wuyts et al. 2006). Similarly, Kirwa and his coworkers (2018) conducted sand assays to elucidate the efficacy of nonvolatile components in tomato root exudates against

Table 5.2 Role of flavonoids against fungi

Flavonoids	Source	Biotic stress	Reported effect	References
Catechin, chrysin, isorhamnetin, naringin, kaempferol and quercetin	*Flourensiamicrophylla*	*Fusarium oxysporum* and *Alternaria alternata*	Inhibited mycelial growth	Carrillo-Lomelí et al. (2022)
Flavonoids	*Commelina erecta* and *Richardia brasiliensis*	*Rhizoctonia solani* and *Macrophomina phaseolina*	Exhibited fungicidal potential against both pathogens	Dilkin et al. (2022)
Polymethoxylated flavonoids	Citrus	*Penicillium italicum*	Disrupted the membrane integrity of fungus by decreasing content of total lipid and ergosterol the cells	Guo et al. (2022)
Glabridin	Synthetic	Ten different fungi	Exhibited fungicidal activity against all the pathogens	Schnarr et al. (2022)
Flavonoids	Microalgae and cyanobacteria	*Botryodiplodia theobromae*, *Fusarium solani* and *Pythium ultimum*	Inhibited fugal growth	Senousy et al. (2022)
Laburnetin	*Retama raetam*	*Stemphylium vesicarium*	Inhibited fungal growth by 55%	Soriano et al. (2022)
Flavonoids	Leaves of *Hyptis spicigera*	*Aspergillus parasitic* and *F. graminearum*	Inhibited growth of both fungi	Adamu et al. (2021)
Flavonoids	*Cupressus atlantica* and *Eucalyptus torquata*	*Mauginiella scaettae*	Inhibited mycelial growth and spore germination	Bouhlali et al. (2021)
Kaempferol, quercetin and hesperidin	Leaf of *Eucalyptus camaldulensis* and *Citrus sinensis* and fruit of *Ficus benghalensis*	*Fusarium* spp.	Increased leaf area and increased yield of zucchini plants	Hassan et al. (2021)
Kaempferol, myricetin and rutin	Leaves of *Conium maculatum* and *Ficus eriobotryoides*, *Acacia saligna* bark and wood of *Schinus terebinthifolius*	*F. eriobotryoides*	All extracts inhibited growth of *Fusarium*	Salem et al. (2021)

Rutin and myricetin	Fruits of *Withania somnifera*	*F. culmorum* and *R. solani*	Inhibited growth of fungal mycelia	El-Hefny et al. (2020)
Flavonoids	*Nicotiana tabacum* and *Ocimum sanctum*	*Aspergillus niger* and *A. flavus*	Suppressed in vitro growth of *A. niger* and *A. flavus*	Naeem et al. (2020)
Flavonoids	*Olea europaea*	*F. proliferatum*	Inhibited germination and growth of fungus	Muzzalupo et al. (2020)
Quercetin	*Mangifera indica*	*Colletotrichum gloeosporioide*	Inhibited conidial germination and fungal growth	Sudheeran et al. (2020)
Catechin, gallate, and procyanidin B1	*Rheum rhabarbarum*	*Blumeria graminis*	Reduced pustules formation by fungus on the leaves of *Hordeum vulgare*	Gillmeister et al. (2019)
Quercetin	Leaves of *Pistacia lentiscus*	*F. oxysporum*	Inhibited fungal growth leading to enhanced growth of infected plants	Hajji-Hedfi et al. (2019)
Rutin	Peels of *Musa paradisiaca*	*R. solani* and *F. culmorum*	Inhibited fungal growth	Behiry et al. (2019)
Medicarpin and 7,4′-dihydroxyflavone	*Medicago sativa*	*F. oxysporum*	Inhibited fungal growth under in-vitro conditions	Gill et al. (2018)
Rutin	*Lycium* spp.	*Harpophora maydis*	Reduced mycelial growth and increased fresh weight of maize plants	Tej et al. (2018)

M. incognita and observed that quercetin attracted the infective stage of nematode at low concentrations, while higher concentration of quercetin acted as repellent for *M. incognita.* The juveniles of *Heteroderazeae* were demonstrated to be killed by rutin, quercetin, patuletin, and patulitrin at different concentrations and exposure times (Faizi et al. 2011). Further, nematode invasion in oats induced generation of defense compounds such as flavone glycosides leading to tolerance in oat plants against *Ditylenchus dipsaci*, *Pratylenchus neglectus,* and *H. avenae* (Soriano et al. 2004). Similarly, glyceollin, a flavonoid that causes resistance in soybean, was produced in response to *M. penetrans* infection to reduce crop damage (Chin et al. 2018). In other studies (Tikoria et al. 2022; Sharma et al. 2022), enhanced flavonoid content was detected in tomato plants infested with nematodes as compared to the control plants. The flavonoids content was observed to be further increased in nematode-infested plants treated with vermicompost and bacterial strain, respectively, indicating the involvement of flavonoids in providing resistance against nematode stress. Bali et al. (2018) found 21.11% increase in flavonoid content in *M. incognita*-treated tomato seedlings. Furthermore, it has been reported that extracts of various plants consist of different flavonoids which help in developing resistance in plants against nematode stress. In a study, aqueous leaf extracts of *Pistacia lentiscus* was reported to exhibit 96% mortality of *M. javanica* at 100% concentration after 72 h. The presence of quercetin flavonoid was revealed by the phytochemical analysis of the extract (Hajji-Hedfi et al. 2019). Presence of quercetin and kaempferol in aqueous extract of *Solanum elaeagnifolium* seeds and leaves also showed nematicidal potential (Balah and AbdelRazek 2020). Bioassays revealed that exogenous application of aqueous extract increased juvenile mortality and reduced the number of galls, egg-masses, and number of females of *M. incognita* which led to increase in growth parameters of nematode-infested eggplants. However, uncertainty surrounds how flavonoids interact with parasitic nematodes, but because they prevent nematode chemotaxis and motility, flavonoids have been discovered to be protective agents and provide defense in plants against nematodes (Table 5.3).

5.5.4 Herbivores

Being sessile creatures, plants serve as a readily available food supply for herbivores like insect pests. Plants have advanced their defense tactics to evade and/or deal with the insect pests by storing and secreting repellent chemicals or producing signals in response to pest invasion (Kortbeek et al. 2019). In reciprocation to chemical cues from injured tissues, the defense system might start in undamaged tissues and deter or enslave insects by secreting phytoalexins. Flavonoids, alkaloids, and other secondary metabolites make up the majority of insect repellent or antifeedant compounds (Adeyemi 2010; Zaynab et al. 2018). Various reports have demonstrated the antagonistic role of flavonoids against insect pests (Table 5.4). For instance, Aboshi and his coworkers (2018) reported the insecticidal potential of *Basella alba* leaf extract, where they found that methanolic leaf extract inhibited larval growth of

Table 5.3 Role of flavonoids against nematodes

Flavonoids	Source	Biotic stress	Reported effect	References
Flavonoids	*Hedysarum coronarium*	*Meloidogyne incognita*	Killed more than 80% of juveniles after 96 h	D'Addabbo et al. (2022)
Flavonoids	*Bidens pilosa*	*M. incognita*	Increased mortality	Kihika-Opanda et al. (2022)
Flavonoids	*Jatropha curcas* seeds	*M. incognita*	Increased mortality of infective juveniles and reduced egg hatching	Ogwudire et al. (2022)
Flavonoids	*Goniothalamus cheliensis*	*M. incognita*	Increased juvenile mortality	Zhao et al. (2022)
Flavonoids	*Allium sativum*	*Meloidogyne* sp.	Exhibited nematicidal and termiticidal activity	Khairan et al. (2021)
Genistein and Daidzein	Synthetic	*Heterodera glycines*	Decreased body length of infective juveniles, increased body fluid leakage and inhibited respiration	Ma et al. (2021)
Flavonoids	*Cucurbita maxima*	*M. incognita*	Suppressed egg hatching and nematode penetration of tomato roots	Sithole et al. (2021)
Lutein	*Tournefortia densiflora*, *Alloispermum integrifolium*, and *Adenophyllum aurantium*	*Nacobbus aberrans*	Increased juvenile immobility	Velasco-Azorsa et al. (2021)
Hydroxyl-3-methoxyflavone and chlorogenic acid	*Solanum elaeagnifolium*	*M. incognita*	Increased juvenile mortality, reduced gall number and egg masses, enhanced growth parameters of eggplant	Balah and AbdelRazek (2020)
Flavonols and flavones	Synthetic and flowers of *Tagetes patula*	*M. incognita*	Increased larval mortality	Bano et al. (2020)
Glycosylated flavonoids	Stem bark of *Piliostigma thonningii*	*M. javanica*	Increased juvenile mortality up to 99%	Dzomba et al. (2020)
Flavonoids	*Chrysanthemum*	*M. incognita*	Increases juvenile mortality in a concentration dependent manner	Jayanthi et al. (2020)

(continued)

Table 5.3 (continued)

Flavonoids	Source	Biotic stress	Reported effect	References
Flavonoids	*Pleurotus floridanus*	*M. incognita*	Inhibited egg hatching and increased juvenile mortality	Tanimola and Adedokun (2020)
Quercetin ad zeatin	*S. lycopersicum*	*M. incognita*	Attracted the juvenile while quercetin acted as repellent at higher concentrations	Dyer et al. (2019)
Quercetin	*S. lycopersicum*	*M. incognita*	Higher concentrations repelled the nematodes	Kirwa et al. (2018)
Daidzin, genistin, glycitin, and glycitein	Synthetic	*H. glycines*	Increased juvenile mortality and reduced egg hatching	Guo et al. (2017)

Spodoptera litura. Phytochemical analysis revealed the presence of vitexin and vitexin-2″-*O*-arabinofuranoside in the methanolic leaf extracts. Further, insecticidal potential of the flavonoids was confirmed by rearing *S. litura* larvae on purified vitexin or vitexin-2″-*O*-arabinofuranoside, where significant impairment of growth of larvae was reported as compared to the larvae raised on control spinach leaves (Aboshi et al. 2018). Furthermore, Kovalikova et al. (2019) investigated the effect of *Pieris brassicae* and *Phyllotreta nemorum* on phenolic profiling of *Brassica oleracea*, where they reported an increased content of flavonoids, particularly quercetin in insect stressed white cabbage plants. Likewise, four isoflavonoids (biochanin A, daidzein, formononetin, and genistein) extracted from two red clover cultivars, were studied for their antifeeding properties against clover root borer (*Hylastinus obscurus*). Insect activity and weight were lowered by all the isoflavonoids whereas, highest antifeeding activity against clover root borer was demonstrated by genistein and formononetin. It was concluded that these outcomes could be used to manage curculionid insect pests (Quiroz et al. 2017). In another study, elevated levels of relative expressions of flavonoid-related genes and flavonoids, especially chrysoeriol7, were observed in rice plants infected with *Nilaparvata lugens*. Chrysoeriol7 was further analyzed for its pest controlling potential in which infected rice seedling were treated with isolated chrysoeriol7, where treatment of seedlings significantly reduced *N. lugens* and enhanced plant growth parameters (Kim et al. 2022). Similar results were obtained on treatment of aphid-infested tomato plants with biopesticide formulation (BPF) prepared from rhizomes of *Allium sativum*. It was found that treatment with BPF reduced the population of *Myzus persicae* and phytochemical analysis demonstrated that pesticidal importance of BPF was because of the existence of flavonoids, arginine, and oligosaccharides (Badar et al. 2022).

Table 5.4 Role of flavonoids against herbivores

Flavonoids	Source	Biotic stress	Reported effect	References
Rutin	Synthetic	*Melanaphis sacchari*	Increased mortality of sugarcane aphid	Aguilar-Marcelino et al. (2022)
Quercetin	*Solidago graminifolia*	*Spodoptera frugiperda*	Exhibited insecticidal potential	Herrera-Mayorg et al. (2022)
Apigenin-7-glucoside catechin, epicatechin and naringin	Spent coffee grounds	*Empoasca fabae* and *Aphis craccivora*	Reduced population density of both insects	Hussein et al. (2022)
Chrysoeriol7	Rice	*Nilaparvata lugens*	Lowered infection on plants and increased growth parameters	Kim et al. (2022)
Flavonoids	Leaves of *Cascabela peruviana*	*Drosophila melanogaster*	Reduced growth and reproduction of fruit flies and induced mortality of second-instar larva	Men et al. (2022)
Chrysoeriol	Leaves and twigs of *Melientha suavis*	*S. litura*	Decreased activity of neurological and detoxification-related enzymes leading to increased mortality of second instar larvae	Ruttanaphan et al. (2022)
Hesperidin	Citrus residues	*Bemisia tabaci and S. frugiperda*	Increased mortality of adults of both the insects	Silva et al. (2022)
Flavonoids	*S. mammosum*	*D. melanogaster*	Reduced rate of pupae formation	Tran et al. (2022)
Flavonols	Neem	*S. frugiperda*	Reduced survival of second stage larvae	Hernandez-Trejo et al. (2021)
Flavonoids	Leaves *Calotropis gigantea*	*Plutella xylostella*	Increased mortality of third instar larvae	Khasanah et al. (2021)
Quercetin	*Euphorbia nivulia*	*Oxycarenus hyalinipennis*	Increased mortality of third instar nymph	Younus et al. (2021)
Apigenin, Luteolin and Luteolin-7-*O*-glucoside	*Cynara cardunculus*	*S. littoralis*	Exhibited antifeeding activity against fourth instar larvae	Abdelaziz (2020)
Flavonoids	*Populus alba*	*Botrytis cinerea* and *Dothiorella gregaria*	Increased resistance to fungal pathogens	Bai et al. (2020)
Flavonoids	*Carica papaya*	*Macrosiphum rosaeformis*	Increased aphid mortality under laboratory and field bioassays	Gupta et al. (2020)

(continued)

Table 5.4 (continued)

Flavonoids	Source	Biotic stress	Reported effect	References
Flavonoids	*Pongamia pinnata*, *Pachyrhizus erosus*, and *Annona squamosa*	*Aphis gossypii*, *Brevicoryne brassicae*, and *Bemisia tabaci*	Reduced population of all insects	Purkait et al. (2019)
Flavonoids	*Moringa oleifera*	*S. litura*	Resulted in weight loss and shrinkage of larvae and reduced food intake by larvae	Tridiptasari et al. (2019)
Vitexin	*Basella alba*	*S. litura*	Impaired growth of larvae	Aboshi et al. (2018)

5.6 Conclusion

In the past decade, flavonoids have gained major attention, and their contribution in stress alleviation has been elucidated. Due to their polyphenolic structure and distinctive chemical nature, flavonoids help plants to survive under different hostile conditions. They accumulate in several plant parts and perform plethora of physiological and biochemical functions there. They perform protective roles and assist plants in dealing with various biotic stressors, thereby acting as environment-friendly patrons in sustainable agriculture. However, most of the studies involving flavonoids as stress mitigators are carried out under in vitro conditions, and much is yet left to be explored and unraveled. The molecular targets of flavonoids and their interactions with cellular signaling systems need further studies for their better and deeper understanding. Furthermore, in order to include them in integrated pest management programs, detailed studies should be done to understand the mechanism behind their stress-ameliorating potential in plants.

References

Abdelaziz S (2020) Insects' deterrent flavonoids from *Cynara cardunculus* for controlling cotton leafworm; *Spodoptera littoralis*. J Plant Protect Pathol 11(6):315–320

Aboshi T, Ishiguri S, Shiono Y, Murayama T (2018) Flavonoid glycosides in Malabar spinach *Basella alba* inhibit the growth of *Spodoptera litura* larvae. Biosci Biotechnol Biochem 82(1): 9–14

Adamu K, Musa H, Aliyu AB, Musa AO (2021) Antifungal activity of *Hyptis spicigera* methanol leaf extract and flavonoid fraction. J Appl Sci Environ Manag 25(7):1167–1172

Adeyemi MH (2010) The potential of secondary metabolites in plant material as deterents against insect pests: a review. Afr J Pure Appl Chem 4(11):243–246

Afendi FM, Okada T, Yamazaki M, Hirai-Morita A, Nakamura Y, Nakamura K, Ikeda S, Takahashi H, Altaf-Ul-Amin M, Darusman LK, Saito K (2012) KNApSAcK family databases:

integrated metabolite-plant species databases for multifaceted plant research. Plant Cell Physiol 53(2):1–12

Agati G, Galardi C, Gravano E, Romani A, Tattini M (2002) Flavonoid distribution in tissues of *Phillyrea latifolia* L. leaves as estimated by microspectrofluorometry and multispectral fluorescence microimaging. Photochem Photobiol 76(3):350–360

Agati G, Matteini P, Goti A, Tattini M (2007) Chloroplast-located flavonoids can scavenge singlet oxygen. New Phytol 174(1):77–89

Aguilar-Marcelino L, Pineda-Alegría JA, Salinas-Sánchez DO, Hernández-Velázquez VM, Silva-Aguayo GI, Navarro-Tito N, Sotelo-Leyva C (2022) In vitro insecticidal effect of commercial fatty acids, β-Sitosterol, and rutin against the sugarcane aphid, *Melanaphis sacchari* Zehntner (Hemiptera: Aphididae). J Food Prot 85(4):671–675

Aherne SA, O'Brien NM (2002) Dietary flavonols: chemistry, food content, and metabolism. Nutrition 18:75–81

Ahmad G, Khan A, Khan AA, Ali A, Mohhamad HI (2021) Biological control: a novel strategy for the control of the plant parasitic nematodes. Antonie Van Leeuwenhoek 114(7):885–912

Allison BJ, Allenby MC, Bryant SS, Min JE, Hieromnimon M, Joyner PM (2017) Antibacterial activity of fractions from three Chumash medicinal plant extracts and in vitro inhibition of the enzyme enoyl reductase by the flavonoid jaceosidin. Nat Prod Res 31(6):707–712

Anderson OM, Jordheim M (2006) The anthocyanins. In: Anderson OM, Markham KR (eds) Flavonoids: chemistry, biochemistry and applications, Separation and quantification of flavonoids. CRC Press/Taylor & Francis Group, Boca Raton, FL, pp 472–551

Arts IC, van de Putte B, Hollman PC (2000) Catechin contents of foods commonly consumed in The Netherlands. 1. Fruits, vegetables, staple foods, and processed foods. J Agric Food Chem 48:1746–1751

Badar R, Ahmed A, Munazir M, Khalique N, Waheed H, Munawar S, Basheer H (2022) Aphicidal effects of organic formulations against tomato aphid *Myzus persicae* (Sulzer) to overcome crop damage/diseases. Pak J Phytopathol 34(1):161–171

Bai Q, Duan B, Ma J, Fen Y, Sun S, Long Q, Lu J, Wan D (2020) Coexpression of PalbHLH1 and PalMYB90 genes from *Populus alba* enhances pathogen resistance in poplar by increasing the flavonoid content. Front Plant Sci 10:1772

Balah MA, AbdelRazek GM (2020) Pesticidal activity of *Solanum elaeagnifolium* Cav. leaves against nematodes and perennial weeds. Acta Ecol Sin 40(5):373–379

Bali S, Kaur P, Sharma A, Ohri P, Bhardwaj R, Alyemeni MN, Wijaya L, Ahmad P (2018) Jasmonic acid-induced tolerance to root-knot nematodes in tomato plants through altered photosynthetic and antioxidative defense mechanisms. Protoplasma 255(2):471–484

Bano S, Iqbal EY, Zik-ur-Rehman S, Fayyaz S, Faizi S (2020) Nematicidal activity of flavonoids with structure activity relationship (SAR) studies against root knot nematode *Meloidogyne incognita*. Eur J Plant Pathol 157:299–309

Baskar V, Venkatesh R, Ramalingam S (2018) Flavonoids (antioxidants systems) in higher plants and their response to stresses. In: Gupta D, Palma J, Corpas F (eds) Antioxidants and antioxidant enzymes in higher plants. Springer, Cham, pp 253–268

Behiry SI, Okla MK, Alamri SA, El-Hefny M, Salem MZ, Alaraidh IA, Ali HM, Al-Ghtani SM, Monroy JC, Salem AZ (2019) Antifungal and antibacterial activities of *Musa paradisiaca* L. peel extract: HPLC analysis of phenolic and flavonoid contents. PRO 7(4):215. https://doi.org/10.3390/pr7040215

Bouhlali EDT, Derouich M, Meziani R, Essarioui A (2021) Antifungal potential of phytochemicals against *Mauginiella scaettae*, the plant pathogen causing inflorescence rot of date palm. Scientifica 2021:1–8

Braidot E, Zancani M, Petrussa E, Peresson C, Bertolini A, Patui S, Macrì F, Vianello A (2008) Transport and accumulation of flavonoids in grapevine (*Vitis vinifera* L.). Plant Signal Behav 3(9):626–632

Buer CS, Muday GK, Djordjevic MA (2007) Flavonoids are differentially taken up and transported long distances in *Arabidopsis*. Plant Physiol 145(2):478–490

Carrillo-Lomelí DA, de Rodríguez DJ, Moo-Huchin VM, Ramón-Canul L, Rodríguez-García R, González-Morales S, Villarreal-Quintanilla JA, Peña-Ramos FM (2022) How does *Flourensia microphylla* extract affect polyphenolic composition, antioxidant capacity, and antifungal activity? Ind Crop Prod 186:115248

Chin S, Behm CA, Mathesius U (2018) Functions of flavonoids in plant–nematode interactions. Plants 7(4):85. https://doi.org/10.3390/plants7040085

Cui Y, Taniguchi S, Kuroda T, Hatano T (2015) Constituents of *Psoralea corylifolia* fruits and their effects on methicillin-resistant *Staphylococcus aureus*. Molecules 20(7):12500–12511

Czemmel S, Heppel SC, Bogs J (2002) R2R3 MYB transcription factors: key regulators of the flavonoid biosynthetic pathway in grapevine. Protoplasma 249(2):109–118

D'Addabbo T, Tava A, Argentieri MP, Biazzi E, Candido V, Avato P (2022) Nematicidal potential of sulla (*Hedysarum coronarium* L.) against the root-knot nematode *Meloidogyne incognita*. Plants 11(19):2550

Del Valle I, Webster TM, Cheng HY, Thies JE, Kessler A, Miller MK, Ball ZT, MacKenzie KR, Masiello CA, Silberg JJ, Lehmann J (2020) Soil organic matter attenuates the efficacy of flavonoid-based plant-microbe communication. Sci Adv 6(5):eaax8254

Dias MC, Pinto DCGA, Silva AMS (2021) Plant flavonoids: chemical characteristics and biological activity. Molecules 26:1–16

Dilkin ERS, Matias R, Oliveira AKM, Corrêa BO (2022) Fungitoxic effect and phytochemical characteristics of Brazilian Cerrado weeds against *Rhizoctonia solani* and *Macrophomina phaseolina* fungi. Braz J Biol 84:1–8

Dixon RA, Xie DY, Sharma SB (2005) Proanthocyanidins–a final frontier in flavonoid research? New Phytol 165(1):9–28

Dyer S, Weir R, Cox D, Cheseto X, Torto B, Dalzell JJ (2019) Ethylene Response Factor (ERF) genes modulate plant root exudate composition and the attraction of plant parasitic nematodes. Int J Parasitol 49(13–14):999–1003

Dzomba P, Mureya C, Gwatidzo L (2020) Nematicidal activity of phytocompounds from *Piliostigma thonningii* stem bark against *Meloidogyne javanica*. Chem Afr 3(4):937–947

El-Hefny M, Salem MZ, Behiry SI, Ali HM (2020) The potential antibacterial and antifungal activities of wood treated with *Withania somnifera* fruit extract, and the phenolic, caffeine, and flavonoid composition of the extract according to HPLC. Processes 8(1):113

Erlund I (2004) Review of the flavonoids quercetin, hesperetin, and naringenin. Dietary sources, bioactivities, bioavailability, and epidemiology. Nutr Res 24:851–874

Faizi S, Fayya S, Bano S, Yawar Iqbal E, Siddiqi H, Naz A (2011) Isolation of nematicidal compounds from *Tagetes patula* L. yellow flowers: structure–activity relationship studies against cyst nematode *Heterodera zeae* infective stage larvae. J Agric Food Chem 59(17): 9080–9093

Fathima A, Rao JR (2016) Selective toxicity of Catechin—a natural flavonoid towards bacteria. Appl Microbiol Biotechnol 100:6395–6402

Fowler ZL, Koffas MAG (2009) Biosynthesis and biotechnological production of flavanones: current state and perspectives. Appl Microbiol Biotechnol 83(5):799–808

Gani U, Vishwakarma RA, Misra P (2021) Membrane transporters: the key drivers of transport of secondary metabolites in plants. Plant Cell Rep 40(1):1–8

Gautam AK, Singh PK, Aravind M (2020) Defensive role of plant phenolics against pathogenic microbes for sustainable agriculture. In: Lone R, Shuab R, Kamili A (eds) Plant phenolics in sustainable agriculture. Springer, Singapore, pp 579–594

Gaxiola RA, Fink GR, Hirschi KD (2002) Genetic manipulation of vacuolar proton pumps and transporters. Plant Physiol 129(3):967–973

Gill US, Uppalapati SR, Gallego-Giraldo L, Ishiga Y, Dixon RA, Mysore KS (2018) Metabolic flux towards the (iso) flavonoid pathway in lignin modified alfalfa lines induces resistance against *Fusarium oxysporum* f. sp. *medicaginis*. Plant Cell Environ 41(9):1997–2007

Gillmeister M, Ballert S, Raschke A, Geistlinger J, Kabrodt K, Baltruschat H, Deising HB, Schellenberg I (2019) Polyphenols from rheum roots inhibit growth of fungal and oomycete phytopathogens and induce plant disease resistance. Plant Dis 103(7):1674–1684

Giusti M, Wrolstad R (2003) Acylated anthocyanins from edible sources and their applications in food systems. Biochem Eng J 14:217–225

Gomez C, Terrier N, Torregrosa L, Vialet S, Fournier-Level A, Verries C, Souquet JM, Mazauric JP, Klein M, Cheynier V, Ageorges A (2009) Grapevine MATE-type proteins act as vacuolar H +-dependent acylated anthocyanin transporters. Plant Physiol 150(1):402–415

Gorniak I, Bartoszewski R, Króliczewski J (2019) Comprehensive review of antimicrobial activities of plant flavonoids. Phytochem Rev 18(1):241–272

Griesbach R (2005) Biochemistry and genetics of flower color. Plant Breed Rev 25:89–114

Guo C-h, Zhu X-f, Duan Y-x, Wang Y-y, Chen L-j (2017) Suppression of different soybean isoflavones on *Heterodera glycines*. Chinese J Oil Crop Sci 39(4):540

Guo L, Li Y, Mao X, Tao R, Tao B, Zhou Z (2022) Antifungal activity of polymethoxylated flavonoids (PMFs)-loaded citral nanoemulsion against *Penicillium italicum* by causing cell membrane damage. J Fungi 8(4):388

Gupta G, Sharma S, Kumar N (2020) Carica papaya aqueous leaf extracts as potential botanical insecticide against rose aphids (*Macrosiphum rosaeformis* D.). J Entomol Zool Stud 8:960–964

Gutiérrez-Grijalva EP, Santos-Zea L, Ambriz-Pérez DL, López-Martínez LX, Heredia JB (2020) Flavones and flavonols: bioactivities and responses under light stress in herbs. In: Lone R, Shuab R, Kamili A (eds) Plant phenolics in sustainable agriculture. Springer, Singapore, pp 91–115

Hajji-Hedfi L, Larayedh A, Hammas NC, Regaieg H, Horrigue-Raouani N (2019) Biological activities and chemical composition of *Pistacia lentiscus* in controlling *Fusarium wilt* and root-knot nematode disease complex on tomato. Eur J Plant Pathol 155:281–291

Hassan S, Mathesius U (2012) The role of flavonoids in root–rhizosphere signalling: opportunities and challenges for improving plant–microbe interactions. J Exp Bot 63(9):3429–3444

Hassan HS, Mohamed AA, Feleafel MN, Salem MZ, Ali HM, Akrami M, Abd-Elkader DY (2021) Natural plant extracts and microbial antagonists to control fungal pathogens and improve the productivity of zucchini (*Cucurbita pepo* L.) *in vitro* and in greenhouse. Horticulturae 7(11):470

Heim KE, Tagliaferro AR, Bobilya DJ (2002) Flavonoid antioxidants: chemistry, metabolism and structure-activity relationships. J Nutr Biochem 13:572–584

Hernandez-Trejo A, Rodríguez-Herrera R, Sáenz-Galindo A, López-Badillo CM, Flores-Gallegos AC, Ascacio-Valdez JA, Estrada-Drouaillet B, Osorio-Hernández E (2021) Insecticidal capacity of polyphenolic seed compounds from neem (*Azadirachta indica*) on *Spodoptera frugiperda* (JE Smith) larvae. J Environ Sci Health B 56(12):1023–1030

Herrera-Mayorg V, Guerrero-Sánchez JA, Méndez-Álvarez D, Paredes-Sánchez FA, Rodríguez-Duran LV, Niño-García N, Paz-González AD, Rivera G (2022) Insecticidal activity of organic extracts of *Solidago graminifolia* and its main metabolites (quercetin and chlorogenic acid) against *Spodoptera frugiperda*: an in vitro and in silico approach. Molecules 27(10):3325

Hussein H, Abouamer W, Ali H, Elkhadragy M, Yehia H, Farouk A (2022) The valorization of spent coffee ground extract as a prospective insecticidal agent against some main key pests of *Phaseolus vulgaris* in the laboratory and field. Plants 11(9):1124

Idris A, Linatoc AC, Bakar MF, Ibrahim ZT, Audu Y (2018) Effect of light quality and quantity on the accumulation of flavonoid in plant species. J Sci Technol 10(3):32–45

Inoue K (2011) Emerging roles of the chloroplast outer envelope membrane. Trends Plant Sci 16(10):550–557

Irani NG, Grotewold E (2005) Light-induced morphological alteration in anthocyanin-accumulating vacuoles of maize cells. BMC Plant Biol 5(1):1–5

Iwashina T (2013) Flavonoid properties of five families newly incorporated into the order Caryophyllales (Review). Bull Natl Mus Nat Sci 39:25–51

Jaiswal P, Kumar P (2014) Antimicrobial screening of free and bound flavonoid from the bark of *Terminalia arjuna*. J Phytopharm 4(6):303–306

Jan R, Asaf S, Numan M, Kim KM (2021) Plant secondary metabolite biosynthesis and transcriptional regulation in response to biotic and abiotic stress conditions. Agronomy 11(5):968

Jayanthi M, Singh K, Chaudhary S (2020) Detection of nematicidal compounds and development of an in vitro mass multiplication protocol in Pusa Centenary *Chrysanthemum*. Indian J Hortic 77(4):704–712

Jha Y (2019) Endophytic bacteria mediated anti-autophagy and induced catalase, β-1,3-glucanases gene in paddy after infection with pathogen *Pyricularia grisea*. Indian Phytopathol 72(1): 99–106

Jucá MM, Cysne Filho FM, de Almeida JC, Mesquita DD, Barriga JR, Dias KC, Barbosa TM, Vasconcelos LC, Leal LK, Ribeiro JE, Vasconcelos SM (2020) Flavonoids: biological activities and therapeutic potential. Nat Prod Res 34(5):692–705

Karak P (2019) Biological activities of flavonoids: an overview. Int J Pharm Sci Res 10(4): 1567–1574

Khairan K, Yusra N, Eriana CN, Bahi M, Syaukani S, Sriwati R, Jacob C (2021) Termiticidal and nematicidal activities of five extracts from garlic (*Allium sativum*). J Phys Conf Ser 1882(1): 012121

Khan MK, Zill-E-Huma, Dangles O (2014) A comprehensive review on flavanones, the major citrus polyphenols. J Food Compos Anal 33(1):85–104

Khasanah N, Martono E, Trisyono YA, Wijonarko A (2021) Toxicity and antifeedant activity of *Calotropis gigantea* L. leaf extract against *Plutella xylostella* L. (Lepidoptcra: Plutellidae). Int J Design Nat Ecodyn 16(6):677–682

Kihika-Opanda R, Tchouassi DP, Ng'ang'a MM, Beck JJ, Torto B (2022) Chemo-ecological insights into the use of the non-host plant vegetable black-jack to protect two susceptible solanaceous crops from root-knot nematode parasitism. J Agric Food Chem 70(22):6658–6669

Kim J, Buell CR (2015) A revolution in plant metabolism: genomeenabled pathway discovery. Plant Physiol 169:1532–1539

Kim EG, Yun S, Park JR, Jang YH, Farooq M, Yun BJ, Kim KM (2022) Bio-efficacy of Chrysoeriol7, a natural chemical and repellent, against brown planthopper in rice. Int J Mol Sci 23(3):1540

Kirwa HK, Murungi LK, Beck JJ, Torto B (2018) Elicitation of differential responses in the root-knot nematode *Meloidogyne incognita* to tomato root exudate cytokinin, flavonoids, and alkaloids. J Agric Food Chem 66(43):11291–11300

Kitamura SA (2006) Transport of flavonoids: from cytosolic synthesis to vacuolar accumulation. In: Grotewold E (ed) The science of flavonoids. Springer, New York, pp 123–146

Kortbeek RW, van der Gragt M, Bleeker PM (2019) Endogenous plant metabolites against insects. Eur J Plant Pathol 154(1):67–90

Kovalikova Z, Kubes J, Skalicky M, Kuchtickova N, Maskova L, Tuma J, Vachova P, Hejnak V (2019) Changes in content of polyphenols and ascorbic acid in leaves of white cabbage after pest infestation. Molecules 24(14):2622

Ku YS, Ng MS, Cheng SS, Lo AW, Xiao Z, Shin TS, Chung G, Lam HM (2020) Understanding the composition, biosynthesis, accumulation and transport of flavonoids in crops for the promotion of crops as healthy sources of flavonoids for human consumption. Nutrients 12(6):1–23

Kuhn BH, Geisler M, Bigler L, Ringli C (2011) Flavonols accumulate asymmetrically and affect auxin transport in *Arabidopsis*. Plant Physiol 156:585–595

Kumar S, Pandey AK (2013) Chemistry and biological activities of flavonoids: an overview. Sci World J 2013:1–16

Kumar S, Bhushan B, Wakchaure GC, Meena KK, Kumar M, Meena NL, Rane J (2020) Plant phenolics under water-deficit conditions: biosynthesis, accumulation, and physiological roles in water stress alleviation. In: Lone R, Shuab R, Kamili A (eds) Plant phenolics in sustainable agriculture. Springer, Singapore, pp 451–465

Lan JE, Li XJ, Zhu XF, Sun ZL, He JM, Zloh M, Gibbons S, Mu Q (2021) Flavonoids from *Artemisia rupestris* and their synergistic antibacterial effects on drug-resistant *Staphylococcus aureus*. Nat Prod Res 35(11):1881–1886

Lichman BR, Godden GT, Buell CR (2020) Gene and genome duplications in the evolution of chemodiversity: perspectives from studies of Lamiaceae. Curr Opin Plant Biol 55:74–83

Liu R, Xu S, Li J, Hu Y, Lin Z (2006) Expression profile of a PAL gene from *Astragalus membranaceus* var. *Mongholicus* and its crucial role in flux into flavonoid biosynthesis. Plant Cell Rep 25(7):705–710

Liu B, Stevens-Green R, Johal D, Buchanan R, Geddes-McAlister J (2022) Fungal pathogens of cereal crops: proteomic insights into fungal pathogenesis, host defense, and resistance. J Plant Physiol 269:153593

Long L, Liu J, Gao Y, Xu FC, Zhao JR, Li B, Gao W (2019) Flavonoid accumulation in spontaneous cotton mutant results in red coloration and enhanced disease resistance. Plant Physiol Biochem 143:40–49

López-Lázaro M (2009) Distribution and biological activities of the flavonoid luteolin. Mini Rev Med Chem 9:31–59

Ma Y, Yuan R, Sikandar A, Zhu X, Duan Y, Wang Y (2021) Genistein and Daidzein effects on the physiological indices of soybean cyst nematodes. Sci Agric 79(2):1–8

Manach C, Scalbert A, Morand C, Remesy C, Jimenez L (2004) Polyphenols: food sources and bioavailability. Am J Clin Nutr 79:727–747

Marinova K, Pourcel L, Weder B, Schwarz M, Barron D, Routaboul JM, Debeaujon I, Klein M (2007) The Arabidopsis MATE transporter TT12 acts as a vacuolar flavonoid/H^+-antiporter active in proanthocyanidin-accumulating cells of the seed coat. Plant Cell 19(6):2023–2038

Mayo-Prieto S, Marra R, Vinale F, Rodríguez-González Á, Woo SL, Lorito M, Gutiérrez S, Casquero PA (2019) Effect of *Trichoderma velutinum* and *Rhizoctonia solani* on the metabolome of bean plants (*Phaseolus vulgaris* L.). Int J Mol Sci 20(3):549

McIntosh CA, Owens DK (2016) Advances in flavonoid glycosyltransferase research: integrating recent findings with long-term citrus studies. Phytochem Rev 15(6):1075–1091

Men TT, Phien HH, Ai TTT, Van Ay N, Hue NTK, Binh TD (2022) The insecticidal capacity of ethanol extract from *Cascabela peruviana* (L.) Lippold against fruit fly. Heliyon 8(4):e09313

Mierziak J, Kostyn K, Kulma A (2014) Flavonoids as important molecules of plant interactions with the environment. Molecules 19:16240–16265

Miranda CL, Maier CS, Stevens JF (2012) Flavonoids. In: Encyclopedia of life sciences. https://doi.org/10.1002/9780470015902.a0003068.pub2

Moellering ER, Muthan B, Benning C (2010) Freezing tolerance in plants requires lipid remodeling at the outer chloroplast membrane. Science 330:226–228

Moriyama Y, Hiasa M, Matsumoto T, Omote H (2008) Multidrug and toxic compound extrusion (MATE)-type proteins as anchor transporters for the excretion of metabolic waste products and xenobiotics. Xenobiotica 38(7–8):1107–1118

Mueller LA, Goodman CD, Silady RA, Walbot V (2000) AN9, a petunia glutathione S-transferase required for anthocyanin sequestration, is a flavonoid-binding protein. Plant Physiol 123(4): 1561–1570

Muzzalupo I, Badolati G, Chiappetta A, Picci N, Muzzalupo R (2020) In vitro antifungal activity of olive (*Olea europaea*) leaf extracts loaded in chitosan nanoparticles. Front Bioeng Biotechnol 8: 151

Nabavi SM, Samec D, Tomczyk M, Milella L, Russo D, Habtemariam S, Suntar I, Rastrelli L, Daglia M, Xiao J, Giampieri F (2020) Flavonoid biosynthetic pathways in plants: versatile targets for metabolic engineering. Biotechnol Adv 38:107316

Naeem Z, Maimoona A, Jabeen K, Khan B, Iqbal S (2020) Evaluation of fungicidal potential of *Ocimum sanctum* and *Nicotiana tabacum* L. against *Aspergillus flavus* and *A. niger*. Pak J Bot 52(6):2223–2228

Ogwudire VE, Agu CM, Ewelike NC, Ojiako FO, Cookey CO, Nwokeji E (2022) Assessment of *Jatropha curcas* L. as alternative nematicide for root knot nematode (*Meloidogyne incognita*) management. Aust J Sci Technol 6(2):65–70

Palanisamy CP, Gunasekaran VP, Dominic S, Xuan TD (2020) Phenolic allelochemicals from crops and weed management. In: Lone R, Shuab R, Kamili A (eds) Plant phenolics in sustainable agriculture. Springer, Singapore, pp 183–199

Panche AN, Diwan AD, Chandra SR (2016) Flavonoids: an overview. J Nutr Sci 5:1–15

Pucker B, Selmar D (2022) Biochemistry and molecular basis of intracellular flavonoid transport in plants. Plants 11(7):963

Purkait A, Biswas S, Saha S, Hazra DK, Roy K, Biswas PK, Ghosh SK, Kole RK (2019) Formulation of plant based insecticides, their bio-efficacy evaluation and chemical characterization. Crop Prot 125:104907

Qiu F, Meng L, Chen J, Jin H, Jiang L (2016) In vitro activity of five flavones from *Scutellaria baicalensisin* combination with Cefazolin against methicillin resistant *Staphylococcus aureus* (MRSA). Med Chem Res 25(10):2214–2219

Quiroz A, Mendez L, Mutis A, Hormazabal E, Ortega F, Birkett MA, Parra L (2017) Antifeedant activity of red clover root isoflavonoids on *Hylastinus obscurus*. J Soil Sci Plant Nutr 17(1): 231–239

Rajendran N, Subramaniam S, Christena LR, Muthuraman MS, Subramanian NS, Pemiah B, Sivasubramanian A (2016) Antimicrobial flavonoids isolated from Indian medicinal plant *Scutellaria oblonga* inhibit biofilms formed by common food pathogens. Nat Prod Res 30(17):2002–2006

Rashad Y, Aseel D, Hammad S (2020) Phenolic compounds against fungal and viral plant diseases. In: Lone R, Shuab R, Kamili A (eds) Plant phenolics in sustainable agriculture. Springer, Singapore, pp 201–219

Rea PA (2007) Plant ATP-binding cassette transporters. Annu Rev Plant Biol 58:347–375

Ruttanaphan T, Thitathan W, Piyasaengthong N, Nobsathian S, Bullangpoti V (2022) Chrysoeriol isolated from *Melientha suavis* Pierre with activity against the agricultural pest *Spodoptera litura*. Chem Biol Technol Agric 9:21. https://doi.org/10.1186/s40538-022-00287-2

Salem MZ, Mohamed AA, Ali HM, Al Farraj DA (2021) Characterization of phytoconstituents from alcoholic extracts of four woody species and their potential uses for management of six *Fusarium oxysporum* isolates identified from some plant hosts. Plants 10(7):1325

Samanta A, Das G, Das SK (2011) Roles of flavonoids in plants. Int J Pharm Sci Technol 6:12–35

Schnarr L, Segatto ML, Olsson O, Zuin VG, Kümmerer K (2022) Flavonoids as biopesticides—systematic assessment of sources, structures, activities and environmental fate. Sci Total Environ 824:153781

Senousy HH, El-Sheekh MM, Saber AA, Khairy HM, Said HA, Alhoqail WA, Abu-Elsaoud AM (2022) Biochemical analyses of ten cyanobacterial and microalgal strains isolated from Egyptian habitats, and screening for their potential against some selected phytopathogenic fungal strains. Agronomy 12(6):1340

Shah A, Smith DL (2020) Flavonoids in agriculture: chemistry and roles in, biotic and abiotic stress responses, and microbial associations. Agronomy 10(8):1–26

Sharma N, Manhas RK, Ohri P (2022) *Streptomyces hydrogenans* strain DH-16 alleviates negative impact of *Meloidogyne incognita* stress by modifying physio-biochemical attributes in *Solanum lycopersicum* plants. Sci Rep 12:15214

Silva DFD, Bomfim J, Marchi RC, Amaral JC, Pinto LS, Carlos RM, Ferreira AG, Forim MR, Fernandes JB, da Silva MF, Bruce TJ (2022) Valorization of hesperidin from citrus residues: evaluation of microwave-assisted synthesis of Hesperidin-Mg Complex and their insecticidal activity. J Braz Chem Soc 33:772–782

Sithole NT, Kulkarni MG, Finnie JF, Van Staden J (2021) Potential nematicidal properties of plant extracts against *Meloidogyne incognita*. S Afr J Bot 139:409–417

Song M, Liu Y, Li T, Liu X, Hao Z, Ding S, Panichayupakaranant P, Zhu K, Shen J (2021) Plant natural flavonoids against multidrug resistant pathogens. Adv Sci 8(15):2100749

Soriano IR, Asenstorfer RE, Schmidt O, Riley IT (2004) Inducible flavone in oats (*Avena sativa*) is a novel defense against plant-parasitic nematodes. Phytopathology 94(11):1207–1214

Soriano G, Petrillo C, Masi M, Bouafiane M, Khelil A, Tuzi A, Isticato R, Fernández-Aparicio M, Cimmino A (2022) Specialized metabolites from the allelopathic plant *Retama raetam* as potential biopesticides. Toxins 14(5):311

Sudheeran PK, Ovadia R, Galsarker O, Maoz I, Sela N, Maurer D, Feygenberg O, Oren Shamir M, Alkan N (2020) Glycosylated flavonoids: fruit's concealed antifungal arsenal. New Phytol 225(4):1788–1798

Sugiyama A (2021) Flavonoids and saponins in plant rhizospheres: roles, dynamics, and the potential for agriculture. Biosci Biotechnol Biochem 85(9):1919–1931

Sultana B, Anwar F (2008) Flavonols (kaempeferol, quercetin, myricetin) contents of selected fruits, vegetables and medicinal plants. Food Chem 108:879–884

Sun ZL, Sun SC, He JM, Lan JE, Gibbons S, Mu Q (2020) Synergism of sophoraflavanone G with norfloxacin against effluxing antibiotic-resistant *Staphylococcus aureus*. Int J Antimicrob Agents 56(3):106098

Tagousop CN, Tamokou JD, Ekom SE, Ngnokam D, Voutquenne-Nazabadioko L (2018) Antimicrobial activities of flavonoid glycosides from *Graptophyllum grandulosum* and their mechanism of antibacterial action. BMC Complement Altern Med 18:1–10

Tak Y, Kumar M (2020) Phenolics: a key defence secondary metabolite to counter biotic stress. In: Lone R, Shuab R, Kamili A (eds) Plant phenolics in sustainable agriculture. Springer, Singapore, pp 309–329

Takanashi K, Shitan N, Yazaki K (2014) The multidrug and toxic compound extrusion (MATE) family in plants. Plant Biotechnol 31:417–430

Tanimola AA, Adedokun OM (2020) Nematicidal potentials of aqueous extracts of two mushroom species and spent mushroom substrate on *Meloidogyne incognita* and *Meloidogyne javanica*. Niger J Agric Food Environ 16(2):110–119

Tej R, Rodríguez-Mallol C, Rodríguez-Arcos R, Karray-Bouraoui N, Molinero-Ruiz L (2018) Inhibitory effect of *Lycium europaeum* extracts on phytopathogenic soil-borne fungi and the reduction of late wilt in maize. Eur J Plant Pathol 152(1):249–265

Tikoria R, Kaur A, Ohri P (2022) Potential of vermicompost extract in enhancing the biomass and bioactive components along with mitigation of *Meloidogyne incognita*-induced stress in tomato. Environ Sci Pollut Res 29:56023–56036

Tofighi Z, Molazem M, Doostdar B, Taban P, Shahverdi AR, Samadi N, Yassa N (2015) Antimicrobial activities of three medicinal plants and investigation of flavonoids of *Tripleurospermum disciforme*. Iran J Pharm Res 14:225–231

Tran TM, Tang HC, Huynh HP, Nguyen YD, Pham TL, Kamei K, Tran DB (2022) Evaluation of the insecticidal activity of *Solanum mammosum* (L.) fruit extract against *Drosophila melanogaster*. J Anim Behav Biometeorol 10(2):2218–2218

Tridiptasari A, Leksono AS, Siswanto D (2019) Antifeedant effect of Moringa oleifera (L.) leaf and seed extract on growth and feeding activity of *Spodoptera litura* (Fab.) (Lepidoptera: Noctuidae). J Exp Life Sci 9(1):25–31

Tsao R (2010) Chemistry and biochemistry of dietary polyphenols. Nutrients 2:1231–1246

Velasco-Azorsa R, Cruz-Santiago H, Cid del Prado-Vera I, Ramirez-Mares MV, Gutiérrez-Ortiz MDR, Santos-Sánchez NF, Salas-Coronado R, Villanueva-Cañongo C, Lira-de León KI, Hernández-Carlos B (2021) Chemical characterization of plant extracts and evaluation of their nematicidal and phytotoxic potential. Molecules 26(8):2216

Weinstein LI, Albersheim P (1983) Host–pathogen interactions: XXIII. The mechanism of the antibacterial action of glycinol, a pterocarpan phytoalexin synthesized by soybeans. Plant Physiol 72:557–563

Wu T, Zang X, He M, Pan S, Xu X (2013) Structure–activity relationship of flavonoids on their anti-*Escherichia coli* activity and inhibition of DNA gyrase. J Agric Food Chem 61(34): 8185–8190

Wu SC, Yang ZQ, Liu F, Peng J, Qu S-Q, Li Q, Song XB, Zhu K, Shen JZ (2019) Antibacterial effect and mode of action of flavonoids from licorice against methicillin-resistant *Staphylococcus aureus*. Front Microbiol 10:2489. https://doi.org/10.3389/fmicb.2019.02489

Wuyts N, Swennen R, De Waele D (2006) Effects of plant phenylpropanoid pathway products and selected terpenoids and alkaloids on the behaviour of the plant-parasitic nematodes *Radopholus similis*, *Pratylenchus penetrans* and *Meloidogyne incognita*. Nematology 8(1):89–101

Xu X, Zhou XD, Wu CD (2012) Tea catechin epigallocatechin gallate inhibits *Streptococcus mutans* biofilm formation by suppressing gtf genes. Archiv Oral Biol 57(6):678–683

Yonekura-Sakakibara K, Higashi Y, Nakabayashi R (2019) The origin and evolution of plant flavonoid metabolism. Front Plant Sci 10:943. https://doi.org/10.3389/fpls.2019.00943

Younus M, Hasan MM, Ali S, Saddq B, Sarwar G, Ullah MI, Maqsood A, Ahmar S, Mora-Poblete F, Hassan F, Darwish H (2021) Extracts of *Euphorbia nivulia* Buch.-Ham. showed both phytotoxic and insecticidal capacities against *Lemna minor* L. and *Oxycarenus hyalinipennis* Costa. PLoS One 16(4):e0250118

Zaynab M, Fatima M, Abbas S, Sharif Y, Umair M, Zafar MH, Bahadar K (2018) Role of secondary metabolites in plant defense against pathogens. Microb Pathog 124:198–202

Zhao J, Dixon RA (2009) MATE transporters facilitate vacuolar uptake of epicatechin 3′-*O*-Glucoside for proanthocyanidin biosynthesis in *Medicago truncatula* and *Arabidopsis*. Plant Cell 21:2323–2340

Zhao J, Dixon RA (2010) The ‘ins’ and ‘outs’ of flavonoid transport. Trends Plant Sci 5(2):72–80

Zhao J, Huhman D, Shadle G, He XZ, Sumner LW, Tang Y, Dixon RA (2011) MATE2 mediates vacuolar sequestration of flavonoid glycosides and glycoside malonates in *Medicago truncatula*. Plant Cell 23(4):1536–1555

Zhao CQ, Li Y, Zhang MF, Zhao HM, Zhou WB, Sang YH, Xu TY, Shen YJ, Li PF, Lv F, Xu JJ (2022) Chemical constituents from *Goniothalamus cheliensis* with nematicidal activity against *Meloidogyne incognita*. Chem Nat Compd 12(6):1340

Zou Z, Xu P, Wu C, Zhu W, Zhu G, He X, Zhang G, Hu J, Liu S, Zeng W, Xu K (2016) Carboxymethyl flavonoids and a chromone with antimicrobial activity from *Selaginella moellendorffii* Hieron. Fitoterapia 111:124–129

Phenolics as Shielding Counterparts from Plants to Combat Biotic Stress Mediated by Microbes and Nematodes

6

Koyel Kar, Kamalika Mazumder, Priyanka Chakraborty, and Sailee Chowdhury

Abstract

The secondary metabolites known as polyphenols, sometimes known as plant phenolics, have a variety of roles in reducing the effects of abiotic (such as temperatures, cold, drought salt, or heavy metals) and biotic (such as bacteria, virus, fungi, insect, or weed) stresses. Phenolic chemicals produced by the typical phenylpropanoid pathway function as signaling molecules and may be used as plant scavengers. Phenolic substances involve in immune responses to infections caused by viruses or bacteria, excessive sun or light exposure, wounds, and stresses involving heavy metals. They are imperative for controlling seed germination as well as working together to control plant growth. One important function of phenolic substances obtained from plant is as protective chemicals against disease-causing organisms such fungi, bacteria, nematodes, weeds, and some other pathogens. The signaling cascade serves as a conduit for the production of the proper physiological or biochemical response to biotic stress.

Keywords

Secondary metabolites · Biotic stress · Polyphenols · Nematodes

K. Kar (✉) · K. Mazumder · S. Chowdhury
Deparment of Pharmaceutical Chemistry, BCDA College of Pharmacy & Technology, Hridaypur, West Bengal, India

P. Chakraborty
Deparment of Pharmacology, BCDA College of Pharmacy & Technology, Hridaypur, West Bengal, India

R. Lone et al. (eds.), *Plant Phenolics in Biotic Stress Management*,
https://doi.org/10.1007/978-981-99-3334-1_6

6.1 Introduction

Plants yield a huge variety of secondary metabolites, one of the significant of which are phenolics. They are the plant kingdom's most firm stuff. They have been known to humans for long, and their role in plant nutrition, fertility, development, and defense has made them compounds of interest and consideration (Kabera et al. 2014). These plant-produced antiherbivore compounds are among the commonest plant allelochemicals in the bionetwork. These types of compounds are distinguished by the presence of one or more hydroxyl groups bound to a six-carbon aromatic ring. These compounds attained prominence due to the resistance properties they possessed (Beninger et al. 2004). There are currently over 8000 phenolic structures known, ranging from simple phenolic acids to highly polymerized substances like tannins (Harborne 1989). Polyphenols and phenolics substances from the plants are the most abundant collections of secondary metabolites, with significant morphological and physiological status.

The compounds, which are aromatic in nature with one or more hydroxyl groups, are produced by the shikimate/phenylpropanoid pathway or the polyketide acetate/malonate pathway, which results in monomeric and polymeric phenols and polyphenols (Chan et al. 2009). The primary function of these phenolic chemicals is in the defensive systems of plants, but they have also been observed to play a part in pollination of crop and camouflage (Acamovic and Brooker 2005; Alasalvar et al. 2001). The majority of these substances are bound to sugar. As they are aromatic in nature, they used to play a significant role in plant communication. With a purpose to keep track of their surroundings, several phenolic compounds in the plant's rhizosphere use quorum sensing (Bjarnsholt and Givskov 2007). Bacteria at the rhizosphere break down the phenolic chemicals, promoting plant development and increasing soil fertility. Additionally, they assist in chelating soil minerals and elements, which increases soil porosity and enhances plants' ability to absorb nutrients (Bais et al. 2006). At the time of stress and invasion of pathogens, the phenolic compounds commonly accumulate in the plant's subepidermal tissues. Numerous factors (internal and external both), like plant physiology, age, developmental phase, environment, and the kind of pathogen assault, have an impact on the synthesis and concentration of the accumulated phenolics (Ozyigit et al. 2007).

Phenolic chemicals are unique because of their capacity to act as both attractants and repellents. The plant either creates phenolic derivatives that act as attractants, such as allelochemicals and chemoattractants, to entice pollinators and symbiotic bacteria, depending on the environment (Ndakidemi and Dakora 2003; Vit et al. 1997), or they act as repellents, such as phenolic derivatives that act to ward off pathogens and pests (Lattanzio et al. 2006). External conditions that negatively impact the growth of a plant, development, or output are stated to as stress in plants (Verma et al. 2013).

Plants have a variety of responses to stress, including changes to their gene expression, metabolic part, parameter of growth and yields of crops, etc. Stresses evolved from plants typically result from certain abrupt variations in the environment. However, exposure to given stress results in adaptation to that specific stress in a time-dependent manner in stress-tolerant plant species (Jones and Dangl 2006). Abiotic stress and biotic stress are the two main types into which plant stress may be separated. Abiotic stress is placed on plants by the environment and can be either physical or chemical, whereas biotic stress is a biological entity like a disease, an insect, etc. that are exposed to crop plants (Mantri et al. 2012).

Living things, in particular viruses, bacteria, fungi, nematodes, insects, arachnids, and weeds, induce biotic stress in plants. Plant death can be the outcome from the agents causing biotic stress straight depriving their host of different nutrients of it. A significant biotic stress can be generated for pre- and postharvest losses. Plants can withstand biotic stressors even if they lack an adaptive immune system by adapting to certain sophisticated techniques (Zhu 2002).

6.2 Influence of Stress in Plants

A variety of biotic stress conditions can affect plants. Once the plant detects stress, a number of molecular and cellular activities are initiated (Rejeb et al. 2014). Plants have developed high end sensory systems to detect biotic invasion and combat the harm it causes to growth, productivity, and survival (Rizhsky et al. 2004; Lamers et al. 2020). Plants have therefore developed an abundance of defense mechanisms to protect themselves from attacks by a wide range of pests and pathogens and insect herbivores (Hammond-Kosack and Jones 2000). To counteract the detrimental effect on their survival, plants often achieve an equilibrium between their reaction and biotic stress (Peck and Mittler 2020).

Both the biotic and abiotic stress each account for 30% and 50%, respectively, of global declines in agricultural productivity. Agriculture is severely impacted by a number of biotic stressors, including viruses, bacteria, fungi, insects, and nematodes, as well as abiotic stressors, including salt, frost, cold, flood, drought, heavy metals, temperature, and nutrient deficit (Dresselhaus and Hückelhoven 2018).

Fungi provide the greatest hazard to plant species out of all the biotic stressors, as they are responsible for 85% of plant illnesses (Behmann et al. 2014). In addition, viruses are considered a severe threat to crop plants. Abundant manifestations in plants, including necrosis, leaf spots, blights, blasts, wilting, mottling, tumor growth, and others, are brought on by these biotic organisms (Saddique et al. 2018). Weeds are another biotic factor that undesirably impacts the growth and output of commercially significant plants by either directly destroying the plant or by escalating competition for nutrients (Melvin et al. 2017) (Fig. 6.1).

6.3 Phenolic Metabolites from Plant Sources

Secondary metabolites of phenylpropanoid metabolism in plant shikimic acid besides pentose phosphate are phenolic compounds (Randhir et al. 2004). They comprise benzene rings by one or else more hydroxyl substituents besides variety after modest phenolic fragments to extremely polymerized substances

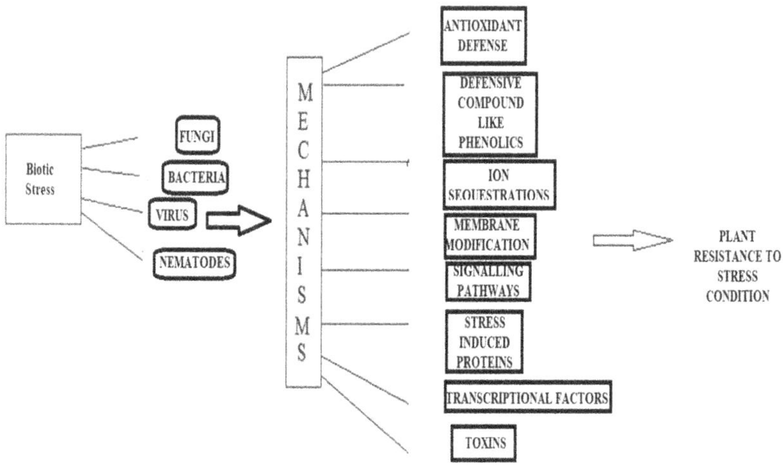

Fig. 6.1 Responses of plants to stress condition

(Velderrain-Rodríguez et al. 2014). The initial stage in the blend of phenolic composites remains to pledge glucose to the pentose phosphate path (PPP) and irreversibly convert glucose-6-phosphate toward ribulose-5-phosphate. Glucose-6-phosphate dehydrogenase pledges the first committed process in the conversion to ribulose-5-phosphate (G6PDH). In one sense, the ribulose-5-phosphate conversion generates reducing peers of nicotinamide adenine dinucleotide phosphate (NADPH) on behalf of cellular anabolic procedures. PPP, contrasted with, generates erythrose-4-phosphate and phosphoenolpyruvate as of glycolysis, remnants afterward cast-off to build phenolic composites by the phenylpropanoid path before actuality funneled toward the shikimic acid path in the direction of yield phenylalanine (Vattem et al. 2005; Lin et al. 2010).

Phenolics stand the maximum noticeable secondary, cutting-edge metabolites floras, and their spreading is observable during the metabolic process. These polyphenols, or phenolic chemicals, contain an extensive variety of components, including phenolic acids, multifaceted flavonoids, and simple flavonoids, besides tinted anthocyanins (Babbar et al. 2014).

These phenolic chemicals often remain allied with plant-defensive responses. Though phenolic metabolites show a noteworthy character in other courses, such as a blend of attractive compounds to advance fertilization, pallor aimed at concealment and protection beside herbivores, besides antibacterial plus antifungal action (Alasalvar et al. 2001; Acamovic and Brooker 2005; Edreva et al. 2008).

Phenolic constituents, specifically trauma-linked phytochemicals, consumed remained related near beneficial effects caused by fruit in addition vegetable diet, notably owing toward their antioxidant possessions (Heima et al. 2002). Balasundram et al. (2006) swotted the presence and potential applications of phenolic chemicals in plants and agricultural by-products. That research indicates that fruits, vegetables, and beverages make up many phenolic chemicals consumed by

Table 6.1 Source of common phenolic compounds from plant sources

Sl. no.	Plant name	Phenolic compounds
1.	*Aspleniumnidus nidus* L.	Quercetin-7-*O*-rutinoside and gliricidin 7-*O*-hexoside (antibacterial)
2.	*Myristica fragrans*	3′,4′,7trihydroxyflavone (antibacterial)
3.	*Pseudomonas aeruginosa* (Schroeter)	2-(3′,4′ dihydroxy-phenyl) 3,5,7-trihydroxy-chromen-4-one (antibacterial)
4.	*Peganum harmala*	Vanillic acid, syringic acid, gallic acid, 3,4-dihydroxybenzoic acid, and 4-hydroxybenzoic acid
5.	*T. vulgaris*	Vanillin
6.	*Rosmarinus officinalis* L.	Rosmarinic acid
7.	*Polygonum cuspidatum*	Piceid, resveratroloside, and piceatannol glucoside
8.	*Rheum tanguticum Maxim. ex Balf.*	*Trans-*, *cis-*, and *trans*-desoxyrhaponticin are three types of rhapontin
9.	*Fraxinus rhynchophylla*	Scopoletin, fraxetin, aesculetin, fraxin, aesculin
10.	*Matricaria chamomilla*	Herniarin
11.	*Phyllanthus amarus*	Niranthin, 5-demethylniranthin, nirtetralin, phyllanthin, hypophyllanthin, virgatusin, and heliobuphthalmin lactone and bursehernin
12.	*Schisandra chinensis*	Schisanchinin A, B, C, and D
13.	*Origanum majorana*	Orientin, luteolin-*O*-glycoside, apigenin-6,8-di-C-hexoside, apigenin-*O*-glucuronide, and luteolin-*O*-glucuronide
14.	*Medicago* spp.	Formononetin, prunetin, biochanin A, daidzin, and glycitein
15.	*Tephrosia vogelii*	Isorhamnetin 3-*O*-galactoside Kaempferol 3-*O*-glucoside
16.	*Curcuma longa*	Curcumin, demethoxycurcumin, and bis-demethoxycurcumin
17.	*Drosera rotundifolia*	7-Methlyjuglone
18.	*Rheum palmatum* and *Rheum hotaoense*	Chrysophanol, physcoin, emodin, aloe-emodin, and rhein

humans. Plant polyphenols can help prevent oxidative damage to human health and disease through their role as dietary antioxidants. Because they act while ordinary antioxidants, phenolic compounds remain frequently originate cutting-edge plant-based foods plus brews and are crucial for human health (Balasundram et al. 2006).

Rendering to numerous professionals, phenolic composites remain the maximum plentiful nutritional antioxidants and cutting-edge regular human diets. A portion of courtesy has recently existed compensated to phenolic compounds because of recent research suggesting that they may have a role in treating or perhaps curing a number of human diseases (Fiorentino et al. 2008; Hoper and Cassidy 2006; Pu et al. 2013). It is well known that frequently expended new and treated fruits, Strawberries, and their related products, such as juices, are similarly rich in phenolic compounds. Fruits and vegetables high in phenolic compounds include apples, pears, raspberries, cranberries, apples, grapes, and jams (Pinto et al. 2007) (Table 6.1).

6.4 Correlation of Phenolics with Biotic Stress

Phenolics operate equally protectants, inhibitors, insecticides, besides normal visceral poisons against a diversity of biotic agents, includes pathogens caused by fungi, viruses, and bacteria, insects, nematodes, and herbivores (Ghosh et al. 2017). Phenolics work similarly to phytoestrogens in mammals as well as to allelochemicals in floras. The most effective allelochemicals are vaporous terpenoids, poisonous water-soluble hydroquinones, hydroxybenzoate, hydroxycinnamates, escopoletins, caffeic acids, besides 5-hydroxynaphthoquinone (Bhattacharya et al. 2010).

Phenolics remain the most effective accepted defensive mediators and may provide a substitute toward natural illness switch on farming harvests. Herbal phenolic composites formed throughout host-pathogen exchanges protect plants through multiple methods (Tripathi 2004). Once a shrub host is attacked by a microbe, the situation begins to accumulate phenolics so a primary reaction, which may prime toward a surge in congregation absorption overall (Mayer et al. 2001).

The two phenolic caffeic acid esters have consistently been found to rise significantly after maize is infected with *Glomerella graminicola* or *C. heterotrophs* (Pusztahely et al. 2017). Even though these polyphenols already else phenolic composites stand nonpoisonous toward fungus, their fast build-up and unexpected decrease in cutting-edge concentration suggested that phenols might turn so a pool for the creation of additional protective chemicals. The ability of the contaminating mediator toward digests the phytoalexins produced through diseased shrub can facilitate communication between the sick herbal and the infecting pathogen (Hardoim et al. 2015).

These chemicals slow the spread of biotic mediator contamination. The bacteriostatic plus bactericidal effects of phytoalexin remain indistinguishable to those of antibiotics (Pina-Pérez and Ferrús Pérez 2018). Phytoalexins inhibit sporulation formation and expansion of plant infective fungus hyphae. Phytoalexin defense against microorganisms or diseases is dependent on the proportion of production besides gathering in plant tissues (Raman et al. 2015). Quorum sensing causes exponential development and normalizes numerous additional actions of cutting-edge microorganisms; polyphenols canister interrupt this progression and limit the propagation of these pathogens, and so phytoalexins show a significant part in limiting the damaging activities of biotic stressors (Duke 2018). Quinones, on behalf of instance, are highly valuable for floras since they engender multifaceted by-products through networking by proteins plus hinder the maceration of proteomic material by herbivores (Easwar Rao et al. 2017).

The occurrence of these molecules in amino acids reduces protein digestion in insects, limiting pest proliferation. Phenols also inhibit the occurrence of ROS such as peroxides, superoxide, besides singlet oxygen, which are engaged in a variety of biotic plus abiotic stressors. Various phenolic chemicals initiate the production of defensive enzymes, thereby shielding plants from these stresses. Phenolics concentrate at the diseased plant site and restrict the overall progress besides expansion of the microbe, which occurs because of cell death caused by a hypersensitive reaction (Lincoln et al. 2018).

L-phenylalanine is converted into lignin through the action of 4-coumaric acid, as well as the CoA-esters of 4-coumaric, ferulic, and synaptic acids towards the conforming alcohols, which are said to polymerize under the action of peroxidase and aid in the cutting-edge defence against biotic stress agents (Varbanova et al. 2011; Peperidou et al. 2017; Bi et al. 2017) Several investigations conducted by different experts reveal that small molecular weight phenols, for example benzoic acids plus phenyl propanoids, are generated in the first reaction to infection. The indication sturdily recommends that the esterification of polyphenols toward cell wall components remains important for plant stress tolerance (Mandal et al. 2010).

Many microorganisms, counting fungi, herbivore insects, bacteria, and viruses pose a hazard to plants (War et al. 2012). Plants have evolved several resistance instruments for defense then existence, many of which remain the result of pathogen attacks.

6.5 Plant Phenolics to Combat Microbial and Nematodal Stress

6.5.1 Bacteria

Plants get hugely infected by bacteria in various mediums like insects, gushing water, or plant parts that are already infected. The above-mentioned mediums infect the different parts of plants via small openings that were fabricated either by damage or naturally (Hirano and Upper 2000). Different secretion systems of bacteria are available but they mainly use the Type-III system which is a bacterial composite structure causing vehemence to the gram-negative pathogen. This system mainly injects effector protein into the cells of plants. The system consists of a plunger system and syringe that enables the injection of effector protein that causes diseases and also activates the defense techniques (Green and Mecsas 2015).

The quorum sensing mechanism is used by bacterial pathogens to ensure gene articulation after reaching a specific size. Mostly all bacterial pathogens cause the spreading of the disease to plants but *Pseudomonas syringae* causes the maximum effect which is a gram-negative bacteria that causes blighting, cankering, wilting, and spotting (Sun et al. 2017).

6.5.1.1 Phenolics in Controlling Bacterial Stress

The Pattern Recognition Receptors (PRRs) receptors help in the recognition of pathogens by the molecular patterns of pathogens causing the triggering of Pathogen-Associated Molecular Patterns (PAMPs) immunity in plants. This phenomenon restricts the spreading of infection by preventing the pathogens to gain complete control over the host (Postel and Kemmerling 2009).

Catechin controls the infection of various bacterial species upon changing the activities and penetrability of the membranes which also helps to release hydrogen peroxide which is considered to be a potent ROS (Wang et al. 2018). The quorum sensing proteins are heterologously expressed by *trans*-cinnamaldehyde and tannic acid in *E. coli*. The quorum-sensing protein in pecto-bacteria is greatly affected by carvacrol and eugenol (Joshi et al. 2016).

6.5.2 Fungus

Fungus is categorized as a eukaryotic microorganism which adsorbs essential food and also digests food ostensibly. Fungal reproduction is executed by hyphae or spores. Hyphae mean vegetative growth composed of tubular cells (Dube 2015).

The fungus causes dead spots or death of the entire leaf when the spores or hyphae pass through the air or soil. So, it also enters plants via roots. Fungi block the water-conduction cells in plants or sometimes kill them leading to wilting. This may also lead to serious damage or even the death of plants (Carris et al. 2012). The fungus responsible for causing rice blast disease is *Magnaporthe oryzae*. The foliar plant tissues have the maximum tendency to be infected but panicle infection may cause loss of grain (Hajano et al. 2011).

6.5.2.1 Phenolics in Controlling Fungal Stress

Resveratrol synthase and resveratrol-methyltransferase escalate malevolence in soybean on exposure to *Rhizoctonia solani*. Cell wall apposition along with formation is appraised as the main defense mechanism for preventing fungal growth (Zernova et al. 2014). Various derivatives of hydroxycinnamic acid, derivatives of flavonol, tyrosol, and oleuropein prevent leaf spot disease in the olive tree caused by *Fusicladium oleagineum* (Talhaoui et al. 2015).

The onion crop gets infected by soil fungus and the infection initiates with the outer layer of the onion. But further, it penetrates within the deeper layers. Both protocatechuic acid and catechol cause resistance of onion to fungus. These two water-soluble compounds readily dispense from the dead outer layer of the bulb to the point of infection, thus resisting perforation as well as germination (Levin 1971).

6.5.3 Virus

Nucleic acid core and protein made the viruses. The infectious virus generates several copies of new virus particles by shedding its protein coat and the nucleic acid and cytoplasmic "bridges" help in their movement (Maccheroni et al. 2005). Transmission of viruses from plant to plant can be also seen by a vector, mechanically, or by seed transmission process (Congdon et al. 2017).

Some vectors are aphids, white flies, nematodes, fungi, and protozoa. Plant viruses are the main cause of photosynthates reallocation and normal cellular processes disruption. The first discovered plant virus was the tobacco mosaic virus (TMV) which attacks mainly the plants of the Solanaceae family and the nightshade members (Tak and Kumar 2020).

Plant phenolics act as a main part of the defense mechanism of plants against viral infections. High concentration of caffeic acid, quercetin, kaempferol, and chlorogenic acid controlled the virus activity (Parr and Bolwell 2000). Phenolic compounds help to defeat the viral infection and damage protein, DNA or RNA to inhibit viral enzyme activities (Kumar and Pandey 2013).

Flavonoids distract viral RNA translation, inhibit viral DNA replication, protein synthesis and also inhibit transcription factors of viral enzymes and genome synthesis. It also interferes with viral structural protein. Coumarins inhibit protease, integrase, and reverse transcriptase and stop viral replication. Thymol and carvacrol, both are monoterpenes, present in essential oil of some plants, showed antiviral activity against cucumber mosaic virus and tobacco mosaic virus. Gramniphenol, a phenolic, acts against the tobacco mosaic virus (Chowdhary et al. 2021).

Fistula flavonoids B and C from the bark and stem of the plant *Cassia fistula* inhibit tobacco mosaic virus (Zhao et al. 2013). Literature showed that the two coumarin derivatives 6-hydroxy-5-methoxy-7-methyl-3-(40-methoxyphenyl)-coumarin and 6-hydroxy-7-methyl-3-(40-methoxyphenyl)-coumarin were isolated from *Nicotiana tabacum* leaves and highly active to defeat tobacco mosaic virus (Liu et al. 2016).

6.5.4 Nematodes

Plant-parasitic nematodes source projected yearly harvest fatalities of $10 billion in the United States besides $125 billion globally. Although producers have access to a range of chemical and other management options, none are optimal in terms of cost, ecological protection, or else effectiveness alongside the altering roundworm biotypes that commonly exist in agricultural soils. The current de-recording of numerous chemical nematodes, as well as the anticipated removal of methyl bromide after the pest-control market, highlights the significance of emerging innovative techniques for managing nematode-persuaded harvest mutilation. Misusing important differences between cutting-edge nematode biology and those of their congregation plants is one strategy for generating novel controls (Sasser and Freckman 1987). The second-stage juvenile worm develops since the plant-scrounging nematode incites and obligation feels a passé of progressive halt. This progression remains functionally comparable toward the inherently well-characterized route cutting-edge the permitted-existing roundworm system that governs the creation of the Dauer, a developmentally stalled state (Bird and Opperman 1998). Dauer trail genetic factor translate a multifaceted system of gesture transduction proteins that perceive nourishment, pheromones, in addition additional ecological inputs besides then progression these indicators by the cellular equal to govern the nematode's metabolic condition as well as reproductive development and lifespan (Riddle and Albert 1997). Plant parasitic nematodes use an oral technique toward feedstuff happening existing plant material. Herbal nematodes introduce enzymes into a host cell to partially break down the cell contents 318 Y (Escobar et al. 2015). Origin-bulge nematodes parasitize herbal source schemes, directly inhibiting water and nutrient

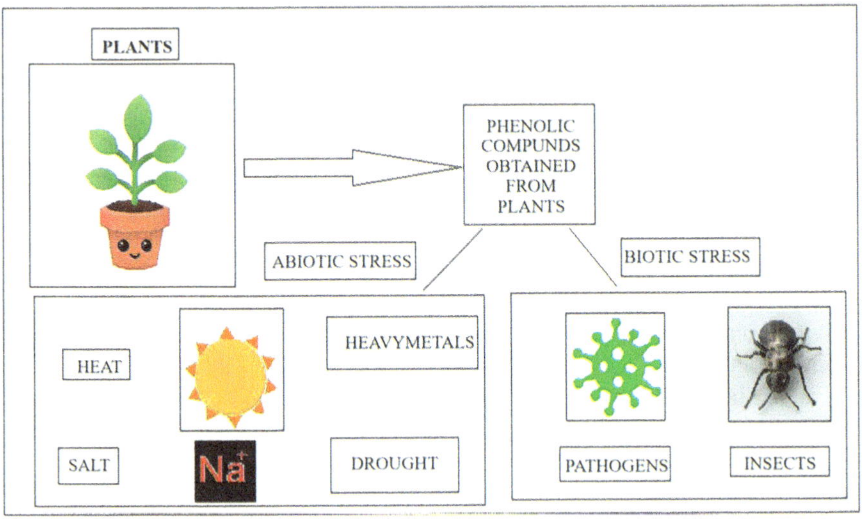

Fig. 6.2 Role of phenolic compounds obtained from plants

consumption required for normal plant progress, expansion, in addition, imitation (Davies and Spiegel 2011). Around 2000 floras universal are vulnerable to origin-bulge nematode contamination, which accounts for about 5% of universal yield damage. Ubiquitous in Europe and North and South America, tuber cystic worms (*Globodera pallida* and *Globodera rostochiensis*) inflict $300 million in losses yearly in Europe (Minnis et al. 2002) (Figs. 6.2 and 6.3).

6.6 Elicitation Strategies of Specific Phenolic Metabolites for Antimicrobial Applications

6.6.1 Physical Elicitation Mechanism

Phenolic metabolites accumulate because of exposure to the stress of temperature, water, saline content, radiation, chemicals, and mechanical reason (Akula and Ravishankar 2011). Physical stress elicitors enhance the content of phenol in plants (Lobiuc et al. 2017; Ravanfar et al. 2018; Zhang and Jiang 2019). Green and red basil show better synthesis of chlorophyll whereas red and blue basil show better synthesis of phenols (Lobiuc et al. 2017). Blue light predominantly increases the quantity of caffeic and rosmarinic derivatives in comparison to normal light due to damage to the membrane (Lobiuc et al. 2017; Gilbert et al. 2018).

Baby spinach leaves when sprayed with caffeic acid enhance their antimicrobial activity (Gilbert et al. 2018). The content of phenol, antioxidant along with antipathogenic activity obtained from *Brassica nigra* L. were investigated. The

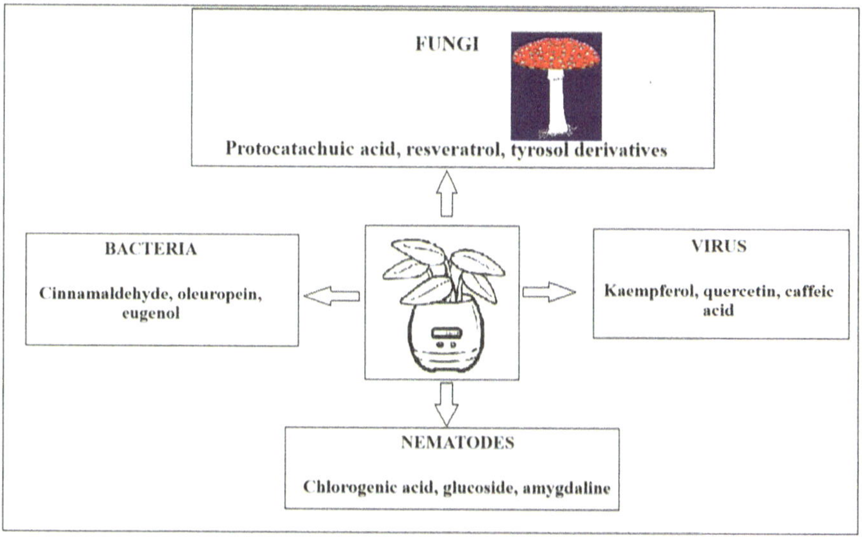

Fig. 6.3 Phenolic compound retaliation to biotic stress

callus grown in white light enacted better properties than those grown in the dark (Hussein et al. 2010).

Radish was exposed to white light blue light, UV radiation, or darkness to analyze the fabrication of anthocyanin. The result depicted that the blue light produced higher phenolic components compared to others (Zhang and Jiang 2019). Radish in presence of blue light also increases the enzymatic activity and also upgrades the genetic expression in the fabrication of anthocyanin (Zhang and Jiang 2019). The antimicrobial activity is also improvised.

Anthocyanins found in berries and fruits showed growth-inhibitory activity against pathogens (Cisowska et al. 2011). Most bacterial pathogens are sensitized to colored light for the accretion of porphyrins and flavins (Yin et al. 2013). Thus, nonwhite light elicitor treatment not only inhibits the maturation of pathogens but also regulates antimicrobial phenolics synthesis in plants providing protection against pathogens. Blue or red-light mechanism is widely utilized in indoor farming. Nonwhite light also intensifies the fabrication of phenolics with antimicrobial properties in plants. Fava beans when stimulated with proline enhanced the phenolic antioxidant enzyme activity (Shetty et al. 2002). UV stress greatly improves the polyphenols in plants. Controlled wounding can also induce a defense mechanism in plants leading to enhanced synthesis of protective phenolics. Three types of carrot pieces were considered for the study—slices, pies, and shreds. The shredded carrots showed increased soluble phenolics, compared to whole carrots (Surjadinata and Cisneros-Zevallos 2012). Ferulic acid was detected in shredded ones (Surjadinata and Cisneros-Zevallos 2012).

Gamma radiation also increases the fabrication of phenolic components which will increase the enzymatic activity associated with phenolic acids (Vardhan and Shukla 2017; Harrison and Were 2007; Maraei and Elsawy 2017; Oufedjikh et al. 2000). Thus, these physical elicitors are used for intensifying the pursuit of stress-induced phenolics, for antimicrobial benefits.

6.6.2 Chemical Elicitation Mechanisms

Examples of chemical elicitors are lipopolysaccharides, plant herbal extracts, peptides, and organic acids. It stimulates phenolic synthesis in seeds leading to increased phenolic phytochemical components and antipathogenic activity (Randhir et al. 2004; McCue and Shetty 2002a, b).

Mung bean seeds grown under a dark environment and treated with oregano combined rosmarinic acid showed higher contents of phenol, in comparison to the seeds grown under light (McCue and Shetty 2002a, b; Akula and Ravishankar 2011). It also showed higher antimicrobial activity (Randhir et al. 2004). There was a subsequent increment in the content of phenol when reacted with a natural elicitor (Randhir et al. 2004). The methanolic extracts of the root of *Plantago lanceolata* when treated with $AgNO_3$ showed maximum antimicrobial activity (Rahamooz-Haghighi et al. 2020).

Ozone treatment is also very effective in grape cultivation. It results in higher phenolic content in the red and white grapes, in comparison to the untreated grapes (Christopher et al. 2018). The ethanolic extracts of grapes from different species also showed antipathogenic activity against various pathogens (Katalinı́c et al. 2010). Roots of hydroponic plants when treated with acetate and chitosan displayed antipathogenic activity that increased doubly after elicitors' treatment (Poulev et al. 2003).

Pineapples and bananas in presence of ozone depicted high phenol content and increased proline enzymatic activity (Alothman et al. 2010). Red cabbage shoots when treated with zeatin showed high content of phenol, and improved antioxidant with cytotoxic activity. COS when sprayed at the preharvest stage of oregano regulates the phenolic content and induces the fabrication of hydrogen peroxide (Yin et al. 2012). Chitosan-treated strawberries have a high quantity of polyphenols with antimicrobial enzyme activity, and the shelf life is also improved (Wang and Gao 2013).

The improved synthesis of stress-inducing phenolics in plants treated with an elicitor provides protection against microbes. Chemical-based elicitor, which is specially bonded with biologically rich elicitors, serves as the safest and most effective technique to improvise antimicrobial phenolics in plants (Hoque et al. 2020).

6.6.3 Microbial Elicitation Mechanism

Microbial biotic elicitors are components that are directly released from the microorganism like xanthino are fabricated by the microorganism when reacts with pectin, or when the enzyme interacts with the microbial cell wall like chitosans. They greatly enhance the fabrication of plant polyphenols (Namdeo 2007). Mung beans when grown in an unlighted environment after reacting with xantham gum, gellan gum, glucan, yeast extract, salicylic, and acetic acid show higher content of polyphenols in different plant parts (McCue and Shetty 2002a, b). The root of sweet basil transformed with microbes like *A. rhizogenes* shows an escalation in the fabrication of rosmarinic acid which shows antimicrobial activity and hence is used effectively against pathogens (Bais et al. 2002).

The extracts are reacted with different elicitors like *Phytophthora cinnamon*, which increases the fabrication of rosmarinic acid (Bais et al. 2002). Soybean seeds when treated with hydrogen peroxide and silver nitrate along with microbial elicitors like *B. subtilis* or Rhizopus improve the quality of phenols showing antimicrobial activity (Kalli et al. 2020). *Coleus blumei* treated with a fungal elicitor or with methyl jasmonate doubly increases the fabrication of rosmarinic acid. The fungal elicitor enhanced the enzymatic activity of rosmarinic acid synthase (Szabo et al. 1999). Venus-fly-trap was elicited using biotic elicitation like *Cronobacter sakazakiito* improve the phenolic compounds (Makowski et al. 2020). This study depicted that the microbial elicitation mechanism increased the fabrication of various acids showing antimicrobial activity.

6.6.4 Phytohormone Elicitation Strategies

Phytohormones promote signaling in plants against bacterial, fungal, and viral pathogens. Examples of phytohormones are jasmonic and salicylic acid, ethylene, abscisic acid, cytokinins, auxins, gibberellins, and brassinosteroids. They help in improving stress-response toward biotic stresses (Bolouri Moghaddam et al. 2016). Salicylic acid largely helps in the initiation of flower, and thermogenesis, and is fabricated by the isochorismate pathway (Dempsey et al. 2011; Raskin 1992).

It behaves like an endogenous signal inducing resistance against bacterial, viral, or fungal attacks. The local acquired resistance shows a hypersensitivity leading to necrosis-causing plant lesions. Plant defense-related genes are expressed during the attack of the pathogen. This strengthens the cell walls of plants fabricating antimicrobial enzymes, antimicrobial peptides, and phytoalexins, displaying a wide antimicrobial activity (Dempsey et al. 1999; Durner et al. 1997).

Jasmonic acid derivatives act as signaling molecules. It activates various transcription factors which bind to gene promoters and activates enzymes with the fabrication of antimicrobial phenolics (Zhou and Memelink 2016). In hydroponic plants when treated with jasmonic acid, antimicrobial activity was noticed after treatment with these elicitors (Poulev et al. 2003). Pea sprouts were grown in a dark environment and treated with acetylsalicylic acid thereafter. The antimicrobial activity was investigated and the treated sprouts attributed high antipathogenic activity than the nontreated ones (Ho et al. 2006).

The wounding stress in carrots was evaluated to elicitate the ROS, ethylene, and jasmonic acid. The carrot shreds were reacted with inhibitors like diphenyleneiodonium chloride, and phenidone (Jacobo-Velázquez et al. 2015). The results depicted that ROS and ET were produced during wounding stress (Jacobo-Velázquez et al. 2015). ROS helps in the signaling of molecules in wound-induced stress and possibly affects the accretion of polyphenols like 3-*O*-caffeoylquinic acid (Jacobo-Velázquez et al. 2015). Plant defense genes get activated resulting in resistance to bacterial, viral, and fungal pathogens (Jacobo-Velázquez et al. 2015; Zhao et al. 2005). Thus, plant hormones played effectively as an elicitor-based treatment mechanism that stimulates the defensive techniques to intensify the antimicrobial phenolics in plants.

6.6.5 Other Mechanisms

Other mechanisms are used sometimes to improve the content of phenols and increase the antimicrobial activity against various pathogens. Plant tissue culture and techniques of micro are used to intensify the stress-inducing polyphenols of various plants. This improves biological, antioxidant, antimicrobial, and cytotoxic activity (Shetty and Wahlqvist 2004; Dias et al. 2016).

Chickpea plants were put through organogenesis and callogenesis, which significantly increased the polyphenols in the callus leading to brown staining for the accretion of phenols (Naz et al. 2008). The callus extract of the cotton plant was examined for its antimicrobial activity. The callus tissues depicted improved antimicrobial activity (Chaturvedi et al. 2010). The alcoholic extracts of *Cichorium pumilum* callus also exhibited antipathogenic activity (Al Khateeb et al. 2012).

Stevia rebaudiana Bertoni leaf extracts in chloroform and methanol were prepared by micropropagation technique. It displays antimicrobial activity pathogens (Debnath 2007). *Tulbaghia violacea* was micropropagated in petroleum ether displaying higher contents of phenol in comparison to the outdoor plants (Ncube et al. 2011). Extract of oregano in phenol, thymol, and carvacrol developed by micropropagation technique showed significant antibacterial activity (Chun et al. 2005; Seaberg et al. 2003). The micropropagation technique helps in developing stress-inducing antimicrobial phenolics (Gutiérrez-Grijalva et al. 2020).

Genetic engineering also helps in improving the quantity of phenol (Verpoorte et al. 2000; Nascimento and Fett-Neto 2010). Genetic engineering targets various factors in the synthetic pathway like reduction of flux and catabolism and regulatory gene expression (Verpoorte et al. 2000).

Genetically engineered plants express bacterial genes that help in enzyme coding along with the fabrication of phenols (Nascimento and Fett-Neto 2010). Plant and pathogenic genes were expressed for enzymes utilized in the carotene biosynthetic pathway which improves the carotenoid content in rice (Ye et al. 2000; Fujisawa et al. 2008). Phenolic biosynthetic pathways involve genetic transcription with the

Table 6.2 Mechanism of elicitation to intensify phenolics and antipathogenic activity

Type of elicitor	Plant types	Targeted pathogens
Blue light	Spinach	*E. coli*
Hydrolysate, extract of oregano and lactoferrin	Mung bean	*H. pylori*
Light or dark callus culture	Black mustard	*K. pneumoniae*, *S. aureus*
$AgNO_3$ and chitosan	*Plantago lanceolata*	*B. cereus*, *P. vulgaris*
Acetate, chitosan, methylsalicylate, and methyljasmonate	Anacardiaceae, Brassicaceae	*S. aureus*, *P. aeruginosa*
Callus culture	Cotton	*S. epidermidis*, *M. smegmatis*
Micropropagation	*Stevia rebaudiana*	*S. mutans*, *B. subtilis*
Callus culture and *ex vitro* plants	Chicory	*B. subtilis*, *S. epidermis*
Micropropagation	*Tulbaghia violacea*	*S. aureus*, *E. coli*
Micropropagation	Oregano	*L. monocytogenes*

help of transcription factors. It initiates the reaction of RNA polymerase II and other proteins (Memelink et al. 2001). Genes coding different enzymes contain different small elements in the promoter regions that are activated by UV radiation, fungal elicitors, or wounding (Dron et al. 1988; Lois et al. 1989; Schulze-Lefert et al. 1989; Feldbrügge et al. 1997).

Genetic engineering of plants improves the quality and hence intensifies the phenolic content involved in antimicrobial activity. Therefore, genetic engineering and micropropagation techniques along with physical, chemical, and microbial elicitation seem to be a unique approach to gratifying the antimicrobial phenolic content (Table 6.2 and Fig. 6.4).

6.7 Future Directions

The increased level of phenol shows its activity as a toxin to insects, pests, microbial and viral growth. It is considered as a natural mechanism of defense developed by a plant. Plants' ability to adapt to harsh environments is increased by the generation of the phenolic compound in plants under biotic stress. The complexity of bacterial contamination and the increased resistance of bacterial pathogens against antibiotics drugs, it is more challenging to get safety solution. Innovations of new antimicrobial strategies like the stimulation of natural defense responses from the host are required. The phenolic content of medicinal plants can be increased which may improve plants' strength against biotic stresses and food safety against bacterial pathogens. The activity of plant phenolic phytochemicals as antimicrobial depends on minimum inhibitory concentration (MIC). Generally, the MIC values are not sufficient as antimicrobials. So, future advancement lies in the combination study of physical, chemical, microbial, tissue culture, and genetic engineering techniques to achieve

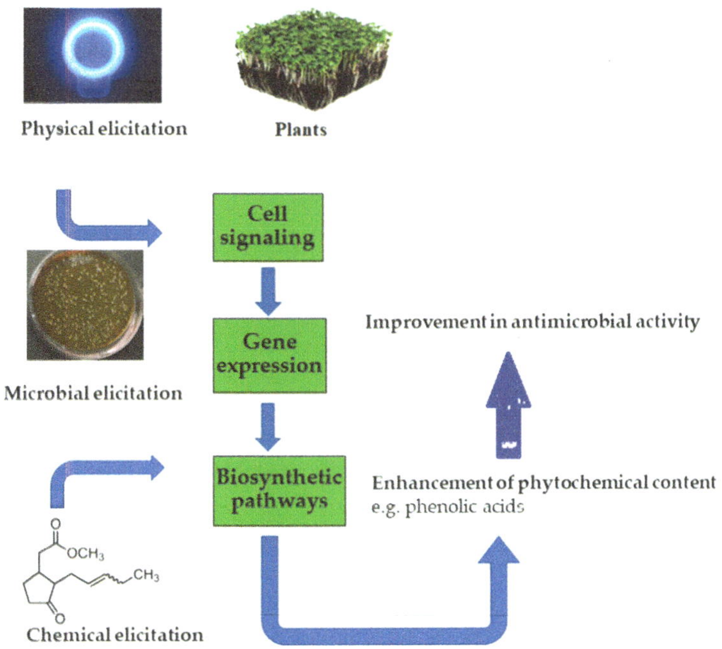

Fig. 6.4 Different elicitation mechanisms to improvise polyphenols for antimicrobial activity

the desired level of phenolic phytochemical content having desired levels of MICs. There are many literature studies focusing on how to increase the phenolic content of plants. The main key point for future development is to understand the molecular mechanism of phenolic compounds and genetic study which help to know the key factor responsible for the biosynthesis of specific phenolic compounds and also its antimicrobial, antiviral, and other mechanisms during each stress.

6.8 Conclusion

Phenolic compounds imparted resistance against various biotic stresses of plants. They showed promising activity to combat microbial, viral, bacterial, and nematode stress. Plants generally synthesized flavonoids and phenolic acids to scavenge ROS generated by the light of UV radiation. To combat temperature stress, phenylalanine ammonia-lyase (PAL) enzyme level is increased that improves phenol accumulation in plants. PAL is a precursor for the biosynthesis of different phenolic compounds. Plants gathered polyphenols, flavonoids, and phenolic acids to resist oxidative stress. Plants can reduce the penetration of microbial pathogens at the site of infection by accumulating phenolic compounds. These naturally synthesized phenolic compounds can recognize microbial pathogens and produce the genetic level of biosynthesized defense metabolites. Plants have to control its systemic-acquired resistance. For that, plants also collect salicylic acid and hydrogen peroxide at the site of infection. Organs of plants accumulate phenolic chemicals that are poisonous or inhibitory to nematodes, insects, etc. The most effective techniques have to be chosen to improvise antimicrobial phenolics in plants. An interactive study was required between salicylic acid and phenolics to know about the biotic stress, oxidative damage, nematodal, viral, and microbial attack.

References

Acamovic T, Brooker JD (2005) Biochemistry of plant secondary metabolites and their effects in animals. Proc Nutr Soc 64:403–412

Akula R, Ravishankar GA (2011) Influence of abiotic stress signals on secondary metabolites in plants. Plant Signal Behav 6:1720–1731

Al Khateeb W, Hussein E, Qouta L, Alu'datt M, Al-Shara B, Abu-Zaiton A (2012) In vitro propagation and characterization of phenolic content along with antioxidant and antimicrobial activities of Cichorium pumilum Jacq. Plant Cell Tissue Organ Cult 110:103–110

Alasalvar C, Grigor JM, Zhang DL, Quantick PC, Shahidi F (2001) Comparison of volatiles, phenolics, sugars, antioxidant vitamins, and sensory quality of different colored carrot varieties. J Agric Food Chem 49:1410–1416

Alothman M, Kaur B, Fazilah A, Bhat R, Karim AA (2010) Ozone-induced changes of antioxidant capacity of fresh-cut tropical fruits. Innov Food Sci Emerg Technol 11:666–671

Babbar N, Oberoi HS, Sandhu SK, Bhargav VK (2014) Influence of different solvents in extraction of phenolic compounds from vegetable residues and their evaluation as natural sources of antioxidants. J Food Sci Technol 51:2568–2575

Bais HP, Walker TS, Schweizer HP, Vivanco JM (2002) Root specific elicitation and antimicrobial activity of rosmarinic acid in hairy root cultures of Ocimum basilicum. Plant Physiol Biochem 40:983–995

Bais HP, Weir TL, Perry LG, Gilroy S, Jorge M (2006) The role of root exudates in rhizosphere interactions with plants and other organisms. Annu Rev Plant Biol 57:233–266

Balasundram N, Sundram K, Samman S (2006) Phenolic compounds in plants and agri-industrial by-products: antioxidant activity, occurrence, and potential uses. Food Chem 99:191–203

Behmann J, Mahlein AK, Rumpf T, Römer C, Plümer L (2014) A review of advanced machine learning methods for the detection of biotic stress in precision crop protection. Precis Agric 16(3):239–260

Beninger CW, Abou-Zaid M, Kistner AL, Hallett RH, Iqbal MJ, Grodzinski B et al (2004) A flavanone and two phenolic acids from Chrysanthemum morifolium with phytotoxic and insect growth regulating activity. J Chem Ecol 30:589–606

Bhattacharya A, Sood P, Citovsky V (2010) The roles of plant phenolics in defence and communication during agrobacterium and rhizobium infection. Mol Plant Pathol 11(5):705–719

Bi B, Tang J, Han S, Guo J, Miao Y (2017) Sinapic acid or its derivatives interfere with abscisic acid homeostasis during Arabidopsis thaliana seed germination. BMC Plant Biol 17(1):99

Bird DM, Opperman CH (1998) Caenorhabditis elegans: a genetic guide to parasitic nematode biology. J Nematol 30:299–308

Bjarnsholt T, Givskov M (2007) Quorum-sensing blockade as a strategy for enhancing host defenses against bacterial pathogens. Philos Trans R Soc B 362:1213–1222

Bolouri Moghaddam MR, Vilcinskas A, Rahnamaeian M (2016) Cooperative interaction of antimicrobial peptides with the interrelated immune pathways in plants. Mol Plant Pathol 17:464–471

Carris LM, Little CR, Stiles CM (2012) Introduction to fungi. The plant health instructor. American Phytopathological Society

Chan YA, Podevels AM, Kevany BM, Thomas MG (2009) Biosynthesis of polyketide synthase extender units. Nat Prod Rep 26(1):90–114
Chaturvedi A, Singh S, Nag TN (2010) Antimicrobial activity of flavonoids from in vitro tissue culture and seeds of Gossypium species. Rom Biotechnol Lett 15:4959–4963
Chowdhary V, Alooparampil S, Pandya RV, Tank JG (2021) Physiological function of phenolic compounds in plant defense system. In: Phenolic compounds - chemistry, synthesis, diversity, non-conventional industrial, pharmaceutical and therapeutic applications. IntechOpen. https://doi.org/10.5772/intechopen.101131
Christopher A, Orwat J, Sarkar D, Hatterman-Valenti H, Shetty K (2018) Ozone elicited phenolic bioactives in grapes and health relevant screening targeted for type 2 diabetes using in vitro assay models. J Med Active Plants 7:1–11
Chun SS, Vattem DA, Lin YT, Shetty K (2005) Phenolic antioxidants from clonal oregano (Origanum vulgare) with antimicrobial activity against Helicobacter pylori. Process Biochem 40:809–816
Cisowska A, Wojnicz D, Hendrich AB (2011) Anthocyanins as antimicrobial agents of natural plant origin. Nat Prod Commun 6:149–156
Congdon BS, Coutts BA, Renton M, Flematti GR, Jones RAC (2017) Establishing alighting preferences and species transmission differences for pea seed-borne mosaic virus aphid vectors. Virus Res 15(241):145–155
Davies K, Spiegel Y (2011) Biological control of plant-parasitic nematodes. Front Microbiol. https://doi.org/10.1007/978-1-4020-9648-8
Debnath M (2007) Clonal propagation and antimicrobial activity of an endemic medicinal plant Stevia rebaudiana. J Med Plant Res 2:45–51
Dempsey DMA, Shah J, Klessig DF (1999) Salicylic acid and disease resistance in plants. Crit Rev Plant Sci 18:547–575
Dempsey DMA, Vlot AC, Wildermuth MC, Klessig DF (2011) Salicylic acid biosynthesis and metabolism. Arabidopsis Book 2011:e0156
Dias MI, Sousa MJ, Alves RC, Ferreira IC (2016) Exploring plant tissue culture to improve the production of phenolic compounds: a review. Ind Crop Prod 82:9–22
Dresselhaus T, Hückelhoven R (2018) Biotic and abiotic stress responses in crop plants. Agron 8: 267
Dron M, Clouse SD, Dixon RA, Lawton MA, Lamb CJ (1988) Glutathione and fungal elicitor regulation of a plant defense gene promoter in electroporated protoplasts. Proc Natl Acad Sci U S A 85:6738–6742
Dube HC (2015) An introduction to fungi, 4th edn. Scientific Publishers, pp 978–998
Duke SO (2018) Interaction of chemical pesticides and their formulation ingredients with microbes associated with plants and plant pests. J Agric Food Chem 66(29):7553–7561
Durner J, Shah J, Klessig DF (1997) Salicylic acid and disease resistance in plants. Trends Plant Sci 2:266–274
Easwar Rao D, Divya K, Prathyusha IVSN, Rama Krishna C, Chaitanya KV (2017) Insect-resistant plants. In: Current developments in biotechnology and bioengineering. Elsevier, Amsterdam, pp 47–74
Edreva A, Velikova V, Tsonev T, Dagnon S, Gürel AL, Aktas L (2008) Stress-protective role of secondary metabolites: diversity of functions and mechanisms. Gen Appl Plant Physiol 34:67–78
Escobar C, Barcala M, Cabrera J, Fenoll C (2015) Overview of root-knot nematodes and Giant cells. In: Plant nematode interactions – a view on compatible interrelationships. Academic, Amsterdam, pp 1–32
Feldbrügge M, Sprenger M, Hahlbrock K, Weisshaar B (1997) PcMYB1, a novel plant protein containing a DNA-binding domain with one MYB repeat, interacts in vivo with a light-regulatory promoter unit. Plant J 11:1079–1093

Fiorentino A, D'Abrosca B, Pacifico S, Mastellone C, Piccolella S, Monaco P (2008) Isolation and structure elucidation of antioxidant polyphenols from quince (Cydonia vulgaris) peels. J Agric Food Chem 56:2660–2667

Fujisawa M, Watanabe M, Choi SK, Teramoto M, Ohyama K, Misawa N (2008) Enrichment of carotenoids in flaxseed (Linum usitatissimum) by metabolic engineering with introduction of bacterial phytoene synthase gene crtB. J Biosci Bioeng 105:636–641

Ghosh P, Bhattacharya A, Char B (2017) Manipulating disease and pest resistance pathways in plants for enhanced crop improvement. Biosci Biotechnol Res Commun 10(4):631–644

Gilbert AR, Alborzi S, Bastarrachea LJ, Tikekar RV (2018) Photo irradiated caffeic acid as an antimicrobial treatment for fresh produce. FEMS Microbiol Lett 365:fny132

Green ER, Mecsas J (2015) Bacterial secretion systems: an overview. In: Virulence mechanisms of bacterial pathogens, 5th edn. ASM Press, Washington, DC, pp 215–239

Gutiérrez-Grijalva EP, Santos-Zea L, Ambriz-Pérez DL, López-Martínez LX, Heredia JB (2020) Flavones and flavonols: bioactivities and responses under light stress in herbs. In: Lone R, Shuab R, Kamili A (eds) Plant phenolics in sustainable agriculture. Springer, Singapore. https://doi.org/10.1007/978-981-15-4890-1_4

Hajano JUD, Pathan MA, Rajput AQ, Lodhi MA (2011) Rice blast-mycoflora, symptomatology and pathogenicity. Int J Agro Vet Med Sci 5(1):53–63

Hammond-Kosack K, Jones JDG (2000) Responses to plant pathogens. Biochem Mol Biol Plants 1: 1102–1156

Harborne JB (1989) General procedures and measurement of total phenolics. In: Dey PM, Harborne JB (eds) Plant phenolics. Academic, London, UK, pp 1–28

Hardoim PR, Van Overbeek LS, Berg G, Pirttilä AM, Compant S, Campisano A, Sessitsch A (2015) The hidden world within plants: ecological and evolutionary considerations for defining functioning of microbial endophytes. Microbiol Mol Biol Rev 79(3):293–320

Harrison K, Were LM (2007) Effect of gamma irradiation on total phenolic content yield and antioxidant capacity of almond skin extracts. Food Chem 102:932–937

Heima KE, Tagliaferroa AR, Bobilya DJ (2002) Flavonoid antioxidants: chemistry, metabolism and structure-activity relationships. J Nutr Biochem 13:572–584

Hirano SS, Upper CD (2000) Bacteria in the leaf ecosystem with emphasis on Pseudomonas syringae-a pathogen, ice nucleus, and epiphyte. Microbiol Mol Biol Rev 64(3):624–653

Ho CY, Lin YT, Labbe RG, Shetty K (2006) Inhibition of Helicobacter pylori by phenolic extracts of sprouted peas (Pisum sativum L.). J Food Biochem 30:21–34

Hoper L, Cassidy A (2006) A review of the health care potential of bioactive compounds. J Sci Food Agric 86:1805–1813

Hoque TS, Sohag AAM, Burritt DJ, Hossain MA (2020) Salicylic acid-mediated salt stress tolerance in plants. In: Lone R, Shuab R, Kamili A (eds) Plant phenolics in sustainable agriculture. Springer, Singapore. https://doi.org/10.1007/978-981-15-4890-1_1

Hussein EA, Taj-Eldeen AM, Al-Zubairi AS, Elhakimi AS, Al-Dubaie AR (2010) Phytochemical screening, total phenolics and antioxidant and antibacterial activities of callus from Brassica nigra L. hypocotyl explants. Int J Pharmacol 6:464–471

Jacobo-Velázquez DA, González-Agüero M, Cisneros-Zevallos L (2015) Cross-talk between signaling pathways: the link between plant secondary metabolite production and wounding stress response. Sci Rep 5:8608

Jones JDG, Dangl JL (2006) The plant immune system. Nature 444:323–329

Joshi JR, Khazanov N, Senderowitz H, Burdman S, Lipsky A, Yedidia I (2016) Plant phenolic volatiles inhibit quorum sensing in pectobacteria and reduce their virulence by potential binding to ExpI and ExpR proteins. Sci Rep 6(1):38126

Kabera JN, Semana E, Mussa AR, He X (2014) Plant secondary metabolites: biosynthesis, classification, function, and pharmacological properties. J Pharm Pharmacol 2:377–392

Kalli S, Araya-Cloutier C, de Bruijn WJ, Chapman J, Vincken JP (2020) Induction of promising antibacterial prenylated isoflavonoids from different subclasses by sequential elicitation of soybean. Phytochemistry 179:112496

Katalinic V, Možina SS, Skroza D, Generalić I, Abramovič H, Miloš M, Ljubenkov I, Piskernik S, Pezo I, Terpinc P et al (2010) Polyphenolic profile, antioxidant properties and antimicrobial activity of grape skin extracts of 14 Vitis vinifera varieties grown in Dalmatia (Croatia). Food Chem 119:715–723

Kumar S, Pandey AK (2013) Chemistry and biological activities of flavonoids: an overview. ScientificWorldJournal 2013:162750

Lamers J, Van Der Meer T, Testerink C (2020) How plants sense and respond to stressful environments. Plant Physiol 182:1624–1635

Lattanzio V, Lattanzio VMT, Cardinali A (2006) Role of phenolics in the resistance mechanisms of plants against fungal pathogens and insects. In: Imperato F (ed) Phytochemistry advances in research. Research Signpost, India, pp 23–67

Levin DA (1971) Plant phenolics: an ecological perspective. Am Nat 105(942):157–181

Lin DR, Hu LJ, You H, Sarkar D, Xing BS, Shetty K (2010) Initial screening studies on potential of high phenolic-linked plant clonal systems for nitrate removal in cold latitudes. J Soils Sediment 10:923–932

Lincoln JE, Sanchez JP, Zumstein K, Gilchrist DG (2018) Plant and animal PR1 family members inhibit programmed cell death and suppress bacterial pathogens in plant tissues. Mol Plant Pathol 19(9):2111–2123

Liu C, Shen Q, Wang Y, Zhang F, He P, Si X, Wang K, Zhu R, Xiang N, Liu Z (2016) Coumarins from the leaves of Nicotiana tabacum and their anti-tobacco mosaic virus activities. Chem Nat Compd 52:992–995

Lobiuc A, Vasilache V, Oroian M, Stoleru T, Burducea M, Pintilie O, Zamfirache MM (2017) Blue and red LED illumination improves growth and bioactive compounds contents in acyanic and cyanic Ocimum basilicum L. microgreens. Molecules 22:2111

Lois R, Dietrich A, Hahlbrock K, Schulz W (1989) A phenylalanine ammonia-lyase gene from parsley: structure, regulation and identification of elicitor and light responsive cis-acting elements. EMBO 8:1641–1648

Maccheroni W, Alegria MC et al (2005) Identification and genomic characterization of a new virus (Tymoviridae family) associated with citrus sudden death disease. J Virol 79(5):3028–3037

Makowski W, Tokarz KM, Tokarz B, Banasiuk R, Witek K, Królicka A (2020) Elicitation-based method for increasing the production of antioxidant and bactericidal phenolic compounds in Dionaea muscipula. J Ellis Tissue Molecules 25:1794

Mandal SM, Chakraborty D, Dey S (2010) Phenolic acids act as signaling molecules in plant microbe symbioses. Plant Signal Behav 5(4):359–368

Mantri N, Patade V, Penna S, Ford R, Pang E (2012) Abiotic stress responses in plants: present and future. In: Abiotic stress responses in plants. Springer, New York, pp 1–19

Maraei RW, Elsawy KM (2017) Chemical quality and nutrient composition of strawberry fruits treated by γ-irradiation. J Radiat Res Appl Sci 10:80–87

Mayer AM, Staples RC, Gil-ad NL (2001) Mechanisms of survival of necrotrophic fungal plant pathogens in hosts expressing the hypersensitive response. Phytochemistry 58(1):33–41

McCue P, Shetty K (2002a) Clonal herbal extracts as elicitors of phenolic synthesis in dark-germinated mungbeans for improving nutritional value with implications for food safety. J Food Biochem 26:209–232

McCue P, Shetty K (2002b) A biochemical analysis of mungbean (Vigna radiata) response to microbial polysaccharides and potential phenolic-enhancing effects for nutraceutical applications. Food Biotechnol 16:57–79

Melvin P, Bankapalli K, D'Silva P, Shivaprasad PV (2017) Methylglyoxal detoxification by a DJ-1family protein provides dual abiotic and biotic stress tolerance in transgenic plants. Plant Mol Biol 94(4–5):381–397

Memelink J, Kijne JW, van der Heijden R, Verpoorte R (2001) Genetic modification of plant secondary metabolite pathways using transcriptional regulators. In: Zhong JJ (ed) Plant cells. Advances in biochemical engineering/biotechnology, vol 72. Springer, Berlin, pp 103–125

Minnis ST, Haydock PPJ, Ibrahim SK, Grove IG, Evans K, Russell MD (2002) Potato cyst nematodes in England and Wales – occurrence and distribution. Ann Appl Biol 140(2):187–195
Namdeo AG (2007) Plant cell elicitation for production of secondary metabolites: a review. Pharmacogn Rev 1:69–79
Nascimento NC, Fett-Neto AG (2010) Plant secondary metabolism and challenges in modifying its operation: an overview. In: do Nascimento NC, Fett-Neto AG (eds) Plant secondary metabolism engineering, Methods in molecular biology (Methods and protocols), vol 643. Humana, Totowa, NJ, pp 1–13
Naz S, Ali A, Iqbal J (2008) Phenolic content in vitro cultures of chickpea (Cicer arietinum L.) during callogenesis and organogenesis. Pak J Bot 40:2525–2539
Ncube B, Ngunge VNP, Finnie JF, Van Staden J (2011) A comparative study of the antimicrobial and phytochemical properties between outdoor grown and micropropagated Tulbaghia violacea Harv. plants. J Ethnopharmacol 134:775–780
Ndakidemi PA, Dakora FD (2003) Review: Legume seed flavonoids and nitrogenous metabolites as signals and protectants in early seedling development. Funct Plant Biol 30:729–745
Oufedjikh H, Mahrouz M, Amiot MJ, Lacroix M (2000) Effect of gamma-irradiation on phenolic compounds and phenylalanine ammonialyase activity during storage in relation to peel injury from peel of Citrus clementina Hort. Ex. Tanaka. J Agric Food Chem 48:559–565
Ozyigit II, Kahraman MV, Ercan O (2007) Relation between explant age, total phenols and regeneration response of tissue cultured cotton (Gossypium hirsutum L). Afr J Biotechnol 6: 003–008
Parr AJ, Bolwell GP (2000) Phenols in the plant and in man. The potential for possible nutritional enhancement of the diet by modifying the phenols content or profile. J Sci Food Agric 80(7): 985–1012
Peck S, Mittler R (2020) Plant signaling in biotic and abiotic stress. J Exp Bot 71:1649–1651
Peperidou A, Pontiki E, Hadjipavlou-Litina D, Voulgari E, Avgoustakis K (2017) Multifunctional cinnamic acid derivatives. Molecules 22(8):1247
Pina-Pérez MC, Ferrús Pérez MA (2018) Antimicrobial potential of legume extracts against foodborne pathogens: a review. Trends Food Sci Technol 72:114–124
Pinto MS, Lajolo FM, Genovese MI (2007) Bioactive compounds and antioxidant capacity of strawberry jams. Plant Foods Hum Nutr 62:127–131
Postel S, Kemmerling B (2009) Plant systems for recognition of pathogen-associated molecular patterns. Semin Cell Dev Biol 20(9):1025–1103
Poulev A, O'Neal JM, Logendra S, Pouleva RB, Timeva V, Garvey AS, Gleba D, Jenkins IS, Halpern BT, Kneer R et al (2003) Elicitation, a new window into plant chemodiversity and phytochemical drug discovery. J Med Chem 46:2542–2547
Pu F, Ren XL, Zhang XP (2013) Phenolic compounds and antioxidant activity in fruits of six *Diospyros kaki* genotypes. Eur Food Res Technol 237:923–932
Pusztahely T, István H, Pócsi IJ (2017) Plant-fungal interactions: special secondary metabolites of the biotrophic, necrotrophic, and other specific interactions. In: Fungal metabolites. Academic, London, pp 1–58
Rahamooz-Haghighi S, Bagheri K, Sharafi A, Danafar H (2020) Establishment and elicitation of transgenic root culture of Plantago lanceolata and evaluation of its anti-bacterial and cytotoxicity activity. Prep Biochem Biotechnol 51:207–224
Raman N, Marchillo K, Lee MR, Rodríguez López A, Andes DR, Palecek SP, Lynn DM (2015) Intraluminal release of an antifungal β-peptide enhances the antifungal and anti-biofilm activities of multilayer-coated catheters in a rat model of venous catheter infection. ACS Biomater Sci Eng 2(1):112–121
Randhir R, Lin YT, Shetty K (2004) Stimulation of phenolics, antioxidant and antimicrobial activities in dark germinated mung bean sprouts in response to peptide and phytochemical elicitors. Process Biochem 39:637–646
Raskin I (1992) Role of salicylic acid in plants. Annu Rev Plant Biol 43:439–463

Ravanfar SA, Karimi E, Mehrabanjoubani P, Ebrahimi M (2018) Enhancement of phenolic and flavonoids compounds, antioxidant and cytotoxic effects in regenerated red cabbage by application of Zeatin. Nat Prod Res 34:898–902
Rejeb IB, Pastor V, Mauch-Mani B (2014) Plant responses to simultaneous biotic and abiotic stress: molecular mechanisms. Plants 3:458–475
Riddle DL, Albert PS (1997) Genetic and environmental regulation of dauer larva development. In: Riddle DL, Blumenthal T, Meyer J, Preiss JR (eds) Caenorhabditis elegans II. Cold Spring Harbor Laboratory Press, Plainview, NY, pp 739–768
Rizhsky L, Liang H, Shuman J, Shulaev V, Davletova S, Mittler R (2004) When defense pathways collide. The response of Arabidopsis to a combination of drought and heat stress. Plant Physiol 134:1683–1696
Saddique M, Kamran M, Shahbaz M (2018) Differential responses of plants to biotic stress and the role of metabolites. In: Plant metabolites and regulation under environmental stress. Elsevier, San Diego, pp 69–87
Sasser JN, Freckman DW (1987) A world perspective on nematology: the role of the society. In: Veech JA, Dickson DW (eds) Vistas on nematology. Society of Nematologists, Hyattsville, MD, pp 7–14
Schulze-Lefert P, Dangl JL, Becker-André M, Hahlbrock K, Schulz W (1989) Inducible in vivo DNA footprints define sequences necessary for UV light activation of the parsley chalcone synthase gene. EMBO 8:651–656
Seaberg AC, Labbe RG, Shetty K (2003) Inhibition of Listeria monocytogenes by elite clonal extracts of oregano (Origanum vulgare). Food Biotechnol 17:129–149
Shetty K, Wahlqvist M (2004) A model for the role of the proline-linked pentose-phosphate pathway in phenolic phytochemical biosynthesis and mechanism of action for human health and environmental applications. Asia Pac J Clin Nutr 13:1–24
Shetty P, Atallah MT, Shetty K (2002) Effects of UV treatment on the proline-linked pentose phosphate pathway for phenolics and L-DOPA synthesis in dark germinated Vicia faba. Process Biochem 37:1285–1295
Sun S, Zhi Y, Zhu Z, Jin J, Duan C, Wu X, Xiaoming W (2017) An emerging disease caused by Pseudomonas syringae pv. phaseolicola threatens mung bean production in China. Plant Dis 101(1):95–102
Surjadinata BB, Cisneros-Zevallos L (2012) Biosynthesis of phenolic antioxidants in carrot tissue increases with wounding intensity. Food Chem 134:615–624
Szabo E, Thelen A, Petersen M (1999) Fungal elicitor preparations and methyl jasmonate enhance rosmarinic acid accumulation in suspension cultures of Coleus blumei. Plant Cell Rep 18:485–489
Tak Y, Kumar M (2020) Phenolics: a key defence secondary metabolite to counter biotic stress. IntechOpen. https://doi.org/10.1007/978-981-15-4890-1_13
Talhaoui N, Taamalli A, Gómez-Caravaca AM, Fernández-Gutiérrez A, Segura-Carretero A (2015) Phenolic compounds in olive leaves: analytical determination, biotic and abiotic influence, and health benefits. Food Res Int 77:92–108
Tripathi P (2004) Exploitation of natural products as an alternative strategy to control postharvest fungal rotting of fruit and vegetables. Postharv Biol Technol 32(3):235–245
Varbanova M, Porter K, Lu F, Ralph J, Hammerschmidt R, Jones AD, Day B (2011) Molecular and biochemical basis for stress-induced accumulation of free and bound p-coumaraldehyde in cucumber. Plant Physiol 157(3):1056–1066
Vardhan PV, Shukla LI (2017) Gamma irradiation of medicinally important plants and the enhancement of secondary metabolite production. Int J Radiat Biol 93:967–979
Vattem DA, Randhir R, Shetty K (2005) Cranberry phenolics-mediated antioxidant enzyme response in oxidatively stressed porcine muscle. Process Biochem 40:2225–2238
Velderrain-Rodríguez GR, Palafox-Carlos H, Wall-Medrano A, AyalaZavala JF, Chen C-YO, Robles-Sanchez M, Astiazaran-García H, Alvarez-Parrilla E, González-Aguilar GA (2014) Phenolic compounds: their journey after intake. Food Funct 5:189–197

Verma S, Nizam S, Verma PK (2013) Biotic and abiotic stress signalling in plants. In: Stress signaling in plants: genomics and proteomics perspective, vol 1, pp 25–49
Verpoorte R, van der Heijden R, Memelink J (2000) Engineering the plant cell factory for secondary metabolite production. Transgenic Res 9:323–343
Vit P, Soler C, Tomás-Barberán FA (1997) Profiles of phenolic compounds of Apis mellifera and Melipona spp. honeys from Venezuela. Z Lebensm Unters Forsch 91:571–577
Wang SY, Gao H (2013) Effect of chitosan-based edible coating on antioxidants, antioxidant enzyme system, and postharvest fruit quality of strawberries (Fragaria x aranassa Duch.). LWT Food Sci Technol 52:71–79
Wang L, Sun R, Zhang Q, Luo Q, Zeng S, Li X, Liu Z (2018) An update on polyphenol disposition via coupled metabolic pathways. In: Expert opinion on drug metabolism & toxicology. Ashley Publications, London, pp 1–15
War AR, Paulraj MG, Ahmad T, Buhroo AA, Hussain B, Ignacimuthu S, Sharma HC (2012) Mechanisms of plant defense against insect herbivores. Plant Signal Behav 7(10):1306–1320
Ye X, Al-Babili S, Klöti A, Zhang J, Lucca P, Beyer P, Potrykus I (2000) Engineering the provitamin A (beta-carotene) biosynthetic pathway into (carotenoid-free) rice endosperm. Science 287:303–305
Yin H, Fretté XC, Christensen LP, Grevsen K (2012) Chitosan oligosaccharides promote the content of polyphenols in Greek oregano (Origanum vulgare ssp. hirtum). J Agric Food Chem 60:136–143
Yin R, Dai T, Avci P, Jorge AES, de Melo WC, Vecchio D, Huang YY, Gupta A, Hamblin MR (2013) Light-based antiinfectives: ultraviolet C irradiation, photodynamic therapy, blue light, and beyond. Curr Opin Pharmacol 13:731–762
Zernova O, Lygin A, Pawlowski M, Hill C, Hartman G, Widholm J, Lozovaya V (2014) Regulation of plant immunity through modulation of phytoalexin synthesis. Molecules 19(6):7480–7496
Zhang W, Jiang W (2019) UV treatment improved the quality of postharvest fruits and vegetables by inducing resistance. Trends Food Sci Technol 92:71–80
Zhao J, Davis LC, Verpoorte R (2005) Elicitor signal transduction leading to production of plant secondary metabolites. Biotechnol Adv 23:283–333
Zhao W, Zeng XY, Zhang T, Wang L, Yang GY, Chen YK et al (2013) Flavonoids from the bark and stems of Cassia fistula and their anti-tobacco mosaic virus activities. Phytochem Lett 6:179–182
Zhou M, Memelink J (2016) Jasmonate-responsive transcription factors regulating plant secondary metabolism. Biotechnol Adv 34:441–449
Zhu JK (2002) Salt and drought stress signal transduction in plants. Annu Rev Plant Biol 53:247–273

Salicylic Acid: A Phytohormone of Antistress and Insecticidal Essence

7

Khursheed Ahmad Wani, Javid Manzoor, Ebru Kafkas, and Junaid Ahmad Malik

Abstract

Salicylic acid has diverse applications due to the presence of phenolic compounds. It has been used as antistress and insecticidal agent. Hence, this phenolic compound is widely studied and has been used by researchers as a signaling phytohormone against biotrophs. It has wide applicability in mitigating abiotic stresses. Similarly, it has been observed that this compound will act as a potent biopesticide. Novel approaches are required to synthesis genes similar to salicylic acid that will survive under harsh conditions to open new ways for the development of biopesticides for future use. The aim of this chapter is to understand the role of salicylic acid as a phytohormone of antistress and insecticidal essence.

Keywords

SA · Stress · Pesticides · Phenolic compounds · Applications

K. A. Wani
Department of Environmental Science, Government Degree College, Thindim Kreeri, Jammu and Kashmir, India

J. Manzoor
Department of Environmental Sciences, Shri JJT University, Jhunjhunu, Rajasthan, India

E. Kafkas
Department of Horticulture, Faculty of Agriculture, University of Çukurova, Adana, Turkey

J. A. Malik (✉)
Department of Zoology, Government Degree College, Kulgam, Jammu and Kashmir, India

R. Lone et al. (eds.), *Plant Phenolics in Biotic Stress Management*,
https://doi.org/10.1007/978-981-99-3334-1_7

7.1 Introduction

Salicylic acid (SA) is a naturally occurring phenolic chemical with seven carbons (C) that is also generated in plants and used as a signaling molecule. Plant phenolics are synthesized through two primary mechanisms; these are the shikimic acid pathway and the malonic acid system. Most phenolic compounds in plants are synthesized through the shikimic acid route. Phenylalanine, the precursor of SA, is produced from simple carbohydrate precursors taken from glycolysis and the pentose phosphate cycle. SA biosynthesis in plants often occurs through the phenylalanine route, although it may also occur along the isochorismate pathway. Many different enzymes are involved in the chain of chemical events that result in SA. SA is synthesized by the hydroxylation of benzoic acid by the enzyme benzoic acid 2-hydroxylase. Both the oxidation of fatty acids and a nonoxidative route are required for cinnamic acid to manufacture benzoic acid. To go from cinnamic acid to coumaric acid, the second step in the process requires the catalytic action of cinnamate 4-hydroxylase. In the presence of phenylalanine ammonialyase, phenylalanine is converted into cinnamic acid. For example, hydroxylation of cinnamic acid yields coumaric acid, oxidation of the side chain yields SA, and still another hydroxylation yields hydroxylated SA. Reports of SA production in plants from shikimic acid through chorismic acid and comaric acid have been researched.

Some plants produce the phenolic molecule salicylic acid (2-hydroxybenzoic acid), which is only one of many such chemicals. Phenolic compounds are those that include a benzene ring with one or more hydroxyl groups. Traditional knowledge argued that plant phenolics, despite their wide distribution, were of little biological significance. Phenolics were formerly thought to have a role in the production of lignin and pigments (Métraux and Raskin 1993), but it has since been shown that they also play a role in allelopathy and the control of responses to abiotic and biotic stressors. For instance, SA is a crucial hormone that regulates thermogenesis, disease resistance, and many other aspects of plant growth and development (Vlot et al. 2009). SA and its acetylated derivate (commonly known as aspirin) are essential pharmacological compounds for human use in addition to their roles in plants. While aspirin is one of the most often used pharmaceuticals for treating pain, fever, edema, and inflammation as well as lowering the risk of heart attack, stroke, and some malignancies, SA is typically used to treat warts, acne, and psoriasis (Weissmann 1991; Antithrombotic Trialists' (ATT) Collaboration 2009; Cuzick et al. 2015).

Multiple abiotic stressors, such as heavy metals, salt, ozone, UV, temperature, and drought, affect plant productivity due to continuous cropping and climate change. Irrespective of whether the stressor is biotic or abiotic, SA regulates the resistance to both (Horvath et al. 2007). It has been shown that SA-mediated most of the process viz., (1) buildup of osmolytes such as glycinebetaine, proline, soluble carbohydrates, and amines, which may assist to maintain osmotic homeostasis to induce abiotic stress tolerance. Increased production of secondary metabolites like terpenes, phenolics, and compounds with nitrogen (alkaloids, cyanogenic glucosides, nonprotein amino acids), and sulfur (glutathione, glucosinolates, phytoalexins, thionins, defensins, and allinin); regulation of other hormone pathways; (2) control of mineral nutrition uptake; (3) increased scavenging activity against reactive oxygen species; salicylic acid is a key component in plant development and growth, serving crucial physiological functions like enhancing the plant's response to stress (both biotic and abiotic) and boosting the plant's system acquired resistance (SAR) through stimulating or altering the endogenous signaling pathways discovered during a dissection of the plant's internal anatomy. The plant is better able to endure salt stress, caused by the especially detrimental sodium chloride compound NaCl, since salicylic acid functions as a stimulant or transmitter of the cell to tolerate these abiotic stresses. It may bind conjugate with certain amino acids, including proline and arginine, which boosts the plant's tolerance to external stressors while also maintaining systemic acquired resistance.

Antioxidant production stimulation is the primary impact of salicylic acid as antioxidant against the damage caused by reactive oxygen species (ROS) free radicals, are also generated during periods of high heat and stress. In addition to its function at the genetic level, drought stress also inhibits the oxidation of algebraic, oxytin, and cytokinein. Specifically, it upregulates the MSDS (manganese superoxide dismutase) gene, which encodes an antioxidant enzyme. By raising its internal concentration, salicylic acid promotes the protective function of pathogenic pathogens, leading to a greater plant response of tolerance and resistance to a wide range of illnesses (Kumar 2014).

Important physiological functions of the SA include regulation of flowering, stomatal opening, ion uptake, nutrient transport, CO_2 gas representation, stomatal closure, stomatal movement, photosynthesis, gas exchange, and protein synthesis. In addition to preventing the representation of ethylene gas and acting counter to the activity of ABA, which is responsible for the demise of the plant leaves, it also speeds up the production of other plant colors like chlorophyll and carotene and increases their levels. It also alters the levels of nucleic and amino acids inside the plant, which plays a significant part in the plant's ability to save energy via alternate routes (Davies 2004). Environmental pressures such as drought, flood, temperature change, and microbial and insect assault are only a few of the many threats that plants confront. Plants' oxidative states have been revealed to be crucial in their defense against biotic and abiotic stressors. To combat these threats, plants use a variety of signaling pathways that ultimately result in the creation of many different types of defensive proteins and nonprotein substances. Several phytohormones are critical components of signaling pathways that contribute to plant defense, including abscisic acid, jasmonic acid (JA), ethylene, and salicylic acid (SA). The phenyl propanoid route is mediated by SA, while the octadecanoid pathway is mediated by JA. The first aids in resistance to diseases, insect pests, and abiotic stressors, whereas the latter is primarily designed to ward off insects and some types of infections. The antioxidative enzyme activities in plants are only one of several physiological, biochemical, and molecular processes that may be manipulated by applying SA and jasmonic acid exogenously. In addition to participating in cross-talk with other pathways regulating plant resistance, SA modulates the components of its own signaling system. Nutrient intake, water relations, stomatal control, and photosynthesis are all thought to be channels via which SA influences plant development under stress. Peroxidase (POD), polyphenol oxidase (PPO), superoxide dismutase (SOD), phenylalanine ammonia lyase (PAL), etc., are all enzymes that it controls, and they are crucial to initiate plant defense against biotic and abiotic stressors.

7.2 History

Historically, the bark and leaves of the willow tree were used by Native Americans, Indians, and Greeks to treat pains and fevers. However, it is known that Hippocrates recommended the compound to alleviate the agony and fever experienced by women during childbirth. This compound is now known as SA. Willow (*Salix* sp.) bark and leaves were utilized as a medicine by the ancient Babylonians, Assyrians, and Chinese, according to historical records. In 1763, the Reverend Edward Stone reported to the Royal Society that willow (*Salix* sp.) bark contained chemicals that successfully eased the symptoms of "ague" (possibly malarial fever). The salicylates, methyl salicylate (MeSA), saligenin (alcohol of SA), and their glycosides were subsequently identified from extracts of willow and other plants, but their active principle remained a mystery until the nineteenth century. In the early nineteenth century, the SA was extracted and refined from willow and meadowsweet (*Filipendula ulmaria*) (Leroux 1830), and in 1860, it was chemically produced by the carboxylation of sodium phenoxide (Kolbe and Lautemann 1860). The main salicylate in willow bark was originally isolated from it by a German chemist, Johann Buchner, in 1928 and given the name "salicin" (the glucoside of salicyl alcohol) (Weissmann 1991). In 1838, Raffaele Piria gave the active constituent of willow (*Salix* sp.) bark the name SA (Cleland and Ajami 1974), which comes from the Latin word salix.

In 1874, Germany was the first country to begin mass producing synthetic SA for commercial use. The Bayer Company created aspirin, a brand name for acetylsalicylic acid, in 1898, and it quickly became one of the world's best-selling medications, displacing SA usage due to its lessening of gastrointestinal discomfort while maintaining its equivalent medical effects.

The twentieth century saw a surge in the use of SA for the treatment of a variety of skin conditions, including acne, psoriasis, warts, and calluses, according to the findings of several scientific studies. It is common practice to use SA as a pore cleanser, skin softener, and for the elimination of dead skin cells, grime, oil, and debris. Salicylates are utilized to treat a wide variety of human ailments, from the common cold to heart attacks, despite the fact that their mechanism of therapeutic action is still up for discussion. Since aspirin may be spontaneously hydrolyzed into SA, the two chemicals have comparable effects on vegetation.

7.3 Salicylic Acid: The Signaling Phytohormone Against Biotrophs

Photosynthesis, transpiration, ion absorption and transport, and defensive responses against pathogens are only some of the functions that SA controls in a growing crop. Seed germination, nodulation, stomatal opening, thermoregulation, respiration, and immunological responses are all controlled by SA. SA also increases blooming and fruit sets and strengthens resilience to environmental stresses including drought, heat, and salt.

When used in conjunction with crop protection products, SA activates the plant's defensive mechanisms, reducing the likelihood that the plant may develop a resistance to the chemicals in the crop protection formula. The interactions of plants with biotrophs, necrotrophs, and hemibiotrophs are likewise regulated by SA.

Infected plant tissues secrete SA, which is then transported throughout the plant to warn its healthy counterparts. The term "Systemic Acquired Resistance" describes this phenomenon (SAR). A considerable decrease in disease pressure may be achieved with the use of extremely low doses of SA prior to the spread of the disease, or in conjunction with monosite fungicides (such as strobilurins). However, SA does not replace fungicides if the disease is already established on the crop. The biological role of salicylic acid in plants is also shown in Fig. 7.1.

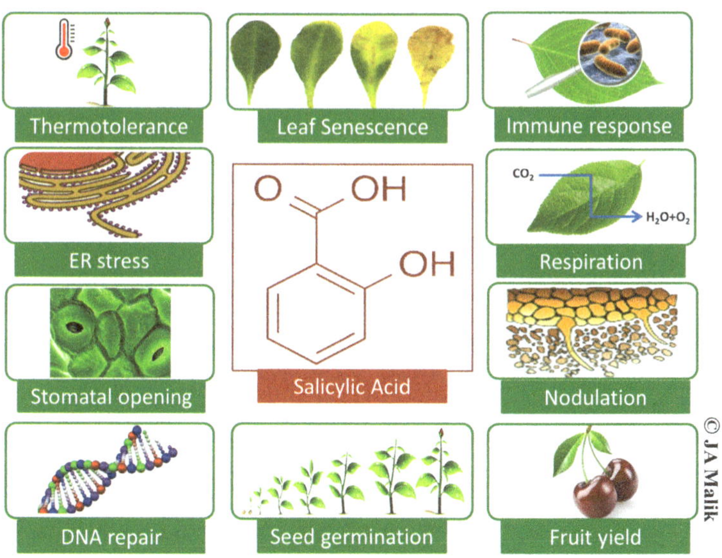

Fig. 7.1 Biological role of salicylic acid in plants

7.4 Functions of SA in Mitigating Abiotic Stresses

7.4.1 Heat

Plant growth and food security are both imperiled by climate change. Heat stress causes sterility and stunted growth in plants, and it may even lead to a decrease in production (Bita and Gerats 2013). Application of exogenous SA increases rice output under high-temperature circumstances, but prevents the synthesis of endogenous SA (Yang et al. 2022). Pea plant thermotolerance was drastically reduced (Pan et al. 2006). Further, numerous plant species, including mustard (Dat et al. 1998), grape (Wang and Li 2006a, b), and melon, creeping bentgrass (Larkindale et al. 2005), showed an increase in SA biosynthesis under heat stress (Widiastuti et al. 2013).

One of the most temperature-sensitive plant systems is photosystem II, an electron transport chain in chloroplasts (Čajánek et al. 1998). Research shows that alfalfa leaves may recover from heat damage to PSII and photosynthetic efficiency if they are sprayed with 0.25 mM SA for 5 days (Wassie et al. 2020). Because SA increases chlorophyll fluorescence and antioxidant capacity, it may be responsible for preserving the thermo-stability of PSII's electron donor and reaction centers (Shi et al. 2006; (Wang et al. 2010). Ion leakage in plant cells is caused by heat stress's interference with osmotic potential and destruction of plasma membranes. The osmoregulation of plant cells relies heavily on free proline, which may be increased with the use of SA.

Several studies have shown this phenomenon in other plant species, including wheat (Afzal et al. 2020), cucumber (Shi et al. 2006), and tomato (Jahan et al. 2019; Khan et al. 2013). It has been shown that stabilizing the activity of proton pumps in membranes, such as H^+- and Ca^{2+}-ATPase, by the application of 100 mM SA to grape leaves represents an additional mechanism for preserving membrane integrity under heat stress (Liu et al. 2008). As a result of SA's role in regulating stomatal aperture and other photosynthetic equipment, including PSII and Rubisco activity, photosynthesis is enhanced under abiotic stresses (Khan et al. 2013).

SA signaling of heat-stress-responsive genes such as NPR1 (nonexpresser of pathogenesis-related), HSPs (heat shock proteins), MBF1c (multiprotein bridging factor 1c), TGA, and PR-1 (pathogenesis-related protein 1) has been discovered by transcriptome study of plants during thermal tolerance (Larkindale et al. 2005; Larkindale and Knight 2002). Heat shock proteins (HSPs) are primarily important for defense against heat stress, and their production is triggered by exogenous administration of SA, as seen in *Arabidopsis thaliana* plants (Clarke et al. 2004), tomato (Cronjé and Bornman 1999; Snyman and Cronjé 2008), and rice (Chang et al. 2007). Contradictory findings were found in a research using transgenic Arabidopsis, suggesting that the molecular process still needs to be studied (Clarke et al. 2004). A lipid-associated enzyme implicated in intracellular signaling, PIP2-phospholipase C of pea was induced in response to heat by endogenous-free SA. Pea plants increase SA synthesis in response to heat stress; this in turn triggers the formation of PIP2-phospholipase C, a lipid-associated enzyme implicated in intracellular signaling (Liu et al. 2006). In response to heat stress, SA also induces the expression of the chitinase-1 gene in melons. In addition, it has been revealed that SA communicates with other plant signalings, including H(2)S, Ca^{2+}, IAA, and ABA (Li et al. 2015; Dinler et al. 2014). Treatment with SA, for instance, enhances the activity of L-cysteine desulfhydrase, a major enzyme in H(2)S production, suggesting that H(2)S may be a downstream signaling component in SA-induced heat tolerance.

7.4.2 Chilling

Damage caused by freezing temperatures is one of the primary factors that impedes the development and production of tropical and subtropical plant species. Numerous plant species, including maize (Janda et al. 1999), mountain rye (Ansari and Sharif-Zadeh 2012), watermelon (Cheng et al. 2016), beans (Gharib and Hegazi 2010), wheat (Ignatenko et al. 2019), and barley, have been shown to have SA play a regulatory role in their resistance to chilling stress. This has been reported in the scientific literature (Mutlu et al. 2013). The buildup of endogenous SA in *Arabidopsis thaliana* and wheat plants was stimulated by low temperatures, which further demonstrated the link between SA and cold stress responses (Scott et al. 2004).

The storage of fruits and vegetables at low temperatures is beneficial, but these temperatures may also cause chilling injuries to the produce. In a broad variety of fruits, the role of SA as a very effective buffering agent against cold stress has been extensively proven. Spraying banana seedlings with 0.5 mM SA, for instance, altered H_2O_2 metabolisms and enhanced the plants' freezing tolerance (Kang et al. 2003). Similar findings have been shown in cut flowers (Aghdam et al. 2016), bamboo shoots (Luo et al. 2012), and fruits including lemons (Siboza et al. 2014), cucumbers (Zhang et al. 2015), bell peppers (Ge et al. 2020), peaches (Wang et al. 2006), and pomegranates (Sayyari et al. 2009; Luo et al. 2011).

7.4.3 Salinity

Plants that are cultivated in saline soils are more likely to suffer from excessive osmotic and ionic stress, which may result in an ion imbalance and toxicity in plant cells. It has been discovered that salt stress can cause a decrease in the content of SA in plants, such as *Iris hexagona* (Wang et al. 2001), tomato (Molina et al. 2002), and soybean (Hamayun et al. 2010), whereas the application of SA increased tolerance to salt toxicity in many plant species, such as pepper, cucumber, and soybean (Farhangi-Abriz and Ghassemi-Golezani 2018).

The inflow and outflow of sodium are both regulated to a significant extent by SA. For instance, the incorporation of SA into the soil reduced the concentration of sodium ions, which in turn relieved the negative effects of salt on maize. Saffron was subjected to saline conditions, and an exogenous foliar injection of 1.5 mM SA resulted in a reduction in osmotic stress and an improvement in the aerial K^+/Na^+ ratio (Feizi et al. 2021). By collecting K^+ and Ca^{2+}, soaking the seeds of *Leymus chinensis* in SA solution reduced the amount of osmotic damage that was done to the plasma membrane. The activation of H^+-ATPase in the membrane (Ghassemi-Golezani and Farhangi-Abriz 2018) may be the cause of SA-signaled K^+ buildup. This activation takes place through guard cells outwardly rectifying K^+ channel (GORK), as was shown in *Arabidopsis thaliana* when it was subjected to salt stress. The salt overly sensitive (SOS) signaling pathway is activated when there is an increase in the amount of Ca^{2+} that enters the cytoplasm. This may cause the activation of the transport system that moves Na^+ and H^+ across the plasma membrane (Sun et al. 2010). In addition, it has been shown that applying SA helps to preserve the membrane's integrity by controlling suitable metabolites like proline and soluble sugars. This helps to ensure that the membrane remains intact. The pepper cells' membrane integrity was maintained while the proline content was raised after irrigation of the soil with the solution containing 1 mM SA. By elevating the levels of proline, soluble carbohydrates, and proteins in soybean leaves, exogenous SA is able to modify the amount of water that is contained inside the cells. It is possible that pretreatment with SA will produce a preadaptive response in the plant in the form of a transitory rise in H_2O_2 level. This transient increase in H_2O_2 level may work as a second messenger to "set up" the plant to protect itself against any subsequent salt stress that may occur. As has been shown in *Leymus chinensis* and *Iris pseudacorus*, pretreatment of plants with SA increases the activities of antioxidant enzymes, which in turn leads to a reduction in the amount of stress-induced oxidative stress (Liu et al. 2021). The signaling function of SA is also cross-linked with ABA, glycinebetaine, and ethylene (ET), all of which are highly connected with the production of stress proteins and the maintenance of leaf water potential. In addition, ET is directly correlated with the signaling role of glycinebetaine (Khan et al. 2014). The role of salicylic acid in plant abiotic stress tolerance is also represented in Fig. 7.2.

7.4.4 Metal Toxicity

Studies on metal phytotoxicity have been extensive in the field of plant biology recently. Roots readily take up heavy metals and carry them to the plant's upper parts, where they may cause growth inhibition, leaf chlorosis, wilting, and even cell death. There are several plant species that have reported SA to be useful in protecting against metal poisoning (Yadav et al. 2021). Using SA on Pb-stressed rice, Cu-stressed *Phaseolus vulgaris*, or Ni-stressed mustard all increased their growth and photosynthesis (Zaid et al. 2019). In more recent times, researchers have also assessed SA's co-reaction with other promoters. To illustrate, exposing maize to both SA and microorganisms that promote plant development mitigated the oxidative damage caused by Cr (Islam et al. 2016). In *Phaseolus vulgaris*, SA, whether combined with kinetin or calcium, reduced plant responses to nickel and lead stress (Khalil et al. 2021). Melatonin and SA supplementation together efficiently

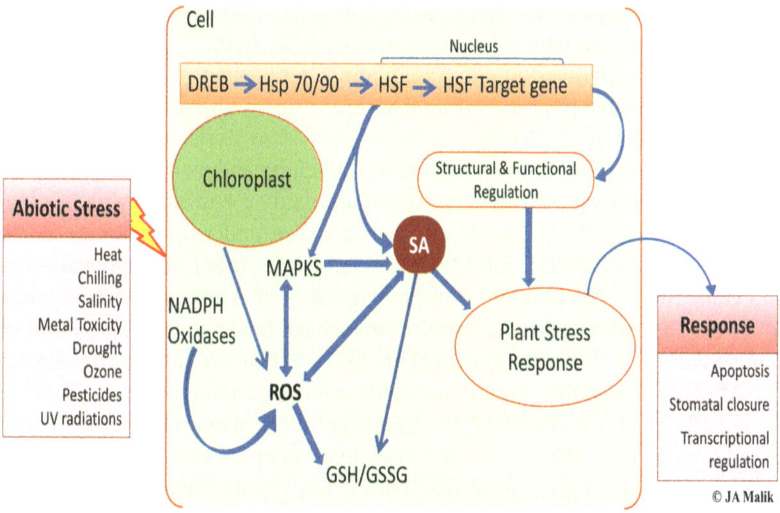

Fig. 7.2 Role of salicylic acid in plant abiotic stress tolerance

detoxified as toxicity in pepper plants by influencing their phytochelatin and nitrogen metabolism (Kaya et al. 2022).

Cadmium is one of the most pervasive and poisonous metals on the earth. It is the prototypical hazardous metal that may cause prototypical plant symptoms such elemental replacement and inactivation, protein structure destruction, and disruption of photosynthesis, respiration, and cell division (Guo et al. 2019). In a broad variety of plant species, SA has been found to play a crucial role in facilitating Cd tolerance throughout a wide range of activities, including but not limited to: plant development, element absorption, Cd translocation, photosynthesis, and senescence (Guo et al. 2019). Therefore, the emphasis of this metal toxicity article is on how SA and Cd interact in plants.

Cadmium's (Cd) phytotoxicity is a key topic of study in modern plant biology. New research shows that Cd stress significantly increases SA production in plants. For instance, the bound SA of maize increased by a factor of 10 after being exposed to 25 M Cd, compared to untreated plants (Krantev et al. 2008). Barley and *Pisum sativum* have both shown signs of this phenomenon, too (Metwally et al. 2005). Many aspects of plant development, absorption, translocation, photosynthesis, and senescence have been found to include SA in enhancing Cd tolerance, according to studies conducted on a broad variety of plant species (Guo et al. 2019).

7.4.5 Other Stresses

7.4.5.1 Drought

Many plant species, including barley (Bandurska 2005), *Phillyrea angustifolia*, and *Salvia officinalis*, show increased SA levels in response to drought stress. Increases in relative water and proline content, as well as regulation of other phytohormones, are all mechanisms by which SA mitigates drought harm (Arif et al. 2020). By upregulating genes involved in redox regulation and proline synthesis, for instance, pretreatment of SA reduced the drought-induced superoxide radical and improved proline content (Lee et al. 2019).

In addition to boosting photosynthetic effectiveness and membrane permeability, SA treatments also increased the activity of antioxidant enzymes, which mitigated the harmful impacts of drought. For instance, maize under drought stress showed significant reductions in ROS and MDA concentrations after foliar application of SA (Shemi et al. 2021). Wheat under drought stress had its membrane stability, chlorophyll content, and photosynthetic rates all improved thanks to the application of SA at a concentration of 100 mM (Ahmad et al. 2021). Wheat seedlings sprayed with SA at 0.5 mM showed significant increases in the activity of antioxidant enzymes (SOD, CAT, and PPO), which protected the plants from the negative effects of drought stress (Khalvandi et al. 2021). It was shown that under water-deficit circumstances, foliar spraying with SA greatly increased sweet basil plant growth and relative water contents. *Rosmarinus officinalis* L. essential oil production was boosted by 2% under moderate drought stress after being sprayed with 2 mM SA (Abbaszadeh et al. 2020). By preserving membrane integrity and carbonic anhydrase activities, which directly impact the rate of photosynthetic CO_2 fixation, SA treatment protected tomato plants against dry stress (Hayat et al. 2008). There seems to be cross-talk between SA and ABA under drought stress, since pretreatment with SA mitigated damage to cell membranes and enhanced ABA levels in barley and maize leaves (Tayyab et al. 2020).

7.4.5.2 Ozone

Reacting with lipids and proteins in plant cells, ozone causes oxidative damage because of its potency as an oxidizing agent (Wedow et al. 2021). Negative effects of ozone on NahG plants' SA insufficiency are mirrored in tobacco, where exposure to the gas increases SA accumulation and buttresses the plant against viral infection. Ozone stress has been demonstrated to increase SA buildup through the ICS route (Ogawa et al. 2007). By regulating cell redox and reducing chlorosis development in leaves, SA suppresses ET generation in ozone-exposed *Salvia officinalis*. But abnormally high or low amounts of SA lead to more ozone damage. Rice under ozone exposure had its stomatal conductance, chlorophyll content, and Mg assimilation reduced after being treated with exogenous SA. An intriguing new investigation has been undertaken to determine if SA, JA, or ethylene regulates O_3-induced cell death. Analysis of transcriptional alterations in single, double, and triple mutants demonstrated that basal SA levels are crucial for plants to guard against ROS-induced cell death, which works in tandem with ethylene and JA signaling (Kittipornkul et al. 2020).

7.4.5.3 Pesticide

Cucumber, pistachio plants (*Pista ciavera* L.), and barley have all shown signs of oxidative damage after being exposed to chemical pesticides like herbicide

(Homayoonzadeh et al. 2020). Paraquat (a herbicide) injury results in a continual production of superoxide in the chloroplasts of plant cells, which in turn encourages redox reaction chains, creates different types of ROS, and ultimately results in oxidative damage. Reduced glutathione (GSH) concentration in transgenic NahG rice plants indicates heightened susceptibility to paraquat, which leads to SA insufficiency. When *Vigna radiata* is treated with fungicide (mancozeb), insecticide (chlorpyrifos), or herbicide (metribuzin), SA dramatically raises enzymatic parameters and photosynthetic pigments (Fatma et al. 2018). Pesticide detoxification enzymes (GSTs: glutathione *S*-transferases; a carbon monoxide-bound enzyme, P450 (absorption band at 450 nm)) activity and expression are stimulated in thiram-treated leaves when they are pretreated with 1 mM SA (Nazish et al. 2022). Cucumbers treated with 1 mM SA prevent the accumulation and facilitate the breakdown of pesticides (Kusumi et al. 2006).

7.4.5.4 Ultraviolet Radiation

UV radiation is an important environmental signal that affects plant growth and development and may lower the prevalence of diseases and pests (Meyer et al. 2021). Excessive exposure, however, may damage cell membranes and proteins and directly trigger the generation of reactive oxygen species (ROS) since it exceeds the capacity of sunlight use in plants. Evidence suggests that through stimulating the production of antioxidant enzymes such POD, APX, CAT, and GR, SA protected pepper against the oxidative stress caused by exposure to ultraviolet radiation of various wavelengths (UV-A, UV-B, and UV-C) (Mahdavian et al. 2008). As a result of UV light activating SA defenses, the JA-deficient genotype of tomato was better able to withstand pathogen assault (Escobar Bravo et al. 2019). UV light, like ozone, causes tobacco to store more SA, and it also increases the activity of benzoic acid 2-hydroxylase, an enzyme necessary for the catalysis of SA production. Many plant species, including blue grass, soybeans, and maize, have had the detrimental effects of UV irradiation reduced with the use of exogenous SA (Liu et al. 2020). Possible functions include boosting photochemical efficiency, antioxidant enzyme activities, and anthocyanin and tocopherol concentration.

7.4.6 Possible Mechanisms of SA in Mitigating Abiotic Stresses

The amount of endogenous SA rises in response to any abiotic stress, and this basic molecule has been implicated in plant stress signaling, according to extensive study (Santisree et al. 2020). Multiple physiological systems, including as embryonic development, photosynthesis, ion assimilation, respiration, the antioxidant system, and interhormonal communication, influence SA's regulatory effects (Rivas-San Vicente and Plasencia 2011). In the initial study of SA signaling, researchers found that it modulates ROS generation and subsequently induces pathogenesis-related (PR1) expression in response to pathogen invasion (Rajendiran and Ramanujam 2003). This finding stimulated more research into the intricate plant signaling network including SA and ROS (Del Valle et al. 2020).

Using SA on tobacco plants has been known to increase their resistance to viral infection and stimulate the production of defense genes since the late 1970s. Upward trends in SA levels were first observed in tobacco and cucumber before the emergence of local and/or systemic disease resistance in 1990, demonstrating SA's importance as an internal indicator for disease resistance. The need of SA for PTI, ETI, and SAR was established by analyzing tobacco and Arabidopsis unable to accumulate SA (due to different mutations or expression of SA-degrading enzymes). Even while SA buildup in the uninfected leaves is necessary for SAR formation, grafting tests employing SA-deficient or wild-type (wt) tobacco showed that SA is not the mobile SAR-inducing signal that moves from the inoculation to the systemic leaves (Kang et al. 2003). The use of SA as a defensive signal has spread to a wide variety of plants. Some monocots and plants that constitutively accumulate large amounts of SA (potato and rice) have different opinions on its function, and its function seems to change depending on the pathogen (Aghdam et al. 2016).

It is important to remember that SA is just one of several plant hormones that communicate resistance to microbial diseases (Zhang et al. 2015). Infection by pathogens that feed on live tissue activates the SA-mediated defensive signaling system. Jasmonic acid (JA) and ethylene control a different defensive system triggered by necro-trophic infections, which feed on dead tissue. Cross-talk between the SA and JA/ethylene-mediated defense pathways is very active, and these interactions are often hostile.

7.5 Future Scope

Growers have been able to greatly improve both production and quality thanks to the advent of synthetic pesticides. Overuse, however, has led to disease resistance, and these substances are typically hazardous. Reducing crop loss by regulation of SAR is an ecologically preferable method. Although SA is a powerful SAR inducer, its phytotoxicity prevents it from being used extensively (Rao and Davis 1999). Defense gene expression and SAR to a wide variety of pathogens can be induced by a number of synthetic compounds, including 2,6-dichloroisonicotinic acid (INA), benzo (1,2,3) thiadiazole-7-carbothioic acid *S*-methyl ester (also called benzothiadiazole; BTH or acibenzolar-*S*-methyl), probenazole (PBZ), and its active metabolite 1,2-benzisothiazol (Xu et al. 2015). SAR is activated by PBZ and BIT, which are SA functional analogs that activate the SA signaling pathway upstream of SA (Dubey et al. 2015; Homayoonzadeh et al. 2021). Plants and suspension cells treated with high concentrations of SA or its functional analogs immediately trigger defenses, whereas low quantities elicit little or no response. In contrast, defenses activate more swiftly and/or forcefully in response to a recurrent infection (Huang et al. 2019). Priming is a phenomenon that happens in systemic leaves of plants that show SAR. A number of molecular processes, including the accumulation of transcripts and/or inactive versions of MAPKs, increased amounts of PRRs, and chromatin remodeling, are presumably involved in priming, although their precise roles have yet to be determined. It is possible that this latter process also helps pass on defensive primed.

Although the SA signaling system for plant disease resistance has been extensively studied, substantial gaps in our understanding persist. To what extent PRRs and R proteins translate pathogen recognition into activation of early cellular responses and SA production is not well understood. Likewise, the function of the ICS vs PAL pathways in many plant species, as well as the enzymes involved in SA production, remains unknown. Additionally, it is important to identify the mechanisms that control ICS1 expression in the nucleus and through retrograde signals from the chloroplast. The identification and evaluation of SA targets/receptors is a critical research direction for understanding the SA signaling system. Together, these studies and efforts detect SA in living organisms (in vivo) should provide light on the many ways in which SA acts. Agriculture stands to gain from a better understanding of the SA signaling pathway, since it will likely lead to the development of techniques for optimizing existing SAR-inducing drugs and new compounds that target as-yet undisclosed system components.

Finally, new possibilities for treating human illnesses have been revealed by the finding of many SA targets in animals (which may have evolved in response to low-level exposure to endogenously generated and/or dietary SA). The anti-inflammatory benefits of low-dose aspirin may be explained by the fact that SA blocks HMGB1 from doing its job. There is great potential for developing SA-based drugs that are even more effective and/or have fewer negative side effects, as evidenced by the discovery of natural and synthetic SA derivatives that suppress disease-associated activities of HMGB1 and GAPDH more effectively than SA (Abreu and Munné-Bosch 2008). In conclusion, there is a substantial possibility that SA's objectives in animals and plants will be substantially similar. Identifying these proteins will provide light on how SA causes its effects across all three kingdoms and may even lead to new approaches to halting harmful processes.

References

Abbaszadeh B, Layeghhaghighi M, Azimi R, Hadi N (2020) Improving water use efficiency through drought stress and using salicylic acid for proper production of Rosmarinus officinalis L. Ind Crop Prod 144:111893

Abreu ME, Munné-Bosch S (2008) Salicylic acid may be involved in the regulation of drought-induced leaf senescence in perennials: a case study in field-grown Salvia officinalis L. plants. Environ Exp Bot 64:105–112

Afzal I, Akram M, Rehman H, Rashid S, Basra S (2020) Moringa leaf and sorghum water extracts and salicylic acid to alleviate impacts of heat stress in wheat. S Afr J Bot 129:169–174

Aghdam MS, Jannatizadeh A, Sheikh-Assadi M, Malekzadeh P (2016) Alleviation of postharvest chilling injury in anthurium cut flowers by salicylic acid treatment. Sci Hortic 202:70–76

Ahmad A, Aslam Z, Naz M, Hussain S, Javed T, Aslam S, Raza A, Ali HM, Siddiqui MH, Salem MZ (2021) Exogenous salicylic acid-induced drought stress tolerance in wheat (Triticum aestivum L.) grown under hydroponic culture. PLoS One 16:e0260556

Ansari O, Sharif-Zadeh F (2012) Does gibberellic acid (GA), salicylic acid (SA) and ascorbic acid (ASc) improve mountain rye (Secale montanum) seeds germination and seedlings growth under cold stress. Int Res J Appl Basic Sci 3:1651–1657

Antithrombotic Trialists' (ATT) Collaboration (2009) Aspirin in the primary and secondary prevention of vascular disease: collaborative meta-analysis of individual participant data from randomised trials. Lancet 373:1849–1860

Arif Y, Sami F, Siddiqui H, Bajguz A, Hayat S (2020) Salicylic acid in relation to other phytohormones in plant: a study towards physiology and signal transduction under challenging environment. Environ Exp Bot 175:104040

Bandurska H (2005) The effect of salicylic acid on barley response to water deficit. Acta Physiol Plant 27:379–386

Bita C, Gerats T (2013) Plant tolerance to high temperature in a changing environment: scientific fundamentals and production of heat stress-tolerant crops. Front Plant Sci 4:273

Čajánek M, Štroch M, Lachetova I, Kalina J, Spunda V (1998) Characterization of the photosystem II inactivation of heat-stressed barley leaves as monitored by the various parameters of chlorophyll a fluorescence and delayed fluorescence. J Photochem Photobiol B Biol 47:39–45

Chang P-FL, Jinn T-L, Huang W-K, Chen Y, Chang H-M, Wang C-W (2007) Induction of a cDNA clone from rice encoding a class II small heat shock protein by heat stress, mechanical injury, and salicylic acid. Plant Sci 172:64–75

Cheng F, Lu J, Gao M, Shi K, Kong Q, Huang Y, Bie Z (2016) Redox signaling and CBF-responsive pathway are involved in salicylic acid-improved photosynthesis and growth under chilling stress in watermelon. Front Plant Sci 7:1519

Clarke SM, Mur LA, Wood JE, Scott IM (2004) Salicylic acid dependent signaling promotes basal thermotolerance but is not essential for acquired thermotolerance in Arabidopsis thaliana. Plant J 38:432–447

Cleland FC, Ajami A (1974) Identification of the flower-inducing factor isolated from aphid honeydew as being salicylic acid. Plant Physiol 54:904–906

Cronjé MJ, Bornman L (1999) Salicylic acid influences Hsp70/Hsc70 expression in Lycopersicon esculentum: dose-and time-dependent induction or potentiation. Biochem Biophys Res Commun 265:422–427

Cuzick J, Thorat MA, Bosetti C, Brown PH, Burn J, Cook NR, Ford LG, Jacobs EJ, Jankowski JA, LaVecchia C, Law M, Meyskens F, Rothwell PM, Senn HJ, Umar A (2015) Estimates of benefits and harms of prophylactic use of aspirin in the general population. Ann Oncol 26:47–57

Dat JF, Foyer CH, Scott IM (1998) Changes in salicylic acid and antioxidants during induced thermotolerance in mustard seedlings. Plant Physiol 118:1455–1461

Davies PJ (2004) The plant hormones: their nature, occurrence and function. In: Davies PJ (ed) Plant hormones: biosynthesis, signal transduction, action, 3rd edn. Kluwer Academic, Dordrecht, pp 1–15

Del Valle JC, Buide ML, Whittall JB, Valladares F, Narbona E (2020) UV radiation increases phenolic compound protection but decreases reproduction in Silene littorea. PLoS One 15: e0231611

Dinler B, Demir E, Kompe Y (2014) Regulation of auxin, abscisic acid and salicylic acid levels by ascorbate application under heat stress in sensitive and tolerant maize leaves. Acta Biol Hung 65:469–480

Dubey P, Mishra AK, Singh AK (2015) Comparative analyses of genotoxicity, oxidative stress and antioxidative defence system under exposure of methyl parathion and hexaconazole in barley (Hordeum vulgare L.). Environ Sci Pollut Res 22:19848–19859

Escobar Bravo R, Chen G, Grosser K, Van Dam NM, Leiss KA, Klinkhamer PG (2019) Ultraviolet radiation enhances salicylic acid-mediated defense signaling and resistance to Pseudomonas syringae DC3000 in a jasmonic acid-deficient tomato mutant. Plant Signal Behav 14:e1581560

Farhangi-Abriz S, Ghassemi-Golezani K (2018) How can salicylic acid and jasmonic acid mitigate salt toxicity in soybean plants? Ecotoxicol Environ Saf 147:1010–1016

Fatma F, Kamal A, Srivastava A (2018) Exogenous application of salicylic acid mitigates the toxic effect of pesticides in Vigna radiata (L.) Wilczek. J Plant Growth Regul 37:1185–1194

Feizi H, Moradi R, Pourghasemian N, Sahabi H (2021) Assessing saffron response to salinity stress and alleviating potential of gamma amino butyric acid, salicylic acid and vermicompost extract on salt damage. S Afr J Bot 141:330–343

Ge W, Zhao Y, Kong X, Sun H, Luo M, Yao M, Wei B, Ji S (2020) Combining salicylic acid and trisodium phosphate alleviates chilling injury in bell pepper (Capsicum annuum L.) through enhancing fatty-acid desaturation efficiency and water retention. Food Chem 327:127057

Gharib F, Hegazi A (2010) Salicylic acid ameliorates germination, seedling growth, phytohormone and enzymes activity in bean (Phaseolus vulgaris L.) under cold stress. J Am Sci 6:675–683

Ghassemi-Golezani K, Farhangi-Abriz S (2018) Foliar sprays of salicylic acid and jasmonic acid stimulate H+-ATPase activity of tonoplast, nutrient uptake and salt tolerance of soybean. Ecotoxicol Environ Saf 166:18–25

Guo B, Liu C, Liang Y, Li N, Fu Q (2019) Salicylic acid signals plant defence against cadmium toxicity. Int J Mol Sci 20:2960

Hamayun M, Khan SA, Khan AL, Shinwari ZK, Hussain J, Sohn E-Y, Kang S-M, Kim Y-H, Khan MA, Lee I-J (2010) Effect of salt stress on growth attributes and endogenous growth hormones of soybean cultivar Hwangkeumkong. Pak J Bot 42:3103–3112

Hayat S, Hasan SA, Fariduddin Q, Ahmad A (2008) Growth of tomato (Lycopersicon esculentum) in response to salicylic acid under water stress. J Plant Interact 3:297–304

Homayoonzadeh M, Moeini P, Talebi K, Roessner U, Hosseininaveh V (2020) Antioxidant system status of cucumber plants under pesticides treatment. Acta Physiol Plant 42:1–11

Homayoonzadeh M, Hosseininaveh V, Haghighi SR, Talebi K, Roessner U, Maali-Amiri R (2021) Evaluation of physiological and biochemical responses of pistachio plants (Pistacia vera L.) exposed to pesticides. Ecotoxicology 30:1084–1097

Horvath E, Szalai G, Janda T (2007) Induction of a biotic stress tolerance by salicylic acid signaling. J Plant Growth Regul 26(3):290–300

Huang Y, Adeleye AS, Zhao L, Minakova AS, Anumol T, Keller AA (2019) Antioxidant response of cucumber (Cucumis sativus) exposed to nano copper pesticide: quantitative determination via LC-MS/MS. Food Chem 270:47–52

Ignatenko A, Talanova V, Repkina N, Titov A (2019) Exogenous salicylic acid treatment induces cold tolerance in wheat through promotion of antioxidant enzyme activity and proline accumulation. Acta Physiol Plant 41:1–10

Islam F, Yasmeen T, Arif MS, Riaz M, Shahzad SM, Imran Q, Ali I (2016) Combined ability of chromium (Cr) tolerant plant growth promoting bacteria (PGPB) and salicylic acid (SA) in attenuation of chromium stress in maize plants. Plant Physiol Biochem 108:456–467

Jahan MS, Wang Y, Shu S, Zhong M, Chen Z, Wu J, Sun J, Guo S (2019) Exogenous salicylic acid increases the heat tolerance in tomato (Solanum lycopersicum L.) by enhancing photosynthesis efficiency and improving antioxidant defense system through scavenging of reactive oxygen species. Sci Hortic 247:421–429

Janda T, Szalai G, Tari I, Paldi E (1999) Hydroponic treatment with salicylic acid decreases the effects of chilling injury in maize (Zea mays L.) plants. Planta 208:175–180

Kang G, Wang C, Sun G, Wang Z (2003) Salicylic acid changes activities of H_2O_2-metabolizing enzymes and increases the chilling tolerance of banana seedlings. Environ Exp Bot 50:9–15

Kaya C, Sarıoglu A, Ashraf M, Alyemeni MN, Ahmad P (2022) The combined supplementation of melatonin and salicylic acid effectively detoxifies arsenic toxicity by modulating phytochelatins and nitrogen metabolism in pepper plants. Environ Pollut 297:118727

Khalil R, Haroun S, Bassyoini F, Nagah A, Yusuf M (2021) Salicylic acid in combination with kinetin or calcium ameliorates heavy metal stress in Phaseolus vulgaris plant. J Agric Food Res 5:100182

Khalvandi M, Siosemardeh A, Roohi E, Keramati S (2021) Salicylic acid alleviated the effect of drought stress on photosynthetic characteristics and leaf protein pattern in winter wheat. Heliyon 7:e05908

Khan MIR, Iqbal N, Masood A, Per TS, Khan NA (2013) Salicylic acid alleviates adverse effects of heat stress on photosynthesis through changes in proline production and ethylene formation. Plant Signal Behav 8:e26374

Khan MIR, Asgher M, Khan NA (2014) Alleviation of salt-induced photosynthesis and growth inhibition by salicylic acid involves glycinebetaine and ethylene in mungbean (Vigna radiata L.). Plant Physiol Biochem 80:67–74

Kittipornkul P, Treesubsuntorn C, Thiravetyan P (2020) Effect of exogenous catechin and salicylic acid on rice productivity under ozone stress: the role of chlorophyll contents, lipid peroxidation, and antioxidant enzymes. Environ Sci Pollut Res 27:25774–25784

Kolbe H, Lautemann E (1860) Uber die constitution und basicitatdersalicylsaure. Justus Liebigs Ann Chem 115:157–206

Krantev A, Yordanova R, Janda T, Szalai G, Popova L (2008) Treatment with salicylic acid decreases the effect of cadmium on photosynthesis in maize plants. J Plant Physiol 165:920–931

Kumar D (2014) Salicylic acid signaling in disease resistance. Plant Sci 228:127–134

Kusumi K, Yaeno T, Kojo K, Hirayama M, Hirokawa D, Yara A, Iba K (2006) The role of salicylic acid in the glutathione-mediated protection against photooxidative stress in rice. Physiol Plant 128:651–661

Larkindale J, Knight MR (2002) Protection against heat stress-induced oxidative damage in Arabidopsis involves calcium, abscisic acid, ethylene, and salicylic acid. Plant Physiol 128: 682–695

Larkindale J, Hall JD, Knight MR, Vierling E (2005) Heat stress phenotypes of Arabidopsis mutants implicate multiple signaling pathways in the acquisition of thermotolerance. Plant Physiol 138:882–897

Lee B-R, Islam MT, Park S-H, Jung H-i, Bae D-W, Kim T-H (2019) Characterization of salicylic acid-mediated modulation of the drought stress responses: reactive oxygen species, proline, and redox state in Brassica napus. Environ Exp Bot 157:1–10

Leroux H (1830) Discovery of salicine. J Chim Med 6:340–432

Li Z-G, Xie L-R, Li X-J (2015) Hydrogen sulfide acts as a downstream signal molecule in salicylic acid-induced heat tolerance in maize (Zea mays L.) seedlings. J Plant Physiol 177:121–127

Liu H-T, Huang W-D, Pan Q-H, Weng F-H, Zhan J-C, Liu Y, Wan S-B, Liu Y-Y (2006) Contributions of PIP2-specific-phospholipase C and free salicylic acid to heat acclimation-induced thermotolerance in pea leaves. J Plant Physiol 163:405–416

Liu Y, Zhang J, Liu H, Huang W (2008) Salicylic acid or heat acclimation pre-treatment enhances the plasma membrane-associated ATPase activities in young grape plants under heat shock. Sci Hortic 119:21–27

Liu T, Yuan C, Gao Y, Luo J, Yang S, Liu S, Zhang R, Zou N (2020) Exogenous salicylic acid mitigates the accumulation of some pesticides in cucumber seedlings under different cultivation methods. Ecotoxicol Environ Saf 198:110680

Liu T, Li T, Zhang L, Li H, Liu S, Yang S, An Q, Pan C, Zou N (2021) Exogenous salicylic acid alleviates the accumulation of pesticides and mitigates pesticide-induced oxidative stress in cucumber plants (Cucumis sativus L.). Ecotoxicol Environ Saf 208:111654

Luo Z, Chen C, Xie J (2011) Effect of salicylic acid treatment on alleviating postharvest chilling injury of 'Qingnai' plum fruit. Postharvest Biol Technol 62:115–120

Luo Z, Wu X, Xie Y, Chen C (2012) Alleviation of chilling injury and browning of postharvest bamboo shoot by salicylic acid treatment. Food Chem 131:456–461

Mahdavian K, Ghorbanli M, Kalantari KM (2008) Role of salicylic acid in regulating ultraviolet radiation-induced oxidative stress in pepper leaves. Russ J Plant Physiol 55:560–563

Métraux JP, Raskin I (1993) Role of phenolics in plant disease resistance. In: Chet I (ed) Biotechnology in plant disease control. Wiley, New York, pp 191–209

Metwally A, Safronova VI, Belimov AA, Dietz K-J (2005) Genotypic variation of the response to cadmium toxicity in Pisum sativum L. J Exp Bot 56:167–178

Meyer P, Van de Poel B, De Coninck B (2021) UV-B light and its application potential to reduce disease and pest incidence in crops. Hortic Res 8:194

Molina A, Bueno P, Marín MC, Rodríguez-Rosales MP, Belver A, Venema K, Donaire JP (2002) Involvement of endogenous salicylic acid content, lipoxygenase and antioxidant enzyme activities in the response of tomato cell suspension cultures to NaCl. New Phytol 156:409–415
Mutlu S, Karadağoğlu Ö, Atici Ö, Nalbantoğlu B (2013) Protective role of salicylic acid applied before cold stress on antioxidative system and protein patterns in barley apoplast. Biol Plant 57: 507–513
Nazish T, Huang Y-J, Zhang J, Xia J-Q, Alfatih A, Luo C, Cai X-T, Xi J, Xu P, Xiang C-B (2022) Understanding paraquat resistance mechanisms in Arabidopsis thaliana to facilitate developing paraquat-resistant crops. Plant Commun 3:100321
Ogawa D, Nakajima N, Tamaoki M, Aono M, Kubo A, Kamada H, Saji H (2007) The isochorismate pathway is negatively regulated by salicylic acid signaling in O_3-exposed Arabidopsis. Planta 226:1277–1285
Pan Q, Zhan J, Liu H, Zhang J, Chen J, Wen P, Huang W (2006) Salicylic acid synthesized by benzoic acid 2-hydroxylase participates in the development of thermotolerance in pea plants. Plant Sci 171:226–233
Rajendiran K, Ramanujam M (2003) Alleviation of ultraviolet-B radiation-induced growth inhibition of green gram by triadimefon. Biol Plant 46:621–624
Rao MV, Davis KR (1999) Ozone-induced cell death occurs via two distinct mechanisms in Arabidopsis: the role of salicylic acid. Plant J 17:603–614
Rivas-San Vicente M, Plasencia J (2011) Salicylic acid beyond defence: its role in plant growth and development. J Exp Bot 62:3321–3338
Santisree P, Jalli LCL, Bhatnagar-Mathur P, Sharma KK (2020) Emerging roles of salicylic acid and jasmonates in plant abiotic stress responses. In: Protective chemical agents in the amelioration of plant abiotic stress. Wiley, Hoboken, NJ, pp 342–373
Sayyari M, Babalar M, Kalantari S, Serrano M, Valero D (2009) Effect of salicylic acid treatment on reducing chilling injury in stored pomegranates. Postharvest Biol Technol 53:152–154
Scott IM, Clarke SM, Wood JE, Mur LA (2004) Salicylate accumulation inhibits growth at chilling temperature in Arabidopsis. Plant Physiol 135:1040–1049
Shemi R, Wang R, Gheith E-S, Hussain HA, Hussain S, Irfan M, Cholidah L, Zhang K, Zhang S, Wang L (2021) Effects of salicylic acid, zinc and glycine betaine on morpho-physiological growth and yield of maize under drought stress. Sci Rep 11:3195
Shi Q, Bao Z, Zhu Z, Ying Q, Qian Q (2006) Effects of different treatments of salicylic acid on heat tolerance, chlorophyll fluorescence, and antioxidant enzyme activity in seedlings of Cucumis sativa L. Plant Growth Regul 48:127–135
Siboza XI, Bertling I, Odindo AO (2014) Salicylic acid and methyl jasmonate improve chilling tolerance in cold-stored lemon fruit (Citrus limon). J Plant Physiol 171:1722–1731
Snyman M, Cronjé M (2008) Modulation of heat shock factors accompanies salicylic acid-mediated potentiation of Hsp70 in tomato seedlings. J Exp Bot 59:2125–2132
Sun J, Wang MJ, Ding MQ, Deng SR, Liu MQ, Lu CF, Zhou XY, Shen X, Zheng XJ, Zhang ZK (2010) H_2O_2 and cytosolic Ca^{2+} signals triggered by the PM H^+-coupled transport system mediate K^+/Na^+ homeostasis in NaCl-stressed Populus euphratica cells. Plant Cell Environ 33:943–958
Tayyab N, Naz R, Yasmin H, Nosheen A, Keyani R, Sajjad M, Hassan MN, Roberts TH (2020) Combined seed and foliar pre-treatments with exogenous methyl jasmonate and salicylic acid mitigate drought-induced stress in maize. PLoS One 15:e0232269
Vlot AC, Dempsey DA, Klessig DF (2009) Salicylic acid, a multifaceted hormone to combat disease. Annu Rev Phytopathol 47:177–206
Wang L-J, Li S-H (2006a) Salicylic acid-induced heat or cold tolerance in relation to Ca^{2+} homeostasis and antioxidant systems in young grape plants. Plant Sci 170:685–694
Wang L-J, Li S-H (2006b) Thermotolerance and related antioxidant enzyme activities induced by heat acclimation and salicylic acid in grape (Vitis vinifera L.) leaves. Plant Growth Regul 48: 137–144

Wang Y, Mopper S, Hasenstein KH (2001) Effects of salinity on endogenous ABA, IAA, JA, and SA in Iris hexagona. J Chem Ecol 27:327–342

Wang L, Chen S, Kong W, Li S, Archbold DD (2006) Salicylic acid pretreatment alleviates chilling injury and affects the antioxidant system and heat shock proteins of peaches during cold storage. Postharvest Biol Technol 41:244–251

Wang L-J, Fan L, Loescher W, Duan W, Liu G-J, Cheng J-S, Luo H-B, Li S-H (2010) Salicylic acid alleviates decreases in photosynthesis under heat stress and accelerates recovery in grapevine leaves. BMC Plant Biol 10:34

Wassie M, Zhang W, Zhang Q, Ji K, Cao L, Chen L (2020) Exogenous salicylic acid ameliorates heat stress-induced damages and improves growth and photosynthetic efficiency in alfalfa (Medicago sativa L.). Ecotoxicol Environ Saf 191:110206

Wedow JM, Ainsworth EA, Li S (2021) Plant biochemistry influences tropospheric ozone formation, destruction, deposition, and response. Trends Biochem Sci 46:992–1002

Weissmann G (1991) Aspirin. Sci Am 264:84–90. https://doi.org/10.1038/scientificamerican0191-84

Widiastuti A, Yoshino M, Hasegawa M, Nitta Y, Sato T (2013) Heat shock-induced resistance increases chitinase-1 gene expression and stimulates salicylic acid production in melon (Cucumis melo L.). Physiol Mol Plant Pathol 82:51–55

Xu E, Vaahtera L, Brosché M (2015) Roles of defense hormones in the regulation of ozone-induced changes in gene expression and cell death. Mol Plant 8:1776–1794

Yadav V, Arif N, Kováč J, Singh VP, Tripathi DK, Chauhan DK, Vaculík M (2021) Structural modifications of plant organs and tissues by metals and metalloids in the environment: a review. Plant Physiol Biochem 159:100–112

Yang J, Duan L, He H, Li Y, Li X, Liu D, Wang J, Jin G, Huang S (2022) Application of exogenous KH_2PO_4 and salicylic acid and optimization of the sowing date enhance rice yield under high-temperature conditions. J Plant Growth Regul 41:1–15

Zaid A, Mohammad F, Wani SH, Siddique KM (2019) Salicylic acid enhances nickel stress tolerance by up-regulating antioxidant defense and glyoxalase systems in mustard plants. Ecotoxicol Environ Saf 180:575–587

Zhang Y, Zhang M, Yang H (2015) Postharvest chitosan-g-salicylic acid application alleviates chilling injury and preserves cucumber fruit quality during cold storage. Food Chem 174:558–563

Role of Plant Phenolics in the Resistance Mechanism of Plants Against Insects

8

Parvaiz Yousuf, Shahid Razzak, Semran Parvaiz, Younis Ahmad Rather, and Rafiq Lone

Abstract

Plant phenolics are secondary metabolites that are produced by plants for numerous reasons. These include several classes of structurally diverse molecules produced from shikimate–phenylpropanoids–flavonoids pathways. The phenolic compounds have diverse roles in plant growth, resistance to pathogens, pigmentation, reproduction, and many other roles. Numerous plant phenolics are helpful in resistance against insects and other pests. The phenolic compounds are adaptations acquired by plants during natural selection. Since plants cannot move physically, so they need a chemical defense against insects and pests. This is where the role of secondary metabolites comes into play. This chapter deals with the classification, synthesis, and role of plant phenolics against insects and other pests.

Keywords

Plant phenolics · Plant defense · Indirect defense · Secondary metabolites · Phenolics

P. Yousuf · S. Razzak · S. Parvaiz
Department of Zoology, Central University of Kashmir, Ganderbal, Jammu and Kashmir, India

Y. A. Rather
Department of Zoology, Government Degree College, Ramban, Jammu and Kashmir, India

R. Lone (✉)
Department of Botany, Central University of Kashmir, Ganderbal, Jammu and Kashmir, India

R. Lone et al. (eds.), *Plant Phenolics in Biotic Stress Management*,
https://doi.org/10.1007/978-981-99-3334-1_8

8.1 Introduction

The family of molecules known as phenolic compounds is considered to be secondary metabolites. They are among the most common and widespread types of molecules that can be discovered in plants. According to Harborne (1989), the chemical definition of a polyphenol or phenolic is a substance with an aromatic ring carrying one (phenol) or more (polyphenol) hydroxyl substituents, as well as functional derivatives (methyl ethers, esters, glycosides, and so on). Polyphenols and phenolics are names that, as a general rule, refer to all natural secondary metabolites that arise biogenetically from a pathway called shikimate–phenylpropanoids–flavonoids pathways. The vast majority of phenolic compounds possess two or more hydroxyl groups, but phenol itself is a natural product that does not require any additions. In general, phenolic compounds found in plants are soluble in polar organic solvents. However, this is not the case if the compounds have been completely esterified, glycosylated, or esterified. The majority of phenolic glycosides are soluble in water, in contrast to their corresponding aglycones, which are not water-soluble. With a few notable exceptions, the solubility in water increases proportionately with the number of hydroxyl groups present in a molecule. Because phenolics are more likely to undergo oxidation in an alkaline environment, the alkaline solvent treatment should either take place in the presence of nitrogen gas or be wholly avoided. Even though they are insoluble in water, phenolics with a low number of hydroxyl groups can be dissolved in other solvents such as ethyl acetate, ether, methanol, chloroform, and ethanol (Van Sumere 1989).

The most common solvents for dissolving phenolic compounds in analytical applications include water, ethanol, methanol, and alcohol–water combinations. Some other widespread solvents are also: Ultraviolet radiation is strongly absorbed by phenolic compounds, especially cultured phenolic compounds. This holds the truest in the ultraviolet part of the spectrum. Different phenolic compounds have different levels of permeability. The peak of absorption for phenolic acids and phenols, for example, is between 250 and 290 nm, whereas those of cinnamic acid derivatives are between 290 and 330 nm, while those of flavonols and flavones are between 250 and 350 nm, while those of chalcones and aurones are above 350 nm. The optical absorption patterns of plant defense 25, phenolics, and betacyanins are strikingly identical to one another (270–275 nm and 535–545 nm, respectively) (Harborne 1964; Mabry et al. 2012).

Plants rely on phenolic compounds for a wide variety of functions, including reproduction, growth, disease resistance, and pigmentation. These compounds make up a sizable portion of a very influential group of secondary metabolites. There are many thousands of unique chemicals, including numerous types of phenolic compounds (dimeric, monomeric, and polymeric) and roughly 8150 different flavonoids. The term "secondary metabolism" is used to describe compounds that are found in a specialized type of cells but are not explicitly fundamental for basic respiratory or photosynthetic metabolism but are thought to be necessary for plants to survive in their environment, in contrast to "basic metabolism," which refers to the anabolic and catabolic processes necessary for cell proliferation and maintenance.

The activities essential for the survival and growth of cells are referred to as "basic metabolism." It is now generally accepted that the raison d'être idea of so-called secondary metabolites are merely waste products of primary metabolism that build up in plant cells due to the absence of an efficient excretory system which fails to explain why plants produce these compounds (Whiting 2001). This is something that has been proven beyond a reasonable doubt in recent times.

Secondary metabolites appear to play a role in defense (against viruses, herbivores, competing plants or viruses) and signal chemicals (to attract seed-dispersing animals or pollination), in addition to protecting the plant from damage caused by oxidants and UV radiation (Kutchan 2001). As a consequence of this, they serve as surrogates for characteristics that have been selected as part of the process through which evolution makes organisms more adaptable. Plants have gathered such a broad chemical arsenal as a result of the necessity for secondary metabolites to execute a wide variety of biological tasks. This desire has caused plants to diversify their chemical composition. The genetic complement of *Drosophila melanogaster* in the fruit fly is significantly smaller than that of plants, with estimates ranging from 20,000 to 60,000 genes in plant genomes. It is estimated that among these genes, only approximately 15% to 20% are responsible for encoding enzymes involved in secondary metabolism (13,601 predicted genes).

It's possible that the difference in genetic and biological complexity can be explained by the different ways in which plants and animals respond to dangers like predation, disease, and environmental stress. Animals are able to avoid potentially hazardous situations because they have developed immune systems and nervous systems that allow them to recognize danger and react appropriately to it. This enables them to escape from potentially hazardous scenarios. Plants, on the other hand, are firmly attached to the soil and are unable to run away from the abiotic and biotic stresses that they face; as a result, they are forced to stay put and adopt preventative measures. The production of chemicals is one method of defense among many. Vincenzo Lattanzio et al. have developed a substance that can be applied to protect against or get rid of hazardous organisms such as pathogens. The distribution of secondary metabolites in any given plant is dynamic and depends on a number of factors, including the stage of development the plant is in, the type of tissue it contains, the function of its organs (for example, the organs responsible for reproduction and survival contain the highest concentrations of secondary metabolites), and the genetic variation that exists both within and between populations (Pichersky and Gang 2000; Lattanzio 2003).

The majority of botanists working during this time period held the view that adaptive and evolutionary explanations were insufficient grounds for accepting the defense hypothesis. In an effort to reduce the complexity of the situation, botanists frequently made the assumption that structural variation was nothing more than a reflection of nature's playful character and that the secondary metabolites were the consequences of primary metabolism. In spite of the fact that the application of plant secondary metabolites in taxonomy has been restricted to the twentieth century, and more especially to the last 40 years, their potential value has been recognized for approximately the previous 200 years. When it comes to determining phylogenetic relationships, the use of secondary compounds has a number of distinct advantages over the use of primary compounds. This is due to the fact that differences in the complement of secondary compounds are qualitative differences rather than quantitative changes that are controlled by environmental and genetic factors. In the presence of a shared pattern of secondary chemicals, one might find more convincing evidence of shared ancestry than morphological similarities attributable to common ancestry or convergent evolution (Perrino et al. 1989; Lattanzio et al. 1996).

8.2 Classification of Plant Phenolics

According to Harborne (1989), these phenolic compounds that can be found in plants can be categorized as belonging to one of the following groups. The fact that C6-simple phenols and benzoquinones both have a single benzene ring can be credited with the beneficial effects they have on one's health. For instance, Embelin, a plant benzoquinone with an action that inhibits the development of sperm, was isolated from the seeds of the Embelia Ribes plant (Kumar et al. 2011).

C6–C1 are phenolic acids with only one carboxylic acid functional group (Table 8.1). There are two distinct carbon frameworks found in these naturally occurring phenolic acids: hydroxybenzoic acid and hydroxycinnamic acid. While both molecules have the same general structure, the position and quantity of hydroxyl groups change between the two types. Cinnamic acid, ferulic acid, coumaric acid, caffeic acid, and sinapic acid are phenolic acids that complement hydroxycinnamic acid effectively. All of the following acids have hydroxybenzoic acid in their structures: gallic acid, benzoic acid, vanillic acid, *p*-hydroxybenzoic acid, syringic acid, protocatechuic acid, gentisic acid, salicylic acid, and veratric acid (Caballero et al. 2003). Foods such as beef, apple, apricot, cheese, and cauliflower contain the phenolic compound C6–C2-acetophenone. It can be found in gum and perfumes. The active auxin phenylacetic acid is a fruit hormone (Wightman and

Table 8.1 Classification of phenolic compounds on the basis of number of carbon atoms

Class	Basic skeleton	Number of carbon atoms
Benzoquinones, simple phenols	C6	6
Phenolic acids	C6–C1	7
Phenylacetic acid	C6–C2	8
Hydroxycinnamic acid, coumarin, polypropene, Isocoumarin	C6–C3	9
Naphthoquinone	C6–C4	10
Xanthone	C6–C1–C6	13
Stilbene, anthrachinone	C6–C2–C6	14
Flavonoids, isoflavonoids	C6–C3–C6	15
Lignans, neolignans	$(C6–C3)_2$	18

Lighty 1982). *Dipteryx odorata* (Sarker and Nahar 2017) possesses a particularly high concentration of the chemical compound C6–C3-Coumarins, which are produced by plants as an antipredator chemical. They are found in nearly all types of grass and clover. Insecticides produced from Calceolaria Andina, notably C6–C4-Naphthoquinones such as naphthazarin and 2-hydroxynaphthoquinones, are effective against tobacco crop pests (Khambay and Jewess 2000).

There is evidence that derivatives of 1,4-naphthoquinone have anticancer and antibacterial properties. Naphthoquinones have demonstrated molluscicidal and larvicidal properties. They are superior to the malaria parasite *Biomphalaria glabrata* and the *Aedes aegypti* mosquito (Ribeiro et al. 2009). The families Clusiaceae and Bonnetiaceae contain C6–C1–C6 Xanthones. Their application is extensive as an ovicide and pesticide against codling moth eggs (Sh 2011).

As a result of stilbenoids' hydroxylation, C6–C2–C6-Stilbenes serve as phytoalexins in plants. Stilbene-structured plant compounds include *trans*-piceid, *trans*-resveratrol, piceatannol, pinosylvin, *trans*-pterostilbene, pinosylvin, rhapontin, and astringin (Chong et al. 2009). Anthraquinones are typically found as glycosides in plants. C6–C3–C6-Flavonoids are often glycosylated derivatives in plants. The fundamental flavonoid structure is comprised of a flavan nucleus. There are numerous types of flavonoids, including flavan-3-ols (epicatechin, catechin, epigallocatechin), flavones (luteolin, apigenin, chrysin), flavonols (fisetin, quercetin, myricetin, kaempferol, galangin), anthocyanidins (cyanin, cyanidin, delphinidin, peonidin, malvidin, and pelargonidin), flavones (luteolin, apigenin, chrysin), and flavanonol (taxifolin) (Caballero et al. 2003). 2-Lignan flavonoids (C6–C3) are a type of phytonutrient with antioxidant effects. Examples of such chemicals are pinoresinol, podophyllotoxin, and steganacin. Both flax and sesame seeds are rich in lignans. (C6–C3–C6) 2-Bioflavonoids are produced by the oxidative coupling of two chalcone units, followed by the modification of the core C3 units. Multiple flowering plants, the gymnosperm order Psilotales and the selaginellalean order Selaginales possess them. They are neither members of the Pinaceae nor the Gnetales. These bioflavonoids hinder insect feeding and impede fungal growth (Iwashina 2003). Polymers of phenolic chemicals include (C6–C3) *n*-lignin and tannins. Lignin is the most prevalent phenolic chemical found within plant cell walls. The three most prevalent kinds of tannin found in plants are ellagitannins, condensed tannins, and gallotannins. The presence of these tannins makes plants less palatable to herbivores. Antioxidant capacity is the essential characteristic of phenols. Thus, the organism is protected from potentially dangerous reactive oxygen species (ROS). Plant-based polyphenols serve as antioxidants, reducing agents, hydrogen donors, and singlet oxygen quenchers, among other functions.

8.3 Role of Phenolic Compounds

The occurrence of phenolic compounds is phylum-specific, yet they are found throughout the plant kingdom. Although the entire spectrum of polyphenols has been detected in vascular plants, bryophytes are regular makers of polyphenols, including flavonoids. The phenolics covered in this article are those that survived natural selection and subsequently evolved in shape and function. In the process of recognizing and distinguishing plant species, taxonomists utilize them frequently. In order to protect themselves from predators, plants have developed a chemical defense consisting of phenolic compounds. It has been demonstrated that plant phenolics serve largely ecological functions, with some performing double or even triple duty. Numerous studies have demonstrated that phenolic compounds and the enzymes responsible for their formation are separated along discrete pathways (Lattanzio 2013).

These phenolics are present in the entire plant. They typically congregate in the core vacuoles of guard cells, epidermal cells, and subepidermal cells in leaves and shoots. Some reside in waxes or on the surface of plant organs, while others are covalently bound to the plant cell wall. Some study indicates that flavonoids are deposited in the nucleus of trees (Falcone Ferreyra et al. 2012). It induces the formation of a molecule between flavonoid and DNA that protects both from oxidative stress. Beninger et al. (2004) performed phenolics synthesized during normal growth of plant tissues, and Harborne (1989) induced phenolics synthesized by plants in response to infection, physical injury, and elicitors such as UV light, temperature, heavy metal salts, etc.

8.3.1 Phenolics in Plant Development

Phenols play a crucial role in plant development by contributing to the production of cell walls. The insoluble or cell wall portion largely comprised ferulic acid and *p*-coumaric acid as esters of hydroxycinnamic acids. Wall-bound phenolic acid pools may be a source of phenylpropanoid building blocks for lignin biosynthesis or the initial phases of the lignification process. These ester populations with numerous attached molecules alter plant growth, turgor pressure, water flux, and cell wall structure as a result of the transduction of light energy. Auxin, a phytohormone, plays a crucial role in the regulation of plant growth (Zetterberg and Öfverholm 1999). It is commonly believed that plants are rooted, stationary organisms that can only move in limited ways. All leguminous plants displayed nyctinasty, a circadian rhythmic leaf movement. It is believed that Schildknecht's turgorins regulate the nyctinastic motions of plants by forcing the pulvini (an organ at the leaf joint) to shrink and swell, causing the leaves to close and open (Schildknecht 1984).

There is a novel class of phytohormones known as turgorins that regulate the turgor of plants. Gentisic acid 5-*O*—D-glucopyranoside and gallic acid 4-*O*-(-D-glucopyranosyl-6′-sulfate), both pulvini-localized in *Mimosa pudica* L., as well as *cis*-*p*-coumaric acid and cinnamic acid—have been identified as phenolic turgorins. Active chemicals have been identified from both *Cassia mimosoides* L. (4-*O*-D-glucopyranoside) and Albizia julibrissin Durazz (*cis*-*p*-coumaroylagmatine). Consequently, phenols also aid plant migration (Ueda et al. 1998). Rapidly germinating seeds exhibit greater quantities of coumaric acid-glucoside, whereas nongerminating seeds, such as those of *Melilotus alba*, have a high concentration of free coumarins

(Van Sumere et al. 1965). Another naturally occurring phenolic compound that inhibits seed germination is ferulic acid. These phenolics inhibit germination because they hinder the delivery of amino acids to the seeds and the formation of proteins (Brglez Mojzer et al. 2016).

8.3.2 Phenolics in Plant Signaling

Allelochemicals are known to have a communication effect between plants. Numerous organic and inorganic nutrients in the vicinity of phenolic compounds are impacted. Complicated is their effects on spore germination, which in turn influence the decomposition rate and the nutrient cycle. Phenols, which are present in all plant parts, are regarded as particularly valuable allelochemicals. *p*-Coumaric acid and *p*-hydroxybenzoic acid are examples of phenolic allelochemicals found in leaves. Quercetin, sorgoleone, catechin, and juglone, are phenolic allelochemicals found in the rhizosphere, bark, and root exudates, respectively (Guntzer et al. 2012).

When polyphenols in vacuoles combine with cytoplasmic proteins, a complex is created called the polyphenols–proteins complex (Lopes da Silva et al. 2007). A complex that promotes the senescence of plant tissues is responsible for the browning of ageing leaves. The analysis of pea root exudates revealed that flavonoids such as apigenin-7-*O*-glucoside and eriodictyol stimulate nod gene expression (Begum et al. 2001). Other isoflavonoids, including hesperidin, naringenin, isoflavones, and chalcones, such as daidzein and genistein, are also released by legume plants and promote nod gene expression. Flowers and fruits obtain their color from phenols in important flavonoids, which also aid in pollination and seed dispersal. Apigenin, kaempferol, luteolin, myricetin, and quercetin, to mention a few, impart a yellowish, whitish, or ivory color to plants where they are found (Brouillard et al. 2010).

8.3.3 Phenolics in Plant Coloration

Flavonoids serve as colors in flowers and fruits, which draws pollinators to blooms and, as a result, attracts animals that consume the fruits and distribute the seeds. Flavonoids are also found in vegetables. Hereditary characteristics account for the vast majority of the factors that influence the coloring of fruit. Environmental elements, including nutrients, temperature, and light, can all have an impact on the final color of the fruit as well as the flavonoid content. Even though anthocyanins are to blame for the purples, blues, and reds that may be seen in plants, the final color is determined by a variety of different factors. It is generally accepted that delphinidin-derived anthocyanins are responsible for the bluish hues, whereas cyanidin- and pelargonidin-derived anthocyanins are present in the mauve and red tissues, respectively. This is because delphinidin is derived from pelargonidin, which is derived from cyanidin. The color that is created when an anthocyanin forms a complex with a so-called copigment can be intensified or changed depending on the nature of the complex. It would appear that the majority of chemical substances, including polyphenols, purines, alkaloids, and metallic cations, are capable of performing the function of copigment. Temperature and pH levels in the vacuolar fluid have also been shown to have an impact on the final color (Brouillard and Dangles 1994; Brouillard et al. 1997).

Two distinct classes of chalcones, flavonoids, and aurones, give certain plants their brilliant yellow color. Yellow carnation contains just chalcone isosalipurposide, while aurone aureusidin, discovered as the 6-glucoside aureusin, is the main carotenoid in snapdragon (*Antirrhinum majus*). Luteolin, apigenin, quercetin, kaempferol, and myricetin are the five most common flavones and flavonols, and their structures provide yellow, white, or ivory tones to the tissues in which they are found, depending on the pattern of hydroxylation they undergo. For instance, the flavone isoetin, a yellow floral color found in various members of the Compositae, is produced by inserting a 2′-hydroxyl group into luteolin. When a hydroxyl group is added to quercetin at either the 6- or 8-position, the resulting compounds (such as quercetagetin, which is found in the flowers of the Coronilla, Primula, Rhododendron species, Lotus, and gossypetin, the pigment of the *Gossypium hirsutum* flowers) are much more yellow in color than quercetin. Lastly, the vast majority of naturally occurring phenolic colors consist of benzoquinones, naphthoquinones, and anthraquinones. Fungi, namely the Hyphomycetes and the Basidiomycetes, contain benzoquinones, but plants possess them infrequently. Priming (6-methoxy-2-*n*-pentylbenzoquinone) is one of the benzoquinones found in the glandular hair of Primula obconica leaves. In contrast, the majority of naphthoquinone pigments are derived from higher plants, such as plumbagin, an orange pigment discovered in *Plumbago capensis* and present in bound form in other members of the Droseraceae, Plumbaginaceae, and Ebenaceae families. The most frequent type of natural quinones, tricyclic anthraquinones, are found in numerous plant families, including Liliaceae, Leguminosae, Polygonaceae, Rubiaceae, Rhamnaceae, and Scrophulariaceae. Anthraquinone, slightly in the past, pigments such as anthragallol (2,3,4-trihydroxyanthraquinone), alizarin (2,3-dihydroxyanthraquinone), and purpurin (1,2,4-trihydroxyanthraquinone) were used in textile colors (Harborne 1980).

8.4 Effect of Plant Phenols on Pests

There is various plant which shows a defense against herbivore damage by boosting their biotic, physical or chemical defenses, and some responses that cannot the plants rest parts from more damage (Karban and Baldwin 1997; Moreira et al. 2018). It is believed that plant phenolics play an important role in the defense mechanism of plants against various herbivores with the help of some specific physiological influences on the herbivore. These defensive substances are treated as digestibility reducers, toxins or even antifeedant agents. These defensive substances generate ROS in the digestive tracts mid-gut of herbivores, where there is a basic pH. The proteins and lipids of herbivores' midgut would get damaged due to the influence of oxidative damage promoted by the ROS elements. With this damage, there occur various problems in the midgut of insects, like oxidized proteins, increased concentration of lipid peroxidation products, and different ion species, which eventually lead to the death of insects. Moreover, the cellular network and pathogen metabolism is disrupted by Phytoalexins. Some practical proofs encompass the rishitin by Solanaceae, medicarpin by *Medicago sativa*, and camalexin by *Arabidopsis thaliana* (Jeandet et al. 2013). In addition, protein inactivation occurs in insects with the help of the binding of tannins to digestive enzymes, such as chymotrypsin and trypsin, and salivary proteins. Those insects which take a greater number of tannins are not gaining sufficient weight, which ultimately leads to their death (Prinz and Lucas 2000). The nature of polymers, i.e., lignins, lends a tough biological boundary as a cell wall toward the herbivore's insects. Furthermore, when the plant gets attacked by any herbivore insects, a group of chemicals known as furanocoumarins gets released by UV rays and gets attached to the herbivore insect's DNA and ultimately leading to the death of an insect (Chen 2008).

8.5 Host Plant Defenses Against Insects

The host plants show some defense mechanisms against the insects. With the help of dynamic and intricate defense mechanisms, plants show responses against the attack of herbivore insects. This defense mechanism encompasses toxic chemicals, the natural enemy attraction of the victim pest and structural barriers, which help a plant to get rid of herbivores' insects (Reise and Waller 2009). As soon as the herbivore insect tries to damage the plant, both these direct and indirect defensive mechanism gets induced or are present constitutively. Induced type of response is one of the significant elements of pest control in plants as far as the agriculture perspective is concerned, and this method has been manipulated to keep the herbivore insect population in check (Reise and Waller 2009; Sharma 2008). In the last few years, the induced response in plants has become an interesting field of ecology and evolutionary biology, and a significant improvement has been earned in studying the role of indices responses against various stresses in plants. Even though there are some metabolic costs of induced response (Agrawal et al. 2002), they are considered very critical as far as lowering the stress of unexpected concern is taken into account, as in response to pest attack, these chemicals get released. Induced defense mechanisms have the quality that, phenotypically, makes the plants very strong and, therefore, reduces the opportunities for insects to attack the plant and to accept these induced chemicals (Reise and Waller 2009; Agrawal 2011). In response to insect attacks on plants, there occurs a modification in the constituents of defensive material, which gets formulated unpredictability in the climate of plants for herbivores (Fig. 8.1). As a result, the behavior and fitness of a herbivore insect get altered (Lin et al. 2014). It will be very beneficial for a plant if there occurs an induced response very early, which modifies the all-around fitness of a plant and disrupt the invasion of pathogen and herbivore insects (Agrawal 2011). Those plants which are having a variety of defensive mechanisms and chemicals can have a better

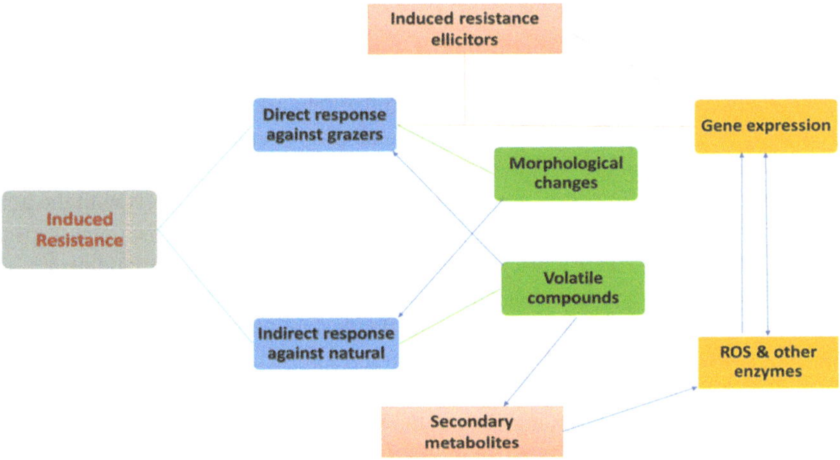

Fig. 8.1 Mechanism of induced response in plants

defense against any pathogen or herbivore insect than those plants that have less variety of defensive chemicals (Lin et al. 2014). The interaction between plants and insects has developed our vision about the evolution of defensive strategies shown by various plants against pathogens, including herbivores insects (Karban 2011), but still, the defense mechanism opted by plants against insects is still not better understood.

Direct defenses. The structural aspects of pants like thorns, trichomes, on the leaf surface, lignification, and thickness of cell wall are the first real boundaries or obstacles for the herbivore insects to feed on plants, and the secondary things like toxins and the things that have an influence on the growth, digestibility and development of insect act as secondary barriers that help the plant to get rid of these attackers (Hanley et al. 2007). Furthermore, the defensive system of a plant against insect attackers gets boosted by the synergic effects of various components of the defense mechanisms. When the phenolics, oxidative enzymes, alkaloids and proteinase inhibitors in tomato are ingested individually, they have less effect or no effect at all, but if they act in combination in a synergistic manner, they can affect the metabolism, digestion as well as ingestion of an insect. The nicotine expression and trypsin proteinase inhibitors of *Nicotiana attenuata* (Torr. ex Watson) provided a more defensive role synergistically in against *Spodoptera exigua* (Steppuhn and Baldwin 2007).

Morphological structures. Against any pathogen, the plant structures serve as the first line of defense and have a crucial role in the HPR of insects. Against herbivore insects or pathogens, the defensive mechanism can be a physical obstacle, either a formation of sticky and waxy cuticles or the formation of trichrome, setae or setae (Hanley et al. 2007). The structural type of defense mainly comprises anatomical and morphological traits that provide more fitness benefits to a plant by preventing the herbivore from damaging the plant directly as an outcome of suberization and lignification, a microscopic modification in the thickness of cell wall occurs, which has a great impact on the prominent protuberances of the plant (Hanley et al. 2007). The structural aspects like trichomes (pubescence), thorns and spines (spinescence), hardened leaves (sclerophylly), divaricated branching, and integration of granular minerals into tissues of plants have the main role in the protection of plants against herbivores (Hanley et al. 2007). The sclerophylly means toughened leaves, and are very important in the defensive mechanism of pants by lessening the tissue's digestibility and deliciousness, therefore making fewer chances for herbivores to damage the insect (Hanley et al. 2007). Spinescence encompasses various configurations like prickles, thorns and spines. These structures are there on the plant to defend it from various insects. Pubescence includes various hair layers known as trichomes that extend from the epidermis of various plant parts above the soil like leaves, stem, or just fruit, and are present in various shapes like a spiral, hooked, straight, glandular or even stellate (Hanley et al. 2007). The plumule, leaf glossiness, and pigmentation of leaf sheath were credible for the resistance of shoot fly *Atherigona soccata* (Rondani) in the case of sorghum *Sorghum bicolor* (L.) (Moench), as reported by Chamarthi et al. (2011).

Trichomes. They have a very prominent function in the defense mechanism of a plant against various pests and implicate deterrents as well as toxic effects (Chamarthi et al. 2011). The density of trichomes has negative impacts on the larval growth, nutrition, development and ovipositional behavior of various insect pests. Moreover, herbivory also gets affected mechanically by the trichome density and checks the insect pests or other arthropod attacker movement on the surface of the plant; hence, the access to the leaf dermis gets reduced for these attackers. The structure of trichomes can be branched, unbranched, hooked, glandular, spiral, nonglandular, or even straight (Hanley et al. 2007). The glandular trichomes release various secondary metabolites such as alkaloids, terpenoids or flavonoids that can trap, repellant, or poisonous for various insects or other organisms as well, thus making a chemical and structural defense combination (Hanley et al. 2007). In many plants, it has been reported that insect damage can induce a response in trichomes (Agrawal 2011). This response can be seen as an increase in the density of plant trichomes and can be identified only in those leaves which get develop during or successive to pest attack, as the trichome density of old leaves remains constant. The damage by *Phratora vulgatissima* L., a leaf beetle in *Salix cinerea* L., induced a more density of trichomes in the newly emerging leaves after the attack, as reported by Dalin and Bjorkman. Similarly, the trichome density gets increased in *S. Cinera* when attacked by coleopteran species, as reported by various studies. It has also been reported in the case of *Lepidium virginicum* L. and *Raphanus raphanistrum* L. that when they get damaged by an insect, an increase in the density of trichomes is observed (Agrawal 2011). When the black mustard is damaged by *Pieris rapae* (L.), there occurs an increased level of glucosinolate and trichome density. Also, when the beetle attacks the *Alnus incana* Moench, the increase in the density of trichomes is seen. In response to the attack of herbivory, the increase in trichome density is usually between 25% and 100%. However, it is also reported that the trichome density gets increases by 500% to 1000% in some cases. After the damage by any damage occurs, the density of trichomes gets changed within days or weeks. Moreover, herbivory is also responsible for inducing the alteration in the relative percentage of glandular and nonglandular trichomes (Agrawal 2011). Several reports have seen a positive relationship between the density of trichomes and the abundance of natural enemies. Exudates from trichomes act as extrafloral nectar (EFN) for the scelonid egg parasitoid of squash bugs, *Gryon pennsylvanicum* Ashmead (Olson and Nechols 1995).

8.6 Phenolics in Plant–Insect Interactions

Since phenolic compounds are widespread in vascular plants and play a central role in hypotheses about the coevolution of plants and their herbivores, ecologists have a particular interest in the effects of phenolics on insect herbivores (Haslam 1989). Schultz investigated the impact of polyphenolics on insects and discovered that herbivores' reactions varied greatly in terms of sensitivity and the aspects of performance that were impacted (Nishizawa et al. 1982). Studies on the effects of polyphenolics on insects indicated that they had either a negative or positive effect on about half of the insects studied. There was a wide range of responses across the different insect groups to phenolics, despite the fact that they typically had negative effects on insects that didn't regularly find them in their host plants and neutral or favorable effects on insects that did. Distinct phenolics may have different impacts on insect herbivores at various stages of the food chain. Polyphenols, for instance, serve as a language understood by both herbivores and pathogenic bacteria. Plant-eating insects may be protected from pathogens including fungi, bacteria, and viruses if they consume phenolics (Nishizawa et al. 1983). Phenolic compounds have been demonstrated to have antibacterial properties; yet, there may be conditions in which they also increase the susceptibility of herbivores to infections. We hypothesize that the varied insect responses to phenolics are due to the oxidative conditions present in the digestive systems of various insect species. Both insect and host plant enzymes and nonenzymatic mechanisms regulate oxidation in the insect and plant guts, respectively. It has two implications for future research into the impact of phenolic compounds on insects. To begin, it is likely that the limited and occasionally incorrect picture of phenolics' function that is gleaned from work with artificial diets devoid of foliar oxidants and enzymes that break down oxygen is the result of working with such diets. Second, the phenolic structure in the leaf may be significantly different from that in the gut due to large differences in redox potential. Therefore, it is possible that the connections between foliar phenolic chemistry and insect behavior don't tell us anything about the function of phenolics in living organisms.

8.6.1 Modes of Phenolic Action in Insects

Insect development retardation by phenolics has been demonstrated numerous times, but the mechanisms involved have received less attention. There was a time when it was assumed that all phenolics would act similarly to prevent insects from feeding on plants once they made their way into an insect. Gross (1992) hypothesized that phenolics prevented insect herbivores from absorbing leaf nutrients by forming hydrogen bonds with protein in the meal and/or digestive enzymes in the gut (Gross 1992). However, circumstances in the digestive systems of many insects prevent hydrogen binding, and phenolic compounds rarely hinder the absorption of food by herbivores (Nishizawa et al. 1982). There is a lot of debate concerning the effects of phenolics on insects, but there have been no *in-vivo* investigations; thus, the evidence is conflicting (Gross 1985). Hydrophobic, hydrogen-based, ionic, and covalent bonds are all possible for phenolics to make with other molecules. Almost all of the theoretical and practical work has been done on hydrogen and covalent bonding. The quantity, structure, and oxidation state of the phenolic, as well as the molecule it may interact with, all have a role in determining the bond type most frequently formed. Insect and host plant characteristics together regulate the pH and redox conditions of the medium (insect gut), which in turn affect the oxidation state (Gross 1985).

There is a complex ecological relationship between insects and plants with various chemical and physical interactions. This connection also gets affected by various factors such as insect factors and plant factors, and also by plant–insect factors like plant resistance to various diseases and hypersensitive reactions. Various situations in the environment can act on the plant, insect, or plant–insect relations and can modify the manifestation of these different factors. Every mechanism, either of plant or insect, can be the outcome of various genetic traits (Painter 1941). So, various queries regarding how plants are chosen by an insect for their food and ovipositing site are yet to be understood.

Various extents of chemoreceptors, either olfactory or gustatory chemo receptive process, are possessed by insects on their mouthparts and antenna, which helps them to determine various kinds of chemical substances even at low concentrations, and insects can easily get the information provided by these chemical compounds. After getting the information, the information is decided with the help of some specific command centers which are present in the central nervous system (Dethier 1970). Various reports on the feeding habits of butterfly larvae on various plants have shown that the main role is played by the secondary metabolites of plants in deducing the utilization patterns. This is not only for butterfly larvae, but almost all phytophagous insects follow the same trend. Chemical compound bio syntheses in the plant tissues are not directly rained to the basic metabolic pathways of the plant but are favorable to their normal development and growth and act to lessen or eliminate the palatability of that plant in which they get elicited (Thorsteinson 1960). The secondary metabolites are the plant material that makes the host unpalatable and is present in enough concentration that it can have unpleasant physiological consequences. Such a plant would get protection from the damage by any phytophagous insects and will likely feel like it had some new adaptive zone. Hence, much more attention has been paid to the plant's secondary metabolites as ultimate and proximate determinants of the host plant in the case of phytophagous insects (Tosh et al. 2003). However, these phytophagous insects can get evolved based on the physiological barrier response. If anyhow any mutation or recombination occurs in a set of insect groups that make them able to feed on protecting a group of plants, then selection will definitely transmit the line into a new adoptive region. When the quantity of the insect-preferred plants becomes shortened, the choice of plant food will be changed at that time and can be favored and can become one of the critical factors for larval survival. Hence, according to Ehrlich and Raven (1964), in a system of coevolution, insects modify themselves to some extent and can show some resistance to the defensive agents that inhibit their feeding so that the same substances that deter their feeding becomes the substance of attractant for insect feeding. At last, it can be said that secondary metabolites of plants can be used by insects to enhance their overall fitness of themselves. It is also reported that the toxic substances in the insect feed are separated by the insects at the time of their feeding in their larval form. These move these separated toxins into the adult stage, and then both adults and larvae generally protect themselves from various bird predators. Apart from the advantages of fitness of separateness in a greatly appreciated species, there are various such reports in which the insect accumulates some chemicals in their cuticle yet are not observed as warningly cultured, and other gets security from the various predators as an outcome of gut content alone (Bernays et al. 1991). Those butterflies (*Polyommatus icarus*) which are raised on the *Vicia villosa* inflorescences (an acceptable kind of plant for the development of butterfly) at their larval stage show various critical characteristics of feed uptake and separation of various important secondary metabolites of the plant as:

- From the host plant, only a few parts of flavonoids are incorporated by larvae, particularly kaempferol and quercetin, while the others are excreted as such.
- There is a positive relationship between the invested number of flavonoids and body mass.
- Metabolism of flavonoids occurs, and their conjugated configurations get stored.

This information provides a clue that in comparison to males, females are best adapted to separate these substances and also, the flavonoids that get accumulated in the female's short wings are proven to be helpful in visual communication, as the females having more flavonoids in their wings gets more attracted toward the males than those females which are flavonoid free (Burghardt et al. 1997). Insects should check and specify the appropriate applicable species of plants before looking for a reasonable host plant. It is obvious that the host plant finding speed will be important. For various reasons, there can be a scarcity of time, and also, various ecological factors can inflict a requirement for speed, like when the reserves are scattered or rare, and there is a risky environment of predators. The exactness of the selection of host plants and the quality of plant assessment is very important for an insect, particularly for those which have a very short spectrum of host plants and specific feeds for their development of larvae. The number of insects has a particular host plant that has specific volatile, phytochemicals and none volatile substances. A combination of these compounds will be of more significance regarding the host plant identification, and also, in a few particulars, the host selection gets dominated by one or more particular chemicals of the host interest (Thorsteinson 1960). The various aspects of a plant, like the cues of thigmotactic, visual, or volatiles, may be implicated in the recognition of a host plant for an insect, and also migration of an insect depends on these things to a host plant. After finding the host, Probing gets initiated as soon as the locomotion gets paused. Now whether the insect remains on the host plant and feeds there depends on the stimuli and deterrents of individual feeding. Also, the rejection or acceptance of a plant by an insect is mainly based on the chemical composition of the host plant, in addition to various physical factors like hairiness, thickness and toughness. Moreover, various other factors play an important role in the acceptance or rejection of host plants such as chemicals inhibitors. The main role in oviposition inhibition is played by the chemical inhibitors and indirectly on the lava growth of an insect and its progeny survival (Harborne 2001). Various studies have been done on the chemical inhibitors' role in the selection of host plants, revealing that various chemicals may have an inhibitory role on the feeding habits of an insect. It is now commonly confirmed that various constituents of plants have an important role in the protection of plants, especially the phenolics, which play a crucial role in protecting the plant species from attackers, particularly insects and herbivory (Todd et al. 1971).

Measuring the content of phenolics in the different tissues of the plant was the traditional method to deduce that the phenolics are engaged in the defensive mechanism in plants against any attackers. While there is no doubt about the relationship between the phenolic content and defense, it has been revealed by various experiences that a specific phenolic subclass or a particular phenolic agent is active against a specific attacker. Moreover, the induced type of defense was one of the significant understandings of the plant–insect relationship. Various plants show different approaches toward the defense against damager as some plants increase the synthesis of specific toxins of phenolics against the response of insect feeding on that plant. Finally, it will be good to say that the main factor in deterrence is the concentration of various products of phenolic toxins, and the feeding barrier for an insect gets boosted by the accumulation of different compounds of phenolic toxins in a particular part of the plant tissues (Castellanos and Espinosa-García 1997). Salicylates in Salix leaves and their contribution to the growth and development of *Operophtera brumata*, polyphagous larvae, is a well-known illustration of a feeding barrier by phenolic toxins in plants. It has been seen that there is a negative relationship between the salicylate content and growth: larvae exposed to those leaves in which these phenolic compounds are present consume less amount of the leaf feed and grow very slowly, and this provides a cue regarding that salicylates can be treated as antifeedants for *O. brumata*. Some researchers have shown that *Hypericum calycinum* flowers that are appeared yellow to humans have a UV structure, likely to be noticeable for insects (Gronquist et al. 2001). Two types of pigments of phenolics are responsible for the UV pattern of this flower, dearomatized isoprenylated phloroglucinols (chinesin II, chinesin I, hypercalin B, hypersaline A (XVIII), and hypercalin C) and flavonoids (I3-II8-biapigenin, a dimeric flavonoid and quercetin-3-*O*-β-D-glucuronide). The presence of isoprenylated phloroglucinols in the ovarian wall and anthers of flowers have increased the chance that these may be involved in the secondary function of defense as well. The saying may not be wrong that developing seeds and pollen require protection, and the discovery that in regard to larvae of *Utetheisa ornatrix*, the hypercalin A was proved to be toxic and a deterrent, indicated that these have a role in defense.

A phenylpropanoid derivative substance, chlorogenic acid, also shows the same antifeeding activity. However, this compound is highly widespread in the entire plant kingdom, so this derivative is now less likely to be used in the case of a defensive mechanism. Hence, various insect species show little tolerance toward this derivative of phenylpropanoid. The impact of real chlorogenic acid on feeding behavior was made by utilizing four different beetles present on the low chlorogenic acid leaves. One among the four species, *Lochmaea capreae* L., was observed as deterred from the willow leaves on which a pure, naturally present concentration of chlorogenic acid was applied. Similarly, *L. capreae* in laboratory feeding trials prefers low chlorogenic willow leaves among the other four species over highly concentrated chlorogenic acid leaves of *Salix myrsinifolia* Salisb and *Salix pentandra* L. When they were present on the *S. phylicifolia* L. leaves, the feeding by *Phratora polaris* Sp.-Schn was also inhibited by pure chlorogenic acid. However, there was no significant effect of chlorogenic acid observed on Ph. Polaris when it existed on the leaves of *S. cinerea* L. Thus, the scientists conclude that the Ph. polaris response toward chlorogenic acid is based on the type of plant species. While as for the other two species, *Plagiodera versicolora* Laich and *Galerucella lineola* F., there was no deterrent effect of chlorogenic acid observed even at unnatural concentrations. From the four different species of common leaf beetles, the effect of chlorogenic acid was only shown by *L. capreae*, while others didn't show any response. Hence, the significance of this phenolic derivative is very low as far as defense agents against the common willow battles are concerned (Ikonen et al. 2001).

Maximum plants have various flavonoids, which differ among the various species, genera and families. The fact that how phytophagous insects are able to differentiate among species, genera or families concludes that flavonoids have some crucial role in the selection of a host. The growth, development and behavior of various insect species get affected by plant flavonoids. For *Anthonomus grandis*, a boll weevil, few flavonoids have the property of feeding impulses (Hedin et al. 1988). For *Papilio xuthus* L., a citrus-feeding butterfly, flavonoids have oviposition impulses, and at last, against some phytophagous insects, flavonoids have antibiotic compounds. Flavonoids not likely to have various secondary metabolites have a very low toxic effect and physiological roles in many insects. Moreover, at low concentrations, various flavonoids act as feeding deterrents against various phytophagous insects. So, in plants, the flavonoid concentration is present more than that is needed for feeding impediment effect on aphid feeding. However, aphids love to

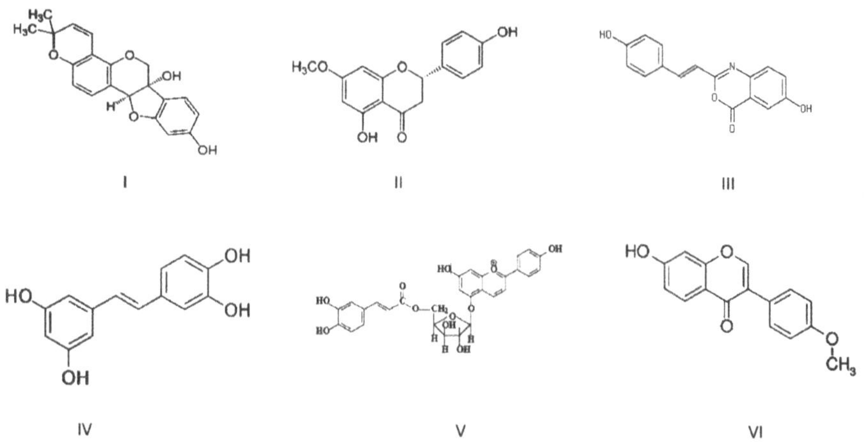

Fig. 8.2 Structure of (I) glyceollin II, (II) sakuranetin, (III) avenalumin I, (IV) piceatannol, (V) caffeic acid ester of arabinosyl 5-*O*-apigeninidin, and (VI) formononetin

feed on those tissue of plants which have a low concentration of flavonoids present line phloem, and thus, they will only experience a high concentration of flavonoids while sapping the phloem, not during phloem feeding (Harborne and Grayer 1993). From the wild chickpea, *Cicer arietinum*, four isoflavonoids were sequestered (maackiain), 2-methoxyjudaicin, judaicin-7-*O*-glucoside, and judaicin, shows a property of deterrent feeding at 100 ppm by larvae of *Heliocoverpa armigera* (Fig. 8.2). All the isolated isoflavonoids show a trend of dose-dependent reduction in function or activity, with maackiain and judaicin retaining their functions at 10 and 50 ppm, respectively. These isolated isoflavonoids were examined in integration and with chlorogenic acid: the mixtures, including maackiain and judaicin, were observed as the highly active combination; the addition of chlorogenic acid boosts the function of all isolated isoflavonoids. The only noctuid that was deterred by all isolated isoflavonoids was *H. armigera*. *Spodoptera frugiperda* and *S. littoralis* were deterred by maackiain and judaicin alone, respectively. However, no iso-flavonoid was effective against the feeding of *S. exigua* and *Heliothis virescens*. When these isolated isoflavonoids were incorporated into the diet, they tended to decrease the weight in the early stages of *H. armigera* larvae more than they do in the late stages, and the most potent ones were judaicin and maackiain. This information includes that these four isolated isoflavonoids, particularly judaicin and maackiain, can have a critical role to play in reducing the chances of Cicer to damage by *H. armigera* (Simmonds and Stevenson 2001).

The wild mustards are the specific and chosen host plant for *Pieris napi oleracea*, an American native butterfly. However, this butterfly cannot show any effect on the *Alliaria petiolata*, garlic mustard, an intrusive weed introduced from Europe. Still, the oviposition occurs in the adults of *P. n. oleracea* on this planet, but its larvae do not get any chance to survive on this garlic mustard. The various feeding bioassays were used with various insect larval stages to regulate the identification and isolation of two important constituents that may clarify the natural resistance of this plant, an alliarinoside, cyanopropenyl glycoside, was found by Renwick et al. (2001), by an apparent post ingestive feedback mechanism, the first instars of insect are strongly inhibited, while isovitexin-6″-D-β-glucopyranoside, a flavone glycoside, serves as a feeding barrier that is distinguished by various taste reports of mouth parts of late instars. Interestingly, the alliarinoside has a very low effect on the late instars, and toward isovitexin-6″-D-β-glucopyranoside, the early instars are not sensitive. Maximum researchers having an interest in the area of insect antifeedants have discussed the effects of these antioxidants on the pests of agriculture to deduce his flavonoids can show their resistance to crops against insect damage. In this regard, the HPLC analysis was performed on wild and cultivated forms of Vigna, a food legume in Latin America and sub-Saharan Africa. The HPLC shows that from a quality point, the cultivated forms of *Vigna unguiculata* L. Walp. are relatively comparable, as it is always showing three aglycones of flavonoids: isorhamnetin, kaempferol, and quercetin. Moreover, a positive relationship was found between the susceptibility/resistance against aphids and the content of glycoside flavonoids. The susceptible lines show a comparatively lower content of glycoside flavonoids than the resistant ones. Among the endogenous flavonoids, isorhamnetin and quercetin have better inhibitors of reproduction rates of aphids, as proven by In vitro bioassays. In comparison to cultivate lines, the HPLC analysis of the wild type of Vigna helped the already present information that there is a presence of various chemotypes of flavonoids in a few species of Vigna. Various kaempferol chemotypes are there, and it is important aglycone detected two isorhamnetin and quercetin chemotypes, where quercetin chemotypes contain only quercetin glycosides. As far as the ecological point is concerned, the most significant phenotypes are some accessions which are of the same species and allow us to examine ceteris paribus, the function of endogenous flavonoids in aphid resistance by a plant. Two chemotypes were found in *V. marina* accessions, *V. marina* var. *oblonga* TVnu 1174, which includes few traces of 2 kaempferol residues and two isorhamnetin glycosides, and *V. marina* var. *marina* TVnu 717, which includes only glycosides of kaempferil. *V. luteola* accessions also have two chemotypes: TVnu 475, which has only quercetin glycosides, and TVnu 172 and TVnu 905, which contains robinin. When the aphid resistance features of different chemotypes of one species were tested, it was displayed that isorhamnetin and quercetin showed the highest resistance in comparison to chemotypes of kaempferol of the same species, thus making it clear that interest in isorhamnetin and quercetin is there in the mechanism of resistance (Lattanzio et al. 2000).

Here it should be noted that the secretion of salivary glands of phytophagous insects had several enzymes which have a crucial role in the digestion of food in the case of piercing-sucking insects. The important enzymes from these salivary enzymes are peroxidases (POD) [E.C. 1.11.1.7], polyphenol oxidases (PPO) [E.C. 1.10.3.1.], oxidoreductases that metabolize the phenolics of the plant. These enzymes have been located in the secretion of the saliva of various species of aphids and are believed to play a crucial role in various functions. Early, It was suggested that these enzymes are believed be implicated in the stylet sheath's chemical stabilization. After that, it was suggested that they are important in neutralizing the effect of phenolics of plants and their secondary derivatives, i.e., they are involved in resisting the defense displayed by a plant. In addition, it has been seen that the damaged cells of plant tissue become partially restored with the saliva secretion of aphids when they aphids try to infiltrate into the plant tissue. Hence, it can be concluded that the metabolites of tissues of plants may be absorbed by the salivary secretion of phytophagous insects. As a result, the secreted saliva secretions PPO gets accumulated with the phenolics of the plant around the stylet sheath of aphids, where oxidation me polymerization occurs to them. So, the naturally present highly concentrated toxins (phenolics) are converted into less effective toxic toxins with the help of these enzymes or salivary secretion (Urbanska et al. 1998). Also, it cannot be denied that the power of phenolics as a factor of resistance to the feeding of insects is boosted by polymer oxidation, which further lessens the nutritional value, palatability and digestibility of the feed. Thus, higher values of PPO, the main enzyme for oxidation in plants, can be associated with the resistance process of plants against the attackers, mostly insects. Polyphenol oxidase, inferred as antiherbivore enzymes, are the enzymes of antinutrition that reduces the value of nutrition of damaged plant tissues by catalyzing the phenolic metabolites oxidation or crosslinking various proteins to polymerizing and reactivating the quinones. In the hybrid poplar, the expression of PPO as inducible ones is examined: after the mechanical wounding, insect damage stimulation, and enhanced activity of PPO in unwounded and wounded leaves on wounded plants starting at 48 h and 24 h, respectively. Also, the expression of PPO is also induced by feeding of forest tent caterpillars. The herbivore and wound Induced PPO expression in poplar hybrid defends the role of defense of this enzyme against various pests and insects. The obtained results deduced that:

- Induced PPO is systemic. This implies that it has a main role in decreasing further damage to herbivores instead of repairing the plant wound.
- The enhanced aggregation of mRNA mediates the PPO induction, which is a combination of various induced defense proteins (Constabel et al. 2000).

For polyphagous insects, various flavonoids are phagostimulatory even at low concentrations, and it is also suggested that they can pertain to novelty attraction for some species with a big advantage of influencing new food. For instance, rutin (quercetin-3-*O*-β-rutinoside), the flavonoid glycosides, serves as phagostimulant even at 10% of the dry weight of disc of glass fiber for *Schistocerca americana* and phagistimulation gets increased when 10% of the dry weight of sucrose gets added by 2% of the dry weight of glass fiber disc. For the other two species of polyphagous insects, *Melanoplus differentialis* (Thomas) and *Schistocerca albolineata* (Thomas), rutin also act as a phagostimulant even at low concentrations. The insects seem to be capable of stabilizing their intake, and after 2 days, rutin ceases to serve as a phagostimulant at higher concentrations. Various amount of rutin in the gut is hydrolyzed and then excreted, while some portion is absorbed and converted to β-3-*O*-glucoside. This remains separated in the portion of the cuticle, where it imparts a yellowish color to the exocuticle and exuviae (Bernays and

Chapman 2000). For *Papilio xuthus* L., a Citrus-feeding swallowtail butterfly, some flavonoids serve as stimulants of oviposition (Nishida et al. 1987). Some components from the leaves of *Citrus unshiu* were isolated, which later on was found to have a crucial role in the behavior of oviposition of *P. xuthus*. The various isolated components are rutin, hesperidin (hesperetin-7-*O*-β-rutinoside) and Narirutin (naringenin-7-*O*-β-rutinoside). Every isolated component do not show any reaction on its own, and even if the mixture of glycosides is prepared in the same concentration that is found in the Citrus leaves, only a weak response is generated from *P. xuthus*. However, suppose the glucosides are varied with other than water-soluble molecules of *C. unshiu* leaves. In that case, the behavior of oviposition of *P. xuthus* gets highly modified and is analogous to that response which is induced by intact leaves, thus concluding that it has a synergistic action.

In East Asia, *Phryma leptostachya* L., a herbaceous perennial plant, was traditionally used as a natural pesticide and grew widely in north-eastern America, temperate Asia and the Himalayas. Various lignans like 3,7-dioxabicyclo-[3.3.0] octanes (furofuran) type have been isolated from the root extract of *Phryma leptostachya* L., which have such insecticidal/pesticide properties against various larvae of lepidopterans insect species. At the concentration of 10 ppm, the lignan, Leptostachyol acetate, extracted from the roots of the same plants, was deleterious for the third instar stage of larvae of three mosquito species, *Aedes aegypti*, *Ocheratatos togoi*, and *Culex pipiens* pallens (Park et al. 2005). Another lignan, magnolol (XXII), extracted from *Magnolia virginiana* leaves also shows the function of defense against the larvae of *Callosomia promethean* (Harborne 2001). Moreover, for phytophagous insects, tannins are commonly treated as deleterious. In three ways, these tannins can influence insect growth:

- Tannins possess an astringent taste that has an influence on palatability and reduces the consumption of a meal.
- The complex of decreased digestibility along with proteins gets formed.
- They serve as inactivators of various enzymes.

In various grains and seeds of plants, tannins are there in a high concentration where they have unfavorable effects on the use of seed and gain for insect feed. The proanthocyanidins, condensed tannins, provide resistance to insect infestation like *Callosobruchus maculatus* (F.) and the cowpea weevil (Lattanzio et al. 2005). During the development and maturation of grains and seeds, an enhanced concentration of proanthocyanidins was analyzed in many different species of plants. During the sorghum grains maturation, an enhanced increase in the polymerization of proanthocyanidins have been noted (Butler 1982). From anthesis to maturity of *Phaseolus vulgaris* seed, the increase in the α-amylase inhibitory activity and proanthocyanidins was observed; both these properties are involved in the defensive mechanism against insects (Coelho and Majolo 1993). I.T. 84E-1-108 showed an increased infestation level (about 30%). However, Vita 7 showed no damage as a result of the cowpea weevil larva. In this concern, two stored cowpea seed accessions that were classified as vulnerable accessions displayed varying bruchid damage during storage and were checked for the contents of α-amylase inhibitors and condensed tannins as a compound of defensive agents against cowpea weevil. The damage by cowpea weevil was recorded to be an increased level of infestation of about 30% in an I.T. 84E-1-108, while no damage for Vita 7 was recorded at all. Moreover, in the Vita 7 cotyledons, no α-amylase inhibitory function was recorded, while a moderate degree of activity of inhibitors fiction was recorded in the case of I.T. 84E-1-108. In contrast to it, the tannin content in the seed coat was found to be 14 times more in nondamaged Vita 7 seeds than in I.T. 84E-1-108-infested seeds. These results are in support the hypothesis that when the grain is stored after harvest, the tannins of the seed coat are much more effective in showing a defensive mechanism against the bruchid larvae that have cowpea seeds as feed (Lattanzio et al. 2005). Even though these tannins shy some defensive mechanism toward the insect attack, recent research has conveyed that the insect shows some sort of resistance by modifying their midgut environment or the structure of the gut. Also, some grasshoppers have the capability to overcome the effect of various gallotannins. Hence, they are resistant to the generated ROS by polyphenol oxidation. Alternatively, they are dependent on comprehensive and quick tannin hydrolysis to prevent any damaging influences (Harborne 1999). With varied feeding patterns, some grasshoppers have large peritrophic envelopes that absorb tannins and prevent tannic acid from passing through, keeping it from reaching the tissues. Additionally, surfactants found in *Schistocerca gregaria*'s midgut environment significantly lower the formation of tannin protein complexes, save for in situations with extremely high tannin concentrations (Bernays and Chapman 2000).

8.7 Conclusion

The phenolics in plants are secondary metabolites that aid in the plant's defense mechanisms against fungal infections and insect herbivores. Plants have evolved complex defense mechanisms against their many natural enemies, including the production of both endogenous and exogenous phenolic chemicals that alter their sensitivity to and resistance to attack. Producing resistance traits may be expensive if limited fitness resources must be allocated to regrowth (tolerance) or the creation of chemicals that directly influence the attacking enemies. Plants selected for increased phenolic content may alter several aspects of their interactions with pests and disease. In order to better understand how increasing the expression of an endogenous phenolic compound or a class/subclass of phenolics in plants can influence plant resistance traits against diseases and phytophagous insects, more research into these interactions is required.

Conflict of Interest The authors declare no conflict of interest.

References

Agrawal AA (2011) Current trends in the evolutionary ecology of plant defence. Funct Ecol 25(2): 420–432

Agrawal AA, Janssen A, Bruin J, Posthumus MA, Sabelis MW (2002) An ecological cost of plant defence: the attractiveness of bitter cucumber plants to natural enemies of herbivores. Ecol Lett 5(3):377–385

Begum AA, Leibovitch S, Migner P, Zhang F (2001) Specific flavonoids induced nod gene expression and pre-activated nod genes of Rhizobium leguminosarum increased pea (Pisum sativum L.) and lentil (Lens culinaris L.) nodulation in controlled growth chamber environments. J Exp Bot 52(360):1537–1543

Beninger CW, Abou-Zaid MM, Kistner AL, Hallett RH, Iqbal MJ, Grodzinski B, Hall JC (2004) A flavanone and two phenolic acids from Chrysanthemum morifolium with phytotoxic and insect growth regulating activity. J Chem Ecol 30(3):589–606

Bernays EA, Chapman RF (2000) Plant secondary compounds and grasshoppers: beyond plant defenses. J Chem Ecol 26(8):1773

Bernays EA, Howard JJ, Champagne D, Estesen BJ (1991) Rutin: a phagostimulant for the polyphagous acridid Schistocerca americana. Entomol Exp Appl 60:19

Brglez Mojzer E, Knez Hrnčič M, Škerget M, Knez Ž, Bren U (2016) Polyphenols: extraction methods, antioxidative action, bioavailability and anticarcinogenic effects. Molecules 21(7):901

Brouillard R, Dangles O (1994) In: Harborne JB (ed) The flavonoids: advances in research since 1986. Chapman and Hall, London, p 565

Brouillard R, Figueiredo P, Elhabiri M, Dangles O (1997) In: Tomás-Barberán F, Robins RJ (eds) Phytochemistry of fruits and vegetables. Oxford University Press, New York, p 29

Brouillard R, Chassaing S, Isorez G, Kueny-Stotz M, Figueiredo P (2010) The visible flavonoids or anthocyanins: from research to applications. In: Recent advances in polyphenol research. Wiley

Burghardt F, Fiedler K, Proksch P (1997) Uptake of flavonoids from Vicia villosa (Fabaceae) by the lycaenid butterfly, Polyommatus icarus (Lepidoptera: Lycaenidae). Biochem Syst Ecol 25:527

Butler LG (1982) Relative degree of polymerization of sorghum tannin during seed development and maturation. J Agric Food Chem 30:1090

Caballero B, Trugo LC, Finglas PM (2003) Encyclopedia of food sciences and nutrition. Academic

Castellanos I, Espinosa-García FJ (1997) Plant secondary metabolite diversity as a resistance trait against insects: a test with Sitophilus granarius (Coleoptera: Curculionidae) and seed secondary metabolites. Biochem Syst Ecol 25(7):591

Chamarthi SK, Sharma HC, Sahrawat KL, Narasu LM, Dhillon MK (2011) Physico-chemical mechanisms of resistance to shoot fly, Atherigona soccata in sorghum, Sorghum bicolor. J Appl Entomol 135(6):446–455

Chen MS (2008) Inducible direct plant defense against insect herbivores: a review. Insect Sci 15(2): 101–114

Chong J, Poutaraud AP, Hugueney (2009) Metabolism and roles of stilbenes in plants. Plant Sci 177:143–155

Coelho JV, Majolo FM (1993) Role of phenolics in the resistance mechanisms of plants against fungal pathogens and insects. Arch Latinoamerican Nutr 43:61

Constabel CP, Yip L, Patton JJ, Christopher ME (2000) Polyphenol oxidase from hybrid poplar. Cloning and expression in response to wounding and herbivory. Plant Physiol 124:285

Dethier VG (1970) In: Sondheimer E, Simeone JB (eds) Chemical ecology. Academic, New York, p 83

Ehrlich PR, Raven PH (1964) Butterflies and plants: a study in coevolution. Evolution 18:586

Falcone Ferreyra ML, Rius SP, Casati P (2012) Flavonoids: biosynthesis, biological functions, and biotechnological applications. Front Plant Sci 3:222

Gronquist M, Bezzerides A, Attygalle A, Meinwald J, Eisner M, Eisner T (2001) Attractive and defensive functions of the ultraviolet pigments of a flower (Hypericum calycinum). Proc Natl Acad Sci 98(24):13745

Gross GG (1985) Biosynthesis and metabolism of phenolic acids and monolignols. Academic, Orlando, FL, pp 229–271
Gross GG (1992) Enzymes in the biosynthesis of hydrolyzable tannins. In: Plant polyphenols. Springer, Boston, MA, pp 43–60
Guntzer F, Keller C, Meunier JD (2012) Benefits of plant silicon for crops: a review. Agron Sustain Dev 32(1):201–213
Hanley ME, Lamont BB, Fairbanks MM, Rafferty CM (2007) Plant structural traits and their role in anti-herbivore defence. Perspect Plant Ecol Evol System 8(4):157–178
Harborne JB (1964) Methods in polyphenol chemistry. Pridham, JB
Harborne JB (1980) Secondary plant products. In: Bell EA, Charlwood BW (eds) Encyclopedia of plant physiology, new series, vol 8. Springer, Berlin, p 329
Harborne JB (1989) Methods in plant biochemistry, Plant phenolics, vol 1. Academic
Harborne JB (1999) Recent advances in chemical ecology. Nat Prod Rep 16:509
Harborne JB (2001) Nat Prod Res 18:361
Harborne JB, Grayer RJ (1993) In: Harborne JB (ed) Flavonoids: advances in researche since 1986. Chapman & Hall, London, p 589
Haslam E (1989) Gallic acid derivatives and hydrolyzable tannins. In: Natural products of woody plants. Springer, Berlin, pp 399–438
Hedin PA, Jenkins JN, Thompson AC, McCarty JC, Smith DH, Parrot WL, Shepherd RL (1988) Effects of bioregulators on flavonoids, insect resistance, and yield of seed cotton. J Agric Food Chem 36:1055
Ikonen A, Tahvanainen J, Roininen H (2001) Chlorogenic acid as an antiherbivore defence of willows against leaf beetles. Entomol Exp Appl 99:47
Iwashina T (2003) Flavonoid function and activity to plants and other organisms. Biol Sci Space 17(1):24–44
Jeandet P, Clément C, Courot E, Cordelier S (2013) Modulation of phytoalexin biosynthesis in engineered plants for disease resistance. Int J Mol Sci 14(7):14136–14170
Karban R (2011) The ecology and evolution of induced resistance against herbivores. Funct Ecol 25(2):339–347
Karban R, Baldwin IT (1997) Induced responses to herbivory. University Press, Chicago
Khambay BP, Jewess P (2000) The potential of natural naphthoquinones as the basis for a new class of pest control agents—an overview of research at IACR-Rothamsted. Crop Prot 19(8–10): 597–601
Kumar AP, Sivashanmugam AT, Umamaheswari M, Subhadradevi V, Jagannath P (2011) Cholesterol esterase enzyme inhibitory and antioxidant activities of leaves of Camellia sinensis (L.) Kuntze. using in vitro models. Int J Pharm Sci Res 2(10):2675
Kutchan TM (2001) Ecological arsenal and developmental dispatcher. The paradigm of secondary metabolism. Plant Physiol 125:58
Lattanzio V (2003) In: Imperato F (ed) Advances in phytochemistry. Research Signpost, Trivandrum, Kerala, p 49. ISBN: 81-271-0018-8
Lattanzio V (2013) Phenolic compounds: introduction. Nat Prod 50:1543–1580
Lattanzio V, Cardinali A, Linsalata V, Perrino P, Ng NQ (1996) A chemosystematic study of the flavonoids of Vigna. Genet Resour Crop Evol 43(6):493
Lattanzio V, Arpaia S, Cardinali A, Di Venere D, Linsalata V (2000) Role of endogenous flavonoids in resistance mechanism of Vigna to aphids. J Agric Food Chem 48(11):5316
Lattanzio V, Terzano R, Cicco N, Cardinali A, Di Venere D, Linsalata V (2005) Seed coat tannins and bruchid resistance in stored cowpea seeds. J Sci Food Agric 85:839
Lin F, Hou Y, Liu Y, Fang Y, Hu P (2014) Internal solitary waves on the southwest shelf of Dongsha Island observed from mooring ADCP. Chin J Oceanol Limnol 32(5):1179–1187
Lopes da Silva F, Escribano-Bailón MT, Pérez Alonso JJ, Rivas-Gonzalo J, Santos-Buelga C (2007) Anthocyanin pigments in strawberry. LWT–Food Sci 40:374
Mabry T, Markham KR, Thomas MB (2012) The systematic identification of flavonoids. Springer Science & Business Media

Moreira X, Galmán A, Francisco M, Castagneyrol B, Abdala-Roberts L (2018) Host plant frequency and secondary metabolites are concurrently associated with insect herbivory in a dominant riparian tree. Biol Lett 14(12):20180281

Nishida R, Ohsugi T, Kokubo S, Fukami H (1987) Oviposition stimulants of a Citrus-feeding swallowtail butterfly. Papilio xuthus L. Experientia 43:342

Nishizawa M, Yamagishi T, Nonaka GI, Nishioka I (1982) Tannins and related compounds. Part 5. Isolation and characterization of polygalloylglucoses from Chinese gallotannin. J Chem Soc Perkin Trans 1:2963–2968

Nishizawa M, Yamagishi T, Nonaka GI, Nishioka I (1983) Tannins and related compounds. Part 9. Isolation and characterization of polygalloylglucoses from Turkish galls (Quercus infectoria). J Chem Soc Perkin Trans 1:961–965

Olson DL, Nechols JR (1995) Effects of squash leaf trichome exudates and honey on adult feeding, survival, and fecundity of the squash bug (Heteroptera: Coreidae) egg parasitoid Gryon pennsylvanicum (Hymenoptera: Scelionidae). Environ Entomol 24(2):454–458

Painter RH (1941) The economic value and biologic significance of insect resistance in plants. J Econ Entomol 34(3):358

Park I-K, Shin S-C, Kim C-S, Lee H-J, Choi W-S, Ahn Y-J (2005) Larvicidal activity of lignans identified in Phryma leptostachya Var. asiatica roots against three mosquito species. J Agric Food Chem 53:969

Perrino P, Maruca G, Linsalata V, Bianco V, Lester RN, Lattanzio V (1989) Flavonoid taxonomic analysis of Vicia species of section Faba. Can J Bot 67:3529

Pichersky E, Gang DR (2000) Genetics and biochemistry of secondary metabolites in plants: an evolutionary perspective. Trends Plant Sci 5(10):439

Prinz JF, Lucas PW (2000) Saliva tannin interactions. J Oral Rehabil 27(11):991–994

Reise SP, Waller NG (2009) Item response theory and clinical measurement. Annu Rev Clin Psychol 5(1):27–48

Renwick JAA, Zhang W, Haribal M, Attygalle AB, Lopez KD (2001) Dual chemical barriers protect a plant against different larval stages of an insect. J Chem Ecol 27(8):1575

Ribeiro KA, de Carvalho CM, Molina MT, Lima EP, López-Montero E, Reys JRM et al (2009) Activities of naphthoquinones against Aedes aegypti (Linnaeus, 1762) (Diptera: Culicidae), vector of dengue and Biomphalaria glabrata (Say, 1818), intermediate host of Schistosoma mansoni. Acta Trop 111(1):44–50

Sarker SD, Nahar L (2017) Progress in the chemistry of naturally occurring coumarins. Prog Chem Org Nat Prod 106:241–304

Schildknecht H (1984) Turgorins—new chemical messengers for plant behaviour. Endeavour 8(3): 113–117

Sh A (2011) Nanotheoretical studies on evaluation of anti-cancer potential on mangosteen plant. Afr J Agric Res 6(19):4661–4670

Sharma HC (2008) Biotechnological approaches for pest management and ecological sustainability. CRC Press

Simmonds MSJ, Stevenson PC (2001) Effects of isoflavonoids from Cicer on larvae of Heliocoverpa armigera. J Chem Ecol 27(5):965

Steppuhn A, Baldwin IT (2007) Resistance management in a native plant: nicotine prevents herbivores from compensating for plant proteinase inhibitors. Ecol Lett 10:499–511

Thorsteinson AJ (1960) Host selection in phytophagous insects. Annu Rev Entomol 5:193

Todd GW, Gethahun A, Cress DC (1971) Resistance in barley to the Greenbug, Schizaphis graminum. 1. Toxicity of phenolic and flavonoid compounds and related substances. Ann Ent Soc Am 64(3):718

Tosh CR, Powell G, Holmes ND, Hardie J (2003) Reproductive response of generalist and specialist aphid morphs with the same genotype to plant secondary compounds and amino acids. J Insect Physiol 49:1173

Ueda M, Sawai Y, Shibazaki Y, Tashiro C, Ohnuki T, Yamamura S (1998) Leaf-opening substance in the nyctinastic plant, Albizzia julibrissin Durazz. Biosci Biotechnol Biochem 62(11): 2133–2137

Urbanska A, Tjallingii WF, Dixon AFG, Leszczynski B (1998) Phenol oxidising enzymes in the grain aphid's saliva. Entomol Exp Appl 86:197

Van Sumere CF (1989) Methods in plant biochemistry. In: Dey PM, Harborne JB (eds) Plant phenolics, vol 1. Academic, London, p 29

Van Sumere CF, Wolf G, Teuchy H, Kint J (1965) A new thin-layer method for phenolic substances and coumarins. J Chromatogr A 20:48–60

Whiting DA (2001) Natural phenolic compounds 1900–2000: a bird's eye view of a century's chemistry. Nat Prod Rep 18:583

Wightman F, Lighty DL (1982) Identification of phenylacetic acid as a natural auxin in the shoots of higher plants. Physiol Plant 55(1):17–24

Zetterberg C, Öfverholm T (1999) Carpal tunnel syndrome and other wrist/hand symptoms and signs in male and female car assembly workers. Int J Ind Ergon 23(3):193–204

Plant Phenolics Role in Bacterial Disease Stress Management in Plants

9

Aadil Farooq War, Subzar Ahmad Nanda, Iqra Bashir, Sumaiya Rehmaan, Ishfaq Ahmad Sheergojri, Ishfaq Ul Rehman, Zafar Ahmad Reshi, and Irfan Rashid

Abstract

Plant phenolic compounds carry out plentiful functions in plant–microbe pathogenic interactions. Some phenolic compounds are produced constitutively by plant on routinely basis while others are induced in response to any tissue damage or pathogen attack. Their vital contribution in constitutive and inducible strategies of plant innate immunity has been supported by robust experimental evidences. Past two decades have witnessed many breakthroughs concerning the biosynthesis, accumulation, transport and perception of phenolic compounds induced upon pathogen attack. These outcomes validate the critical role of phenolic compounds in both local and systemic acquired resistance (SAR). This review discusses the strategies of plant immunity against bacterial diseases in the context of phenolic compounds viz. role in bacterial disease resistance via phytoalexins and phytoanticipins, rapid alterations in cell wall to protect bacterial entry and highlighting the crucial role of salicylic acid in inducing SAR. This study might be important in understanding dynamic nature of phenolics in disease resistance against pathogens.

Keywords

Phenolics · Biotic stress · Bacterial disease · SAR · Resistance

A. F. War (✉) · S. A. Nanda · I. Bashir · S. Rehmaan · I. A. Sheergojri · I. U. Rehman · Z. A. Reshi · I. Rashid
Department of Botany, University of Kashmir, Srinagar, Jammu and Kashmir, India
e-mail: Aadilwar.scholar@kashmiruniversity.net

R. Lone et al. (eds.), *Plant Phenolics in Biotic Stress Management*,
https://doi.org/10.1007/978-981-99-3334-1_9

9.1 Introduction

The plants regularly come across the pathogenic microorganisms, both below and above the ground. Despite the uninterrupted encounters with potential pathogens, most of the pathogens fail to successfully colonize the plant. This failure of potential pathogens to infect most of the plants demonstrates either the lack of instigation of pathogenicity functions or sturdiness of immune system of plants, whose primary molecular mechanisms have been largely deciphered in the last decade (Chisholm et al. 2006; Boller and Felix 2009; Jones and Dangl 2006). Major advances which lead to understand innate immunity of plants include decrypting the processes underneath signal transduction, pathogen recognition and subsequent activation of defense-related genes (Dodds and Rathjen 2011). In addition, experimental proof advocates that strengthening of cell wall, biosynthesis of low-molecular-weight secondary metabolites (particularly phenolics) and antimicrobial peptides make their mark to restraint infection development (Miedes et al. 2014; Maróti et al. 2011; Bednarek 2012; Voigt 2014). Contrary to these path breaking discoveries, there is still a lack of knowledge about the processes that actually restrict the entry of pathogen, its growth and successful establishment inside plants.

Plant phenolics are structurally diverse group of compounds, produced in plants from numerous primary metabolites or their biosynthetic intermediates. These compounds are having a benzene ring with one or more hydroxyl groups, some of which are carboxylic acids and glycosides soluble either in water or in organic solvents, while others are insoluble polymers (Taiz et al. 2015). They are biosynthesized from the shikimate-phenylpropanoids-flavonoid pathways, generating monomeric and polymeric phenols and polyphenols (Van Sumere 1989). Phenolic compounds have been known to play an important role in interaction of plant with other species for a long time (Hartmann 2008). Their potential role in plant responses to microbial diseases has been a subject of much research and is considered as an important mechanism of plant immunity (Link et al. 1929; Nicholson and Hammerschmid 1992). For instance, the phenylpropanoid pathway intermediates, including ferulic acid, sinapic acid, *p*-coumaric acid, and caffeic acid and pathway derivatives, including glycosides and flavonoid aglycones, have been verified experimentally to possess antibiotic activities against the potential pathogens (Scalbert 1991; Hamilton-Miller 1995).

Phenolics are rare in fungi, bacteria and algae; however, they are not known to produce flavonoids. Although bryophytes produce a limiting range of flavonoids, the full range of flavonoids is produced by vascular plants including ferns, gymnosperms and angiosperms (Harborne 1998). Phenolics play significant roles in plant development, pigmentation, cell wall formation, reproduction, pathogen resistance and variety of other functions (McDowell et al. 1995; Vogel 2008; Humphry et al. 2010). Some of the phenolic compounds directly act as defense compounds against microbes, viruses and herbivores, while some act as signal compounds to induce the production of defense compounds (Swain 1977; Kutchan 2001). Hence, they are regarded as adaptive characteristics that get selected through natural selection during plant evolution. Therefore, the plants are well equipped with the phenolic diversity and the defense mechanisms to deal with the biotic stresses, as they cannot move and change position to escape from their parasites and predators (Al-Khayri et al. 2023). In this review, discussion of the role of these phenolic compounds in providing resistance to the plant has been presented in detail.

9.2 Phenolics Diversity and Importance

Phenols are universally biosynthesized by plants and are known to be synthesized by three main pathways: shikimic/phenylpropanoid acid pathway, iso-prenoid pathway and malonate-acetate pathway (Fig. 9.1; Macheroux et al. 1999; Dewick 2002; Vickery and Vickery 1981). The shikimic acid pathway being the main contributor yields three different aromatic amino acids (tyrosine, tryptophan and phenylalanine) that serves as precursors of phenolic compounds (Tzin and Galili 2010). Phenylalanine, the important phenolic precursor as far as the pathogen resistance is concerned, is gradually converted into hydroxycinnamic acids like salicylic acid, *p*-coumaric acid, cinnamic acid, *p*-hrdoxybenzoic acid, sinapic acid, caffeic acid and ferulic acid (Cheynier et al. 2013). Some of these compounds need further processing so that to become biologically active, for instance, *p*-coumaric acid gets further modified to coumarins to perform diverse biological functions including antiviral, antimicrobial, antioxidant, anti-inflammatory and enzyme inhibiting activity (Cushnie and Lamb 2011; Li et al. 2014; Fang et al. 2016). The malonate-acetate pathway, also called the polyketide pathway functions somewhat similar to the fatty acid biosynthesis in

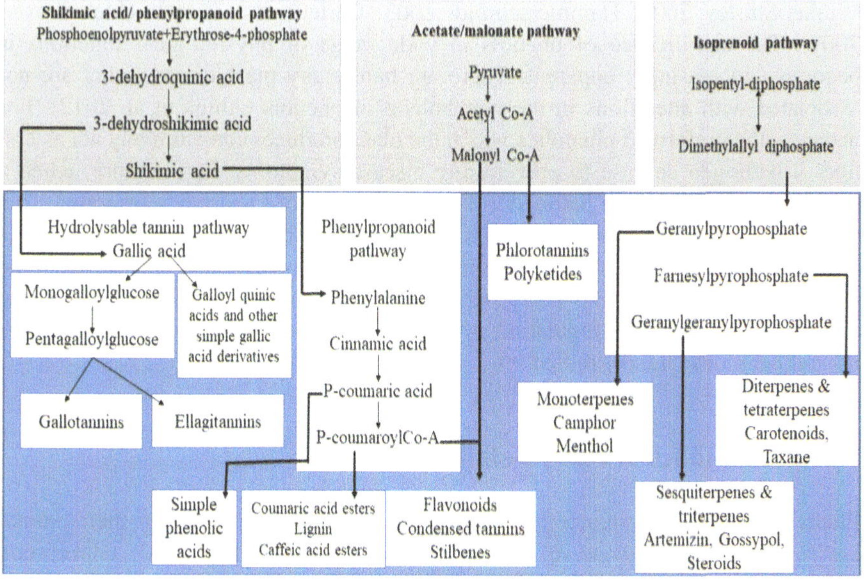

Fig. 9.1 Summarizes the pathways responsible for the synthesis of different phenolic compounds

which malonyl-CoA are added one at a time to form a polyketide chain (Vickery and Vickery 1981). This polyketide chain undergoes a cyclization process to form a benzene ring structure. The intermediates of polyketide and phenylpropanyl pathway like acetyl-CoA, malonyl-CoA and *p*-coumaric acid give rise to chalcone. This compound undergoes several modifications and leads to the production of miscellany of flavanones, isoflavones, flaven-3-ols, flavones and anthocyanidins (Vickery and Vickery 1981). The isoprenoid pathway although underrated produces some important phenolics like artimisin, gossypol and stigmasterol. The classification of phenolic compounds is typically based on their carbon skeleton. Examples include simple phenols and benzoquinones (C6), acetophenones and phenylacetic acids (C6–C2), phenolic acids (C6–C1), phenylpropanes, coumarins, hydroxycinnamic acids and chromones (C6–C3), xanthones (C6–C1–C6), naphthoquinones (C6–C4), stilbenes and anthraquinones (C6–C2–C6), lignans and neolignans $(C6–C3)_2$, flavonoids and isoflavonoids (C6–C3–C6), biflavonoids $(C6–C3–C6)_2$, lignins $(C6–C3)_n$, catechol melanins $(C6)_n$ and condensed tannins $(C6–C3–C6)_n$. The phenolic compounds with the carbon skeleton ranging from C6 to C6–C3 come under the broad category of simple phenolic compounds, whereas all other phenolics including most complex polymers lignin ($(C6–C3)_n$) and tannins $(C6–C6)_n$ constitute complex phenolic compounds (Cheynier et al. 2013; Aoki et al. 2000; Hättenschwiler and Vitousek 2000; Iwashina 2000; Whiting 2001; Lattanzio and Ruggiero 2003). Prominent functions of these phenols include defense through the production of phytoalexins, lignin production for structural support, antioxidant activity to counteract peroxidants generated during pathogen or environmental stresses, pigmentation through anthocyanins and antimicrobial properties through isoflavonoids (Table 9.1) (Wink 1997; Nicholson and Hammerschmidt 1992; Winkel-Shirley 2002; Hammerschmidt 2003; Carletti et al. 2003; Hückelhoven 2007). The significance of phenols in wide range of physiological functions is becoming increasingly apparent. There are barely any plant diseases that are not associated with alterations in the metabolism of phenols (Ahuja et al. 2012). For instance, the preformed phenolics which the plant produces constitutively act as first line of pathogen defense to prevent any disease symptoms. Furthermore, when a plant is under attack by a pathogen, it triggers an induced resistance response. This response leads to the synthesis and accumulation of phenolic compounds such as antimicrobial compounds, enzymes and structural reinforcement, which can limit the growth of the pathogen (Al-Khayri et al. 2023). Despite some notable reports, the function of phenolics in regulating interactions between plants and pathogens has not yet been thoroughly studied.

9.2.1 Phenolics in Plant Defense

Plants come across innumerable attacks by pathogens and pests in their natural environment that may cause number of diseases (Mengiste 2012; Glazebrook 2005; Lai and Mengiste 2013). In order to survive the attack by these pathogens, the plants respond in such a way that may lead to either tolerance or resistance

Table 9.1 Phytoanticipins present in different plants with antibacterial properties against different pathogenic bacteria

Phytoanticipin	Phenolic class	Producing species	Bacteria affected	References
Amygdalin	Cyanogenic glycoside	*Prunus armenica* *Prunus persica* *Malas domestica*	*Pseudomonas aeruginosa*, *Streptococcus pyogenes*, *Serratia marcescens*, *Staphylococcus aureus*	Saleh et al. (2010)
Catechol	Simple phenols	*Ipomoea batatas* *Camellia sinensis*	*Pseudomonas putida*, *Pseudomonas pyocyanea*	Halder et al. (1998), Kocaçalışkan et al. (2006)
Pyrogallol	Phenolic acid	*Myriophyllum spicatum*	*Pseudomonas putida*, *Pseudomonas pyocyanea*	Kocaçalışkan et al. (2006)
Juglone	Naphthoquinones	*Juglanas nigra*	*Bacillus cereus*, *B. subtilis*, *Escherichia coli*, *Klebsiella pneumoniae*, *Pseudomonas aeruginosa*, *Staphylococcus aureus*	Clark et al. (1990), Pereira et al. (2007)
Protocatechuic acid	Phenolic acid	*Hibiscus sabdariffa* *Boswellia dalzielii* *Diospyros melanoxylon*	*Staphylococcus aureus*, *Klebsiella pneumoniae*, *Acinetobacter baumannii*, *Pseudomonas aeruginosa*	Liu et al. (2005), Mallavadhani and Mahapatra (2005), Alemika et al. (2006)
Vanillic acid	Phenolic acid	*Angelica sinensis* *Jatropha curcus*	*Bacillus subtilis*, *Escherichia coli*, *Klebsiella pneumonia*, *Pseudomonas aeruginosa*, *Staphylococcus aureus*	Duke (1992), Namuli et al. (2011)
Chlorogenic acid	Hydrocinnamic acid	*Arctium lapa* *Phyllostachys edulis* *Calluna vulgaris*	*Bacillus subtilis*, *Escherichia coli*, *Salmonella Typhimurium*, *Staphylococcus aureus*, *Stenotrophomonas maltophilia*, *Streptococcus pneumoniae*, *Shigella dysenteriae*	Kweon et al. (2001), Lou et al. (2010), Lou et al. (2011)
Rutin	Flavonol	*Carpobrotus edulis* *Castanea sativa* etc.	*Enterobacter cloacae*, *Enterobacter aeroge*, *Klebsiella pneumoniae*, *Escherichia coli*, *Staphylococcus aureus*, *Pseudomonas aeruginosa*, *Proteus vulgaris*, *Salmonella typhi*	Basile et al. (2000), Kreft et al. (1999)

(continued)

Table 9.1 (continued)

Phytoanticipin	Phenolic class	Producing species	Bacteria affected	References
p-Coumaric acid	Hydroxycinnamic acid	*Allium cepa* *Dacus carrota* *Lycopersicum aesculentum* *Ocimum basilum* *Pisum sativum*	*Bacillus subtilis*, *Escherichia coli*, *Salmonella typhimurium Shigella dysenteriae*, *Streptococcus pneumoniae*	Mussatto et al. (2007), Lou et al. (2012)
Naringin	Flavonoid	Citrus fruits	*Listeria monocytogenes*, *Staphylococcus aureus*	Peterson et al. (2006), Celiz et al. (2011)
Oleuropein	Terpene	*Olea europaea*	*Haemophilus influenzae*, *Moraxella catarrhalis*, *Salmonella typhi*, *Staphylococcus aureus*, *Vibrio cholerae*, *Vibrio parahaemolyticus*, *Vibrio parahaemolyticus*	Bisignano et al. (1999)
Hispidulin and Nepetin	Flavonoid	*Salvia fruticose* *Tarconanthus camphoratus* *Tamarix ramosissima*	*Listeria monocytogenes*, *Staphylococcus castellani*, *Staphylococcus aureus*	Osman et al. (2014), Ren et al. (2019)
Genistein	Isoflavonoid	*Flemingia vestita* *F. macrophylla* *Genista tinctoria* *Glycine max* *Lupinus* spp.	*Escherichia coli*, *Klebsiella pneumonia*, *Shigella sonnei*, *Staphylococcus aureus*	Rao (1991), Kaufman et al. (1997), Hong et al. (2006)
Quercetin	Flavonoid	Broadly present in fruits and vegetables	*Bacillus cereus*, *Escherichia coli*, *Klebsella pneumoniae*, *Staphylococcus aureus*, *Pseudomonas aeruginosa*	Bravo and Anacona (2001), "USDA Database for the Flavonoid Content of Selected Foods, Release 3" U.S. Department of Agriculture (2011)

strategies. Resistance strategies involve traits that limit or inhibit the disease caused by pathogen, whereas tolerance strategies does not limit or inhibit the disease but lower the consequences on plant fitness by buffering the effects of disease. The tolerance strategies often adjust the physiology of the plant to recompense for the disease damage by increasing nutrient uptake, leaf size and number, chlorophyll pigment content, advancing plant phenology etc. (Paul et al. 2000; Taylor et al. 2004; Dietrich et al. 2005). Resistance strategies encompass chemical and physical barriers and the underlying mechanisms to swiftly blowout infection (hypersensitive response), localized cell death to limit the spread of disease within the host and to encourage the systemic acquired resistance (SAR) (Jones and Dangl 2006). The plants in order to carry-out the resistance process produce diverse phenolic compounds either against the pathogen infection (phytoalexins) or normally as a part of their normal development (phytoanticipins) (Jones and Dangl 2006; Dakora and Phillips 1996; Lattanzio et al. 2006). These compounds function as signal transducers, inhibitors, pesticides, natural toxicants against range of invaders like nematodes, herbavores, phytophagous insects and bacterial and fungal pathogens (Akhtar and Malik 2000; Lattanzio et al. 2006). However, both the resistance and tolerance strategies come with the cost of host resources, which the plant reallocates for the biosynthesis and accumulation of the products needed for either strategy. The plants can reduce this cost by synthesizing defense chemicals only after exposure to the pathogen, though this could be risky as the infection process may be too fast for the plant to initiate effective defense response. However, if the plant learns about its pattern of infection, it can program its defense strategy likewise to minimize the cost, e.g., the plants frequently infected by pathogens may be better off spending in constitutive production of defense-related products, while plants which are rarely infected may depend mostly on induced defenses (Morrissey and Osbourn 1999; Purrington 2000; Wittstock and Gershenzon 2004; Koricheva et al. 2004). Moreover, it has been found that plants by default rely on both phytoanticipins and phytoalexins, and they do not care much about the cost which they bear to produce these compounds. However, it is clear that plant amends its phenolic status so that to protect itself from pathogens before and after the pathogen attack at minimized cost.

9.2.2 Phytoanticipins (Troops Ready) vs. Phytoalexins (Troops Arriving)

Plants possess an innate immunity to halt the spread of bacterial diseases, by the production of antimicrobial phenolic compounds that operate at different lines of defense. The compounds that act as first line of defense are called phytoanticipins and others are induced by pathogen elicitors and are called phytoalexins. Phytoanticipins are "low molecular weight compounds, antimicrobial in nature that are present in plants before challenge by plant pathogens, or are produced from pre-existing precursors after encountering pathogen" (Osbourn 1996). Some phytoanticipins are found to be recruited either at the surface, while others are stored in organelles and vacuoles as preformed compounds and released by a hydrolyzing enzyme after encountering pathogen. Although these stored compounds are released after the exposure to the pathogen, they are not considered as phytoalexins because the enzyme used as liberation molecule is not formed de novo. However, the phytoalexins are described as "low molecular weight compounds antimicrobial in nature that the produced and accumulated in plants only after exposure to pathogens" (Bednarek and Osbourn 2009; Grayer and Kokubun 2001). Therefore, biosynthesis of phytoalexins entails not only translational or transcriptional activity in the plant after exposure to pathogens but also the mechanisms that involve trafficking and their secretion at the site of infection (Bednarek and Osbourn 2009). The definitions of phytoanticipins and phytoalexins are sometimes arbitrary as they are based on the deviations in the time of synthesis without concerning much about their chemical composition. Since the same chemical can serve as phytoanticipin in one plant and phytoalexin in other, even in the same plant, a chemical can be phytoalexin in one organ and phytoanticipin in another (Donnez et al. 2009; Edmands et al. 2013). The antimicrobial compound maackiain, an isoflavonoid derivative, presents the fine example of above problem. The maackiain in the roots of red clove act as phytoanticipin as they are stored as aglycone of preformed glycoside in the vacuoles and are released during tissue decompartmentalization by preformed glucosidase (McMurchy and Higgins 1984). However, maackiain is also produced as phytoalexin when the plant is under pathogen attack or the same plant's defense system is induced by MAMP (microbe-associated molecular patterns) (Higgins and Smith 1972; Dewick 1975). Furthermore, the phytoalexins being more evolved and effective than phytoanticipins represent an important component of an array of induced defense systems. In addition to phytoanticipins and phytoalexins, mechanisms like production of oxidizing agents, chitinase and glucanase enzymes, pathogen-related proteins and cell wall reinforcement, all are believed to have evolved with time to serve as the basis for plant defense mechanisms (van der Hoorn and Jones 2004; Torres et al. 2002; Hückelhoven 2007).

9.2.3 Phytoanticipins Are Preformed Resistance Factors Against Bacterial Pathogens

Plants produce phytoanticipins as the normal part of their growth and development in the expectation of serving the plant at the first line of defense as chemical barriers to potential pathogens (Wang et al. 2014). These preformed antibiotic compounds are distributed in the tissue-specific manner within plants. The tendency of the phytoanticipins to get concentrated in the exterior layer of plant organs suggest that their primary function is to act as deterrents to pests and pathogens and/or inhibit their growth (Bennett and Wallsgrove 1994). It is evident from the studies that phytoanticipins like catechol and proto-catechuic acid effect growth of pathogens at the plant surface. Moreover, it is evident that these preformed compounds are sequestered in the organelles and vacuoles; therefore, the concentration that an invading pathogen encounters depends upon the extent of the site damage caused by that pathogen. The type and nature of preformed phenolics encountered by any pathogen may also vary depending upon age, host genotype and environmental factors (Price et al. 1987). Scientists were of the opinion that these variations in preformed compounds against any particular pathogen may be natural, but all the attempts failed to reveal any positive correlation. There is huge diversity of phytoanticipins reported to be produced by plants in the genotype-specific manner (Table 9.1). Some well-known examples include saponins, cyanogenic glycosides, alkylated resorcinols, glucosinolates, dienes etc. (Bennett and Wallsgrove 1994; Kuc 1995; Osbourn 1996). However, to determine their potential role in plant defense against bacterial infections, only a limited subset of phenolic chemicals have been thoroughly examined.

Saponins being the most studied phytoanticipins are plant glycosides having surfactant properties (Osbourn et al. 2011; Faizal and Geelen 2013). They are usually made up of glycon (sugar) and aglycon (nonsugar) part linked together by glycosidic linkage. The aglycone part, sometimes called sapoginin is either steroid or a triterpinoid which is attached to one or more glycone moieties particularly pentoses and hexoses. The saponin molecules are stored in vacuoles as inactive precursors and are activated by hydrolase enzymes following wounding or pathogen infection (Ökmen et al. 2013; Nielsen et al. 2004). They possess many biological activities which is mostly the function of their chemical structure, such as anti-inflammatory, antiviral, antifungal and antibacterial (Osbourn 2003; Morrissey and Osbourn 1999). Saponins first bind the cell membrane, function like detergents in increasing the permeability of bacterial cell membranes, which in turn facilitates the influx of antibiotics through the cell membranes and destroys the functioning of the bacterial cell (Ahuja et al. 2012; Bednarek 2012). Khan et al. (2018) while working on saponins derived from *Camellia sinensis* confirmed the antibacterial effects against both Gram-positive and Gram-negative bacteria. These results confirm that saponins are involved in protecting the plants against bacterial diseases. Saponins are not only phytoanticipins that may target plant pathogens but are actually a member of class comprising of numerous such compounds with similar inhibitory functions. Although these inhibitory compounds are effective against broad range of pathogens, some pathogens successfully circumvent their antibiotic effect by tolerating or detoxifying them (Bennett and Wallsgrove 1994; Osbourn 1996).

9.2.4 Induced Disease Resistance by Phenolic Compounds

When a pathogen detoxifies the phytoanticipins or somehow manages to escape from constitutive defense barriers it may be recognized at the cell surface. This recognition of an invariant pathogen-associated molecular pattern (PAMP), also known as elicitors that represent an entire class of microbial species, likely activates the induced plant defense (Mackey and Mcfall 2006; Boller and Felix 2009). These PAMPs activate an immune response known as PAMP-triggered immunity when they are detected by pattern recognition receptors (PRR) on cell surfaces (Monaghan and Zipfel 2012; Jones and Dangl 2006). The PAMPs are actually the molecules important for microbial survival, although they activate the plants immune system, they do not necessarily have a role in pathogenicity. These PAMPs include a variety of structural compounds including polysaccharides, lipids, lipopolysaccharides, glycoproteins, flagellin, oligosaccharides and even enzymes (Zipfel and Felix 2005; Felix et al. 1999). Some plant species have extended the capacity of their recognition system over time for additional microbial molecules. For example, members of Brassicacae and Solanaceae recognize EF-Tu (bacterial elongation factor) and bacterial cold shock proteins respectively as pattern recognition molecules in order to activate the induced immunity responses (Kunze et al. 2004; Boller and Felix 2009; Furukawa et al. 2014). PAMP sensing systems initiate downstream signaling cascades, the primary goal of which is to activate "PAMP triggered immunity" in natural plant–pathogen encounters (Jones and Dangl 2006). PAMP-activated immunity induces broad-spectrum innate immune responses to react to infections that function both locally at pathogen infection site and systemically in uninfected tissue. Typical early PAMP responses are ion fluxes across the plasma membrane, the production of salicylic acid, NO, ROS, ethylene. The early responses in turn trigger the late responses which include cell wall strengthening and synthesis of phytoalexins (Boller and Felix 2009). Furthermore, the salicylic acid produced as the early PAMP response has been found occupying the central position in many plant defense strategies particularly in the establishment of SAR (Boller and Felix 2009; Tsuda et al. 2008). PAMPs also cause the production of calcium-dependent protein kinases (CDPK), the activation of mitogen-activated protein kinase (MAPK) cascades, and alterations in the transcription of numerous defense-related genes (Boller and Felix 2009; Rasmussen et al. 2012; Wurzinger et al. 2011).

9.3 Salicylic Acid Mediated Systemic Acquired Resistance

SAR can be described as whole plant resistance response that comes in place succeeding a previous localized exposure to a pathogen (Walters et al. 2014; Durrant and Dong 2004). SAR is found to function analogous to innate immune system present in animals, and both are believed to be evolutionary conserved. Bernard and Nobecourt were the first to report that plants surviving pathogen attack develop systemic resistance that protects the plant against subsequent infections. This was supported by Kuc (1995) who experimentally proved that inoculating one cucumber leaf with pathogen, develops resistance not only against this pathogen but other pathogens as well. Systemic acquired resistance provides protection against diverse pathogens, not just one that induced the response. By this means systemic acquired resistance primarily differs from the specific antigen-antibody mediated immunity of animals.

The process for inducing SAR entails the formation of slowly developing necrotic lesions. Plants respond the pathogen attack with the release of alarm signals like jasmonic acid (JA), salicylic acid (SA) and ethylene (ET) that greatly varies in timing and chemical composition. These alarm signals also called signal signatures that contributes to the specific plant's primary induced defense response (Bezemer and van Dam 2005; De Vos et al. 2005; Stout et al. 2006). SAR is believed to be induced with the commencement of localized responses like phytoalexin accumulation, hypersensitive cell death, deposition of papilla and lignin. After the appearance of necrotic lesion, a signal originates from the lesion and is translocated systematically in the phloem. This signal is active until the plant begun to flower and is graft transmissible not genus, species or cultivar specific. This systemic signal then guides the whole plant to take further action against pathogen challenge. The plant then responds with the rapid and intense expression of ROS species, phytoalexins accumulation, cell death and increased expression of pathogen-related proteins to tackle the pathogen infection (Chong et al. 1999; Nishimura et al. 2003; Tsuda et al. 2008; Feys and Parker 2000).

The precise identification of the signal involved in triggering SAR is still the topic of debate. However, there are several molecules that own the features to induce the SAR, such as, nitric oxide, salicylic acid, phosphorylated sugar glyceraol-3-phosphate (G3P), B-ionone, pipecolic acid, ROS and jasmonic acid (Ryals et al. 1996; Chanda et al. 2011; Mandal et al. 2012; Yu et al. 2013; Wang et al. 2014; El-Shetehy et al. 2015; Gao et al. 2015). However, systemic resistance is induced by crosstalk between many complex signal transducing pathways controlled by many stress signals (Tada et al. 2008; Lindermayr et al. 2010; Wang et al. 2014). Salicylic acid is a key player in the plant's defense mechanisms, whose biosynthesis and accumulation in plants take place via contribution of two distinct sub-branches of the shikimic acid pathway, viz.; isochorismate synthase (ICS) and phenylalanine ammonia-lyase (PAL)-derived pathway. Both these subpathways use chorismate as the common precursor (Shah 2003; Kachroo and Kachroo 2009; Chen et al. 2009; Yu et al. 2010; Dempsey et al. 2011; An and Mou 2011; Singh et al. 2013; Fig. 9.1). The first step of the PAL-derived pathway involves converting phenylalanine, a byproduct of the shikimic acid pathway, into *trans*-cinnamic acid. This reaction is catalyzed by the enzyme phenylalanine ammonia lyase, which is induced in response to pathogen attack (Huang et al. 2010). The experimental support for this came from the work of Umesha (2006) on tomato, where she found that tomato plants when inoculated with *Clavibacter michiganasis*, the resistant genotypes, showed increased expression of enzyme PAL as compared to susceptible plants. This increased expression of PAL in resistant genotypes can be well correlated with the degree of SAR induction conferred by this enzyme. On the other hand, isochromate synthase pathway converts the chorismate to isochorismate by enzyme isochromate synthase. The isochromate is then converted into salicylic acid by the enzyme isochorismate pyruvate lyase (IPL). The Arabidopsis contains two isoforms of ICS (ICS1 and ICS2). The ICS1 also called SID2 is responsible for synthesis of 95% of the pathogen-induced salicylic acid (Strawn et al. 2007; Garcion et al. 2008). The relative contribution of ICS and PAL toward SA biosynthesis is different in different plant species, for instance, in Arabidopsis, ICS pathway appears to be accountable for majority for the pathogen-induced synthesis of salicylic acid. A mutation in either PAL or ICS impairs SAR, suggesting that both pathways are critical for the establishment of SAR in plants (Wildermuth et al. 2001; Wang et al. 2014). In transgenic plants, the compromised systemic acquired resistance (SAR) phenotype that expresses the bacterial enzyme salicylate hydroxylase (NahG) further highlights the significance of salicylic acid (SA) in SAR, as this enzyme converts SA into catechol, making it unavailable for SAR signaling (Vernooij et al. 1994). However, it is still unclear that what guides the specific recruitment of ICS or PAL for SA biosynthesis and needs to be addressed.

Once synthesized, salicylic acid (SA) binds to two proteins located in the cell membrane of plants. One of these proteins has catalase activity and upon binding to SA gets inhibited which then leads to the localized accumulation of hydrogen peroxide, causing various changes in the plant cell membrane. The second salicylic acid-binding protein is directly involved in the increased gene expression of many pathogen-related proteins and eventually increases the resistance to hosts to pathogens (Ryals et al. 1996; Tsuda et al. 2008; Feys and Parker 2000). Research has shown that salicylic acid levels increase rapidly around necrotic lesions and remain elevated in plants that have acquired systemic resistance. Additionally, some synthetic analogs of salicylic acid, such as dichloroisonicotinic acid (INA), benzothiazoles and benzothiazole, benzo (1,2,3) thiadiazole-7-carbothioic acid *S*-methyl ester (BTH), have been found to induce similar resistance responses to those stimulated by salicylic acid. However, the disease protectant like dichloroisonicotinic acid (INA) may prove effective in inducing resistance under field and green house trails, but it may sometimes prove phytotoxic multiple studies have supported the role that exogenous application of either SA or its synthetic analogs play in inducing disease resistance in plants (Conrath 2009; Görlach et al. 1996; Bovie et al. 2004).

9.4 Cell Wall Strengthening

The first challenge microbes face is to get through the cell wall of the host, while avoiding being caught by the immune receptors. Some of these receptors recognize conserved microbial patterns; others sense the physical changes in cell wall caused by pathogen attack (Hématy et al. 2009). Cell walls of plants are tough and dynamic in nature, which get assembled by joining complex sugars, i.e., cellulose with diverse cross-links of hemicellulose followed by extensive postsynthetic modifications to fulfill the demands of the prevailing requirements (Scheller and Ulvskov 2010; Davidsson et al. 2013; Nuhse 2012; Pauly et al. 2013). In addition, the primary walls also contain pectin polysaccharides, while secondary walls contain less pectin but more lignin polysaccharides (Endler and Persson 2011). In order to penetrate this tough structured cell wall, microbes are well equipped with the collection of wall degrading enzymes which determine the strategic virulence factors (Davidsson et al. 2013; Nuhse 2012). The pathogen attack triggers several changes in cell wall composition and integrity. Most of these changes have been described well, while there are many more which are still unclear and need to be described.

The first response of the cell wall against attempted pathogen penetration is often increased in cytoplasmic streaming and buildup of cytoplasm near the site of attempted penetration in order to transport the cellular apparatus to carry out the synthesis of cell wall fortifications (Osbourn 1996; Hückelhoven 2007). The cell wall fortifications produced in response to call wall penetration varies with the pathogen type. Some pathogens may induce papilla deposition along callose (P-1,3 glucan) with lignin, silicon and proteins (Nuhse 2012; Endler and Persson 2011). These reinforcements are deposited just under the penetration site between the plasma membrane and cell interface. The deposition of papillae is a common response of many plants particularly against cell wall penetration of epidermal cells by phytopathogens. In addition, a polymer of β-1,3 glucan called callose is found to be ubiquitously present in cell wall reinforcements. Studies found that when callose synthase was knocked down through RNAi in leaves of *Citrus lemon*, the susceptibility of plant toward *Xanthomans citri* got increased (Underwood 2012). Furthermore, *Xanthomonas compestris* has been found to inhibit callose deposition in *Arabidopsis* and *Nicotiana benthamiana* by producing exopolysaccharide called xanthan, thus suggesting a possible role of callose deposition in limiting the excess of *Xanthomonas* to the host cells (Yun et al. 2006). Other kinds of cell wall reinforcement include hydroxyproline-rich glycoproteins, which are structural proteins required for the assembly of secondary wall thickenings. The products of the genes that are responsible for the synthesis of hydroxyproline-rich glycoproteins are kept ready in anticipation of pathogen invasion. The onset of pathogen leads to the production of hydrogen peroxide, which in turn leads to extensive cross linking between the proteins that are high in hydroxyproline-rich glycoproteins and other components of the wall (Albersheim et al. 2010). This cross-linking directs the deposition of lignin on plant cell walls, thereby making the cell wall more resistant to digestion by cell wall degrading microbial enzymes like pectinases and cellulases (Sederoff et al. 1999; Davin and Lewis 2000; Boerjan et al. 2003; Sattler and Funnell-Harris 2013). In addition to providing strength to the cell wall, lignin entraps bacterial cells and restricts bacterial growth and movement. Lignin also blocks the nutrient uptake and water by the pathogen and diffusion of pathogen enzyme and toxin into the host cells. Furthermore, the free radicals and the precursors produced during lignin biosynthesis are found to be toxic to microbes and deactivate their enzymes, elicitors, toxins and suppressors. The antimicrobial effects of lignin become multifold in the presence of ROS along with the activation of phenol oxidase enzymes. These phenol oxidase enzymes convert the phenols into more toxic phenolic polymers and quinones during the induced defense response (Sattler and Funnell-Harris 2013). Thus, lignin could be considered as most important cell wall reinforcements because in addition to mechanical barrier, they provide many essential services like production of toxic precursors, free radicals, diffusion blockage to and from the host, pathogen entrapment in order to increase resistance of plant against pathogens.

9.5 Phytoalexins

When a pathogen detoxifies the phytoanticipins or somehow manages to escape from constitutive defense barriers, it is destroyed by the plants active defense systems (Andersen et al. 2018). Plants possess a range of active defense responses in addition to passive defense bestowed by phytoanticipins. Phytoalexins represent one of the active responses whose synthesis and accumulation are induced elicitors only after the onset of pathogenic signal (Akhtar and Malik 2000; Lattanzio et al. 2006; Boller and Felix 2009). In natural settings, the presence of microorganisms provides the stimulus, and the host's perception of it starts a chain of events that leads to the synthesis of phytoalexins (Jones and Dangl 2006). After synthesis, the phytoalexins buildup at infection sites where they halt the bacterial growth. As a result, it is appropriate to regard phytoalexins as trustworthy plant-defense molecules that combat bacterial infections (Table 9.2 Nicaise et al. 2009). The phytoalexin success depends upon the time taken for the response, location of the synthesis and magnitude of response. There is decent experimental relationship between the trio in most host-pathogen interactions. Studies have shown that resistant cultivars buildup phytoalexin faster and in higher concentrations than susceptible plants, as transcription commences within 1 h of microbial pattern recognition and phyalexins concentrations peak within 24 h after pathogen attack (Hammerschmidt 1999). Moreover, the synthesis of phytoalexins is localized in cells that are proximately surrounding the lesion and do not get dispersed throughout the plant. Experiments using techniques such as laser microprobe analysis, histochemistry, radioimmunoassay, hybridization and immunocylochemistry have confirmed that there is a localized action in several host–pathogen interactions during the biosynthesis pathway (Kuc 1995). However, it has been found that if biosynthesis of phytoalexins is inhibited, disease resistance is lost. Furthermore, exogenous application of phytoalexins or the expression of novel phytoalexins in transgenic plants is found to increase disease resistance (Kuc 1995). The hydroxycoumarin scopoletin (6-methoxy-7-hydroxycoumarin) is a form of phytoalexin discovered in tobacco plants that accumulates in response to infections and elicitors and has antimicrobial properties (Valle et al. 1997; Chong et al. 2002; Matros and Mock 2004). Rishitin is another phytoalexin possessing antagonistic activity against *Erwinia atroseptica* (Lyon and Bayliss 1975). Resveratrol also possess antibacterial activity against variety of bacteria including *Helicobacter*, *Chlamydia*, *Staphylococcus*, *Pseudomonas*, *Enterococcus* and *Neisseria* (Langcake and Pryce 1976; Sobolev et al. 1955). There are some other extensively studied phytoalexins including pisatin in pea; phaseollin in bean; alfalfa and clover; glyceollin in soybean rishitin in potato; capsidiol in pepper, gossypol in cotton and sakuranetin from rice (Va VanEtten and Bateman 1971; Perrin and Pisatin 1961; Ebel et al. 1976; He and Dixon 2000; Hasegawa et al. 2014). Even though the contribution of innumerable number of phenolic compounds in the plant immunity has been explicitly confirmed, we are still far off from understanding the mechanism of their function.

9.6 Conclusion

During the co-evolution of bacteria and plant hosts, plants have developed numerous strategies to protect themselves from microbial invaders. Phenols have long been accompanying in active and passive defense responses of plants. Phytoanticipins

Table 9.2 Phytoalexins produced from higher plants against diverse pathogenic bacteria

Scopoletin	Coumarin	*Solanum nigrum* *Scopolia carniolica* *Mallotus resinosus* *Scopolia japonica* *Artemisia scoparia* *Urtica dioica* *Brunfelsia* *Viburnum prunifolium* *Kleinhovia hospita*	*Staphylococcus aureus* *Pseudomonas aeruginosa* *Klebsiella* sp. *Acinetobacter baumannii* *Stenotrophomonas maltophila* *Escherichia coli* *Salmonella choleraesuis* *Enterococcus faecium*	Ji et al. (2004), Zhao et al. (2010), Buathong et al. (2019)
Camalexin	Simple alkaloid	*Brassicaceae*	*Pseudomonas syringae* *Pseudomonas lingam* *Acinetobactor junii* *Agrobacterium tumefaciens* *Bacillus pumilus* *Bacillus subtilis* *Escherichia coli*	Pedras et al. (1998), Browne et al. (1991), Kumar et al. (2017)
Coumestrol	Isoflavonoid	*Leguminosae*, *Trifolium* *Medicago sativa*	*Xanthomonas* *Achromobacter* *Pseudomonas*	Jeand et al. (2013), Wyman and Van Etten (1978)
Wyerone	Furanoacetylene	*Leguminosae*	*Micrococcus lysodeikticus* *Bacillus megaterium* *Corynebacterium betae* *Corynebacterium fascians* *Mycobacterium phlei* *Streptomyces scabies*	Tanaka et al. (1983), Gnanamanickam and Mansfield (1981)
Kiavetone	Isoflavonoid	*Leguminosae*	*Micrococcus lysodeikticus* *Bacillus megaterium* *Corynebacterium betae* *Corynebacterium fascians* *Mycobacterium phlei* *Streptomyces scabies*	Gnanamanickam and Mansfield (1981), Goossens and Vendrig (1982)

(continued)

Table 9.2 (continued)

Gossypol	Terpenoid	*Gossypium*	*Enterococcus faecium* *Enterococcus faecalis* *Staphylococcus aurens*	Przybylski et al. (2009)
Medicarpin	Isoflavonoid	*Medicago sativa* *Swartzia madagascariensis.*	*Pseudomonas aeruginosa* *Escherichia coli* *Bacillus subtilis* *Staphylococcus aureus*	Rufatto et al. (2018)
Resveratrol	Stilbene	*Vitaceae* *Rosaceae* *Arachis hypogea*	*Staphylococcus aureus* *Escherichia coli* *Acinetobacter baumannii* *Staphylococcus epidermidis*	Jasiński et al. (2013), Sangkanu et al. (2017)
Brassinin	Sulfur containing phytoalexin	*Brassicaceae*	*Pseudomonas cichorii* *Staphylococcus aureus* *Bacillus subtilis* *Pseudomonas aeruginosa* *Escherichia coli*	Pedras et al. (1998), Browne et al. (1991)
Sakuranetin	Flavonoid	*Rosaceae* *Asteraceae*	*Cryptococcus* *Saccharomyces cerevisiae*	Domínguez and Roehll de la Fuente (1973), Teresa et al. (1984), Grecco et al. (2014), Stompor (2020)
Nomilactone B	Terpenoid	*Poaceae*	*Pseudomonus ovalis* *Bacillus cereus* *Bacillus pumilus*	Poloni and Schirawski (2014), Schmelz et al. (2014), Fukuta et al. (2007)
Coumarin	Terpenoid	*Solanaceae*	*Bacillus subtilis* *Escherichia coli Staphylococcus aureus Pseudomonas aeruginosa*	de Souza et al. (2005), Nitiema et al. (2012)
Falcarinol	Polyacetylene	*Umbelliferae* *Daucus carota* *Panax ginseng*	*Bacillus subtilis Entereococcus faecialis* *Escherichia coli* *Staphylococcus aureus*	Brantner and Grein (1994)

responsible for passive immunity act at first line of defense and not only blocks pathogen penetration but also inhibits their growth and spread. When a pathogen bypasses the plant's passive defense mechanisms, receptors in the plant activate signaling pathways that lead to the expression of PTI (PAMP-triggered immunity). The defense mechanisms of plants rely on the ability of PPRs (pattern recognition receptors) to recognize PAMPs (microbial-associated molecular patterns), transmit signals and activate defense responses through pathways involving genes and their products. Nonetheless, there is huge reservoir of constitutive and inducible phenolic compounds with known antimicrobial activity, but our knowledge about their induced biosynthesis and mechanisms that contribute to plant immunity is scare and needs to be investigated. Furthermore, phenolic compounds have become epicenter of current research on plant disease research, which will uncover additional resistance regulation responses and mechanisms in future years. Therefore, recognizing proper response molecules and mechanisms behind plant immunity and pathogen resistance will significantly assist agricultural productivity by decreasing crop loss, and also help in understanding fundamentals of co-evolution and molecular interactions that underpins this field and other biological systems.

References

Ahuja I, Kissen R, Bones AM (2012) Phytoalexins in defense against pathogens. Trends Plant Sci 17(2):73–90

Akhtar M, Malik A (2000) Roles of organic soil amendments and soil organisms in the biological control of plant-parasitic nematodes: a review. Bioresour Technol 74(1):35–47

Albersheim P, Darvill A, Roberts K, Sederoff R, Staehelin A (2010) Plant cell walls. Garland Science

Alemika TE, Onawunmi GO, Olugbade TA (2006) Antibacterial phenolics from Boswellia dalzielii. Nigerian J Nat Prod Med 10:108–110

Al-Khayri JM, Rashmi R, Toppo V, Chole PB, Banadka A, Sudheer WN, Nagella P, Shehata WF, Al-Mssallem MQ, Alessa FM, Almaghasla MI (2023) Plant secondary metabolites: the weapons for biotic stress management. Metabolites 13(6):716

An C, Mou Z (2011) Salicylic acid and its function in plant immunity F. J Integr Plant Biol 53(6): 412–428

Andersen EJ, Ali S, Byamukama E, Yen Y, Nepal MP (2018) Disease resistance mechanisms in plants. Genes 9(7):339

Aoki T, Akashi T, Ayabe SI (2000) Flavonoids of leguminous plants: structure, biological activity, and biosynthesis. J Plant Res 113(4):475

Basile A, Sorbo S, Giordano S, Ricciardi L, Ferrara S, Montesano D, Cobianchi RC, Vuotto ML, Ferrara L (2000) Antibacterial and allelopathic activity of extract from Castanea sativa leaves. Fitoterapia 71:110–116

Bednarek P (2012) Chemical warfare or modulators of defence responses–the function of secondary metabolites in plant immunity. Curr Opin Microbiol 15(4):407–414

Bednarek P, Osbourn A (2009) Plant-microbe interactions: chemical diversity in plant defense. Science 324(5928):746–748

Bennett RN, Wallsgrove RM (1994) Secondary metabolites in plant defence mechanisms. New Phytol 127(4):617–633

Bezemer TM, van Dam NM (2005) Linking aboveground and belowground interactions via induced plant defenses. Trends Ecol Evol 20(11):617–624

Bisignano G, Tomaino A, Cascio RL, Crisafi G, Uccella N, Saija A (1999) On the in-vitro antimicrobial activity of oleuropein and hydroxytyrosol. J Pharm Pharmacol 51(8):971–974
Boerjan W, Ralph J, Baucher M (2003) Lignin biosynthesis. Annu Rev Plant Biol 54(1):519–546
Boller T, Felix G (2009) A renaissance of elicitors: perception of microbe-associated molecular patterns and danger signals by pattern-recognition receptors. Annu Rev Plant Biol 2(60): 379–406
Bovie C, Ongena M, Thonart P, Dommes J (2004) Cloning and expression analysis of cDNAs corresponding to genes activated in cucumber showing systemic acquired resistance after BTH treatment. BMC Plant Biol 4(1):1–12
Brantner A, Grein E (1994) Antibacterial activity of plant extracts used externally in traditional medicine. J Ethnopharmacol 44(1):35–40
Bravo A, Anacona JR (2001) Metal complexes of the flavonoid quercetin: antibacterial properties. Transit Med Chem 26(1):20–23
Browne LM, Conn KL, Ayer WA, Tewari JP (1991) The camalexins: new phytoalexins produced in the leaves of Camelina sativa (Cruciferae). Tetrahedron 47(24):3909–3914
Buathong R, Chamchumroon V, Schinnerl J, Bacher M, Santimaleeworagun W, Kraichak E, Vajrodaya S (2019) Chemovariation and antibacterial activity of extracts and isolated compounds from species of Ixora and Greenea (Ixoroideae, Rubiaceae). PeerJ 7(7):e6893
Carletti P, Masi A, Wonisch A, Grill D, Tausz M, Ferretti M (2003) Changes in antioxidant and pigment pool dimensions in UV-B irradiated maize seedlings. Environ Exp Bot 50(2):149–157
Celiz G, Daz M, Audisio MC (2011) Antibacterial activity of naringin derivatives against pathogenic strains. J Appl Microbiol 111(3):731–738
Chanda B, Xia Y, Mandal MK, Yu K, Sekine KT, Gao QM, Selote D, Hu Y, Stromberg A, Navarre D, Kachroo A (2011) Glycerol-3-phosphate is a critical mobile inducer of systemic immunity in plants. Nat Genet 43(5):421–427
Chen Z, Zheng Z, Huang J, Lai Z, Fan B (2009) Biosynthesis of salicylic acid in plants. Plant Signal Behav 4(6):493–496
Cheynier V, Comte G, Davies KM, Lattanzio V, Martens S (2013) Plant phenolics: recent advances on their biosynthesis, genetics, and ecophysiology. Plant Physiol Biochem 72:1–20
Chisholm ST, Coaker G, Day B, Staskawicz BJ (2006) Host-microbe interactions: shaping the evolution of the plant immune response. Cell 124(4):803–814
Chong J, Baltz R, Schmitt C, Beffa R, Fritig B, Saindrenan P (1999) An early salicylic acid-, pathogen- and elicitor-inducible tobacco glucosyltransferase: role in compartmentalization of phenolics and H_2O_2 metabolism. FEBS Lett 458(2):204–208
Chong J, Baltz R, Schmitt C, Beffa R, Fritig B, Saindrenan P (2002) Downregulation of a pathogen-responsive tobacco UDP-Glc: phenylpropanoid glucosyltransferase reduces scopoletin glucoside accumulation, enhances oxidative stress, and weakens virus resistance. Plant Cell 1495: 1093–1107
Clark, Alice M., Tannis M. Jurgens, and Charles D Hufford (1990). Antimicrobial activity of juglone. Phytother Res 411–14
Conrath U (2009) Priming of induced plant defense responses. Adv Bot Res 51:361–395
Cushnie TT, Lamb AJ (2011) Recent advances in understanding the antibacterial properties of flavonoids. Int J Antimicrob Agents 38(2):99–107
Dakora FD, Phillips DA (1996) Diverse functions of isoflavonoids in legumes transcend antimicrobial definitions of phytoalexins. Physiol Mol Plant Pathol 49(1):1–20
Davidsson PR, Kariola T, Niemi O, Palva T (2013) Pathogenicity of and plant immunity to soft rot pectobacteria. Front Plant Sci 4:191
Davin LB, Lewis NG (2000) Dirigent proteins and dirigent sites explain the mystery of specificity of radical precursor coupling in lignan and lignin biosynthesis. Plant Physiol 123(2):453–462
De Vos M, Van Oosten VR, Van Poecke RM, Van Pelt JA, Pozo MJ, Mueller MJ, Buchala AJ, Métraux JP, Van Loon LC, Dicke M, Pieterse CM (2005) Signal signature and transcriptome changes of Arabidopsis during pathogen and insect attack. Mol Plant Microb Interact 18(9): 923–937

Dempsey DM, Vlot AC, Wildermuth MC, Klessig DF (2011) Salicylic acid biosynthesis and metabolism. The Arabidopsis Book 9:e0156
de Souza SM, Delle Monache F, Smânia A (2005) Antibacterial activity of coumarins. Zeitschrift fuer Naturforschung C 60(9-10):693–700
Dewick PM (1975) Pterocarpan biosynthesis: chalone and isoflavone precursors of demethylhomopterocarpin and maackiain in Trifolium pratense. Phytochemistry 14(4):979–982
Dewick PM (2002) Medicinal natural products: a biosynthetic approach. Wiley
Dietrich R, Ploss K, Heil M (2005) Growth responses and fitness costs after induction of pathogen resistance depend on environmental conditions. Plant Cell Environ 28(2):211–222
Dodds PN, Rathjen JP (2011) Plant immunity: towards an integrated view of plant–pathogen interactions. Nat Rev Genet 11(8):539–548
Domínguez XA, Roehll de la Fuente E (1973) Sakuranetin and pulcherryl acetate from Eupatorium havanense. Phytochemistry 12:2060
Donnez D, Jeandet P, Clément C, Courot E (2009) Bioproduction of resveratrol and stilbene derivatives by plant cells and microorganisms. Trends Biotechnol 27(12):706–713
Duke JA (1992) Database of phytochemical constituents of GRAS herbs and other economic plants. CRC Press
Durrant WE, Dong X (2004) Systemic acquired resistance (2004). Annu Rev Phytopathol 42:185–209
Ebel J, Ayers AR, Albersheim P (1976) Host-pathogen interactions: XII. Response of suspension-cultured soybean cells to the elicitor isolated from Phytophthora megasperma var. sojae, a fungal pathogen of soybeans. Plant Physiol 57(5):775–779
Edmands WM, Gooderham NJ, Holmes E, Mitchell SC (2013) S-Methyl-L-cysteine sulphoxide: the Cinderella phytochemical. Toxicol Res 2(1):11–22
El-Shetehy M, Wang C, Shine MB, Yu K, Kachroo A, Kachroo P (2015) Nitric oxide and reactive oxygen species are required for systemic acquired resistance in plants. Plant Sig Behav 10(9): e998544
Endler A, Persson S (2011) Cellulose synthases and synthesis in Arabidopsis. Mol Plant 4(2): 199–211
Faizal A, Geelen D (2013) Saponins and their role in biological processes in plants. Phytochem Rev 12(4):877–893
Fang Y, Lu Y, Zang X, Wu T, Qi X, Pan S, Xu X (2016) 3D-QSAR and docking studies of flavonoids as potent Escherichia coli inhibitors. Sci Rep 6(1):1–3
Felix G, Duran JD, Volko S, Boller T (1999) Plants have a sensitive perception system for the most conserved domain of bacterial flagellin. Plant J 18(3):265–276
Feys BJ, Parker JE (2000) Interplay of signaling pathways in plant disease resistance. Trends Genet 16(10):449–455
Fukuta M, Dang Xuan T, Deba F, Tawata S, Dang Khanh T, Min Chung I (2007) Comparative efficacies in vitro of antibacterial, fungicidal, antioxidant, and herbicidal activities of momilatones A and B. J Plant Interact 2(4):245–251
Furukawa T, Inagaki H, Takai R, Hirai H, Che FS (2014) Two distinct EF-Tu epitopes induce immune responses in rice and Arabidopsis. Mol Plant Microbe Interact 27(2):113–124
Gao QM, Zhu S, Kachroo P, Kachroo A (2015) Signal regulators of systemic acquired resistance. Front Plant Sci 6:228
Garcion C, Lohmann A, Lamodière E, Catinot J, Buchala A, Doermann P, Métraux JP (2008) Characterization and biological function of the ISOCHORISMATE SYNTHASE2 gene of Arabidopsis. Plant Physiol 147(3):1279–1287
Glazebrook J (2005) Contrasting mechanisms of defense against biotrophic and necrotrophic pathogens. Annu Rev Phytopathol 43:205–227
Gnanamanickam SS, Mansfield JW (1981) Selective toxicity of wyerone and other phytoalexins to gram-positive bacteria. Phytochemistry 20(5):997–1000
Goossens JF, Vendrig JC (1982) Effects of abscisic acid, cytokinins, and light on isoflavonoid phytoalexin accumulation in Phaseolus vulgaris L. Planta 154(5):441–446

Görlach J, Volrath S, Knauf-Beiter G, Hengy G, Beckhove U, Kogel KH, Oostendorp M, Staub T, Ward E, Kessmann H, Ryals J (1996) Benzothiadiazole, a novel class of inducers of systemic acquired resistance, activates gene expression and disease resistance in wheat. Plant Cell 8(4): 629–643

Grayer RJ, Kokubun T (2001) Plant–fungal interactions: the search for phytoalexins and other antifungal compounds from higher plants. Phytochemistry 56(3):253–263

Grecco DS, Dorigueto AC, Landre IM, Soares MG, Martho K, Lima R, Pascon RC, Vallim MA, Capello TM, Romoff P, Sartorelli P (2014) Structural crystalline characterization of sakuranetin—an antimicrobial flavanone from twigs of Baccharis retusa (Asteraceae). Molecules 19(6):7528–7542

Halder J, Tamuli P, Bhaduri AN (1998) Isolation and characterization of polyphenol oxidase from Indian tea leaf (Camellia sinensis). J Nutr Biochem 9(2):75–80

Hamilton-Miller JM (1995) Antimicrobial properties of tea (Camellia sinensis L.). Antimicrob Agents Chemother 39(11):2375–2377

Hammerschmidt R (1999) Phytoalexins: what have we learned after 60 years? Annu Rev Phytopathol 37(1):285–306

Hammerschmidt R (2003) Phytoalexin accumulation: response or defense. Physiol Mol Plant Pathol 3(62):125–126

Harborne AJ (1998) Phytochemical methods a guide to modern techniques of plant analysis. Springer Science & Business Media

Hartmann T (2008) The lost origin of chemical ecology in the late 19th century. Proc Natl Acad Sci 105(12):4541–4546

Hasegawa M, Mitsuhara I, Seo S, Okada K, Yamane H, Iwai T, Ohashi Y (2014) Analysis on blast fungus-responsive characters of a flavonoid phytoalexin sakuranetin; accumulation in infected rice leaves, antifungal activity and detoxification by fungus. Molecules 19(8):11404–11418

Hättenschwiler S, Vitousek PM (2000) The role of polyphenols in terrestrial ecosystem nutrient cycling. Trends Ecol Evol 15(6):238–243

He XZ, Dixon RA (2000) Genetic manipulation of isoflavone 7-O-methyltransferase enhances biosynthesis of 4′-O-methylated isoflavonoid phytoalexins and disease resistance in alfalfa. Plant Cell 12(9):1689–1702

Hématy K, Cherk C, Somerville S (2009) Host–pathogen warfare at the plant cell wall. Curr Opin Plant Biol 12(4):406–413

Higgins VJ, Smith DG (1972) Separation and identification of two pterocarpanoid phytoalexins produced by red clover leaves. Phytopathology 62(2):235–238

Hong H, Landauer MR, Foriska MA, Ledney GD (2006) Antibacterial activity of the soy isoflavone genistein. J Basic Microbiol 46(4):329–335

Huang J, Gu M, Lai Z, Fan B, Shi K, Zhou YH, Yu JQ, Chen Z (2010) Functional analysis of the Arabidopsis PAL gene family in plant growth, development, and response to environmental stress. Plant Physiol 153(4):1526–1538

Hückelhoven R (2007) Cell wall–associated mechanisms of disease resistance and susceptibility. Annu Rev Phytopathol 45:101–127

Humphry M, Bednarek P, Kemmerling B, Koh S, Stein M, Göbel U, Stüber K, Piślewska-Bednarek M, Loraine A, Schulze-Lefert P, Somerville S (2010) A regulon conserved in monocot and dicot plants defines a functional module in antifungal plant immunity. Proc Natl Acad Sci 107(50):21896–21901

Iwashina T (2000) The structure and distribution of the flavonoids in plants. J Plant Res 113(3):287

Jasiński M, Jasińska L, Ogrodowczyk M (2013) Resveratrol in prostate diseases–a short review. Cent Eur J Urol 66(2):144

Ji M, Shannon HJ, Hecht SM (2004) A Coumarin from Mallotus r esinosus that mediates DNA cleavage. J Nat Prod 67:1614–1616

Jones JD, Dangl JL (2006) The plant immune system. Nature 444(7117):323–329

Kachroo A, Kachroo P (2009) Fatty acid–derived signals in plant defense. Annu Rev Phytopathol 47:153–176

Kaufman PB, Duke JA, Brielmann H, Boik J, Hoyt JE (1997) A comparative survey of leguminous plants as sources of the isoflavones, genistein and daidzein: implications for human nutrition and health. J Altern Complete Med 3(1):7–12
Khan MI, Ahhmed A, Shin JH, Baek JS, Kim MY, Kim JD (2018) Green tea seed isolated saponins exerts antibacterial effects against various strains of gram positive and gram-negative bacteria, a comprehensive study in vitro and in vivo. Evid Based Complement Altern Med 2018:1
Kocaçalışkan I, Talan I, Terzi I (2006) Antimicrobial activity of catechol and pyrogallol as allelochemicals. Zeitschrift für Naturforschung C 61(9–10):639–642
Koricheva J, Nykänen H, Gianoli E (2004) Meta-analysis of trade-offs among plant antiherbivore defenses: are plants jacks-of-all-trades, masters of all. Am Nat 163(4):64–75
Kreft S, Knapp M, Kreft I (1999) Extraction of rutin from buckwheat (Fagopyrum esculentum Moench) seeds and determination by capillary electrophoresis. J Agric Food Chem 47(11): 4649–4652
Kuc J (1995) Phytoalexins, stress metabolism, and disease resistance in plants. Annu Rev Phytopathol 33(1):275–297
Kumar K, Gupta SM, Arya MC, Nasim M (2017) In vitro antimicrobial and antioxidant activity of camelina seed extracts as potential source of bioactive compounds. Proc Natl Acad Sci India B Biol Sci 87(2):521–526
Kunze G, Zipfel C, Robatzek S, Niehaus K, Boller T, Felix G (2004) The N terminus of bacterial elongation factor Tu elicits innate immunity in Arabidopsis plants. Plant Cell 16(12):3496–3507
Kutchan TM (2001) Ecological arsenal and developmental dispatcher. The paradigm of secondary metabolism. Plant Physiol 125(1):58–60
Kweon MH, Hwang HJ, Sung HC (2001) Identification and antioxidant activity of novel chlorogenic acid derivatives from bamboo (Phyllostachys edulis). J Agric Food Chem 49(10): 4646–4655
Lai Z, Mengiste T (2013) Genetic and cellular mechanisms regulating plant responses to necrotrophic pathogens. Curr Opin Plant Biol 16(4):505–512
Langcake P, Pryce R (1976) The production of resveratrol by Vitis vinifera and other members of the Vitaceae as a response to infection or injury. Physiol Plant Pathol 9(1):77–86
Lattanzio V, Ruggiero P (2003) Biochimica Agraria. Patron Editore, Bologna, p 631
Lattanzio V, Lattanzio VM, Cardinali A (2006) Role of phenolics in the resistance mechanisms of plants against fungal pathogens and insects. Phytochem Adv res 661(2):23–67
Li AN, Li S, Zhang YJ, Xu XR, Chen YM, Li HB (2014) Resources and biological activities of natural polyphenols. Nutrients 6(12):6020–6047
Lindermayr C, Sell S, Müller B, Leister D, Durner J (2010) Redox regulation of the NPR1-TGA1 system of Arabidopsis thaliana by nitric oxide. Plant Cell 22(8):2894–2907
Link KP, Angell HR, Walker JC (1929) The isolation of protocatechuic acid from pigmented onion scales and its significance in relation to disease resistance in onions. J Biol Chem 1(81):369–375
Liu KS, Tsao SM, Yin MC (2005) In vitro antibacterial activity of roselle calyx and protocatechuic acid. Phytother Res 11:942–945
Lou Z, Wang H, Lv W, Ma C, Wang Z, Chen S (2010) Assessment of antibacterial activity of fractions from burdock leaf against food-related bacteria. Food Control 21(9):1272–1278
Lou Z, Wang H, Zhu S, Ma C, Wang Z (2011) Antibacterial activity and mechanism of action of chlorogenic acid. J Food Sci 76(6):398–403
Lou Z, Wang H, Rao S, Sun J, Ma C, Li J (2012) p-Coumaric acid kills bacteria through dual damage mechanisms. Food Control 25(2):550–554
Lyon G, Bayliss C (1975) The effect of rishitin on Erwinia carotovora var. atroseptica and other bacteria. Physiol Plant Pathol 6(2):177–186
Macheroux P, Schmid J, Amrhein N, Schaller A (1999) A unique reaction in a common pathway: mechanism and function of chorismate synthase in the shikimate pathway. Planta 207(3): 325–334
Mackey D, McFall AJ (2006) MAMPs and MIMPs: proposed classifications for inducers of innate immunity. Mol Microbiol 61(6):1365–1371

Mallavadhani UV, Mahapatra A (2005) A new aurone and two rare metabolites from the leaves of Diospyros melanoxylon. Nat Prod Res 19(1):91–97

Mandal MK, Chandra-Shekara AC, Jeong RD, Yu K, Zhu S, Chanda B, Navarre D, Kachroo A, Kachroo P (2012) Oleic acid–dependent modulation of NITRIC OXIDE ASSOCIATED1 protein levels regulate nitric oxide–mediated defense signaling in Arabidopsis. Plant Cell 24(4):1654–1674

Maróti G, Kereszt A, Kondorosi É, Mergaert P (2011) Natural roles of antimicrobial peptides in microbes, plants and animals. Res J Microbiol 4:363–374

Matros A, Mock H-P (2004) Ectopic expression of a UDP-glucose: phenylpropanoid glucosyltransferase leads to increased resistance of transgenic tobacco plants against infection with Potato Virus Y. Plant Cell Physiol 45(9):1185–1193

McDowell I, Taylor S, Gay C (1995) The phenolic pigment composition of black tea liquors—Part I: Predicting quality. J Sci Food Agric 69(4):467–474

McMurchy RA, Higgins VJ (1984) Trifolirhizin and maackiain in red clover: changes in Fusarium roseum "Avenaceum"-infected roots and in vitro effects on the pathogen. Physiol Plant Pathol 25(2):229–238

Mengiste T (2012) Plant immunity to necrotrophs. Annu Rev Phytopathol 50:267–294

Miedes E, Vanholme R, Boerjan W, Molina A (2014) The role of the secondary cell wall in plant resistance to pathogens. Front Plant Sci 5(5):358

Monaghan J, Zipfel C (2012) Plant pattern recognition receptor complexes at the plasma membrane. Curr Opin Plant Biol 15(4):349–357

Morrissey JP, Osbourn AE (1999) Fungal resistance to plant antibiotics as a mechanism of pathogenesis. Microbiol Mol Biol Rev 63(3):708–724

Mussatto SI, Dragone G, Roberto IC (2007) Ferulic and p-coumaric acids extraction by alkaline hydrolysis of brewer's spent grain. Ind Crop Prod 25(2):231–237

Namuli A, Abdullah N, Sieo CC, Zuhainis SW, Oskoueian E (2011) Phytochemical compounds and antibacterial activity of Jatropha curcas Linn. extracts. J Med Plant Res 5(16):3982–3990

Nicaise V, Roux M, Zipfel C (2009) Recent advances in PAMP-triggered immunity against bacteria: pattern recognition receptors watch over and raise the alarm. Plant Physiol 150(4): 1638–1647

Nicholson RL, Hammerschmidt R (1992) Phenolic compounds and their role in disease resistance. Annu Rev Phytopathol 30(1):369–389

Nielsen KA, Gotfredsen CH, Buch-Pedersen MJ, Ammitzbøll H, Mattsson O, Duus JØ, Nicholson RL (2004) Inclusions of flavonoid 3-deoxyanthocyanidins in Sorghum bicolor self-organize into spherical structures. Physiol Mol Plant Pathol 65(4):187–196

Nishimura MT, Stein M, Hou BH, Vogel JP, Edwards H, Somerville SC (2003) Loss of a callose synthase results in salicylic acid-dependent disease resistance. Science 301(5635):969–972

Nitiema LW, Savadogo A, Simpore J, Dianou D, Traore AS (2012) In vitro antimicrobial activity of some phenolic compounds (coumarin and quercetin) against gastroenteritis bacterial strains. Int J Microbiol Res 3(3):183–187

Nuhse TS (2012) Cell wall integrity signaling and innate immunity in plants. Front Plant Sci 3:280

Ökmen B, Etalo DW, Joosten MH, Bouwmeester HJ, de Vos RC, Collemare J, de Wit PJ (2013) Detoxification of α-tomatine by C ladosporium fulvum is required for full virulence on tomato. New Phytol 198(4):1203–1214

Osbourn AE (1996) Preformed antimicrobial compounds and plant defense against fungal attack. Plant Cell 8(10):1821

Osbourn AE (2003) Saponins in cereals. Phytochemistry 62(1):1–4

Osbourn A, Goss RJ, Field RA (2011) The saponins–polar isoprenoids with important and diverse biological activities. Nat Prod Rep 28(7):1261–1268

Osman KA, Al-Humaid AI, Al-Redhaiman KN, El-Mergawi RA (2014) Safety methods for chlorpyrifos removal from date fruits and its relation with sugars, phenolics and antioxidant capacity of fruits. J Food Sci Technol 51:1762–1772

Paul ND, Hatcher PE, Taylor JE (2000) Coping with multiple enemies: an integration of molecular and ecological perspectives. Trends Plant Sci 5(5):220–225

Pauly M, Gille S, Liu L, Mansoori N, de Souza A, Schultink A, Xiong G (2013) Hemicellulose biosynthesis. Planta 238(4):627–642

Pedras MS, Khan AQ, Taylor JL (1998) The phytoalexin camalexin is not metabolized by Phoma lingam, Alternaria brassicae, or phytopathogenic bacteria. Plant Sci 139(1):1–8

Pereira JA, Oliveira I, Sousa A, Valentão P, Andrade PB, Ferreira IC, Ferreres F, Bento A, Seabra R, Estevinho L (2007) Walnut (Juglans regia L.) leaves: phenolic compounds, antibacterial activity and antioxidant potential of different cultivars. Food Chem Toxicol 45(11):2287–2295

Perrin DR, Pisatin BW (1961) An antifungal substance from Pisum sativum L. Nature 191(4783): 76–77

Peterson JJ, Beecher GR, Bhagwat SA, Dwyer JT, Gebhardt SE, Haytowitz DB, Holden JM (2006) Flavanones in grapefruit, lemons, and limes. A compilation and review of the data from the analytical literature. J Food Compost Anal 19:74–80

Poloni A, Schirawski J (2014) Red card for pathogens: phytoalexins in sorghum and maize. Molecules 19(7):9114–9133

Price KR, Johnson IT, Fenwick GR, Malinow MR (1987) The chemistry and biological significance of saponins in foods and feeding stuffs. Crit Rev Food Sci Nutr 26(1):27–135

Przybylski P, Pyta K, Stefańska J, Ratajczak-Sitarz M, Katrusiak A, Huczyński A, Brzezinski B (2009) Synthesis, crystal structures and antibacterial activity studies of aza-derivatives of phytoalexin from cotton plant–gossypol. Eur J Med Chem 44(11):4393–4403

Purrington CB (2000) Costs of resistance. Curr Opin Plant Biol 3(4):305–308

Rao HS (1991) Soflavones from Flemingia vestita. Fitoterapia 63:458

Rasmussen MW, Roux M, Petersen M, Mundy J (2012) MAP kinase cascades in Arabidopsis innate immunity. Front Plant Sci 3:169

Ren X, Bao Y, Zhu Y, Liu S, Peng Z, Zhang Y, Zhou G (2019) Isorhamnetin, hispidulin, and cirsimaritin identified in Tamarix ramosissima barks from southern Xinjiang and their antioxidant and antimicrobial activities. Molecules 24(3):390

Rufatto LC, Luchtenberg P, Garcia C, Thomassigny C, Bouttier S, Henriques JA, Roesch-Ely M, Dumas F, Moura S (2018) Brazilian red propolis: chemical composition and antibacterial activity determined using bioguided fractionation. Microbiol Res 214:74–82

Ryals JA, Neuenschwander UH, Willits MG, Molina A, Steiner HY, Hunt MD (1996) Systemic acquired resistance. Plant cell 8(10):1809

Saleh MA, Clark S, Woodard B (2010) Antioxidant and free radical scavenging activities of essential oils. Ethn Dis 20:78–82

Sangkanu S, Rukachaisirikul V, Suriyachadkun C, Phongpaichit S (2017) Evaluation of antibacterial potential of mangrove sediment-derived actinomycetes. Microb Pathog 112:303–312

Sattler SE, Funnell-Harris DL (2013) Modifying lignin to improve bioenergy feedstocks: strengthening the barrier against pathogens? Front Plant Sci 5(4):70

Scalbert A (1991) Antimicrobial properties of tannins. Phytochemistry 30(12):3875–3883

Scheller HV, Ulvskov P (2010) Hemicelluloses. Annu Rev Plant Biol 61:263–289

Schmelz EA, Huffaker A, Sims JW, Christensen SA, Lu X, Okada K, Peters RJ (2014) Biosynthesis, elicitation and roles of monocot terpenoid phytoalexins. Plant J 79(4):659–678

Sederoff RR, MacKay JJ, Ralph J, Hatfield RD (1999) Unexpected variation in lignin. Curr Opin Plant Biol 2(2):145–152

Shah J (2003) The salicylic acid loop in plant defense. Curr Opin Plant Biol 6(4):365–371

Singh V, Roy S, Giri MK, Chaturvedi R, Chowdhury Z, Shah J, Nandi AK (2013) Arabidopsis thaliana FLOWERING LOCUS D is required for systemic acquired resistance. Plant Microbe Interact 26(9):1079–1088

Sobolev VS, Cole RJ, Dorner JW, Yagen B (1955) Isolation, purification, and liquid chromatographic determination of stilbene phytoalexins in peanuts. JAOAC Int 78(5):1177–1182

Stompor M (2020) A review on sources and pharmacological aspects of sakuranetin. Nutrients 12(2):513
Stout MJ, Thaler JS, Thomma BP (2006) Plant-mediated interactions between pathogenic microorganisms and herbivorous arthropods. Annu Rev Entomol 51:663–689
Strawn MA, Marr SK, Inoue K, Inada N, Zubieta C, Wildermuth MC (2007) Arabidopsis isochorismate synthase functional in pathogen-induced salicylate biosynthesis exhibits properties consistent with a role in diverse stress responses. J Biol Chem 282(8):5919–5933
Swain T (1977) Secondary compounds as protective agents. Annu Rev Plant Physiol 28(1): 479–501
Tada Y, Spoel SH, Pajerowska-Mukhtar K, Mou Z, Song J, Wang C, Zuo J, Dong X (2008) Plant immunity requires conformational charges of NPR1 via S-nitrosylation and thioredoxins. Science 321(5891):952–956
Taiz L, Zeiger E, Møller IM, Murphy A (2015) Plant physiol and dev (No. Ed. 6). Sinauer Associates Incorporated
Tanaka H, Uegaki R, Fujimori T, Kato K (1983) Antibacterial activity of sesquiterpenoids from tobacco leaves elicited by Pseudomonas solanacearum and tobacco mosaic virus. Jap J Phytopathol 49(4):501–507
Taylor JE, Hatcher PE, Paul ND (2004) Crosstalk between plant responses to pathogens and herbivores: a view from the outside in. J Exp Bot 55(395):159–168
Teresa JD, González MS, Muriel MR, Bellido IS (1984) Phenolic derivatives from Artemisia campestris subsp. glutinosa. Phytochemistry 23(8):1819–1821
Torres MA, Dangl JL, Jones JD (2002) Arabidopsis gp91phox homologues AtrbohD and AtrbohF are required for accumulation of reactive oxygen intermediates in the plant defense response. Proc Natl Acad Sci 99(1):517–522
Tsuda K, Sato M, Glazebrook J, Cohen JD, Katagiri F (2008) Interplay between MAMP-triggered and SA-mediated defense responses. Plant J 53(5):763–775
Tzin V, Galili G (2010) New insights into the shikimate and aromatic amino acids biosynthesis pathways in plants. Mol Plant 3(6):956–972
Umesha S (2006) Note: Phenylalanine ammonia lyase activity in tomato seedlings and its relationship to bacterial canker disease resistance. Phytoparasitica 34(1):68–71
Underwood W (2012) The plant cell wall: a dynamic barrier against pathogen invasion. Front Plant Sci 3:85
Van Sumere CF (1989) Phenols and phenolic acids. In: Methods in plant biochemistry, vol 1. Academic Press, pp 29–73
Va VanEtten HD, Bateman DF (1971) Studies on the mode of action of the phytoalexin phaseollin. Phytopathology 1(61):1363–1372
Valle T, López JL, Hernández JM, Corchete P (1997) Antifungal activity of scopoletin and its differential accumulation in Ulmus pumila and Ulmus campestris cell suspension cultures infected with Ophiostoma ulmi spores. Plant Sci 125(1):97–101
van der Hoorn RA, Jones JD (2004) The plant proteolytic machinery and its role in defence. Curr Opin Plant Biol 7(4):400–407
Vernooij B, Friedrich L, Morse A, Reist R, Kolditz-Jawhar R, Ward E, Uknes S, Kessmann H, Ryals J (1994) Salicylic acid is not the translocated signal responsible for inducing systemic acquired resistance but is required in signal transduction. Plant Cell 6(7):959–965
Vickery ML, Vickery B (1981) Polyketides. In: Secondary plant metabolism. Palgrave, London, pp 88–111
Vogel J (2008) Unique aspects of the grass cell wall. Curr Opin Plant Biol 11(3):301–307
Voigt CA (2014) Callose-mediated resistance to pathogenic intruders in plant defense-related papillae. Front Plant Sci 28(5):168
Walters DR, Newton AC, Lyon GD (2014) Induced resistance for plant defense: a sustainable approach to crop protection. Wiley
Wang C, El-Shetehy M, Shine MB, Yu K, Navarre D, Wendehenne D, Kachroo A, Kachroo P (2014) Free radicals mediate systemic acquired resistance. Cell Rep 7(2):348–355

Whiting DA (2001) Natural phenolic compounds 1900–2000: a bird's eye view of a century's chemistry. Nat Prod Rep 18(6):583–606

Wildermuth MC, Dewdney J, Wu G, Ausubel FM (2001) Isochorismate synthase is required to synthesize salicylic acid for plant defence. Nature 414(6863):562–565

Wink M (1997) Compartmentation of secondary metabolites and xenobiotics in plant vacuoles. Adv Bot Res 25:141–169

Winkel-Shirley B (2002) Biosynthesis of flavonoids and effects of stress. Curr Opin Plant Biol 5(3): 218–223

Wittstock U, Gershenzon J (2004) Constitutive plant toxins and their role in defense against herbivores and pathogens. Curr Opin Plant Biol 5(4):300–307

Wurzinger B, Mair A, Pfister B, Teige M (2011) Cross-talk of calcium-dependent protein kinase and MAP kinase signaling. Plant Signal Behav 6(1):8–12

Wyman JG, Van Etten HD (1978) Antibacterial activity of selected isoflavonoids. Physiol Biochem 68:583–589

Yu J, Gao J, Wang XY, Wei Q, Yang LF, Qiu K, Kuai BK (2010) The pathway and regulation of salicylic acid biosynthesis in probenazole-treated Arabidopsis. J Plant Biol 53:417–424

Yu K, Soares JM, Mandal MK, Wang C, Chanda B, Gifford AN, Fowler JS, Navarre D, Kachroo A, Kachroo P (2013) A feedback regulatory loop between G3P and lipid transfer proteins DIR1 and AZI1 mediates azelaic-acid-induced systemic immunity. Cell Rep 3:1266–1278

Yun MH, Torres PS, Oirdi ME, Rigano LA, Gonzalez-Lamothe R, Marano MR, Castagnaro AP, Dankert MA, Bouarab K, Vojnov AA (2006) Xanthan induces plant susceptibility by suppressing callose deposition. Plant Physiol 141(1):178–187

Zhao Y, Liu F, Lou HX (2010) Studies on the chemical constituents of Solanum nigrum. Zhong yao cai= J Chinese med material 33(4):555–556

Zipfel C, Felix G (2005) Plants and animals: a different taste for microbes? Curr Opin Plant Biol 8: 353–360

Polyphenol Phytoalexins as the Determinants of Plant Disease Resistance

10

Ashutosh Sharma, Aditi Sharma, Ajay Sharma, Yogesh Kumar, Pooja Sharma, Renu Bhardwaj, and Indu Sharma

Abstract

Polyphenols is a large group of phytochemicals known for diverse roles in plants metabolism, particularly under biotic and abiotic stress environments. These secondary metabolites may provide the biochemical shield to the sessile plants, against different classes of plant pathogens. Some of these polyphenols are accumulated in plants upon the infestation of pathogens. The defense molecules synthesized in plants in response to a particular plant pathogen are referred to as, phytoalexins. Although the phytoalexins are a group of chemically heterogeneous group of compounds, but out of them, the polyphenolic phytoalexins are of

A. Sharma
Faculty of Agricultural Sciences, DAV University, Jalandhar, Punjab, India

A. Sharma
College of Horticulture and Forestry, Thunag, Mandi, Dr. Y. S. Parmar University of Horticulture and Forestry, Solan, Himachal Pradesh, India

A. Sharma
Department of Chemistry, Career Point University, Hamirpur, Himachal Pradesh, India

Y. Kumar
Department of Botany, Central University of Jammu, Jammu, Jammu and Kashmir, India

P. Sharma
Department of Botanical and Environmental Sciences, Guru Nanak Dev University, Amritsar, Punjab, India

Department of Microbiology, DAV University, Jalandhar, Punjab, India

R. Bhardwaj
Department of Botanical and Environmental Sciences, Guru Nanak Dev University, Amritsar, Punjab, India

I. Sharma (✉)
Department of Life Sciences, Sant Baba Bhag Singh University, Jalandhar, Punjab, India

R. Lone et al. (eds.), *Plant Phenolics in Biotic Stress Management*,
https://doi.org/10.1007/978-981-99-3334-1_10

particular interest. The present chapter highlights the antimicrobial/ antipathogenic nature of these polyphenol phytoalexins against a number of phytopathosystems. Since the considerable details of their biosynthetic pathway and associated enzymes have been worked out, the expression of these polyphenol phytoalexins may be altered as per the requirement to engineer future crops with better plant disease resistance. Further, an attempt has also been made to discuss their role in human health, since many of them have been known to possess bioactive properties. This chapter is intended to review their chemical groups and subgroups, role in providing disease resistance in plants, biosynthesis and its regulation, role in protection against mycotoxins and in promoting human health, and an attempt has also been made to discuss the possibility to use them as food preservatives.

Keywords

Phenylpropanoid pathway · Shikimic acid pathway · Flavonoids · Germ tube · Biosynthetic pathway · Plant disease · Transgenic plants

10.1 Introduction

Polyphenols is a class of plant secondary metabolites consisting of a wide range of compounds (like phenolic acids, flavonoids, stilbenes, lignans etc.), that differing in their structure, physiochemical properties and distribution among different plants (Chiriac et al. 2021; El Gharras 2009). Plant polyphenols have been a constituent in variety of human diets. The dietary polyphenols are known for their ability to improve the composition of gut microbiota and function (Singh et al. 2019). The gut microbes have also developed multifaced interaction with gut microbiota. The intestinal microbiota also plays a key role in mediating the physiological functions of these dietary polyphenols (Kawabata et al. 2019). The plant phenolics possess a wide array of pharmaceutically important properties like antioxidant (Bors and Michel 2002), hepatoprotective (Yang et al. 2010), immunomodulatory (Yahfoufi et al. 2018), nephroprotective (Al-Sayed et al. 2015), anti-inflammatory (Yahfoufi et al. 2018), antimicrobial (Daglia 2012), antitumor (Juli et al. 2019), antiaging (Maleki et al. 2020) and neuroprotective (Silva and Pogačnik 2020) properties. Therefore, the dietary phenolics are important nutraceuticals that can be added in the day-today life for promoting human health (Piccolella et al. 2019; Visioli et al. 2011). The detailed review of the bioactivities of dietary polyphenols is available elsewhere (Luca et al. 2020). Due to the antimicrobial properties of some dietary polyphenols, they have also been studied as a potential candidate as natural food preservatives; however, more extensive studies are required in this regard to make their extraction cost effective and to optimize their dosage to minimize any side effect (Ofosu et al. 2020). The increased biosynthesis of phenol and polyphenol compounds is generally associated with the ability of plants to cope with the different abiotic and biotic stresses conditions (Tuladhar et al. 2021). Out of which, the biotic stress is particularly important for the plant pathologists and agricultural entomologists, worldwide. The occurrence of phenolics in plants either acts as the preformed resistance factor or is synthesized in plants upon the invasion by a plant pathogen (Farkas and Kiraly 1962; Ohri and Pannu 2010).

Plants being sedentary have evolved the various structural and biochemical arrangements to save themselves from the invading pathogens (Agrios 2005). Out of which, the biochemical defense molecules are particularly considered very important by the plant pathologists that includes two major and distinct classes of biochemicals, i.e., pathogenesis-related (PR) proteins (Sharma et al. 2021a) and the phytoalexins (Sharma et al. 2021b). PR proteins are the inducible defense-related proteins that are synthesized upon pathogen invasion and include chitinases, defensins, glucanases, lipid transfer proteins, osmotin-like proteins, oxalate oxidase (or oxalate oxidase-like) proteins, thaumatin-like proteins, thionins etc., which have been classified in 17 major families (from PR1 to PR17) based on their physiochemical properties (Ali et al. 2018; Sharma et al. 2021a), whereas the phytoalexins are the low molecular weight plant secondary metabolites, belonging to the diverse chemical classes that are produced by the plants in response to the invading pathogens (inducible molecules) possessing antimicrobial properties (Hammerschmidt 1999; Sharma et al. 2021b). Out of different chemically diverse classes of phytoalexins, a lot of information has been accumulated regarding the polyphenolic phytoalexins. Further, the role of polyphenolic phytoalexins in human health is also being studied in details due to their various bioactive properties.

The chemical classification of polyphenolic phytoalexins is discussed in details in the subsequent sections and the specificity of each major subgroup of these compounds has been separately discussed under different sections. The present chapter therefore focuses on the broad classification of phenolic and polyphenolic phytoalexins, their antimicrobial properties against plant pathogens, role in plant disease resistance, biosynthesis, regulation of biosynthesis and their role in promoting human health. Further, their role as a potential food preservative and the safety concerns thereof have also been addressed.

10.2 Different Chemical Groups and Subgroups of Polyphenolic Phytoalexins

The polyphenolic compounds are mainly classified into two main groups, viz., flavonoids and nonflavonoids (Libro et al. 2016). Flavonoids include various subgroups like flavonols, flavones, flavanols, flavanones, isoflavones, anthocyanins, pro-anthocyanidins etc., whereas the nonflavonoids may mainly be classified into the subgroups like lignans, stilbenes, phenolic acids etc. (Dixon et al. 2005; Libro et al. 2016; Rentzsch et al. 2009). The phenolics in plants are mainly synthesized via a shikimate/phenylpropanoid pathway (Sharma et al. 2019). Some phenylpropanoid guaiacols like coniferyl alcohol have also been identified as the phytoalexins, which may have a possible involvement in enhanced lignification mediated strengthening of plant defense (Keen and Littlefield 1979). The representatives of different subgroups of phytoalexins have been presented in Fig. 10.1, and their antimicrobial activities against the phytopathogens have been presented under relevant heading in subsequent sections.

10.3 Antimicrobial Properties and Role in Enhancing Disease Resistance in Plants

10.3.1 Flavonoids as Natural Versatile Molecules with Promising Antimicrobial Activity

Flavonoids are a versatile group of low molecular weight secondary metabolites (like flavonols, flavones, flavanols, flavanones, isoflavones, pro-anthocyanidins etc.) with polyphenolic structure that are generally involved in plethora of different physiological functions and also show significant protection against biotic and abiotic stress (Brunetti et al. 2013; Chalker 1999; Chen et al. 2019; El Gharras 2009; Wang et al. 2011). Some flavonoids also play important role in plant rhizobia association by inducing *nod* genes and in symbiotic association of arbuscular mycorrhizal fungi (Liu and Murray 2016; Singla and Garg 2017). These flavonoids are compartmentalized in different cellular and subcellular locations and are chemically diverse in nature. The flavonoid biosynthesis is most extensively studied utilizing different molecular biology techniques (Baskar et al. 2018).

Flavonoids are a highly characterized group of phytoalexins that plays key role in plant protection (Shah and Smith 2020). These antimicrobial flavonoids have been reported in many plants including soybean, vegetables, green tea, grape seeds, red pepper, apple, citrus fruits, berries, peaches, fruit juices, mulberry etc. (Grayer and Harborne 1994). Pathogen resistance due to these secondary metabolites also depends on the genotype of plant, pathogen virulence and the existing environmental factors. Generally, during fungal infections, flavonoids inhibit the spore development and hyphal elongation, dis-organize fungal cell contents, regulate the levels reactive oxygen species (ROS) and can modulate hormonal responses in plants (Baskar et al. 2018; Blount et al. 1992; McLay et al. 2020; Wiesel et al. 2014). Flavonoids have been classified into six major subgroups (flavonols, flavanones, isoflavones, flavones, flavan-3-ols and anthocyanins) according to their basic structure and polyphenolic phytoalexins of each of these subgroups have the antimicrobial activities against the phytopathogens.

A yellow lupine isoflavone compound called genistein has been found to inhibit the infections caused by many pathogenic fungi including *Aspergillus flavus*, *Fusarium oxysporum* and *Sclerotinia sclerotiorum* (Morkunas et al. 2005). Flavan-3-ols significantly inhibited spore germination and hyphal growth *Melampsora larici-populina* that causes rust disease poplar and other economic plants (Ullah et al. 2017). Yuxing and Shanfa (2017) have recently reported the antifungal properties of several flavonoids against different species of *Colletotrichum* and *Cochliobolus*. In another experiment, it has been found that the overexpression of *isoflavone synthase* in *M. truncatula* resulted in isoflavonoid, medicarpin accumulation in plant during

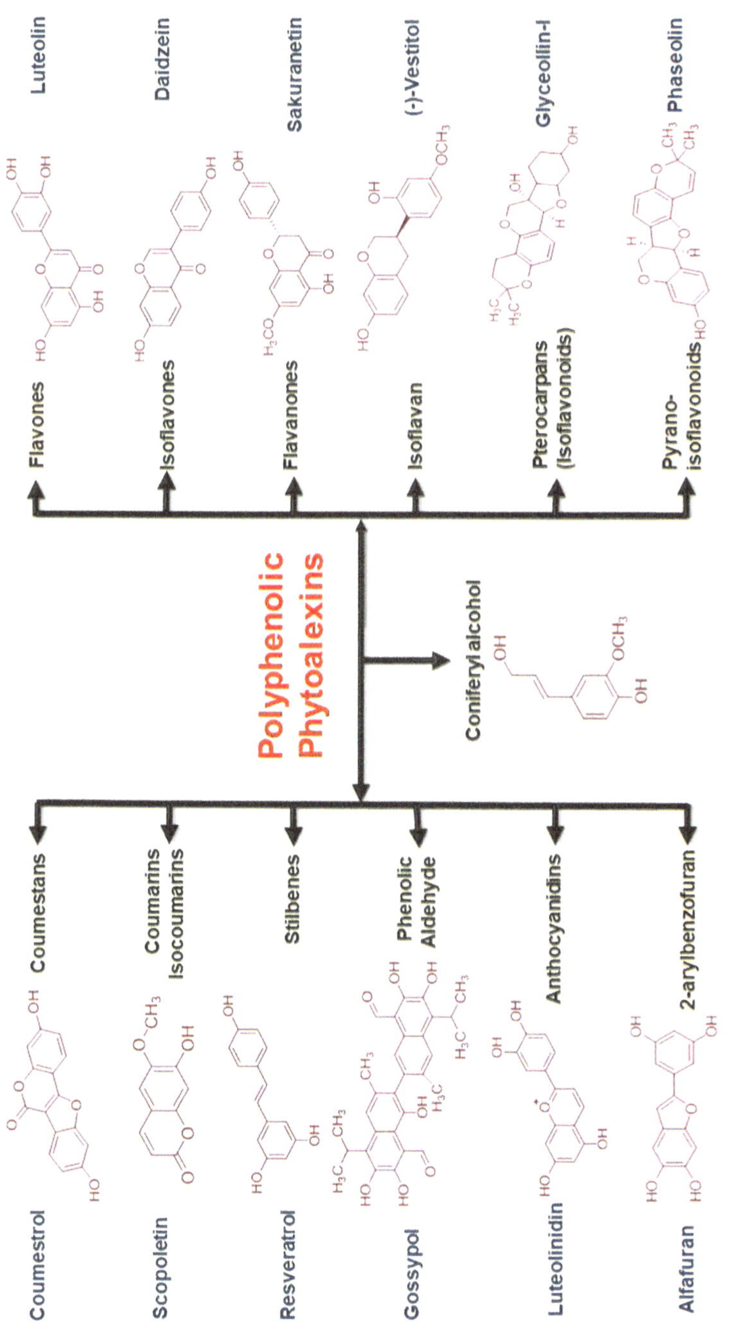

Fig. 10.1 Chemical classification of polyphenolic phytoalexins into different subgroups

the defense response toward *Rhizoctonia solani* (Liu et al. 2017). Sakuranetin, a rice flavanone, has been reported to possess substantial inhibitory activity against the important fungal pathogens of rice, i.e., *Pyricularia oryzae* and *Rhizoctonia solani* (Park et al. 2014; Katsumata et al. 2018).

Further Bilska and co-workers (2018) reported decreased mycotoxin accumulation in *Fusarium culmorum* and *F. graminearum* after exposure to flavonoids such as luteolin, kaempferol, naringenin and apigenin. Flavonoid accumulation in cotton has been reported to be more in red mutant S156 that significantly increases resistance to *Verticillium dahlia* causing wilt that poses serious threat to cotton production worldwide (Long et al. 2019). Recent report advocated that exogenously applied luteolin could maintain the fruit quality and also inhibit fungal pathogens in sweet cherry. The data suggested that Luteolin at a concentration of 100 or 200 mg/L inhibited growth of *Penicillium expansum*, reduced mycotoxin production and maintained better organoleptic quality by activating phenylpropanoid pathway (Liu et al. 2021). Induction of flavonoid biosynthesis pathway has also been reported by transcriptome and metabolome profiling during the interaction of *Stylosanthes* with *Colletotrichum gloeosporioides*. In vitro study showed mycelial growth inhibition by phloretin and naringenin. In addition to these flavonoids, apigenin, daidzein, quercetin and kaempferol suppressed conidial germination of the *Colletotrichum* strain (Jiang et al. 2021a). These compounds are also reported to significantly inhibit growth of bacterial pathogens after infection. A flavonoid compound rutin was found effective against most devastating bacteria causing blight, wilt and leaf spot in rice, tomato and tobacco. Rutin mediates salicylic acid-based resistance in plants and also stimulates the defense genes in plants (Farahani and Taghavi 2018; Yang et al. 2016). Secondary metabolites (flavonoids and polyphenols) from mulberry leaves have shown antibacterial activity against *Pseudomonas syringae*. These flavonoids inhibited pathogen growth by changing cell morphology, damaging structure and unbalancing of osmotic pressure in the bacterial cell (Wang et al. 2017).

Flavonoids and phenolic compound accumulation in citrus leaves has been correlated with increased citrus tolerance to *Candidatus liberibacter asiaticus* transmitted by *Diaphorina citri* (Hijaz et al. 2020). Recent research focused on identifying and characterizing the natural bioactive molecules against pathogens has come up with an introduction of new compounds against biotic stress. Findings on antiviral agents have suggested that glyceollidin and moracin are potent antiviral compounds that inhibit RAP-1 and RAP-2 (replication associated proteins) and can be employed as new promising anticucumber mosaic virus (CMV) candidates (Kumar et al. 2021). Besides the reports discussed above, several other workers have also suggested the role of the flavonoid phytoalexins in combating biotic stress in plants. The recent reports on antimicrobial action of flavonoids on different pathogens including the mode of action of these compounds have been presented as Table 10.1.

Table 10.1 Antimicrobial activity of flavonoids against pathogenic microbes

Flavonoid (compound name)		Host	Pathogen	Disease	Mode of action	Reference
Antifungal	Quercetin 3-*O*-rhamnoside and syringetin rhamnoside	Blueberries cultivar 'Elliott'	*Colletotrichum acutatum*	Anthracnose fruit rot	Growth inhibition	Miles et al. (2013)
	Medicarpin and 7,4′-dihydroxyflavone	Alfalfa (*Medicago sativa* L.)	*Fusarium oxysporum* f.sp. *medicaginis*	Wilt	Mycelial growth inhibition	Gill et al. (2018)
	Myricetin, and naringenin	*Musa paradisiaca*	*Agrobacterium tumefaciens*, *Dickeya solani*, *Erwinia amylovora and Rhizoctonia solani*	Crown gall, blackleg and soft rot, black scurf	Growth inhibition	Behiry et al. (2019)
	5,7-Dihydroxy-6 methylflavanone, 5,7-Dihydroxy-6,8-dimethylflavanone, 5-Hydroxy-7-methoxy-6-methylflavanone	*Agathis australis*	*Phytophthora agathidicida*	Dieback	Inhibited zoospore germination	Lawrence et al. (2019)
	Quercetin and kaempferol aglycones	*Dianthus caryophyllus*	*Fusarium oxysporum* f.sp. *dianthi*	Vascular wilt	Resistance response and growth inhibition	Romero-Rincón et al. (2021)
	Luteolin	Sweet cherry	*Botrytis cinerea* and *Penicillium expansum*	Rots	Inhibitory efficacy on mycelial growth and mycotoxin production	Liu et al. (2021)
Antibacterial	Rutin	Rice, tobacco and *Arabidopsis thaliana*	*Xanthomonas oryzae* pv. *oryzae*, *Ralstonia solanacearum, and Pseudomonas syringae* pv. *tomato*	Blight, wilt and spot	Salicylic acid mediated resistance	Yang et al. (2016)

(continued)

Table 10.1 (continued)

Flavonoid (compound name)		Host	Pathogen	Disease	Mode of action	Reference
	Rutin	Tomato	*Xanthomonas perforans*	Bacterial spot	Stimulation defense genes	Farahani and Taghavi (2018)
	Luteolin	Citrus	*Candidatus Liberibacter asiaticus*	Citrus greening	Significant reduction in gene expression	Zuo et al. (2019)
	Naringenin	*Arabidopsis*	*Pseudomonas syringae*	Bacterial spot	Monomerization and nuclear translocation of nonexpressor of pathogenesis-related genes 1 (NPR1)	An et al. (2021)
	Kaempferol and quercetin	Rice	*Xanthomonas Oryzae* pv. *oryzae*	Bacterial leaf blight	Induction of defense genes and hormones	Jan et al. (2021)
Antiviral	Flavones: 8-hydroxy-7-(3-hydroxypropyl)-2′-methoxyflavone (1) and 8-hydroxy-7-(2-hydroxyethyl)-2′-methoxyflavone	*Cassia pumila*	Tobacco mosaic virus	Mosaic	Antiviral	Kong et al. (2018)

10.3.2 Coumarins as Antimicrobial Compounds Against Plant Pathogens

Coumarins comprises of another family of plant-derived secondary metabolites that are produced via the phenylpropanoid pathway. These coumarins were named after the name of a plant *Coumarouna odorata* (syn., *Dipteryx odorata*) (Borges et al. 2005). Coumarins are generally produced by different plant families such as Rutaceae, Leguminosae, Umbelliferae, Compositae, Moraceae, Thymelaeaceae, Brassicaceae and Oleaceae (Bourgaud et al. 2006; Matos et al. 2015). Different studies on accumulation of coumarins in response to pathogen attack have been advocated by many researchers throughout the globe. Coumarins accumulation has been reported in tobacco mosaic virus-resistant cultivar *Nicotiana tabacum* cv. *xanthi* (Tanguy and Martin 1972). Similarly, accumulation of coumarin called scopoletin in young leaves of *Corchorus olitorius* in response to inoculation of plants with spores of *Helminthosporium turcicum* has been reported by Abou Zeid (2002). Scopoletin was also found to be highly toxic against different pathogens such as *Phytophthora parasitica var. nicotianae*, *Alternaria alternate*, *Fusarium graminearum*, *Sclerotinia sclerotiorum*, *Ophiostoma ulmi*, *Cercospora nicotianae*, *Botrytis cinerea*, *Pseudomonas syringae* pv. *tabaci*, *P. syringae* pv. *syringae* and tobacco mosaic virus (El Oirdi et al. 2010; Goy et al. 1993; Sun et al. 2014; Venugopala et al. 2013).

Scopoletin from the roots of cassava plant exhibited antifungal activity against *Aspergillus flavus* and *Aspergillus niger* at a minimal inhibitory concentration from 0.07 ± 0.00 µg/mL and 0.15 ± 0.00 µg/mL (Njankouo Ndam et al. 2020). Similarly, a coumarin derivative, i.e., 5′-hydroxy-auraptene inhibited conidial germination and aflatoxin production in *Aspergillus flavus* (Ali et al. 2021). Coumarin derivatives were also found effective against *Macrophomina phaseolina* and *Sclerotinia sclerotiorum* (Rastija et al. 2021). These antimicrobial coumarins have also exhibited strong antibacterial activities against clinical (e.g., *Staphylococcus aureus*, *Escherichia coli* and *Pseudomonas aeruginosa*) and plant pathogenic bacteria (e.g., *R. solanacearum* and *Pseudomonas syringae*) (Chen et al. 2016; Goy et al. 1993; Souza et al. 2005). Benzo-4-methyl coumarin, a coumarin derivative, has shown antibacterial activity against *Erwinia amylovora* and *Ralstonia solanacearum* (Desheesh et al. 2017). One basic coumarin from *Brassica oleracea* has shown resistance against *Xanthomonas campestris* pv. *campestris* (Tortosa et al. 2018). Recent data elucidated antibacterial action of coumarins against *Xanthomonas euvesicatoria* pv. *euvesicatoria* in tomato (Schiavi et al. 2021). Besides these reports, the antiviral effect of scopoletin was shown by metabolomic and transcriptomic profiling of tomato plants in response to tomato yellow leaf curl virus infection (Sade et al. 2015). Another coumarin called osthole has been recently proven as potential antiviral agent against tobacco mosaic virus (Chen et al. 2020). Antiviral activity of these coumarins includes inhibition of different enzymes such as reverse transcriptase, protease and integrase (Shokoohinia et al. 2014). The activity of coumarins and its derivatives against major plant pathogens infecting crops have been summarized in Table 10.2.

Table 10.2 Antimicrobial activity of coumarins against pathogenic microbes

Coumarin (compound name)		Host	Pathogen	Disease	Mode of action	Reference
Antifungal	*Coumarin derivatives*: pyranocoumarin, libanorin, furanocoumarin, and disenecioyl khellactone, pyranocoumarin	*Angelica* spp. and *Peucedanum decursivum*	*Sclerotinia sclerotiorum* and *Botrytis cinerea*	Rots	Mycelial growth inhibition	Song et al. (2017)
	Scopoletin	*Arabidopsis thaliana*	*Fusarium oxysporum* and *Verticillium dahliae*	Wilt	Inhibits spore germination	Stringlis et al. (2018)
	3-Hydroxycoumarin	*Theobroma cacao* L.	*Moniliophthora perniciosa*	Witches' broom	Inhibited the production of chitin syntheses in fungi	de Andrade et al. (2019)
	Scopoletin	*Manihot esculenta*	*Aspergillus flavus* and *Aspergillus niger*	Food spoilage fungi	Inhibited mycelium growth	Njankouo Ndam et al. (2020)
	Coumarin derivative: 5′-hydroxy-auraptene	*Lotus lalambensis*	*Aspergillus flavus*	Postharvest rot	Inhibited conidial germination and antiaflatoxigenic activity	Ali et al. (2021)
	6-bromo3-2,2-dibromoacetyl-2*H*-chromen-2-one	Solanaceous crops	*Fusarium foetens*, *F. pseudoanthophilum* and *F. fujikuroi*	Wilt	Grown inhibition	Amobonye et al. (2021)
Antibacterial	Coumarin	Tobacco	*Ralstonia solanacearum*	Wilt	Damaging bacterial cell membrane and preventing swarming motility and biofilm formation	Chen et al. (2016)
	Umbelliferone	Tobacco	*Ralstonia solanacearum*	Wilt	Reduces biofilm formation, transcription of type III secretion system regulators and effectors	Yang et al. (2017)

	Hydroxycoumarins: umbelliferone, esculetin and daphnetin	Tobacco	*Ralstonia pseudosolanacearum*	Wilt	Growth inhibition	Yang et al. (2018)
	6-Methylcoumarin	Tobacco	*Ralstonia solanacearum*	Wilt	Bacterial cell elongation, disrupts cell division, and suppresses the expression of the bacterial division protein coding genes *ftsZ*.	Yang et al. (2021)
	Coumarin	Tomato	*Xanthomonas euvesicatoria* pv. *euvesicatoria*	Bacterial spot	Inhibited bacterial growth	Schiavi et al. (2021)
Antiviral	Scopoletin	Tomato	Tomato yellow leaf curl virus	Leaf curl	Inhibitory activity	Sade et al. (2015)
	Osthole	*Cnidium monnieri*	Tobacco mosaic virus	Mosaic	Inhibitory activity	Chen et al. (2020)

10.3.3 Stilbenes as Antimicrobial Compounds Against Plant Pathogens

Stilbenes are nonflavonoid polyphenols and have been identified in different plant families including Fabaceae, Moraceae, Gnetaceae, Vitaceae, Polygonaceae and Pinaceae (Chong et al. 2009; El Khawand et al. 2018). Viniferins, resveratrol and other stilbenes are generally accumulated in *Vitis vinifera* in response to infections caused by *Plasmopara viticola*, *Rhizopus stolonifer*, *Uncinula necator* and *Botrytis cinerea* (Jeandet et al. 2002; Pezet et al. 2003). Stilbene derivatives display potent fungicidal activities against *Colletotrichum lagenarium* and *Pseudoperonospora cubensis* and are responsible for membrane disruption and respiration inhibition of fungal mycelium (Jian et al. 2015). Resveratrol is a natural phytoprotective stilbenoid compound synthesized by various plants and possess anticancer, antioxidant, antiageing and antimicrobial activities against human and plant pathogens (Fibach et al. 2012; Bostanghadiri et al. 2017). Antifungal activity of resveratrol has been tested against *Botrytis cinerea* (Caruso et al. 2011). Xu and co-workers (2018) reported that combination of resveratrol and pyrimethanil has shown synergistic effects against gray mold (*B. cinerea*) infection on grapes (Xu et al. 2018). Resveratrol has also shown inhibitory action against blight causing *Xanthomonas oryzae* pv. *oryzae* (Xoo) in rice. Lou and co-workers reported that double bond of resveratrol contributes to inhibitory action against Xoo by creating oxidative stress to bacterial cells and could disturb the metabolic pathways including energy, purine, amino acid synthesis (Luo et al. 2020). Recently, Song et al. (2021) reported systematic study on resveratrol and its derivatives that provide strong evidence of antitobacco mosaic activity of these compounds. The antimicrobial action of these stilbene compounds against plant pathogens has been summarized in Table 10.3.

10.3.4 Phenolic Acids, Tannins and Lignans as Natural Versatile Molecules with Promising Antimicrobial Activity

Phenolic acids are naturally occurring phenolic compounds with one carboxylic acid group that are found in plant-based foods, viz., seeds, skins of fruits, leaves of vegetables etc. Phenolic acids are mainly divided into two subgroups: hydroxybenzoic (e.g., *p*-hydroxybenzoic, protocatechuic, vanillic and syringic acids) and hydroxycinnamic acid (e.g., chlorogenic acid, ferulic, caffeic, *p*-coumaric and sinapic acids) (Pereira et al. 2009; Mandal et al. 2010). Different phenolic compounds from *Cymbopogon citratus* (such as gentisic acid, chlorogenic acid, sinapic acid, protocatechuic acid, caffeic acid and *p*-couramic acid) inhibited the in vitro growth of *Fusarium graminearum* and *Fusarium oxysporum* f.sp. *tulipae* (Kouassi et al. 2017).

Chlorogenic acid is basically (an ester of caffeic acid and quinic acid) a phernolic secondary metabolite that falls under the category of hydroxycinnamic acid generally synthesized by plants via the phenylpropanoid pathway (Lallemand et al. 2012). It has shown complete inhibition of spore germination *Botrytis cinerea, Sclerotinia*

Table 10.3 Antimicrobial activity of stilbenes against pathogenic microbes

Stilbene (compound name)	Host	Pathogen	Disease	Mode of action	Reference
Fluorine-containing stilbene derivatives	Cucumber	*Colletotrichum lagenarium and Pseudoperonospora cubensis*	Anthracnose and downy mildew	Membrane-disruption	Jian et al. (2015)
r-Viniferin, hopeaphenol, r2-viniferin	*Vitis vinifera*	*Plasmopara viticola*	Downey mildew	Growth inhibition	Gabaston et al. (2017)
Resveratrol and isorhamnetin 3 *O* rutinoside	*Solanum linnaeanum*	*Fusarium oxysporum* f.sp. *radicis-lycopersici*	Crown and root rot	Reduced mycelial growth	Nefzi et al. (2018)
Resveratrol	*Vitis vinifera*	*Uncinula necator*	Powdery mildew	Overexpression of stilbene synthase *VqSTS6*	Liu et al. (2019b)
Resveratrol	Rice	*Xanthomonas oryzae* pv. *oryzae*	Blight	Inhibited flagellum growth	Luo et al. (2020)
Resveratrol	*Vitis vinifera*	*Pseudomonas syringae*	Bacterial spot	Programmed cell death and accumulation of reactive oxygen species	Jiang et al. (2021b)
Resveratrol and its derivatives	Tobacco	Tobacco mosaic virus	Mosaic	Inactivation of virus	Song et al. (2021)

sclerotiorum, *Cercospora sojina*, *Fusarium solani* and *Verticillium dahlia* (Martínez et al. 2017). Phenolic acids such as *p*-coumaric acid could act as uncoupler of oxidative phosphorylation due to their acidic, nonpolar and planar characteristics. These structural and chemical characteristics favor for their antifungal activity against *Botrytis cinerea* (Morales et al. 2017). Vanillic acid, *p*-hydroxybenzoic acid from *Vanilla* juice showed a fungistatic effect against *Alternaria alternata* (Romero-Cortes et al. 2019). Ferulic acid from infected blackberry fruits in combination with other natural antifungal agents has shown activity against *Aspergillus japonicus* and *Gilbertella persicaria* with a fractional inhibitory concentration index of 0.281 (Liu et al. 2019a). Recent data on antifungal phenolic acids have shown that syringic acid, *p*-hydroxybenzoic, *p*-coumaric acids and vanillin are able to protect the crops naturally against fungal contamination and also avoid aflatoxin production in soybean (Silva et al. 2020).

Vanillic acid and protocatechuic acid have shown antifungal activity against *Macrophomina phaseolina* (Javed et al. 2021). Hernández et al. (2021) investigated the antifungal activity of orange peel polyphenolic extract (OPE) against postharvest fungal pathogens, i.e., *Botrytis cinerea*, *Monilinia fructicola* and *A. alternata*. Ferulic acid and *p*-coumaric acid displayed significant inhibitory activity in synthetic medium at a concentration of 1.5 g/L by reducing mycelial growth and conidial germination of target fungi. These small phenolic molecules such as cinnamic acid, ferulic acid, caffeic acid and *p*-coumaric acid have shown significant reduction in disease symptoms caused by *Xanthomonas oryzae* pv. *oryzae*. These compounds resulted in virulence attenuation of bacterial blight pathogen and can be efficiently used for its control (Talreja and Nerurkar 2018). The cinnamic acid derivative, *p*-coumaric acid mediates JA-signaling-based resistance response to *Xanthomonas campestris* pv. *campestris* (*Xcc*) in *Brassica napus* (Islam et al. 2019). Kang et al. (2020) experimentally advocated the potential antivirulence activity of 4-hydroxybenzoic acid and vanillic from the root extract of *Sedum middendorffianum* against *Pseudomonas syringae* pv. *tomato*. These phenolic acids significantly suppressed the expression of *hopP1*, *hrpA*, and *hrpL* in the *hrp/hrc* gene cluster at a concentration of 2.5 mM.

Tannins are the most complex group of polyphenolic compounds with high molecular weight (Crozier et al. 2006). They are known to display antimicrobial activities against human and plant pathogenic agents. Antifungal mechanisms of action generally comprise of inhibition of oxidative phosphorylation, extracellular enzyme inhibition, metal complexation and protein in-solubilization (Ogawa and Yazaki 2018). Condensed tannins extracted from conifers showed antifungal activity against brown-rot, white-rot and soft-rot fungi in the liquid cultures (Anttila et al. 2013). Tannins can also be used to control leaf blight disease on *Toona sureni* caused by *Rhizoctonia* sp. (Firmansyah et al. 2015). Antifungal activity of tannic acid was observed against *Penicillium digitatum* causing postharvest losses in citrus, and the mechanism of action of tannic acid was cell wall disruption and the leakage of the intracellular contents of the fungi (Zhu et al. 2019). Antibacterial activity of tannins isolated from *Sapium baccatum* extract was also recorded against *Ralstonia solanacearum* (Vu et al. 2017).

Likewise, lignans are polyphenols or diphenolic compounds, which are found in flaxseeds, sesame-seeds, legumes, whole grains, fruits and vegetables. Lignans isolated from *Larrea divaricata* exhibited strong antifungal activity against *Fusarium graminearum* that led to the reduced disease incidence of seedling blight and crown rot in wheat (Vogt et al. 2013). These compounds also inhibited trichothecene biosynthesis in *Fusarium graminearum* (Kulik et al. 2014). Li et al. (2017) reported a novel dirigent gene, *GmDIR22* in soybean that promotes lignan biosynthesis and enhances resistance to a *Phytophthora* root and stem rot. Collectively, these data suggest that these secondary metabolites enhance disease resistance in different plants and can be developed to be utilized as various formulations in plant protection. The antimicrobial activity of phenolic acids against pathogenic microbes has been presented in Table 10.4.

10.4 Regulation of the Biosynthesis

The polyphenolic phytoalexins are mainly synthesized by phenylpropanoid pathway or shikimate (or shikimic acid) metabolic pathway (Dinelli et al. 2019; Kumar et al. 2020; Shrestha et al. 2019; Tak and Kumar 2020; Valletta et al. 2021). In bacteria, plants and fungi, maximum phenolic compounds are derived from the shikimic acid pathway, whereas relatively less phenolic compounds are derived from the malonic acid pathway in higher plants (Santos-Sánchez et al. 2019; Tak and Kumar 2020). Both shikimic acid and chorismic acid are the common precursors in the shikimic acid pathway for the synthesis of aromatic amino acids as well as structurally diverse phenolic compounds (Santos-Sánchez et al. 2019). Further report by Ahmed et al. (2020) highlighted that the phenolics may be derived from three biosynthetic pathways, viz., (1) shikimate (or chorismate or succinylbenzoate) pathway which produces derivatives of phenylpropanoids; (2) malonate (acetate or polyketide) pathway which produces phenylpropanoids elongated with side-chain; and (3) mevalonate (or acetate) pathway that produces aromatic terpenoids.

Burgeoning studies emphasized that shikimate pathway generates L-phenylalanine (an aromatic amino acid) which acts as the primary starting branch point for the phenylpropanoid pathway; and from the phenylpropanoid biosynthetic pathway, phenolic phytoalexins such as coumarins, stilbenes, stilbenoids etc., are synthesized (Valletta et al. 2021). Recently, a lot of information have been emerged in the scientific literature about the transcriptional regulation of the phenolic/polyphenolic phytoalexins. Different families of transcription factors (TFs), viz., WRKY, MYC, MYB, bHLH, NAC, NAM, bHLH, AP2/ERF, bZIP etc., have been found to regulate the expression of biosynthetic pathway genes for different types of various plant metabolites including phenolic compounds/phenolic phytoalexins at the level of transcription (Devi et al. 2017; Jahan et al. 2019, 2020; Thiel and Rössler 2017; Xu et al. 2004, 2014, 2015; Valletta et al. 2021).

It was found that GaWRKY1 participates in regulation of sesquiterpene phytoalexin (like gossypol) biosynthesis in cotton, via targeting CAD_1-A (a member of cadinene synthase (CAD_1), a sesquiterpene cyclase gene family, catalyzing a

Table 10.4 Antimicrobial activity of phenolic acids against pathogenic microbes

Phenolic acids (compound name)		Host	Pathogen	Disease	Mode of action	Reference
Antifungal	Protocatechuic acid	Strawberry	*Botrytis cinerea*	Gray mold	Growth inhibition	Nguyen et al. (2015)
	Gallic acid, chlorogenic acid, catechin, caffeic acid, *p*-coumaric acid and ferulic acid	Cherry tomato	*Alternaria alternata*	Soft rot	Suppression of fungus	Pane et al. (2016)
	Ferulic acid	Grapes	*Botrytis cinerea*	Rot	Growth inhibition	Patzke and Schieber (2018)
	p-Coumaric acid and Cinnamic acid	Soybean	*Phytophthora sojae*	Blight	Growth inhibition	Zhang et al. (2020)
	Caffeic acid, *p*-coumaric acid, ferulic acid and sinapic acid	Date palm	*Fusarium oxysporum* f.sp. *albedinis*	Bayoud or fungal wilt	Inhibition of fungal biomass production, sporulation and hydrolytic enzymes	El Hassni et al. (2021)
Antibacterial	Chlorogenic acid, caffeic acid, coumaric acid, and ferulic acid	Chinese cabbage	*Pectobacterium carotovorum* subsp. *carotovorum*	Soft rot	Inactivation of bacteria	Kang et al. (2018)
	Ortho-coumaric acid	Rice	*Xanthomonas oryzae* pv. *oryzae*	Blight	Inhibitor of type III secretion system	Fan et al. (2019)
	Vanillic acid, benzoic acid, chlorogenic acid	*Moringa oleifera* seeds	*Agrobacterium tumefaciens*, *Erwinia amylovora*, and *Pectobacterium atrosepticum*	Gall and soft rot	Growth inhibition	Salem et al. (2021)

branch-point step leading to the biosynthesis of sesquiterpenes) gene (Xu et al. 2004). It was found that the biosynthesis of the rice phytoalexin, sakuranetin is highly dependent on jasmonic acid (JA) signaling that induces the expression of *OsNOMT* gene (encodeing naringenin 7-*O*-methyltransferase, a key enzyme for its production) through JA-inducible basic helix-loop-helix transcriptional factor, OsMYC2 that drastically upregulates their activity of OsNOMT promoter. Further, the two interacting collaborators of OsMYC2 viz., OsMYC2-like protein 1 and 2 (OsMYL1 and OsMYL2) that acts in synergy with OsMYC2 for the OsNOMT activation, thereby leading to inductive production of sakuranetin during defense reaction (Ogawa et al. 2017).

In an attempt to identify the regulatory network of glyceollin (an iso-flavonoid phytoalexin), biosynthesis in soybean GmNAC42-1 protein was found to be interacting with the promoters of its two biosynthetic genes, namely Isoflavone synthase 2 and Glycinol 4-dimethylallyltransferase in the yeast one hybrid assay. The overexpression of *GmNAC42*-1 increased the expressions of the gene transcripts in soybean hairy roots upon treatment with *Phytophthora sojae* WGE (wall glucan elicitors) (Jahan et al. 2019). In another experiment to identifying glyceollin biosynthesis-related TFs, the transcriptomic studies were made on two soybean varieties responding to *P. sojae* WGE and two soybean R2R3-type MYB TFs were identified with the contrasting roles in its regulation. Further, it was also recorded that the silencing *GmMYB29A2* in hairy roots decreased the expressions of *GmNAC42*-1, glyceollin biosynthesis gene transcripts also with the corresponding decline in glyceollin, whereas its overexpression led to the increased their levels (Jahan et al. 2020). Further, evidences are now emerging that the flavonoid biosynthesis is also affected by the transcriptional regulation mediated by MBW (MYB-bHLH-WDR) protein complexes (Xu et al. 2014, 2015).

The involvement of the plant growth regulators (PGRs), viz., ethylene, kinetin, jasmonic acid (JA), gibberlins and salicylic acid, in the regulation of phenolic phytoalexins during pathogen infestation have also been emerging. During the RNAseq analysis of the moderately resistant *Medicago truncatula* accession A17 and highly susceptible *skl* mutant (defective in ethylene sensing) revealed that the possible role of ethylene-mediated accumulation of isoflavonoid phytoalexins like medicarpin, during the defense against *Rhizoctonia solani* (Liu et al. 2017). Shimizu et al. (2012) recorded that the sakuranetin production is induced by JA treatment. Further, it was also found that the coronatine induced the accumulation of sakuranetin in rice, under the control of kinetin and ascorbic acid (AsA). It was also suggested that coronatine might interact at the same active site(s) as JA, leading to phytoalexin production (Tamogami and Kodama 2000). Later, the involvement of JA signaling was suggested to be involved in the sakuranetin biosynthesis in rice by the induction of *OsNOMT* gene through OsMYC2 TF (Ogawa et al. 2017). Previously, the role of amino acid conjugates of JA in eliciting the production of sakuranetin in rice was also recorded and was suggested as later component in the signaling transduction cascade of rice defense reaction (Tamogami et al. 1997). In another experiment, involving two contrasting genotypes of rice (resistant and susceptible against bakanae disease caused by *Fusarium fujikuroi*), the involvement of three PGRs, gibberellin, abscisic acid and JA was observed in the sakuranetin production (Siciliano et al. 2015). However, the involvement of different PGRs in the regulation of the biosynthesis of phenolic phytoalexins has been reported by several others, yet more detailed studies are required to precisely know the exact mechanism of action of each PGR in regulation of biosynthesis of phenolic phytoalexins.

10.5 Role in Human Health

Phytoalexins have been evolved as the molecules with antimicrobial action in plant defense, leading to the perception that they might have a similar possible role in protecting humans from microbes and may have some other bioactive properties, which could be pharmaceutically important. Some important roles played by the phytoalexins have been presented elsewhere (Sharma et al. 2021b; Jeandet et al. 2014). Out of chemically diverse phytoalexins, polyphenolic phytoalexins are particularly important. Plant polyphenols have already been described as the molecules with human health promoting properties (Di Lorenzo et al. 2021; Tresserra-Rimbau 2020; Cory et al. 2018). In the subsequent paragraphs, the role of crucial polyphenol phytoalexins in promoting human health is documented.

The rice flavanone phytoalexin, sakuranetin is one of the well-characterized, pharmaceutically active phytoalexin that possesses antioxidant, antimicrobial, anti-inflammatory, antiparasitic, antimutagenic and antiallergic properties (Stompor 2020). Sakuranetin has been reported to suppress the rhinoviruses, which cause common cold and chronic inflammatory respiratory diseases in human beings (Choi 2017). The antirhinoviral (against human rhinovirus 3) potential of sakuranetin, which was isolated from *Sorbus commixta* Hedl., in human epithelioid carcinoma cervix (HeLa) cells was evaluated in the clinical setting. Sakuranetin significantly inhibited the replication of human rhinovirus 3 in HeLa cells and exhibited in vitro antiviral activity without cytotoxicity, thereby inhibiting the viral adsorption. In rice plants or cherry trees, sakuranetin is reported to be biosynthesized from naringenin (Jeong et al. 2018). In human liver microsomes, sakuranetin is revealed to modulate the drug metabolizing enzymes, and in human hepatocarcinoma cells, it induced transactivation of cytochrome P450 3A4 gene, whereas UDP-glucuronosyltransferases (UGT) 1A9 was inhibited. The chemopreventive potential of sakuranetin for influenza virus-related diseases has been emphasized by Kwon et al. (2018). It inhibited the replication of influenza B/Lee/40 virus, viral RNA synthesis and altered the entry, attachment and postentry of the virus. Yamauchi et al. (2019) reported that sakuranetin downregulated the expression of inducible nitric oxide synthase through modulation of expression of interleukin-1 receptor and CCAAT/enhancer-binding protein β.

The antibacterial activity against *Salmonella typhi* of scopoletin (a coumarin) which was isolated from stem bark of *Aleurites moluccana* (candlenut) was observed (Prabowo and Agustina 2020). Another study by Alkorashy et al. (2020) at transcriptomic level highlighted the phagocytic potential of scopoletin in

U937-derived human macrophages. Scopoletin along with daidzein and fisetin has been reported to inhibit aggregation of α-synuclein (Rane et al. 2021). The aggregation of α-synuclein is associated with various neurological disorders such as Parkinson's disease. In vitro studies on human adenocarcinoma HT-29 cell lines revealed the protective roles of resveratrol and pterostilbene, two compounds of stilbenes family (Spevakova et al. 2020). The antioxidant and anti-inflammatory potential of resveratrol along with curcumin and other dietary nitric oxide supplementation on human cardiovascular health have been well documented (Banez et al. 2020). Further study by Medrano-Padial et al. (2021) assessed the antimutagenic and antigenotoxic potential of stilbene extract. Resveratrol and pterostilbene have been reported to inhibit the SARS-CoV-2 (COVID-19) replication in air–liquid interface-cultured human primary bronchial epithelial cells (Ter Ellen et al. 2021).

The neuroprotective potential of luteolin, a glycosylated flavonoid, against lead acetate-induced cortical and neuronal damage in rats has been identified (Baty et al. 2020). It emphasizes that neuronal damage may be induced by luteolin via inhibition of neuroinflammation, oxidative damage, and the cortical cell death. Another study revealed health promoting effects of luteolin and emphasized that low doses of luteolin can activate the Nrf2 in mice liver and its aglycon form may lead to activation of the Nrf2 pathway at ZT12 in the mice liver (Kitakaze et al. 2020). Apigeninidin and luteolinidin were reported from the seed head and stalk dermal layer of sweet sorghum (*Sorghum bicolor* L. Moench) which were reported to have potential for human health protection (Vanamala et al. 2018). The role of stilbenes, 3-Deoxyanthocyanins such as yellow apigeninidin and the orange luteolinidin, phenolamides and other phytochemicals present in sorghum has been documented by Dykes (2019).

It was found that a polyphenolic phytoalexin, polydatin can help in reducing the prion pathogenesis and may serve as a therapeutic agent targeting amyloidogenic transition in prions. Polydatin binds with moderate affinity to the recombinant protease resistant core of human prion protein, encompassing the $rPrP^{res}$ sequence (amino acid residues at 90–231 position, corresponding to the protease-resistant fragment originating after the truncation of amino-terminal of the mature human prion protein) and inhibits its conversion into the highly neurotoxic forms (Sirohi et al. 2021). The paraquat (herbicide)-induced human MRC-5 fibroblast injury was alleviated by polydatin (a resveratrol glycoside) via inhibition of the NLRP3 inflammasome activation (Fu et al. 2020). It also reduced that paraquat induced inflammatory responses and improved the antioxidant stress capacity of human embryonic lung fibroblast MRC-5 cells. The protective effects of polydatin against bleomycin-induced idiopathic pulmonary fibrosis in Sprague-Dawley rats have been reported by Liu et al. (2020). Polydatin administration resulted in inhibition of TGF-β1 (transforming growth factor-β1)-induced phenotypic transformation, reduction in α-smooth muscle actin and collagen I gene expression levels. Further polydatin also ameliorated the fiber production in lung tissues, reduced the levels of tumor necrosis factor-α, interleukin-6 and 13. In human osteosarcoma MG-63 cells, polydatin is reported to induce cell apoptosis and trigger cell autophagy whereas enhanced cell viability (Jiang et al. 2020). The MG-63 cell death was induced through the stimulation of STAT3 signaling and thus, polydatin might be employed as a potential future clinical drug for the treatment of osteosarcoma. Bonucci et al. (2020) suggested the dietary supplementation with polydatin for the nutritional management of individuals against SARS-CoV-2 infection (COVID-19). Recently, Tang (2021) documented the protective role of polydatin against dementia-related disorders such as alcohol-related dementia, vascular dementia, Lewy body dementias and Alzheimer's disease in human populations.

Coumestrol (belongs to coumestans) has been reported to protect skin from photoaging, improve skin elasticity, maintain skin thickness in postmenopausal women and recent experiments on Wistar rats experimental model revealed their wound healing potential (Bianchi et al. 2018). Besides anti-inflammatory, neuroprotective, antiadipogenesis, having potential role in skin protection and wound healing, coumestrol is also known to promote the cell death in the choriocarcinoma cells through regulation of ERK1/2 MAPK and JNK MAPK signaling pathways (Lim et al. 2017). Further, it depolarizes the mitochondrial membrane potential, disrupts the homeostasis Ca^{2+} and reactive oxygen species (ROS), suppresses the proliferation of human placental choriocarcinoma cells and induces Bax and Bak (pro-apoptotic proteins) that enhance the apoptosis in JAR and JEG3 cells. Reger et al. (2018) reported that dietary intake of two phytoestrogens like coumestrol and isoflavones may modulate the risk of ovarian, colorectal, lung and prostate cancer in the prostate in human populations because of the structural similarity of these phytoestrogens to 17β-estradiol. Coumestrol is also reported to inhibit the proliferation of breast cancer MCF-7 cells by targeting the cellular copper to induce apoptosis in malignant cells, redox cycling of Cu(II), mediating the DNA damage through neocuproine and ROS scavengers, and thereby inducing activation of caspases, arrest of G1/S phase and stimulating p53/p21 expression (Zafar et al. 2018).

The role of coumestrol in attenuation of brain mitochondrial dysfunction has also been documented in Wistar rat models (Anastacio et al. 2019). Neonatal hypoxia-ischemia causes permanent brain damage. When coumestrol was administrated, it blocked the late reactive astrogliosis after hypoxia-ischemia injury through the maintenance of mitochondrial function and thereby resulting in neuroprotection in rat neonatal hypoxia-ischemia. Coumestrol is also reported to ameliorate the doxorubicin-induced cell damage, cell apoptosis and cardiotoxicity through activation of 5′ AMP-activated protein kinase alpha (Wu et al. 2020). Another study revealed the potential role of coumestrol against human skin cancer through activation of caspase and thus, stimulation of apoptosis, inhibition of expression of MMP-2 and MMP-9 as well as blocking m-TOR/PI3K/AKT signaling pathway in SKEM-5 cells (Kuang et al. 2021). Of late, Xu et al. (2021) emphasized that diabetes-induced oxidative stress, cell apoptosis and retinal cell inflammation through activation of SIRT1 (sirtuin 1) was mitigated by coumestrol in a rat model.

10.6 Polyphenolic Phytoalexins: A Way Ahead

Being well established as antimicrobial substances, one of the possible roles of naturally occurring phytoalexins has been anticipated as food preservations due to their widespread presence in plants. But only few attempts have been made to study their role in food preservation (Ejike et al. 2013; Sharma et al. 2021b). However, at present, the major limiting factors in their applications in food industry are the lack of adequate safety evaluation data, difficulty in the enhanced accumulation in plant tissue and shelf-life-related issues in processed foods. In addition to these, the detoxification of phytoalexins by some microbes can affect their usefulness to act as natural preservatives (Jeandet 2015). Being antimicrobial against phytopathogenic fungi, the possible role of enhanced level of phenolic phytoalexin in decreasing the mycotoxins in food stuffs was also anticipated and needed experimental support. In a study made in soybean, the clear-cut relation between the accumulation of glyceollin in infected plants was correlated to prevent aflatoxin accumulation (Song and Karr 1993). However, no clear-cut association was established in the other study by Zorzete et al. (2011) in peanuts, indicating the involvement of some other factors also in mycotoxin production. Phytoalexins may also protect humans from the ill effects of mycotoxins. In a toxicological investigation on in human embryonic kidney (HEK293) cells, it was found that resveratrol ameliorates the toxicity caused by a nephrotoxic mycotoxin, Ochratoxin A, which is produced by the fungal genera like *Aspergillus* and *Penicillium* (Raghubeer et al. 2015).

Besides their role in plant protection, phytoalexins are helpful in promoting the human health due to a number of health benefits associated with them (Sharma et al. 2021b). Since they are also commonly present in some plant tissues, that are part of human nutrition, they are commonly referred as nutraceuticals. Word 'nutraceuticals' refers to a food/food component playing a significant role in maintaining normal physiological functions to maintain human health, and has been derived from the amalgamation of two separate words 'nutrition' and 'pharmaceutical' (Das et al. 2012). Some polyphenolic phytoalexins like resveratrol of stilbene subgroup have already been described as nutraceutical compounds (Machado et al. 2019).

The detailed understanding of the complicated regulatory network of phenolic phytoalexin will help us engineer the future crops with better diseases resistance. Further, the engineering of phytoalexin gene regulatory networks could serve as a tool for producing phytoalexins in the nutraceuticals for promoting human health (Ahmed and Kovinich 2021). Besides the increase in phytoalexin production, the efforts can also be made to apply phytoalexins or their chemical analogs under field conditions and efforts may be done for their target-specific and need-based delivery.

Acknowledgment Authors are thankful to both DAV University and Sant Baba Bhag Singh University administration for the continuous encouragement and infrastructural support.

Conflicts of Interests Authors declare that they do not have any conflict of interests.

References

Abou Zeid AHS (2002) Stress metabolites from *Corchorus olitorius* L. leaves in response to certain stress agents. Food Chem 76:187–195

Agrios GN (2005) Plant pathology, 5th edn. Elsevier Academic Press, Burlington, MA

Ahmed S, Kovinich N (2021) Regulation of phytoalexin biosynthesis for agriculture and human health. Phytochem Rev 20:483–505

Ahmed A, Tariq A, Habib S (2020) Interactive biology of auxins and phenolics in plant environment. In: Plant phenolics in sustainable agriculture. Springer, Singapore, pp 117–133

Ali S, Ganai BA, Kamili AN, Bhat AA, Mir ZA, Bhat JA, Tyagi A, Islam ST, Mushtaq M, Yadav P, Rawat S (2018) Pathogenesis-related proteins and peptides as promising tools for engineering plants with multiple stress tolerance. Microbiol Res 212:29–37

Ali EM, Alkuwayti MA, Aldayel MF, Abdallah BM (2021) Coumarin derivative, 5′-hydroxy-auraptene, extracted from Lotus lalambensis, displays antifungal and anti-aflatoxigenic activities against Aspergillus flavus. J King Saud Univ Sci 33(1):101216

Alkorashy AI, Doghish AS, Abulsoud AI, Ewees MG, Abdelghany TM, Elshafey MM, Elkhatib WF (2020) Effect of scopoletin on phagocytic activity of U937-derived human macrophages: insights from transcriptomic analysis. Genomics 112(5):3518–3524

Al-Sayed E, Abdel-Daim MM, Kilany OE, Karonen M, Sinkkonen J (2015) Protective role of polyphenols from *Bauhinia hookeri* against carbon tetrachloride-induced hepato- and nephro-toxicity in mice. Ren Fail 37(7):1198–1207

Amobonye A, Bhagwat P, Ranjith D, Mohanlall V, Pillai S (2021) Characterisation, pathogenicity and hydrolytic enzyme profiling of selected *Fusarium* species and their inhibition by novel coumarins. Arch Microbiol 203:3495–3508

An J, Kim SH, Bahk S, Vuong UT, Nguyen NT, Do HL, Chung WS (2021) Naringenin induces pathogen resistance against *Pseudomonas syringae* through the activation of NPR1 in Arabidopsis. Front Plant Sci 12:672552

Anastacio JB, Sanches EF, Nicola F, Odorcyk F, Fabres RB, Netto CA (2019) Phytoestrogen coumestrol attenuates brain mitochondrial dysfunction and long-term cognitive deficits following neonatal hypoxia–ischemia. Int J Dev Neurosci 79:86–95

Anttila AK, Pirttilä AM, Häggman H, Harju A, Venäläinen M, Haapala A, Julkunen-Tiitto R (2013) Condensed conifer tannins as antifungal agents in liquid culture. Holzforschung 67(7):825–832

Banez MJ, Geluz MI, Chandra A, Hamdan T, Biswas OS, Bryan NS, Von Schwarz ER (2020) A systemic review on the antioxidant and anti-inflammatory effects of resveratrol, curcumin, and dietary nitric oxide supplementation on human cardiovascular health. Nutr Res 78:11–26

Baskar V, Venkatesh R, Ramalingam S (2018) Flavonoids (antioxidants systems) in higher plants and their response to stresses. In: Antioxidants and antioxidant enzymes in higher plants. Springer, Cham, pp 253–268

Baty RS, Hassan KE, Alsharif KF, El-Hennamy RE, Elmahallawy EK, Hafez MM, Moneim AA, Kassab RB (2020) Neuroprotective role of luteolin against lead acetate-induced cortical damage in rats. Hum Exp Toxicol 39(9):1200–1212

Behiry SI, Okla MK, Alamri SA, El-Hefny M, Salem MZ, Alaraidh IA, Salem AZ (2019) Antifungal and antibacterial activities of Musa paradisiaca L. peel extract: HPLC analysis of phenolic and flavonoid contents. Processes 7(4):215

Bianchi SE, Machado BE, da Silva MG, da Silva MM, Dal Bosco L, Marques MS, Horn AP, Persich L, Geller FC, Argenta D, Teixeira HF (2018) Coumestrol/hydroxypropyl-β-cyclodextrin association incorporated in hydroxypropyl methylcellulose hydrogel exhibits wound healing effect: in vitro and in vivo study. Eur J Pharm Sci 119:179–188

Bilska K, Stuper-Szablewska K, Kulik T, Buśko M, Załuski D, Jurczak S, Perkowski J (2018) Changes in phenylpropanoid and trichothecene production by *Fusarium culmorum* and *F. graminearum* sensu stricto via exposure to flavonoids. Toxins 10(3):110

Blount JW, Dixon RA, Paiva NL (1992) Stress responses in alfalfa (Medicago sativa L.) XVI. Antifungal activity of medicarpin and its biosynthetic precursors; implications for the genetic manipulation of stress metabolites. Physiol Mol Plant Pathol 41:333–349

Bonucci M, Raggi R, Vacca RA (2020) Polydatin and its potential protective effect on COVID-19. Clin Nutr (Edinburgh, Scotland) 39(12):3850

Borges F, Roleira F, Milhazes N, Santana L, Uriarte E (2005) Simple coumarins and analogues in medicinal chemistry: occurrence, synthesis and biological activity. Curr Med Chem 12:887–916

Bors W, Michel C (2002) Chemistry of the antioxidant effect of polyphenols. Ann N Y Acad Sci 957(1):57–69

Bostanghadiri N, Pormohammad A, Chirani AS, Pouriran R, Erfanimanesh S (2017) Comprehensive review on the antimicrobial potency of the plant polyphenol resveratrol. Biomed Pharmacother 95:1588–1595

Bourgaud F, Hehn A, Larbat R, Doerper S, Gontier E, Kellner S (2006) Biosynthesis of coumarins in plants: a major pathway still to be unravelled for cytochrome P450 enzymes. Phytochem Rev 5:293–308

Brunetti C, Di FM, Fini A, Pollastri S, Tattini M (2013) Flavonoids as antioxidants and developmental regulators: relative significance in plants and humans. Int J Mol Sci 14:3540–3555

Caruso F, Mendoza L, Castro P, Cotoras M, Aguirre M, Matsuhiro B, Isaacs M, Rossi M, Viglianti A, Antonioletti R (2011) Antifungal activity of resveratrol against Botrytis cinerea is improved using 2-furyl derivatives. PLoS One 6:e2542110

Chalker SL (1999) Environmental significance of anthocyanins in plant stress responses. Photochem Photobiol 70:1–9

Chen J, Yu Y, Li S, Ding W (2016) Resveratrol and coumarin: novel agricultural antibacterial agent against *Ralstonia solanacearum* in vitro and in vivo. Molecules 21(11):1501

Chen S, Wu F, Li Y, Qian Y, Pan X, Li F, Wang Y, Wu Z, Fu C, Lin H, Yang A (2019) NtMYB4 and NtCHS1 are critical factors in the regulation of flavonoid biosynthesis and are involved in salinity responsiveness. Front Plant Sci 10:178

Chen YH, Guo DS, Lu MH, Yue JY, Liu Y, Shang CM, Zhao MM (2020) Inhibitory effect of osthole from *Cnidium monnieri* on tobacco mosaic virus (TMV) infection in *Nicotiana glutinosa*. Molecules 25(1):65

Chiriac ER, Chiţescu CL, Geană EI, Gird CE, Socoteanu RP, Boscencu R (2021) Advanced analytical approaches for the analysis of polyphenols in plants matrices-a review. Separations 8(5):65

Choi HJ (2017) *In vitro* antiviral activity of sakuranetin against human rhinovirus 3. Osong Public Health Res Perspect 8(6):415

Chong J, Poutaraud A, Hugueney P (2009) Metabolism and roles of stilbenes in plants. Plant Sci 177:143–155

Cory H, Passarelli S, Szeto J, Tamez M, Mattei J (2018) The role of polyphenols in human health and food systems: a mini-review. Front Nutr 5:87

Crozier A, Jaganath IB, Clifford MN (2006) Phenols, polyphenols and tannins: an overview. In: Plant secondary metabolites: occurrence, structure and role in the human diet, vol 1. Wiley, pp 1–25

Daglia M (2012) Polyphenols as antimicrobial agents. Curr Opin Biotechnol 23(2):174–181

Das L, Bhaumik E, Raychaudhuri U, Chakraborty R (2012) Role of nutraceuticals in human health. J Food Sci Technol 49(2):173–183

de Andrade Gonçalves P, dos Santos Junior MC, do Sacramento Sousa C, Góes-Neto A, Luz EDMN, Damaceno VO, de Assis SA (2019) Study of sodium 3-hydroxycoumarin as inhibitors in vitro, in vivo and in silico of Moniliophthora perniciosa fungus. Eur J Plant Pathol 153(1): 15–27

Desheesh MA, El-Zemity SR, Kadous EA, Fahmy MM, Tawfeek EE (2017) Antimicrobial activities of synthesized 3-acetyl coumarin and benzo-4-methyl coumarin. Alexandria Sci Exchange J 38:515–520

Devi EL, Kumar S, Singh TB, Sharma SK, Beemrote A, Devi CP, Chongtham SK, Singh CH, Yumlembam RA, Haribhushan A, Prakash N (2017) Adaptation strategies and defence mechanisms of plants during environmental stress. In: Medicinal plants and environmental challenges. Springer, Cham, pp 359–413

Di Lorenzo C, Colombo F, Biella S, Stockley C, Restani P (2021) Polyphenols and human health: the role of bioavailability. Nutrients 13(1):273

Dinelli G, Bonetti A, Elementi S, Antuono LF, Catiozone P (2019) Biosynthesis of polyphenol phytoestrogens in plants. In: Phytoestrogens in functional foods. CRC Press, pp 19–80

Dixon RA, Xie DY, Sharma SB (2005) Proanthocyanidins–a final frontier in flavonoid research? New Phytol 165(1):9–28

Dykes L (2019) Sorghum phytochemicals and their potential impact on human health. Sorghum 1931:121–140

Ejike CE, Gong M, Udenigwe CC (2013) Phytoalexins from the Poaceae: biosynthesis, function and prospects in food preservation. Food Res Int 52(1):167–177

El Gharras H (2009) Polyphenols: food sources, properties and applications - a review. Int J Food Sci Technol 44(12):2512–2518

El Hassni M, Laadouzaa H, El Hadrami A, Dihazi A, Rakibi Y, Lemjiber N, Naamani K (2021) An in vitro evaluation of the effect of hydroxycinnamic acids on the growth and hydrolytic enzyme production in *Fusarium oxysporum* f. sp. albedinis. Arch Phytopathol Plant Protect 54:17–18

El Khawand T, Courtois A, Valls J, Richard T, Krisa S (2018) A review of dietary stilbenes: sources and bioavailability. Phytochem Rev 17:1007–1029

El Oirdi M, Trapani A, Bouarab K (2010) The nature of tobacco resistance against Botrytis cinerea depends on the infection structures of the pathogen. Environ Microbiol 12:239–253

Fan S, Tian F, Fang L, Yang CH, He C (2019) Transcriptional responses of *Xanthomonas oryzae* pv. oryzae to type III secretion system inhibitor ortho-coumaric acid. BMC Microbiol 19(1): 1–12

Farahani AS, Taghavi SM (2018) Rutin promoted resistance of tomato against Xanthomonas perforans. Eur J Plant Pathol 151(2):527–531

Farkas GL, Kiraly Z (1962) Role of phenolic compounds in the physiology of plant diseases and disease resistance. J Phytopathol 44:105–150

Fibach E, Prus E, Bianchi N, Zuccato C, Breveglieri G, Salvatori F, Finotti A, Lipucci di Paola M, Brognara E, Lampronti I, Borgatti M (2012) Resveratrol: antioxidant activity and induction of fetal hemoglobin in erythroid cells from normal donors and β-thalassemia patients. Int J Mol Med 29(6):974–982

Firmansyah MA, Soekarno BP, Witarto AB (2015) Effects of tannin to control leaf blight disease on Toona sureni Merr. caused by two isolates of Rhizoctonia sp. Plant Pathol J (Faisalabad) 14(3): 148–152

Fu Y, Yan M, Xie C, Hu J, Zeng X, Hu Q (2020) Polydatin relieves paraquat-induced human MRC-5 fibroblast injury through inhibiting the activation of the NLRP3 inflammasome. Ann Transl Med 8(12):765

Gabaston J, Cantos-Villar E, Biais B, Waffo-Teguo P, Renouf E, Corio-Costet MF, Mérillon JM (2017) Stilbenes from Vitis vinifera L. waste: a sustainable tool for controlling *Plasmopara viticola*. J Agric Food Chem 65(13):2711–2718

Gill US, Uppalapati SR, Gallego-Giraldo L, Ishiga Y, Dixon RA, Mysore KS (2018) Metabolic flux towards the (iso) flavonoid pathway in lignin modified alfalfa lines induces resistance against *Fusarium oxysporum* f. sp. medicaginis. Plant Cell Environ 41(9):1997–2007

Goy PA, Signer H, Reist R, Aichholz R, Blum W, Schmidt E, Kessmann H (1993) Accumulation of scopoletin is associated with the high disease resistance of the hybrid *Nicotiana glutinosa* X *Nicotiana debneyi*. Planta 191:200–206

Grayer RJ, Harborne JB (1994) A survey of antifungal compounds from higher plants, 1982–1993. Phytochemistry 37:19–42

Hammerschmidt R (1999) Phytoalexins: what have we learned after 60 years? Annu Rev Phytopathol 37:285–306

Hernández A, Ruiz-Moyano S, Galván AI, Merchán AV, Nevado FP, Aranda E, Martín A (2021) Anti-fungal activity of phenolic sweet orange peel extract for controlling fungi responsible for post-harvest fruit decay. Fungal Biol 125(2):143–152

Hijaz F, Al-Rimawi F, Manthey JA, Killiny N (2020) Phenolics, flavonoids and antioxidant capacities in Citrus species with different degree of tolerance to Huanglongbing. Plant Signal Behav 15(5):1752447

Islam MT, Lee BR, Lee H, Jung WJ, Bae DW, Kim TH (2019) p-Coumaric acid induces jasmonic acid-mediated phenolic accumulation and resistance to black rot disease in Brassica napus. Physiol Mol Plant Pathol 106:270–275

Jahan MA, Harris B, Lowery M, Coburn K, Infante AM, Percifield RJ, Ammer AG, Kovinich N (2019) The NAC family transcription factor GmNAC42–1 regulates biosynthesis of the anti-cancer and neuroprotective glyceollins in soybean. BMC Genomics 20(1):1–21

Jahan MA, Harris B, Lowery M, Infante AM, Percifield RJ, Kovinich N (2020) Glyceollin transcription factor GmMYB29A2 regulates soybean resistance to *Phytophthora sojae*. Plant Physiol 183(2):530–546

Jan R, Aaqil Khan M, Asaf S, Park JR, Lee IJ, Kim KM (2021) Flavonone 3-hydroxylase relieves bacterial leaf blight stress in rice via overaccumulation of antioxidant flavonoids and induction of defense genes and hormones. Int J Mol Sci 22(11):6152

Javed S, Mahmood Z, Khan KM, Sarker SD, Javaid A, Khan IH, Shoaib A (2021) Lupeol acetate as a potent antifungal compound against opportunistic human and phytopathogenic mold *Macrophomina phaseolina*. Sci Rep 11(1):1–11

Jeandet P (2015) Phytoalexins: current progress and future prospects. Molecules 20:2770–2774

Jeandet P, Douillet-Breuil AC, Bessis R, Debord S, Sbaghi M, Adrian M (2002) Phytoalexins from the Vitaceae: biosynthesis, phytoalexin gene expression in transgenic plants, antifungal activity, and metabolism. J Agric Food Chem 50(10):2731–2741

Jeandet P, Hébrard C, Deville MA, Cordelier S, Dorey S, Aziz A, Crouzet J (2014) Deciphering the role of phytoalexins in plant-microorganism interactions and human health. Molecules 19(11): 18033–18056

Jeong H, Lee J, Kim S, Yeo YY, So H, Wu H, Song YS, Jang CY, Kim HD, Kim MJ, Chang M (2018) Hepatic metabolism of Sakuranetin and its modulating effects on cytochrome P450s and UDP-glucuronosyltransferases. Molecules 23(7):1542

Jian W, He D, Xi P, Li X (2015) Synthesis and biological evaluation of novel fluorine-containing stilbene derivatives as fungicidal agents against phytopathogenic fungi. J Agric Food Chem 63(45):9963–9969

Jiang CQ, Ma LL, Lv ZD, Feng F, Chen Z, Liu ZD (2020) Polydatin induces apoptosis and autophagy via STAT3 signaling in human osteosarcoma MG-63 cells. J Nat Med 74(3): 533–544

Jiang L, Wu P, Yang L, Liu C, Guo P, Wang H, Wang S, Xu F, Zhuang Q, Tong X, Liu P (2021a) Transcriptomics and metabolomics reveal the induction of flavonoid biosynthesis pathway in the interaction of Stylosanthes-Colletotrichum gloeosporioides. Genomics 113:2702–2716

Jiang C, Wang D, Zhang J, Xu Y, Zhang C, Zhang J, Wang Y (2021b) VqMYB154 promotes polygene expression and enhances resistance to pathogens in Chinese wild grapevine. Hortic Res 8(1):1–17

Juli G, Oliverio M, Bellizzi D, Gallo Cantafio ME, Grillone K, Passarino G, Colica C, Nardi M, Rossi M, Procopio A, Tagliaferri P (2019) Anti-tumor activity and epigenetic impact of the polyphenol oleacein in multiple myeloma. Cancers 11(7):990

Kang M, Kim SJ, Lee JY, Yoon SR, Kim SH, Ha JH (2018) Inactivation of *Pectobacterium carotovorum* subsp. carotovorum on Chinese cabbage (*Brassica rapa* L. subsp. pekinensis) by wash treatments with phenolic compounds. LWT 93:229–236

Kang JE, Jeon BJ, Park MY, Kim BS (2020) Inhibitory activity of *Sedum middendorffianum* - derived 4-hydroxybenzoic acid and vanillic acid on the type III secretion system of *Pseudomonas syringae* pv. tomato DC3000. Plant Pathol J 36(6):608

Katsumata S, Toshima H, Hasegawa M (2018) Xylosylated detoxification of the rice flavonoid phytoalexin sakuranetin by the rice sheath blight fungus *Rhizoctonia solani*. Molecules 23(2): 276

Kawabata K, Yoshioka Y, Terao J (2019) Role of intestinal microbiota in the bioavailability and physiological functions of dietary polyphenols. Molecules 24(2):370

Keen NT, Littlefield LJ (1979) The possible association of phytoalexins with resistance gene expression in flax to *Melampsora lini*. Physiol Plant Pathol 14(3):265–280

Kitakaze T, Makiyama A, Yamashita Y, Ashida H (2020) Low dose of luteolin activates Nrf2 in the liver of mice at start of the active phase but not that of the inactive phase. PLoS One 15(4): e0231403

Kong WS, Xing HH, Li J, Ye L, Liu X, Li YP, Rao GX, Zhou M, Hu QF, Li YK, Li XM (2018) Two new flavones from Cassia pumila and their anti-tobacco mosaic virus activity. Chem Nat Compd 54(6):1048–1051

Kouassi EK, Coulibaly I, Rodica P, Pintea A, Ouattara S, Odagiu A (2017) HPLC phenolic compounds analysis and antifungal activity of extract's from *Cymbopogon citratus* (DC) Stapf against *Fusarium graminearum* and *Fusarium oxysporum* sp tulipae. J Sci Res Rep 14:1–11

Kuang W, Hu W, Ren H, Shao Y, Liu B (2021) Plant derived coumestrol phytochemical targets human skin carcinoma cells by inducing mitochondrial-mediated apoptosis, cell cycle arrest, inhibition of cell migration and invasion and modulation of m-TOR/PI3K/AKT signalling pathway. Saudi J Biol Sci 28(5):2739–2746

Kulik T, Buśko M, Pszczółkowska A, Perkowski J, Okorski A (2014) Plant lignans inhibit growth and trichothecene biosynthesis in Fusarium graminearum. Lett Appl Microbiol 59(1):99–107

Kumar S, Abedin MM, Singh AK, Das S (2020) Role of phenolic compounds in plant-defensive mechanisms. In: Plant phenolics in sustainable agriculture. Springer, Singapore, pp 517–532

Kumar M, Anantapur R, Peter A, Chaitra HV (2021) Computational investigation of phytoalexins as potential antiviral RAP-1 and RAP-2 (replication associated proteins) inhibitor for the management of cucumber mosaic virus (CMV): a molecular modeling, in silico docking and MM-GBSA study. J Biomol Struct Dyn. https://doi.org/10.1080/07391102.2021.1968500

Kwon DH, Ji JH, Yim SH, Kim BS, Choi HJ (2018) Suppression of influenza B virus replication by sakuranetin and mode of its action. Phytother Res 32(12):2475–2479

Lallemand LA, Zubieta C, Lee SG, Wang Y, Acajjaoui S, Timmins J, McCarthy AA (2012) A structural basis for the biosynthesis of the major chlorogenic acids found in coffee. Plant Physiol 160(1):249–260

Lawrence SA, Burgess EJ, Pairama C, Black A, Patrick WM, Mitchell I, Gerth ML (2019) Mātauranga-guided screening of New Zealand native plants reveals flavonoids from kānuka (*Kunzea robusta*) with anti-Phytophthora activity. J R Soc N Z 49:137–154

Li N, Zhao M, Liu T, Dong L, Cheng Q, Wu J, Zhang S (2017) A novel soybean dirigent gene GmDIR22 contributes to promotion of lignan biosynthesis and enhances resistance to *Phytophthora sojae*. Front Plant Sci 8:1185

Libro R, Giacoppo S, Rajan TS, Bramanti P, Mazzon E (2016) Natural phytochemicals in the treatment and prevention of dementia: an overview. Molecules 21(4):518

Lim W, Yang C, Jeong M, Bazer FW, Song G (2017) Coumestrol induces mitochondrial dysfunction by stimulating ROS production and calcium ion influx into mitochondria in human placental choriocarcinoma cells. MHR Basic Sci Reprod Med 23(11):786–802

Liu CW, Murray JD (2016) The role of flavonoids in nodulation host-range specificity: an update. Plants (Basel) 5:33

Liu Y, Hassan S, Kidd BN, Garg G, Mathesius U, Singh KB, Anderson JP (2017) Ethylene signaling is important for isoflavonoid-mediated resistance to *Rhizoctonia solani* in roots of *Medicago truncatula*. Mol Plant Microbe Interact 30(9):691–700

Liu H, Zhao H, Lyu L, Huang Z, Fan S, Wu W, Li W (2019a) Synergistic effect of natural antifungal agents for postharvest diseases of blackberry fruits. J Sci Food Agric 99(7): 3343–3349

Liu M, Ma F, Wu F, Jiang C, Wang Y (2019b) Expression of stilbene synthase VqSTS6 from wild Chinese *Vitis quinquangularis* in grapevine enhances resveratrol production and powdery mildew resistance. Planta 250(6):1997–2007

Liu YL, Chen BY, Nie J, Zhao GH, Zhuo JY, Yuan J, Li YC, Wang LL, Chen ZW (2020) Polydatin prevents bleomycin-induced pulmonary fibrosis by inhibiting the TGF-β/Smad/ERK signaling pathway. Exp Ther Med 20(5):1

Liu X, Cui X, Ji D, Zhang Z, Li B, Xu Y, Chen T, Tian S (2021) Luteolin-induced activation of the phenylpropanoid metabolic pathway contributes to quality maintenance and disease resistance of sweet cherry. Food Chem 342:128309

Long L, Liu J, Gao Y, Xu FC, Zhao JR, Li B, Gao W (2019) Flavonoid accumulation in spontaneous cotton mutant results in red coloration and enhanced disease resistance. Plant Physiol Biochem 143:40–49

Luca SV, Macovei I, Bujor A, Miron A, Skalicka-Woźniak K, Aprotosoaie AC, Trifan A (2020) Bioactivity of dietary polyphenols: the role of metabolites. Crit Rev Food Sci Nutr 60(4): 626–659

Luo HZ, Guan Y, Yang R, Qian GL, Yang XH, Wang JS, Jia AQ (2020) Growth inhibition and metabolomic analysis of Xanthomonas oryzae pv. oryzae treated with resveratrol. BMC Microbiol 20:1–13

Machado ND, Fernández MA, Häring M, Saldías C, Díaz DD (2019) Niosomes encapsulated in biohydrogels for tunable delivery of phytoalexin resveratrol. RSC Adv 9(14):7601–7609

Maleki M, Khelghati N, Alemi F, Bazdar M, Asemi Z, Majidinia M, Sadeghpoor A, Mahmoodpoor A, Jadidi-Niaragh F, Targhazeh N, Yousefi B (2020) Stabilization of telomere by the antioxidant property of polyphenols: anti-aging potential. Life Sci 24:118341

Mandal SM, Chakraborty D, Dey S (2010) Phenolic acids act as signaling molecules in plant-microbe symbioses. Plant Signal Behav 5(4):359–368

Martínez G, Regente M, Jacobi S, Del Rio M, Pinedo M, de la Canal L (2017) Chlorogenic acid is a fungicide active against phytopathogenic fungi. Pestic Biochem Physiol 140:30–35

Matos MJ, Santana L, Uriarte E, Abreu OA, Molina E, Yordi EG (2015) Coumarins-an important class of phytochemicals. In: Phytochemicals-isolation, characterisation and role in human health, vol 25. IntechOpen, pp 533–538

McLay ER, Pontaroli AC, Wargent JJ (2020) UV-B induced flavonoids contribute to reduced biotrophic disease susceptibility in lettuce seedlings. Front Plant Sci 11:1691

Medrano-Padial C, Prieto AI, Puerto M, Pichardo S (2021) In vitro assessment of the mutagenic and genotoxic potential of a pure stilbene extract. Food Chem Toxicol 150:112065

Miles TD, Vandervoort C, Nair MG, Schilder AC (2013) Characterization and biological activity of flavonoids from ripe fruit of an anthracnose-resistant blueberry cultivar. Physiol Mol Plant Pathol 83:8–16

Morales J, Mendoza L, Cotoras M (2017) Alteration of oxidative phosphorylation as a possible mechanism of the antifungal action of p-coumaric acid against *Botrytis cinerea*. J Appl Microbiol 123(4):969–976

Morkunas I, Marczak L, Stachowiak J, Stobiecki M (2005) Sucrose-induced lupine defense against *Fusarium oxysporum*: sucrose-stimulated accumulation of isoflavonoids as a defense response of lupine to *Fusarium oxysporum*. Plant Physiol Biochem 43(4):363–373

Nefzi A, Abdallah RAB, Jabnoun-Khiareddine H, Ammar N, Medimagh-Saïdana S, Haouala R, Daami-Remadi M (2018) Management of Fusarium crown and root rot of tomato by *Solanum linnaeanum* L. extracts. Sci Hortic 238:204–214

Nguyen XH, Naing KW, Lee YS, Moon JH, Lee JH, Kim KY (2015) Isolation and characteristics of protocatechuic acid from Paenibacillus elgii HOA73 against Botrytis cinerea on strawberry fruits. J Basic Microbiol 55(5):625–634

Njankouo Ndam Y, Nyegue MA, Mounjouenpou P, Kansci G, Kenfack MJ, Eugène EE (2020) LC-MS quantification of scopoletin in cassava (*Manihot esculenta* Crantz) varieties, local derived foods, and activity on some food spoilage fungi. J Food Process Preserv 44(4):e14387

Ofosu FK, Daliri EB, Elahi F, Chelliah R, Lee BH, Oh DH (2020) New insights on the use of polyphenols as natural preservatives and their emerging safety concerns. Front Sustain Food Syst 4:223

Ogawa S, Yazaki Y (2018) Tannins from *Acacia mearnsii* De wild. Bark: tannin determination and biological activities. Molecules 23(4):E837

Ogawa S, Miyamoto K, Nemoto K, Sawasaki T, Yamane H, Nojiri H, Okada K (2017) OsMYC2, an essential factor for JA-inductive sakuranetin production in rice, interacts with MYC2-like proteins that enhance its transactivation ability. Sci Rep 7(1):1–1

Ohri P, Pannu SK (2010) Effect of phenolic compounds on nematodes-a review. J Appl Nat Sci 2(2):344–350

Pane C, Fratianni F, Parisi M, Nazzaro F, Zaccardelli M (2016) Control of *Alternaria* post-harvest infections on cherry tomato fruits by wild pepper phenolic-rich extracts. Crop Prot 84:81–87

Park HL, Yoo Y, Hahn TR, Bhoo SH, Lee SW, Cho MH (2014) Antimicrobial activity of UV-induced phenylamides from rice leaves. Molecules 19:18139–18151

Patzke H, Schieber A (2018) Growth-inhibitory activity of phenolic compounds applied in an emulsifiable concentrate-ferulic acid as a natural pesticide against Botrytis cinerea. Food Res Int 113:18–23

Pereira DM, Valentão P, Pereira JA, Andrade PB (2009) Phenolics: from chemistry to biology. Molecules 14:2202–2211

Pezet R, Perret C, Jean-Denis JB, Tabacchi R, Gindro K, Viret O (2003) δ-Viniferin, a resveratrol dehydrodimer: one of the major stilbenes synthesized by stressed grapevine leaves. J Agric Food Chem 51(18):5488–5492

Piccolella S, Crescente G, Candela L, Pacifico S (2019) Nutraceutical polyphenols: new analytical challenges and opportunities. J Pharm Biomed Anal 175:112774

Prabowo WC, Agustina R (2020) Antibacterial activity of scopoletin from stem bark of *Aleurites moluccana* against *Salmonella typhi*. J Trop Pharm Chem 5(1):29–32

Raghubeer S, Nagiah S, Phulukdaree A, Chuturgoon A (2015) The phytoalexin resveratrol ameliorates ochratoxin A toxicity in human embryonic kidney (HEK293) cells. J Cell Biochem 116(12):2947–2955

Rane AR, Paithankar H, Hosur RV, Choudhary S (2021) Modulation of α-synuclein fibrillation by plant metabolites, daidzein, fisetin and scopoletin under physiological conditions. Int J Biol Macromol 182:1278–1291

Rastija V, Vrandečić K, Ćosić J, Majić I, Šarić GK, Agić D, Molnar M (2021) Biological activities related to plant protection and environmental effects of coumarin derivatives: QSAR and molecular docking studies. Int J Mol Sci 22(14):7283

Reger MK, Zollinger TW, Liu Z, Jones JF, Zhang J (2018) Dietary intake of isoflavones and coumestrol and the risk of prostate cancer in the prostate, lung, colorectal and ovarian cancer screening trial. Int J Cancer 142(4):719–728

Rentzsch M, Wilkens A, Winterhalter P (2009) Non-flavonoid phenolic compounds. In: Wine chemistry and biochemistry. Springer, New York, pp 509–527

Romero-Cortes T, Pérez España VH, López Pérez A, Rodríguez-Jimenes GDC, Robles-Olvera VJ, Aparicio Burgos JE, Cuervo-Parra JA (2019) Antifungal activity of vanilla juice and vanillin against *Alternaria alternata*. CyTA 17(1):375–383

Romero-Rincón A, Martínez ST, Higuera BL, Coy-Barrera E, Ardila HD (2021) Flavonoid biosynthesis in *Dianthus caryophyllus* L. is early regulated during interaction with *Fusarium oxysporum* f. sp. *dianthi*. Phytochemistry 192:112933

Sade D, Shriki O, Cuadros-Inostroza A, Tohge T, Semel Y, Haviv Y, Willmitzer L, Fernie AR, Czosnek H, Brotman Y (2015) Comparative metabolomics and transcriptomics of plant response to tomato yellow leaf curl virus infection in resistant and susceptible tomato cultivars. Metabolomics 11:81–97

Salem MZ, Ali HM, Akrami M (2021) *Moringa oleifera* seeds-removed ripened pods as alternative for papersheet production: antimicrobial activity and their phytoconstituents profile using HPLC. Sci Rep 11(1):1–13

Santos-Sánchez NF, Salas-Coronado R, Hernández-Carlos B, Villanueva-Cañongo C (2019) Shikimic acid pathway in biosynthesis of phenolic compounds. In: Plant physiological aspects of phenolic compounds, vol 1. IntechOpen

Schiavi D, Francesconi S, Bischetti G, Giovanale G, Fortunati E, Balestra GM (2021) Antibacterial activity of coumarin as an innovative organic control strategy for *Xanthomonas euvesicatoria* pv. euvesicatoria. J Plant Dis Protect 30:1–7

Shah A, Smith DL (2020) Flavonoids in agriculture: chemistry and roles in, biotic and abiotic stress responses, and microbial associations. Agronomy 10(8):1209

Sharma A, Shahzad B, Rehman A, Bhardwaj R, Landi M, Zheng B (2019) Response of phenylpropanoid pathway and the role of polyphenols in plants under abiotic stress. Molecules 24(13):2452

Sharma A, Sharma A, Kumar R, Sharma I, Vats AK (2021a) PR proteins: key genes for engineering disease resistance in plants. In: Crop improvement. CRC Press, pp 81–98

Sharma I, Thakur A, Sharma A, Singh N, Kumar R, Sharma A (2021b) Phytoalexins: implications in plant defence and human health. In: Plant secondary metabolites: phytochemical properties and therapeutic applications. Springer. https://doi.org/10.1007/978-981-16-4779-6_10

Shimizu T, Lin F, Hasegawa M, Okada K, Nojiri H, Yamane H (2012) Purification and identification of naringenin 7-O-methyltransferase, a key enzyme in biosynthesis of flavonoid phytoalexin sakuranetin in rice. J Biol Chem 287(23):19315–19325

Shokoohinia Y, Sajjadi SE, Gholamzadeh S, Fattahi A, Behbahani M (2014) Antiviral and cytotoxic evaluation of coumarins from Prangos ferulacea. Pharm Biol 52(12):1543–1549

Shrestha A, Pandey RP, Sohng JK (2019) Biosynthesis of resveratrol and piceatannol in engineered microbial strains: achievements and perspectives. Appl Microbiol Biotechnol 103(7):2959–2972

Siciliano I, Amaral Carneiro G, Spadaro D, Garibaldi A, Gullino ML (2015) Jasmonic acid, abscisic acid, and salicylic acid are involved in the phytoalexin responses of rice to *Fusarium fujikuroi*, a high gibberellin producer pathogen. J Agric Food Chem 63(37):8134–8142

Silva RF, Pogačnik L (2020) Polyphenols from food and natural products: neuroprotection and safety. Antioxidants 9(1):61

Silva B, Souza MM, Badiale-Furlong E (2020) Antioxidant and antifungal activity of phenolic compounds and their relation to aflatoxin B1 occurrence in soybeans (*Glycine max* L.). J Sci Food Agric 100(3):1256–1264

Singh AK, Cabral C, Kumar R, Ganguly R, Rana HK, Gupta A, Lauro MR, Carbone C, Reis F, Pandey AK (2019) Beneficial effects of dietary polyphenols on gut microbiota and strategies to improve delivery efficiency. Nutrients 11(9):2216

Singla P, Garg N (2017) Plant flavonoids: key players in signaling, establishment, and regulation of rhizobial and mycorrhizal endosymbioses. In: Mycorrhiza-function, diversity, state of the art. Springer, Cham, pp 133–176

Sirohi PR, Kumari A, Admane N, Somvanshi P, Grover A (2021) The polyphenolic phytoalexin polydatin inhibits amyloid aggregation of recombinant human prion protein. RSC Adv 11(42):25901–25911

Song DK, Karr AL (1993) Soybean phytoalexin, glyceollin, prevents accumulation of aflatoxin B1 in cultures of *Aspergillus flavus*. J Chem Ecol 19(6):1183–1194

Song PP, Zhao J, Liu ZL, Duan YB, Hou YP, Zhao CQ, Wu M, Wei M, Wang NH, Lv Y, Han ZJ (2017) Evaluation of antifungal activities and structure–activity relationships of coumarin derivatives. Pest Manag Sci 73(1):94–101

Song P, Yu X, Yang W, Wang Q (2021) Natural phytoalexin stilbene compound resveratrol and its derivatives as anti-tobacco mosaic virus and anti-phytopathogenic fungus agents. Sci Rep. https://doi.org/10.21203/rs.3.rs-458237/v1

Souza SM, Monache FD, Smânia A (2005) Antibacterial activity of coumarins. Zeitschr Naturforschung C 60:693–700

Spevakova I, Fernandez-Cruz ML, Tokarova K, Greifova H, Capcarova M (2020) The protective effect of stilbenes resveratrol and pterostilbene individually and combined with mycotoxin citrinin in human adenocarcinoma HT-29 cell line in vitro. J Environ Sci Health A 56(1):75–88

Stompor M (2020) A review on sources and pharmacological aspects of sakuranetin. Nutrients 12(2):513

Stringlis IA, Yu K, Feussner K, De Jonge R, Van Bentum S, Van Verk MC, Pieterse CM (2018) MYB72-dependent coumarin exudation shapes root microbiome assembly to promote plant health. Proc Natl Acad Sci 115(22):E5213–E5222

Sun H, Wang L, Zhang B, Ma J, Hettenhausen C, Cao G, Sun G, Wu J, Wu J (2014) Scopoletin is a phytoalexin against *Alternaria alternata* in wild tobacco dependent on jasmonate signalling. J Exp Bot 65(15):4305–4315

Tak Y, Kumar M (2020) Phenolics: a key defence secondary metabolite to counter biotic stress. In: Plant phenolics in sustainable agriculture. Springer, Singapore, pp 309–329

Talreja SS, Nerurkar AS (2018) Small molecules cause virulence attenuation of Xanthomonas oryzae pv. oryzae, the pathogen causing bacterial blight of rice. Eur J Plant Pathol 151(1): 229–241

Tamogami S, Kodama O (2000) Coronatine elicits phytoalexin production in rice leaves (*Oryza sativa* L.) in the same manner as jasmonic acid. Phytochemistry 54(7):689–694

Tamogami S, Rakwal R, Kodama O (1997) Phytoalexin production by amino acid conjugates of jasmonic acid through induction of naringenin-7-O-methyltransferase, a key enzyme on phytoalexin biosynthesis in rice (*Oryza sativa* L.). FEBS Lett 401(2–3):239–242

Tang KS (2021) Protective effects of polydatin against dementia-related disorders. Curr Neuropharmacol 19(2):127–135

Tanguy J, Martin C (1972) Phenolic compounds and hypersensitivity reaction in *Nicotiana tabacum* infected with tobacco mosaic virus. Phytochemistry 11:19–28

Ter Ellen BM, Kumar ND, Bouma EM, Troost B, van de Pol DP, van der Ende-Metselaar HH, Apperloo L, van Gosliga D, van den Berge M, Nawijn MC, van der Voort PH (2021) Resveratrol and pterostilbene inhibit SARS-CoV-2 replication in air-liquid interface cultured human primary bronchial epithelial cells. Viruses 13(7):1335

Thiel G, Rössler OG (2017) Resveratrol regulates gene transcription via activation of stimulus-responsive transcription factors. Pharmacol Res 117:166–176

Tortosa M, Cartea ME, Rodriguez VM, Velasco P (2018) Unraveling the metabolic response of Brassica oleracea exposed to *Xanthomonas campestris* pv. campestris. J Sci Food Agric 98: 3675–3683

Tresserra-Rimbau A (2020) Dietary polyphenols and human health. Nutrition 12:2893

Tuladhar P, Sasidharan S, Saudagar P (2021) Role of phenols and polyphenols in plant defense response to biotic and abiotic stresses. In: Biocontrol agents and secondary metabolites. Elsevier, pp 419–441

Ullah C, Unsicker SB, Fellenberg C, Constabel CP, Schmidt A, Gershenzon J, Hammerbacher A (2017) Flavan-3-ols are an effective chemical defense against rust infection. Plant Physiol 175(4):1560–1578

Valletta A, Iozia LM, Leonelli F (2021) Impact of environmental factors on stilbene biosynthesis. Plants 10:90

Vanamala JK, Massey AR, Pinnamaneni SR, Reddivari L, Reardon KF (2018) Grain and sweet sorghum (*Sorghum bicolor* L. Moench) serves as a novel source of bioactive compounds for human health. Crit Rev Food Sci Nutr 58(17):2867–2881

Venugopala KN, Rashmi V, Odhav B (2013) Review on natural coumarin lead compounds for their pharmacological activity. BioMed Res Int 2013:1. https://doi.org/10.1155/2013/963248

Visioli F, Lastra CA, Andres-Lacueva C, Aviram M, Calhau C, Cassano A, D'Archivio M, Faria A, Favé G, Fogliano V, Llorach R (2011) Polyphenols and human health: a prospectus. Crit Rev Food Sci Nutr 51(6):524–546

Vogt V, Cifuente D, Tonn C, Sabini L, Rosas S (2013) Antifungal activity in vitro and in vivo of extracts and lignans isolated from *Larrea divaricata* Cav. against phytopathogenic fungus. Ind Crop Prod 42:583–586

Vu TT, Kim H, Tran VK, Vu HD, Hoang TX, Han JW, Choi YH, Jang KS, Choi GJ, Kim JC (2017) Antibacterial activity of tannins and use for control of tomato bacterial wilt. PLoS One 12(7): e0181499

Wang Y, Chen S, Yu O (2011) Metabolic engineering of flavonoids in plants and microorganisms. Appl Microbiol Biotechnol 91:949–956

Wang N, Yanliang C, Yurui P (2017) Extraction of flavonoids and polyphenols from coronatine-treated mulberry leaves and evaluation of their antibacterial activity against *Pseudomonas syringae*. Agric Biotechnol 6(4):51–62

Wiesel L, Newton AC, Elliott I, Booty D, Gilroy EM, Birch PRJ, Hein I (2014) Molecular effects of resistance elicitors from biological origin and their potential for crop protection. Front Plant Sci 5:655

Wu ZZ, Rao M, Xu S, Hu HY, Tang QZ (2020) Coumestrol ameliorates doxorubicin-induced cardiotoxicity via activating AMPKα. Free Radic Res 54(8–9):629–639

Xu Y-H, Wang J-W, Wang S, Wang J-Y, Chen X-Y (2004) Characterization of GaWRKY1, a cotton transcription factor that regulates the sesquiterpene synthase gene (+)-δ-cadinene synthase-A. Plant Physiol 135(1):507–515

Xu W, Grain D, Bobet S, Le Gourrierec J, Thévenin J, Kelemen Z, Lepiniec L, Dubos C (2014) Complexity and robustness of the flavonoid transcriptional regulatory network revealed by comprehensive analyses of MYB–b HLH–WDR complexes and their targets in *Arabidopsis* seed. New Phytol 202(1):132–144

Xu W, Dubos C, Lepiniec L (2015) Transcriptional control of flavonoid biosynthesis by MYB–bHLH–WDR complexes. Trends Plant Sci 20(3):176–185

Xu D, Yu G, Xi P, Kong X, Wang Q, Gao L, Jiang Z (2018) Synergistic effects of resveratrol and pyrimethanil against Botrytis cinerea on grape. Molecules 23(6):1455

Xu Y, Zhang Y, Liang H, Liu X (2021) Coumestrol mitigates retinal cell inflammation, apoptosis, and oxidative stress in a rat model of diabetic retinopathy via activation of SIRT1. Aging (Albany NY) 13(4):5342

Yahfoufi N, Alsadi N, Jambi M, Matar C (2018) The immunomodulatory and anti-inflammatory role of polyphenols. Nutrients 10(11):1618

Yamauchi Y, Okuyama T, Ishii T, Okumura T, Ikeya Y, Nishizawa M (2019) Sakuranetin downregulates inducible nitric oxide synthase expression by affecting interleukin-1 receptor and CCAAT/enhancer-binding protein β. J Nat Med 73(2):353–368

Yang J, Li Y, Wang F, Wu C (2010) Hepatoprotective effects of apple polyphenols on CCl4-induced acute liver damage in mice. J Agric Food Chem 58(10):6525–6531

Yang W, Xu X, Li Y, Wang Y, Li M, Wang Y, Ding X, Chu Z (2016) Rutin-mediated priming of plant resistance to three bacterial pathogens initiating the early SA signal pathway. PLoS One 11(1):e0146910

Yang L, Li S, Qin X, Jiang G, Chen J, Li B, Ding W (2017) Exposure to umbelliferone reduces Ralstonia solanacearum biofilm formation, transcription of type III secretion system regulators and effectors and virulence on tobacco. Front Microbiol 8:1234

Yang L, Wu L, Yao X, Zhao S, Wang J, Li S, Ding W (2018) Hydroxycoumarins: new, effective plant-derived compounds reduce *Ralstonia pseudosolanacearum* populations and control tobacco bacterial wilt. Microbiol Res 215:15–21

Yang L, Wang Y, He X, Xiao Q, Han S, Jia Z, Ding W (2021) Discovery of a novel plant-derived agent against *Ralstonia solanacearum* by targeting the bacterial division protein FtsZ. Pestic Biochem Physiol 2021:104892

Yuxing D, Shanfa L (2017) Biosynthesis and regulation of phenylpropanoids in plants. Crit Rev Plant Sci 36(4):257–290

Zafar A, Singh S, Satija YK, Saluja D, Naseem I (2018) Deciphering the molecular mechanism underlying anticancer activity of coumestrol in triple-negative breast cancer cells. Toxicol In Vitro 46:19–28

Zhang H, Yang Y, Mei X, Li Y, Wu J, Li Y, Liu Y (2020) Phenolic acids released in maize rhizosphere during maize-soybean intercropping inhibit Phytophthora blight of soybean. Front Plant Sci 11:886

Zhu C, Lei M, Andargie M, Zeng J, Li J (2019) Antifungal activity and mechanism of action of tannic acid against *Penicillium digitatum*. Physiol Mol Plant Pathol 107:46–50

Zorzete P, Reis TA, Felício JD, Baquião AC, Makimoto P, Corrêa B (2011) Fungi, mycotoxins and phytoalexin in peanut varieties, during plant growth in the field. Food Chem 129(3):957–964

Zuo R, de Oliveira A, Bullita E, Torino MI, Padgett-Pagliai KA, Gardner CL, Lorca GL (2019) Identification of flavonoids as regulators of YbeY activity in *Liberibacter asiaticus*. Environ Microbiol 21(12):4822–4835

Phytochemicals of *Withania somnifera* and Their Perspective on Plant Defense Against Stress

11

Manali Singh, Kuldeep, Parul Chaudhary, Shruti Bhasin, Anshi Mehra, and Shivani Bhutani

Abstract

Plants produce a variety of phytochemicals in response to the different biotic or abiotic stress conditions. These metabolites not provide protection to the plant and help in its survival. But are also exploited by humans for their own needs. Secondary metabolites of Ashwagandha have great therapeutic importance. The plants herbal rejuvenative properties have made plant exceptionally esteemed in pharmacology and naturopathy. The active biochemical constituents of *W. somnifera* are steroidal lactones, flavonoids, alkaloids, sitoindosides, glycowithaolides, withanosides, amino acids, saponins etc., all help in plant defense. Withanolides also act in plant defense mechanism actively protecting the plant from pathogens and in withstanding biotic and a biotic stress conditions. Sterols released by WS also having protective functioning in plants and improve the phenotypic and metabolic activities in plants under stress conditions. The plants response to these stress conditions by generation of stress-tolerating enzymes such as superoxide dismutase and glutathione reductase. Increase in production of secondary metabolite Withaferin (terpenoids) protects plants from

M. Singh (✉)
Department of Life Sciences, Parul Institute of Applied Sciences, Parul University, Vadodara, Gujarat, India

Kuldeep
Department of Agricultural & Food Engineering, IIT Kharagpur, Kharagpur, West Bengal, India

P. Chaudhary
Department of Animal Biotechnology, Animal Genomics Lab, NDRI, Karnal, India

S. Bhasin
Department of Biotechnology, Banasthali Vidyapith, Aliyabad, Rajasthan, India

A. Mehra · S. Bhutani
Department of Biotechnology, Invertis Village, Bareilly, Uttar Pradesh, India

R. Lone et al. (eds.), *Plant Phenolics in Biotic Stress Management*,
https://doi.org/10.1007/978-981-99-3334-1_11

cold stress via production of antioxidant enzymes. Terpenoids act as signaling molecules in plants and help in defense response toward stress conditions (biotic/ abiotic stress). The phytochemical constituent of the plant contributes in drugs as well as assists with getting the physiological property for the treatment of various illnesses. Because of its significance in herbal drugs, many types of exploration are proceeding to determine new substance compounds and furthermore distinguishing the degree of toxicity.

Keywords

Withania somnifera · Stress · Phytochemicals · Withanolides

11.1 Introduction

Herbal and plants' preparations have been used as scent, flavor and medicine. Medicinal plants show low toxicity therefore uses from last few decades. Research on medicinal plants has turned in the major pharmaceutical population because of its safety and efficacy. *Withania somnifera* (L.) Dunal is also known as Indian ginseng, Ashwagandha, winter cherry. Ashwagandha belongs to Solanaceae family that has been used in indigenous and Ayurvedic medicine for more than 3000 years (Umadevi et al. 2012). *Withania somnifera* (WS) is from Plantae kingdom, Tracheoionta subkingdom, Spermatophyta superdivision, Magnoliopsida subclass, Solanales order, Solanaceae family and Withania genus. Local name of *Withania somnifera* are Ashwagandha (Bengali), Amkulang (Tamil), Pulivendram (Telugu), Ghodaasoda (Gujarati) and Punir (Hindi) (Pratibha et al. 2013). Ashwagandha plant grows abundantly in India, Pakistan, Sri Lanka and Africa. Numerous pharmacological examinations have been completed to portray various organic properties of *W. somnifera* and results got from these investigations demonstrate that it is additionally valuable for various treatments (Kulkarni and Dhir 2008). This species is branched, erect and short branched with a grayish color. Its height is 30–150 cm tall with tuberose roots along with tomentose branches. Leaves are dull green color with simple petiolate, alternate, broadly ovate of size 4–10 cm long and 2–7 cm broad (Mir et al. 2012).

Ashwagandha is suitable for both the sexes and of all the ages without any side effects. Its root contains a group of steroidal lactone like withanolides which attributed to pharmacological effects. Roots are used to prepare tonic which augmenting defense against infectious disease, arresting the aging process and promote longevity. From the leaves of *Withania somnifera*, many withanolide steroidal lactones have been extracted which show antitumor, antifungal and antibacterial properties (John 2014). Its leaves show the properties of insect repellent, and fruits contained considerable amount of saponins (Schmelze et al. 2008). Extract of the herb shows excellent immunomodulatory affect by activation of nonspecific lymphocytes, natural killer cells, granulocytes and macrophages. It also protects against different pathogen like viruses, fungi etc. by generating various effecter molecules by activated cells (Rizvi Faizi et al. 2016). Ashwagandha is widely used in treatment of several diseases like inflammatory disease, bronchitis, ulcer, asthma and stomach problems (Narinderpal et al. 2013). The biochemical constituents of *W. somnifera* are steroidal lactones, flavonoids, alkaloids, sitoindosides, glycowithaolides, withanosides, amino acids, saponins and many more (Elsakka et al. 1990; Mishra and Singh 2000). Various clinical researches have extensively work on toxicology of *W. somnifera* and obtained data that in wide range of practical doses of this plant are nontoxic (Kulkarni and Dhir 2008; Prabu et al. 2013; Sharada et al. 1993). But regular consumption can lead to arrest senescence, increase fertility, rejuvenate the reproductive organs and rectify abnormalities of the sense organs (Kumar et al. 2013).

Pharmacology and chemistry information of *W. somnifera* is now available that extracts obtained from its different parts have been reviewed that if it combines with other plant appropriately then could be used for cure and prevention of chronic diseases and diverse disorders. Steroidal lactones and steroidal alkaloids are the major constituents in root of Ashwagandha. Moreover, because of excessive usage of plant has become endangered. So we expect to develop the plant in laboratory by in vitro procedures with enhance withanolide content to fulfill commercial demands (Singh and Singh 2016). *Withania somnifera* is an exceptionally utilized therapeutic plant of Ayurveda. This is one of the handfuls chosen significant therapeutic plant, which is in great request in global and public market. Ashwagandha has been perceived as the therapeutic plant in the assembling of more than 100 traditional therapeutic herbal grown formulation thus this plant is exceptionally esteemed in pharmacology area. *W. somnifera* shows pharmacological significance as it has life-prolonging and rejuvenating properties. It is used to make tonic to improve sleep, increase stamina, reduce the weakness and calm the mind. It provides antioxidant protection and increase the production of cells, semen, blood, lymph and activation of the immune system (Nadkarni 1954). The normal prescriptions have less or no secondary effect as contrast with manufactured drugs. From the hundreds of years, Ashwagandha is effectively used for treatment of numerous medical issues. Because of its significance in herbal drugs, many types of exploration are proceeding to determine new substance compounds and furthermore distinguishing the degree of toxicity. The quantitative assessment of dynamic biological component is possible by UV spectrometer strategy. In spite of the fact that *Withania somnifera* is a multipurpose medication, it additionally has a few limitations which can be overwhelmed by performing more clinical trials. For the helpful use, the result of biochemical mixtures and their activities is important (Bhasin et al. 2019).

Analysts showed that phytoconstituents profile of the plant is basically impacted by abiotic and biotic factors. These elements modulate the guard system of the plant by balancing the secondary metabolites through biosynthetic qualities (Hugueney et al. 1996; Zulak et al. 2007). Elicitation is a productive tool which incites the plant to incorporate advantageous phytochemical (Kim et al. 2007) as well as nutraceuticals (Zhao et al. 2002). A few factors, for example, development stages, kind of elicitor, concentration and so on, are liable for the upgrade of secondary metabolites. It has been accounted by different analysts that the molecules inciting the development of secondary metabolites to guarantee their endurance in plants are known as "elicitors" (Namdeo 2007). For instance, chitosan has additionally been accounted to inspire the production of different secondary metabolites in the in vitro culture of cell (Ferri and Tassoni 2011).

11.2 Phytochemicals of *Withania somnifera* and Their Pharmacological Significance

Withania somnifera (WS) is known as Ashwagandha and is used in traditional treatment for cardioprotection, neurological illnesses and skin illnesses. The quality of medicinal plants depends on the plant part used, its collection time and its phytochemicals (Tetali et al. 2020). There are various plant parts of *W. somnifera* such as root, steam, leaves and fruit which are rich in bioactive phytochemicals such as saponins, flavonoids, carbohydrates, scopoletin, withanine, withananine alkaloids, phenols and lactones having the pharmacological actions as reported by Saleem et al. (2020). WS plant can be used as dietary supplement as it has a variety of nutrients and phytochemicals. Roots are described to contain alkaloids (0.13–0.31%), amino acids and steroids. Numerous phytochemicals have been isolated from WS using diverse techniques like gas chromatography mass spectroscopy (GC-MS), thin layer, column and high-performance chromatography. There are more than 12 alkaloids, withanolides (40) and various sitoindosides found in numerous parts of WS and assessed using phytochemical examines. Withanolides are steroids (C_{28}) with an ergostane skeleton and derived from phytosterol pathway using enzyme cytochrome P450 having role in biotic stress management (Singh et al. 2014; Shilpashree et al. 2022). The metabolites such as Withanolide A, B, D, F and withanosides (glycosylated steroids) are secreted by WS having neuroprotective, anticancer, hepatoprotective, antiaging, diuretic, antipogenic, hemopoietic, immunomodulatory functions and antioxidant activities (Huang et al. 2009; Kim et al. 2019). Sitoindosides and acylsterylglucosides metabolites in WS are used against stress conditions (antistress) as reported by Bhattacharya et al. (1987). Recently, it was observed that Withaferin A having therapeutic potential protects from COVID-19 infection to mitigate the virus-made cardiovascular disease (Straughn and Kakar 2020). Withasomniferol D was isolated from WS having antiadipogenic properties which is used in obesity treatment as reported by Lee et al. (2021). Withanolides also have antimicrobial effect on pathogenic microbes such as *Salmonella*, *Staphylococcus aureus*, *Pseudomonas aeruginosa*, *Escherichia coli* and *Candida albicans* (Jain and Varshney 2011). Girme et al. (2020) reported different withanosides and withanolides from WS roots using ultra high-performance chromatography and mass fragmentation studies. Sterols released by WS also have protective functioning in plants and improve the phenotypic and metabolic activities in plants under stress conditions. It was observed that sterol glycosyltransferase (SGTL1) gene in WS improved the growth of plants and protect from cold stress (abiotic) and biotic stresses due to *Spodoptera litura* (Saema et al. 2016). Mishra et al. (2021) also observed that expression of SGT gene from WS in *Arabidopsis* improved the agronomic traits and help in salt stress tolerance through modulation in sterols. Qadir et al. (2021) reported that WS tolerated the fly ash contamination and removed their pollution via generation of stress-tolerating enzymes such as superoxide dismutase and glutathione reductase. Increase in production of secondary metabolite Withaferin (terpenoids) protects plants from cold stress via production of antioxidant enzymes. Terpenoids act as signaling molecules in plants and help in defense response toward stress conditions (biotic/abiotic stress). Sabir et al. (2012) observed that WS plant grows better in presence of sodium chloride (50 mM) by releasing the withanolides which protect plants from salt stress. Increase in secondary metabolites such as phenols released by WS plant also increases the antioxidant enzymes, which help in copper stress tolerance at polluted sites (Khatun et al. 2008). Withanolides are also synthesized via shoot cultures, which depend on growth hormones applied in culture media. Withanolide D was identified via shoot and root culture supplemented with sucrose which provide protection to the plants. Application of biotic and abiotic elicitors (chitosan and aluminum chloride) increases the production of withanolide (Sivanandhan et al. 2012). Withaferin A level was increased in root cultures when WS plant was inoculated with *A. rhizogens* during biotic stress due to *Aspergillus* sp. (Varghese et al. 2014). Production of secondary metabolites such as Withaferin and withanolide A/B produced by WS via inoculation of microbes showed protection from *Alternaria alternata* infection (Mishra et al. 2018). Application of different media, tissue culture techniques and genetic engineering approaches can enhance the maximum production of different withanolides in roots of WS.

11.3 Physiological and Biochemical Changes in *Withania somnifera* Under Different Types of Stress Conditions (Biotic, Abiotic) Like Soil, Water, pH, Drought, Air Etc.

Withania coagulans is a kind of withania. Dunal is a medicinal plant that belongs to the Solanaceae family and includes a variety of physiologically active compounds such as milk coagulating enzyme, steroids, alkaloids, tannin, saponin, amino acids, organic acids and pharmaceutically active withanolide. Cultivation in a natural setting increases the species' natural conservation potential as well as its therapeutic worth. However, several abiotic variables such as drought stress, salt stress, severe temperature, cold and heavy metals all have an impact on plant development and output.

11.3.1 Drought Stress

Drought stress has a significant impact on plant morphology, physiology and biochemical activity in recent years, resulting in a decrease in agricultural output (Hsiao 1973). Drought survival strategies are developed by plants in three ways—(1) drought relief, (2) tolerance to drought and (3) drought avoidance. Drought resistance is the term for the emergence of these states. Due to stomatal restrictions, drought produced metabolic and physiological reactions as well as a significant drop in CO_2 absorption. As a result, the intake of reduced equivalents ($NADPH_2^+$) for CO_2 absorption via the Calvin cycle decreases, resulting in increased oxidative stress and an excess of antioxidant enzymes. Drought causes metabolic processes to shift toward biosynthetic activities, which utilize reduction equivalents. As a result, decreased chemical production (isoprenoids, phenols and alkaloids) is boosted (Singh et al. 2018).

11.3.2 Moisture Stress

Despite its drought resistance, a protracted soil moisture deficit limits the crop's growth and development, lowering crop yield potential value and plant productivity and efficiency. The impact of long-term moisture stress on active principle quality content and root quality will aid functional characterization and biochemical incorporation of molecular and genetic data.

11.3.3 Salt Stress

Salinity stress reduced the proportion of seeds that germinated seedling growth and chlorophyll content. Similarly, antioxidants such as ascorbic acid (AA), reduced glutathione (GSH), and -tocopherol (-toc) in all plant sections were badly impacted (root, stem and leaf). Furthermore, antioxidant enzymes such as superoxide dismutase (SOD), peroxidase (POX), polyphenol oxidase (PPO) and catalase had their activity altered (CAT) (Jaleel et al. 2008).

11.3.4 Cold Stress

According to Mir et al. after 7 days of cold therapy, the antioxidative enzyme activities (SOD, CAT, APX and GR) as well as the quantity of marker withanolides increased significantly. Under the experimental conditions, there was a favorable association between the induction of antioxidant enzymes and the synthesis of withanolides in *W. somnifera*. According to our findings, AGB002 had less oxidative stress in response to cold therapy than AGB001 (Fig. 11.1 and Table 11.1).

11.4 Prospects and Perspectives of Biosynthetic Pathway Engineering

Withanolides are triterpenoids, which are 30-carbon molecules. Triterpenoid backbone, like other terpenoid molecules, is produced through a metabolic mechanism that uses isoprene units as precursors (isopentenyln pyrophosphate; IPP and

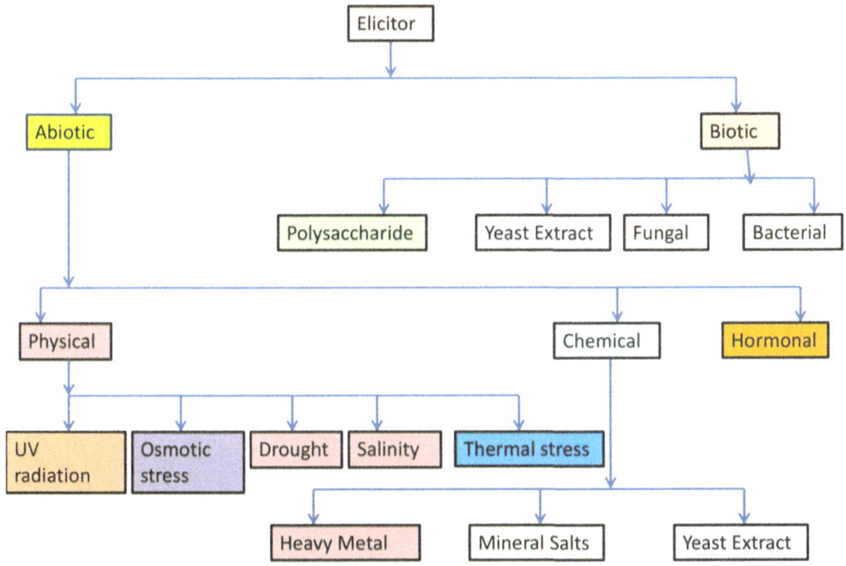

Fig. 11.1 Schematic representation of the different types of elicitors that affect the production of withanolides in *Withania somnifera*

dimethyl allyl pyrophosphate; DMAPP). As a result, isoprenogenesis might be one of the important upstream metabolic processes controlling the flux of isoprene units for the synthesis of triterpenoid pathway metabolic intermediate(s) dedicated to withanolide biosynthesis (Bhatia et al. 2013). In plant cells, two independent routes for isoprenoid precursor biosynthesis exist: the traditional cytosolic mevalonic acid (MVA) pathway and the alternate plastidial methylerythritol phosphate (MEP) pathway.

11.4.1 Tissue-Specific Pathway Engineering

The tremendous therapeutic potential of *W. somnifera* motivates researchers to investigate all possible way of acquiring chemotypes with favored enhanced chemoprofiles, including recombinant DNA methods and tools and in vitro methods. Although recombinant DNA or cell fusion procedures are promising options, they are limited by the availability of genetic and biochemical expertise about secondary metabolite production. As a result, urgent biotechnology improvements are required to improve yield in a shorter time frame. In vitro approaches, on the other hand, give a viable option for the creation of these therapeutically useful molecules in the hunt for alternatives. Tissue culture techniques provide an endless, steady, and renewable supply of valuable plant medicines, which are used in large-scale plant cell culture for secondary metabolite extraction. There has been a lot of study done on *W. somnifera* utilizing various in vitro methodologies, with the major focus on

Table 11.1 Withanolide production in response to stress condition

S. no.	Abiotic/ biotic stress	Stress condition	Plant part involved	Name of metabolite	References
1	Biotic	Hormone	Leaves	27-deoxy-16-en-Withaferin A, 2,3-dihydro3b-hydroxywithanone, 27-acetoxy-3-oxo-witha-1,4,24-trienolide	Misra et al. (2005)
2		Enzyme	Roots	Withaferin A, withanolide A, 12-deoxywithastramonolide, and 20-deoxywithanolide A	Subbaraju et al. (2006)
3		Terpenoids	Leaves, roots	27-hydroxywithanone, 17-hydroxy Withaferin A, 17-hydroxy-27-deoxy Withaferin A, Withaferin A, withanolide D, 27-hydroxy withanolide B, withanolide A, withanone and 27-deoxywithaferin A	Chaurasiya et al. (2008), Akhoon et al. (2016), Shilpashree et al. (2022)
4		Biotic (microorganism)	Differential plant parts	Withaferin A, Withanone and Withanolide A, squalene	Dhar et al. (2016)
5		Biotic (hormone)	Multiple shoots, teratoma	Withaferin A, withanolide A	Sangwan et al. (2007)
6		Biotic (hormone)	Multiple shoots	Glycowithanolides; withanolides	Ahuja et al. (2009)
7		Biotic (chitosan)	Adventitious roots from semi-friable callus of leaves	Withanolide B, Withaferin A, withanosides IV and V	Sivanandhan et al. (2012b)
8		Biotic (protein)	Plantlet	Withaferin A, withanolide A	Rana et al. (2013)
9		Biotic (hormone, microbe)	Cell suspension culture	Withanolide A, withanolide B, Withaferin A, withanone	Sivanandhan et al. (2013b)
10		Biotic (cytokines, methyl jasmonate, salicylic acid, cadmium stress tolerance by releasing phenols, flavonoids and lignin)	Multiple shoot cultures	Withanolide A, withanolide B, Withaferin A, withanone	Sivanandhan et al. (2013c), Turrini et al. (2016), Mishra and Singh Sangwan (2019)

11		Biotic (chitosan, squalene, salicylic signaling)	Cell suspension culture	Withanolide A, withanolide B, Withaferin A, withanone, 12 deoxywithanstramonolide, withanoside IV, withanoside V	Sivanandhan et al. (2014a), Dhuley (2008), Ghosh et al. (2014)
12		Biotic (seaweed)	Shoot suspension culture	Withanolide A, withanolide B, Withaferin A, withanone	Sivanandhan et al. (2014b)
13		Biotic (vermicompost)	Whole plant/plant parts	Withaferin A, Withanolide D	Raja and Veerakumari (2013)
14		Biotic (bacteria)	Hairy roots	Withanolide D	Ray et al. (1996)
15		Biotic (bacteria)		Withaferin A, Withanolide D	Bandyopadhyay et al. (2007)
		Biotic (fungal protein)		Withaferin A and withanolide A	Sil et al. (2015)
16		Biotic (bacteria)	Teratoma	Withaferin A, Withanolide D	Ray and Jha (1999)
17		Biotic (antioxidant enzyme activities)	Leaves	Quercetin	Bhargavi and Shankar (2021)
18		Abiotic (cold stress tolerance)	Leaves	Withaferin A	Hahm et al. (2011), Mir et al. (2015)
19	Abiotic	Squalene	Leaves	Withanolide A	Grover et al. (2013)
20		Methyl jasmonate, salicylic acid		Withaferin A, Withanone and Withanolide A	Sivanandhan et al. (2013a)
21		Soil	Leaves, roots	Withaferin A, Withanone and Withanolide A	Gupta et al. (2016)
22		Carbon and nitrogen sources	Flowers, fruits	Withanolide A, withanolide B, Withaferin A, withanone	Sivanandhan et al. (2015)
23		pH	Adventitious root culture	Withanolide A	Murthy and Praveen (2013)
24		pH	Whole plant/plant parts	Withanolide-A	Praveen and Murthy (2010)
25		Ethanol, methanol	Dried roots/leaves	Withanone, Withastramonolide, 27-hydroxywithanone, Physagulin-D, Withanolide A, Withaferin A	Dhar et al. (2006)

(continued)

Table 11.1 (continued)

S. no.	Abiotic/ biotic stress	Stress condition	Plant part involved	Name of metabolite	References
26		Sucrose	Whole plant	6α,7α-epoxy-3β,5α,20β-trihydroxy-1-oxowitha-24-enolide, 5β,6β-epoxy-4β,17α,27-trihydroxy-1-oxowitha-2,24-dienolide, Withaferin A, 2,3-dihydrowithaferin A, 6α,7α-epoxy-5α,20β-dihydroxy-1-oxowitha-2,24-dienolide, 5β,6β-epoxy-4β-hydroxy-1-oxowitha-2,14,24-trienolide	Choudhary et al. (2004)
27		Methanol	Leaves, roots	Acetoacetate, alanine, arginine, aspartic acid, choline, citric acid, citrulline, GABA, glutamine, glutamic acid, lactic acid, leucine, lysine, malic acid, methanol, myoinositol, ornithine, proline, succinic acid, threonine, valine, fructose, glucose (Glc), galactose and sucrose, adenosine, formic acid, fumaric acid, phenylalanine, trigonelline, tyrosine, uracil, uridine	Bharti et al. (2011)
28		Drugs	Roots	Withaferin A, Withanone and Withanolide A	Kumar and Kalonia (2007)
29		Methanol	Leaves	Withaferin A	Furmanowa et al. (2001)
30		Methanol	Leaves, roots, seedling	Withaferin A	Johny et al. (2015)

manipulating plant development regulator adjuvants and cultural environment for withanolide accumulation.

11.4.2 Mevalonate Pathway

For terpenoid biosynthesis, the MVA route comprises seven enzymes for the production of precursor molecules, including IPP and DMAPP. The enzyme AcAc-CoA thiolase converts two molecules of acetyl-CoA onto acetoacetyl (AcAc)-CoA in the first step, 3-hydroxy-3-methylglutaryl-coenzyme is formed (HMG-CoA) HMG-CoA synthase facilitates the condensation of AcAc-CoA with one molecule of acetyl-CoA to generate HMG-CoA in the second phase. Following that, production of mevalonate from HMG-CoA is catalyzed by HMG-CoA reductase (HMGR), a nicotinamide adenine dinucleotide (phosphate)-dependent (NAD (P)H) enzyme that catalyzes a twofold reduction step requiring four electron transfers. Two phosphorylations and decarboxylation activities that involve mevalonate kinase, phosphomevalonate kinase and mevalonate diphosphate decarboxylase enzymes are involved in the conversion of mevalonate to IPP. Furthermore, isopentenyl diphosphate isomerase, a divalent, metal ion-requiring enzyme, acts on IPP generated from the cytosolic MVA route to create DMAPP.

11.4.3 Methylerythritol Phosphate Pathway

1-Deoxy-D-xylulose 5-phosphate synthase (DXS) catalyzes the first step in the MEP pathway, converting pyruvate and glyceraldehyde 3-phosphate to 1-deoxy-Dxylulose 5-phosphate (DXP). DXP reductoisomerase converts DXP to MEP, which is then transformed to 1-hydroxy-2-methyl-2-(*E*)-butenyl 4-diphosphate by the enzymes 2-C-methyl-D-erythritol 4-phosphate cytidylyltransferase, 4-diphosphocytidyl-2-C-methyl-D-erythritol kinase, 2-C-methyl-(HMBPP). The last step is the conversion of HMBPP to IPP and DMAPP, which is performed by a single enzyme, (*E*)-4-hydroxy-3-methyl but-2-enyl diphosphate reductase (HDR). Although HDR in the MEP route generates both IPP and DMAPP in an 85:15 ratio, the plastid-localized isopentenyl diphosphate isomerase catalyzes IPP isomerization, which aids in substrate optimization. The synthesis of farnesyl pyrophosphate is caused by the head-to-tail condensation of IPP (FPP). The major precursor for triterpenoids is FPP. Condensation of IPP with DMAPP to generate 10-C intermediate geranyl diphosphate (GPP) and condensation of GPP with another molecule of IPP to form 15-C FPP are the two stages of the FPPS-catalyzed process.

Squalene is a metabolic step in the manufacture of a variety of triterpenoids. Its production is catalyzed by the squalene synthase enzyme, which catalyzes the NADPH-dependent head-to-head condensation of two molecules of FPP to create squalene. Squalene epoxidase epoxidizes one of the squalene's terminal double bonds, resulting in squalene 2,3-epoxide. A ring closure process mediated by the cycloartenol synthase enzyme on squalene 2,3-epoxide results in the biosynthesis of cycloartenol, which may then be transformed into a range of steroidal triterpenoidal skeletons (Dhar et al. 2014).

11.4.4 De Novo Tissue-Specific Withanolide Biosynthesis

The qualitative overlap of leaf and root withanolides in *W. somnifera* is supported by phytochemical data collected throughout time. The presence of a gradient of withanolide concentrations in leaves and roots, as well as radiotracer experiments employing 24-methylene cholesterol as a precursor, suggest that withanolides may be imported from leaves to roots. Nonetheless, in vitro grown roots and native/orphan roots of *W. somnifera* show 14C incorporation from [2–14C]-acetate and [U-14C]-glucose into WS-1. This study found that these main metabolites were incorporated into WS-1, indicating that root-specific WS-1 is synthesized from primary isoprenogenic precursors instead of being imported from leaves (Sangwan et al. 2007).

11.4.5 Gene Elucidation of Withanolide Biosynthetic Pathway

The stereochemical ring closure, prevalence of chiral centers, stiff *trans*-lactone groups and highly energetic epoxy ring make chemical synthesis of withanolides difficult. As a result, due to low yields and high costs, synthetic manufacturing is economically unviable. As a result, effective options for the commercial manufacturing of withanolides in great numbers are required. Genetic modification of genes encoding enzymes engaged in withanolide production appears to be a promising method for creating genotypes of *W. somnifera* having higher amounts of withanolides, as well as alternate microbial/yeast hosts that produce withanolides. This method necessitates the expression of metabolic circuits in heterologous host (s). At the molecular level, the withanolide biosynthesis pathway is still in its hypothetical stage. As a result, elucidating the withanolide biosynthetic route, as well as the regulatory elements of their production, is of basic scientific relevance.

11.4.5.1 3-Hydroxy-3-methylglutaryl Coenzyme A Reductase

Coenzyme 3-hydroxy-3-methylglutaryl in the isoprenoid biosynthesis pathway, a reductase (3-HMGR) is a NADH-dependent rate restricting enzyme engaged in the conversion of 3-hydroxy-3-methylglutaryl coenzyme A (HMG-CoA) into mevalonic acid, which is the primary precursor of IPP and DMAPP. Plant HMGR has been found in subcellular organelles such as the endoplasmic reticulum (ER), mitochondria and plastids, with catalytic domains in the cytoplasm. This enzyme is a subject of numerous cholesterol-lowering medicines such as statins because of its rate-limiting nature. It has been functionally described for both mammalian species and plant systems. HMGR is encoded by a multigene family in plants, with a wide range of temporal and geographic expression patterns. HMGR-1 and HMGR-2-encode HMGR isozymes in Arabidopsis. Dwarf phenotype, sterility and senescence result from the loss of HMGR-1 function, which corresponds to decreased sterol production.

11.4.5.2 1-Deoxy-D-xylulose-5-phosphate Reductoisomerase and 1-Deoxy-D-xylulose-5-phosphate Synthase

Withanolides, *W. somnifera*'s hallmark secondary metabolites, are likely biosynthesized by a metabolic departure from the sterol route at the level of 24-methylene cholesterol. MVA and MEP pathways provide isoprenoid precursors for the same, with MEP being a plastid-derived alternative path for isoprenoid biosynthesis. The MEP route contributes around 30% of the production of withanolide precursor isoprenoids, with the first step being pyruvate condensation with D-glyceraldehyde-3-phosphate to generate 1-deoxy-D-xylulose-5-phosphate (DXP), which is mediated by DXP synthase (DXS). DXP is a precursor for the biosynthesis of IPP and DMAPP. DXP reductoisomerase then catalyzes the conversion of DXP to MEP (DXR). Though DXR is the first enzyme in the MEP route for terpenoid biosynthesis, DXS, the first enzyme in this pathway, is also important for isoprenoid biosynthesis in a variety of species, including bacteria and plants.

11.4.5.3 Farnesyl Diphosphate

The MVA and MEP pathways are responsible for the majority of the bioactive compounds produced in Withania. Farnesyl diphosphate (FPP), a substrate for the first committed reaction of many branching pathways, is generated in two stages by the enzyme farnesyl diphosphate synthase (FPPS) in several biosynthetic pathways. To begin, 10-C intermediate geranyl diphosphate is formed via condensation of IPP with DMAPP structures (GPP). GPP is further condensed with another molecule of IPP to generate FPP. In the early stages of triterpenoid precursor generation linked to withanolide biosynthesis, FPPS also plays a crucial role.

11.4.5.4 Squalene Synthase

Squalene synthase (SQS) catalyzes one of the first enzymatic steps in the phytosterol biosynthesis pathway, allowing two farnesyl pyrophosphate radicals to condense into squalene. SQS directs carbon flow from the isoprenoid pathway to phytosterol biosynthesis, resulting in the creation of braassinosteroids, withanolides and triterpenoids as end products. Despite the abundance of evidence for the function of SQS in phytosterol biosynthesis, little is known about the withanolide biosynthetic pathway, genes involved in withanolide biosynthesis and regulatory components of the promoter region directing gene expression in *W. somnifera*. Bhat et al. studied the importance of squalene synthase in withanolide production for pathway intensification resulting to increased withanolide deposition in *W. somnifera*.

11.4.5.5 Cytochrome P450 Reductase and Monooxygenases

Cytochrome P450 enzymes are a component of a functionally varied protein superfamily. It plays an important role in a number of metabolic molecular circuits. P450s are heme thiolate-proteins that catalyze a wide range of reactions, including hydroxylations, dealkylations, sulfoxidations, epoxidations, reductive dehalogenations, peroxidations and various types of isomerization, to produce a variety of primary and secondary metabolites required for plant growth and development. P450 monooxygenases are a substrate-specific group of enzymes with a high degree of regio and stereo specificity. Cytochrome P450s make up roughly 1% of all genes in the vegetation genome, according to gene annotation. P450s are found in the ER, and their catalytic action is dependent on the availability of electrons via NADPH cytochrome P450 reductase.

Furthermore, P450 monooxygenases are a strongly regio- and stereo-specific group of static substrate-specific enzymes that serve a vital role in secondary metabolism, primarily assisting in the functionalization of core structures of compounds such as withanolides. These are potential targets for industrial biocatalysis because to their regio- and stereo-specific catalytic versatility. P450s have proved beneficial in the pharmaceutical sector for testing novel medicines, pharmaceuticals and xenobiotics.

11.4.5.6 Glucosyltransferases

Glycosylation is a typical modifying process in plant metabolism that is inextricably linked to secondary metabolism. Uridine diphosphate glycosyltransferases (UGTs) are enzymes that catalyze the synthesis of glucosides. They belong to the glycosyltransferase superfamily, which encompasses over 80 enzyme families, and their role entails transferring an UDP-activated glucose to an analogous acceptor molecule. UGTs utilize UDP-activated sugars as donors, distributing their sugar moiety to a large number of acceptors. The glycosylation of excess bioactive natural chemicals is catalyzed by plant family 1 UGTs. This is often the last stage in the biosynthesis of numerous natural products, in order to increase their solubility and stability, as well as to make storage and accumulation in plant cells easier.

The first investigation on *W. somnifera* sterol glucosyltransferases (SGTs) identified three distinct (SGTs) with conserved SGT family domains: SGTL1, SGTL2 and SGTL3. Among these, the full-length SGTL1 gene (DQ356887) was cloned and confirmed to be expressed in all regions of the plant. SGTL1 has transmembrane domains and prefers membrane sterol glucosylation, according to its deduced amino acid sequence. Furthermore, partly purified recombinant SGT was shown to be selective for sterols containing a hydroxyl group at the C-3 position. Stress has also been linked to the functional activation of SGTL1 in response to environmental challenges.

11.5 Tissue-Specific Transcriptomal Analysis

After the EST research, it became discovered that the transcriptomic and genomic statistics of *W. somnifera* are very limited and handiest 742 ESTs are to be had inside the national center for biotechnology statistics (NCBI) database. The in-depth generation of EST databases of Withania may additionally offer statistics approximately all the expressed regions of genome and may be used to symbolize styles of gene expression in unique tissues but after the development of subsequent generation sequencing approach the transcriptomic evaluation of leaf and root tissues became performed to increase our information associated with withanolide (most important and vital secondary metabolite in *Withania* genus) biosynthesis. Those transcriptomic records showed the expression profiling of differentially expressed methyltransferases, cytochrome p450s, glycosyltransferase and transcription elements which can be concerned in withanolide biosynthesis in leaf and root tissues. These data will help us in growing techniques of metabolic engineering and to grow the manufacturing of specific withanolides. Further to gene identification for pathway elucidation, molecular markers were additionally diagnosed the use of transcriptome facts from leaf and root on the way to be useful for marker-assisted selection and breeding program (Gupta et al. 2013). Further to the above-noted leaf versus root transcriptome, analysis to perceive the gene worried in secondary metabolite biosynthesis in *W. somnifera* the leaf transcriptome of *W. somnifera* after exogenous application of SA turned into additionally finished to apprehend the signaling mechanism concerned in the manufacturing of secondary metabolites. SA is the key hormone throughout biotic protection response and tiers of SA and its glycosylated conjugate (SAG) in tissues is regarded to considerably gather both locally and systemically after pathogen infection. Exogenously software of SA changed into stated to decorate sickness resistance and induce pathogenesis-associated (PR) gene expression in a huge form of plant species. So this transcriptomic study centered the analysis of transcripts expressed in salicylic acid-handled leaf tissues. The production of secondary metabolites beneath in vitro condition is pronounced to be more desirable by way of exogenous application of elicitors (biotic and abiotic) in lifestyle media and methyl jasmonate and salicylic acid are broadly pronounced to induce production of secondary metabolites under subculture conditions. The transcriptomic have a look at recorded the boom in manufacturing of three predominant metabolites of *W. somnifera* such as withanoside V, Withaferin A and withanolide A in leaf tissues, subsequent to exogenous utility of SA (Ghosh et al. 2014). The development of excessive throughput sequencing strategies and their analysis no longer most effective broadens our know-how associated with plant molecular and biochemical studies and metabolic engineering however additionally precious for drug development and breeding program. In plants, this approach has expanded the knowledge of complicated transcriptional styles and has furnished measurements of gene expression in extraordinary tissues or at different stages of development.

11.6 Survey of Proteomics Studies in Members of Solanaceae Family

Proteomic analysis in Solanaceae individuals began over three decades ago. In the past, even prior to the advent of MS for analysis of larger biomolecules like proteins. At the beginning, Nover and Scharf used tomato suspension subculture cells for identity of heat shock responsive proteins (HSPS) the use of a 2-DE and N-terminal sequencing approach (Nover and Scharf 1984). Following this take a look at, Solanaceae proteomics has progressed at a fast phase, majorly inside the beyond decade, to establish proteomes of various tissues, organs and organelles for the duration of ordinary and damaging environmental situations. One of the first huge-scale studies of gel-loose proteomics even before the genome sequence finishing touch changed into applied to a potato breeding trial in discipline research analyzing 12 different potato cultivars (Hoehenwarter et al. 2008, 2011). The use of a novel aggregate of records processing, alignment and multivariate records referred to as mass accuracy precursor alignment (MAPA), sturdy cultivar-and trait-unique protein marker was diagnosed in discipline grown potato tubers and the biggest tuber proteomed at a set was assembled (Hoehenwarter et al. 2008, 2011). This study showed for the primary time the possibility to use big-scale shotgun proteomics in discipline conditions for the identification of robust and trait-particular protein marker which may be applied for in addition breeding procedures. These days, finishing touch of genome sequencing of potato in 2011 (P.G.S. Consortium 2011), tomato in 2012 (T.G. Consortium 2012) and pepper in 2014 (Kim et al. 2014a, b) affords a platform for big-scale proteome evaluation and identification of proteins with better accuracy. Similarly, availability of draft genomes for *Nicotiana benthamiana*, *Nicotiana tabacum*, *Solanum pimpinellifolium*, *Petunia axillaris* and *Petunia inflate* offers a big benefit for proteome evaluation to recognize the biology of these tremendously unexplored plant life of the Solanaceae family. Both gel-based totally and gel-unfastened proteomic technologies have been used to establish the proteome of various tissues, organs and organelles and PTM-proteome. Maintaining in view their financial significance, distinct tissues in one-of-a-kind vegetation had been preferentially explored for proteomic studies to become aware of key gamers, which may be used as ability objectives for crop upgrades, e.g., fruits in tomato and pepper; leaves in tobacco and Ashwagandha; tubers in potato and flora in ornamental vegetation.

11.7 Proteomics Studies of Solanaceae Plant Response to Abiotic Stresses

To evaluate plant response to pressure situations, it is far important to set up a preferred and sturdy system that may mimic real abiotic pressure conditions encountered by the flora inside the area (Ghatak et al. 2016). Plant reaction to abiotic pressure is inspired not simplest with the aid of its intensity but additionally by the duration and increase stage of the plant. Proteomics studies related to abiotic stress conditions had undertaken the use of each gel-based totally and gel-free proteomic techniques to decide global adjustments at proteome level (Gupta et al. 2015). Similarly, to decide molecular occasions occurring due to protein regulatory mechanisms, it is far important to perceive and quantify proteins and PTMs that might act as stress sensors and chargeable for version of plant life to abiotic stresses (Vanderschuren et al. 2013). Usually, proteomics studies aim to recognize the complexity and diversity in context of regulation underneath abiotic stresses in model plant species, e.g., *Arabidopsis thaliana*, *Oryza sativa* and *Thellungiella halophile* (Hossain et al. 2012; Kosová et al. 2011). Right here, we offer information on proteomic studies in Solanaceae plant species and their response to abiotic stresses.

11.7.1 Salinity Stress

Salinity influences ~6% of global land and thereby influences the agriculture productiveness (Parihar et al. 2015). Further to modulating gene expression pattern and disturbance of the protein homeostasis in a plant mobile, soil salinity reasons other important consequences: (a) nutritional imbalance because of saline ions and low water uptake and distribution and (b) high accumulation of Na^+ and Cl^- ions in the cytoplasm (Amini et al. 2007). Amini et al. investigated salt-induced proteome adjustments in tomato (*L. esculentum* mill., cv. shirazy) seedlings, which were supplemented with 0, 40, 120 and 160 mm of NaCl for 24 days earlier than subjecting it for two-DE proteomics approach. This examine led to detection of 400 spots, of which 18 spots showed big changes underneath salt pressure. Apparently, epidermal increase element receptor protein (EGFR), a member of tyrosine kinase superfamily, showed multiplied ranges underneath salt stress condition, which is also regarded to be a precursor for sign transduction occasions via phosphorylation. A few reports advocate that hydrogen peroxide (H_2O_2) additionally induces EGF-receptor phosphorylation. Beneath salt stress, reactive oxygen species (ROS) inclusive of H_2O_2 are fashioned as by using-products of metabolism, consequently up-law of pressure protein EGFR can also be related to detoxification of H_2O_2. In addition, M2D3, three, RAV-like B domain DNA-binding protein and salt-tolerance protein (molecular weight 25 kDa) were also suggested to be multiplied in abundance beneath salt-strain conditions. Further, tomato seedlings of cv. moneymaker had been salt confused for 4 days at specific concentrations of NaCl (50 mm, 100 mm, and 150 mm) and a couple of days with 200 mm, in a hydroponic answer and subjected to second-dige. As a result, 17 differential spots had been diagnosed such as a ferredoxin-NADP(+) reductase and UDP-glucose pyrophosphorylase. Ferredoxin-NADP(+) reductase was caused by 1.2-fold beneath salt pressure, which can be predicted because plant chloroplast thioredoxin system uses ferredoxin reductase to detoxify peroxidases with the intention to preserve cellular homeostasis (Parihar et al. 2015). Similarly, UDP-glucose pyrophosphorylase, which acts as an crucial precursor for sucrose (Meng et al. 2007), cell wall polysaccharides (Ordin and Hall 1968) and starch biosynthesis (Viola et al. 2001), become also diagnosed. Juan et al. found an enhanced level of sucrose in salt-tolerant tomato cultivars compared to touchy cultivars. On the opposite, ribulose-1,5-bisphosphate carboxylase/oxygenase (rubisco) activase, which converts rubisco from inactive to energetic state, turned into observed at lower degrees, indicating an ordinary suppression of general rubisco activity and much less CO_2 assimilation (Juan et al. 2005). Manaa et al. verified adjustments in physiology and molecular behavior of special cultivars of tomato (cv. roma and cv. supermarmande) under salt stress condition. Careworn flowers confirmed a decrease in increase and leaf osmotic ability, while an increase within the lipid peroxidation (MDA content) becomes located. In comparison to the previous look at, wherein rubisco activase become found to be reduced, here, an improved stage of rubisco activase at the side of different photosynthesis-related proteins like rubisco huge subunit (LSU), pyruvate dehydrogenase, glucose-6-phosphate dehydrogenase and malate dehydrogenase become observed beneath salt stress situation (Manaa et al. 2013a, b).

11.7.2 Metal Toxicity Stress

Silicon (Si) is the second maximum considerable element in soil and is to be had as silicic acid to vegetation (Ma and Yamaji 2006). Its miles discovered that silicon has various effects on flora; it improves photosynthesis and remediates nutrient imbalances in flowers (Datnoff et al. 1997; Gong et al. 2006; Romero-Aranda et al. 2006); however, at better concentrations, it will increase metallic toxicities and mineral deficiencies (Kidd et al. 2001; Neumann and Zur Nieden 2001). Young seedlings (15-day vintage) of tomato were used to analyze the effect of Si under salt stress. The 2-DE evaluation of root proteome brought about the identification of 40 proteins differentially expressed beneath silicon and salt-treated plant life, of which 24 showed differential expressions (induction or suppression) by way of Si remedy in salt-pressured roots. Si transporter genes such as lelsi-1, lelsi-2 and lelsi-3 were appreciably expressed in roots of the flowers beneath silicon treatment, at the same time as the expression changed into low in control flora (Muneer et al. 2014). Subsequently, it was concluded that silicon has green deposition in shoots and roots via root permeability (Gong et al. 2006; Kim et al. 2014b). Further, a mixed effect of salinity and calcium became analyzed on tomato end result the use of 2-de (Manaa et al. 2013a, b). Proteins related to carbon and strength metabolism, salt stress, oxidative pressure and ripening technique have been located to be differentially modulated in reaction to salt and calcium stress (Manaa et al. 2013a, b).

11.7.3 Heat Stress

Excessive temperature past a threshold stage causes irreversible damage to plant increase, photosynthetic interest and yield (Wahid et al. 2007). Severe heat stress blended with drought strain reasons permanent impairment to agricultural plants (Berry and Bjorkman 1980). Tomato demonstrates a drastic poor reaction to extreme temperatures, which have mentioned effects on its reproductive growth, specifically at some point of pollen improvement (Fragkostefanakis et al. 2016). Nevertheless, various effects of high temperature additionally depend upon the genotypes (i.e., warmth tolerant or warmness sensitive cultivars). Chaturvedi et al. used a shotgun proteomic method (gel-ltq-orbitrap MS) to evaluate warmness careworn proteome of two pollen developmental levels (put up—meiotic and mature) in tomato cv. hazera

3017, a heat-sensitive cultivar, which brought about the identification of N2000 proteins in total (Chaturvedi et al. 2015). In this look at, the diagnosed peptides have been quantified via making use of a novel tMAPA strategy considering "proteotypic peptides" that resulted in detection of 51 precise proteins as warmness responsive applicants. In submit-meiotic degree, proteins like late embryogenesis abundant (LEA) protein, bloodless surprise protein 1, small HSP20 (sHSP) and 22 chaperone protein HTPG had been diagnosed. LEA is a critical protein marker for heat stress, as it participates in the right folding and conformation of the proteins (both structural and practical). LEA protein is expressed in all developmental stages with no tissue specificity under stress circumstance. In addition, sHSP20 and 22 are known to play a shielding function during postmeiotic ranges (microspore and polarized microspore) beneath warmth treatment. Apparently, these proteins were additionally recognized in pollen mother cells (microsporocytes) below manipulate situation in tomato (Chaturvedi et al. 2013) and also showed excessive expression stages in a transcriptome evaluation of growing microspores of tomato (Frank et al. 2009). In mature pollen, proteins corresponding to TCA cycle confirmed accelerated ranges below warmth strain and included ATP synthase, mitochondrial ATP synthase, citrate synthase, ATP-citrate lyase A-2. The presynthesis of proteins in mature pollen acts as a reservoir for the imminent electricity hard manner and pollen tube germination. This phenomenon was termed as developmental priming (Chaturvedi et al. 2013), that is mentioned elaborately in a recent evaluation article (Chaturvedi et al. 2016).

11.7.4 Drought Stress

Drought is every other principal proscribing thing affecting plant molecular and physiological tactics (Stebbins 1952; Bohnert et al. 1995). A comparative root proteomes evaluation between mesophytic tomato cultivar (walter la3465) and dehydration-tolerant cultivar (s. Chilense cv. La1958) identified proteins that play an important role in tolerance mechanism (Zhou et al. 2013). In *S. chilense* cv. La1985, 170 proteins had been diagnosed, of which 106 proteins had been repressed and 64 proteins were prompted under stress circumstance. Similarly, in *S. lycopersicum* cv. walter la3465, of 130 diagnosed proteins, 104 proteins were suppressed and 26 proteins had been caused extensively underneath stress condition. In *S. chilense* la1958, a LEA and an abscisic acid (ABA), stress- and ripening-prompted (ASR) protein had been brought on. Wherelse "walter" la3465, a dehydrin and a water-pressure inducible protein 3 were brought about. Further, calmodulin was brought about inside the tolerant tomato cultivar but repressed in the inclined cultivar. Proteins for folding nascent and denatured proteins and proteins mediating protein processing and localization in er and mitochondrial organelles were affected in another way in the species. Xei et al. investigated drought-precipitated proteomic adjustments in leaves of cultivated tobacco (cv. Honghuadajingyuan), using an ITRAQ method. In overall, 5570 proteins were identified, and most of differentially regulated proteins had been related to photosynthesis, metabolism, stress and defense (Xie et al. 2016). Tiers of ABA had been appreciably more suitable underneath drought strain in tobacco leaves. Similarly, HSP82, an HSP90-like protein, confirmed very high abundance underneath drought-stress situation, followed with the aid of warmness shock 70 kDa protein 17 and warmth shock cognate 70 kDa protein 2-like protein. Thioredoxins (TRXs) showed reduced levels underneath drought strain situation. The TRXs are well-conserved disulfide reductases in flowers which manage the redox repute of the target proteins and also play vital roles in plant tolerance to oxidative strain (Vieira Dos Santos and Rey 2006).

11.7.5 Cold Stress

Approach was utilized in tobacco (Jin et al. 2011). Five-week-vintage tobacco seedlings were handled at 4°C for 4 h, which led to phenotypic alterations like smoothing and shallowing leaves and expanded relative electrolyte leakage, in comparison to control. The identified differential proteins have been binned into practical category of photosynthesis, protein processing, redox homeostasis, RNA processing, sign transduction, translation, cellular department/cycle and metabolisms of carbon and energy. Protein applicants like N enolase, rcbL, rcbA, ascorbate peroxidase, and HSPs have been diagnosed along with several novel proteins such as 2-cys peroxiredoxin, armadillo repeat-containing protein, lr1-like protein rga-1 and putative nascent polypeptide-associated complex α chain. To improve the cryopreservation protocols for potato, apical shoots from 3-week-old in vitro plantlets from two species (*S. commersonii* dun and *S. tuberosum* L.) had been given osmotic observed via cold pressure (Folgado et al. 2013). The shoot period, the variety of leaves and the water content were decreased considerably in *S. commersonii* species. For proteomic analysis, 2-dige with maldi-tof/tof was used. In total, 310 spots were notably exclusive, of which 94 spots/proteins were recognized in this look at. *S. commersonii* is called the most frost tolerant species of potato collectively with *S. acaule* (Gusta and Wisniewski 2013; Seppänen et al. 1998), whereas cultivated potato *S. tuberosum* is touchy to freezing (Chen and Li 1980). Proteins associated with photosynthesis had been differentially expressed in reaction to osmotic and cold strain. All proteins related to playstation I and ps II subunit confirmed a decrease in awareness which include ferredoxin reductases. Annexin p34 and germin-like protein confirmed down-law in both species (Folgado et al. 2013). Folgado et al. studied 3-week-vintage shoots from in vitro plantlets of the cultivated *S. tuberosum* and its frost-resistant relative *S. commersonii* beneath osmotic stress (imposed via sucrose) and chilling (6°C). The use of 2d-dige mixed with maldi-tof/tof, a complete of numerous protein spots, had been diagnosed which includes HSPS/chaperones which have been abundant in each species under strain conditions (Folgado et al. 2014). Those proteins help in protein folding, assembly, translocation and degradation machinery (Wang et al. 2014; Boston et al. 1996). Effect of preculture remedies (sucrose pretreatment medium and cold-culturing throughout weeks) on donor vegetation of four potato species become studied in

S. commersonii, *S. juzepcukii*, *S. ajanhuiri* and *S. tuberosum* (Folgado et al. 2015). Proteome of donor plantlets after the pretreatments was analyzed by means of second-dige, which confirmed that carbon metabolism at the side of stress-response and oxidative-homeostasis was the main elegance of proteins that modified in abundance after pretreatments (Folgado et al. 2015).

11.7.6 Other Abiotic Stress Conditions

Hattrup et al. studied the proteomic sample of tomato (variety better boy) leaves in color avoidance response. On this examine, flora had been grown at once under sunlight in ambient temperature in addition to in-shade material environment. Leaf proteins have been investigated in the usage of 2-De and nesi-lc-ms/ms. A complete of 59 proteins had been identified with differential expression as part of the shade avoidance mechanism. Those identified proteins were binned into purposeful categories like respiration, Calvin cycle, glycolysis, biosynthesis, mobile maintenance, strain response, breathing, light harvesting, cell maintenance, strain response and unknown (Hattrup et al. 2007). In addition, Ahsan et al. completed comparative proteomic analysis of tomato (cv. Koma) leaves in reaction to waterlogging strain (Ahsan et al. 2007). To advantage higher resolution, total proteins had been extracted from leaves, observed with the aid of polyethylene glycol (PEG) fractionation and its 2-de analysis. Formerly, efficacy of PEG fractionation changed into shown in depletion of rubisco and enrichment of low abundance proteins in leaf samples (Kim et al. 2001). From this evaluation, a novel institution of proteins become recognized; Pto disorder resistance proteins, salt-tolerance protein, putative hydrogenase protein, ef-tu, cysteine protease have been diagnosed. It was also observed that waterlogging strain produces a high amount of Ros in tomato leaves, which blocks the pastime of rubisco subunit and rubisco activase. It additionally inhibits protein biosynthesis and protease inter-EST that plays a primary position in programmed mobile loss of life, which may be a possible cause of leaf senescence beneath waterlogging stress (Ahsan et al. 2007). Expression of cyanobacterial flavodoxin (fld) in tobacco flowers confirmed more advantageous tolerance to a wide variety of abiotic stresses inclusive of drought, temperature and UV. In every other similar have a look at, adjustments in proteome pattern of fld-expressing tobacco (cv. Petit havana) vegetation in reaction to drought pressure had been analyzed with the aid of high-resolution 2-de. In overall, 930 protein spots have been recognized of which 52 proteins spots showed good-sized abundance below drought in transgenic and/or wild-kind flowers (Gharechahi et al. 2015). The usage of blended maldi-tof/tof and esi-q/Tof evaluation, 39 (24 in wild-type, 11 in transgenic and 4 in each) drought-responsive proteins were identified. Some of these proteins such as remurin, ferredoxin-NADP reductase, chloroplast manganese stabilizing protein-II, phosphoglycerate mutase and GST confirmed reduced degrees underneath drought harassed fld-tobacco even as some others like *s*-formylglutathione hydrolase and pyridoxine biosynthesis protein confirmed expanded degrees. Quantitative proteomics observe become also performed in potato to evaluate the reaction to contrasting fertilization regimes (Rempelos et al. 2013). The results of changing mineral with composted livestock manure fertilizer input and excluding pesticide-based totally crop protection on *S. tuberosum* tuber yield, leaf and tuber mineral nutrient content material and leaf protein profiles were investigated. Dehydroascorbate reductase and GST proteins were expanded in abundance beneath mineral fertilization, possibly associated with better cd composition. Nitrogen and phosphorus compositions in leaf that were the primary drivers inside the protein expression had been improved (Rempelos et al. 2013). The study changed into done particularly to broaden purposeful molecular marker to improve breeding in potato nutrient use efficiency (Rempelos et al. 2013).

11.8 Conclusion

For more than 2500 years an ancient Indian medicine, Rasayana herb made of *Withania somnifera* is used. In India and other countries, currently Ayurvedic formulations are commercialized in which Ashwagandha is an active ingredient as it is prescribed for treatment of various diseases that affect health of humans. Various parts of plant have been studied for their pharmacological activities like antitumor, anti-inflammatory, antiparkinsonia, memory enhancing, adaptogen, antidepressant, anxiolytic, antioxidant, neuroprotective, anticancer properties and cardiovascular protection, antibacterial, hypolipidemic, hypoglycemic and immunomodulation. The most active ingredient of *W. somnifera* is withanolides which has potential in treating different health disorder. For therapeutic applications in human, there is need of purification, isolation and commercial formation of withanolides. Different parts of plant show natural antioxidants such as flavonoids; in clinical trials, it acts as antiaging agent. However, interface of pharmacology and chemistry is required to research toward lower toxicity and enhance its activity in drug development of *W. somnifera*.

References

Ahsan N, Lee DG, Lee SH, Lee KW, Bahk JD, Lee BH (2007) A proteomic screen and identification of waterlogging-regulated proteins in tomato roots. Plant Soil 295:31–57

Ahuja A, Kaur D, Sharada M, Kumar A, Suri KA, Dutt P (2009) Glycowithanolides accumulation in *in vitro* shoot cultures of Indian ginseng (*Withania somnifera* Dunal). Nat Prod Commun 4: 479–482

Akhoon BA, Pandey S, Tiwari S, Pandey R (2016) Withanolide A offers neuroprotection, ameliorates stress resistance and prolongs the life expectancy of Caenorhabditis elegans. Exp Gerontol 78:47–56

Amini F, Ehsanpour A, Hoang Q, Shin J (2007) Protein pattern changes in tomato under in vitro salt stress. Russ J Plant Physiol 54:464–471

Bandyopadhyay M, Jha S, Tepfer D (2007) Changes in morphological phenotypes and withanolide composition of Ri-transformed roots of *Withania somnifera*. Plant Cell Rep 26:599–609. https://doi.org/10.1007/s00299-006-0260-0

Berry J, Bjorkman O (1980) Photosynthetic response and adaptation to temperature in higher plants. Annu Rev Plant Physiol 31:491–543

Bhargavi S, Shankar SRM (2021) Dual herbal combination of Withania somnifera and five Rasayana herbs: a phytochemical, antioxidant, and chemometric profiling. J Ayurveda Integr Med 12:283–293. https://doi.org/10.1016/j.jaim.2020.10.001

Bharti SK, Bhatia A, Tewari SK, Sidhu OP, Roy R (2011) Application of HR-MAS NMR spectroscopy for studying chemotype variations of *Withania somnifera* (L.) Dunal. Magn Reson Chem 49:659–667. https://doi.org/10.1002/mrc.2817

Bhasin S, Singh M, Singh D (2019) Review on bioactive metabolites of *Withania somnifera* (L.) Dunal and its pharmacological significance. J Pharmacognosy Phytochem 8(3):3906–3909

Bhatia A, Bharti SK, Tewari SK, Sidhu OP, Roy R (2013) Metabolic profiling for studying chemotype variations in *Withania somnifera* (L.) Dunal fruits using GC–MS and NMR spectroscopy. Phytochemistry 93:105–115. https://doi.org/10.1016/j.phytochem.2013.03.013

Bhattacharya SK, Goel RK, Kaur R, Ghosal S (1987) Anti-stress activity of sitoindosides VII and VIII, new acylsterylglucosides from Withania somnifera. Phytother Res 1:32–37

Bohnert HJ, Nelson DE, Jensen RG (1995) Adaptations to environmental stresses. Plant Cell 7: 1099

Boston RS, Viitanen PV, Vierling E (1996) Molecular chaperones and protein folding in plants. Plant Mol Biol 32:191–222. https://doi.org/10.1007/BF00039383

Chaturvedi P, Ischebeck T, Egelhofer V, Lichtscheidl I, Weckwerth W (2013) Cell-specific analysis of the tomato pollen proteome from pollen mother cell to mature pollen provides evidence for developmental priming. J Proteome Res 12:4892–4903. https://doi.org/10.1021/pr400197p

Chaturvedi P, Doerfler H, Jegadeesan S, Ghatak A, Pressman E, Castillejo MA, Wienkoop S, Egelhofer V, Firon N, Weckwerth W (2015) Heat-treatment-responsive proteins in different developmental stages of tomato pollen detected by targeted Mass Accuracy Precursor Alignment (tMAPA). J Proteome Res 14:4463–4471. https://doi.org/10.1021/pr501240n

Chaturvedi P, Ghatak A, Weckwerth W (2016) Pollen proteomics: from stress physiology to developmental priming. Plant Reprod 29:119–132. https://doi.org/10.1007/s00497-016-0283-9

Chaurasiya ND, Uniyal GC, Lal P, Misra L, Sangwan NS, Tuli R et al (2008) Analysis of withanolides in root and leaf of *Withania somnifera* by HPLC with photodiode array and evaporative light scattering detection. Phytochem Anal 19:148–154. https://doi.org/10.1002/pca.1029

Chen HH, Li PH (1980) Characteristics of cold acclimation and deacclimation in tuberbearing Solanum species. Plant Physiol 65:1146–1148. https://doi.org/10.1104/pp.65.6.1146

Choudhary MI, Yousuf S, Nawaz SA, Ahmed S, Atta-ur-Rahman (2004) Cholinesterase inhibiting withanolides from *Withania somnifera*. Chem Pharm Bull 52:1358–1361. https://doi.org/10.1248/cpb.52.1358

Datnoff LE, Deren CW, Snyder GH (1997) Silicon fertilization for disease management of rice in Florida. Crop Prot 16:525–531. https://doi.org/10.1016/S0261-2194(97)00033-1

Dhar RS, Verma V, Suri KA, Sangwan RS, Satti NK, Kumar A et al (2006) Phytochemical and genetic analysis in selected chemotypes of *Withania somnifera*. Phytochemistry 67:2269–2276. https://doi.org/10.1016/j.phytochem.2006.07.014

Dhar N, Rana S, Razdan S, Bhat WW, Hussain A, Dhar RS et al (2014) Cloning and functional characterization of three branch point oxidosqualene cyclases from *Withania somnifera* (L.) dunal. J Biol Chem 289:17249–17267. https://doi.org/10.1074/jbc.M114.571919

Dhar RS, Khan S, Khajuria RK, Bedi YS (2016) Dynamics of squalene content in different tissues of Ashwagandha (*Withania somnifera* L. Dunal) during its growth phases. Ind Crop Prod 84: 375–380. https://doi.org/10.1016/j.indcrop.2016.02.003

Dhuley JN (2008) Adaptogenic and cardioprotective action of ashwagandha in frogs and rats. J Ethnopharmacol 70:57–63

Elsakka M, Grigorescu E, Stănescu U, Stănescu U, Dorneanu V (1990) New data referring to chemistry of *Withania somnifera* species. Rev Med Chir Soc Med Nat Iasi 94:385–387

Ferri M, Tassoni A (2011) Chitosan as elicitor of health beneficial secondary metabolites in in vitro plant cell cultures. In: Mackay RG, Tait JM (eds) Handbook of chitosan research and applications. Nova Science Publishers, New York, pp 389–413

Folgado R, Panis B, Sergeant K, Renaut J, Swennen R, Hausman JF (2013) Differential protein expression in response to abiotic stress in two potato species: Solanum commersonii Dun and Solanum tuberosum L. Int J Mol Sci 14:4912–4933. https://doi.org/10.3390/ijms14034912

Folgado R, Sergeant K, Renaut J, Swennen R, Hausman JF, Panis B (2014) Changes in sugar content and proteome of potato in response to cold and dehydration stress and their implications forcryopreservation. J Proteome 98:99–111. https://doi.org/10.1016/j.jprot.2013.11.027

Folgado R, Panis B, Sergeant K, Renaut J, Swennen R, Hausman J-F (2015) Unravelling the effect of sucrose and cold pretreatment on cryopreservation o of potato through sugar analysis and proteomics. Cryobiology 71:432–441

Fragkostefanakis S, Mesihovic A, Simm S, Paupière MJ, Hu Y, Paul P, Mishra SK, Tschiersch B, Theres K, Bovy A, Schleiff E, Scharf K-D (2016) HsfA2 controls the activity of developmentally and stress-regulated heat stress protection mechanisms in tomato male reproductive tissues. Plant Physiol 170:2461–2477. https://doi.org/10.1104/pp.15.01913

Frank G, Pressman E, Ophir R, Althan L (2009) Transcriptional profiling of maturing tomato (Solanum lycopersicum L.) microspores reveals the involvement of heat shock proteins, ROS scavengers, hormones, and sugars in the heat stress response. J Exp Bot 60:3891

Furmanowa M, Gajdzis-Kuls D, Ruszkowska J, Czarnocki Z, Obidoska G, Sadowska A et al (2001) *In vitro* propagation of *Withania somnifera* and isolation of withanolides with immunesuppressive activity. Planta Med 67:146–149. https://doi.org/10.1055/s-2001-11494

Gharechahi J, Hajirezaei MR, Salekdeh GH (2015) Comparative proteomic analysis of tobacco expressing cyanobacterial flavodoxin and its wild type under drought stress. J Plant Physiol 175: 48–58. https://doi.org/10.1016/j.jplph.2014.11.001

Ghatak A, Chaturvedi P, Nagler M, Roustan V, Lyon D, Bachmann G, Postl W, Schröfl A, Desai N, Varshney RK, Weckwerth W (2016) Comprehensive tissue-specific proteome analysis of drought stress responses in Pennisetum glaucum (L.) R. Br. (pearl millet). J Proteome 143: 122–135. https://doi.org/10.1016/j.jprot.2016.02.032

Ghosh Dasgupta M, George BS, Bhatia A, Sidhu OP (2014) Characterization of Withania somnifera leaf transcriptome and expression analysis of pathogenesis – related genes during salicylic acid signaling. PLoS One 9(4):e94803. https://doi.org/10.1371/journal.pone.0094803

Girme A, Saste G, Pawar S, Balasubramaniam AK, Musande K, Darji B, Satti NK, Verma MK, Anand R, Singh R, Vishwakarma RA, Hingorani L (2020) Investigating 11 withanosides and withanolides by UHPLC–PDA and mass fragmentation studies from Ashwagandha (Withania somnifera). ACS Omega 43:27933–27943. https://doi.org/10.1021/acsomega.0c03266

Gong HJ, Randall DP, Flowers TJ (2006) Silicon deposition in the root reduces sodium uptake in rice (Oryza sativa L.) seedlings by reducing bypass flow. Plant Cell Environ 29:1970–1979. https://doi.org/10.1111/j.1365-3040.2006.01572.x

Grover A, Samuel G, Bisaria VS, Sundar D (2013) Enhanced withanolide production by overexpression of squalene synthase in *Withania somnifera*. J Biosci Bioeng 115:680–685. https://doi.org/10.1016/j.jbiosc.2012.12.011

Gupta P, Goel R, Pathak S, Srivastava A, Singh SP, Sangwan RS, Asif MH, Trivedi PK (2013) De novo assembly, functional annotation and comparative analysis of Withania somnifera leaf and root transcriptomes to identify putative genes involved in the withanolides biosynthesis. PLoS One 8(5):e62714

Gupta R, Wang Y, Agrawal GK, Rakwal R, Jo IH, Bang KH, Kim ST (2015) Time to dig deep into the plant proteome: a hunt for low-abundance proteins. Front Plant Sci 6:22

Gupta M, Srivastava PK, Shikha NA, Tewari SK (2016) Use of a bioaugmented organic soil amendment in combination with gypsum for *Withania somnifera* growth on sodic soil. Pedosphere 26:299–309. https://doi.org/10.1016/S1002-0160(15)60044-3

Gusta LV, Wisniewski M (2013) Understanding plant cold hardiness: an opinion. Physiol Plant 147:4–14. https://doi.org/10.1111/j.1399-3054.2012.01611.x

Hahm ER, Moura MB, Kelley EE, Van Houten B, Shiva S, Singh SV (2011) Withaferin A-induced apoptosis in human breast cancer cells is mediated by reactive oxygen species. PLoS One 2011(6):1–12
Hattrup E, Neilson KA, Breci L, Haynes PA (2007) Proteomic analysis of shade-avoidance response in tomato leaves. J Agric Food Chem 55:8310–8318. https://doi.org/10.1021/jf0713049
Hoehenwarter W, van Dongen JT, Wienkoop S, Steinfath M, Hummel J, Erban A, Sulpice R, Regierer B, Kopka J, Geigenberger P (2008) A rapid approach for phenotypescreening and database independent detection of cSNP/protein polymorphism using mass accuracy precursor alignment. Proteomics 8:4214–4225
Hoehenwarter W, Larhlimi A, Hummel J, Egelhofer V, Selbig J, van Dongen JT, Wienkoop S, Weckwerth W (2011) MAPA distinguishes genotype-specific variability of highly similar regulatory protein isoforms in potato tuber. J Proteome Res 10:2979–2991
Hossain Z, Nouri MZ, Komatsu S (2012) Plant cell organelle proteomics in response to abiotic stress. J Proteome Res 11:37–48. https://doi.org/10.1021/pr200863r
Hsiao TC (1973) Plant responses to water stress. Annu Rev Plant Physiol 24(1):519–570
Huang CF, Ma L, Sun LJ, Ali M, Arfan M, Liu JW, Hu LH (2009) Studies of Ashwagandha (Withania somnifera Dunal). Int J Pharm Biol Arch 7:1–11
Hugueney P, Bouvier F, Badillo A, Quennemet J, d'Harlingue A, Camara B (1996) Developmental and stress regulation of gene expression for plastid and cytosolic isoprenoid pathways in pepper fruits. Plant Physiol 111(2):619–626
Jain P, Varshney R (2011) Antimicrobial activity of aqueous and methanolic extracts of Withania somnifera (Ashwagandha). J Chem Pharmaceut Res 3(3):260–263
Jaleel CA, Lakshmanan GMA, Gomathinayagam M, Panneerselvam R (2008) Triadimefon induced salt stress tolerance in Withania somnifera and its relationship to antioxidant defense system. S Afr J Bot 74(1):126–132
Jin Y, Zhang C, Yang H, Yang Y, Huang C, Tian Y, Lu X (2011) Proteomic analysis of cold stress responses in tobacco seedlings. Afr J Biotechnol 5:11900–18991. https://doi.org/10.5897/AJB11.900
John J (2014) Therapeutic potential of *Withania somnifera*: a report on phytopharmacological properties. Int J Pharm Sci Res 5:2131–2148
Johny L, Conlan X, Cahill D, Adholeya A (2015) *In vitro* and *in situ* screening systems for morphological and phytochemical analysis of *Withania somnifera* germplasms. Plant Cell Tissue Organ Cult 120:1191–1202. https://doi.org/10.1007/s11240-014-0673-3
Juan M, Rivero RM, Romero L, Ruiz JM (2005) Evaluation of some nutritional and biochemical indicators in selecting salt-resistant tomato cultivars. Environ Exp Bot 54:193–201. https://doi.org/10.1016/j.envexpbot.2004.07.004
Khatun S, Babar M, Ali EJ, Hahn KYP (2008) Copper toxicity in Withania somnifera: growth and antioxidant enzymes responses of in vitro grown plants. Environ Exp Bot 64:279–285. https://doi.org/10.1016/j.envexpbot.2008.02.004
Kidd PS, Llugany M, Poschenrieder CH, Gunse B, Barcelo J (2001) The role of root exudates in aluminium resistance and silicon-induced amelioration of aluminium toxicity in three varieties of maize (Zea mays L.). J Exp Bot 52:1339–1352
Kim ST, Cho KS, Jang YS, Kang KY (2001) Two-dimensional electrophoretic analysis of rice proteins by polyethylene glycol fractionation for protein arrays. Electrophoresis 22:2103–2109
Kim OT, Bang KH, Shin YS, Lee MJ, Jung SJ, Hyun DY, Kim YC, Seong NS, Cha SW, Hwang B (2007) Enhanced production of asiaticoside from hairy root cultures of *Centella asiatica* (L.) urban elicited by methyl jasmonate. Plant Cell Rep 26(11):1941–1949
Kim S, Park M, Yeom S-I, Kim Y-M, Lee JM, Lee H-A, Seo E, Choi J, Cheong K, Kim K-T (2014a) Genome sequence of the hot pepper provides insights into the evolution of pungency in Capsicum species. Nat Genet 46:270–278

Kim Y-H, Khan AL, Kim D-H, Lee S-Y, Kim K-M, Waqas M, Jung H-Y, Shin J-H, Kim J-G, Lee I-J (2014b) Silicon mitigates heavy metal stress by regulating P-type heavy metal ATPases, Oryza sativa low silicon genes, and endogenous phytohormones. BMC Plant Biol 14:1

Kim S, Yu JS, Lee JY, Choi SU, Lee J, Kim KH (2019) Cytotoxic withanolides from the roots of Indian ginseng (Withania somnifera). J Nat Prod 82:765–773. https://doi.org/10.1021/acs.jnatprod.8b00665

Kosová K, Vítámvás P, Prášil IT, Renaut J (2011) Plant proteome changes under abiotic stress – contribution of proteomics studies to understanding plant stress response. J Proteome 74:1301–1322. https://doi.org/10.1016/j.jprot.2011.02.006

Kulkarni SK, Dhir A (2008) *Withania somnifera*: an Indian ginseng. Prog Neuropsychopharmacol Biol Psychiatry 32:1093–1105

Kumar A, Kalonia H (2007) Protective effect of *Withania somnifera* Dunal on the behavioral and biochemical alterations in sleep-disturbed mice (Grid over water suspended method). Indian J Exp Biol 45:524–528

Kumar R, Singh JK, Nath A, Ali M, Kumar A, Sinha P (2013) Effect of *Withania somnifera* on estrogen, cholesterol and sub-cellular structure of ovary of chlorpyrifos exposed mice. Caribbean J Sci Technol 1:061–069

Lee BS, Yoo MJ, Kang H, Lee SR, Kim S, Yu JS, Kim J-C, Jang TS, Pang C, Kim KH (2021) Withasomniferol D, a new anti-adipogenic withanolide from the roots of ashwagandha (Withania somnifera). Pharmaceuticals 14:1017. https://doi.org/10.3390/ph14101017

Ma JF, Yamaji N (2006) Silicon uptake and accumulation in higher plants. Trends Plant Sci 11: 392–397. https://doi.org/10.1016/j.tplants.2006.06.007

Manaa A, Mimouni H, Wasti S, Gharbi E, Aschi-Smiti S, Faurobert M, Ben Ahmed H (2013a) Comparative proteomic analysis of tomato (Solanum lycopersicum) leaves under salinity stress. Plant Omics 6:268–277

Manaa A, Faurobert M, Valot B, Bouchet J-P, Grasselly D, Causse M, Ben Ahmed H (2013b) Effect of salinity and calcium on tomato fruit proteome. OMICS 17:338–352

Meng M, Geisler M, Johansson H, Mellerowicz EJ, Karpinski S, Kleczkowski LA (2007) Differential tissue/organ-dependent expression of two sucrose-and cold-responsive genes for UDP-glucose pyrophosphorylase in Populus. Gene 389:186–195

Mir BA, Khazir J, Mir NA, Hasan T, Koul S (2012) Botanical, chemical and pharmacological review of *Withania somnifera* (Indian ginseng): an ayurvedic medicinal plant. Indian J Drugs Dis 1(6):2278–2958

Mir BA, Mir SA, Khazir J, Tonfack LB, Cowan DA, Vyas D, Koul S (2015) Cold stress affects antioxidative response and accumulation of medicinally important withanolides in Withania somnifera (L.) Dunal. Ind Crop Prod 74:1008–1016

Mishra LC, Singh BB (2000) Scientific basis for the therapeutic use of *Withania somnifera* (ashwagandha): a review. Altern Med Rev 5:334–346

Mishra B, Singh Sangwan N (2019) Amelioration of cadmium stress in Withania somnifera by ROS management: active participation of primary and secondary metabolism. Plant Growth Regul 87:403–412. https://doi.org/10.1007/s10725-019-00480-8

Mishra A, Singh SP, Mahfooz S, Bhattacharya A, Mishra N, Shirke PA, Nautiyal CS (2018) Bacterial endophytes modulates the withanolide biosynthetic pathway and physiological performance in Withania somnifera under biotic stress. Microbiol Res 212–213:17–28. https://doi.org/10.1016/j.micres.2018.04.006

Mishra MK, Tiwari S, Misra P (2021) Overexpression of WssgtL3.1 gene from Withania somnifera confers salt stress tolerance in Arabidopsis. Plant Cell Rep 40:2191–2204. https://doi.org/10.1007/s00299-021-02666-9

Misra L, Lal P, Sangwan RS, Sangwan NS, Uniyal GC, Tuli R (2005) Unusually sulfated and oxygenated steroids from *Withania somnifera*. Phytochemistry 66:2702–2707. https://doi.org/10.1016/j.phytochem.2005.10.001

Muneer S, Park YG, Manivannan A, Soundararajan P, Jeong BR (2014) Physiological and proteomic analysis in chloroplasts of Solanum lycopersicum L. under silicon efficiency and salinity stress. Int J Mol Sci 15:21803–21824. https://doi.org/10.3390/ijms151221803

Murthy HN, Praveen N (2013) Carbon sources and medium pH affects the growth of *Withania somnifera* (L.) Dunal adventitious roots and withanolide A production. Nat Prod Res 27:185–189. https://doi.org/10.1080/14786419.2012.660691

Nadkarni KM (1954) Indian materia medica, 3rd edn. Popular Prakshan Limited, Bombay, India

Namdeo AG (2007) Plant cell elicitation for production of secondary metabolites: a review. Pharmacognosy Rev 1(1):69–79

Narinderpal K, Junaid N, Raman B (2013) A review on pharmacological profile of *Withania somnifera* (Ashwagandha). Res Rev J Bot Sci 2:6–14

Neumann D, Zur Nieden U (2001) Silicon and heavy metal tolerance of higher plants. Phytochemistry 56:685–692. https://doi.org/10.1016/S0031-9422(00)00472-6

Nover L, Scharf K (1984) Synthesis, modification and structural binding of heat-shock proteins in tomato cell cultures. Eur J Biochem 139:303–313

Ordin L, Hall MA (1968) Cellulose synthesis in higher plants from UDP glucose. Plant Physiol 43: 473

P.G.S. Consortium (2011) Genome sequence and analysis of the tuber crop potato. Nature 475: 189–195

Parihar P, Singh S, Singh R, Singh VP, Prasad SM (2015) Effect of salinity stress on plants and its tolerance strategies: a review. Environ Sci Pollut Res 22:4056–4075. https://doi.org/10.1007/s11356-014-3739-1

Prabu PC, Panchapakesan S, Raj CD (2013) Acute and sub-acute oral toxicity assessment of the hydroalcoholic extract of *Withania somnifera* roots in Wistar rats. Phytother Res 27:1169–1178

Pratibha C, Madhumati B, Akash P (2013) Therapeutic properties and significance of different parts of Ashwagandha-a medicinal plant. Int J Pure Appl Biosci 1(6):94–101

Praveen N, Murthy HN (2010) Production of withanolide-A from adventitious root cultures of *Withania somnifera*. Acta Physiol Plant 32:1017–1022. https://doi.org/10.1007/s11738-010-0489-7

Qadir SU, Raja V, Siddiqui WA et al (2021) Foliar concentrations of selected elements, assessment of oxidative stress markers and role of antioxidant defense system is associated with fly ash stress tolerance in Withania somnifera. J Plant Growth Regul 40:1450–1465. https://doi.org/10.1007/s00344-020-10200-6

Raja G, Veerakumari L (2013) Influence of vermicomposts on the yield and alkaloid content of *Withania somnifera* Dunal. Int J Adv Biol Res 3:223–226

Rana S, Lattoo SK, Dhar N, Razdan S, Bhat WW, Dhar RS et al (2013) NADPH-cytochrome P450 reductase: molecular cloning and functional characterization of two paralogs from *Withania somnifera* (L.) dunal. PLoS One 8:e57068. https://doi.org/10.1371/journal.pone.0057068

Ray S, Jha S (1999) Withanolide synthesis in cultures of *Withania somnifera* transformed with *Agrobacterium tumefaciens*. Plant Sci 146:1–7. https://doi.org/10.1016/S0168-9452(99)00077-1

Ray S, Ghosh B, Sen S, Jha S (1996) Withanolide production by root culture of *Withania somnifera* transformed with *Agrobacterium rhizogenes*. Planta Med 62:571–573. https://doi.org/10.1055/s-2006-957977

Rempelos L, Cooper J, Wilcockson S, Eyre M, Shotton P, Volakakis N, Orr CH, Leifert C, Gatehouse AMR, Tétard-Jones C (2013) Quantitative proteomics to study the response of potato to contrasting fertilisation regimes. Mol Breed 31:363–378

Rizvi Faizi T, Razauddin M, URRahman Saif M, Jahan T, Naz Z, Kumar R, Kumar A, Ali M (2016) Immunomodulatory effect of Ashwagandha against doxorubicin toxicity. Eur J Pharm Med Res 3(7):463–467

Romero-Aranda MR, Jurado O, Cuartero J (2006) Silicon alleviates the deleterious salt effect on tomato plant growth by improving plant water status. J Plant Physiol 163:847–855. https://doi.org/10.1016/j.jplph.2005.05.010

Sabir F, Sangwan RS, Kumar R et al (2012) Salt stress-induced responses in growth and metabolism in callus cultures and differentiating in vitro shoots of Indian ginseng (Withania somnifera Dunal). J Plant Growth Regul 31:537–548. https://doi.org/10.1007/s00344-012-9264-x

Saema S, Rahman LU, Singh R, Niranjan A, Ahmad IZ, Mishra P (2016) Ectopic overexpression of WsSGTL1, a sterol glucosyltransferase gene in Withania somnifera, promotes growth, enhances glycowithanolide and provides tolerance to abiotic and biotic stresses. Plant Cell Rep 35:195–211. https://doi.org/10.1007/s00299-015-1879-5

Saleem S, Muhammad G, Hussain MA, Altaf M, Bukhari SNA (2020) Withania somnifera L.: insights into the phytochemical profile, therapeutic potential, clinical trials, and future prospective. Iran J Basic Med Sci 23(12):1501–1526. https://doi.org/10.22038/IJBMS.2020.44254.10378

Sangwan RS, Chaurasiya ND, Lal P, Misra L, Uniyal GC, Tuli R et al (2007) Withanolide A biogeneration in *in vitro* shoot cultures of ashwagandha (*Withania somnifera* DUNAL), a main medicinal plant in Ayurveda. Chem Pharm Bull 55:1371–1375. https://doi.org/10.1248/cpb.55.1371

Schmelze GH, Gurib-Fakim A, Arroo RRJ, Bosch CH, de Ruijter A, Simmonds MSJ (2008) Plant resources of tropical Africa 11(1)–medicinal plants. Backhuys Publishers, Wageningen, p 630

Seppänen MM, Majaharju M, Somersalo S, Pehu E (1998) Freezing tolerance, cold acclimation and oxidative stress in potato. Paraquat tolerance is related to acclimation but is a poor indicator of freezing tolerance. Physiol Plant 102:454–460

Sharada AC, Emerson Solomon F, Uma Devi P (1993) Toxicity of *Withania somnifera* root extract in rats and mice. Pharm Biol 31:205–212

Shilpashree HB, Sudharshan SJ, Shasany AK et al (2022) Molecular characterization of three CYP450 genes reveals their role in withanolides formation and defense in Withania somnifera, the Indian Ginseng. Sci Rep 12:1602. https://doi.org/10.1038/s41598-022-05634-9

Sil B, Mukherjee C, Jha S, Mitra A (2015) Metabolic shift from with asteroid formation to phenylpropanoid accumulation in cryptogein-cotransformed hairy roots of *Withania somnifera* (L.) Dunal. Protoplasma 252:1097–1110. https://doi.org/10.1007/s00709-014-0743-8

Singh M, Singh R (2016) *Withania somnifera* in enhancement of withanolides via rhizogenesis. In: Advances in bio and medico sciences. https://doi.org/10.13140/RG.2.1.1421.9127

Singh S, Pal S, Shanker K, Chanotiya CS, Gupta MM, Dwivedi N, Shasany AK (2014) Sterol partitioning by HMGR and DXR for routing intermediates toward withanolide biosynthesis. Physiol Plant 152:617–633. https://doi.org/10.1111/ppl.12213

Singh R, Gupta P, Khan F, Singh SK, Sanchita, Mishra T, Kumar A, Dhawan SS, Shirke PA (2018) Modulations in primary and secondary metabolic pathways and adjustment in physiological behaviour of Withania somnifera under drought stress. Plant Sci 272:42–54. https://doi.org/10.1016/j.plantsci.2018.03.029. Epub 2018 Apr 5. PMID: 29807605

Sivanandhan G, Arun M, Mayavan S, Rajesh M, Mariashibu TS, Manickavasagam M, Selvaraj N, Ganapathi A (2012) Chitosan enhances withanolides production in adventitious root cultures of Withania somnifera (L.) Dunal. Ind Crop Prod 3:124–129

Sivanandhan G, Arun M, Mayavan S, Rajesh M, Mariashibu TS, Manickavasagam M et al (2012b) Chitosan enhances withanolides production in adventitious root cultures of *Withania somnifera* (L.) Dunal. Ind Crop Prod 37:124–129. https://doi.org/10.1016/j.indcrop.2011.11.022

Sivanandhan G, Dev GK, Jeyaraj M, Rajesh M, Arjunan A, Muthuselvam M et al (2013a) Increased production of withanolide A, withanone, and withaferin A in hairy root cultures of *Withania somnifera* (L.) Dunal elicited with methyl jasmonate and salicylic acid. Plant Cell Tissue Organ Cult 114:121–129. https://doi.org/10.1007/s11240-013-0297-z

Sivanandhan G, Dev GK, Jeyaraj M, Rajesh M, Muthuselvam M, Selvaraj N et al (2013b) A promising approach on biomass accumulation and withanolides production in cell suspension culture of *Withania somnifera* (L.) Dunal. Protoplasma 250:885–898. https://doi.org/10.1007/s00709-012-0471-x

Sivanandhan G, Rajesh M, Arun M, Jeyaraj M, Dev GK, Arjunan A et al (2013c) Effect of culture conditions, cytokinins, methyl jasmonate and salicylic acid on the biomass accumulation and production of withanolides in multiple shoot culture of *Withania somnifera* (L.) Dunal using liquid culture. Acta Physiol Plant 35:715–728. https://doi.org/10.1007/s11738-012-1112-x

Sivanandhan G, Selvaraj N, Ganapathi A, Manickavasagam M (2014a) Enhanced biosynthesis of withanolides by elicitation and precursor feeding in cell suspension culture of *Withania somnifera* (L.) Dunal in shake-flask culture and bioreactor. PLoS One 9:e104005. https://doi.org/10.1371/journal.pone.010400

Sivanandhan G, Selvaraj N, Ganapathi A, Manickavasagam M (2014b) Improved production of withanolides in shoot suspension culture of *Withania somnifera* (L.) Dunal by seaweed extracts. Plant Cell Tissue Organ Cult 119:221–225. https://doi.org/10.1007/s11240-014-0521-5

Sivanandhan G, Theboral J, Dev GK, Selvaraj N, Manickavasagam M, Ganapathi A (2015) Effect of carbon and nitrogen sources on *in vitro* flower and fruit formation and withanolides production in *Withania somnifera* (L.) Dunal. Indian J Exp Biol 53:177–183

Stebbins GL (1952) Aridity as a stimulus to plant evolution. Am Nat 132:87–106

Straughn AR, Kakar SS (2020) Withaferin A: a potential therapeutic agent against COVID-19 infection. J Ovarian Res 13(1):79. https://doi.org/10.1186/s13048-020-00684-x

Subbaraju GV, Vanisree M, Rao CV, Sivaramakrishna C, Sridhar P, Jayaprakasam B et al (2006) Ashwagandhanolide, a bioactive dimeric thiowithanolide isolated from the roots of *Withania somnifera*. J Nat Prod 69:1790–1792. https://doi.org/10.1021/np060147p

T.G. Consortium (2012) The tomato genome sequence provides insights into fleshy fruit evolution. Nature 485:635–641

Tetali SD, Acharya S, Ankari AB, Nanakram V, Raghavendra AS (2020) Metabolomics of Withania somnifera (L.) Dunal: advances and applications. J Ethnopharmacol 267:113469. https://doi.org/10.1016/j.jep.2020.113469

Turrini E, Calcabrini C, Sestili P, Catanzaro E, De Gianni E, Diaz RA, Hrelia P, Tacchini M, Guerrini A, Canonico B, Papa S, Valdrè G, Fimognari C (2016) Withania somnifera induces cytotoxic and cytostatic effects on human T leukemia cells. Toxins 8(5):147. https://doi.org/10.3390/toxins8050147

Umadevi M, Rajeshwari R, Rahale Sharmila C, Selvavenkadesh S, Pushpa R, Kumar Sampath KP, Bhowmik D (2012) Traditional and medicinal uses of *Withania somnifera*. Pharma Innov 9: 102–110

Vanderschuren H, Lentz E, Zainuddin I, Gruissem W (2013) Proteomics ofmodel and crop plant species: status, current limitations and strategic advances for crop improvement. J Proteome 93: 5–19. https://doi.org/10.1016/j.jprot.2013.05.036

Varghese S, Keshavachandran R, Baby B, Nazeem PA (2014) Genetic transformation in ashwagandha (Withania somnifera (L.) Dunal) for hairy root induction and enhancement of secondary metabolites. J Trop Agric 52:47–53

Vieira Dos Santos C, Rey P (2006) Plant thioredoxins are key actors in the oxidative stress response. Trends Plant Sci 11:329–334. https://doi.org/10.1016/j.tplants.2006.05.005

Viola R, Nyvall P, Pedersén M (2001) The unique features of starch metabolism in red algae. Proc R Soc Lond B Biol Sci 268:1417–1422

Wahid A, Gelani S, Ashraf M, Foolad MR (2007) Heat tolerance in plants: an overview. Environ Exp Bot 61:199–223. https://doi.org/10.1016/j.envexpbot.2007.05.011

Wang P, Sun X, Xie Y, Li M, Chen W, Zhang S, Liang D, Ma F (2014) Melatonin regulates proteomic changes during leaf senescence in Malus hupehensis. J Pineal Res 57:291–307

Xie H, Yang D-H, Yao H, Bai G, Zhang Y-H, Xiao B-G (2016) iTRAQ-based quantitative proteomic analysis reveals proteomic changes in leaves of cultivated tobacco (Nicotiana tabacum) in response to drought stress. Biochem Biophys Res Commun 469:768–775

Zhao J, Nakamura N, Hattori M, Kuboyama T, Tohda C, Komatsu K (2002) Withanolide derivatives from the roots of *Withania somnifera* and their neurite outgrowth activities. Chem Pharm Bull (Tokyo) 50(6):760–765

Zhou S, Palmer M, Zhou J, Bhatti S, Howe KJ, Fish T, Thannhauser TW (2013) Differential root proteome expression in tomato genotypes with contrasting drought tolerance exposed to dehydration. J Am Soc Hortic Sci 138:131–141. http://journal.ashspublications.org/content/138/2/131.short

Zulak KG, Cornish A, Daskalchuk TE, Deyholos MK, Goodenowe DB, Gordon PM, Klassen D, Pelcher LE, Sensen CW, Facchini PJ (2007) Gene transcript and metabolite profiling of elicitor-induced opium poppy cell cultures reveals the coordinate regulation of primary and secondary metabolism. Planta 225(5):1085–1106

Interactive Role of Silicon and Phenolics in Biotic Stress Regulation in Plants and Expression of Phenylpropanoid Pathway Genes

12

Naveed Gulzar, Rafiq Lone, Abdullah Mohammed Al-Sadi, and Abdul Azeez

Abstract

Because of the multiple benefits of silicon (Si) and phenolics to plants, experts are very interested in their application in agriculture. The usage of silicon and phenolics has lowered the severity of various ailments in commercially significant crops. In this study, the association between silicon and phenolics in reducing biotic stress was examined. The current analysis emphasizes the role silicon and phenolics in reducing biotic stress in various plants, as well as the regulation of phenylpropanoid biosynthetic pathway. By combining the information, as stated in this study, a deeper understanding of the association between phenolics and silicon treatments has improved plant resistance, and a reduction in the severity of fungal infections could be achieved.

Keywords

Phenolics · Silicon · Biotic stress · Resistance · Phenylpropanoid pathway genes

N. Gulzar
Center of Research for Development, University of Kashmir, Srinagar, Jammu and Kashmir, India

R. Lone (✉)
Department of Botany, Central University of Kashmir, Ganderbal, Jammu and Kashmir, India

A. M. Al-Sadi
Department of Crop Sciences, College of Agricultural & Marine Science, Sultan Qaboos University, AlKhoud, Oman

A. Azeez
Institute of Biological Chemistry, Washington State University, Pullman, WA, USA

R. Lone et al. (eds.), *Plant Phenolics in Biotic Stress Management*,
https://doi.org/10.1007/978-981-99-3334-1_12

12.1 Introduction

Various types of environmental stress influence development and growth of plants; these stresses form important restrictions for long-term agricultural output in terms of fertility and soil health (Kumar and Verma 2018). Higher agricultural productivity losses are due to biotic stress, abiotic stress, and other factors around the world. Biological stresses such as viruses, bacteria, fungus, insects, and nematodes have a significant and detrimental impact on agriculture (Dresselhaus and Hückelhoven 2018). Biological stresses are the most common and among them, fungi are the most dangerous to plant species, as fungus or fungal-like organisms are responsible for 85% of plant illnesses (Behmann et al. 2014). Viruses are also considered to be a potentially harmful agent capable of harming crop plants. Wilting, mottling, chlorosis, necrosis, blights, and blasts and tumor development are all symptoms caused by these biotic organisms in plants (Saddique et al. 2018). Weeds are another biotic factor that has an impact on plant growth and productivity by either directly damaging them or by raising competition for resources and space (Melvin et al. 2017). Phenolics and silicon augmentation were found to be efficient in reducing the effects of multiple stresses in plants and also enhance growth and development in plants.

12.2 Phenolics and Their Biosynthesis in Plants

Phenolics are chemicals that are important in plant protection and are required to combat a variety of stressors. An aromatic ring is coupled to one or more hydroxyl groups in the structure of phenols. Plants require them in order to grow and flourish. They are useful as they help to make lignin and pigments as well as protecting the plants from external illnesses. Many microbes are repelled by phenolic phytoalexins, which are emitted by plants that have been harmed or otherwise damaged (Asgari Lajayer et al. 2017). Plant phenolics can be divided into two groups.

(a) Plant tissues produce preformed phenolics as a natural component of their development and growth.
(b) When a cell is injured physically or mechanically or when it is attacked by a microbe or pathogen, or when it is exposed to heavy metals, salt, UV irradiation, or extreme temperatures, induced phenolics are generated.

In the initial stage in phenolic biosynthesis, glucose participates in the hexose monophosphate shunt and is converted to glucose-6-phosphate, which is subsequently irreversibly turned into ribulose-5-phosphate. Glucose-6-phosphate dehydrogenase is involved to carry out the initial step in phenolic biosynthesis (Lin et al. 2016). On the other hand, end products of glycolysis are involved in the formation of phenolic compounds before being transported for the formation of phenylalanine through shikimic acid pathway. In plants, microbes, and fungus, the majority of phenolic compounds are formed by the shikimic acid pathway. In higher plants, the route of malonic acid is less significant than in fungi and bacteria in the synthesis of phenolic acids. After phenolic acid biosynthesis, it becomes a part of the plant's cell wall to reimburse for biotic stress with augmented unpredictability by producing cinnamic and benzoic acid derivatives (Ledesma-Escobar et al. 2019).

12.3 Phenolics Regulating Biotic Stress

Pathogens like bacteria, viruses, fungi, and herbivores all contribute to biotic stress. These pathogens are responsible for a significant amount of yield loss around the world. Bacteria are the microorganisms that can obstruct plant's capacity to transmit nutrients and water to other areas of the plant, leading to the plant's death. Bacterial infections are indicated by leaf and fruit spots, early blights on leaves, tissue death, stem lesions, and root rots in plants (Jones et al. 2004; Ludwig-Müller 2015). Plants infected by viruses exhibit a variety of physiological and metabolic problems. The infection lesions or wounded plant portions provide direct pathway for viral genetic material to enter plant cells and the infection spreads through them (Culver and Padmanabhan 2007). Mottling, mosaic, leaf spots, leaf curling, and puckering are the symptoms of virally infected plants that result in the stunted growth and yield decrease. These symptoms could be induced by physiologic effects on the host, whether direct or indirect (Thakur et al. 2018). Another type of biotic stress factor involves fungi having up to 1.5 million species. Fungi can be divided into three groups: biotrophs, necrotrophs, and hemibiotrophs. Fungi have evolved many mechanisms to collect food from either living or nonliving creatures because they are nonphotosynthetic and unable to synthesis it (Lattanzio et al. 2006).

As a defense mechanism against biotic stress, plant machinery produces a variety of metabolites during different stages of development. Typically, these metabolites are phenolic chemicals that are generated by a particular metabolite (Guo et al. 2018). Secondary metabolites include phenols; they are comprised of one of the most typical and extensively dispersed chemical families discovered in plants. When pathogens attack plants, the phenylpropanoid pathway produces these chemicals, which are then concentrated in the subepidermal layer of plant tissue. Furthermore, plant cell walls are covalently linked to phenolics and other compounds found at different locations in plant organs (Vishwanath et al. 2014). Natural factors of the environment involving temperature and moisture content have profound impact on phenolic biosynthesis and accumulation (Mukherjee et al. 2018). Phenolics are produced as a first defense mechanism by the pathogen as this may cause an overall increase in host metabolism. It has been observed that following infection of maize with *Glomerella graminicola*, two phenolic caffeic acid esters considerably rise (Pusztahely et al. 2017). Their quick accumulation and abrupt decrease in concentration suggested that these phenols could be used as a starting point for the production of other protective chemicals. In plant pathogenic fungus, phenolics stop hyphae from growing and developing. The amount of production and accumulation of phytoalexins in plants determine their ability to defend against bacteria or diseases (Duke 2018). Peroxides, superoxides, and singlet oxygen, all of which have

Table 12.1 Mitigation of biotic stress by phenolics

Phenolics	Function	References
Secoiridoids oleuropein Verbascoside catechol	Alter plasma membrane characteristics and permeability, and enhance the formation of peroxide (H_2O_2)	Buonaurio et al. (2015)
Pterostilbene and resveratrol	Stilbene causes the infectious structure to grow in the penetrated epidermal cell, resulting in resistance	Schnee et al. (2008)
Chlorogenic acid and Rutin	From the amine tyramine and the thioester p-coumaroyl-CoA, the THT gene catalyzes the formation of p-coumaroyltyramine and phenylalanine ammonia-lyase	Mandhania et al. (2018)
Chlorogenic acid (leaves and roots), transcinnamic acid	Browning and the delayed establishment of extensive necrosis in migrating nematode-affected plants	Ahmed et al. (2009) Ohri and Pannu (2010)
Cinnamic acid, naringin and Rutin	Amylase activity in weeds is reduced by phenolic compounds in herbal preparations, which limits their growth. The hydrolysis of starch, and germination of seeds	Mendes and Rezende (2014)
Chlorogenic acid (caffeic acid, quinic acid), catechin	Plant polyphenol oxidase can oxidize chlorogenic acid to produce reactive quinone species, which cross-link cell wall components and inhibit microbial growth and tissue damage	Lee et al. (2017)

been linked to biotic and abiotic stresses, are also reduced by phenols. These phenolic compounds protect plants from these threats by triggering the production of defense enzymes (Saddique et al. 2018). Phenolics get concentrated at the infected locations in plants, limiting the microorganism's overall growth and development through hypersensitive reaction that causes cell death (Lincoln et al. 2018). Phenolics alleviate the negative effects of biotic stressors (Table 12.1). Phenols such as benzoic acids and phenyl propanoids are produced in the earliest reaction to infection. The evidence strongly suggests that phenol esterification to cell wall components is important for plants to cope with varied stressors (Mandal et al. 2010).

Bacterial pathogens such as *Pseudomonas aeruginosa*, *Salmonella choleraesuis*, *Escherichia coli*, *Bordetella Pertusis*, and *Staphylococcus aureus* are all reduced by phenolics such as catechins through affecting plasma membrane permeability and characteristics. Additionally, reactive oxygen species such hydrogen peroxide, hydroxyl, and superoxide anions are formed (Wang et al. 2018). Appearance of quorum sensing proteins after exogenous infusion of trans-cinnamaldehyde and tannic acid (LasI and RhlI) was discovered in *E. coli*. In *Pectobacteria*, the quorum-sensing regulator is particularly inhibited by eugenol and carvacrol (Joshi et al. 2016).

The antagonism of olive trees to *Fusicladium oleagineum* leaf spot is mostly due to various derivatives as flavonol, flavinones, monoglucosides, and tyrosol (Talhaoui et al. 2015). Fungus is resistant to several onion cultivars and this resistance is linked to catechol, protocatechuic acid, and the color of the dead outer bulb scales, which can be red or yellow (Levin 1971).

Viruses are dangerous pathogens that destroy agricultural crops far more than fungi all over the world. Phenolics reduce the infection of viral pathogens in plants. Virus activity in several host plants has been associated with increase in caffeic acid, quercetin, kaempferol, and chlorogenic acid concentrations (Parr and Bolwell 2000). Plant parasites directly attack the vascular system of host plants, thereby limiting water, nutrient, and carbohydrate intake and, as a result, drastically reducing biomass and seed production. Gossypol formation in leaves and flower buds is thought to protect plants against mammalian herbivores, and it has also been found to protect plants from multiple fungal infestations (Levin 1971). A phenolic chemical known as Juglone is generated by *Carya ovata* that is disliked by bark beetle *Scolytus multistriatus* (Byers 1995).

12.4 Silicon and Biotic Stress

Even though silicon is not measured among the essential elements larger plants, it is a typical useful substance that not only promotes plant growth and development but also increases a plant's tolerance to a variety of challenges. Its favorable benefits are more pronounced in stressed plants because it acts as a protective element, allowing plants to withstand both biotic and abiotic stress elements. Plants fed with Si have more resistant to both fungal and bacterial infections. Pathogens packed with biotrophic, necrotrophic, and hemibiotrophic stress factors have been demonstrated to have negative impacts on plants. Plants have developed various inherent defensive systems, which Si regulates via physical, molecular, and biochemical pathways. Si is an effective method for controlling illnesses in a variety of plant species caused by both bacterial and fungal pathogens (Rodrigues and Datnof 2015). Silicon has been discovered to be engaged in the multiple pathways by establishing effective barriers to disease entry into plants. These pathways involve physical fortifications to strengthen the cells, preventing pathogen translocation by promoting systemic acquired resistance (SAR) and antimicrobial substances are produced in plants, and numerous defense signaling pathways are activated (Table 12.1) (Rodrigues and Datnof 2015; Vivancos et al. 2015).

12.5 Silicon as a Physical Barrier

The fact that silicon strengthens cells by the creation of a physical barrier that stops infections from entering the cells demonstrates its effectiveness as a physical barrier. This greater resistance in plants due to the number of silicified epidermal cells has increased. The double cuticle silicon layer, which is formed by a dense layer of silica behind the cuticle, provides mechanical strength to plants driven by silicon application (Ma and Yamaji 2006, 2008). Papilla development thickens the cellulosic membrane, allowing it to interact with organic compounds contained within the epidermal cell wall, inhibiting enzyme breakdown caused by fungus penetration into the membrane (Fauteux et al. 2005; Datnoff et al. 2007). Pectins and polyphenols, which are the principal elements of the cell wall, become complexed with silicon, enhancing the elasticity of the cell wall throughout plant growth (Emadian and Newton 1989). When silicon is sprayed to the leaf blades of rice plants, it confers resistance to *M. grisea* blast disease, as silicon produces a physical barrier that prevents pathogen access (Kim et al. 2002). Silicon creates a physical barrier that protects cucumbers from powdery mildew and wheat leaves from the hypheal fungus *Pyricularia oryzae* (Sousa et al. 2013). When silicon is delivered to the roots, it induces systemic acquired resistance (SAR) in plants, preventing infection from spreading to healthy plant sections. Thus, silicon performs a protective role in a variety of ways, including limiting disease access into plants. In plants, silicon supplies defense systems via biochemical and molecular mechanisms. Activities of defense-related enzymes such as phenylalanine ammonia-lyases, peroxidases, polyphenol oxidases, and gluconases are increased in plants after exposure to silicon. Plants with silicon treatment have increased all antimicrobial substances as phenols, phytoalexins, flavonoids, pathogenesis-related proteins (PR), and lignin. Silicon therapy enhances biochemical resistance in plants by increasing systemic signal regulation such as salicylic acid (SA), jasmonic acid (JA), and ethylene (ET) regulation.

12.6 Silicon Enhancing Defense-Related Enzyme Activities

Van et al. (2013) observed significant stimulation of defense enzyme activity on silicon application. Phenylalanine ammonia-lyase in plants is implicated as secondary antimicrobial compounds that are important in plant disease resistance response. Higher action of phenylalanine ammonia-lyase following silicon treatment increases more buildup of phenolic and lignin thiogilcolic acid in coffee plant's leaves, providing resistance to diseases (Fortunato et al. 2012). In Si-treated perennial rye grass, the higher expression of phenylalanine ammonia-lyase (PALa and PALb), as well as lipoxygenase, was increased, resulting in the repression of gray leaf spot (Rahman et al. 2015). Quarta et al. (2013) reported that polyphenol oxidase enzyme exists in free and bound forms in the cytoplasm, chloroplasts, and other subcellular organelles, and is mainly engaged in pathogen resistance. Polyphenoloxidase and cross-linking of cell wall proteins also help to strengthen the cell wall in plants (Brisson et al. 1994). Silicon upon addition through roots in Pythum species increases the activity of chitinases, peroxidases, and polyphenoloxidases (Liang et al. 2005). Chitinase is a pathogenesis-related protein that aids in the hydrolysis of pathogenic fungi's cell walls. Application of Si to the roots of a cucumber plant infected with Pythium spp. inhibits the pathogen infection by enhancing defense-related enzyme activity (Liang et al. 2005). Increased chitinase and 1,3-glucanase activity reduces the prevalence of *Mycosphaerella pinodes* in pea seeds after silicon application as potassium silicate (Dann and Muir 2002). The infection of wheat leaves with powdery mildew caused by *Blumeria graminis* is reduced by silicon treatment, which increases peroxidase activity in the leaves (Yang et al. 2003). When rice is treated with Si, the activity of chitinase and peroxidase rises, making it resistant to the brown spot disease (*Bipolaris oryzae*) (Dallagnol et al. 2011). The harshness of pink rot in Chinese cantaloupe is reduced by using sodium silicate, which increases peroxidase and phenylalanine ammonia-lyase activity (Guo et al. 2007). Higher peroxidase and polyphenol oxidase activities in perennial rye grass prevent *Magnaporthe oryzae* infection (Rahman et al. 2015). In silicon-treated leaves of soybean, target spot infection is reduced by increased activity of chitinases, peroxidases, and phenylalanine ammonia-lyases (Corynespora cassiicola) (Fortunato et al. 2015). On silicon treatment, sheath blight lesions in rice can be minimized by increasing the activities of phenylalanine ammonia-lyases, polyphenol oxidases, peroxidases, and chitinases (Schurt et al. 2014).

12.7 Silicon Increasing Activities of Antimicrobial Compounds

Activities of antimicrobial compounds get increased considerably after the use of silicon, giving protection to plants against infections (Van et al. 2013). Silicon stimulates the activity of phenols, flavonoids, and phytoalexins, which are antibacterial substances. The formation of lignin and flavonoids is attributed to silicon application, which results in increased PAL activity (Hao et al. 2011). Shetty et al. (2012) observed that silicon treatment causes the induction of another phenolic chemical known as flavonoid, which confers resistance to *Podosphaera pannosa* in rose plants. During *Podosphaera xanthii* infection, the presence of Si promotes the formation of flavonoid in cucumber plants (Fawe et al. 1998). Silicon addition confers resistance to *Pyricularia oryzae* in wheat and protects cucumber plants from damping off (Silva et al. 2010). Silicon suppresses the formation of target spots in soybean, brown spot, and sheath blight on rice (Zhang et al. 2013; Fortunato et al. 2015). It was found that silicon when applied to perennial ryegrass (*Magnaporthe oryzae*) pathosystems, there was an increase in flavonoids, phenolics, and defense genes that encode PAL and lipoxygenase and all of which are involved in gray spot disease resistance (Rahman et al. 2015). Silicon treatment resulted in a higher buildup of fungitoxic phenolic compounds, and it protected Arabidopsis from *Erysiphe cichoracearum's* powdery mildew (Fauteux et al. 2005). Higher production of phytoalexin was discovered to be the cause of blast resistance (*M. grisea*) in rice (Rodrigues et al., 2005). Silicon-induced increases in phenolic metabolism boosted a sensitive rice cultivar's susceptibility to sheath blight in the rice—*Rhizoctonia solani* pathosystem (Zhang et al. 2013). Dallagnol et al. (2011) discovered that increased buildup of lignin and phenolics resulted in a lower level of rice brown spot. Wheat resistance to blast produced by *Pyricularia oryzae* was boosted when high doses of lignin-thioglycolic acid derivatives were applied (Filha et al. 2011). Fortunato et al. (2015) discovered that enhanced activity of total soluble phenolics resulted in a decreased incidence of target spot in leaves of silicon-supplied soybean plants (*Corynespora cassiicola*).

12.8 Silicon Regulating the Levels of Systemic Signals

Plants face a constant barrage of biotic threats. Plants have developed a sophisticated defense mechanism system to cope up with severe situations as a result of their sessile nature. The network of signal transduction pathways has controlled a wide range of inducible and constitutive defensive mechanisms established by plants to defend against pathogen infection (Grant et al. 2013). SA, JA, and ET play significant role in the systemic signaling cascade in plants, improving their immunity to pathogen invasion and adjusting defense responses (Devadas et al. 2002). SA targets biotrophic and hemibiotrophic pathogens, whereas JA and ET primarily target necrotrophic pathogens (Pieterse et al. 2012). Rice plants infected with *Magnaporthe oryzae* following Si treatment stimulates JA and ET, and improves blast disease resistance (Brunings et al. 2009). Si causes the leaves to produce SA, JA, and ET, which protects them from powdery mildew infection (Ghareeb et al. 2011). SA regulation is mediated by the genes EDS5 and SID2, and SA synthesis is mediated by the genes EDS1 and PAD4. NPR1 dramatically upregulates NPR1 transcription in tomato plants, and WRKY proteins favorably upregulate NPR1 expression in response to SA (Ghareeb et al. 2011). In wild-type plants, NPR1 inhibits the formation of SA. NPR1 reduces SA/JA cross-talk during herbivore attack, allowing JA-mediated resistance against herbivore attack to begin, supporting NPR1's regulatory role. By functioning as a priming factor for JA pathway, Si enhances defensive response expression and increases intensity of the JA-mediated defense response. After silicon addition, there occurs rise in the expression of defense-related enzymes and proteins that were implicated in JA signaling.

By enhancing Si deposition, JA increases overall leaf silicification and the production of phytolith-containing silica cells (Ye et al. 2013). In Arabidopsis, JA regulates the expression of PDF1,2 gene, and in tomato, JA regulates the expression of proteinase inhibitor I and II. JAZ1, the repressor of the JA signaling pathway, gets degraded by E3 ubiquitin ligase involved in the fine modification of JA-related genes (Thines et al. 2007). As ubiquitin protein ligase action was enhanced in pathogen-infected plants, ubiquitin protein ligase contributed to defense reaction signaling (Dreher and Callis 2007). Genes JERF3, TSRF1, and ACCO are found in plants, and in response to ET and JA signals, JERF3 is activated. TSRF1 is an ET-responsive transcription factor, and ACCO is involved in ethylene synthesis (Pirrello et al. 2012). Following Si therapy for tomato, *R. solanacearum* infection, enhanced expression of JERF3, TSRF1, and ACCO genes was discovered by Ghareeb et al. (2011), demonstrating that silicon-induced resistance was transmitted through the ET and JA pathways. PDF1 expression was increased in Arabidopsis during *Botrytis cinerea* infection, demonstrating its role as a modulator of a signaling pathway involved in plant fungal defense (Cabot et al. 2013). In cells infected with pathogens, the primary reaction occurs. The secondary reaction is evoked by elicitors and is confined to cells surrounding the initial site of infection, whereas the tertiary response is the systemic acquired response (SAR), which is hormonally distributed throughout the plant.

12.9 Silicon Mediates Molecular Mechanism in Plants

Studies on transcriptomic and proteomic levels have been carried out to reveal how silicon is playing a definite role in defense responses in distinct pathosystems (Ghareeb et al. 2011). Vivancos et al. (2015) showed that when Si is applied to plants, they develop resistance to pathogens thus preventing disease onset. When silicon is added to cucumber plants, it leads to induce systemic acquired resistance (SAR) and upregulates gene expression levels involved in generating an unique proline-rich protein linked to cell wall strengthening at places where fungi attempts to penetrate into epidermal cells (Kauss et al. 2003). Fauteux et al. (2005) used molecular techniques such as subtractive cDNA libraries and microarrays to evaluate the expression of defense genes in control and infected plants. Their microarray conclusion exposed that Si-treated plants upregulate defense and pathogenesis-related gene in response to pathogen inoculation. It has been confirmed that after Si application to pathogen-inoculated plants, multiple transcriptional changes get overturned as compared to the control and only Si supplemented plants. Arabidopsis–*Erysiphe cichoracearum* pathosystem resulted in alterations in expression of approximately 4000 genes and no significant variations in the quantity and expression of upregulated defensive genes. The magnitude of the genes that were downregulated was connected to primary metabolism and was diminished after Si treatment (Fauteux et al. 2005). Si swiftly alters 26 proteins in a tomato plant that has been downregulated after inoculation with *R. solanacearum*, demonstrating that the wide change in the level of proteins in tomato plants is linked to Si-supplemented resistance against the fungal disease (Chen et al. 2014). Silicon plays an important role in both local and systemic resistance, as well as acts as a second messenger (Fauteux et al. 2005; Bockhaven et al. 2012) wide range of defense-related genes were detected in tomato, rice, and Arabidopsis cultivated in silicon-enriched soil compared to nonamended control plants. According to genome-wide studies, In host plants diseases resistance was enhanced by silicon supplementation (Table 12.2).

12.10 Phenylpropanoid Biosynthetic Pathway

Phenylpropanoids are natural compounds that are formed by phenylalanine ammonia-lyase deamination of the amino acid phenylalanine (PAL). Phenyl propane unit is condensed with a component made from acetate using malonyl coenzyme A, produces more complex phenylpropanoids. Not all plant species contain all types of phenylpropanoid chemicals. Although hydroxycinnamic acid and flavonoids are found in abundance in higher plants, isoflavonoids and stilbenes are phenylpropanoid groups that are restricted to specific plant families. Isoflavonoids are predominantly found in the Leguminosae subfamily Papilionoideae. Phenylpropanoids are quantified as and are associated with biotic and abiotic factor defense. Because several fast-expressed sequence tag (EST) libraries have been created, the National Center for Biotechnology Information's (NCBI) dbEST database has been used to catalog the EST sequence. In order to find genes, a large

Table 12.2 Si application enhances disease resistance

Host	Pathogen	Effect	Disease	References
Rose	*Sphaerotheca pannosa* *Podosphera pannosa*	+	Powdery mildew	Shetty et al. (2012)
Wheat	*Septoria nodorum*	+	Septoria leaf blotch	Rodgers-Gray and Shaw (2004)
	Pyriclaria oryzae	+	Leaf blast	Silva et al. (2015)
	Oculimacula vallundae	+	Eye spot	Rodgers-Gray and Shaw (2004)
	Pyricularia grisea	+	Blast	Filha et al. (2011)
	Xanthomonas translucens	+	Leaf streak	Silva et al. (2010)
	Bipolaris sorokiniana	+	Spot blotch	Domiciano et al. (2010)
Arabidopsis	*Erysiphe cichoracearum*	+	Powdery mildew	Vivancos et al. (2015)
Asparagus	*Phomopsis asparagi*	+	Stem blight	Lu et al. (2008)
Banana	*Mycosphaerella fijiensis* *Xanthomonas campestris*	+	Black sigatoka Fusarium wilt	Kablan et al. (2012) Mburu et al. (2015)
Barley	*Alternaria spp.*	+	Black point	Kunoh and Ishizaki (1975)
Bean Bellepepper	*Blumeria graminis f. sp. hordei* *Pseudocercospora griseola* *Phytophthora capsici*	+ + +	Powdery mildew Angular leaf spot Phytophthora blight	Wiese et al. (2005) Rodrigues et al. (2010) French-Monar et al. (2010)
Bitter gourd	*Erysiphe sp.*	+	Powdery mildew	Ratnayake et al. (2016)
Bermuda grass	*Biolaris cynodontis*	+	Leaf spot	Datnoff et al. (2007)
Capsicum	*Colletotrichum gloeosporioides*	+	Anthracnose	Jayawardana et al. (2016)
Cherry	*Penicillium expansum*	+	Fruit decay	Qin and Tian (2005)
Chinese cantaloupe	*Fusarium spp.*	+	Fusarium root rot	Liu et al. (2009)
	Trichothecium roseum	+	Pink rot	Guo et al. (2007)
Coffee	*Hemileia vastatrix*	+	Leaf rust	Carré-Missio et al. (2014)
	Meloidogyne exigua	+	Root-knot	Silva et al. (2010)
Cotton	*Fusarium oxysporum*	+	Fusarium wilt	Whan et al. (2016)

(continued)

Table 12.2 (continued)

Host	Pathogen	Effect	Disease	References
Corn	*Fusarium graminearum*	+	Corn smut	Sun et al. (1994)
Creeping, turf grass	*Rhizoctonia solani*	+	Brown patch	Zhang et al. (2006)
Cucumber	*Fusarium oxysporum*	+	Fusarium wilt	Miyake and Takahashi (1983)
	Sphaerotheca fuliginea	+	Powdery mildew	Liang et al. (2005)
	Colletrichum orbiculare		Anthracnose	Kantoo (2002)
Hami melons	*Alternaria alternata*	+	Decay	Bi et al. (2006)
Lettuce	*Bremia lactucae*	+	Downy mildew	Garibaldi et al. (2011)
Melon	*Acidovorax citrulli*	+	Fruit blotch	Conceição et al. (2014)
	Podosphaera xanthii	+	Powdery mildew	Dallagnol et al. (2015)
Muskmelon	*Trichothecium roseum*	+	Pinkrot disease	Li et al. (2011)
	Sphaerotheca fuliginea	+	Powdery mildew	Menzies et al. (1992)
Oil palm	*Ganoderma boninense*	+	Basal stem rot	Najihah et al. (2015)
Pea	*Mycosphaerella pinodes*	+	Brown spot	Dann and Muir (2002)
Pearl millet	*Sclerospora graminicola*	+	Downy mildew	Deepak et al. (2008)
Perennial ryegrass	*Microdochium nivale* *Gray leaf spot*	+ +	Fusarium patch *Magnaporthe oryzae*	McDonagh and Hunter (2010) Rahman et al. (2015)
Potato	*Fusarium sulphureum* *Phytopthora infestans*	+ +	Dry hot Late blight	Li et al. (2009), Soratto et al. (2012)
Pumpkin	*Podosphaera xanthii*	+	Powdery mildew	Lepolu Torlon et al. (2016)
	Sphaerotheca xanthii	+	Powdery mildew	Heckman et al. (2003)
Peach	*Monilinia fructicola*	+	Brown rot	Yang et al. (2010)
Rice	*Pyricularia oryzae, Magnaporthe oryzae*	+	Blast	Hayasaka et al. (2008), Brunings et al. (2009), Domiciano et al. (2015)
	Bipolaris oryzae	+	Brown spot	Dallagnol et al. (2013)
	Rhizoctonia solani	+	Sheath blight	Zhang et al. (2013)

(continued)

Table 12.2 (continued)

Host	Pathogen	Effect	Disease	References
	Pyricularia oryae	+	Leaf and panicle blast	Cacique et al. (2013)
Sorghum	*Colletotrichum sublineolum*	+	Anthracnose	Resende et al. (2013)
Soybean	*Phytophthora sojae*	+	*Phytophthora stem and root rot*	Guérin et al. (2014)
	Phakopsora pachyrhizi	+	Rust	Cruz et al. (2014)
	Phakopsora pachyrhizi	+	Asian soybean rust	Lemes et al. (2011), Arsenault-Labrecque et al. (2012)
Strawberry	*Sphaerotheca macularis f sp. Macularis*	+	Powdery mildew	Kanto et al. (2006)
	Pestalotia longisetula	+	Pestalotia leaf spot	Carre-Missio et al. (2014)
	Botrytis cinerea	+	Gray mold	Lopes et al. (2015)
Sugarcane	*Puccinia melanocephala*	+	Brown rust	Ramouthar et al. (2015)
Tobacco	*Tobacco mosaic virus* *Belladonna mottle virus*	+	Viral infection	Zellner et al. (2011)
Tomato	*Pythium aphanidermatum*	+	Root rot	Heine et al. (2007)
	Odiopsis sicula	+	Powdery mildew	Garibaldi et al. (2011)
Zucchini squash	*Erysiphe cichoracearum*, *Podosphaera xanthii*	+	*Erysiphe cichoracearum*	Savvas et al. (2009)

number of ESTs have been produced. Phenylalanine ammonia-lyase, 4-coumarate coenzyme-A (CoA) ligase, and chalcone synthase are all significant enzymes in the phenylpropanoid pathway. The aromatic amino acid phenylalanine is the first step in phenylpropanoid production. PAL catalyzes the exchange of phenylalanine to cinnamic acid which is then converted into p-coumaroyl-CoA that is catalyzed by Cinnamate 4-hydroxylase and 4-coumarate coenzyme-A (CoA). P-coumaroyl-CoA forms the precursor to a number of phenylpropanoid chemicals. Flavonols, anthocyanins, and isoflavonoids are produced from p-coumaroyl-CoA via a convoluted phenylpropanoid pathway. Following next-generation sequencing of many plant species, for phenylpropanoid biosynthesis, seven genes (PAL1,2, C4H, 4CL1,2, CHS, and DFR) were found. The leaf had the highest levels of PAL1, C4H, CL1, CHS, and DFR expression; however, CL2 expression was equivalent in the bud, leaf, and stem. The expression patterns of PAL2 and CL2 were similar. PAL1 transcript levels were found greater in leaves as compared to the bud, stem, flowers, and roots. When compared to PAL2, PAL1 had a higher level of expression. Three isoforms of PAL were identified and studied in *Scutellaria baicalensis*, with distinct expression pattern in various organs. Elevated levels of PAL1, PAL2, and PAL3 transcripts were found in root, stem, and leaf, respectively (Xu et al. 2010). The leaf had much transcript level of C4H than the flower, bud, and root. The expression level of PAL1 was comparable to CL1. Transcriptional establishment of a wide range of biosynthetic pathway genes is widely regarded to be the reason for the foundation of phenylpropanoid metabolites through plant response to infection. Following infection elicitation, microarray nuclear transcript run-on tests have discovered multiple incidences of increased transcription rates of phenylpropanoid pathway genes (Rushton and Somssich 1999). The types of transcription factors that control gene expression in plants, including those in the phenylpropanoid pathway, have been described in several reviews (Liu et al. 1999). Different groups of transcription factors appear in the overall regulation of phenylpropanoid biosynthesis, among them myb factors receiving more attention. WRKY, Ntmyb2, PAP1-D, Ntlim1, G/HBF-1, and KAP2 regulate the genes and pathways as phenylpropanoids/ PR proteins, PAL/defense response genes, D phenylpropanoid pathway, PAL, 4CL, and CAD in tobacco, CHS in soybean (Nesi et al. 2001; Lindsay et al. 2002). Si-mediated stimulation of a cascade of events via the phenylpropanoid pathway facilitates the production and storage of defense chemicals such as flavonoids and phenols against infections. The expression of phenylalanine ammonia-lyase genes, cinnamyl alcohol dehydrogenase, and chalcone synthases in phenylpropanoid pathway are effected by silicon exposure. There was a definite relationship between enhanced phenolic synthesis and greater PAL and CHS transcript levels, especially with Si treatments. Si application resulted in the upregulation of PAL and CHS and phenolic biosynthesis in Arabidopsis, wheat, and in rice plants after infected with various biotrophic fungal pathogens (Chain et al. 2009). Similarly, silicon supplementation causes activation of genes involved in the phenylpropanoid biosynthesis pathway in rose plants infected with the *P. pannosa* pathogen (Shetty et al. 2012). However, transcriptomic analysis revealed that the gene expression of the phenylpropanoid pathway is not significantly affected by silicon (Fauteux et al. 2005). Silicon is supposed to enhance antimicrobial defense in plants by upregulating the levels of phenylpropanoid pathway genes, phenols, and flavonoids synthesis (Shetty et al. 2012).

12.11 Conclusion

By gaining the information on the host–pathogen interaction mediated by phenolics and silicon, it has been concluded that phenolics and Si are found to have significant role in plants against various biotic stress factors. Phenolics as secondary metabolites impart protection to wide range of biotic stresses in plants. Plants produce phenolics as an early response to biotic stress factors, which might lead to elevate other defense-related compounds in hosts against multiple stresses. Si acts as a physical barrier that develops resistance to a wide range of diseases by forming preformed defense barriers before pathogens are introduced. Si-induced biochemical resistance entails the activation of defense-related enzymes, the stimulation of antimicrobial compound synthesis, and the regulation of a complex network of signal pathways. Finally, Si may control the expression of defense-related genes at the molecular level.

Acknowledgment The authors would like to acknowledge Prof. Azra N Kamili (Central University of Kashmir) for significant analysis of the manuscript and many helpful suggestions.

Conflicts of Interest No conflict of interest is declared.

References

Ahmed N, Abbasi MW, Shaukat SS, Zaki MJ (2009) Physiological changes in leaves of mungbean plants infected with Meloidogyne javanica. Phytopathol Mediterr 48:262–268

Arsenault-Labrecque G, Menzies JG, Belanger RR (2012) Effect of silicon absorption on soybean resistance to *Phakopsora pachyrhizi* in different cultivars. Plant Dis 96:37–42

Asgari Lajayer B, Ghorbanpour M, Nikabadi S (2017) Heavy metals in contaminated environment: Destiny of secondary metabolite biosynthesis, oxidative status and phytoextraction in medicinal plants. Ecotoxicol Environ Saf 145:377–390

Behmann J, Mahlein AK, Rumpf T, Römer C, Plümer L (2014) A review of advanced machine learning methods for the detection of biotic stress in precision crop protection. Precis Agric 16(3):239–260

Bi Y, Tian SP, Guo YR, Ge YH, Qin GZ (2006) Sodium silicate reduces postharvest decay on Hami melons: induced resistance and fungistatic effects. Plant Dis 90:279–283. https://doi.org/10.1094/pd-90-0279

Bockhaven JV, De Vleesschauwer D, Höfte M (2012) Towards establishing broad-spectrum disease resistance in plants: silicon leads the way. J Exp Bot 64:1281–1293

Brisson LF, Tenhaken R, Lamb C (1994) Function of oxidative crosslinking of cell wall structural proteins in plant disease resistance. Plant Cell 6:1703–1712. https://doi.org/10.2307/3869902

Brunings AM, Datnoff LE, Ma JF, Mitani N, Nagamura Y, Rathinasabapathi B et al (2009) Differential gene expression of rice in response to silicon and rice blast fungus *Magnaporthe oryzae*. Ann Appl Biol 155:161–170. https://doi.org/10.1111/j.1744-7348.2009.00347.x

Buonaurio R, Moretti C, da Silva DP, Cortese C, Ramos C, Venturi V (2015) The olive knot disease as a model to study the role of interspecies bacterial communities in plant disease. Front Plant Sci 6:434

Byers JA (1995) Host-tree chemistry affecting colonization in bark beetles. Chem Ecol Insect 2: 154–213

Cabot C, Gallego B, Martos S, Barceló J, Poschenrieder C (2013) Signal cross talk in Arabidopsis exposed to cadmium, silicon, and Botrytis cinerea. Planta 237:337–349. https://doi.org/10.1007/s00425-012-1779-7

Cacique IS, Domiciano GP, Moreira WR, Rodrigues FA, Cruz MFA, Serra NS, Catala AB (2013) Effect of root and leaf applications of soluble silicon on blast development in rice. Bragantia 72: 304–309

Carré-Missio V, Rodrigues FA, Schurt DA, Resende RS, Souza NFA, Rezende DC et al (2014) Effect of foliar-applied potassium silicate on coffee leaf infection by Hemileia vastatrix. Ann Appl Biol 164:396–403. https://doi.org/10.1111/aab.12109

Chain F, Côté-Beaulieu C, Belzile F, Menzies JG, Bélanger RR (2009) A comprehensive transcriptomic analysis of the effect of silicon on wheat plants under control and pathogen stress conditions. Mol Plant Microbe Interact 22:1323–1330

Chen Y, Liu M, Wang L, Lin W, Fan X, Cai K (2014) Proteomic characterization of silicon-mediated resistance against Ralstonia solanacearum in tomato. Plant Soil 387:425–440. https://doi.org/10.1007/s11104-014-2293-4

Conceição CS, Felix KCS, Mariano RLR, Medeiros EV, Souza EB (2014) Combined effect of yeast and silicon on the control of bacterial fruit blotch in melon. Sci Hortic 174:164–170. https://doi.org/10.1016/j.scienta.2014.05.027

Culver JN, Padmanabhan MS (2007) Virus-induced disease: altering host physiology one interaction at a time. Annu Rev Phytopathol 45(1):221–243

Cruz MFA, Rodrigues FA, Diniz APC, Alves Moreira M, Barros EG (2014) Soybean resistance to Phakopsora pachyrhizi as affected by acibenzolar-S-methyl, jasmonic acid and silicon. J Phytopathol 162:133–136

Dallagnol LJ, Rodrigues FA, Damatta FM, Mielli MV, Pereira SC (2011) Deficiency in silicon uptake affects cytological, physiological, and biochemical events in the rice—Bipolaris oryzae interaction. Phytopathology 101:92–104. https://doi.org/10.1094/PHYTO-04-10-0105

Dallagnol L, Rodrigues F, Pascholati S, Fortunato A, Camargo L (2013) Comparison of root and foliar applications of potassium silicate in potentiating post-infection defences of melon against powdery mildew. Plant Pathol 64:1085–1093. https://doi.org/10.1111/ppa.12346

Dallagnol L, Rodrigues F, Pascholati S, Fortunato A, Camargo L (2015) Comparison of root and foliar applications of potassium silicate in potentiating post-infection defences of melon against powdery mildew. Plant Pathol 64:1085–1093. https://doi.org/10.1111/ppa.12346

Dann EK, Muir S (2002) Peas grown in media with elevated plant-available silicon levels have higher activities of chitinase and b-1, 3-glucanase, are less susceptible to a fungal leaf spot pathogen and accumulate more foliar silicon. Australas Plant Pathol 31:9–13. https://doi.org/10.1071/AP01047

Datnoff LE, Elmer WH, Huber DM (2007) Mineral nutrition and plant disease. The American Phytopathological Society, St. Paul

Deepak S, Manjunath G, Manjula S, Niranjan-Raj S, Geetha N, Shetty HS (2008) Involvement of silicon in pearl millet resistance to downy mildew disease and its interplay with cell wall proline/hydroxyproline-rich glycoproteins. Australas Plant Pathol 37:498–504. https://doi.org/10.1071/AP08047

Devadas SK, Enyedi A, Raina R (2002) The Arabidopsis hrl1 mutation reveals novel overlapping roles for salicylic acid, jasmonic acid and ethylene signalling in cell death and defence against pathogens. Plant J 30:467–480. https://doi.org/10.1046/j.1365-313X.2002.01300.x

Domiciano GP, Rodrigues FA, Vale FXR, Filha MSX, Moreira WR, Andrade CCL et al (2010) Wheat resistance to spot blotch potentiated by silicon. J Phytopathol 158:334–343. https://doi.org/10.1111/j.1439-0434.2009.01623.x

Domiciano G, Cacique I, Freitas C, Filippi M, DaMatta FM, Vale F, Rodrigues F (2015) Alterations in gas exchange and oxidative metabolism in rice leaves infected by Pyricularia oryzae are attenuated by silicon. Phytopathology 105:738–747

Dreher K, Callis J (2007) Ubiquitin, hormones and biotic stress in plants. Ann Bot 99:787–822. https://doi.org/10.1093/aob/mcl255

Dresselhaus T, Hückelhoven R (2018) Biotic and abiotic stress responses in crop plants. Agronomy 8:267

Duke SO (2018) Interaction of chemical pesticides and their formulation ingredients with microbes associated with plants and plant pests. J Agric Food Chem 66(29):7553–7561

Emadian SF, Newton RJ (1989) Growth enhancement of lob-lolly pine (Pinus taeda L.) seedlings by silicon. J Plant Physiol 134:89–103

Fauteux F, Remus-Borel W, Menzies JG, Belanger RR (2005) Silicon and plant disease resistance against pathogenic fungi. FEMS Microbiol Lett 249:1–6. https://doi.org/10.1016/j.femsle.2005.06

Fawe A, Abou-Zaid M, Menzies J, Bélanger R (1998) Silicon-mediated accumulation of flavonoid phytoalexins in cucumber. Phytopathology 88:396–401. https://doi.org/10.1094/PHYTO.1998.88.5.396

Filha MSX, Rodrigues FA, Domiciano GP, Oliveira HV, Silveira PR, Moreira WR (2011) Wheat resistance to leaf blast mediated by silicon. Australas Plant Pathol 40:28–38. https://doi.org/10.1007/s13313-010-0010-1

Fortunato AA, Rodrigues F, Do Nascimento KJ (2012) Physiological and biochemical aspects of the resistance of banana plants to fusarium wilt potentiated by silicon. Phytopathology 102:957–966. https://doi.org/10.1094/PHYTO-02-12-0037-R

Fortunato AA, Debona D, Bernardeli AMA, Rodrigues FA (2015) Defence-related enzymes in soybean resistance to target spot. J Phytopathol 163:731–742. https://doi.org/10.1111/jph.12370

French-Monar RD, Rodrigues FA, Korndörfer GH, Datnoff LE (2010) Silicon suppresses Phytophthora blight development on bell pepper. J Phytopathol 158:554–560. https://doi.org/10.1111/j.1439-0434.2009.01665.x

Garibaldi A, Gilardi G, Cogliati EE, Gullino ML (2011) Silicon and increased electrical conductivity reduce downy mildew of soilless grown lettuce. Eur J Plant Pathol 132:123–132. https://doi.org/10.1007/s10658-011-9855-6

Ghareeb H, Bozsó Z, Ott PG, Repenning C, Stahl F, Wydra K (2011) Transcriptome of silicon-induced resistance against Ralstonia solanacearum in the silicon non-accumulator tomato implicates priming effect. Physiol Mol Plant Pathol 75:83–89. https://doi.org/10.1016/j.pmpp.2010.11.004

Grant MR, Kazan K, Manners JM (2013) Exploiting pathogens' tricks of the trade for engineering of plant disease resistance: challenges and opportunities. Microb Biotechnol 6:212–222. https://doi.org/10.1111/1751-7915.12017

Guérin V, Lebreton A, Cogliati EE, Hartley SE, Belzile F, Menzies JG et al (2014) A zoospore inoculation method with Phytophthora sojae to assess the prophylactic role of silicon on soybean cultivars. Plant Dis 98:1632–1638. https://doi.org/10.1094/pdis-01-14-0102-re

Guo Y, Liu L, Zhao J, Bi Y (2007) Use of silicon oxide and sodium silicate for controlling Trichothecium roseum postharvest rot in Chinese cantaloupe (Cucumis melo L.). Int J Food Sci Techol 42:1012–1018. https://doi.org/10.1111/j.1365-621.2006.01464.x

Guo Q, Major IT, Howe GA (2018) Resolution of growth–defense conflict: mechanistic insights from jasmonate signaling. Curr Opin Plant Biol 44:72–81

Hao Z, Wang L, He Y, Liang J, Tao R (2011) Expression of defense genes and activities of antioxidant enzymes in rice resistance to rice stripe virus and small brown planthopper. Plant Physiol Biochem 49:744–751

Hayasaka T, Fujii H, Ishiguro K (2008) The role of silicon in preventing appressorial penetration by the rice blast fungus. Phytopathology 98:1038–1044. https://doi.org/10.1094/PHYTO-98-9-1038

Heckman JR, Johnston S, Cowgill W (2003) Pumpkin yield and disease response to amending soil with silicon. Hortic Sci 38(4):552–554

Heine G, Tikum G, Horst WJ (2007) The effect of silicon on the infection by and spread of Pythium aphanidermatum in single roots of tomato and bitter gourd. J Exp Bot 58:569–577. https://doi.org/10.1093/jxb/erl232

Jayawardana HARK, Weerahewa HLD, Saparamadu MDJS (2016) The mechanisms underlying the Anthracnose disease reduction by rice hull as a silicon source in capsicum (*Capsicum annuum* L.) grown in simplified hydroponics. Procedia Food Sci 6:147–150

Jones DL, Hodge A, Kuzyakov Y (2004) Plant and mycorrhizal regulation of rhizodeposition. New Phytol 163(3):459–480

Joshi JR, Khazanov N, Senderowitz H, Burdman S, Lipsky A, Yedidia I (2016) Plant phenolic volatiles inhibit quorum sensing in pectobacteria and reduce their virulence by potential binding to ExpI and ExpR proteins. Sci Rep 6:38126

Kablan L, Lagauche A, Delvaux B, Legrève A (2012) Silicon reduces black Sigatoka development in banana. Plant Dis 96:273–278. https://doi.org/10.1094/pdis-04-11-0274
Kantoo T (2002) Research of silicate for improvement of plant defense against pathogens in Japan. Abstract of Second Silicon in Agriculture Conference, p. 22-26
Kanto T, Miyoshi A, Ogawa T, Maekawa K, Aino M (2006) Suppressive effect of liquid potassium silicate on powdery mildew of strawberry in soil. J Gen Plant Pathol 72:137–142
Kauss H, Kai S, Franke R, Gilbert S, Dietrich RA, Kröger N (2003) Silica deposition by a strongly cationic proline-rich protein from systemically resistant cucumber plants. Plant J 33:87–95. https://doi.org/10.1046/j.1365-313X.2003.01606.x
Kim SG, Kim KW, Park EW, Choi D (2002) Silicon-induced cell wall fortification of rice leaves: a possible cellular mechanism of enhanced host resistance to blast. Phytopathology 92:1095–1103
Kumar A, Verma JP (2018) Does plant—microbe interaction confer stress tolerance in plants: a review? Microbiol Res 207:41–52
Kunoh H, Ishizaki H (1975) Silicon levels near penetration sites of fungi on wheat, barley, cucumber and morning glory leaves. Physiol Plant Pathol 5:283–287
Lattanzio V, Lattanzio VMT, Cardinali A (2006) Role of phenolics in the resistance mechanisms of plants against fungal pathogens and insects. Phytochem Adv Res 81:23–67
Ledesma-Escobar CA, Priego-Capote F, Luque de Castro MD (2019) Relevance and analysis of citrus flavonoids. In: Polyphenols in plants. Elsevier, London, pp 133–150
Lee G, Joo Y, Kim SG, Baldwin IT (2017) What happens in the pith stays in the pith: tissue -localized defense responses facilitate chemical niche differentiation between two spatially separated herbivores. Plant J 92:414–425
Lemes EM, Mackowiak CL, Blount A, Marois JJ, Wright DL, Coelho L et al (2011) Effects of silicon applications on soybean rust development under greenhouse and field conditions. Plant Dis 95:317–324. https://doi.org/10.1094/pdis-07-10-0500
Lepolu Torlon J, Heckman J, Simon J, Wyenandt C (2016) Silicon soil amendments for suppressing powdery mildew on pumpkin. Sustainability 8:293. https://doi.org/10.3390/su8040293
Levin DA (1971) Plant phenolics: an ecological perspective. Am Nat 105(942):157–181
Li YC, Bi Y, Ge YH, Sun XJ, Wang Y (2009) Antifungal activity of sodium silicate on *Fusarium sulphureum* and its effect on dry rot of potato tubers. J Food Sci 74:M213–M218
Li W, Bi Y, Ge Y, Li Y, Wang J, Wang Y (2011) Effects of postharvest sodium silicate treatment on pink rot disease and oxidative stress-antioxidative system in muskmelon fruit. Eur Food Res Technol 234:137–145. https://doi.org/10.1007/s00217-011-1611-9
Liang YC, Sun W, Si J, Römheld V (2005) Effects of foliar- and root-applied silicon on the enhancement of induced resistance to powdery mildew in Cucumis sativus. Plant Pathol 54: 678–685. https://doi.org/10.1111/j.1365-3059.2005.01246.x
Lin D, Xiao M, Zhao J, Li Z, Xing B, Li X, Chen S (2016) An overview of plant phenolic compounds and their importance in human nutrition and management of type 2 diabetes. Molecules 21(10):1374
Lincoln JE, Sanchez JP, Zumstein K, Gilchrist DG (2018) Plant and animal PR1 family members inhibit programmed cell death and suppress bacterial pathogens in plant tissues. Mol Plant Pathol 19(9):2111–2123
Lindsay WP, McAlister FM, Zhu Q, He X-Z, Droge-Laser W, Hedrick S, Doerner P, Lamb C, Dixon RA (2002) KAP-2, a protein that binds to the H-box in a bean chalcone synthase promoter, is a novel plant transcription factor with sequence identity to the large subunit of human Ku autoantigen. Plant Mol Biol 49:503–514
Liu L, Guo Y, Bi Y, Li M, Zhao J, Zhao H (1999) Inhabited mechanisms of silicon compounds against Fusarium rot (Fusarium spp.) of postharvest Chinese cantaloupe. J Food Process Preserv 33:187–202. https://doi.org/10.1111/j.1745-4549.2008.00292.x
Liu L, Guo Y, Bi Y, Li M, Zhao J, Zhao H (2009) Inhabited mechanisms of silicon compounds against Fusarium rot (Fusarium spp.) of postharvest Chinese cantaloupe. J Food Process Preserv 33:187–202

Lopes MR, Klein MN, Ferraz LP et al (2015) *Saccharomyces cerevisiae*: a novel and efficient biological control agent for *Colletotrichum acutatum* during pre-harvest. Microbiol Res 175: 93–99

Ludwig-Müller J (2015) Bacteria and fungi controlling plant growth by manipulating auxin: balance between development and defense. J Plant Physiol 172:4–12

Lu G, Jian W, Zhang J, Zhou Y, Cao J (2008) Suppressive effect of silicon nutrient on Phomopsis stem blight development in asparagus. Hortic Sci 43(3):811–817

Ma JF, Yamaji N (2006) Silicon uptake and accumulation in higher plants. Trends Plant Sci 11: 392–397. https://doi.org/10.1016/j.tplants.2006.06.007

Ma JF, Yamaji N (2008) Functions and transport of silicon in plants. Cell Mol Life Sci 65:3049–3057. https://doi.org/10.1007/s00018-008-7580-x

Mandal SM, Chakraborty D, Dey S (2010) Phenolic acids act as signaling molecules in plant-microbe symbioses. Plant Signal Behav 5(4):359–368

Mandhania S, Sangwan RS, Siwach SS, Pundir SR, Sangwan O, Janu A (2018) Role of biochemical constituents and minerals against cotton leaf curl disease in cotton. J Environ Biol 39:221–227

Mburu K, Oduor R, Mgutu A, Tripathi L (2015) Silicon application enhances resistance to xanthomonas wilt disease in banana. Plant Pathol 65:807–818. https://doi.org/10.1111/ppa.12468

McDonagh D, Hunter A (2010) Effect of silicon application on Lolium perenne development and Fusarium control. Paper presented at the XXVIII international horticultural congress on science and horticulture for people (IHC2010), Lisbon, pp 195–201

Melvin P, Bankapalli K, D'Silva P, Shivaprasad PV (2017) Methylglyoxal detoxification by a DJ-1 family protein provides dual abiotic and biotic stress tolerance in transgenic plants. Plant Mol Biol 94(4–5):381–397

Mendes IDS, Rezende MOO (2014) Assessment of the allelopathic effect of leaf and seed extracts of Canavalia ensiformis as post emergent bioherbicides: a green alternative for sustainable agriculture. J Environ Sci Health B 49(5):374–380

Menzies J, Bowen P, Ehret D, Glass AD (1992) Foliar applications of potassium silicate reduce severity of powdery mildew on cucumber, muskmelon, and zucchini squash. J Am Soc Hortic Sci 117:902–905

Miyake Y, Takahashi E (1983) Effect of silicon on the growth of solution-cultured cucumber plant. Soil Sci Plant Nutr 29:71–83. https://doi.org/10.1080/00380768.1983.10432407

Mukherjee S, Kutty NN, Bera P, Mitra A (2018) Impact of light and sucrose supplementation on cellular differentiation, metabolic shift and modulation of gene expression in hairy roots of Daucus carota. In: Plant cell, tissue and organ culture. Springer, Berlin, pp 1–15

Najihah NI, Hanafi MM, Idris AS, Hakim MA (2015) Silicon treatment in oil palms confers resistance to basal stem rot disease caused by Ganoderma boninense. Crop Prot 67:151–159. https://doi.org/10.1016/j.cropro.2014.10.004

Nesi N, Jond C, Debeaujon I, Caboche M, Lepiniec L (2001) The Arabidopsis TT2 gene encodes an R2R3 MYB domain protein that acts as a key determinant for proanthocyanidin accumulation in developing seed. Plant Cell 13:2099–2114

Ohri P, Pannu SK (2010) Effect of phenolic compounds on nematodes—a review. J Appl Nat Sci 2(2):344–350

Parr AJ, Bolwell GP (2000) Phenols in the plant and in man. The potential for possible nutritional enhancement of the diet by modifying the phenols content or profile. J Sci Food Agric 80(7): 985–1012

Pieterse CM, Van DDD, Zamioudis C, Leonreyes A, Van Wees SC (2012) Hormonal modulation of plant immunity. Cell Dev Biol 28:489–521. https://doi.org/10.1146/annurev-cellbio-092910-154055

Pirrello J, Prasad BN, Zhang W, Chen K, Mila I, Zouine M et al (2012) Functional analysis and binding affinity of tomato ethylene response factors provide insight on the molecular bases of plant differential responses to ethylene. BMC Plant Biol 12:190. https://doi.org/10.1186/1471-2229-12-190

Pusztahely T, István H, Pócsi IJ (2017) Plant-fungal interactions: special secondary metabolites of the biotrophic, necrotrophic, and other specific interactions. In: Fungal metabolites. Academic, London, pp 1–58

Qin GZ, Tian SP (2005) Enhancement of biocontrol activity of Cryptococcus laurentii by silicon and the possible mechanisms involved. Phytopathology 95:69–75

Quarta A, Mita G, Durante M, Arlorio M, De PA (2013) Isolation of a polyphenol oxidase (PPO) cDNA from artichoke and expression analysis in wounded artichoke heads. Plant Physiol Biochem 68:52–60. https://doi.org/10.1016/j.plaphy.2013.03.020

Rahman A, Wallis CM, Uddin W (2015) Silicon-induced systemic defense responses in perennial ryegrass against infection by Magnaporthe oryzae. Phytopathology 105:748–757. https://doi.org/10.1094/PHYTO-12-14-0378-R

Ramouthar PV, Caldwell PM, Mcfarlane SA (2015) Effect of silicon on the severity of brown rust of sugarcane in South Africa. Eur J Plant Pathol 145:53–60. https://doi.org/10.1007/s10658-015-0812-7

Ratnayake RMRNK, Daundasekera WAM, Ariyarathne HM, Ganehenege MYU (2016) Some biochemical defense responses enhanced by soluble silicon in bitter gourd-powdery mildew pathosystem. Australas Plant Pathol 45:425–433. https://doi.org/10.1007/s13313-016-0429-0

Resende RS, Rodrigues F, Costa RV, Silva DD (2013) Silicon and fungicide effects on anthracnose in moderately resistant and susceptible sorghum lines. J Phytopathol 161:11–17. https://doi.org/10.1111/jph.12020

Rodgers-Gray B, Shaw M (2004) Effects of straw and silicon soil amendments on some foliar and stem-base diseases in pot-grown winter wheat. Plant Pathol 53:733–740

Rodrigues FA, Datnof LE (2015) Silicon and plant diseases. Springer, Berlin. https://doi.org/10.1007/978-3-319-22930-0

Rodrigues F, Duarte HSS, Rezende DC, Filho JAW, Korndörfer GH, Zambolim L (2010) Foliar spray of potassium silicate on the control of angular leaf spot on beans. J Plant Nutr 33:2082–2093. https://doi.org/10.1080/01904167.2010.519082

Rushton PJ, Somssich IE (1999) Transcriptional regulation of plant genes responsive to pathogens and elicitors. In: Stacey G, Keen NT (eds) Plant–microbe interactions, vol 4. American Phytopathological Society, St Paul, pp 251–274

Saddique M, Kamran M, Shahbaz M (2018) Differential responses of plants to biotic stress and the role of metabolites. In: Plant metabolites and regulation under environmental stress. Elsevier, San Diego, pp 69–87

Savvas D, Giotis D, Chatzieustratiou E, Bakea M, Patakioutas G (2009) Silicon supply in soilless cultivations of zucchini alleviates stress induced by salinity and powdery mildew infections. Environ Exp Bot 65:11–17. https://doi.org/10.1016/j.envexpbot.2008.07.004

Schnee S, Viret O, Gindro K (2008) Role of stilbenes in the resistance of grapevine to powdery mildew. Physiol Mol Plant Pathol 72(4–6):128–133

Schurt DA, Cruz MF, Nascimento KJ, Filippi MC, Rodrigues FA (2014) Silicon potentiates the activities of defense enzymes in the leaf sheaths of rice plants infected by *Rhizoctonia solani*. Trop Plant Pathol 39:457–463. https://doi.org/10.1590/S1982-56762014000600007

Shetty R, Jensen B, Shetty NP, Hansen M, Hansen CW, Starkey KR et al (2012) Silicon induced resistance against powdery mildew of roses caused by Podosphaera pannosa. Plant Pathol 61:120–131. https://doi.org/10.1111/j.1365-3059.2011.02493.x

Silva R, Oliveira R, Nascimento K, Rodrigues F (2010) Biochemical responses of coffee resistance against Meloidogyne exigua mediated by silicon. Plant Pathol 59:586–593. https://doi.org/10.1111/j.1365-3059.2009.02228.x

Silva RV, de Lima Oliveira RDA, da Silva FP, Castro DB, Rodrigues FA (2015) Effects of silicon on the penetration and reproduction events of Meloidogyne exigua on coffee roots. Bragantia 74:196–199. https://doi.org/10.1590/1678-4499.360

Soratto RP et al (2012) Leaf application of silicic acid to white oat and wheat. Rev Bras Cienc Solo 36(5):1538–1544

Sousa RS, Rodrigues FA, Schurt DA, Souza NFA, Cruz MFA (2013) Cytological aspects of the infection process of Pyricularia oryzae on leaves of wheat plants supplied with silicon. Trop Plant Pathol 38:472–477. https://doi.org/10.1590/S1982-56762013000600002

Sun X, Sun Y, Zhang C, Song Z, Chen J, Bai J, Cui Y, Zhang C (1994) The mechanism of corn stalk rot control by application of potassic and siliceous fertilizers. Acta Phytophysiol Sin 21:102–108

Talhaoui N, Taamalli A, Gómez-Caravaca AM, Fernández-Gutiérrez A, Segura-Carretero A (2015) Phenolic compounds in olive leaves: analytical determination, biotic and abiotic influence, and health benefits. Food Res Int 77:92–108

Thakur H, Jindal SK, Sharma A, Dhaliwal MS (2018) Chilli leaf curl virus disease: a serious threat for chilli cultivation. J Plant Dis Protect 125(3):239–249

Thines B, Katsir L, Melotto M, Niu Y, Mandaokar A, Liu G et al (2007) JAZ repressor proteins are targets of the SCFCOI1 complex during jasmonate signalling. Nature 448:661–665. https://doi.org/10.1038/nature05960

Van BJ, De Vleesschauwer D, Hofte M (2013) Towards establishing broad-spectrum disease resistance in plants: silicon leads the way. J Exp Bot 64:1281–1293. https://doi.org/10.1093/jxb/ers329

Vishwanath SJ, Delude C, Domergue F, Rowland O (2014) Suberin: biosynthesis, regulation, and polymer assembly of a protective extracellular barrier. Plant Cell Rep 34(4):573–586

Vivancos J, Labbe C, Menzies JG, Belanger RR (2015) Silicon-mediated resistance of Arabidopsis against powdery mildew involves mechanisms other than the salicylic acid (SA)-dependent defence pathway. Mol Plant Pathol 16:572–582. https://doi.org/10.1111/mpp.12213

Wang L, Sun R, Zhang Q, Luo Q, Zeng S, Li X, Liu Z (2018) An update on polyphenol disposition via coupled metabolic pathways. In: Expert opinion on drug metabolism & toxicology. Ashley Publications, London, pp 1–15

Whan JA, Dann EK, Aitken EA (2016) Effects of silicon treatment and inoculation with Fusarium oxysporum f. sp. vasinfectum on cellular defences in root tissues of two cotton cultivars. Ann Bot 118:219–226. https://doi.org/10.1093/aob/mcw095

Wiese J, Wiese H, Schwartz J, Schubert S (2005) Osmotic stress and silicon act additively in enhancing pathogen resistance in barley against barley powdery mildew. J Plant Nutr Soil Sci 168:269–274. https://doi.org/10.1002/jpln.200420490

Xu H, Park NI, Li X, Kim YK, Lee SY, Park SU (2010) Molecular cloning and characterization of phenylalanine ammonia-lyase, cinnamate 4-hydroxylase and genes involved in flavone biosynthesis in Scutellaria baicalensis. Bioresour Technol 101:9715–9722

Yang YF, Liang YC, Lou YS, Sun WC (2003) Influences of silicon on peroxidase, superoxide dismutase activity and lignin content in leaves of wheat *Tritium aestivum* L. and its relation to resistance to powdery mildew. Sci Agric Sin 36:813–817

Yang YF, Liang YC, Lou YS, Sun WC (2010) Influences of silicon on peroxidase, superoxide dismutase activity and lignin content in leaves of wheat Tritium aestivum L. and its relation to resistance to powdery mildew. Sci Agric Sin 36:813–817

Ye M, Song YY, Long J, Wang RL, Baerson SR, Pan ZQ et al (2013) Priming of jasmonate-mediated antiherbivore defense responses in rice by silicon. Proc Natl Acad Sci U S A 110: 3631–3639. https://doi.org/10.1073/pnas.1305848110

Zellner W, Frantz J, Leisner S (2011) Silicon delays tobacco ringspot virus systemic symptoms in Nicotiana tabacum. J Plant Physiol 168:1866–1869. https://doi.org/10.1016/j.jplph.2011.04.002

Zhang Q, Fry J, Lowe K, Tisserat N (2006) Evaluation of calcium silicate for brown patch and dollar spot suppression on turfgrasses. Crop Sci 46:1635–1643. https://doi.org/10.2135/cropsci2005.04-0002

Zhang G, Cui Y, Ding X, Dai Q (2013) Stimulation of phenolic metabolism by silicon contributes to rice resistance to sheath blight. J Plant Nutr Soil Sci 176:118–124. https://doi.org/10.1002/jpln.201200008

Plants' Fungal Diseases and Phenolics Response

13

Luis A. Cabanillas-Bojórquez, Cristina A. Elizalde-Romero, Erick P. Gutiérrez-Grijalva, and J. Basilio Heredia

Abstract

Plants are susceptible to pathogens like fungi, which reduce plant survival and cause losses in yield and fruit quality. However, pathogens are inhibited as plants defend various mechanisms such as increased lignin biosynthesis, hydrolysis of cell walls, and the production of phenolic compounds. Phenolic compounds synthetized as second metabolites are accumulated in plants by fungi such as *Fusarium*, *Alternaria alternata*, and others. Furthermore, phenolic compounds possess antifungal activity, which could be attributed to several mechanisms. In this sense, investigations show that phenolic compounds affect fungal growth due to the plasma membrane disruption, as well as inhibited fungi cell wall and inhibited fungi cell division. Also, phenolic compounds induce the mitochondrial dysfunction of fungi and inhibit RNA and protein synthesis. In this work, we present the most relevant report on plants affected by fungi and its relation to the phenolic compounds' response.

Keywords

Plants · Phenolics · Pathogens · Fungi · Resistance

L. A. Cabanillas-Bojórquez · C. A. Elizalde-Romero · J. B. Heredia (✉)
Post-Doct CONAHCyT-Centro de Investigación en Alimentación y Desarrollo, Culiacán, Sinaloa, México
e-mail: jbheredia@ciad.mx

E. P. Gutiérrez-Grijalva
Cátedras CONAHCyT-Centro de Investigación en Alimentación y Desarrollo, Culiacán, Sinaloa, México

R. Lone et al. (eds.), *Plant Phenolics in Biotic Stress Management*,
https://doi.org/10.1007/978-981-99-3334-1_13

13.1 Introduction

Plants are sensitive to environmental conditions and stress due to their sessile condition, which generated high amounts of reactive oxygen species (ROS). These species cause oxidation of proteins, nucleic acids, and lipids, compromising plant survival (Ganugi et al. 2021; Khare et al. 2020). Plants synthetized bioactive compounds by stress conditions which are accumulated in different parts of plant, such as cuticles and walls of aerial plant organs (Khare et al. 2020; Simaan et al. 2020). This metabolism involves various enzymatic and nonenzymatic molecules (Bartwal et al. 2013).

Different groups of bioactive compounds with different structure could protect the plant from biotic and abiotic stresses, also, the bioactive compounds accumulation and quantity could affect the protect effect (Beeby et al. 2020; Ganugi et al. 2021; Khare et al. 2020; Zamljen et al. 2021). Secondary metabolites are classified in groups due to their characteristics; the most abundant compounds belong to the phenolic group. Phenolic compounds are obtained by two pathways (shikimic and malonic acid) (Khare et al. 2020). The structure of phenolic compounds is characterized by at least one hydroxyl (–OH) in an aromatic ring (Vuolo et al. 2019). Plant stress is classified as biotic, involving other organisms, and abiotic (Fig. 13.1), which refers to plants' exposure to environmental factors or conditions.

Biotic stress is caused by herbivory and pathogenic microorganisms such as fungi and bacteria. Therefore, plants need to avoid being eaten, which is achieved by producing substances (SM) that could cause the death of herbivores or confer repellent flavors. These SM can also repel oviposition herbivores and attract their enemies. These toxins or repellents have been thoroughly studied mainly because of their strong insecticide characteristics (Ganugi et al. 2021; Mithöfer and Boland 2012).

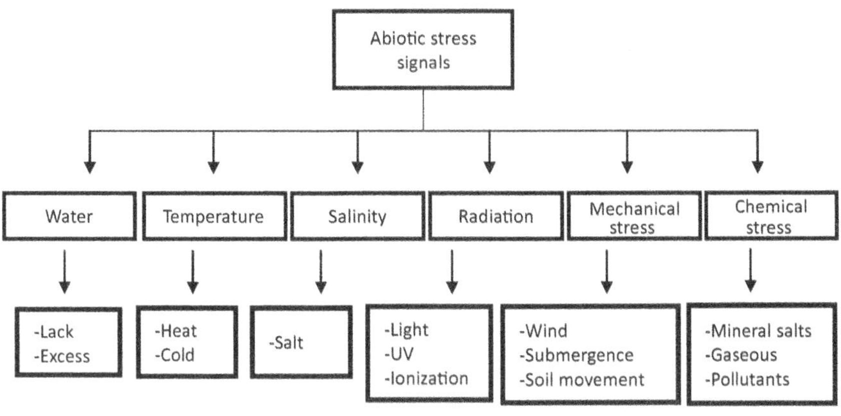

Fig. 13.1 Abiotic stress in plants (Adapted from Khare et al. 2020; Mahajan and Tuteja 2005; Zhu 2016)

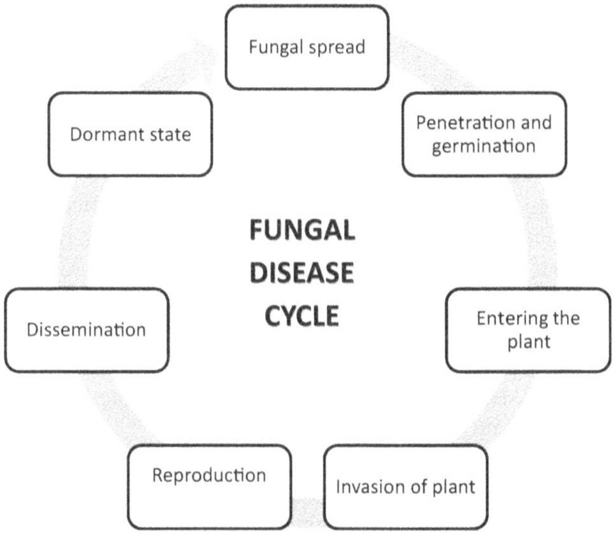

Fig. 13.2 Fungal disease cycle in plants: (1) Fungal spread and contact with host/plant. (2) Penetration and germination involve attachment to the host and consequently triggering defense signals. (3) Fungi penetrate directly into the plant through wounds. (4) Spores invasion which infect plant cells, starting symptoms. (5) Reproduction of several fungal spores. (6) Dissemination of spore to different surfaces or possible hosts. (7) Dormancy allows the pathogen to survive until favorable conditions are available again (own making) (Ray et al. 2017; Zeilinger et al. 2015)

Biotic stress represents a menace the plant will be continuously exposed to during its life span, forcing it to develop defenses that include constitutive and inducible, direct, and indirect mechanisms (Mithöfer and Boland 2012). Even though SM protect the plant against pathogens growth linked by a structural protection mechanism of plants. Leaves can present lignified cell walls and waxes, which act repelling water from the leaves' surface, reducing exposure to infections. Cell walls are also protected by lignin or suberin, and they work by stopping the pathogen from getting into the cell content (Daayf et al. 2012; Rodrigues and Furlong 2022).

Fungal diseases in plants have been studied for different purposes, mainly because they represent 70% to 80% of plant diseases and because they provoke commercial repercussions as it reduces the quality and yield of products. The fungal diseases could be inhibited by early stages detection (Arora et al. 2022; Ray et al. 2017; Rodrigues and Furlong 2022).

Fungi have a great capacity for adaptation because they can survive environmental changes using signals to activate specific stimuli (Simaan et al. 2019). The disease cycle can be summarized in seven states (Fig. 13.2).

13.2 Fungal Diseases in Plants

Fungi are eukaryotes and heterotrophs microorganism capable of digesting food externally and absorbing the nutrients it needs through its cell walls. Fungi produce spores that replicate to obtain energy, nutrients, or carbon from living organism to survive in different environments (Carris et al. 2012).

Along with being the most common type of pathogens to affect plants, fungi can survive in a dormant state. Around 20,000 species of fungi have been reported to be pathogenic to plants (Aboody and Mickymaray 2020). Spores are also an important aspect of their pathogenicity. This is due to their ability to spread easily by wind, animals, soil, and other plants. It is the main reason fungi can rapidly affect an entire harvest. Due to these aspects, fungi disease control represents a big challenge. Besides, fungicide resistance is common, as well as the fungi's capacity to persist in the plant once it is introduced (Ray et al. 2017).

The plant's immune system can be innate and acquired or developed as it adapts to particular conditions or is exposed to different pathogens. The plant response to pathogens can occur only 5 min after interacting with the two entities. This response involves signaling transduction, such as ion fluxes or protein phosphorylation (Shen et al. 2017). Plants are exposed to different stresses simultaneously; both can activate pathways to control these stresses. Plants with OsMPK5 gene overexpression have shown an increased tolerance to stress. In addition, studies have shown that phosphatases and protein kinases are involved in activating resistant responses of fungi diseases (Rejeb et al. 2014).

Plant protection against fungi start with the spore's detection by plant cells, then plants fortifying their cell walls structurally and chemically. However, if the plant is already infected, plant synthetized compounds such as gums, gels, and tyloses. Plant responses are related to the emergence of ROS, and fungal infections intensify the manufacture of these along with the synthesis of pathogenesis-related proteins. A higher ROS production by fungi could damage the plant cells and cause cell death. Damaged cells could liberate toxic compounds (phytoalexins) by cellular decompartmentalization and inhibit pathogen dissemination (Dehgahi et al. 2016). Another important aspect in the early stages of plant–pathogen interaction is the mechanical protection provided by trichomes and stomata; stoma is a major route for pathogens to invade, and plants can react to fungal toxins causing stomatal closure as an immune response (Dehgahi et al. 2016; Rejeb et al. 2014).

Chemical protection plays an important role, too; reports have shown that plants produce specific compounds against fungi. As shown in Table 13.1, different parts of the plants can be under the attacks of specific fungus, leading to a production or increased production of certain phenolics. Some symptoms may show when fungus overcomes the plant's first barrier and becomes infected. First, it is important to identify and quantify the pathogen correctly to determine the ideal approach or management. Diagnosis has conventional methods, including (1) visual examination, (2) culturing and planting methods, and (3) isozyme analysis is explained in Table 13.2 (Ray et al. 2017; Truong et al. 2017).

Table 13.1 Plat defense against fungi

Plant (part)	Fungus	Compound produced	References
Barley roots	*Fusarium*	t-cinnamic acid	Lanoue et al. (2010)
Cereal grains	*F. graminearum*	Mycotoxins	Ponts et al. (2011)
Chili plants	*Fusarium oxysporum*	Phenolics	Arora et al. (2022)
Grapevine plant	*Alternaria spp.*	Gallic acid, stilbenes, flavanols	Rusjan et al. (2017)
Maize leaves	*U. maydis*	Hydroxycinnamic acid	Doehlemann et al. (2008)
Pineapple fruit	*Fusarium ananatum*	Coumaroyl-isocitric and caffeoyl-isocitric acids	Barral et al. (2017)
Raspberry cane	*Didymella applanata* and *Leptosphaeria coniothyrium*	Flavonols, glycosides of quercetin and ellagic acid	Mikulic-Petkovsek et al. (2014)
Settatia marcescens plant	*Phytophthora nicotianae*	Phenolics	Lavania et al. (2006)
Tomato plant roots	*Fusarium oxysporum*	Phenolics	Michielse et al. (2012)
Tomato plant	*Fusarium oxysporum*	Phenolics	Arora et al. (2022)
Wheat leaves and spikes	*Pyricularia oryzae*	Phenolics	Silva et al. (2019)

Table 13.2 Fungal disease detection methods

Method	Description	Limitations
Visual examination	Interpretation of visual symptoms such as spots, galls, cankers, wilts, and damping-off, among others. Followed by the isolation of the pathogen and microscopy techniques	They are subjected to an individual's experience Not all results are conclusive For visual, the identification procedure expertise microbiologist is needed
Culturing and planting methods	Isolation of fungi and cultured at artificial media, with different specific conditions. This is followed by microscopy techniques. The pathogen diagnosis is based on its morphological characteristics	Depends on the skill and experience of the evaluator It is time-consuming
Isozyme analysis	It is used to differentiate and identify different species, providing an efficient tool for revealing genetic variability between fungi	There is low variability among the different fungi varieties

Ray et al. (2017), Zamljen et al. (2021)

There are current methods for fungal disease detection. Direct methods are enzyme-linked immunosorbent assay (ELISA), flow cytometry, and serologically specific electron microscopic (SSEM). Indirect methods include stressed-based disease detection techniques (such as fluorescence, x-ray, or nuclear magnetic resonance) and biomarker-based detection techniques based on plant metabolites (Ray et al. 2017; Rodrigues and Furlong 2022; Sankaran et al. 2010).

13.2.1 Classification of Fungal Diseases

Fungi are responsible for several plant diseases, which cause severe injuries and may also cause plant death. For example, it has been reported that the spur blight, cane blight, esca, and blast fruitlet core rot are dangerous to plant diseases caused by fungi (Barral et al. 2017; Mikulic-Petkovsek et al. 2014; Rusjan et al. 2017; Silva et al. 2019).

13.2.1.1 Spur Blight and Cane Blight

Spur blight and cane blight have been reported as diseases caused in fruits such as raspberry by fungi (*Didymella applanata* and *Leptosphaeria coniothyrium*). These diseases can invade both the internal tissue and the bark, cause leaf falls, and damage the branches, therefore inducing a decrease in fruit yield (Mikulic-Petkovsek et al. 2014; Shternshis et al. 2006; Williamson and Jennings 1992).

13.2.1.2 Esca

Esca is a disease associated with fungi such as *Alternaria alternate*, *Botryospaeriaceae* sp., *Aureobasidium pullulans*, *Fomitiporia mediterranea*, and *Phaeomoniella chlamydospora* observed in grapevine plants which are related to wood deterioration, discoloration, and decay (Rusjan et al. 2017).

13.2.1.3 Blast

Blast is a fungal disease caused by the fungus *Pyricularia oryzae* that affects cereals such as wheat. Although the disease affects the root of the plants, causing a lower uptake of nutrients and a decrease in the transport of them to the grains and leaves, it also induces the death of the spikes, as well as wilting leaves (Ceresini et al. 2018; Rodrigues et al. 2017; Silva et al. 2019).

13.2.1.4 Fruitlet Core Rot

Another disease produced by fungi (such as the *Fusarium* genus) is fruitlet core rot, also called black spots. This disease affects some fruits during ripening. It has been reported that pathogen penetrates the plant and reaches during fruit growth, producing physiological changes to the fruit such as darkening, necrosis, and drier rot (Barral et al. 2017).

13.3 Plant Defense Against Fungal

Pathogenic fungi are considered important biotic stress that causes death of plants (Olufolaji and Ajayi 2019). Plants protect themselves from pathogenic fungi through different strategies such as increased lignin biosynthesis, hydrolysis of cell walls, and phenolic compounds production (Sharma et al. 2011; Silva et al. 2019). Likewise, biotic stress, like fungi infection, triggers a systemic response in plants as a defense mechanism to enhance the production of metabolites. This signaling cascade is involved with enzymes like cytochrome P450, among others pathways (Pandian et al. 2020).

Phenolic compounds are a widespread group characterized by a phenol ring attached to a hydroxyl group (–OH). These compounds are obtained from the shikimate pathway and can be classified as flavonoids and phenolic acids (Gutiérrez-Grijalva et al. 2018; Andrés-Lacueva et al. 2009; Croteau et al. 2015).

It has been demonstrated that phenolic compounds can have an antifungal effect, and the enhanced biosynthesis of these molecules might affect other metabolites of physiological and commercial importance. For instance, phenolic enhancement as a result of fungi infection by *Verticillium dahlia* might decrease the production of volatile compounds because of increased synthesis of compounds with a C6 backbone rather than C5 backbone precursor molecules responsible for the aroma of virgin olive oil (Landa et al. 2019). In this sense, fungi infections can also downregulate the phenylpropanoid pathway, from which phenolics are produced, decreasing its content in infected plants and fruits (Kong et al. 2020).

13.3.1 Plant Phenolics Synthesized and Fungal Diseases

Plants under biotic stress activate a defense system, in this sense, plants synthetized phenolic compounds for inhibited fungi attack. It has been found that flavonols, dihydrochalcones, and simple phenolics inhibited the fungi grown in plants inoculated with fungi. Likewise, if the infection predominates in all parts of the plant, many phenolic compounds are synthesized at the infection site (Arora et al. 2022; Martini et al. 2009; Mikulic-Petkovsek et al. 2014; Rodrigues and Furlong 2022).

Similarly, other studies demonstrated that leaves under fungal diseases have an increase in phenolic compounds since leaves are deteriorated by fungi growth, and therefore, the production of fruits decreases (Ganugi et al. 2021; Xu et al. 2009; Zamljen et al. 2021). Furthermore, it has been related those phenolic compounds reduced the fungi disease on grape leaves due to the increase of phenolic compounds in leaves with high disease severity (Dai et al. 1994; Vagiri et al. 2017). In this sense, Vagiri et al. (2017) reported that black currant leaves under fungal disease synthetized different phenolic compounds. Furthermore, they showed the synthesis of flavonols on black currant leaves with disease and reported a decrease in leaf fungal disease with increased accumulation of phenolic compounds.

Similarly, Arici et al. (2014) studied different apple cultivar's leaves on the apple scab caused by *Venturia inaequalis*. They found that phenolic compounds and shikimic acid accumulated on leaves of apple cultivars that are considered moderate or resistance susceptible were higher than apple leaves cultivars considered susceptible. Also, an accumulation of phenolic compounds such as hydroxycinnamic acid and anthocyanins are found in infected maize leaves by fungi (*Ustilago maydis*, *Euchlaena mexicana*, *Cochliobolus heterostrophus*) in response to injury (Condon et al. 2013; Doehlemann et al. 2008; Simaan et al. 2020). Mincuzzi et al. (2020) concluded that these compounds could be related to plant disease reduction on pomegranate leaves from *Pilidiella granati*.

Plant roots, spikes, and wood are susceptible to the attack of fungi. In this sense, Rusjan et al. (2017) demonstrated that phenolic compounds of grapevine wood are increased by interaction with fungi (*Alternaria alternate*, *Botryospaeriaceae* sp. and *Aureobasidium pullulans*, *Fomitiporia mediterranea*, and *Phaeomoniella chlamydospora*) that cause the esca disease. Also, Silva et al. (2019) reported a high the total soluble phenolic of wheat spikes (up to 35%) and flag leaves (up to 16%) by the fungus *Pyricularia oryzae*. They found that the most significant increase was in wheat spikes than flag leaves after 72 h postinoculation by fungus.

One of the most important phenolic compounds is flavonoids, which are linked to inhibited fungi growth in diseased plants (Aboody and Mickymaray 2020).

The major phenolic compounds in maize kernels infected with *Fusarium graminearum* were caffeoylquinic acid, flavonoids, hydroxycinnamic, and hydroxybenzoic acids. These compounds also demonstrated potent antifungal activity in an in vitro test (Atanasova-Penichon et al. 2012; Barral et al. 2017).

In addition, high hydroxycinnamic acids were found in pineapple fruit inoculated with *Fusarium ananatum*. The hydroxycinnamic acids identified in infected fruitlet were coumaroyl isocitrate and caffeoyl isocitrate with high levels compared to control fruits. The higher level of p-coumaroyl-isocitric acid was an increase of 150 times, as well as an increase of 250 times of caffeoyl-isocitric acid compared with healthy fruitlet. Besides, both hydroxycinnamic acids had antifungal activity against *Fusarium ananatum* in an in vitro test (Barral et al. 2017).

13.3.2 Mode of Action

Phenolic compounds have antifungal activity, and have been demonstrated that increase in fungal plant diseases. Different authors have reported that phenolic compounds create aggregate complexes with plant proteins that reduce the enzymatic function of fungi, thereby decreasing the spread of the fungus in plants (McAllister et al. 1994; Mikulic-Petkovsek et al. 2014; Schwalb and Feucht 1999).

The phenolic compounds production in plants by pathogen infection (principally by fungi) contributes to slowing penetration and resistance of the same into the plant wall (Barral et al. 2017). In this sense, the accumulation of phenolic compounds on fungal diseases from pomegranates leaves could be related to phenolic stimulation production by polyphenol degradation as well as the existence of some enzymes that catalyzed phenolic compounds as ellagitannase (Ascacio-Valdés et al. 2016; Fischer et al. 2011; Mincuzzi et al. 2020).

Shalaby et al. (2016) found that plant phenolic acids could program cell death of fungi. The authors reported that phenolic compounds could cause different effects at the molecular level on fungi. It was also shown that ferulic acid could modify gene expression (dispersion of histone H1:GFP as well as hyphal shrinkage) and consequently damage the fungal cell membrane. Likewise, the phenolic acids caffeic and ferulic were activated by fungal stress. Besides, it has been found that these phenolic acids reduced pathogenicity and cell integrity at the molecular level causing the fungi death.

The antifungal mechanism of flavonoids could be related to various mechanisms that disrupted the fungi membrane, cell division inhibition and cell wall formation, also induction of mitochondrial dysfunction, and decrease RNA and protein synthesis (Aboody and Mickymaray 2020).

Moreover, inoculated with fungi, hydroxycinnamic acid accumulated in pineapple fruit was reported with antifungal activities. In this sense, caffeoylquinic acid has been reported with ability to disrupt the cell membrane of some fungi such as *Candida albicans* and may inhibit mycelial growth of *Fusarium ananatum*, also reduce the growth of black spot disease (Barral et al. 2017; Sung and Lee 2010).

13.4 Perspectives

Recent studies have demonstrated that phenolic compounds have antifungal activity and these compounds are related with a decrease of plant diseases caused by fungi (Vuolo et al. 2019; Aboody and Mickymaray 2020; Zamljen et al. 2021; Arora et al. 2022). The induction of enzymatic activity and phenolic compound accumulation on plants infected by fungi causes a reduction of the disease. Nevertheless, some fungi could survive and infect aerials parts of plants (Silva et al. 2019; Simaan et al. 2019). Moreover, some fungi are drug-resistant; therefore, a potential solution could be the application of synergetic combination therapies of phenolic compounds as well as other compounds synthesized by plants and the addition of external medicinal plant extract, which has a high concentration of phenolic compounds (Hassan et al. 2020; Mincuzzi et al. 2020; Seepe et al. 2020). The application of medicinal plant extracts (tea tree, garlic, and black walnut, among others) rich in phenolic compounds in the early stages of plant growth seems to be a promising strategy for the prevention and reduction of diseases caused by pathogenic fungi (Sepahvand et al. 2017; Seepe et al. 2020). Studies are still needed to know, and the efficacy against disease-causing strains of extracts with antifungal potential, the synergistic effect of phenolic compounds in the stress response caused by pathogenic fungi, and its safety are still unknown.

13.5 Conclusions

Plant infections caused by fungi are a problem that has been increasing. New species of fungi have been reported to affect different plants and cause severe injuries and, in some cases, death. Therefore, studies are still needed to know the plants response against biotic stress caused by pathogens. As a result of biotic stress, the biosynthesis of phytochemicals like phenolic compounds may have an antagonistic effect on fungi diseases. However, further studies are needed to elucidate the antifungal mechanism of action of phenolic compounds, the potential synergistic effects of these molecules may have, and the development of antifungal products derived from phenolic compounds.

References

Aboody MSA, Mickymaray S (2020) Anti-fungal efficacy and mechanisms of flavonoids. Antibiotics (Basel) 9(2):45. https://doi.org/10.3390/antibiotics9020045

Andrés-Lacueva C, Medina-Remon A, Llorach R, Urpi-Sarda M, Khan N, Chiva-Blanch G, Zamora-Ros R, Rotches-Ribalta M, Lamuela-Raventós RM (2009) Phenolic compounds: chemistry and occurrence in fruits and vegetables. In: de la Rosa LA, Alvarez-Parrilla E, González-Aguilar GA (eds) Fruit and vegetable phytochemicals—chemistry, nutritional value and stability

Arici SE, Kafkas E, Kaymak S, Koc NK (2014) Phenolic compounds of apple cultivars resistant or susceptible to Venturia inaequalis. Pharm Biol 52(7):904–908. https://doi.org/10.3109/13880209.2013.872674

Arora H, Sharma A, Poczai P, Sharma S, Haron FF, Gafur A, Sayyed RZ (2022) Plant-derived protectants in combating soil-borne fungal infections in tomato and chilli. J Fungi (Basel) 8(2):213

Ascacio-Valdés JA, Aguilera-Carbó AF, Buenrostro JJ, Prado-Barragán A, Rodríguez-Herrera R, Aguilar CN (2016) The complete biodegradation pathway of ellagitannins by Aspergillus niger in solid-state fermentation. J Basic Microbiol 56(4):329–336. https://doi.org/10.1002/jobm.201500557

Atanasova-Penichon V, Pons S, Pinson-Gadais L, Picot A, Marchegay G, Bonnin-Verdal M-N, Ducos C, Barreau C, Roucolle J, Sehabiague P, Carolo P, Richard-Forget F (2012) Chlorogenic acid and maize ear rot resistance: a dynamic study investigating Fusarium graminearum development, deoxynivalenol production, and phenolic acid accumulation. Int Soc Mol Plant Microbe Interact 25(12):1605–1616. https://doi.org/10.1094/mpmi-06-12-0153-r

Barral B, Chillet M, Minier J, Léchaudel M, Schorr-Galindo S (2017) Evaluating the response to Fusarium ananatum inoculation and antifungal activity of phenolic acids in pineapple. Fungal Biol 121(12):1045–1053. https://doi.org/10.1016/j.funbio.2017.09.002

Bartwal A, Mall R, Lohani P, Guru SK, Arora S (2013) Role of secondary metabolites and Brassinosteroids in plant defense against environmental stresses. J Plant Growth Regul 32(1):216–232. https://doi.org/10.1007/s00344-012-9272-x

Beeby E, Magalhães M, Poças J, Collins T, Lemos MFL, Barros L, Ferreira ICFR, Cabral C, Pires IM (2020) Secondary metabolites (essential oils) from sand-dune plants induce cytotoxic effects in cancer cells. J Ethnopharmacol 258:112803. https://doi.org/10.1016/j.jep.2020.112803

Carris LM, Little CR, Stiles CM (2012) Introduction to fungi. The plant health instructor. The American Phytopathological Society. https://doi.org/10.1094/PHI-I-2012-0426-01

Ceresini PC, Castroagudín VL, Rodrigues FÁ, Rios JA, Aucique-Pérez CE, Moreira SI, Alves E, Croll D, Maciel JLN (2018) Wheat blast: past, present, and future. Annu Rev Phytopathol 56(1):427–456. https://doi.org/10.1146/annurev-phyto-080417-050036

Condon BJ, Leng Y, Wu D, Bushley KE, Ohm RA, Otillar R, Martin J, Schackwitz W, Grimwood J, MohdZainudin N, Xue C, Wang R, Manning VA, Dhillon B, Tu ZJ, Steffenson BJ, Salamov A, Sun H, Lowry S, LaButti K, Han J, Copeland A, Lindquist E, Barry K, Schmutz J, Baker SE, Ciuffetti LM, Grigoriev IV, Zhong S, Turgeon BG (2013) Comparative genome structure, secondary metabolite, and effector coding capacity across Cochliobolus pathogens. PLoS Genet 9(1):e1003233. https://doi.org/10.1371/journal.pgen.1003233

Croteau R, Kutchan TM, Lewis NG (2015) Natural products (secondary metabolites). In: Buchanan B, Gruissem W, Jones R (eds) Biochemistry & molecular biology of plants. American Society of Plants, Rockville, pp 1250–1318

Daayf F, El Hadrami A, El-Bebany AF, Henriquez MA, Yao Z, Derksen H, El-Hadrami I, Adam LR (2012) Phenolic compounds in plant defense and pathogen counter-defense mechanisms. In: Recent advances in polyphenol research, vol. 3, pp 191–208. https://doi.org/10.1002/9781118299753.ch8

Dai GH, Andary C, Cosson-Mondolot L, Boubals D (1994) Polyphenols and resistance of grapevines to downy mildew. In: International Society for Horticultural Science (ISHS), Leuven, Belgium, pp 763–766. https://doi.org/10.17660/ActaHortic.1994.381.110

Dehgahi R, Subramaniam S, Zakaria L, Joniyas A, Beiki F, Haghnama K, Razinataj M (2016) Review of research on fungal pathogen attack and plant defense mechanism against pathogen. Int J Sci Res Agric Sci 2:197–208. https://doi.org/10.12983/ijsras-2015-p0197-0208

Doehlemann G, Wahl R, Horst RJ, Voll LM, Usadel B, Poree F, Stitt M, Pons-Kühnemann J, Sonnewald U, Kahmann R (2008) Reprogramming a maize plant: transcriptional and metabolic changes induced by the fungal biotroph Ustilago maydis. Plant J 56(2):181–195

Fischer UA, Carle R, Kammerer DR (2011) Identification and quantification of phenolic compounds from pomegranate (Punica granatum L.) peel, mesocarp, aril and differently produced juices by HPLC-DAD–ESI/MSn. Food Chem 127(2):807–821. https://doi.org/10.1016/j.foodchem.2010.12.156

Ganugi P, Martinelli E, Lucini L (2021) Microbial biostimulants as a sustainable approach to improve the functional quality in plant-based foods: a review. Curr Opin Food Sci 41:217–223. https://doi.org/10.1016/j.cofs.2021.05.001

Gutiérrez-Grijalva EP, Picos-Salas MA, Leyva-López N, Criollo-Mendoza MS, Vazquez-Olivo G, Heredia JB (2018) Flavonoids and phenolic acids from oregano: occurrence. Biol Activity Health Benefits Plants 7(1):2. https://doi.org/10.3390/plants7010002

Hassan O, Chang T, Hossain A (2020) Changes in the secondary compounds of persimmon leaves as a defense against circular leaf spot caused by Plurivorosphaerella nawae. PLoS One 15(3): e0230286. https://doi.org/10.1371/journal.pone.0230286

Khare S, Singh NB, Singh A, Hussain I, Niharika K, Yadav V, Bano C, Yadav RK, Amist N (2020) Plant secondary metabolites synthesis and their regulations under biotic and abiotic constraints. J Plant Biol 63(3):203–216. https://doi.org/10.1007/s12374-020-09245-7

Kong Q, An P, Xu Z, Zhang R, Qi J, Ren X (2020) New insights into the alleviating role of Melaleuca alternifolia oil on metabolites pathway disorder of grapes caused by Aspergillus niger, verified by corresponding key genes expression. Food Chem 327:127083. https://doi.org/10.1016/j.foodchem.2020.127083

Landa BB, Perez AG, Luaces P, Montes-Borrego M, Navas-Cortes JA, Sanz C (2019) Insights into the effect of Verticillium dahliae defoliating-pathotype infection on the content of phenolic and volatile compounds related to the sensory properties of virgin olive oil. Front Plant Sci 10:12. https://doi.org/10.3389/fpls.2019.00232

Lanoue A, Burlat V, Henkes GJ, Koch I, Schurr U, Röse US (2010) De novo biosynthesis of defense root exudates in response to Fusarium attack in barley. New Phytol 185(2):577–588

Lavania M, Chauhan PS, Chauhan S, Singh HB, Nautiyal CS (2006) Induction of plant defense enzymes and phenolics by treatment with plant growth–promoting rhizobacteria Serratia marcescens NBRI1213. Curr Microbiol 52(5):363–368

Mahajan S, Tuteja N (2005) Cold, salinity and drought stresses: an overview. Arch Biochem Biophys 444(2):139–158

Martini S, D'Addario C, Colacevich A, Focardi S, Borghini F, Santucci A, Figura N, Rossi C (2009) Antimicrobial activity against Helicobacter pylori strains and antioxidant properties of blackberry leaves (Rubus ulmifolius) and isolated compounds. Int J Antimicrob Agents 34(1): 50–59. https://doi.org/10.1016/j.ijantimicag.2009.01.010

McAllister TA, Bae HD, Yanke LJ, Cheng KJ, Muir A (1994) Effect of condensed tannins from birdsfoot trefoil on endoglucanase activity and the digestion of cellulose filter paper by ruminal fungi. Can J Microbiol 40(4):298–305. https://doi.org/10.1139/m94-048

Michielse CB, Reijnen L, Olivain C, Alabouvette C, Rep M (2012) Degradation of aromatic compounds through the β-ketoadipate pathway is required for pathogenicity of the tomato wilt pathogen Fusarium oxysporum f. sp. lycopersici. Mol Plant Pathol 13(9):1089–1100

Mikulic-Petkovsek M, Schmitzer V, Stampar F, Veberic R, Koron D (2014) Changes in phenolic content induced by infection with Didymella applanata and Leptosphaeria coniothyrium, the causal agents of raspberry spur and cane blight. Plant Pathol 63(1):185–192. https://doi.org/10.1111/ppa.12081

Mincuzzi A, Ippolito A, Brighenti V, Marchetti L, Benvenuti S, Ligorio A, Pellati F, Sanzani SM (2020) The effect of polyphenols on pomegranate fruit susceptibility to Pilidiella granati provides insights into disease tolerance mechanisms. Molecules 25(3):515. https://doi.org/10.3390/molecules25030515

Mithöfer A, Boland W (2012) Plant defense against herbivores: chemical aspects. Annu Rev Plant Biol 63:431–450

Olufolaji DB, Ajayi AM (2019) Management of plant biotic stress with botanicals and antagonistic fungi in the tropics. In: Ansari RA, Mahmood I (eds) Plant health under biotic stress, vol 1. Organic strategies. Springer Singapore, Singapore, pp 59–72. https://doi.org/10.1007/978-981-13-6043-5_3

Pandian BA, Sathishraj R, Djanaguiraman M, Prasad PVV, Jugulam M (2020) Role of cytochrome P450 enzymes in plant stress response. Antioxidants (Basel) 9(5):454. https://doi.org/10.3390/antiox9050454

Ponts N, Pinson-Gadais L, Boutigny A-L, Barreau C, Richard-Forget F (2011) Cinnamic-derived acids significantly affect Fusarium graminearum growth and in vitro synthesis of type B trichothecenes. Phytopathology 101(8):929–934

Ray M, Ray A, Dash S, Mishra A, Achary KG, Nayak S, Singh S (2017) Fungal disease detection in plants: traditional assays, novel diagnostic techniques, and biosensors. Biosens Bioelectron 87: 708–723. https://doi.org/10.1016/j.bios.2016.09.032

Rejeb IB, Pastor V, Mauch-Mani B (2014) Plant responses to simultaneous biotic and abiotic stress: molecular mechanisms. Plants 3(4):458–475

Rodrigues MHP, Furlong EB (2022) Fungal diseases and natural defense mechanisms of tomatoes (Solanum lycopersicum): a review. Physiol Mol Plant Pathol 122:101906. https://doi.org/10.1016/j.pmpp.2022.101906

Rodrigues FÁ, Rios JA, Debona D, Aucique-Pérez CE (2017) Pyricularia oryzae-wheat interaction: physiological changes and disease management using mineral nutrition and fungicides. Trop Plant Pathol 42(3):223–229. https://doi.org/10.1007/s40858-017-0130-z

Rusjan D, Persic M, Likar M, Biniari K, Mikulic-Petkovsek M (2017) Phenolic responses to esca-associated fungi in differently decayed grapevine woods from different trunk parts of 'Cabernet Sauvignon'. J Agric Food Chem 65(31):6615–6624. https://doi.org/10.1021/acs.jafc.7b02188

Sankaran S, Mishra A, Ehsani R, Davis C (2010) A review of advanced techniques for detecting plant diseases. Comput Electron Agric 72(1):1–13

Schwalb P, Feucht W (1999) Changes in the concentration of phenolic substances in the bark during the annual development of the cherry tree (Prunus avium L.). Adv Hortic Sci 13(2):71–75

Seepe HA, Amoo SO, Nxumalo W, Adeleke RA (2020) Sustainable use of thirteen South African medicinal plants for the management of crop diseases caused by Fusarium species—An in vitro study. S Afr J Bot 130:456–464. https://doi.org/10.1016/j.sajb.2020.01.038

Sepahvand A, Ezatpour B, Tarkhan F, Bahmani M, Khonsari A, Rafieian-Kopaei M (2017) Phytotherapy in fungi and fungal disease: a review of effective medicinal plants on important fungal strains and diseases. Int J Pharm Sci Res 8(11):4473–4495. https://doi.org/10.13040/ijpsr.0975-8232.8(11).4473-95

Shalaby S, Larkov O, Lamdan NL, Goldshmidt-Tran O, Horwitz BA (2016) Plant phenolic acids induce programmed cell death of a fungal pathogen: MAPK signaling and survival of Cochliobolus heterostrophus. Environ Microbiol 18(11):4188–4199. https://doi.org/10.1111/1462-2920.13528

Sharma N, Sharma KP, Gaur RK, Gupta VK (2011) Role of chitinase in plant defense. Asian J Biochem 6(1):29–37. https://doi.org/10.3923/ajb.2011.29.37

Shen Y, Liu N, Li C, Wang X, Xu X, Chen W, Xing G, Zheng W (2017) The early response during the interaction of fungal phytopathogen and host plant. Open Biol 7(5):170057. https://doi.org/10.1098/rsob.170057

Shternshis MV, Beljaev AA, Shpatova TV, Duzhak AB, Panfilova ZI (2006) The effect of chitinase on Didymella applanata, the causal agent of raspberry cane spur light. BioControl 51(3):311–322. https://doi.org/10.1007/s10526-005-1034-2

Silva ET, Rios JA, Araujo MUP, Silveira PR, Rodrigues FA (2019) Defence responses in flag leaves and spikes of common wheat Triticum aestivum cultivars with contrasting levels of basal resistance to blast caused by Pyricularia oryzae. Plant Pathol 68(4):645–658. https://doi.org/10.1111/ppa.12979

Simaan H, Lev S, Horwitz BA (2019) Oxidant-sensing pathways in the responses of fungal pathogens to chemical stress signals. Front Microbiol 10:567. https://doi.org/10.3389/fmicb.2019.00567

Simaan H, Shalaby S, Hatoel M, Karinski O, Goldshmidt-Tran O, Horwitz BA (2020) The AP-1-like transcription factor ChAP1 balances tolerance and cell death in the response of the maize pathogen Cochliobolus heterostrophus to a plant phenolic. Curr Genet 66(1):187–203. https://doi.org/10.1007/s00294-019-01012-7

Sung WS, Lee DG (2010) Antifungal action of chlorogenic acid against pathogenic fungi, mediated by membrane disruption. J Pure Appl Chem 82(1):219. https://doi.org/10.1351/PAC-CON-09-01-08

Truong T, Dinh A, Wahid K (2017) An IoT environmental data collection system for fungal detection in crop fields. In: 2017 IEEE 30th Canadian conference on electrical and computer engineering (CCECE), 30 April–3 May 2017, pp 1–4. https://doi.org/10.1109/CCECE.2017.7946787

Vagiri M, Johansson E, Rumpunen K (2017) Phenolic compounds in black currant leaves—an interaction between the plant and foliar diseases? J Plant Interact 12(1):193–199. https://doi.org/10.1080/17429145.2017.1316524

Vuolo MM, Lima VS, Junior MRM (2019) Phenolic compounds: structure, classification, and antioxidant power. In: Bioactive compounds. Elsevier, pp 33–50

Williamson B, Jennings DL (1992) Resistance to cane and foliar diseases in red raspberry (Rubus idaeus) and related species. In: Johnson R, Jellis GJ (eds) Breeding for disease resistance. Springer, Dordrecht, pp 59–70. https://doi.org/10.1007/978-94-017-0954-5_5

Xu XM, Robinson JD, Berrie AM (2009) Infection of blackcurrant leaves by Drepanopeziza ribis in relation to weather conditions and leaf position. J Phytopathol 157(5):280–286. https://doi.org/10.1111/j.1439-0434.2008.01488.x

Zamljen T, Medič A, Veberič R, Hudina M, Štampar F, Slatnar A (2021) Apple fruit (Malus domestica Borkh.) metabolic response to infestation by invasive Brown Marmorated stink bug (Halyomorpha halys Stal.). Horticulturae 7(8):212

Zeilinger S, Gupta VK, Dahms T, Silva R, Singh H, Upadhyay R, Gomes E, Tsui C, S CN (2015) Friends or foes? Emerging insights from fungal interactions with plants. FEMS Microbiol Rev 40:fuv045. https://doi.org/10.1093/femsre/fuv045

Zhu J-K (2016) Abiotic stress signaling and responses in plants. Cell 167(2):313–324. https://doi.org/10.1016/j.cell.2016.08.029

Fungal Control Through Plant Phenolics: A Biotic Constraint

14

Sagnik Nag, Rafiq Lone, Mahima Praharaju, Prattusha Khan, and Arsalan Hussain

Abstract

Agriculture is a prime necessity at this given point in time. Fifty percent of global activities revolve around agriculture. Studies from the Food and Agriculture Organization (FAO) show that 20–40% of crops are destroyed globally due to pests. They are known to cause a major loss to the economy leading to global poverty. Fungal pestilence among them is a burning issue as 85% of plant diseases are caused by fungi. They are a source of biotic stress in plants. Biotic stress includes any live flora or fauna that is known to be pathogenic to plants. To counteract the effects of such pathogenic fungi, plants tend to neutralize the pestilence effect through plant secondary metabolites or polyphenols. Polyphenols or phenolic substances can be explained as the naturally occurring chemical compounds in plants that succor them from biotic and abiotic stress. They have formed from the amalgamation of shikimate; phenylpropanoids; and flavonoid pathways. These derivatives can nurture and endorse many developmental aspects of the plant uniquely counteracting the stress constraints

S. Nag (✉)
Department of Bio-Sciences, School of Bio-Sciences & Technology (SBST), Vellore Institute of Technology (VIT), Vellore, Tamil Nadu, India

R. Lone
Department of Botany, Central University of Kashmir, Ganderbal, Jammu and Kashmir, India

M. Praharaju
Department of Environmental Sciences, IGNOU University, Hyderabad, India

P. Khan
Department of Microbiology, St. Xavier's College, Kolkata, West Bengal, India

A. Hussain
Department of Life Science, Presidency University, Calcutta University, Kolkata, West Bengal, India

R. Lone et al. (eds.), *Plant Phenolics in Biotic Stress Management*,
https://doi.org/10.1007/978-981-99-3334-1_14

simultaneously. Most polyphenols play a key role as antioxidants against oxidative stress inducers. Hampering the biotic stress restraints using these secondary phytochemicals has been a major milestone in the respective field of plant biology. In this book chapter, we weighed up the pros and cons of the thwarting response of plant-derived polyphenols to the fungal stress constraints, furthermore providing insight into the biosynthesis and biochemistry of the phenolic compounds, their counteract mechanism against fungal toxins, and the aftermath of fungal pestilence on crops and *in vitro* in a long run.

Keywords

Agriculture · Biotic stress · Fungal pestilence · Polyphenols

14.1 Introduction

Being sessile, plants are prone to a diverse range of stress conditions which causes adverse effects on the growth of the plants, limiting development and leading to reduced productivity (Mittler and Blumwald 2010). These environmental stresses are broadly subdivided into biotic (fungal, bacterial, viral, insects, herbivores, aphids, etc.) and abiotic stresses (temperature, pH, salinity, light intensity, availability of nutrients, etc.) (Bailey-Serres and Voesenek 2008).

There are several pathogens, pests, and parasites that are responsible for causing plant infections and instigating biotic stress. Nematodes, for example, feed on various plant parts and are the primary causative agent of soil-borne plant diseases which leads to wilting, stunted growth, and nutrient deficiency (Bekal et al. 2011). Similarly, viruses lead to local as well as systemic damage to plants which results in stunted growth and chlorosis (Pallas and García 2011). Insects and mites cause impairment of plants by their feeding habits (like sucking, piercing) and they might as well play the role of carrier organism for bacteria and viruses. However, phytopathogenic strains of fungi are regarded as a major biotic constraint that substantially bestows massive loss in plant yield (Rashad et al. 2020). It is estimated that around 10–20% of crops (both food and cash) are damaged by the activity of phytopathogenic fungi (Hewitt 2000). Several generic fungi are known to degrade the quality of yield in respect of shelf life, nutritional value, and organoleptic characteristics. Fungal pathogens induce leaf spots, cankers, and vascular wilts (Sobiczewski et al. 2017). To combat the various stress factors, plants have evolved a complex immune system and have physical barriers like trichomes, wax, and cuticles to avert pests and pathogens (Saijo and Loo 2020). Moreover, plants being bound to the ground via root systems are unable to escape their stressors and hence have to protect themselves by staying at their position (Osbourn et al. 2003). This gives rise to the need to produce chemical compounds that would kill or deter pathogens and pests. In agriculture and for commercial production of fungal pathogen-free crops, mostly synthetic fungicides have been used as a control strategy against these fungi but they are found to be less efficacious and are accompanied by harmful environmental impacts (Cabanillas-Bojórquez et al. 2020; Paulitz and Bélanger 2001). In addition to this, the heightened demand for environment-friendly alternate agricultural strategies has encouraged research on the use of plant secondary metabolites as antifungal agents (Fiori et al. 2000). Secondary metabolites produced by plants have a diverse span of biological activity and are a very important defense mechanism of plants against several pathogenic fungi (Einhellig 1994). These secondary metabolites are nontoxic, environment friendly, and biodegradable and have the potential of being used by the farmers as an agrochemical for pest management (Kim et al. 2004) Secondary metabolites are nonessential for plants to sustain life but play crucial roles in survival of the producers (Hadacek 2002). Studies show, in comparison to animals, plants are known to produce a greater variety of secondary metabolites as they are incapable of using physical mobility as a means of escape from the predators which lead them to evolve chemical defense strategy for self-protection. Polyphenols are a major class of secondary metabolites produced by plants. These phenolic compounds are needed by plants for growth, pigmentation, reproduction, and for providing resistance against pathogens (Whiting 2001). Secondary metabolites not only provide defense against biotic stressors like fungi, viruses, other microbes, and herbivores but also play a vital role in the protection of plants from oxidants and ultraviolet radiation (Kutchan 2001).

However, the conditions in the field vary from those in the laboratory because under natural field conditions more than one stressor act in combination, with a heightened negative impact on the productivity of the crop. Changes in climatic conditions are known to play a pivotal role in influencing the natural habitat of plant pathogens (fungi, bacteria, nematodes, and viruses) and pests (Gautam et al. 2020). Studies suggest that the plant defense mechanism is weakened by abiotic stresses which enhance plant susceptibility toward pathogens (Atkinson and Urwin 2012). For example, cold stress impairs the silencing of genes which is a potent defense mechanism of plants against viral pathogens (Szittya et al. 2003). Simultaneous occurrence of these stresses leads to the higher complexity of the plant responses which are controlled by several signaling pathways which might interact and even cause inhibition of each other (Prasch and Sonnewald 2013). On sensing stress, plants trigger their various cellular and molecular processes (Lamers et al. 2020). For the identification and overcoming of biotic invasions, plants have evolved complicated sensory mechanisms (Lamers et al. 2020). In this book chapter, we will be discussing biotic stress in detail, specifically fungi as abiotic constraint, and how phenolics produced by plants can provide resistance against fungal diseases.

14.1.1 Biotic Stress and Fungal Pestilence

An adverse condition that hampers the normal growth and development of a plant is a state of stress to them and the conditions can be either abiotic, which is environmental factors, or biotic, which is caused by plant pathogens (Mahalingam 2015; Pandey et al. 2015). The plant pathogens which grow either inside the plant tissue or on it are bacteria, fungi, viruses, nematodes, etc., and cause diseases like rotting,

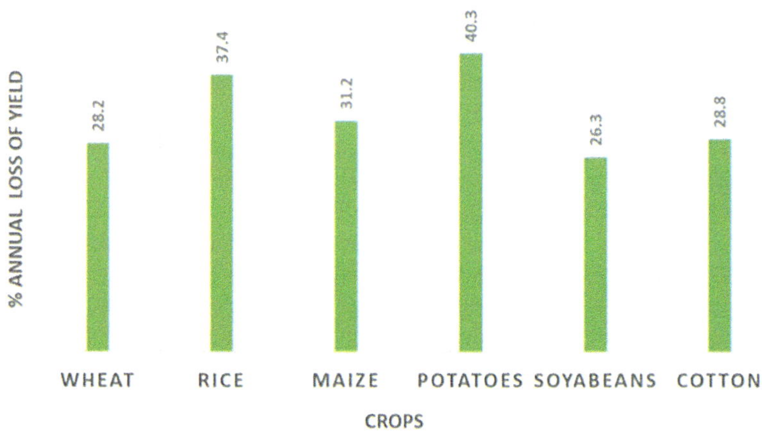

Fig. 14.1 Percent loss in crop yields due to biotic stresses

chlorosis, lesions, and stunting. The combined effect of these pathogens can cause severe disease to the plants and is very commonly seen rather than the effect of an individual pathogen causing the disease to the plant (Pandey et al. 2017). The biotic factors lead to a decrease in photosynthesis by the crops and thus a huge loss in the yield of the crops. According to a report from 2013, the yield loss of different crops due to biotic factors is shown in Fig. 14.1 (Wang et al. 2013). The abiotic factors that are mostly responsible for the loss of plant growth and yield are temperature fluctuations, humidity, salinity, and drought and also influence the biotic factors in the spread of the disease (Peters et al. 2014; Tomar et al. 2023). If all of these crops get affected together, then about 61% of the population would be without food (Fisher et al. 2012).

Fungal pathogens are the main source of causing serious diseases to plants and crops, among other biotic factors (Doehlemann et al. 2017). Among all the species, there are reports of 8000 species of fungi and oomycetes that cause disease to the plants (Fisher et al. 2020). These diseases lead to a significant loss in the quantity, yield, and quality of various agricultural systems that are economically important (Godfray et al. 2016; Shuping and Eloff 2017; Rodriguez-Moreno et al. 2018; Asibi et al. 2019). Fungal infections are estimated to destroy 125 million tons of food crops (wheat, rice, potato, maize, and soybean). The estimated global loss is around 60 billion US dollars (Fisher et al. 2012). Seventy percent to 80% of the loss in the world agricultural population is due to phytopathogenic fungi (Moore 2020). The frequently occurring diseases caused by phytopathogenic fungi are a blight, wilt, scab, leaf spot or necrosis, root rot, pustules, canker, anthracnose, gall, powdery mildew, and gall (Iqbal et al. 2018; Hussain and Usman 2019; Jain et al. 2019). The fungi that are responsible for the loss of yield of rice and wheat upon which the economy of several countries depend is *Magnaporthe oryzae*. The next important fungi for the economic devastation are *Botrytis cinerea* which has a broad host range including wheat and *Puccinia* sp. (Dean et al. 2012). Other important crops that get affected by fungal pathogens are mangoes, bananas, nuts, spices, coffee, cocoa, etc. (Drenth and Guest 2016).

The phytopathogenic fungi produce cell wall degrading enzymes (CWDEs) which help them to invade the plant cell walls and cause infection. In turn, they derive nutrition from the plants for their growth and further infection. The fungi can be categorized into three types based on the way by which they derive nutrition. They are—the biotrophs, which derive nutrition from living plant tissue, the necrotrophs, which derive nutrition by killing the plant cells and the hemibiotrophic, which initially depend on the living plant cells for their nutrition but later derive nutrition by degrading the plant cells. The four major enzymes that help in the degradation of the main constituents of the plant cell wall are—cellulose (endoglucanase, cellobiohydrolase, and β-glucosidase), and pectinase, amylase, and laccase (Gurjar and Kanade 2020). A few more enzymes released by the phytopathogenic fungi that aid in the degradation of the plant cells and contribute to their virulence are lipase, xylanase, and cutinase. All of the hydrolytic enzymes together play a role in the host cell pathogenesis (Ramos et al. 2016).

The plant pathogens also secrete host selective toxins which are secondary metabolites of low molecular weight that are responsible for the pathogenesis, but as the name suggests they are specific to particular host plants and the total number of host selective toxins are also very low in number. There are also some nonspecific toxins released by the phytopathogenic fungus which help them in their invasion and infection. Toxins make the plant cell membrane more permeable, leading to easier leakage of the cell electrolytes and damage to the metabolism, leading to its loss of functions and cell death (Lyu et al. 2015).

The important and the most predominant phytopathogenic fungi are *Magnaporthe oryzae, Botrytis cinerea, Fusarium oxysporum, Fusarium graminearum, Ustilago maydis, Melampsora lini, Rhizoctonia solani, Blumeria graminis, Mycosphaerella graminicola,* and *Phakopsora pachyrhizi* (Nazarov et al. 2020). The worldwide trend of emerging fungal diseases shows that plant infection is predominant all over the world and the extinction of species leading to plant diseases is also lesser as compared to the animal disease-causing fungal pathogens (Fisher et al. 2012). *Botrytis cinerea* and *Sclerotinia sclerotiorum* are the necrotrophic plant pathogens that have the most devastating effect and have a broad host range and *Magnaporthe oryzae* is a hemibiotrophic ascomycete that is responsible for the loss of the staple food crop of India, rice and is responsible for 30% loss of its yield (Talbot 2003). *Bipolaris sorokiniana,* a hemibiotrophic plant pathogen causing the spot blotch disease in wheat, is also responsible for a huge annual loss in the yield of wheat (Aich et al. 2017).

Biotic stress imposes a great threat to the nutritional status of people all around the world; thus, proper management to overcome and control biotic and abiotic stress is an absolute necessity. The use of pesticides and fungicides (both chemical and biological control agents) is the most common practice to fight biotic stress but they have their drawbacks (Majeed et al. 2018; Monteiro et al. 2022). Genetic engineering of the crops has shown much better results in this aspect. But before this, the plant activates its defense mechanism to fight against any of these infections. The first line of defense mechanism starts with the pattern recognition receptors or the PRRs. The receptors identify the pathogen-associated molecules or the PAMPs and activate the defense mechanism against those molecules. One such hypersensitive response is the production of phenolic compounds which helps the plant fight against various stresses (Thakur et al. 2019; Abdul Malik et al. 2020). The phenolic compounds benefit the plants in both ways; as they act as an attractant to various organisms that are beneficial to the plants and act as repellents to harmful pathogens (Pratyusha 2022). Thus, a detailed study of these plant phenolics will help in understanding the plant defense mechanism against fungal pathogens.

14.1.2 Plant-Based Polyphenol Biosynthesis and Biochemistry

Polyphenols are groups of natural or synthetic organic chemicals whose basic structural composition is the presence of an aromatic ring with a hydroxyl (OH) group attached to it. They have a wide distribution with more than 8000 known structures and can be either a single molecule or a complex polymeric structure such as tannins (Evans 1991). Polyphenolic compounds have several bioactive properties which have beneficial roles as shown in Fig. 14.2. Most of the phenolic compounds are plant-based; however, synthetic analogs are now also being synthesized and are extensively used in the cosmetics, food and beverages, and plastic industries.

Plants are the most abundant source of phenolic compounds which they use for their growth and development. They also use these secondary metabolites to fight against several abiotic and biotic stresses in their ways. Examples of plants that are rich in phenolic compounds are tea and coffee, nuts, berries, flax seeds, herbs, vegetables, spices, olives, cocoa powder, etc. (Velderrain-Rodríguez et al. 2014; Lin et al. 2016). The most common form of plant-based polyphenols is condensed tannins. The major classification of plant phenolic compounds is shown in Fig. 14.3.

Synthetic polyphenols have made their position in industrial applications, especially in the food industries. The food products can easily get oxidized due to many environmental factors which decrease the shelf life and quality of the food. The use of synthetic polyphenols as antioxidants in such cases can help increase the shelf life of the food without compromising the nutritional benefits. Synthetic polyphenols are easily available, highly stable, and cost less compared to natural ones (Saad et al.

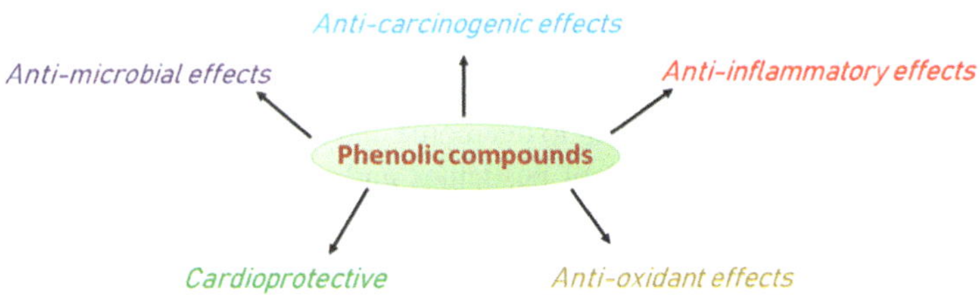

Fig. 14.2 Bioactive properties of phenolic compounds

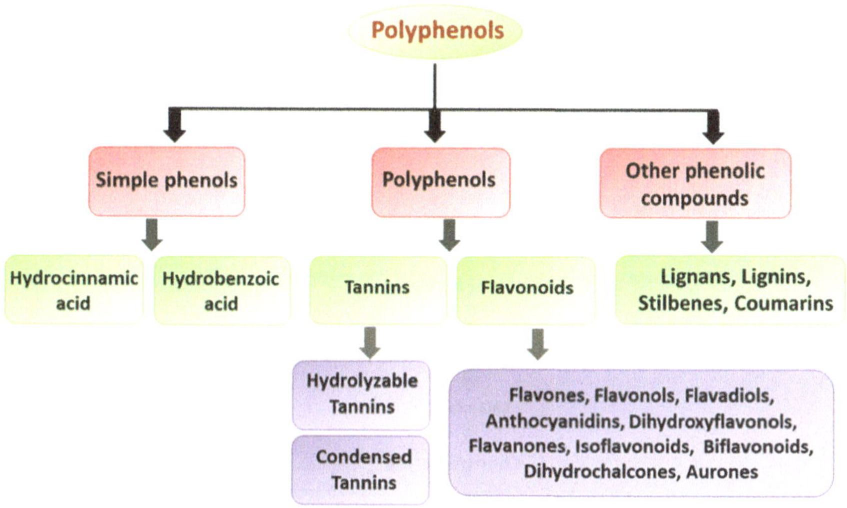

Fig. 14.3 Classification of plant phenolic compounds

2007; Xiu-Qin et al. 2009; Mishra et al. 2023). Examples of the commonly used synthetic polyphenols are Propyl Gallate (PG), 4-Phenylphenol (OPP), Butylated Hydroxyanisole (BHA), 2-Naphthol (2-NL), Tert-Butyl Hydroquinone (TBHQ), 2,4-Dichlorophenoxyacetic acid (2,4-DA), and Butylated Hydroxytoluene (BHT) (Xiu-Qin et al. 2009). But the synthetic polyphenols pose several health issues like increased risk of cancer, liver problems, skin allergies, gastrointestinal problems, and DNA damage (Botterweck et al. 2000; Randhawa and Bahna 2009; Engin et al. 2011; Kornienko et al. 2019).

The chemical constituents of different types of phenols are the following:

Phenolic acids—All phenolic compounds are constituted of a basic aromatic ring with an attached hydroxyl group (–OH). These compounds have one or multiple aromatic side chains, with one or more hydroxyl groups and have structures varying from simple molecules to complex polymers. Phenolic acids can be divided into two major categories, benzoic acid or cinnamic acid derivatives based on their C1–C6 and C3–C6 backbones, respectively.

Flavonoids—Flavonoid structures are based upon 15-carbon skeletons with two benzene rings (A and B) linked via a heterocyclic pyrene ring in the middle. Flavonoids are further classified depending on the distribution of –OH groups and glycosylation and prenylation into flavanones, flavonols, flavones, isoflavones, flavan-3-ols, anthocyanidins, etc.

Tannins—These are the most complex of the polyphenols with molecular weight ranging from 0.5 to 3 kDa. These are complex units of n number of C6–C3–C6. They are of two types—hydrolyzable, which has carbohydrate units at the center, and condensed, which has flavan-3-of units at the center (Tresserra-Rimbau et al. 2018; Kumar and Goel 2019).

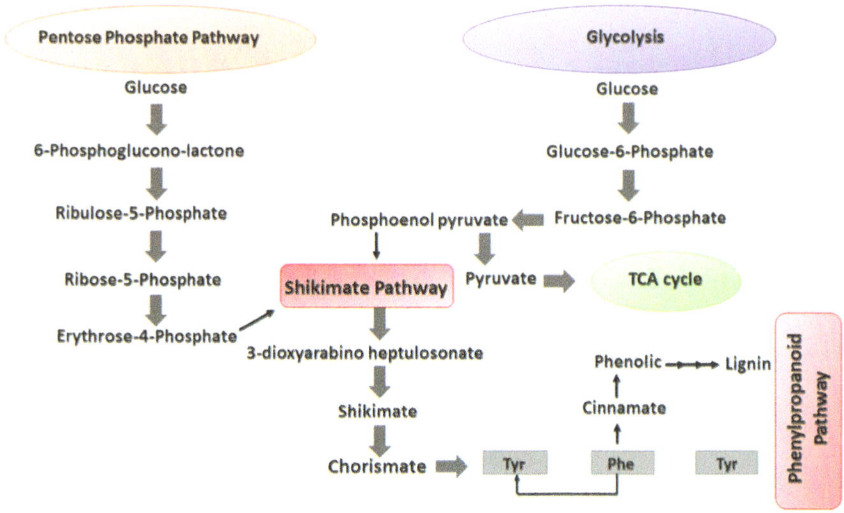

Fig. 14.4 Polyphenol biosynthesis

Polyphenol biosynthesis includes two metabolic pathways—Shikimic acid and Phenylpropanoid pathways (Wang et al. 2019). The biosynthesis of the phenolic compounds is shown in a schematic in Fig. 14.4 (Vattem et al. 2005; Lin et al. 2010).

Few of the identified polyphenols from the plant source and the genes involved in their synthesis are tabulated in Table 14.1.

The different types of phenolic compounds have distinct ways of acting as antifungal agents.

(a) Simple phenols—At infection, the antifungal molecules start accumulating and the oxidative enzymes also increase, leading to the oxidation of phenols to quinones. This leads to oxidative stress on the pathogens and disrupts their metabolism and causes DNA damage which leads to cell death. More is the complexity of the structure, more is the number of –OH groups and thus more is the oxidative stress of the pathogens (Shabana et al. 2008; Daayf et al. 2012).
(b) Phenolic acids—The plants defend themselves by accumulating signal molecules on being attacked by a pathogen. Salicylic acid is a phenolic acid which is one such signaling molecule that activates the plants' defense mechanism. Moreover, the biotic stress leads to increased synthesis of cinnamic acid and benzoic acid which further strengthens the cell wall (de Ascensao and Dubery 2003; Mandal et al. 2010).
(c) Flavonoids—The toxic effects of flavonoids on plant pathogens include disruption of the cell wall, damage to the cytoplasmic membrane, induction of cell death, enzymatic inhibition, metal ions chelation, and/or soluble or extracellular protein binding (Mierziak et al. 2014a, b).

Table 14.1 Common polyphenols from plant sources

Polyphenol	Plant source	Function	Characterization	References
Maslinic acid	Olives	Helps in weight loss	LC-MS	Guodong et al. (2019)
Proanthocyanidins	Grapes	Cardioprotective, antioxidant, anticancer	HPLC-MS	Xia et al. (2010)
Catechins	Tea	Antidiabetic, anticarcinogenic, antioxidant	MALDI, EI, iTRAQ	Li et al. (2022)
Hydrocinammic acid, gallic acid	Coffee	Diminishing risk of Alzheimer's and Parkinson's, type II diabetes, and liver disease	Folin-Ciocalteu	Dybkowska et al. (2017)
Tannin and catechin	Cocoa beans	Decrease in LDL oxidation, improved endothelial dysfunction	RP LC-MS	Natsume et al. (2000)
Tannins, lignans, naphthoquinones, stilbenes	Nuts	Antiproliferative, antiviral, anti-inflammatory, hypocholesterolemic	Folin-Ciocalteu	Bolling et al. (2010)
Gallic acid, ellagic acid, punicalin A and B	Pomegranate	Antioxidant, anti-inflammatory, therapeutic agent for cancer	HPLC-DAD	Li et al. (2016)
Phenolic acid, flavonoids, lignans, stilbenes	Herbs and spices	Neuroprotective, anti-inflammatory, anticancer, antiasthmatic, antimicrobial	HPLC	Opara and Chohan (2014)

(d) Lignin—Lignin acts as a physical barrier, as thickening of the cell wall has been reported in pathogen-resistant crops. Lignin suppresses fungal invasion, and also diffusion of their toxins is repressed (Xie et al. 2018).
(e) Tannins—The main antifungal action of tannins is inhibition of the extracellular enzymes (laccase, cellulase, xylanase, pectinase, etc.) and inhibition of oxidative phosphorylation (Ogawa and Yazaki 2018).

Polyphenols are the largest group of plant metabolites that help plants fight against abiotic and biotic stress. The molecules show different effects on plant growth and development according to their structure, and therefore, it is important to know how these polyphenols help to fight against these adverse conditions. A better understanding of these metabolites further helps in biotechnological applications toward genetically modified pathogen-resistant crops (Tuladhar et al. 2021).

14.2 Fungi as a Biotic Constraint

Fungi are known to cause a manifold of plant diseases as compared to any other plant pests (Williams and Myers 1984). Fungi are a eukaryotic group of versatile heterotrophs that are found in most natural habitats. Phytopathogenic groups of fungi are the major cause of the significant loss of crop yield and hence are regarded as a serious economic hurdle among plant breeders, farmers, and scientists (Hewitt 2000). Fungal pathogens have evolved several strategies for invading and colonizing plant tissue, optimal growth, and propagation in plants. Fungal parasites mainly invade via natural openings like cuts or wounds. However, many phytopathogenic groups of fungi have developed mechanisms of actively traversing structural barriers like cuticles by secretion of several hydrolytic enzymes which include cellulose, cutinase, protease, and pectinase. In living plants, aerial parts are covered by cuticle which has to undergo piercing to set up effective pathogenic mechanisms (Ryan et al. 1995). Thus, cutin degradation by enzymes is considered crucial in the penetration and pathogenesis of fungi. It has been studied that during the prepenetration process, cutinase can alter the cuticle's adhesive property and hence facilitate the attachment of fungi to the surface of the plant and can also release signal molecules that are necessary for the early development of fungi on plants (Nicholson and Epstein 1991). In this section, we have discussed some of the major fungal pathogens to which plants are prone to infection and how they are a cause of concern worldwide.

The most important of all is *Magnaporthe oryzae* which is a filamentous ascomycetes group of fungi that is the cause of one of the major diseases of rice in the world called the rice blast disease (Ou 1980). A huge population of the world consumes rice as a part of their staple diet which highlights the exact agronomic significance of this crop (Khush 2005). *M. oryzae* develops appressorium which is a specialized organ for penetration and is necessary for infection development (Howard and Valent 1996). Another significant fungus, *Botrytis cinerea, also* called the grey mold is a typical necrotroph that is capable of infecting greater than 200 species of plants (Van Baarlen et al. 2007). The cost of damage caused by *Botrytis* is difficult to estimate due to the occurrence of damage at different production stages (Billard et al. 2012). Sometimes this fungus stays in the quiescent stage for a considerable amount of time before it causes the rotting of the plant tissues during favorable environmental conditions (Williamson et al. 2007). *Puccinia* sp. is another group of significant fungi that is responsible for causing rust disease in wheat plants—black and brown rust of stems and leaves, respectively (Chen 2005). The most damaging and notorious of these is stem rust (Jin et al. 2010). They are a basidiomycetes group of fungi that are obligately biotrophic (Jin et al. 2010). *Puccinia* sp. is known to form haustoria which are specialized feeding structures through which they efficiently obtain nutrients from plant cells (Voegele and Mendgen 2011).They also possess specialized structures for infection via which they suppress the defense mechanisms of the host. In 1935, black stem rust of wheat by *Puccinia* sp. destroyed around 60% of the total wheat crop in South Dakota and Minnesota (Peterson 2018). *Fusarium graminearum* is an ascomycetes group of fungi that is an immensely destructive pathogen that affects all cereal species causing head blight disease (Keller et al. 2014). This disease causes a reduction in the quality of grain rather than a reduction in yield and gives rise to mycotoxin-contaminated grain which is unsafe for consumption by humans and animals (Simpson et al. 2001). If postharvest, the infected cereal is transported or stored under high moisture conditions then it leads to the growth of fungus, and the level of mycotoxin increases (Magan et al. 2010). *Fusarium oxysporum* on the other hand is a soil-borne phytopathogen that causes progressive vascular wilting, leaf epinasty, vascular browning, stunting, and even death of plants (Di et al. 2017). It is known to infect greater than 100 varieties of plants leading to a severe loss of crops including cotton, tomato, melon, banana, etc. (Michielse and Rep 2009). It also causes invasive infectious diseases in immunocompromised humans (Nucci and Anaissie 2007). Another ascomycete, *Blumeria graminis* is the causal agent of powdery mildew disease of grasses which include barley and wheat which are among the agriculturally important crops in the world (Takamatsu 2004). Mildew infection immensely reduces the yield of grains and should be kept under control to ensure the economic viability of the crops (Hückelhoven 2005). *B. graminis* is of particular importance due to its central and persistent role as the causative agent of disease in cereal crops great significance and as a model for studying other mildew infections (Murray and Brennan 2010). One of the major economic constraints on the productivity of wheat especially in the temperate regions of the world is *Mycosphaerella graminicola* which causes STB (*Septoria tritici* blotch) disease (Orton et al. 2011). Infection is marked by hyphal extension on the surface of leaves and via stomatal penetration without appressorium formation (Kema 1996). There is a long symptomless period of intercellular colonization which is followed by necrotic leaf lesion formation (Keon et al. 2007). In the study of population dynamics of pathogen and evolution, *Mycosphaerella graminicola* serves as a model organism and is affected by climatic changes (Bearchell et al. 2005). *Colletotrichum* spp. is a postharvest phytopathogenic fungus as it causes latent infection before harvest that does not become active until storage (Prusky 1996). Almost every crop around the world is prone to infection by one or more than one species of *Colletotrichum* (De Silva et al. 2017). These fungi lead to blights and anthracnose spot formation in aerial parts of the plants and also cause postharvest rotting (Chen et al. 2006). This fungus causes severe loss of economically significant crops, specifically vegetables, fruits, ornamental plants, and even staple food crops (including sorghum, cassava, and bananas) (De Silva et al. 2017). Corn smut caused by *Ustilago maydis* is not a very economically devastating disease but is characterized by dramatic symptoms with mostly local infections which do not spread (Gold et al. 1997). *Phytophthora infestans* which causes late blight diseases of potatoes is a historically significant fungus that was responsible for the Irish famine leading to starvation, mass migration, and death (Haverkort et al. 2009). Other than food crops, cash crops like coffee are also prone to fungal infections like coffee rust caused by *Hemileia vastatrix* which is the most destructive coffee disease characterized by powdery orange-yellow spots on the lower surface of leaves (Vandermeer et al. 2009). Hence, a vast range of high-value crops are being infected by pathogenic fungi and leading to disastrous consequences economically, agriculturally, and also leading to an ecological imbalance (Dean et al. 2012).

Several strategies have been designed to control the growth of these economically significant fungi (Cook 1993). Disease-resistant varieties of plants and several fungicides have been developed over the years but need continuous revision and update due to the emergence of fungicide resistance and the constant evolution of the fungal strains. Other than good cultural and agronomic practices farmers usually rely on chemical pesticides. The use of excessive agrochemicals causes environmental pollution, but in recent years, people's attitude has changed toward the usage of chemicals in agricultural practices and there are certain strict regulations on selling them in the market (Cook 1993; Heydari and Pessarakli 2010). This highlights the need of developing an alternative method of controlling fungal diseases in plants. Biocontrol methods of managing fungal diseases in plants are now a viable alternative. Some phenolic compounds produced by plants are known to possess antimicrobial properties (Cook 1993). Plant phenolics could also be used in combination with clinical antifungals which would increase the effectiveness of the phenolics as well as can reduce the dosage of the clinical antifungals (Heydari and Pessarakli 2010). This strategy could work against the problem of the development of resistance of the fungal pathogens toward the antifungals due to the combined action of phenolics and antifungals with distinct mechanisms and targets. Antifungal phenolics are usually found conjugated with glycosides in the vacuole or other organelles of healthy plants (Beckman 2000). It has been reported that protocatechuic acid and catechol from onion scales can reduce the germination of *Colletotrichum circinans* spores and hence protect onion plants from smudge diseases (Walker and Stahmann 1955). Similarly, chlorogenic acid at an inadequate level has the potential to provide resistance against *Phytophthora infestans* and protects potato tubers (Johnson and Schaal 1952). Additionally, several flavanones and flavones show activity against fungal pathogens like *B. cinerea*, *F. oxysporum*, and *Aspergillus* sp. that arise during the storage of vegetables and fruits (Picman et al. 1995). The activity of plant polyphenols as antifungal agents has been described extensively in the following sections of this chapter.

14.3 Plant Polyphenols Against Fungal Constraints

As mentioned earlier polyphenols are the phytochemicals or secondary metabolites that are both structurally and biogenetically multifarious compared to natural products (Mayer 2006). A wide range of secondary products can be produced by plants for their defense as they are not physically mobile like animals that can escape from predators (Lattanzio and Ruggeiro 2003). Physiological and ecological facets can alter the internal parts of the plant in a unique way that is quite clear at various developmental phases indicating the plant's complexity (Mayer and Harel 1979). Propitious use of secondary metabolites can help develop lineal associations compared to primary products, showing concentration differences (Lattanzio et al. 2015). The secondary products show qualitative differences while the primary ones exhibit

Table 14.2 Plant-based polyphenols against fungal species

Antifungal phenolics	Fungus	References
Benzaldehyde, ethyl benzoate	*Botrytiscinerea, Momilinia fructicola*	Wilson and Wisniewski (1989)
Salicylic acid, Vanillic acid, 4-Hydroxybenzoic acid, Chlorogenic acid	*Eutypa lata, Phytophthora infestans, Verticillium albo-atrum, Phytophthora vagabunda*	Amborabé et al. (2002), Janzen (1981), Lee and LeTorneau (1958), Valle (1957)
Chlorogenic acid, Rutin p-Coumaric acid, Cyanidin 3- and 7-Hydroxyflavone	*Fusarium oxysporum, Gloeosporium perenas, Pencillium glabrum, Cladosporum herbarum*	Carrasco et al. (1978), Hulme and Edney (1960), Martini et al. (1997)
Oleuropin, Nobiletin, Genistein, Biochannin	*Phytophthora spp., Phoma tracheiphila, Monilinia fructicola, Cercospora bieticola*	Del Río (2003), McClure (1979), Johnson et al. (1976)
Hordatine A and B	*Helminthosporium sativum*	Overeem (1976)

quantitative differences that can further restrict environmental and genetic aspects (Pichersky and Gang 2000). In general, comprehensive reflexes occur due to the host–parasite interaction in plants. It might either be a tolerance mechanism or a resistance mechanism (Constabel et al. 2000; Lattanzio et al. 2015). Tolerance effects do not limit the pathogen attack but they can either mitigate or neutralize it adapting to the plant's function to counteract the pathogens (Kutchan 2001). Resistance mechanisms can limit or inhibit the attack through traits. A large quantum of secondary products is produced as a part of the plant growth or under adverse conditions that are noxious to the pathogens (Seigler 1998).

Fungi cannot possibly synthesize their food since photosynthesis is absent. Instead, they rely on plants, and that often turned out to be a fruitful association with added advantages (Osbourn et al. 2003). However, a fraction of them broke the mutual barriers to become fungal pathogens. Due to the biotic stress exerted by these pathogens, plants play out the resistance mechanism to prevent infection. This phenomenon is known as nonhost or species-specific resistance (Perrino et al. 1989). Prebuilt compounds like phenolics and polyphenols are pervasive and can play out their strengths in the species-specific resistance against the filamentous fungi. These compounds are termed "phytoanticipins" to stand out from the section of "phytolexins" that are known to be the remote precursors formed in response to the attack (Swain 1977). The antifungal phenolics are listed in Table 14.2.

Recent studies show that enzymes called Polyphenol oxidases (PPO) or tyrosinases might play a key role under both biotic and abiotic stress conditions. They are reportedly said to have diverse functions other than those that are inventoried (Lee 2014). Nevertheless, the primary function of PPOs is mechanisms like quinone toxicity that can form structural barricades and safeguard the plant from pathogens (Wink 1997). On the other hand, they are also found to coexist with proteins of photosystem II in the thylakoids of plant chloroplasts where they are metabolically active (Webb et al. 2014). Owing to this fact, one cannot rule out the role of PPOs in chloroplastic functions. The PPOs in both plants and fungi can exist in both active and dormant states. Despite that, the reason behind their latency has not been determined (Fuerst et al. 2014). But a few authors anticipated that this might happen due to the relation between these enzymes and the hemocyanins that are responsible for oxygen transport in invertebrates (Baydar et al. 2006). The function of the PPOs in fungi is different from that of plants. Nonetheless, both are known to exhibit defensive properties against pathogens (Nürnberger and Lipka 2005).

In-vitro and *in-vivo* studies on the formation of the polyphenolic compounds procured from the olive mill wastes (OMWW) are found to counteract the fungal as well as food-borne pathogens (Obied et al. 2005). Experimental studies of their concentrations have been done in three stages. The zone of inhibition of fungal pathogens was determined using different techniques like disk diffusion and well diffusion assays constituting the primary stage (Walker and Stahmann 1955). The effect of polyphenols against 14 fungal pathogens was evaluated in the second stage to estimate the microbicidal concentration (MIC/MFC). The OMWW were examined for liquid polyphenols (LFP) which are known to be the natural biochemicals against some fungal species and were evaluated *in vivo* on tomato plants (Overeem 1976; Mahanil et al. 2008). The resultant MIC/MFC depicted that the fungus *Aspergillus flavus* was highly resistant to the LFP concentration, showing potential counteractivity against the pathogens (Leontopoulos et al. 2015).

Polyphenols are notorious copper-binding enzymes occurring in plants. They act as catalysts in the oxidation of polyphenols to highly active quinones (Tran et al. 2012). Being encoded by a multigene family, gene expression of PPOs is quite crucial for plant growth and development (Mayer and Harel 1979). They protect the plant under biotic as well as abiotic stress conditions, regardless of which it might also be a pathogenic factor during the fungal attack against other organisms (Mayer 2006). PPOs under standard conditions occur during the vegetative, and reproductive phases, in riposte to damage and defensive functional analysis. With that being said, PPO genes at times might be overexpressed or reduced, leading to a delay in the fungal infection process in strawberry fruit or causing browning in potatoes respectively. The suppression of the standard StuPPO gene in transgenic tomatoes (Li and Steffens 2002) induced proneness of insects while the overexpression of the hybrid PtdPPO gene resulted in hindering the susception effect.

Considering the two gene expression scenarios, we reviewed a few articles related to both overexpression and suppression of the PPO genes. After careful consideration, a comparative account of both scenarios was drawn. Overexpression of the PPO gene in the strawberry fruit was analyzed for the change in polyphenol contents during the fruit development and also changes in the PPOs (Preuß et al. 2014). It was observed that, as the fruit develops, the polyphenolic content descends (Jia et al. 2016). They found four PPO encoding genes in strawberries with various ranges of expression levels in tissues. As discussed, induced PPO activity in the strawberry fruit due to the overexpression of FaPPO1 results in the retardation of the fungal infection process (Chai et al. 2011), bringing about several changes that can affect the pathogenic gene expressions such as Phenylalanine ammonia-lyase (PAL), Superoxide dismutase (SOD), Peroxidase (POD), Butyrophilin genes (BG), and

Chitinase genes (Jia et al. 2016) (Ding et al. 2002). It is also noted that fruit damage can increase the overexpression of the FaPPO1 gene where the ascorbic acid and methyl jasmonic acid were regulatory in function (Concha et al. 2013). On the other hand, reduced PPO activity and low browning of potatoes can be obtained by the suppression of StuPPO2 to StuPPO4 genes (Aziz et al. 2019). However, the suppression of all four genes (StuPPO1 to StuPPO4) resulted in a greater reduction of PPO activity and potato browning (Chi et al. 2014).

14.4 Mechanism of Polyphenols in Countering Fungal Toxin

Most phenolic compounds own a very strong binding affinity toward biomolecules like proteins or glycoproteins (Simonetti et al. 2020). Many polyphenols belonging to the flavonoid family possess antioxidant properties that cause inhibition of biofilm formation (H. Lee et al. 2018) and lipid peroxidation (El Moussaoui et al. 2019) thereby altering the potential and generating disturbances in the cell membrane (Dai et al. 2009). This may result in a reduction in the cellular size of pathogens (Lee et al. 2018) and leakage of most intracellular components. Moreover, many polyphenols belonging to the flavonoid family such as Quercetin, catechin, or epigallocatechin have been reported to target the fatty acid synthase enzyme which is an essential enzyme in the fungal cell membrane for endogenous fatty acid synthesis (Bitencourt et al. 2013; da Silva et al. 2014; Li et al. 2002). Recent studies on gene expression microarrays of fungal biofilm-forming pathogen *Candida albicans* on exposure to Pterostilbene showed downregulation in filamentation genes about the Ras/cyclic AMP pathways. Alteration in gene expression of ergosterol biosynthesis genes and heat shock proteins was seen on *in-vitro* treatments with pterostilbene (Li et al. 2014). Another important polyphenolic compound Resveratrol is known to induce metacaspase-regulated apoptosis of Candida albicans. The compound mediates its activity by lowering mitochondrial membrane potential via the effect of ROS (reactive oxygen species) and hydroxy free radicals resulting in defective mitochondrial functions (Lee and Lee 2015). Another important polyphenol Gallic acid shows inhibitory effects on a broad spectrum against filamentous fungal pathogens like *Trichophyton rubrum* and other strains. The compound targets 14alpha-demethylase P450 which is an essential enzyme in sterol biosynthesis (Li et al. 2017). Ellagic acid, a dilactone of hexahydroxy diphenic acid also shows similar effects against the ergosterol biosynthesis pathway of the T. Rubrum membrane (Li et al. 2015). The mechanism of action of another rare polyphenol epigallocatechin-3-gallate (EGCG) against various fungal pathogens in combination with azole antifungals *in vitro* has been elucidated. EGCG inhibits folic acid metabolism by targeting the dihydrofolate reductase enzyme of Candida (Navarro-Martinez 2006). Isocitrate lyase (ICL1), an important enzyme of the glyoxylate pathway having no human ortholog is one of the most fungal pathogens like C. albicans which is inhibited by caffeic acid (Cheah et al. 2014). Since ergosterol are major components that maintain cell membrane integrity of most fungal pathogens, disruption of ergosterol biosynthesis pathway components remains a vital mechanism to counter such biotic stress (Ghannoum and Rice 1999).

Apart from cell membrane disruptive action of polyphenols, antifungal mechanisms of these compounds also include inhibition of beta-glucans and chitin which comprise the primary components of the fungal cell walls (Lagrouh et al. 2017). Deformations in the fungal cell wall are mainly promoted by isoflavones like Glabridin and pedaling which promote membrane depolarization, reduction in membrane fluency, and enhancement of genes responsible for chromatin disorganization and DNA fragmentation (Liu et al. 2014) which eventually leads to apoptosis and cell death. Another important mechanism followed by some classes of flavonoids to overcome fungal stress is inhibiting mitotic spindle formation and consequently inhibiting cell division (Lagrouh et al. 2017). Biophysical analyses of hyphal transition on the action of honey flavonoids showed a reduction in the G0/G1 phase and increment in G2/M transitions in phenotypes of *Candida albicans*, thereby reducing infection rates (Canonico et al. 2014).

Most antifungal compounds enter cells via active transport and inhibit either protein synthesis in the cytosol, or enter the nucleus to disrupt replication or transcription. The mechanism of inhibition of the translation process in fungal cells is most studied and a prominent antifungal target (Lagrouh et al. 2017). Apigenin interferes with the protein synthesis machinery driven by the internal ribosome entry site (IRES) in various disease-causing fungal pathogens (Qian et al. 2015). *In-vitro* studies involving Gallic acid extracted from *Paeonia rockii* were shown to inhibit protein synthesis in Candida albicans. Such inhibition was followed by a reduction in the hyphal size and germinal tube of the pathogenic fungi (Picerno et al. 2011). Western blot analysis of hypha-specific gene expression on *in-vitro* treatment with polyphenols like catechin was shown to inhibit MAPK and cyclic adenosine 3,5-monophosphate pathways, consequently affecting the nucleic acid synthesis (Al Aboody and Mickymaray 2020). Similarly, carvacrol, which is a plant-derived polyphenol of the flavonoid family, also inhibits the synthesis of nucleic acid and is known to induce apoptosis by disrupting cytoplasmic membranes (Gabriela et al. 2014).

14.5 Future Prospects of Polyphenols Against Fungal Pestilence

The role of polyphenols as explained in previously published literature is mostly in terms of their antioxidative properties. However, advancements in research strategies have thrown light on their emergence as regulatory, signaling, and interactive compounds against both biotic and abiotic factors. The prospects for practical applications of polyphenols and flavonoids are wide both in the agricultural, health sectors, medicines, and pharmacy, mostly because it has minimal adverse effects and is naturally occurring. Polyphenols can be employed as ideal constituents of fungicides, weedicides, and biopesticides. Genetic engineering of transgenic flax plants where phenylpropanoid and glucosyltransferase led to resistance against pathogenic strains of *Fusarium* (Mierziak et al. 2014a, b). Many such cross-breeding experiments have led to the development of new plants resistant to such phytopathogens. Fusarium head blight of wheat caused by *Gibberella zea*, overexpressing for naringenin and kaempferol showed resistance against the pathogen (Bollina et al. 2010). Similar antimicrobial flavonoids also showed pathogenic activities against *Xanthomonas oryzae*, infecting *Oryza sativa* (Padmavati et al. 1997). Polyphenol-based biopesticides that may show allelopathic or phytotoxic properties can be used as a potential natural defense against weeds. 5,7,4′-Trihydroxy-3′,5′-dimethoxyflavone, extracted from an allelopathic variety of rice, showed a phytotoxic effect on weed species of *Echinochloa* and *Cyperus* (Kong 2004). Many classes of polyphenols are characterized by their antibacterial, antiviral, and antifungal properties, not only against phytopathogens but also against pathogens infecting humans. Flavonoid compounds originating from vegetables are an important component of the human diet and have a potential antiallergic/inflammatory, anticancer, and antidiabetic role (Kritas et al. 2013; Middleton et al. 2000). Some of these compounds are highly active against drug-resistant strains of fungus and bacteria (Saleem et al. 2010).

Another implication of our knowledge about plant phenolics and their mode of action against a fungal pathogen can be used to translate in research to counter fungal pathogenesis in humans using combinatorial drug strategies (Al Aboody and Mickymaray 2020). Such strategies are recognized to be highly effective against microbial resistance along with the aforementioned fungal infections in humans, enhancing the degree and rate of microbial elimination (Mukherjee et al. 2005). The drug resistance process of fungal pathogens can be reduced by combinatorial strategies since each drug or compound has a diverse mechanism of action, therefore using two drugs can be effective against diverse targets; hence, multitargeting can be achieved (Wagner and Ulrich-Merzenich 2009). *In-vitro* studies diminished minimum inhibitory concentrations (MIC) of polyphenols along with other antifungals against pathogenic strains of fungi. This has also led to reduced toxicity and intolerance of the engaged drugs in use (Gallucci et al. 2014; Fuentefria et al. 2018). A study on *Candida albicans* showed that bioactive phenolics triggered other antifungals and potentiated their mode of action by increasing intracellular concentrations by blocking efflux pumps (Moraes et al. 2017). Thus, polyphenols and flavonoids in general are competent and proficient in effective synergistic drug therapies and can be supportive in finding novel therapeutic approaches against fungal pathogens.

14.6 Conclusion

Biotic stress imposes a great threat to the nutritional status of people all around the world; thus, proper management to overcome and control biotic and abiotic stress is an absolute necessity. The use of pesticides and fungicides (both chemical and biological control agents) is the most common practice to fight biotic stress, but they have their drawbacks. The study of phenolic compounds and their biological activity has been a major field of research for a long time. The use of these phytochemicals or polyphenols must be seen as a distinguished opportunity for the betterment of plant health and good yield. Plant phenolic compounds in a nutshell seem simple, yet have a vast range of possibilities with multiple genes encoding their traits. Both biotic and abiotic factors can affect plant growth and development in several ways. This book chapter puts forward a detailed review of fungi as biotic stress constraints and the role of secondary metabolites called polyphenols or phenolic compounds. Polyphenols promote plant growth, preventing them from the fungal pathogens synchronously. The fascinating biochemistry of polyphenol synthesis helped the veterans of the research field to perform various *in-vitro* and *in-vivo* experiments that resulted in a lot of probabilities. In contempt of huge damage and economic losses, a major set of fungal diseases remain unstudied. The antifungal compounds that are identified have less effect than the anticipatory results, while some are still being developed across the globe. Considering the few antifungal drugs, they were often used for diseases that with time developed resistance against them. As a consequence of this, the outlook on efficient treatment of fungal diseases can remain unpredictable in near future. This chapter thus tends to provide an insight into the basic understanding and conceptualization of plant-based polyphenols that can be used as phytochemical weapon to combat the disastrous ill-effects of biotic fungal pestilence.

References

Abdul Malik NA, Kumar IS, Nadarajah K (2020) Elicitor and receptor molecules: orchestrators of plant defense and immunity. Int J Mol Sci 21:963

Aich S, Singh RK, Kundu P, Pandey SP, Datta S (2017) Genome-wide characterization of cellulases from the hemibiotrophic plant pathogen, Bipolaris sorokiniana, reveals the presence of a highly stable GH7 endoglucanase. Biotechnol Biofuels 10:135

Al Aboody MS, Mickymaray S (2020) Anti-fungal efficacy and mechanisms of flavonoids. Antibiotics (Basel) 9(2):45

Amborabé B-E, Fleurat-Lessard P, Chollet J-F, Roblin G (2002) Antifungal effects of salicylic acid and other benzoic acid derivatives towards Eutypa lata: structure-activity relationship. Plant Physiol Biochem 40(12):1051–1060

Asibi AE, Chai Q, Coulter JA (2019) Rice blast: a disease with implications for global food security. Agronomy 9:451

Atkinson NJ, Urwin PE (2012) The interaction of plant biotic and abiotic stresses: from genes to the field. J Exp Bot 63(10):3523–3543

Aziz E, Batool R, Akhtar W, Rehman S, Gregersen PL, Mahmood T (2019) Expression analysis of the polyphenol oxidase gene in response to signaling molecules, herbivory and wounding in antisense transgenic tobacco plants. 3 Biotech 9(2):55

Bailey-Serres J, Voesenek LACJ (2008) Flooding stress: acclimations and genetic diversity. Annu Rev Plant Biol 59:313–339

Baydar NG, Sagdic O, Ozkan G, Cetin S (2006) Determination of antibacterial effects and total phenolic contents of grape (Vitis vinifera L.) seed extracts. Int J Food Sci Technol 41(7): 799–804

Bearchell SJ, Fraaije BA, Shaw MW, Fitt BDL (2005) Wheat archive links long-term fungal pathogen population dynamics to air pollution. Proc Natl Acad Sci U S A 102(15):5438–5442

Beckman CH (2000) Phenolic-storing cells: keys to programmed cell death and periderm formation in wilt disease resistance and in general defence responses in plants? Physiol Mol Plant Pathol 57(3):101–110

Bekal S, Domier LL, Niblack TL, Lambert KN (2011) Discovery and initial analysis of novel viral genomes in the soybean cyst nematode. J Gen Virol 92(8):1870–1879

Billard A, Fillinger S, Leroux P, Lachaise H, Beffa R, Debieu D (2012) Strong resistance to the fungicide fenhexamid entails a fitness cost in Botrytis cinerea, as shown by comparisons of isogenic strains. Pest Manag Sci 68(5):684–691

Bitencourt TA, Komoto TT, Massaroto BG, Miranda CES, Beleboni RO, Marins M, Fachin AL (2013) Trans-chalcone and quercetin down-regulate fatty acid synthase gene expression and reduce ergosterol content in the human pathogenic dermatophyte Trichophyton rubrum. BMC Complement Altern Med 13(1):229

Bollina V, Kumaraswamy GK, Kushalappa AC, Choo TM, Dion Y, Rioux S et al (2010) Mass spectrometry-based metabolomics application to identify quantitative resistance-related metabolites in barley against Fusarium head blight. Mol Plant Pathol 11(6):769–782

Bolling BW, McKay DL, Blumberg JB (2010) The phytochemical composition and antioxidant actions of tree nuts. Asia Pac J Clin Nutr 19:117–123

Botterweck AAM, Verhagen H, Goldbohm RA, Kleinjans J, van den Brandt PA (2000) Intake of butylated hydroxyanisole and butylated hydroxytoluene and stomach cancer risk: results from analyses in the Netherlands cohort study. Food Chem Toxicol 38:599–605

Cabanillas-Bojórquez LA, Gutiérrez-Grijalva EP, Contreras-Angulo LA, Aviles-Gaxiola S, Heredia JB (2020) Biotechnology for extraction of plant phenolics. In: Plant phenolics in sustainable agriculture. Springer, Singapore, pp 39–67

Canonico B, Candiracci M, Citterio B, Curci R, Squarzoni S, Mazzoni A, Papa S, Piatti E (2014) Honey flavonoids inhibit Candida albicans morphogenesis by affecting DNA behavior and mitochondrial function. Future Microbiol 9(4):445–456

Carrasco A, Boudet AM, Marigo G (1978) Enhanced resistance of tomato plants to Fusarium by controlled stimulation of their natural phenolic production. Physiol Plant Pathol 12(2):225–232

Chai Y-M, Jia H-F, Li C-L, Dong Q-H, Shen Y-Y (2011) FaPYR1 is involved in strawberry fruit ripening. J Exp Bot 62(14):5079–5089

Cheah H-L, Lim V, Sandai D (2014) Inhibitors of the glyoxylate cycle enzyme ICL1 in Candida albicans for potential use as antifungal agents. PLoS One 9(4):e95951

Chen XM (2005) Epidemiology and control of stripe rust [Puccinia striiformis f. sp. tritici] on wheat. Can J Plant Pathol 27(3):314–337

Chen C, Ha YS, Min JY, Memmott SD, Dickman MB (2006) Cdc42 is required for proper growth and development in the fungal pathogen Colletotrichum trifolii. Eukaryot Cell 5(1):155–166

Chi M, Bhagwat B, Lane WD, Tang G, Su Y, Sun R, Oomah BD, Wiersma PA, Xiang Y (2014) Reduced polyphenol oxidase gene expression and enzymatic browning in potato (Solanum tuberosum L.) with artificial microRNAs. BMC Plant Biol 14(1):62

Concha CM, Figueroa NE, Poblete LA, Oñate FA, Schwab W, Figueroa CR (2013) Methyl jasmonate treatment induces changes in fruit ripening by modifying the expression of several ripening genes in Fragaria chiloensis fruit. Plant Physiol Biochem 70:433–444

Constabel CP, Yip L, Patton JJ, Christopher ME (2000) Polyphenol oxidase from hybrid poplar. Cloning and expression in response to wounding and herbivory. Plant Physiol 124(1):285–296

Cook RJ (1993) Making greater use of introduced microorganisms for biological control of plant pathogens. Annu Rev Phytopathol 31:53–80

da Silva CR, de Andrade Neto JB, de Sousa Campos R, Figueiredo NS, Sampaio LS, Magalhães HIF, Cavalcanti BC, Gaspar DM, de Andrade GM, Lima ISP, de Barros Viana GS, de Moraes MO, Lobo MDP, Grangeiro TB, Nobre Júnior HV (2014) Synergistic effect of the flavonoid Catechin, quercetin, or epigallocatechin Gallate with fluconazole induces apoptosis in Candida tropicalis resistant to fluconazole. Antimicrob Agents Chemother 58(3):1468–1478

Daayf F, El Hadrami A, El-Bebany AF, Henriquez MA, Yao Z, Derksen H, El-Hadrami I, Adam LR (2012) Phenolic compounds in plant defense and pathogen counter-defense mechanisms. In:

Cheynier V, Sarni-Manchado P, Quideau S (eds) Recent advances in polyphenol research. Wiley-Blackwell, Oxford, pp 191–208
Dai BD, Cao YY, Huang S, Xu YG, Gao PH, Wang Y, Jiang YY (2009) Baicalein induces programmed cell death in candida albicans. J Microbiol Biotechnol 19(8):803–809
de Ascensao ARFDC, Dubery IA (2003) Soluble and wall-bound phenolics and phenolic polymers in Musa acuminata roots exposed to elicitors from Fusarium oxysporum f. sp. cubense. Phytochemistry 63:679–686
De Silva DD, Crous PW, Ades PK, Hyde KD, Taylor PWJ (2017) Life styles of Colletotrichum species and implications for plant biosecurity. Fungal Biol Rev 31(3):155–168
Dean R, Van Kan JA, Pretorius ZA, Hammond-Kosack KE, Di Pietro A, Spanu PD, Rudd JJ, Dickman M, Kahmann R, Ellis J (2012) The top 10 fungal pathogens in molecular plant pathology. Mol Plant Pathol 13:414–430
Del Río J (2003) Enhancement of phenolic compounds in olive plants (Olea europaea L.) and their influence on resistance against Phytophthora sp. Food Chem 83(1):75–78
Di X, Gomila J, Takken FLW (2017) Involvement of salicylic acid, ethylene and jasmonic acid signalling pathways in the susceptibility of tomato to Fusarium oxysporum. Mol Plant Pathol 18(7):1024–1035
Ding C-K, Wang C, Gross K, Smith D (2002) Jasmonate and salicylate induce the expression of pathogenesis-related-protein genes and increase resistance to chilling injury in tomato fruit. Planta 214(6):895–901
Doehlemann G, Ökmen B, Zhu W, Sharon A (2017) Plant pathogenic fungi. Microbiol Spectr. 5(1)
Drenth A, Guest DI (2016) Fungal and oomycete diseases of tropical tree fruit crops. Annu Rev Phytopathol 54:373–395
Dybkowska E, Sadowska A, Rakowska R, Dębowska M, Swiderski F, Świąder K (2017) Assessing polyphenols content and antioxidant activity in coffee beans according to origin and the degree of roasting. Rocz Panstw Zakl Hig 68:347–353
Einhellig FA (1994) Allelopathy: current status and future goals, pp 1–24
El Moussaoui A, Jawhari FZ, Almehdi AM, Elmsellem H, Fikri Benbrahim K, Bousta D, Bari A (2019) Antibacterial, antifungal and antioxidant activity of total polyphenols of Withania frutescens.L. Bioorg Chem 93:103337
Engin AB, Bukan N, Kurukahvecioglu O, Memis L, Engin A (2011) Effect of butylated hydroxytoluene (E321) pretreatment versus l-arginine on liver injury after sub-lethal dose of endotoxin administration. Environ Toxicol Pharmacol 32:457–464
Evans CS (1991) J. B. Harborne (ed.) Methods in plant biochemistry: 1. Plant phenolics. Academic Press, New York and London, 1990. £65. Phytochem Anal 2:48–48
Fiori ACG, Schwan-Estrada KRF, Stangarlin JR, Vida JB, Scapim CA, Cruz MES, Pascholati SF (2000) Antifungal activity of leaf extracts and essential oils of some medicinal plants against Didymella bryoniae. J Phytopathol 148(7–8):483–487
Fisher MC, Henk D, Briggs CJ, Brownstein JS, Madoff LC, McCraw SL, Gurr SJ (2012) Emerging fungal threats to animal, plant and ecosystem health. Nature 484:186–194
Fisher MC, Gurr SJ, Cuomo CA, Blehert DS, Jin H, Stukenbrock EH, Stajich JE, Kahmann R, Boone C, Denning DW (2020) Threats posed by the fungal kingdom to humans, wildlife, and agriculture. mBio 11:e00449–e00420
Fuentefria AM, Pippi B, Dalla Lana DF, Donato KK, de Andrade SF (2018) Antifungals discovery: an insight into new strategies to combat antifungal resistance. Lett Appl Microbiol 66(1):2–13
Fuerst EP, Okubara PA, Anderson JV, Morris CF (2014) Polyphenol oxidase as a biochemical seed defense mechanism. Front Plant Sci 5:689
Gabriela N, Rosa AM, Catiana ZI, Soledad C, Mabel OR, Esteban SJ, Veronica B, Daniel W, Ines IM (2014) The effect of Zuccagnia punctata, an argentine medicinal plant, on virulence factors from candida species. Nat Prod Commun 9(7):933–936
Gallucci MN, Carezzano ME, Oliva MM, Demo MS, Pizzolitto RP, Zunino MP, Zygadlo JA, Dambolena JS (2014) In vitro activity of natural phenolic compounds against fluconazole-

resistant Candida species: a quantitative structure-activity relationship analysis. J Appl Microbiol 116(4):795–804
Gautam AK, Singh PK, Aravind M (2020) Defensive role of plant phenolics against pathogenic microbes for sustainable agriculture. In: Plant phenolics in sustainable agriculture. Springer, Singapore, pp 579–594. https://doi.org/10.1007/978-981-15-4890-1_25
Ghannoum MA, Rice LB (1999) Antifungal agents: mode of action, mechanisms of resistance, and correlation of these mechanisms with bacterial resistance. Clin Microbiol Rev 12(4):501–517
Godfray HCJ, Mason-D'Croz D, Robinson S (2016) Food system consequences of a fungal disease epidemic in a major crop. Philos Trans R Soc B Biol Sci 371:20150467
Gold SE, Brogdon SM, Mayorga ME, Kronstad JW (1997) The Ustilago maydis regulatory subunit of a cAMP-dependent protein kinase is required for gall formation in maize. Plant Cell 9(9): 1585–1594
Guodong R, Jianguo Z, Xiaoxia L, Ying L (2019) Identification of putative genes for polyphenol biosynthesis in olive fruits and leaves using full-length transcriptome sequencing. Food Chem 300:125246
Gurjar G, Kanade M (2020) Analysis of phytopathogenic fungi isolated from some important crop plants using morpho-molecular tools—foldscope and ITS region sequencing. Mycol Prog 19: 1475–1493
Hadacek F (2002) Secondary metabolites as plant traits: current assessment and future perspectives. CRC Crit Rev Plant Sci 21(4):273–322
Haverkort AJ, Struik PC, Visser RGF, Jacobsen E (2009) Applied biotechnology to combat late blight in potato caused by Phytophthora infestans. Potato Res 52(3):249–264
Hewitt G (2000) New modes of action of fungicides. Pestic Outlook 11(1):28–32
Heydari A, Pessarakli M (2010) A review on biological control of fungal plant pathogens using microbial antagonists. J Biol Sci 10(4):273–290
Howard RJ, Valent B (1996) Breaking and entering: host penetration by the fungal rice blast pathogen Magnaporthe grisea. Annu Rev Microbiol 50:491–512
Hückelhoven R (2005) Powdery mildew susceptibility and biotrophic infection strategies. FEMS Microbiol Lett 245(1):9–17
Hulme AC, Edney KL (1960) Phenolic in plant and disease. Pridham, JB
Hussain F, Usman F (2019) Fungal biotic stresses in plants and its control strategy. In: Abiotic and biotic stress plants
Iqbal Z, Khan MA, Sharif M, Shah JH, ur Rehman MH, Javed K (2018) An automated detection and classification of citrus plant diseases using image processing techniques: a review. Comput Electron Agric 153:12–32
Jain A, Sarsaiya S, Wu Q, Lu Y, Shi J (2019) A review of plant leaf fungal diseases and its environment speciation. Bioengineered 10:409–424
Janzen DH (1981) Secondary plant products, Edited by E. A. Bell & B. V. Charlwood. (Encyclopedia of plant physiology, new series, Vol. 8.) Springer-Verlag, Berlin—Heidelberg—New York: xvi + 674 pp., illustr., 23 × 16 × 3.5 cm, DM. 198 (approx. US$ 110.90), 1980. Environ Conserv 8(1):79
Jia H, Zhao P, Wang B, Tariq P, Zhao F, Zhao M, Wang Q, Yang T, Fang J (2016) Overexpression of polyphenol oxidase gene in strawberry fruit delays the fungus infection process. Plant Mol Biol Report 34(3):592–606
Jin Y, Szabo LJ, Carson M (2010) Century-old mystery of Puccinia striiformis life history solved with the identification of berberis as an alternate host. Phytopathology 100(5):432–435
Johnson G, Schaal LA (1952) Resistance in potatoes. 1:7–9
Johnson G, Maag DD, Johnson DK, Thomas RD (1976) The possible rôle of phytoalexins in the resistance of sugarbeet (Beta vulgaris) to Cercospora beticola. Physiol Plant Pathol 8(3): 225–230
Keller MD, Bergstrom GC, Shields EJ (2014) The aerobiology of Fusarium graminearum. Aerobiologia 30(2):123–136
Kema (1996) Phyto86n07_777.PDF

Keon J, Antoniw J, Carzaniga R, Deller S, Ward JL, Baker JM, Beale MH, Hammond-Kosack K, Rudd JJ (2007) Transcriptional adaptation of Mycosphaerella graminicola to programmed cell death (PCD) of its susceptible wheat host. Mol Plant Microbe Interact 20(2):178–193

Khush GS (2005) What it will take to feed 5.0 billion Rice consumers in 2030. Plant Mol Biol 59(1):1–6

Kim YM, Lee CH, Kim HG, Lee HS (2004) Anthraquinones isolated from Cassia tora (Leguminosae) seed show an antifungal property against phytopathogenic fungi. J Agric Food Chem 52(20):6096–6100

Kong C (2004) Two compounds from allelopathic rice accession and their inhibitory activity on weeds and fungal pathogens. Phytochemistry 65(8):1123–1128

Kornienko JS, Smirnova IS, Pugovkina NA, Ivanova JS, Shilina MA, Grinchuk TM, Shatrova AN, Aksenov ND, Zenin VV, Nikolsky NN, Lyublinskaya OG (2019) High doses of synthetic antioxidants induce premature senescence in cultivated mesenchymal stem cells. Sci Rep 9: 1296

Kritas SK, Saggini A, Varvara G, Murmura G, Caraffa A, Antinolfi P, Toniato E, Pantalone A, Neri G, Frydas S, Rosati M, Tei M, Speziali A, Saggini R, Pandolfi F, Cerulli G, Theoharides TC, Conti P (2013) Luteolin inhibits mast cell-mediated allergic inflammation. J Biol Regul Homeost Agents 27(4):955–959

Kumar N, Goel N (2019) Phenolic acids: natural versatile molecules with promising therapeutic applications. Biotechnol Rep (Amst) 24:e00370

Kutchan TM (2001) Ecological arsenal and developmental dispatcher. The paradigm of secondary metabolism. Plant Physiol 125(1):58–60

Lagrouh F, Dakka N, Bakri Y (2017) The antifungal activity of Moroccan plants and the mechanism of action of secondary metabolites from plants. J Mycol Med 27(3):303–311

Lamers J, Van Der Meer T, Testerink C (2020) How plants sense and respond to stressful environments. Plant Physiol 182(4):1624–1635

Lattanzio V, Ruggeiro P (2003) Biochemica Agraria. Patron Editore, Bologna, p 631

Lattanzio V, Lattanzio VMT, Cardinali A (2015) Role of phenolics in the resistance mechanisms of plants against fungal pathogens and insects. Phytochemistry 661(2)

Lee MRF (2014) Forage polyphenol oxidase and ruminant livestock nutrition. Front Plant Sci 5:694

Lee J, Lee DG (2015) Novel antifungal mechanism of resveratrol: apoptosis inducer in Candida albicans. Curr Microbiol 70(3):383–389

Lee S, LeTorneau DJ (1958) Chlorogenic acid content and Verticillium wilt resistance of potatoes. Phytopathology 48:268–274

Lee H, Woo ER, Lee DG (2018) Apigenin induces cell shrinkage in Candida albicans by membrane perturbation. FEMS Yeast Res, 18(1)

Leontopoulos SV, Giavasis I, Petrotos K, Kokkora M, Makridis C (2015) Effect of different formulations of polyphenolic compounds obtained from OMWW on the growth of several fungal plant and food borne pathogens. Studies in vitro and in vivo. Agric Agric Sci Procedia 4: 327–337

Li L, Steffens J (2002) Overexpression of polyphenol oxidase in transgenic tomato plants results in enhanced bacterial disease resistance. Planta 215(2):239–247

Li X-C, Joshi AS, ElSohly HN, Khan SI, Jacob MR, Zhang Z, Khan IA, Ferreira D, Walker LA, Broedel SE, Raulli RE, Cihlar RL (2002) Fatty acid synthase inhibitors from plants: isolation, structure elucidation, and SAR studies. J Nat Prod 65(12):1909–1914

Li D-D, Zhao L-X, Mylonakis E, Hu G-H, Zou Y, Huang T-K, Yan L, Wang Y, Jiang Y-Y (2014) In vitro and in vivo activities of Pterostilbene against Candida albicans biofilms. Antimicrob Agents Chemother 58(4):2344–2355

Li Z-J, Guo X, Dawuti G, Aibai S (2015) Antifungal activity of Ellagic acid in vitro and in vivo. Phytother Res 29(7):1019–1025

Li R, Chen XG, Jia K, Liu ZP, Peng HY (2016) A systematic determination of polyphenols constituents and cytotoxic ability in fruit parts of pomegranates derived from five Chinese cultivars. Springerplus 5:914

Li Z-J, Liu M, Dawuti G, Dou Q, Ma Y, Liu H-G, Aibai S (2017) Antifungal activity of gallic acid in vitro and in vivo. Phytother Res 31(7):1039–1045

Li J, Wang Y, Suh JH (2022) Multi-omics approach in tea polyphenol research regarding tea plant growth, development and tea processing: current technologies and perspectives. Food Sci Human Wellness 11:524–536

Lin D, Hu L, You H, Sarkar D, Xing B, Shetty K (2010) Initial screening studies on potential of high phenolic-linked plant clonal systems for nitrate removal in cold latitudes. J Soils Sediments 10:923–932

Lin D, Xiao M, Zhao J, Li Z, Xing B, Li X, Kong M, Li L, Zhang Q, Liu Y, Chen H, Qin W, Wu H, Chen S (2016) An overview of plant phenolic compounds and their importance in human nutrition and management of type 2 diabetes. Molecules 21:1374

Liu W, Li LP, Zhang JD, Li Q, Shen H, Chen SM, He LJ, Yan L, Xu GT, An MM, Jiang YY (2014) Synergistic antifungal effect of Glabridin and fluconazole. PLoS One 9(7):e103442

Lyu X, Shen C, Fu Y, Xie J, Jiang D, Li G, Cheng J (2015) Comparative genomic and transcriptional analyses of the carbohydrate-active enzymes and secretomes of phytopathogenic fungi reveal their significant roles during infection and development. Sci Rep 5:1–16

Magan N, Aldred D, Mylona K, Lambert RJW (2010) Limiting mycotoxins in stored wheat. Food Addit Contam Part A Chem Anal Control Expo Risk Assess 27(5):644–650

Mahalingam R (2015) Consideration of combined stress: a crucial paradigm for improving multiple stress tolerance in plants. In: Combined stresses in plants. Springer, pp 1–25

Mahanil S, Attajarusit J, Stout MJ, Thipyapong P (2008) Overexpression of tomato polyphenol oxidase increases resistance to common cutworm. Plant Sci 174(4):456–466

Majeed A, Muhammad Z, Ahmad H (2018) Plant growth promoting bacteria: role in soil improvement, abiotic and biotic stress management of crops. Plant Cell Rep 37:1599–1609

Mandal SM, Chakraborty D, Dey S (2010) Phenolic acids act as signaling molecules in plant-microbe symbioses. Plant Signal Behav 5:359–368

Martini H, Weidenbörner M, Adams S, Kunz B (1997) Increased antifungal activity of 3- and 7-hydroxyflavone against Cladosporium herbarum and Penicillium glabrum through ester formation. Mycol Res 101(8):920–922

Mayer AM (2006) Polyphenol oxidases in plants and fungi: going places? A review. Phytochemistry 67(21):2318–2331

Mayer AM, Harel E (1979) Polyphenol oxidases in plants. Phytochemistry 18(2):193–215

McClure JW (1979) Recent advances in phtochemistry and biochemistry of plant phenolics. Plenum Press, New York, p 525

Michielse CB, Rep M (2009) Pathogen profile update: Fusarium oxysporum. Mol Plant Pathol 10(3):311–324

Middleton E, Kandaswami C, Theoharides TC (2000) The effects of plant flavonoids on mammalian cells: implications for inflammation, heart disease, and cancer. Pharmacol Rev 52(4): 673–751

Mierziak J, Kostyn K, Kulma A (2014a) Flavonoids as important molecules of plant interactions with the environment. Molecules 19:16240–16265

Mierziak J, Wojtasik W, Kostyn K, Czuj T, Szopa J, Kulma A (2014b) Crossbreeding of transgenic flax plants overproducing flavonoids and glucosyltransferase results in progeny with improved antifungal and antioxidative properties. Mol Breed 34(4):1917–1932

Mishra UN, Dey P, Singhal RK, Sahu C, Jena D, Nanda SP, Chauhan J (2023) Plant phenolics: as antioxidants and potent compounds under multiple stresses. In: Plant phenolics in abiotic stress management. Springer Nature Singapore, Singapore, pp 215–234

Mittler R, Blumwald E (2010) Genetic engineering for modern agriculture: challenges and perspectives. Annu Rev Plant Biol 61(1):443–462

Monteiro E, Gonçalves B, Cortez I, Castro I (2022) The role of biostimulants as alleviators of biotic and abiotic stresses in grapevine: a review. Plants 11:396

Moore D (2020) 21st century guidebook to fungi, 2nd edn. Reviews and contents

Moraes RC, Carvalho AR, Lana AJD, Kaiser S, Pippi B, Fuentefria AM, Ortega GG (2017) In vitro synergism of a water insoluble fraction of Uncaria tomentosa combined with fluconazole and terbinafine against resistant non-Candida albicans isolates. Pharm Biol 55(1):406–415
Mukherjee PK, Sheehan DJ, Hitchcock CA, Ghannoum MA (2005) Combination treatment of invasive fungal infections. Clin Microbiol Rev 18(1):163–194
Murray GM, Brennan JP (2010) Estimating disease losses to the Australian barley industry. Australas Plant Pathol 39(1):85–96
Natsume M, Osakabe N, Yamagishi M, Takizawa T, Nakamura T, Miyatake H, Hatano T, Yoshida T (2000) Analyses of polyphenols in cacao liquor, cocoa, and chocolate by normal-phase and reversed-phase HPLC. Biosci Biotechnol Biochem 64:2581–2587
Navarro-Martinez MD (2006) Tea polyphenol epigallocatechin-3-gallate inhibits ergosterol synthesis by disturbing folic acid metabolism in Candida albicans. J Antimicrob Chemother 57(6): 1083–1092
Nazarov PA, Baleev DN, Ivanova MI, Sokolova LM, Karakozova MV (2020) Infectious plant diseases: etiology, current status, problems and prospects in plant protection. Acta Nat 12:46–59
Nicholson RL, Epstein L (1991) Adhesion of fungi to the plant surface. In: The fungal spore and disease initiation in plants and animals, pp 3–23
Nucci M, Anaissie E (2007) Fusarium infections in immunocompromised patients. Clin Microbiol Rev 20(4):695–704
Nürnberger T, Lipka V (2005) Non-host resistance in plants: new insights into an old phenomenon. Mol Plant Pathol 6(3):335–345
Obied HK, Allen MS, Bedgood DR, Prenzler PD, Robards K, Stockmann R (2005) Bioactivity and analysis of biophenols recovered from olive mill waste. J Agric Food Chem 53(4):823–837
Ogawa S, Yazaki Y (2018) Tannins from Acacia mearnsii De Wild. Bark: tannin determination and biological activities. Molecules 23:837
Opara EI, Chohan M (2014) Culinary herbs and spices: their bioactive properties, the contribution of polyphenols and the challenges in deducing their true health benefits. Int J Mol Sci 15:19183–19202
Orton ES, Deller S, Brown JKM (2011) Mycosphaerella graminicola: from genomics to disease control. Mol Plant Pathol 12(5):413–424
Osbourn AE, Qi X, Townsend B, Qin B (2003) Dissecting plant secondary metabolism—constitutive chemical defences in cereals. New Phytol 159(1):101–108
Ou SH (1980) Pathogen variability and host resistance in Rice blast disease. Annu Rev Phytopathol 18(1):167–187
Overeem JC (1976) Biochemical aspects of plant parasite relationships. Academic, London, p 195
Padmavati M, Sakthivel N, Thara KV, Reddy AR (1997) Differential sensitivity of rice pathogens to growth inhibition by flavonoids. Phytochemistry 46(3):499–502
Pallas V, García JA (2011) How do plant viruses induce disease? Interactions and interference with host components. J Gen Virol 92(12):2691–2705
Pandey P, Ramegowda V, Senthil-Kumar M (2015) Shared and unique responses of plants to multiple individual stresses and stress combinations: physiological and molecular mechanisms. Front Plant Sci 6:723
Pandey P, Irulappan V, Bagavathiannan MV, Senthil-Kumar M (2017) Impact of combined abiotic and biotic stresses on plant growth and avenues for crop improvement by exploiting physio-morphological traits. Front Plant Sci 8:537
Paulitz TC, Bélanger RR (2001) Biological control in greenhouse systems. Annu Rev Phytopathol 39(1):103–133
Perrino P, Maruca G, Linsalata V, Bianco VV, Lester RN, Lattanzio V (1989) Flavonoid taxonomic analysis of Vicia species of section Faba. Can J Bot 67(12):3529–3533
Peters K, Breitsameter L, Gerowitt B (2014) Impact of climate change on weeds in agriculture: a review. Agron Sustain Dev 34:707–721
Peterson PD (2018) The barberry eradication program in Minnesota for stem rust control: a case study. Annu Rev Phytopathol 56(June):203–223

Picerno P, Mencherini T, Sansone F, Del Gaudio P, Granata I, Porta A, Aquino RP (2011) Screening of a polar extract of Paeonia rockii: composition and antioxidant and antifungal activities. J Ethnopharmacol 138(3):705–712
Pichersky E, Gang DR (2000) Genetics and biochemistry of secondary metabolites in plants: an evolutionary perspective. Trends Plant Sci 5(10):439–445
Picman AK, Schneider EF, Picman J (1995) Effect of flavonoids on mycelial growth of Verticillium albo-atrum. Biochem Syst Ecol 23(7–8):683–693
Prasch CM, Sonnewald U (2013) Simultaneous application of heat, drought, and virus to Arabidopsis plants reveals significant shifts in signaling networks. Plant Physiol 162(4): 1849–1866
Pratyusha S (2022) Phenolic compounds in the plant development and defense: an overview. In: Physiology. IntechOpen
Preuß A, Augustin C, Figueroa CR, Hoffmann T, Valpuesta V, Sevilla JF, Schwab W (2014) Expression of a functional jasmonic acid carboxyl methyltransferase is negatively correlated with strawberry fruit development. J Plant Physiol 171(15):1315–1324
Prusky D (1996) Pathogen quiescence in postharvest diseases. Annu Rev Phytopathol 34(1): 413–434
Qian S, Fan W, Qian P, Zhang D, Wei Y, Chen H, Li X (2015) Apigenin restricts FMDV infection and inhibits viral IRES driven translational activity. Viruses 7(4):1613–1626
Ramos AM, Gally M, Szapiro G, Itzcovich T, Carabajal M, Levin L (2016) In vitro growth and cell wall degrading enzyme production by Argentinean isolates of Macrophomina phaseolina, the causative agent of charcoal rot in corn. Rev Argent Microbiol 48:267–273
Randhawa S, Bahna SL (2009) Hypersensitivity reactions to food additives. Curr Opin Allergy Clin Immunol 9:278–283
Rashad Y, Aseel D, Hammad S (2020) Phenolic compounds against fungal and viral plant diseases. In: Plant phenolics in sustainable agriculture. Springer, Singapore, pp 201–219. https://doi.org/10.1007/978-981-15-4890-1_9
Rodriguez-Moreno L, Ebert MK, Bolton MD, Thomma BP (2018) Tools of the crook-infection strategies of fungal plant pathogens. Plant J 93:664–674
Ryan C, Ryan C, Lamb CJ, Lamb CJ, Jagendorf AT, Jagendorf AT, Kolattukudy PE, Kolattukudy PE (1995) Signaling pathogenesis. Genes Dev 92(May):4080–4087
Saad B, Sing YY, Nawi MA, Hashim N, Mohamed Ali AS, Saleh MI, Sulaiman SF, Talib KM, Ahmad K (2007) Determination of synthetic phenolic antioxidants in food items using reversed-phase HPLC. Food Chem 105:389–394
Saijo Y, Loo EPI (2020) Plant immunity in signal integration between biotic and abiotic stress responses. New Phytol 225(1):87–104
Saleem M, Nazir M, Ali MS, Hussain H, Lee YS, Riaz N, Jabbar A (2010) Antimicrobial natural products: an update on future antibiotic drug candidates. Nat Prod Rep 27(2):238–254
Seigler DS (1998) Flavonoids. In: Plant secondary metabolism. Springer US, pp 151–192
Shabana YM, Abdel-Fattah GM, Ismail AE, Rashad YM (2008) Control of brown spot pathogen of rice (Bipolaris oryzae) using some phenolic antioxidants. Braz J Microbiol 39:438–444
Shuping D, Eloff JN (2017) The use of plants to protect plants and food against fungal pathogens: a review. Afr J Tradit Complement Altern Med 14:120–127
Simonetti G, Brasili E, Pasqua G (2020) Antifungal activity of phenolic and polyphenolic compounds from different matrices of Vitis vinifera L. against human pathogens. Molecules 25(16):3748
Simpson DR, Weston GE, Turner JA, Jennings P, Nicholson P (2001) Differential control of head blight pathogens of wheat by fungicides and consequences for mycotoxin contamination of grain. Eur J Plant Pathol 107(4):421–431
Sobiczewski P, Iakimova ET, Mikiciński A, Węgrzynowicz-Lesiak E, Dyki B (2017) Necrotrophic behaviour of Erwinia amylovora in apple and tobacco leaf tissue. Plant Pathol 66(5):842–855
Swain T (1977) Secondary compounds as protective agents. Annu Rev Plant Physiol 28(1): 479–501

Szittya G, Silhavy D, Molnár A, Havelda Z, Lovas Á, Lakatos L, Bánfalvi Z, Burgyán J (2003) Low temperature inhibits RNA silencing-mediated defence by the control of siRNA generation. EMBO J 22(3):633–640

Takamatsu S (2004) Phylogeny and evolution of the powdery mildew fungi (Erysiphales, Ascomycota) inferred from nuclear ribosomal DNA sequences. Mycoscience 45(2):147–157

Talbot NJ (2003) On the trail of a cereal killer: exploring the biology of Magnaporthe grisea. Annu Rev Microbiol 57:177–202

Thakur A, Verma S, Reddy VP, Sharma D (2019) Hypersensitive responses in plants. Agric Rev

Tomar M, Kumar M, Seva Nayak D (2023) Role of phenolic metabolites in salinity stress management in plants. In: Plant phenolics in abiotic stress management. Singapore, Springer Nature Singapore, pp 353–368

Tran LT, Taylor JS, Constabel CP (2012) The polyphenol oxidase gene family in land plants: lineage-specific duplication and expansion. BMC Genomics 13(1):395

Tresserra-Rimbau A, Lamuela-Raventos RM, Moreno JJ (2018) Polyphenols, food and pharma. Current knowledge and directions for future research. Biochem Pharmacol 156:186–195

Tuladhar P, Sasidharan S, Saudagar P (2021) Role of phenols and polyphenols in plant defense response to biotic and abiotic stresses. In: Biocontrol agents and secondary metabolites. Elsevier, pp 419–441

Valle G (1957) On anti-fungal factors in potato leaves. Acta Chem Scand 11(2):395–397

Van Baarlen P, Woltering EJ, Staats M, Van Kan JAL (2007) Histochemical and genetic analysis of host and non-host interactions of Arabidopsis with three Botrytis species: an important role for cell death control. Mol Plant Pathol 8(1):41–54

Vandermeer J, Perfecto I, Liere H (2009) Evidence for hyperparasitism of coffee rust (Hemileia vastatrix) by the entomogenous fungus, Lecanicillium lecanii, through a complex ecological web. Plant Pathol 58(4):636–641

Vattem DA, Randhir R, Shetty K (2005) Cranberry phenolics-mediated antioxidant enzyme response in oxidatively stressed porcine muscle. Process Biochem 40:2225–2238

Velderrain-Rodríguez GR, Palafox-Carlos H, Wall-Medrano A, Ayala-Zavala JF, Chen C-YO, Robles-Sánchez M, Astiazaran-García H, Alvarez-Parrilla E, González-Aguilar GA (2014) Phenolic compounds: their journey after intake. Food Funct 5:189–197

Voegele RT, Mendgen KW (2011) Nutrient uptake in rust fungi: how sweet is parasitic life? Euphytica 179(1):41–55

Wagner H, Ulrich-Merzenich G (2009) Synergy research: approaching a new generation of phytopharmaceuticals. Phytomedicine 16(2–3):97–110

Walker JC, Stahmann MA (1955) Chemical nature of disease resistance in plants. Annu Rev Plant Physiol 6(1):351

Wang M, Zheng Q, Shen Q, Guo S (2013) The critical role of potassium in plant stress response. Int J Mol Sci 14:7370–7390

Wang J, Xu J, Gong X, Yang M, Zhang C, Li M (2019) Biosynthesis, chemistry, and pharmacology of polyphenols from Chinese salvia species: a review. Molecules 24:E155

Webb KJ, Cookson A, Allison G, Sullivan ML, Winters AL (2014) Polyphenol oxidase affects normal nodule development in red clover (Trifolium pratense L.). Front Plant Sci 5:700

Whiting DA (2001) Natural phenolic compounds 1900–2000: a bird's eye view of a century's chemistry. Nat Prod Rep 18(6):583–606

Williams KS, Myers JH (1984) Oecologia for fall webworm larvae. Oecologia 63:166–170

Williamson B, Tudzynski B, Tudzynski P, Van Kan JAL (2007) Botrytis cinerea: the cause of grey mould disease. Mol Plant Pathol 8(5):561–580

Wilson CL, Wisniewski ME (1989) Biological control of postharvest diseases of fruits and vegetables: an emerging technology. Annu Rev Phytopathol 27(1):425–441

Wink M (1997) Compartmentation of secondary metabolites and xenobiotics in plant vacuoles, pp 141–169

Xia E-Q, Deng G-F, Guo Y-J, Li H-B (2010) Biological activities of polyphenols from grapes. Int J Mol Sci 11:622–646

Xie M, Zhang J, Tschaplinski TJ, Tuskan GA, Chen J-G, Muchero W (2018) Regulation of lignin biosynthesis and its role in growth-defense tradeoffs. Front Plant Sci 9:1427

Xiu-Qin L, Chao J, Yan-Yan S, Min-Li Y, Xiao-Gang C (2009) Analysis of synthetic antioxidants and preservatives in edible vegetable oil by HPLC/TOF-MS. Food Chem 113:692–700

Momilactone B and Potential in Biological Control of Weeds

15

Truong Ngoc Minh and Tran Dang Xuan

Abstract

Rice has been known to release allelochemical(s) into the surroundings and has been demonstrated by numerous studies conducted within both lab-scale and field-scale. Many potential allelochemicals in rice include fatty acids, phenolic acids, hydroxamic acids, phenylalkanoic acids, terpenes, momilactones, and indoles. In this review, the role of labdane-related diterpenoid momilactones, typically momilactone B (MB), is highlighted as the principle allelochemical in rice plants. During the plant wheel of life, momilactone B is excreted from the roots to inhibit weeds. Structurally, momilactone B is also a phenol which differs from momilactone A (MA) and other momilactones by an oxygen bridge, where the –OH is attached to the C3 of the benzene ring. Basically, the amount of MB is less than MB in all plant parts in rice, but its biological activity, especially weed suppression level, is more excellent than MA. The synthesis of function group from MB is potential for development of novel herbicides in agricultural production.

Keywords

Phenolics · Momilactone B · Biocontrol · Herbicide: Weeds

T. N. Minh
Center for Research and Technology Transfer, Vietnam Academy of Science and Technology (VAST), Hanoi, Vietnam

T. D. Xuan (✉)
Transdisciplinary Science and Engineering Program, Graduate School of Advanced Science and Engineering, Hiroshima University, Hiroshima, Japan
e-mail: tdxuan@hiroshima-u.ac.jp

R. Lone et al. (eds.), *Plant Phenolics in Biotic Stress Management*,
https://doi.org/10.1007/978-981-99-3334-1_15

15.1 Introduction

Weed infestation, to a certain extent, is termed as the assault of other disturbing plants and putting a significant stress on the global agricultural sector. This kind of plant competes with crops to obtain nutrients, water, and space, directly affecting the crop yields (Shahzad et al. 2021). Moreover, high levels of weed infestation interfere with automated harvesting techniques. Invasive weeds have the highest potential for production loss worldwide (34%), more than animal pests or pathogens (18% and 16%, respectively) (Oerke 2006). Rice is one of many crops to be disturbed by weed such as barnyard grass (*Echinochloa crus-galli*), and farmers must flood their rice fields to restrict the invasion of weed (Bayer 1991). Nevertheless, it was only efficient with certain species. This situation led to the rise of synthetic herbicide, which initially was expected to solve the problem but in fact still insufficient to the weed inhibition, expensive, environmentally hazardous and more important, created herbicide-resistant weeds. In that situation, allelopathy, which is a capability of plant to induce specialized (or secondary) allelochemicals to the soil to limit the development of the neighboring plants, is a potential aspect that experts and scientists can focus to find the real solution (Tukey 1969; Latif et al. 2017; Putnam and Duke 1978). Notably, momilactone B is a widely-recognized allelochemical that can sense the neighboring plants and synthesized by rice (Kato-Noguchi and Ino 2003). The biosynthesis of this allelochemical has been demonstrated but a deep insight of its mechanism as well as biosynthesis regulation is still limited (Belz 2007). Therefore, further study of molecular mechanism of momilactone B and its regulation can fulfill the overall picture of the chemical for the fight against weed infestation on rice fields as well as other crop plants.

15.1.1 The Invasion of Weeds to Rice Fields

There are plenty of weeds that invade and threaten the boundary of rice colony. The most invasive two species within the genus *Echinochloa* are *E. crus-galli* (barnyard grass) and *E. colona*. These two species are widely observed in rice production sector of all rice-exporting countries (Jankulovski and Khan 2022). Noticeably, it was found that *E. crus-galli* is one of the most 10 concerning weeds that can be resistant to herbicides due to its evolution to withstand at least eight different acts of herbicide. The widespread application of herbicide on rice fields and the long-time coexistence has allowed many weeds species to adapt and thrive (Heap 2014). Notably, 51 weed species were reported of their tolerance against the long-term usage of herbicide in rice fields (Peterson et al. 2018). It is likely to state that the effect of herbicide on the invasive weed is becoming less effective in the future. Another problematic fact is that the seedling morphology of this grass species is very resemble to that of rice, making the discrimination between crops and weeds become difficult to farmers, who occasionally call this phenomenon as "Vavilovian mimicry" or "crop mimicry" (Ye et al. 2019). All of the concerning problems mentioned above indicate that there should be an alternative approach to deal with the "disturbing" weeds in a more sustainable way, which is affordable, environment friendly, and can inhibit the resistance within weeds (Gomiero et al. 2011; Mtenga et al. 2019; Weis et al. 2012).

Although in general, the attributes of rice weeds are diverse in several aspects such as anatomy, biology and physiology (Vaughan et al. 2001) and the weeds usually share the similar outlook with rice plants, the differences between them are not impossible to recognize. The weedy rice plants are more resemble to rice plants in their seedling stage and starts to grow differently from their tillering stage as the tillers are more abundant, longer and slenderer. Moreover, in this phase, stiff bristles are observed in both sides of leaves surface, pigmentation can be identified in several plant parts; the grains have awns, the pericarp is red and the seeds are shattering (Kwon et al. 1992; Shu et al. 1997). Espinoza et al. (2005) and Chauhan (2014) also indicated the distinctive features of weedy rice plants to the cultivated rice including the more brittle and rounder stem, the softer and spongier leaf sheath surface of the former in compared with the latter. Notably, anthocyanin is the pigmentation that mostly found in the first leaf and auricles of the weedy rice plants.

Similar to rice plants, the germination of weedy rice seeds is restricted in saturation soil; however, in the mature stage, weedy rice seeds are likely to shatter and exhibit a diverse range of dormancy (Perreto et al. 1993). Famers usually eliminate the weedy seeds lately due to they grow up rapidly and shatter right after to construct and solidify the seed bank. Sastry and Seetharaman (1973) have found that in conditions of dominant homozygosis (Sh Sh) or heterozygosis (sh Sh), the gene Sh, regulates the typical early shattering of weedy rice seeds. Nine days after the flowering stage was the time for seed shattering, recorded and investigated by Ferrero and Vidotto (1998) and Ferrero and Vidotto (2010). The first two authors revealed that the germination reach 65% of the total grains after 30 days, while in the other, author claimed that germination rate was up to 85% after only 12 days after flowering. It was reported that the trait of weedy rice is tend to be dependent on environmental factors during its seed development, storage, and germination (Nair et al. 1994; Gu et al. 2006). Footitt and Cohn (1992) indicated that the shrink in pH value of the embryo tissue is the signal for the breaking of weedy rice dormancy. The depth of soil for sowing seed is another notable factor influencing to the survival rate of weedy rice seeds as it declined by 6% after one year and 5% after 2 years of burial, employing ploughing technique in loamy soil (Ferrero and Vidotto 1998). However, although numerous studies on seed dormancy were conducted, the insight of mechanism regulating the germinability remained obscure (Foley 2001; Koornneef et al. 2002). Veasey et al. (2004) reported that favorable condition for germination and seedling viability is high moisture of the environment since the duration of seed dormancy is longer in dry region than wet region. The factors that govern seed longevity were analyzed by Noldin (1995), consisting of burial depth, category of soil, moisture, cultivation procedure, the scale of seed production and dormancy intensity. Besides, the texture of soil, amount of water and the burial depth are the main regulators for the emergence of rice weeds (Ferrero and Finassi 1995). The temperature of at least 10 °C can assure the normal germination of weedy rice while the optimum depth for good germination is 4 cm. It is necessary to bring seeds on to the surface in subsequent ploughing practices if farmers want to break the dormancy of the seeds.

Yield reduction is frequently observed as a consequence of weedy rice infestation and is controlled by numerous factors including (1) season, (2) weed varieties and density, (3) rice species, growth rate and density. It was reported that yield reduction of 60% is the consequence of weedy rice infestation of 35% and the loss can be up to 74% by the act of direct seeded rice (Watanabe et al. 1996). As an increase in the number of weedy seed per meter square (10, 100 and 1000), the yield of weedy rice declined from 4.05 down to 2.75, and 0.43 t/ha (Chin and Siddiqui 2000). In fact, shorter rice species can tolerate weedy rice weaker than the tall varieties (Kwon et al. 1991). Climate change can worsen the problem due to the incline of CO_2 content in the atmosphere, which can trigger the competition ability of weedy rice plants in rice production system.

There are four elements such as land preparation, water management, varietal selection and fertilizer management that need to focus in terms of adopting weed control methods of nonchemicals. Methods to prevent the colonization of weedy rice regardless of chemical utilization should aim for two targets: (1) minimize the germination possibility of weedy rice seeds at crop establishment; (2) avert the seed to drop back to the soil from surviving mature plants. The first approach consists of appropriate tillage to clean out the seedbeds, utilization of rotational methods for crop establishment (transplanting, water and wet-direct seeding) and selection of varieties for good water management. However, these kinds of approach can lead to the reduction of pant survival rate. Tough problems in the case of red rice have been tested by the used of minimum tillage systems. Notably, subsequently to the land preparation, the growing area was intentionally abandoned for the growth of both red rice and weed to form a mulching cover. The application of nonselective herbicides was used before germination procedures (drilling or water seedlings) and as soon as the emergence of the plants, the growing area should be flooded to maintain the degree of weed control. An investigation of Chauhan (2014) reported the difference between two methods of crop establishment (transplanting and broadcasting), revealing the decline in weedy rice seed production from 96–98% to 71–87% as compared to direct-seeding approach. Other methods for the control of weedy rice were the utilization of suppressing and submergence tolerant species. It was observed that the taller and longer cycle varieties tend to compete greater than the others (modern early and semi dwarf). The selection of glufosinate, nonselective herbicides, for herbicide tolerant crop cultivars is the alternative and novel strategy to deal with the chemical issue (Sankula et al. 1997). The most effective methods to reduce weed infestation can be considered as rotational crop growth in temperate climate, introducing other crop cultivar to grow in cycle such as soybean, maize, wheat, sunflower, sorghum, etc. Watanabe et al. (1998) reported that the rotational cultivation of mung bean in Vietnam obtained a promising result as the practice reduced the growth of weedy rice plants and other species significantly.

An alternative approach that has been considered to be encouraging is to focus on the innate ability of crops so-called allelopathy, which has been developed by several plants, allowing them to constraint the growth of other plants by the group of compounds known as allelochemicals (Wezel et al. 2014). The definition of allelopathy was first developed by Molisch (1937), describing the direct and indirect effects of biochemical compounds transmitted between two different plants. Thereafter, it has attracted interests of scientists to study on the possibility of allelopathy to become the better approach in replace of synthetic herbicides (Macias et al. 2003). In fact, for a more sustainable agriculture, criteria have been set to direct the target study on allelopathy: (1) The inhibition effects of allelochemicals in crops need to be strong and selective to the objective plants (Zhao et al. 2021); (2) The endogenous allelochemicals should not affect the crop plants themselves; moreover, the plants should selectively drive the production of these substances, especially when they can sense the presence of the unwanted weeds in their territory (Prasad et al. 2014); (3) The impact range of the allelochemicals need to be no further than the "target plants," otherwise, it can put stresses to the surround environment and other harmless organism (Kong et al. 2014; Habben et al. 2014). Studies indicated that several rice species exhibit allelopathy that fits at least the two former requirements but research on the impact of rice allelopathy on fungi and animals are limited (Toyomasu et al. 2014; Dilday et al. 1994; Olofsdotter et al. 2002).

15.1.2 Momilactones Are the Main Allelochemicals in Rice

With the motivation to identify the target allelochemicals due to the rise of many reports indicated allelopathic potential in rice species, some researches were conduct (Dilday et al. 1994; Olofsdotter et al. 2002). At first, phenolic acid was considered the possible allelochemicals of rice plants (Hartley and Whitehead 1985). However, the content of these compounds was not adequate to inhibit the growth of the "target plants" root (Lovett and Ryuntyu 1992); thus, the attention to the main allelochemicals of rice was moved to the other substances (Kato et al. 1977). Eventually, the question has been answered when the Japanese scientists have been successfully isolated momilactone A and B from rice husks, which are both named after the meaning of the husk in Japanese "*momi*" (Kato et al. 1977). It was found that *E. crus galli*, *Lactuca sativa*, and other lowland weed species tended to grow closer to rice plants lacked of momilactone than the other wild type rice plants (Kodama et al. 1988). Rice shoots, roots, and root exudates were the other areas containing momilactone, but at micromolar levels (Kato-Noguchi et al. 2007, 2012; Hasegawa et al. 2010). In comparison between the activity of the two mentioned allelopathic substances, momilactone B was far stronger than momilactone A as the former substance can inhibit the development of *Arabidopsis thaliana* 17 times greater than the latter compound (Kato-Noguchi et al. 2012). However, it was revealed that rice tolerance against fugal pathogens has been more exhibited by the act of momilactone A (Kato-Noguchi et al. 2012; Helliwell and Yang 2013; Kato et al. 1973).

Generally, in this book chapter, we illustrate the structure of momilactone B, revise the techniques used to isolate and purify momilactone B, its content in rice and some of other plants, the main bioactivity on weeds and other potential significant effects before elaborating its relevance within the scope of agriculture sector. To further study on this aspect of Momilactone B for the success application on practice, a broad and deep understanding of allelopathy in rice is necessary.

15.2 Chemical Structure of Momilactone B

Momilactone B: Colorless crystalline compound; R_f 0.42 ($CHCl_3$:MeOH; 9.5:0.5); m.p. 240 °C; IR ν_{max}: 2920, 1737, 1662, 1637, 1461, 1296, 992, 916; ^1H NMR ($CDCl_3$; 500 MHz): δ 1.99 (m, H 1α), 2.13–2.06 (m, complex H 2, H 14), 2.20 (dd, $J = 6.5, 2.0$, H 5), 4.97 (t, $J = 4.5$, H 6), 5.68 (d, $J = 5.0$, H 7), 1.72–1.64 (m, H 9, H 11α), 1.30 (m, H 11β), 1.56–1.51 (m, complex, H 1β, H 12), 5.82 (dd, $J = 17.0, 11.0$, H 15), 4.93 (d d, $J = 10.0$ & 1, H 16), 0.87 (s, H 17), 1.43 (s, H 18), 3.58, 4.07 (dd, 9.0, 3.1 7 9.0, 3.5). ^{13}C NMR ($CDCl_3$; 125 MHz): δ 28.81 (C 1), 26.44 (C 2), 96.60 (C 3), 50.35 (C 4), 42.97 (C 5), 73.76 (C 6), 114.00 (C 7), 146.70 (C 8), 44.68 (C 9), 30.74 (C 10), 24.79 (C 11), 37.22 (C 12), 39.99 (C 13), 47.42 (C 14), 148.83 (C 15), 110.23 (C 16), 21.86 (C 17), 18.99 (C 18), 180.48 (C 19), 72.72 (C 20); HPLC PDA MS ESI$^+$: 331 $[M + H]^+$ ($C_{20}H_{27}O_4$); ESI$^-$: 329 $[M - H]^-$ ($C_{20}H_{25}O_4$); HRMS 330.1905 $[M + H]^+$ (calc for $C_{20}H_{27}O_4$, 315.1909) (Cartwright et al. 1981; Kim et al. 2007; Mennan et al. 2012) (Fig. 15.1).

15.3 Isolation and Purification of Momilactone B

Rice husks was the material used for the first isolation and identification of momilactone B, which was initially considered as an inhibitor to the germination and root elongation of rice seedlings. Rice leaves and straws were the other two parts that contain momilactone B, which was then identified as phytoalexins. For the quantification of momilactone B, many techniques were employed such as thin layer chromatography and flame ionization detection (TLC/FID), thin layer

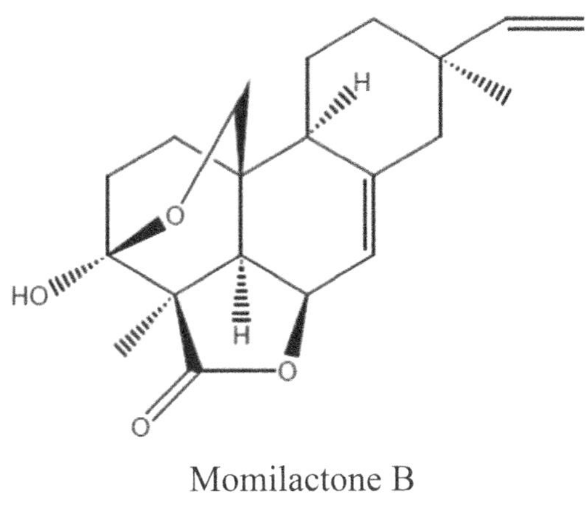

Fig. 15.1 Chemical structures of momilactone B

chromatography (TLC), reverse phase high-performance liquid chromatography (RP/HPLC), high-performance liquid chromatography tandem mass spectrometry (HPLC/MS/MS), gas chromatography mass spectrometry (GC/MS), and chromatography mass spectrometry in the selected ion monitoring mode (GC/MS/SIM). Cold percolation and different solvent systems of methanol or methanol:water (8: 2 *w*/*w*) were examined to separate MB, but the result was not very promising as the actual yields of MB were low, revealing the value of 0.5 to 10 μg/g husks. It was reported that rice varieties, growing stage, and extraction methods are the driving factor of MB yield. An approach to enrich the yields of natural products was given, suggesting a combination between the higher temperature and relevant extracting solvents (Minh et al. 2018).

15.4 Content of Momilactone B in Rice and Moss Plants

MB was isolated from rice husks as plant growth inhibitors. The chemical structure of MB was determined as 3,20 epoxy 3α hydroxy synpimara 7,15 dien 19,6β olide, $C_{20}H_{26}O_4$. Secretion of momilactone B was later confirmed for other rice cultivars as well. MB was further found in various rice cultivars and origins. To date, the biological activities of MB have been limited to allelopathy, antioxidant, antifungal, and antimicrobial activities, although the cytotoxic and antitumor activity of MB on human colon cancer cell was reported. The medicinal and pharmaceutical properties of MB have not been much known, as the isolation and purification of MB are complicated and laborious. There are only several laboratories worldwide have worked on MB, and thus, no standards of MB can be purchased. This fact has prevented us from understanding the physiological roles of this compound in rice plants and moss plants (Minh et al. 2018).

Table 15.1 summarized procedures and techniques (HPLC, LC/MS/MS, GC/MS, and GC/MS/SIM) employed so far to isolate and identify MB in different rice parts such as seedlings, and root exudates. Although reasons for the distinct taxonomies between rice and *H. plumaeforme* were concealed, MB was still detected in *H. plumaeforme* and effectively identified by HPLC and TLC. The content of purified MB identified by CC was 0.7 μg/g DW, which was in line with the same experiments (using CC for MB isolation) conducted by Kato et al. (1973) and Takahashi et al. (1976), revealing the value of MB content of 0.8 and 10 μg/g DW, respectively. The correlation among these mentioned results is explained by the usage of the same rice varieties (cv. Koshihikari). Chung et al. (2005) revealed that the content of MB in their study was approximately ten times higher (10.0 μg/g DW) than the previous studies. However, the husks they used as the experimental objective were originated from South Korea, which significantly contributed to the difference between other results. This study indicated that the yields of MB prepared by using both dried (100 °C, 1 h) and boiled (100 °C, 2 h) husks followed by MeOH (100%) extraction can be much higher than others studies (up to 104.4 μg/g DW). Besides, it was noted that the content of purified MB obtained by column chromatography was much less than the amounts of MB detected by GC MS. Indeed,

Table 15.1 Contents of MB in rice husks and moss plants

MB (μg/g DW)	Materials	Instruments	Extraction protocols
10.0	Rice husks	CC	Hexane:EtOAc (8:2)
3.1	Rice husks	CC	Hexane:EtOAc (8:2)
3.0	Rice straw	LC/MS/MS	EtOAc
2.9	Rice husks	GC/MS	EtOAc:H_2O (1:1)
95.0	Whole rice plants	HPLC	EtOAc
64.4–114.1	Seedling*	GC/MS	EtOAc
0.5	Rice husks	CC	$CHCl_2$ + EtOH
245.0	Rice shoots	HPLC	MeOH
64.1	Rice roots	HPLC	MeOH
Nd	Rice seeding	HPLC	MeOH: H_2O (8:2)
0.8	Rice husks	CC	Benzene:EtOAc (10:1)
9.6	UV irradiated rice leaves	GC/MS/SIM	MeOH
37.8	Rice husks	GC/MS	EtOAc:H_2O (1:1)
23.4	*Hypnum plumaeforme* L.	HPLC	EtOAc
4.2	*Hypnum plumaeforme* L.	TLC	EtOAc
0.7	Rice husks	CC	Hexane:EtOAc (8:2)
3.0–104.4	Rice husks**	HPLC	EtOAc; MeOH; Temperature; Pressure

nd not detected, *DW* dry weight, *CC* column chromatography, *HPLC* high-performance liquid chromatography, *GC/MS* gas chromatography mass spectrometry, *GC/MS/SIM* gas chromatography mass ion monitoring

* Differences in quantities were observed across rice origins and subtypes; ** Levels of MA and MB exhibited variation and enrichment among different extractions

37.8 μg/g DW is the value of MB that Chung et al. found in his study while the actual number dropped to only 3.1 μg/g DW when column chromatography was employed (Chung et al. 2006). Meanwhile, in this study, the theoretical level of MB quantified by HPLC (104.4 μg/g DW) were approximately 50 times higher than the actual yields of MB purified by CC (0.7 μg/g DW).

It can be said that hexane, MeOH, EtOAC, and water are the most favorable system of solvents for the extract of MB even though benzene and $CHCl_3$ are also within the list since the use of the last two solvents can be hazardous and toxic. In regard to the reduction of hazard within the extraction protocol of MB, the use of EtOH can be considered instead of MeOH. MeOH $\leq$50% or distilled water only could not yield the optimized level of purified MB. The content of MB was enriched and found at high level μg/g DW but it was lower than phenolic and fatty acids, two other allelochemicals detected in rice husks (Minh et al. 2018).

The quantity enrichment of MB was demonstrated to be regulated by the temperature modification (Tables 15.1 and 15.2). It was shown that the change in MB can be observed by the procedure of 100 °C in 3 h. The measurement by HPLC was

Table 15.2 Yields of ethyl acetate (EtOAC) extract MB in different extractions

Methods	EtOAc crude extract (g)	MB (μg/g DW)
Controls (standards MA and MB by column chromatography)	0.17	0.70 ± 0.03 h
MeOH 100%	0.50	42.80 ± 8.76 cdef
MeOH 70%	0.46	67.81 ± 4.76 bc
MeOH 50%	0.12	nd
MeOH 30%	0.10	nd
MeOH 10%	0.10	nd
Distilled water (room temperature)	0.04	nd
Distilled water (100 °C)	0.01	nd
Distilled water (100 °C, 30 min) + MeOH 100%	0.67	63.80 ± 5.33 cd
Distilled water (100 °C, 30 min) + EtOAc 100%	0.50	49.63 ± 2.49 cde
Distilled water (100 °C, 1 h) + MeOH 100%	0.45	3.02 ± 0.02 g
Distilled water (100 °C, 2 h) + MeOH 100%	0.50	102.23 ± 5.32 ab
Distilled water (100 °C, 3 h) + MeOH 100%	0.50	45.65 ± 2.62 cdef
Distilled water (100 °C, 4 h) + MeOH 100%	0.50	30.65 ± 1.38 ef
Dried (100 °C, 1 h) + MeOH 100%	0.45	104.43 ± 6.44 a
Dried (100 °C, 1 h) + MeOH 70%	0.40	29.68 ± 1.89 ef
Dried (100 °C, 1 h) + MeOH 50%	0.17	nd
Dried (100 °C, 1 h) + MeOH 30%	0.06	nd
Dried (100 °C, 1 h) + MeOH 10%	0.20	nd
Dried (100 °C, 1 h) + distilled water (room temperature)	0.16	nd
Dried (100 °C, 1 h) + distilled water (100 °C)	0.62	nd
Dried (100 °C, 1 h), distilled water (100 °C, 1 h) + MeOH 100%	0.31	14.04 ± 0.33 f
Dried (100 °C, 2 h), distilled water (100 °C, 2 h) + MeOH 100%	0.50	35.19 ± 2.43 def
Dried (100 °C, 3 h), distilled water (100 °C, 3 h) + MeOH 100%	0.60	53.03 ± 2.45 cde
Dried (100 °C, 4 h), distilled water (100 °C, 4 h) + MeOH 100%	0.40	20.26 ± 1.51 f
Distilled water (100 °C, 120 kPa) + MeOH 100%	0.40	40.78 ± 2.82 cdef
Dried (100 °C, 120 kPa) + MeOH 100%	0.50	71.00 ± 6.14 bc

Values with similar letters in each column are not significantly different ($p < 0.05$); Values are means ± SD (standard deviation)
EtOAc ethyl acetate, *MB* momilactones B, *nd* not detected

conducted after the extraction treatment and showed the high yield of MB (Minh et al. 2018). Nevertheless, the variation of MB in rice husks due to different treatments needs further researches. Furthermore, Germain and Deslongchamps noted that high yields of MB might need the temperature above 150 °C (Germain and Deslongchamps 2002). An oxygen bridge in a lactone ring of MB might be the answer for the employment of higher temperature to isolate MB; however, these hypotheses need more evidence. It was stated that rice cultivars and origins are the important factors explaining the deviation content of MB and thus the amounts of MB in Kato et al. (1973) and Takahashi et al. (1976), and this study was not much different since the Koshihikari cultivar was employed to conduct the experiments. The purification of MB, utilizing column chromatography in this study resulted in the value of 0.7 μg/g DW. Meanwhile, the content of purified MB in the work of Kato et al. (2002) and Takahashi et al. (1976) was 0.5 and 0.8 μg/g DW, respectively. Although pressure did not play a vital role in yielding the highest content of MB, 40.78–71.0 μg/g DW were still considered as the encouraging number in this context (Minh et al. 2018).

15.5 Role of Momilactone B as a Phytoalexin

Plants use a complex defensive system, which includes phytoalexins as the major factor to resist pests and pathogens. These groups of compounds are low in molecular weight and produced by and compiled in plants to fight against biotic and abiotic stresses. Müller and Börger initiated in defining the idea of phytoalexins 70 years ago while observing the significant repulsion of potato infection against the second strain of *Phytophthora* by exhibiting hypersensitive reactions. Since then, the principle of phytoalexin was associated with the hypersensitive reaction of the plant cells.

The understanding of phytoalexins mostly originated from thorough studies of several plant families including *Fabaceae, Solanaceae, Orchidaceae, Chenopodiaceae, Compositae, Convolvulaceae, Ginkgoaceae,* and *Poaceae*. Recently, plant families such as *Poaceae* (maize and rice), *Vitaceae* and *Malvaceae* (cotton) were the main subjects for phytoalexins studies. Notably, the production of Camalexin in *Arabidopsis thaliana* as well as its related biosynthetic pathway and regulatory networks has attracted attentions of many scientists. However, it is necessary to conduct more research of the diversity of phytoalexins within the plant kingdom (Jeandet et al. 2010; Sunilkumar et al. 2006).

Substances generated from remote precursors limit the activities of phytoalexins via an enzymatic synthesis. This peculiarity brings difficulties to the decipherment of its biosynthesis as well as the regulation mechanisms (Jeandet et al. 2013). It is considered that controllers of the adjustment in producing phytoalexins as well as inhibiting pathogen include phosphorylation cascades, defensive marker genes, sensors and elicitors of calcium and hormone signaling. As a result, the understanding of phytoalexin accumulation can be seen as the foundation of the genetic adjustment of those substances in engineered plants for the improvement in tolerance capability against diseases (Jeandet et al. 2013).

Groups of compounds including zealexins, kauralexins, momilactones, oryzalexins, phytocassanes, flavanones, 3 deoxyanthocyanidins, and phenylamides are the major phytoalexins of Poaceae (maize, rice and sorghum). Researches on luteolinidin and apigeninidin in sorghum as well as zealexins and kauralexins in maize were studied recently, contributing further to the current knowledge of phytoalexins (Delaunois et al. 2014).

Allelopathy in rice was demonstrated when cultivating closely with plants as various genus of rice can suppress the development of several plant species, in both conditions of field and laboratory conditions (Dilday et al. 1998). Momilactone B found in root exudates was then identified as the active substances which represent the inhibition effect (Kato-Noguchi et al. 2002). Authors noted that the release of momilactone B accumulated in roots led to an increase in its level in the growing medium, exerting the suppressive effects on plant (Kong et al. 2004). Thereafter, rice secretory fluid was the other part identified to contain momilactone B (Kato-Noguchi et al. 2008). Thus, it can be said that momilactone B is a key factor to represent inhibitory effects as well as allelopathy in rice.

In is notable that rice tissues or roots were not the part to accumulate or release momilactones habitually. In fact, momilactone secretion or exudation were seem to be dependent on external factors since the significant increase in the level of momilactone B was observed in rice plants grown beside barnyard grass or treated with barnyard grass exudates. Furthermore, the stimulation of momilactone production and/or exudation in rice plant was also detected with the presence of other weed species *Eclipta prostrata* and *Leptochloa chinensis* (Yang and Kong 2017). Thus, this finding can be considered the basis for these three ideas: first, the nearby plants can play as a primary factor to stimulate the allelopathic responses based on the activity of momilactone B; second, substances in root exudates of several weeds, so-called inducers, can trigger the yield and excretion of momilactone in rice; and finally, inducers in the exudation process are likely cumulative rather than an outcome of the interaction between rice plants and weeds (Chagas et al. 2018). Besides, it seems that the presence of a substance, so-called a repressor, excreted by rice can be seen in this context because it is considered related to a sophisticated stoichiometric regulation of the response. Other rice plants in the population can detect this kind of substance, which can negate the inducer by inhibiting its signaling response. An increase in the concentration momilactone is observable under the impact of induction but it cannot be confirmed that the rising content is an output of a more productive biosynthesis or due to the activation of a novel derivative of momilactone B (Fig. 15.2) (Sørensen and Møller 2021).

An incline in the momilactone B level and exudation is detected with the growth of the unfriendly neighbors. A certain and unknown signal, possibly a chemical compound, of the root exudates can be identified by the rice plant. As such, the key factor for the increase of momilactone B level is still there to be unveiled. Downstream signaling could act as a transcriptional activation of biosynthetic process, leading to a more effective production. Another circumstance is that the biosynthetic

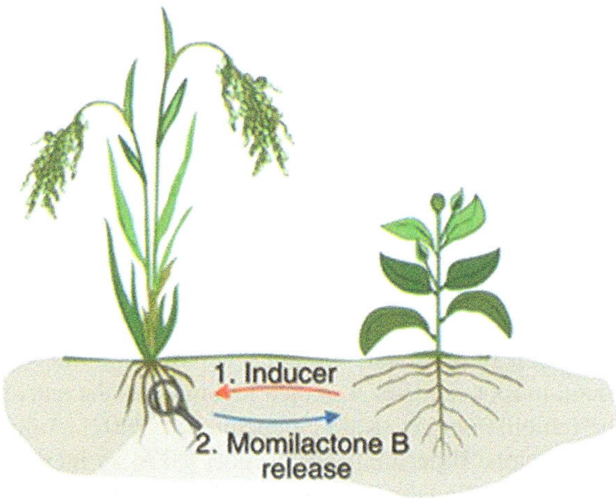

Fig. 15.2 Momilactone B release in the soil

enzymes is kept inactivated until the induction occurs, where its state are switched and the enzymes itself are free, leading to an increase in the synthesis rate. Thus, there are two possible ways to regulate momilactone B: an activated momilactone B is stimulated by the induction and momilactone B can be stored in the cell, e.g., momilactone B is kept in the oil bodies of *C. plumiforme* (Murakami and Kudo 2004).

15.6 Inhibitory Activity of Momilactone B on Weeds

For some parasitic plants, the development of associations to the host plants is highly dependent on compounds secreted from their roots. These compounds would play the role of chemical cross talk between the host and the parasite, thus providing better understanding in plant-to-plant interactions, which include the location of parasite germination and the development of physical connections between both entities (Palmer et al. 2004; Callaway et al. 2002). Recently, some scientists have reported that allelopathic activity in rice was increased by the presence of barnyard grass. Three allelochemicals have also been determined, including momilactone B (Zhao et al. 2005). However, it was still not clear if the reason for such increase in allelopathic activity and allelochemicals were chemical mediated interactions with barnyard grass, or it was simply due to plant-to-plant interference reactions, which in turn were caused by competitive relationship between plants for essential resources such as nutrients, light, water, and even chemical interactions. For example, when incubated under nitrogen or phosphate limited conditions, rice was observed to exhibit elevated allelopathic activity, suggesting that nutrient starvation played an important role in the increase of their allelopathic activity. In particular, Chung et al. (2002) reported significant increase in allelopathic activity of aqueous extracts from rice plants cultivated in nutrient competitive conditions with barnyard grass, and such increases were even greater with higher density of barnyard grass. Therefore, it is reasonable to believe that allelopathic activity of rice was mostly caused by nutrient competitive conditions, while the role of chemical mediated interaction between neighboring plants was not quite apparent in this case.

Due to their common occurrence in a wide range of plants, phenolic acids were routinely investigated as potential putative allelochemicals in various researches regarding the subject (Inderjit 1996). However, it is unlikely that they are responsible for allelopathic activity in rice, because the concentration of phenolic acids in rice root exudates never has never been found to be reaching phytotoxic levels (Wu et al. 2019). In contrast, momilactone B have been isolated from root exudates of various rice cultivars, including the rice cultivar Koshihikari (Kato-Noguchi et al. 2002), at concentrations greater than 1 μmol/L, which could effectively inhibit the growth of typical rice weeds such as *Echinochloa colonum* or barnyard grass.

Others also observed the secretion of momilactone B from the roots into the rhizosphere over the entire life cycle in various rice cultivars, thus making it a potential allelochemical in allelopathic research. In rice agriculture, especially in southern areas of China, barnyard grass (*Echinochloa crus-galli*) is often considered a major noxious weed; therefore, their presence is not welcomed. At concentrations higher than 1 μM, momilactone B has been found to be able to inhibit up to 59–82% of the growth of shoots and roots in barnyard grass, with an IC_{50} value of around 6.5 μM (Scognamiglio et al. 2013). Additionally, momilactone B has also been found to be able to inhibit the growth of roots and shoots in other monocots and dicots, albeit at different concentrations, including *Echinochloa colonum*, *Arabidopsis thaliana,* timothy (*Phleum pratense*), alfafa (*Medicago sativa*), lettuce (*Lactusa sativa* cv. Santanasu), ryegrass (*Lodium multiflorum*), cress (*Lepidium sativum*), Chinese cabbage (*Brassica rapa* cv. *Harumaki ichigou*), and crabgrass (*Digitaria sanguinalis*) (Kato-Noguchi et al. 2010). However, momilactone A has been determined to have much less allelopathic potency compared to momilactone B.

The growth inhibitory of momilactone compounds in barnyard grass has been linked to various different mechanisms, but mostly the altered expressions of miRNA relevant to nucleotide excision repair, the p53 signaling pathway, peroxisome proliferator activated receptor pathway (PPAR pathway), and plant hormone signal transduction (Kato-Noguchi and Kitajima 2015). When treated with momilactone A or B, the expressions of cruciferina, cruciferin, and cruciferin storage proteins which play pivotal role in providing nitrogen for seed germination—in Arabidopsis—were found to be up regulated. In other words, through inhibiting the degradation process of cruciferina and cruciferins, momilactone compounds could effectively inhibit the germination of Arabidopsis seeds. Under similar mechanism, momilactone B has been found to be capable of inhibiting the growth of Arabidopsis seedlings, through inhibiting the degradation process of subtilisins such as malate synthase, amyrin synthase LUP2, serine protease, and β-glucosidase—important proteins associated with the metabolic process, and the production of intermediate compounds necessary for the development of cell structures in plants (Kato-Noguchi and Ota 2013).

Additionally, momilactone B could also induce the production of 1 cysteine peroxiredoxin and glutathione S transferase—two proteins that play important roles in the defense responses of plants (Kato-Noguchi and Ota 2013). Interestingly, momilactone A and momilactone B have also shown inhibitory effects on the root and shoot growth of rice seedlings. However, such effects were mostly minimal, with positive results only recorded at concentrations greater than 100 and 300 μM, respectively. In other words, at concentration levels mostly harmless to rice seedlings, momilactone A and momilactone could still exhibit high cytotoxic activity to a wide range of common weeds (Özkara et al. 2016).

15.7 Potential of Biological Control of Weeds by Momilactone B in Agricultural Practice

Concerning issues such as environmental pollution, tolerant agricultural crops and risks on human health are the consequences of the widespread utilization of herbicides. The management of weed can be controllable by adopting distinctive properties of allelopathic plant and understanding the volatilization, leaching, and root exudations during the breakdown of plant residues can regulate the effect between plants and microorganisms. Allelopathic effects of various plant species have become an interesting topic for many scientists and it have been expected as the solution for weed control in rice ecosystems. A study covered nearly 12,000 rice accessions has been conducted and unveiled that the effects of rice allelopathy worked on duck salad, which is among the most common aquatic weeds of rice in the United States. Thereafter, the attention on allelopathic studies on rice has broadened to the boundary of Japan and Korea. Chung and Ahn have collected barnyard grass (*E.crus galli* Beauv. var. *oryzicola* Ohwi) as the subject for their allelopathic research on rice, preparing extracts from leaves, straw, and hulls (Chung et al. 2001). The higher allelopathic effects were observed in Korean rice varieties, at the middle age of growth and in those having hull and awn color. Moreover, temperature was another factor that controls the allelopathy of plant as it can directly affect the activity and saturation of allelochemical. The warm extracts exert more strains on the germination, seedling growth, weight, and caloric content of barnyard grass as compared to hot extracts. It is noted that rice cultivars were not only main factor causing the deviation of allelopathic potential in rice but different parts of rice plant and extract conditions (temperature, solvent, and time) also involved in the variation. According to the result in this study, it can be said that extraction time and temperature had the most important influence on allelochemical analysis results (Minh et al. 2018).

It was found that allelopathy in rice plant was likely dependent on the activity of several secondary compounds including terpenoids, steroids, phenols, coumarins, alkaloids, flavonoids, diterpenoid substances, momilactones, long chain hydrocarbons, and fatty acids. Among those, most of studies focus on the activity of phenolic compounds, which are considered phytotoxins. Employing thin layer chromatography, six allelopathic compounds similar to *p*-hydroxybenzoic acid were isolated successfully from rice straw. Moreover, over 90% of barnyard grass germination was constrained by 10^3 M of *p* coumaric acid, *m* coumaric acid, ferulic acid, and *p* hydroxybenzoic acid, which were derived from rice hull extracts. Song et al. also reported that long chain fatty acids, phenols, phthalic acids, and benzene derivatives were isolated from root exudates of rice varieties such as Sathi, Jinmi, and Nongan at the yellow ripening stage and that root exudates containing these chemicals significantly inhibited the height, dry weight, and number of barnyard grass tillers (Song et al. 2004). However, it was not evident for the link between phenolic compounds and allelopathy as the concentration of soil was always insignificant to exert toxicity.

In this context, momilactones, especially Momilactone B, turned out to be the reliable compound for the application of allelopathic properties on the metabolism of crop-friendly pesticide, the evolution of rice and moss defense as well as the sustainability of green agriculture. In fact, patent has been issued for the activity of momilactone A and B as herbicides and anticancer agents. However, supply of momilactone for the practical use is always limited on a large scale because it is still assumed as impractical for the economy as well as unfriendly to the environment despite momilactone A was successfully synthesized many years ago. Bioengineering is suggested as the alternative approach but investigations on this aspect have been limited. Another possible method is to shift the attention on momilactone-like molecules derived from natural products such as *Icacina*, *Casimirella,* and their affinitive genera. These kind of molecules in *Icacina* and *Casimirella* are quite different than momilactone found in rice and moss, which are metabolites inducing stress at limited quantities; they present within the plants' body at high content, with high structural diversity and can open a gate for further studies of promising lead compounds as well as precursors for semi synthesis (Kato-Noguchi and Kitajima 2015). Additionally, apart from their potential as crop-friendly herbicides, antifungal, antibacterial, and antitumor agents, further properties are there to discover such as anti-herpes as well as antiviral activity (Kato-Noguchi and Kitajima 2015). These new attributes of momilactone-like molecules (members of (9β H) pimarane lactones) are expected to inspire further research interests of scientists and experts (Olofsdotter et al. 1995).

15.8 Future Prospect

It is stated that the reaction to unrelated species have been the objective for researches on "kin recognition" of plants. The operation process interpreting kin and nonkin recognition has remained unclear so far. Restating the mentioned explanation, the response to unrelated neighbor plants is activated by several substances in their root exudates rather than by the direct interaction between the roots or by volatile compounds. Thus, there are three directions need to be aim for future researches:

- Identification of "non-kin" compounds detected by rice plants
- Explanation of the methods to recognize those compounds
- Determination of the downstream signaling pathways which activate and aid to produce allelochemical

Metabolomic techniques are required to tighten the ranges of compounds in charge of nonkin recognition. Profiles of root exudate substances from the relative plants of rice and other species has been compared by employing a technique of untargeted metabolomics; however, the study turned out impossible due to the vast range of bioactive compounds, which can make confusion for the authors. Besides, fractionation bioassays can become a candidate for the identification of allelochemical due to its continuous fractionation procedures, turning samples of root exudates into smaller fractions until the range of target compounds are narrowed. Kong et al. (2018) have applied this approach successfully as the potent inducers activating the DIMBOA production in wheat in more than 100 species have been addressed (Olofsdotter et al. 1995).

Forthcoming researches requires an insight in the genetic aspects, e.g., membrane-bound or intracellular receptors, to discover the mechanism of rice to recognize the inducers and to call out the associated downstream signaling pathways for the information transmit which cause the rising content of momilactone (Hoang Anh et al. 2022). The identification of unknown genetic compositions is still effective when using forward genetics screens despite its laborious procedures. Besides, genetics techniques, which can quantify wide range of germplasm, are potential for the identification of target alleles linked to momilactone inducibility. Challenge for these two techniques is the requirement of a feasible approach to determine efficiency to generate momilactone as well as its concentration in a wide range of plants. Reporter lines or high-efficiency metabolomics are the two suggested approaches to overcome the challenge.

Allelopathic response in rice is one of their natural defense mechanisms against noxious weeds, thus improving such response could in turn reduce the requirement for herbicide and manual removal of weeds during cultivation, while also potentially increasing production yield. In fact, generating rice varieties with increased allelopathic capacity had already been one of the main targets of breeding efforts in the recent past. However, such an approach had also encountered many difficulties caused by unknown factors, which significantly hampered its efficiency.

Aside from that, researchers are also motivated to study and understand the mechanism of allelopathic response in rice, because it would consequently provide them with more knowledge regarding the mechanistic basis of autotoxicity resistance/tolerance in rice. In particular, an ideal rice cultivar is the one that exhibits high allelopathic response against noxious weeds, but at the same time adequately tolerant to momilactone B. In another word, there should be no yield losses caused by allelopathic response, or the losses should at least be significantly less than potential losses due to the infestation of noxious weeds. Additionally, the increase in energetic and resource cost for cultivating highly allelopathic rice variations should also be taken into account, since such investments could potentially lead to negative economic results.

Resistance against allelopathic compounds in donor plants could be achieved through selected mutagenesis. In particular, if the targets of an allelopathic compound—such as momilactone B—were known, then we could generate resistant variants through selected mutagenesis, so that the resistant variants are highly insensitive to the allelopathic compound. Alternatively, resistance or increased tolerance could be screened by other methods such as mutagenized population, or germplasm collection. However, currently, there have been no systematic analyses on the relationships between important factors regarding this problem, including autotoxicity, strength of inducibility, and constitutive momilactone levels. Since it is safe to assume that at least some highly allelopathic rice cultivars should also have high autotoxicity tolerance, the authors believe it would be worth the efforts to search for rice cultivars that are both low cost and capable of producing high levels of momilactone B. In fact, over 20 years ago, such efforts were initiated by Maria Olofsdotter and her colleagues (Olofsdotter et al. 1995, 1999). However, after her death, the research had been largely abandoned. Now, while high-throughput metabolomics can be achieved at affordable costs thanks to scientific advances, it should be more feasible than ever to resume such efforts in screening for the ideal allelopathic rice cultivar.

15.9 Conclusion

The topic regarding momilactone B illustrates its division into two different paths of development: (1) The wide recognition of allelopathic effects of momilactone B has been affirmed by numerous reliable studies, while (2) the secret of genetic basis of the mechanism to constrain the growth of foreign weeds as well as the process of biosynthesis and inducibility is still unclear. Successes on the study of rice allelopathy open a wide gate for the breeding of rice cultivars to aim for novel species capable of withstanding and resisting weeds efficiently and sustainably. In addition, the clarification of function groups in the chemical structure of momilactone B on weed suppression is indispensable. Subsequently, the synthesis of the functional groups which are responsible for weed suppression is promising to develop novel herbicides. This chapter is expected to motivate more scientists and experts for the extensive study regarding this topic.

References

Bayer DE (1991) Weed management. In: Rice. Springer, Boston, pp 287–309

Belz RG (2007) Allelopathy in crop/weed interactions—an update. Pest Manag Sci 63(4):308–326

Callaway RM, Brooker RW, Choler P, Kikvidze Z, Lortie CJ, Michalet R, Paolini L, Pugnaire FI, Newingham B, Aschehoug ET, Armas C, Kikodze D, Cook BJ (2002) Positive interactions among alpine plants increase with stress. Nature 417(6891):844–848

Cartwright DW, Langcake P, Pryce RJ, Leworthy DP, Ride JP (1981) Isolation and characterization of two phytoalexins from rice as momilactones A and B. Phytochemistry 20(3):535–537
Chauhan BS, Abeysekera ASK, Wickramarathe MS, Kulatunga SD, Wickrama UB (2014) Effect of rice establishment methods on weedy rice (Oryza sativa L.) infestation and grain yield of cultivated rice (O. sativa L.) in Sri Lanka. Crop Prot 55:42–49
Chagas FO, de Cassia Pessotti R, Caraballo-Rodríguez AM, Pupo MT (2018) Chemical signaling involved in plant–microbe interactions. Chem Soc Rev 47(5):1652–1704
Chin OC, Siddiqui KM (2000) Characteristics of some biomass briquettes prepared under modest die pressures. Biomass Bioenergy 18(3):223–228
Chung IM, Ahn JK, Yun SJ (2001) Assessment of allelopathic potential of barnyard grass (*Echinochloa crus-galli*) on rice (*Oryza sativa* L.) cultivars. Crop Prot 20(10):921–928
Chung IM, Kim KH, Ahn JK, Chun SC, Kim CS, Kim JT, Kim SH (2002) Screening of allelochemicals on barnyardgrass (*Echinochloa crus-galli*) and identification of potentially allelopathic compounds from rice (*Oryza sativa*) variety hull extracts. Crop Prot 21(10): 913–920
Chung I-M, Hahn S-J, Ahmad A (2005) Confirmation of potential herbicidal agents in hulls of rice, *Oryza sativa*. J Chem Ecol 31(6):1339–1352. https://doi.org/10.1007/s10886-005-5290-5
Chung IM, Kim JT, Kim S-H (2006) Evaluation of allelopathic potential and quantification of momilactone A, B from rice hull extracts and assessment of inhibitory bioactivity on paddy field weeds. J Agric Food Chem 54(7):2527–2536
Delaunois B, Farace G, Jeandet P, Clément C, Baillieul F, Dorey S, Cordelier S (2014) Elicitors as alternative strategy to pesticides in grapevine? Current knowledge on their mode of action from controlled conditions to vineyard. Environ Sci Pollut Res 21:4837–4846
Dilday RH, Lin J, Yan W (1994) Identification of allelopathy in the USDA-ARS rice germplasm collection. Aust J Exp Agric 34(7):907
Dilday RH, Yan WG, Moldenhauer KAK, Gravois KA (1998) Allelopathic activity in rice for controlling major aquatic weeds. In: Olofsdotter M (ed) Allelopathy in rice. International Rice Research Institute, Manila, pp 7–26
Espinoza F, Diaz Y, Argenti P, Quintana H, Leon L (2005) Use of leucaena meal (Leucaena leucocephala) on the post-weaning lamb feeding during dry season. Revista de la Facultad de Agronomia, Universidad del Zulia 22(1):42–53
Ferrero A, Finassi A (1995) Viability and soil distribution of red rice (Oryza sativa var. Silvatica) seeds [wild rice, a weed in most rice growing areas in the world]. Mededelingen-Faculteit Landbouwkundige en Toegepaste Biologische Wetenschappen Universiteit Gent (Belgium)
Ferrero A, Vidotto F (1998) Germinability after flowering, shattering ability and longevity of red rice seeds. Pages 205–211
Ferrero A, Vidotto F (2010) 11 History of rice in Europe. Rice: origin, antiquity and history, 341
Footitt S, Cohn MA (1992) Seed dormancy in red rice: VIII. embryo acidification during dormancy-breaking and subsequent germination. Plant Physiol 100(3):1196–1202
Foley ME (2001) Seed dormancy: an update on terminology, physiological genetics, and quantitative trait loci regulating germinability. Weed Sci 49(3):305–317
Germain J, Deslongchamps P (2002) Total synthesis of (±)-momilactone A. J Org Chem 67(15): 5269–5278
Gomiero T, Pimentel D, Paoletti MG (2011) Is there a need for a more sustainable agriculture? CRC Crit Rev Plant Sci 30(1–2):6–23
Gu XY, Kianian SF, Foley ME (2006) Dormancy genes from weedy rice respond divergently to seed development environments. Genetics 172(2):1199–1211
Habben JE, Bao X, Bate NJ, DeBruin JL, Dolan D, Hasegawa D, Helentjaris TG, Lafitte RH, Lovan N, Mo H, Reimann K, Schussler JR (2014) Transgenic alteration of ethylene biosynthesis increases grain yield in maize under field drought-stress conditions. Plant Biotechnol J 12(6): 685–693

Hartley RD, Whitehead DC (1985) Phenolic acids in soils and their influence on plant growth and soil microbial processes. In: Soil organic matter and biological activity. Springer, Dordrecht, pp 109–149

Hasegawa M, Mitsuhara I, Seo S, Imai T, Koga J, Okada K, Yamane H, Ohashi Y (2010) Phytoalexin accumulation in the interaction between rice and the blast fungus. Mol Plant Microbe Interact 23(8):1000–1011

Heap I (2014) Herbicide resistant weeds. In: Integrated pest management. Springer, Dordrecht, pp 281–301

Helliwell EE, Yang Y (2013) Molecular strategies to improve rice disease resistance. Methods Mol Biol 956:285–309

Hoang Anh L, Van Quan N, Quang Lam V, Takami A, Dang Khanh T, Dang Xuan T (2022) Rice momilactones and phenolics: expression of relevant biosynthetic genes in response to UV and chilling stresses. Agronomy 12(8):1731

Inderjit (1996) Plant phenolics in allelopathy. Bot Rev 62(2):186–202

Jankulovski N, Khan UE (2022) Asian rice exports and the European demand: an application of the modified panel gravity model. J Agric Sci 14(8):18

Jeandet P, Delaunois B, Conreux A, Donnez D, Nuzzo V, Cordelier S, Clément C, Courot E (2010) Biosynthesis, metabolism, molecular engineering, and biological functions of stilbene phytoalexins in plants. Biofactors 36(5):331–341

Jeandet P, Clément C, Courot E, Cordelier S (2013) Modulation of phytoalexin biosynthesis in engineered plants for disease resistance. Int J Mol Sci 14(7):14136–14170

Kato T, Kabuto C, Sasaki N, Tsunagawa M, Aizawa H, Fujita K, Kato Y, Kitahara Y, Takahashi N (1973) Momilactones, growth inhibitors from rice, *Oryza sativa* L. Tetrahedron Lett 14(39): 3861–3864

Kato T, Tsunakawa M, Sasaki N, Aizawa H, Fujita K, Kitahara Y, Takahashi N (1977) Growth and germination inhibitors in rice husks. Phytochemistry 16(1):45–48

Kato-Noguchi H, Ino T, Sata N, Yamamura S (2002) Isolation and identification of a potent allelopathic substance in rice root exudates. Physiol Plant 115(3):401–405

Kato-Noguchi H, Ino T (2003) Rice seedlings release momilactone B into the environment. Phytochemistry 63(5):551–554

Kato-Noguchi H, Kitajima S (2015) Momilactone sensitive proteins in Arabidopsis thaliana. Nat Prod Commun 10(5):729–732

Kato-Noguchi H, Ota K (2013) Biological activities of rice allelochemicals momilactone A and B. Rice Res Open Access 1:108

Kato-Noguchi H, Kujime H, Ino T (2007) UV-induced momilactone B accumulation in rice rhizosphere. J Plant Physiol 164(11):1548–1551

Kato-Noguchi H, Ino T, Ota K (2008) Secretion of momilactone A from rice roots to the rhizosphere. J Plant Physiol 165(7):691–696

Kato-Noguchi H, Hasegawa M, Ino T, Ota K, Kujime H (2010) Contribution of momilactone A and B to rice allelopathy. J Plant Physiol 167(10):787–791

Kato-Noguchi H, Ota K, Kujime H (2012) Absorption of momilactone A and B by *Arabidopsis thaliana* L. and the growth inhibitory effects. J Plant Physiol 169(15):1471–1476

Kim S-J, Park H-R, Park E, Lee S-C (2007) Cytotoxic and antitumor activity of momilactone B from rice hulls. J Agric Food Chem 55(5):1702–1706

Kodama O, Suzuki T, Miyakawa J, Akatsuka T (1988) Ultraviolet-induced accumulation of phytoalexins in rice leaves. Agric Biol Chem 52(10):2469–2473

Kong C, Liang W, Xu X, Hu F, Wang P, Jiang Y (2004) Release and activity of allelochemicals from allelopathic rice seedlings. J Agric Food Chem 52(10):2861–2865

Kong F, Deng Y, Zhou B, Wang G, Wang Y, Meng Q (2014) A chloroplast-targeted DnaJ protein contributes to maintenance of photosystem II under chilling stress. J Exp Bot 65(1):143–158

Kong C-H, Zhang S-Z, Li Y-H, Xia Z-C, Yang X-F, Meiners SJ, Wang P (2018) Plant neighbor detection and allelochemical response are driven by root-secreted signaling chemicals. Nat Commun 9(1):3867

Koornneef M, Bentsink L, Hilhorst H (2002) Seed dormancy and germination. Curr Opin Plant Biol 5(1):33–36

Kwon SL, Smith RJ, Talbert RE (1991) Interference of red rice (Oryza sativa) densities in rice (O. sativa). Weed Sci 39(2):169–174

Kwon SL, Smith RJ, Talbert RE (1992) Comparative growth and development of red rice (Oryza sativa) and rice (O. sativa). Weed Sci 40:57–62

Latif S, Chiapusio G, Weston LA (2017) Allelopathy and the role of allelochemicals in plant defence. In: Advances in botanical research. Elsevier, pp 19–54

Lovett J, Ryuntyu M (1992) Allelopathy: broadening the context. In: Allelopathy. Springer, Dordrecht, pp 11–19

Macias FA, Marin D, Oliveros-Bastidas A, Varela RM, Simonet AM, Carrera C, Molinillo JM (2003) Allelopathy as a new strategy for sustainable ecosystems development. Biol Sci Space 17(1):18–23

Mennan H, Ngouajio M, Sahin M, Isik D, Kaya Altop E (2012) Quantification of momilactone B in rice hulls and the phytotoxic potential of rice extracts on the seed germination of *Alisma plantago-aquatica*: allelopathic potential of rice extracts. Weed Biol Manag 12(1):29–39

Minh TN, Xuan TD, Ahmad A, Abdelghany Elzaawely A, Teschke R, Van TM (2018) Momilactones A and B: optimization of yields from isolation and purification. Separations 5:28

Molisch H (1937) Einfluss einer pflanze auf die andere, allelopathie

Mtenga NC, Tarimo TM, Ndakidemi PA, Mbega ER (2019) Carrot-weed: a noxious plant that threatens biodiversity in Africa. Am J Plant Sci 10(03):433–446

Murakami M, Kudo I (2004) Recent advances in molecular biology and physiology of the prostaglandin E2-biosynthetic pathway. Prog Lipid Res 43(1):3–35

Nair MG, Chandra A, Thorogood DL, Ammermann E, Walker N, Kiehs K (1994) Gopalamicin, an antifungal macrodiolide produced by soil actinomycetes. J Agric Food Chem 42(10):2308–2310

Noldin JA (1995) Characterization, seed longevity, and herbicide sensitivity of red rice (Oryza sativa L.) ecotypes, and red rice control in soybeans (Glycine max (L.) Merr.). Texas A&M University

Oerke E-C (2006) Crop losses to pests. J Agric Sci 144(1):31–43

Olofsdotter M, Navarez D, Moody K (1995) Allelopathic potential in rice (*Oryza sativa* L.) germplasm. Ann Appl Biol 127(3):543–560

Olofsdotter M, Navarez D, Rebulanan M, Streibig JC (1999) Weed-suppressing rice cultivars: does allelopathy play a role? Weed Res 39:441–454

Olofsdotter M, Jensen LB, Courtois B (2002) Improving crop competitive ability using allelopathy—an example from rice. Plant Breed 121(1):1–9

Özkara A, Akyıl D, Konuk M (2016) Pesticides, environmental pollution, and health. In: Larramendy ML, Soloneski S (eds) Environmental health risk—hazardous factors to living species. IntechOpen, London, pp 3–27

Palmer M, Bernhardt E, Chornesky E, Collins S, Dobson A, Duke C, Gold B, Jacobson R, Kingsland S, Kranz R, Mappin M, Martinez ML, Micheli F, Morse J, Pace M, Pascual M, Palumbi S, Reichman OJ, Simons A, Townsend A, Turner M (2004) Ecology for a crowded planet. Science 304(5675):1251–1252

Perreto EL, Peske ST, Galli J (1993) Avaliação de sementes e plantas de arroz daninho. Revista Brasileira de Sementes 15:49–54

Peterson MA, Collavo A, Ovejero R, Shivrain V, Walsh MJ (2018) The challenge of herbicide resistance around the world: a current summary: herbicide resistance around the world. Pest Manag Sci 74(10):2246–2259

Prasad R, Kumar V, Prasad KS (2014) Nanotechnology in sustainable agriculture: present concerns and future aspects. Afr J Biotechnol 13(6):705–713

Putnam AR, Duke WB (1978) Allelopathy in agroecosystems. Annu Rev Phytopathol 16(1): 431–451

Sankula S, Braverman MP, Jodari F, Linscombe SD, Oard JH (1997) Evaluation of glufosinate on rice (Oryza sativa) transformed with the BAR gene and red rice (Oryza sativa). Weed Technol 11(1):70–75

Sastry MVS, Seetharaman R (1973) Inheritance of grain shattering and lazyhabit and their interrelationship in rice. Indian J Genet PlantBreedinf 38:318–321

Scognamiglio M, D'Abrosca B, Esposito A, Pacifico S, Monaco P, Fiorentino A (2013) Plant growth inhibitors: allelopathic role or phytotoxic effects? Focus on Mediterranean biomes. Phytochem Rev 12(4):803–830

Shahzad A, Ullah S, Dar AA, Sardar MF, Mehmood T, Tufail MA, Shakoor A, Haris M (2021) Nexus on climate change: agriculture and possible solution to cope future climate change stresses. Environ Sci Pollut Res Int 28(12):14211–14232

Shu HS, Sato YI, Morishima H (1997) Genetic characterization of weedy rice (Oryza sativa L.) based on morpho-physiology, isozymes, and RAPD markers. Theor Appl Genet 94:316–321

Song Z, Lu BR, Chen J (2004) Pollen flow of cultivated rice measured under experimental conditions. Biodivers Conserv 13:579–590

Sørensen M, Møller BL (2021) Metabolic engineering of photosynthetic cells—in collaboration with nature. In: Nielsen J, Stephanopoulos G, Lee SY (eds) Metabolic engineering—concepts and applications. Wiley, pp 803–857

Sunilkumar G, Campbell LM, Puckhaber L, Stipanovic RD, Rathore KS (2006) Engineering cottonseed for use in human nutrition by tissue-specific reduction of toxic gossypol. Proc Natl Acad Sci U S A 103(48):18054–18059

Takahashi N, Kato T, Tsunagawa M, Sasaki N, Kitahara Y (1976) Mechanisms of dormancy in rice seeds: II. New growth inhibitors, momilactone-A and-B isolated from the hulls of rice seeds. Jpn J Breed 26(2):91–98

Toyomasu T, Usui M, Sugawara C, Otomo K, Hirose Y, Miyao A, Hirochika H, Okada K, Shimizu T, Koga J, Hasegawa M, Chuba M, Kawana Y, Kuroda M, Minami E, Mitsuhashi W, Yamane H (2014) Reverse-genetic approach to verify physiological roles of rice phytoalexins: characterization of a knockdown mutant of OsCPS4 phytoalexin biosynthetic gene in rice. Physiol Plant 150(1):55–62

Tukey HB (1969) Implications of allelopathy in agricultural plant science. Bot Rev 35(1):1–16

Vaughan LK, Ottis BV, Prazak-Havey AM, Sneller C, Chandler JM, Park WD (2001) Is all red rice found in commercial rice really Oryza sativa? Weed Sci 49:468–476

Veasey EA, Karasawa MG, Santos PP, Rosa MS, Mamani E, Oliveira GCX (2004) Variation in the loss of seed dormancy during after-ripening of wild and cultivated rice species. Ann Bot 94 (6):875–882

Watanabe T, Shimbo S, Moon CS, Zhang ZW, Ikeda M (1996) Cadmium contents in rice samples from various areas in the world. Sci Total Environ 184(3):191–196

Watanabe H, Zoli M, Changeux JP (1998) Promoter analysis of the neuronal nicotinic acetylcholine receptor α4 gene: methylation and expression of the transgene. Eur J Neurosci 10(7):2244–2253

Weis M, Keller M, Rueda V (2012) Herbicide reduction methods. In: Herbicides—environmental impact studies and management approaches. InTech

Wezel A, Casagrande M, Celette F, Vian J-F, Ferrer A, Peigné J (2014) Agroecological practices for sustainable agriculture. A review. Agron Sustain Dev 34(1):1–20

Wu L-M, Fang Y, Yang H-N, Bai L-Y (2019) Effects of drought-stress on seed germination and growth physiology of quinclorac-resistant *Echinochloa crusgalli*. PLoS One 14(4):e0214480

Yang X-F, Kong C-H (2017) Interference of allelopathic rice with paddy weeds at the root level. Plant Biol (Stuttg) 19(4):584–591

Ye C-Y, Tang W, Wu D, Jia L, Qiu J, Chen M, Mao L, Lin F, Xu H, Yu X, Lu Y, Wang Y, Olsen KM, Timko MP, Fan L (2019) Genomic evidence of human selection on Vavilovian mimicry. Nat Ecol Evol 3(10):1474–1482

Zhao H, Li H, Kong C, Xu X, Liang W (2005) Chemical response of allelopathic rice seedlings under varying environmental conditions. Allelopathy J 15(1):105–110
Zhao L-X, Wang Z-X, Zou Y-L, Gao S, Fu Y, Ye F (2021) Phenoxypyridine derivatives containing natural product coumarins with allelopathy as novel and promising proporphyrin IX oxidase-inhibiting herbicides: design, synthesis and biological activity study. Pestic Biochem Physiol 177:104897

Plant Phenolics and Their Versatile Promising Role in the Management of Nematode Stress

16

Younis Ahmad Hajam, Diksha, Rajesh Kumar, and Rafiq Lone

Abstract

Biotic stress which is caused by various insects, viruses, fungi, bacteria, etc., and abiotic stress which is caused by various stressors for instance, drought, flood, elevated temperature, light, heavy metals etc. These stressors influence the growth, development and productivity of plant at large extent. Biotic stresses influence photosynthesis because chewing insects decreases the leaf area and viral infection decreases the photosynthesis rate per leaf area. Like sponges, corals, and anemones, plants also act in response to various stress situations with the alterations in the expression of genes, protein patterns that can control metabolite biosynthesis concerned with the relations between a plant and its surroundings. Polyphenols are the most important class of special types of metabolites that shows vital physiological role during the life cycle of plant which includes responses to stress conditions. Therefore, in this chapter we have discussed about the functions of various polyphenols like "phenolic acids, flavonoids, lignans and stilbenoids" in the management of biotic stress. Therefore, this chapter was aimed to focus mainly on biotic stress, and its management by plant phenolics.

Y. A. Hajam (✉)
Department of Life Sciences and Allied Health Sciences, Sant Baba Bhag Singh University, Jalandhar, Punjab, India

Diksha
Division Zoology, Department of Biosciences, Career Point University, Hamirpur, Himachal Pradesh, India

R. Kumar
Department of Biosciences, Himachal Pradesh University, Shimla, Himachal Pradesh, India

R. Lone
Department of Botany, Central University of Kashmir, Ganderbal, Jammu and Kashmir, India

R. Lone et al. (eds.), *Plant Phenolics in Biotic Stress Management*,
https://doi.org/10.1007/978-981-99-3334-1_16

Keywords

Biotic stress · Secondary metabolites · Plants · Production

16.1 Introduction

During the past decades, it has been found by various scientists that changes in the climate and atmosphere can fastly change biological diversity as well as the production of crops globally (Zavaleta et al. 2003; Kumar 2016). Due to these changes environment can face various stresses. These changes include any change or alteration in the state of growth inside the plant or its natural habitat that further changes its metabolic homeostasis it is known as environmental stress.

Generally, environmental stress can be of two types: Biotic stress which is caused by various insects, viruses, fungi, bacteria, etc., and abiotic stress which is caused by various stressors for instance, drought, flood, elevated temperature, light, heavy metals etc. These stressors influence the growth, development and productivity of plant at large extent (Meena et al. 2020). Therefore, this chapter was aimed to focus mainly on biotic stress, and its management by plant phenolics. There are number of living things like viruses, bacteria, fungi, nematodes, arachnids and weeds which are responsible for causing biotic stress in plants. All the necessary nutrients are taken up by these stressors from the host plant and finally cause the death of that plant. Due to the losses during the pre-harvesting and post-harvesting of crops, biotic stress may become a major issue. In spite of the adaptive immune system is lacking in plants, they can work against biotic stress by developing themselves to certain complicated approaches. Plants have genetic code stored inside them which can act as a defense system for these plants and can counteract these stressors. In plants, these resistant genes are encoded in hundreds that can work against these stressors. Abiotic stress is totally varied from biotic stress. Various kinds of non-living factors are there which cause abiotic stress on plants and negative impacts as well. The climatic conditions in which the crop plant lives, decide the type of stress to be imposed on crop plants. These climatic conditions also decide the capability of these crop plants to show resistance to that specific type of stress. Various types of biotic stresses are there which influence photosynthesis because chewing insects decreases the leaf area and viral infection decreases the photosynthesis rate per leaf area. Like sponges, corals, and anemones, plants also act in response to various stress situations with the alterations in the expression of genes, protein patterns that can control metabolite biosynthesis concerned with the relations between a plant and its surroundings (Tak and Kumar 2020).

Under different adverse environmental conditions phenylpropanoid biosynthetic pathway gets activated and cause accretion of many polyphenols (Sharma et al. 2019; Linić et al. 2019). Plants also come across various agents cause biotic stress concurrently with the agents cause abiotic stress. The effect of various environmental factors on the plant ailments commonly called as the *disease triangle*. It has for all time been a vital consideration for plant pathologists. There are various studies that showed the impact of drought and salinity causing vulnerability of plants to species of *Puccinia* which is a causal agent of rust, species of *Verticillium* which is a causative agent of verticillium wilt, species of *Fusarium* which cause Fusarium wilt, species of *Pythium* which cause root rot, and species of *Erysiphe* which cause powdery mildew. Cooccurrence of drought and elevated temperature or cold stress increases competitiveness of weed plants more than other crop plants.

Polyphenols are the most important class of special types of metabolites that shows vital physiological role during the life cycle of plant which includes responses to stress conditions. The largest and mostly studied group of plant secondary metabolites is polyphenols. These include above 8000 molecules (González-Sarrías et al. 2020). The pathway for the synthesis of these polyphenols is the shikimate or phenylpropanoid pathway. In this pathway there is a formation of a broad range of monomeric and polymeric polyphenols (Sharma et al. 2019). There is a wide difference between the structures of these phenolic compounds, though they show a similar character.

Generally, the presence of phenolic acids might be in free forms, but frequently these are present in conjugated forms. The primary characteristic of phenolic acids is that a carboxyl group is attached to an aromatic ring. These are attached to various compounds like cellulose, proteins or lignin of plant cell; other small organic molecules like quinic acid, cis-butenedioic acid, or dihydroxybutane dioic acid, and glucose; or other larger molecules like flavonoids, or terpenes etc. with the help of ether, acetal or ester bonds (Andreasen et al. 2000; Lam et al. 2001). On the basis of the structure of phenolic acids these can be sub-divided into following categories:. (1) Based on the structure of polyphenols these can be categorized into two classes: (2) derivatives of benzene carboxylic acid includes hydroxybenzoic acids and derivatives of cinnamic acid include "hydroxycinnamic acid". One or more carbohydrate residues attached by "beta glycosidic bonds to a hydroxyl group or a C atom" of benzene ring in their conjugated forms. The related carbohydrate compounds may be monosaccharides, or disaccharides, or polysaccharides (González-Sarrías et al. 2020). These polyphenols are considered as an assembly of molecules having same biological activities. There are various studies conducted on the total polyphenolic content as a "marker of biological activity". Nevertheless, the structure of phenolic compounds considerably affects their activity and function in various biological processes, and as a result the involvement of polyphenols in plant stresses responses (Hajam et al. 2020). Therefore, in this chapter we have discussed about the functions of various polyphenols like phenolic acids, flavonoids, lignans and stilbenoids in the management of biotic stress.

16.2 Different Plants Stresses

The growth and development of plants are affected by a number of biotic stressors present there in the environment. These stressors greatly affect soil productivity and fertility and acts as the main restrictions for the sustainable agricultural productivity (Kumar and Verma 2018). Approximately 30–50% loss in the agricultural productivity is only because of abiotic and biotic stress. An extensive range of biotic stressors such as bacteria, viruses, fungi, nematode, insects etc. and abiotic stressors like frost, salinity, temperature, drought, toxic metals etc., cause an enormous harmful effecton agriculture (Dresselhaus and Hückelhoven 2018). Keeping in mind all these biotic factors, the huge negative impact to the plant species is caused by fungi. It has been reported by Behmann et al. (2015) that fungi or fungi like other living things are responsible for causing approximately 86% of the diseases in plants. Moreover, the serious factors that are responsible for the destruction of crop plants are viruses. Viruses cause a number of diseases like necrosis, spots on the leaves, blights, blasts, wilting, mottling, development of tumors in plants (Saddique et al. 2018a, 2018b). Weeds are the other biotic stressors those also shows their effect on the development and yield of plants that are economically very important by directly destructing the plant or by causing struggle for the availability of nutrients and area (Melvin et al. 2017).

16.3 Biotic Stress and Plant Polyphenols

Several secondary metabolites are present in plants which acts against biotic stressors and protects the plants. These secondary metabolites may be polyphenols, or these are biosynthesized from a particular type of primary metabolite of plants (Guo et al. 2018). Polyphenols are secondary metabolites of plants that are categorized into one of the main and an extensive group of compounds in plants. The synthesis of these metabolites occurs through the phenylpropanoid pathway and when biotic stressors attack the plants these are present in the sub-epidermal layer of plant tissue. In addition to this, the presence of these polyphenols is also found in plant cells where these are covalently attached and the presence of other components is detected in waxes or on the peripheral side of plant organ (Vishwanath et al. 2015). Besides this, the accumulation and biosynthesis of polyphenols might be affected by various natural factors like the type of soil, moisture content, temperature, and UV rays etc. (Mukherjee et al. 2019).

16.4 Biotic Stressors and Their Effect on Plants

Commonly, biotic stressors like nematods, insects, pests (rodents), lead to 6–25% loss in crop respectively. The capacity of the plant to transport water and nutrients from one part to another is greatly affected by bacteria (prokaryotic microorganisms) which cause death of the plant (Jones et al. 2004). In bacterial infection plants show symptoms like mosaic and these symptoms are alike virus causing disease and may also cause a huge number of plant abnormalities like galls or tumors (Ludwig-Müller 2015). These also affect the other organs of plants like leaves, and fruits cause blights or deadening of leaves tissues, stems or trunks of trees etc. Due to this infection caused by viruses, plants also show a number of physiological and metabolic deformities (Culver and Padmanabhan 2007). Lesions or other injured parts of plants provide a pathway to the viral RNA/DNA to take entry inside the cells of the plants and leads to the infection. The plants that are affected by viruses shows various symptoms such as mottling, mosaic, spots on the leaves, curling of leaves, puckering of leaves and stunting and decrease in the yield. All these symptoms can cause various effects to the plants directly or indirectly and interrupt the whole physiology of the affected plant (Thakur et al. 2018).

Fungi are also the biotic stressors and have almost 1.6 million of species. Fungi can be categorized into two classes such as biotrophs and necrotrophs (Lattanzio et al. 2006). Because fungi are not undergone photosynthesis reaction means they cannot make their own food. These are evolved in such a way that they can take an approach to take their food from either living or non-living organisms. Biotrophic fungi take nutrients from the host tissues of living organism through special type of cells called haustoria. Haustoria are form inside host cells. Necrotrophic fungi produce toxins and kill host tissues and take nutrients "from dead cells (Delaye et al. 2013).

Nematodes are also the biotic stressors and called root knot nematodes. These cause the damage to the plant below ground and cause root knot by feeding on plant roots and decrease the capability of plant to get nutrients and water (Bais et al. 2006). It has been reported by Chitwood (2003) that invasion of nematode cause approximately 15% of global loss accounting upto $100 billion. These are the plant parasites having stylets with these stylets, these cause mechanical damage to the host plant (Holbein et al. 2016). Evidently, nematodes can also secrete various digesting enzymes due to which there is occurrence of various intracellular and intercellular movement of endoparasites and because of these digestive enzymes nematodes can digest various solid compounds of cells. Therefore, digestion of internal matter of plants also done by nematodes and ultimately kills those plant cells (Kyndt et al. 2017). The distinctive symptoms of nematodes attack on plants are root knots, lesions on roots, extreme root branching, root tips are injured and short root system. Wilting is also shown by plants even with plenty soil humidity, yellow leaves and short growth (Marahatta 2018).

Insects are the other biotic factors that cause injury to the plants by two main ways such as indirect way and direct way. The direct damage that is caused by insects to the plants is that these directly feed on plants. And indirect damage is caused by transmitting an infection that can be bacterial, fungal, or viral into crop plants (Fried et al. 2017). The symptoms that are shown by the plants after the attack of insects are flower deformity, irregular shoot branching, chlorotic or yellow mosaic pattern or short growth of plant. Amongst crop plants, the fall in the total global potential because of the pests ranging from approximately 51% in wheat crop to more than 81% in the production of cotton (Ueda et al. 2019).

The potential loss i.e., 35% in the agricultural crops is because of the weeds. Weeds are having a number of species that cause the loss in yield almost each time and these species can vary significantly in competitive capability. Weeds having different growth habits and adaptation, and because of these factors, they can grow well in a lot of ecological niches where the growth of other plants is impossible. Weeds compete with themselves and they also have the potential to compete with

Fig. 16.1 Various biotic stressors and their effect on plants

other agricultural or horticultural crop plants (Wallace et al. 2017). These cause the impact on the growth and yield of plant by causing direct damage to the crop or by causing competition for the availability of nutrient and area (Fig. 16.1).

16.5 Plant Polyphenols

The secondary metabolites of plant are plant polyphenols. These play very significant role in the plant protection and essential to fight against various stressors (Wani et al. 2016). Benzene ring is present in polyphenols that are attached to one or more hydroxyl groups. Polyphenols have an important role in the development and growth of plant because these are essential in the biosynthesis of pigments and lignin and provide protection to the plant from external infection (Asgari Lajayer et al. 2017). The structural and scaffolding reliability to plants is also provided by these polyphenols. Phytoalexins is the plant phenolic which is secreted by most of the plants that are infected or damaged. It repels a lot of microorganisms but there are such pathogens which can prevent themselves from these protections or even use it for their own benefit (Dar et al. 2017). Plant polyphenols can be classified into two categories: Preformed phenolics, or Induced phenolics. The formation of preformed phenolic occurs at the time of normal development of tissues of plant. And on the other hand, the formation of induced phenolics occurs during the physical or mechanical injury, attack by the microbes or pathogens or when attacked by abiotic factors.

16.5.1 Biosynthesis of Plant Polyphenols

Plant polyphenols are common plant secondary metabolites but these are not ordinary amongst bacteria, algae and fungi (Ma et al. 2015). The first step comes in the biosynthesis of polyphenols is the involvement of glucose in the hexose monophosphate shunt (HMP) pathway in which conversion of glucose is takes place into glucose-6-phosphate and after that irreversible conversion into ribulose-5-phosphate. The very first step in the transformation of ribulose-5-phosphate is the formation of Glucose-6-phosphate (Lin et al. 2016). On the other hand, HMP shunt pathway generates erythrose-4-phosphate along with phosphoenolpyruvate from glycolytic pathway, which is later used via phenylpropanoid pathway to produce polyphenolic components after being transported to the shikimic acid pathway to generate phenylalanine amino acid. The majority of phenolic compounds are formed in the Shikimic acid pathway in various fungi, plants, and bacteria (Santos-Sánchez et al. 2019). The one more pathway is malonic acid pathway, this pathway is less involved in the generation of polyphenolic compounds in higher plants than fungi and bacteria (Cheynier et al. 2013).

The major group of polyphenols is the flavonoids. These are composed of 15 carbon compounds and two phenolic rings bounded by three units of carbon. Flavonoids are derived from phenolic acids. These are biosynthesized from malonyl CoA and p-coumaroyl CoA, resulting from acetate and shikimate (Patel et al. 2017). Cinnamic acid is formed from phenylalanine amino acid and a key enzyme phenylalanine ammonia lyase is present and it is a key step primary and secondary metabolism (Saltveit 2017). This is the significant regulatory step in the generation of phenols. After biosynthesis of phenolics, these are incorporated in the plants cell wall to reimburse for biotic stress with high unpredictability through the formation of cinnamic acid and benzoic acid and defend the plant against biotic stressors (Ledesma-Escobar et al. 2019).

16.5.2 Functions of Phenolic Acids in the Management of Biotic Stress

Phenolic compounds are the protecting agents, "inhibitors, pesticides and natural animal toxicants antagonistic" to a number of biotic stressors such as weeds, nematods, insects, fungi, bacteria, virus etc. (Ghosh et al. 2017). The most efficient allelochemical compounds are terpenoids, the harmful hydroquinones, escopoletins, hydroxybenzoates, Caffeic acids, hydrocinnamates and the 5-hydroxynapthoquinones (Malik et al. 2020; Bhattacharya et al. 2010). These are the best natural agents provides protection against biotic stressors and also acts as an alternate to the chemical control of stressors on crops. The production of these phenolic compounds occurs during the host-pathogen interactions by various mechanisms in the defense of plants (Tripathi and Dubey 2004). Phenolic compounds get accumulated, show primary response and may cause normal enhancement in the metabolism of host, once the host plant is infected by biotic stressor (Mayer et al. 2001). There is a considerable enhancement in the two phenolic Caffeic acid ester in the maize crop after maize plant gets infected with *Glomerella graminicola* or *Cochliobolus heterostrophus* (Pusztahelyi et al. 2017). Though, these polyphenolics does not show any toxicity to fungi, the fast accumulation and unexpected reduction in the concentration of these phenolic compounds showed that these polyphenols can act as pool for the formation of other defensive compounds. There is a buildup of intercommunication strength between the host plant and pathogen due to the ability of pathogen to metabolize the phytoalexins that are formed by the host plant (Hardoim et al. 2015). These polyphenolics are responsible for reducing the distribution of infection caused by biotic agent. Phytoalexins show similar bacteriostatic and bactericidal properties as antibiotics have (Pina-Pérez and Pérez 2018). Spores' development and hyphae growth in fungi host plant is limited by phytoalexins. The synthesis rate and accumulation of phytoalexins in the tissues of plants is responsible for the defense of plant against microbial infection (Duke 2018). The exponential growth is caused due to quorum sensing and it also regulates various activities in bacteria; phenolic compounds have the potential to interrupt this process and limit the growth of these biotic agents, and thus phytoalexins show a vital function in the control of destructive actions of the biotic agents (Raman et al. 2016). Quinines are also derived from polyphenols. Quinines play a very important role in plants due to their ability of forming complex derivatives by joining with proteins and hinder the maceration of proteomic compound by the herbivores (Easwar Rao et al. 2017). The amount of proteomics in amino acids causes reduction in the digestion of protein in that insects obstruct the growth of pest. Polyphenols also cause the reduction in the occurrence of ROS like hydrogen peroxides, superoxide ion and singlet oxygen, these are involved in a chain of abiotic and biotic stresses. Polyphenols responsible for the increase in the antioxidant enzymatic defense system and therefore shield the plant from these stressors (Saddique et al. 2018a, 2018b). Phenolic compounds are present at the infected place of plant and hinder the development of the pathogen which takes place as the result of cell death because of the hypersensitive reaction (Lincoln et al. 2018). To defeat the biotic stress, lignin biosynthesis starts from phenylalanine through 4-coumaric acid and the CoA-esters of 4-coumaric, ferulic and sinapic acids to the analogous alcohols which evidently polymerize under the action of peroxidase and assist in fight against biotic stressors (Varbanova et al. 2011; Peperidou et al. 2017; Bi et al. 2017). It has been reported by various scientists that polyphenols those are having "low molecular weight such as benzoic acids and phenylpropanoids are" produced in the primary response to pathogenic infection. There is evidence which strongly recommended that esterification of polyphenols to plant cell wall is particularly for managing of the plants against a number of stressors (Mandal et al. 2010). Several microbes like bacteria, virus, fungi, and insects are the main threat of damage to plants (War et al. 2012) (Fig. 16.2).

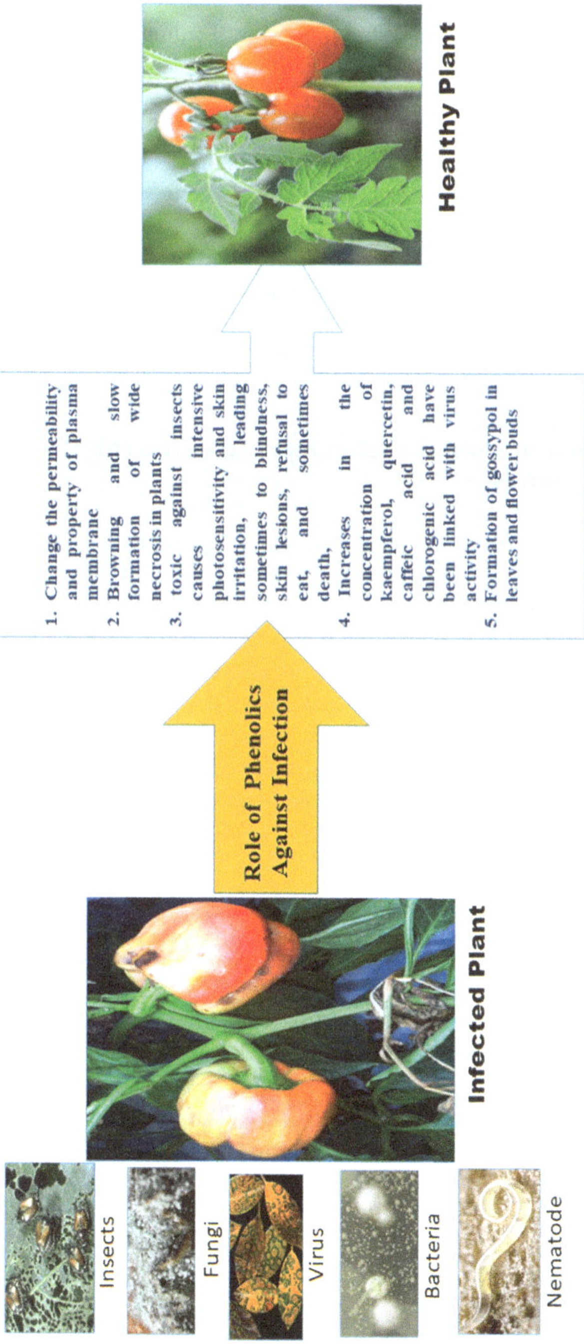

Fig. 16.2 Protective roles of polyphenols against biotic stress

16.6 Bacteria

Bacteria have the ability to infect the plant and distribute in various ways such as insects, washing water or through infected parts of plants. Bacteria find their way through the small pores either through destruct or cuts (Hirano and Upper 2000). A Type III secretion system is there in bacteria, it is a complex structure of bacteria that gives virulence to gram −ve bacteria to insert chemicals or protein directly into the cells of plants. This system in bacteria shows similarity with a syringe and plunger system that bacteria utilize to insert proteins that affect the plant and cause diseases and activate protective response in plants (Green and Mecsas 2016). Amongst all the pathogens of bacteria that cause infection in plants, *Pseudomonas syringae* bacteria are the well-known ones. It is a gram −ve bacterium that cause a variety of diseases in plants like blights, cankers, spots on the leaves (Sun et al. 2017).

16.6.1 Protection from Bacterial Infection Through Phenolic Compounds

The recognition receptors in the plant distinguish capable biotic agents by conserved pathogen associated molecular patterns (PAMP) because of the innate immunity of plants and they cause PAMP-triggered immunity. Consequently, the growth of infection is obstructed long before the biotic agent put on absolute hold of the host plant. Phenolic compounds like catechins are able to influence a number of bacterial species such as *Pseudomonas aeruginosa*, *Serratia marcescens*, *Salmonella choleraesuis*, *Escherichia coli*, *Bordetella bronchiseptica*, *Klebsiella pneumoniae*, *Bacillus subtilis* and *Staphylococcus aureus*. Catechins change the permeability and property of cell membrane and affect the generation of ROS like H_2O_2 (Wang et al. 2019). Other phenolic compounds like eugenol and carvacrol particularly obstruct the quorum sensing regulator in *pectobacterium brasiliense* (Joshi et al. 2016).

16.7 Fungi

Fungi come under eukaryotes which have the capability to digest food externally and absorbs nutrients directly to its cell wall. Spores and hyphae are there which are responsible for the reproduction process in fungi. A vegetative growth phase is also occurring in which a thallus is there which contains microscopic tubular cells. Fungi attack the plants through its spores. Air is responsible for the passage of these spores and then these spores cause dead spots on plant leaves and sometimes cause the death of the entire leaf. It also causes blockage in the water-conducting cells and leads to wilting in plant. In most of the cases it causes severe damage to the plant and can also cause death of host plant. In the worldwide a disease named rice blast disease is also caused by a filamentous ascomycetes fungus named *Magnaporthe oryzae*. Globally, this disease is the most disturbing disease of the *Oryza sativa* plant. Due to this fungus, all foliar tissues of the fungus infected plant is infected; thus, panicle infection causes entire loss of grains (Rashad et al. 2020; Jamal-u-ddinHajano et al. 2011).

16.7.1 Protection from Fungal Infection Through Phenolic Compounds

Polyphenols like resveratrol-O-methyltranferse and resveratrol synthase 3 enhance the antagonism to soybean when soyabean plant was exposed to *Rhizoctonia solani* pathogen (Zernova et al. 2014). These phenolic compounds are capable in halting the growth of fungi, relative rate of formation and quality of cell wall oppositions considered as the first line of the complementary Defence mechanisms. It has been also reported that derivatives of hydroxycinnamic acids, oleuropein, flavanols monoglucosides and tyrosol are the polyphenols which protect the olive tree against leaf spot disease caused by *Fusicladium oleagineum* (Talhaoui et al. 2015). Soil saprophyte is a fungus that affects the onion crop. This fungus functions as a parasite and affects the dead outer scales of onion bulb. Afterward, it punctures the lower succulent scales and profound into the onion bulb. Some of the varieties of onion crop show resistance to the fungus, the fight being associated with catechol, protocatechuic acid and red or yellow pigment of the dead outer scales of onion bulb (Levin 1971).

16.8 Virus

Viruses are the biotic agents that cause infection to the plants and made up of nucleic acids and proteins. Once the virus infects the host cell, it releases its proteins and nucleic acids and then leads to the development of a number of copies and associated proteins mainly the formation of new virus particle (Maccheroni et al. 2005) The movement of virus inside the plants takes place through cell-to-cell or by cytoplasmic bridges. Vectors are mainly responsible for the transmission of plant viruses but other transmissions like mechanical and seed transmission are also noted there (Congdon et al. 2017). Various nematodes, insects, fungi and protozoa can act as vectors of plant virus. These viruses can cause diseases in various parts of plant. These cause reallocation of photosynthetic reaction and destruction of general cell responses as they replicate (Burch-Smith and Zambryski 2016).

16.8.1 Protection from Viral Infection Through Phenolic Compounds

Several polyphenols such as kaempferol, Quercetin, Caffeic acids and chlorogenic acid all are associated with various activities of virus in different plants infected with virus (Parr and Bolwell 2000). In *Matthiola incana* phenolic compound like kaempferol, its level was enhanced, while anthocyanin level was reduced (Baskar et al. 2018).

16.9 Nematode

The most critical plant pathogens are the nematodes. Globally, loss due to nematodes is more than $125 billion yearly (Chitwood 2003). These nematodes rupture the plant cells and take nutrients from that host plant. These also inject various digesting enzymes into host plant cell for the partial digestion (Escobar et al. 2015). Root-knot nematodes parasitize roots of the plants and takes water and nutrients that are needed for the typical growth of that plant (Davies and Spiegel 2011). Globally, root-knot nematodes affect approximately 2000 plants and cause 6% of worldwide crop loss. In Europe and North and South America there are potato cysts nematodes widely affects the potato plant and cause 305-million-dollar worth of destruction in Europe each year (Minnis et al. 2002).

16.9.1 Protection from Nematode Infection Through Phenolic Compounds

Plant polyphenols protect the plant from nematodes by various mechanisms such as (1) browning of plants and steady production of broad necrosis in plants (2) fast browning and appearance of non-expandable necrosis (3) the inhibiton of IAA-oxidase cause accumulation of auxin and consequently leads to the generation of galls or large cell (4) the stimulation of IAA-oxidase cause the decomposition of auxin and of ultimate death in plants (Giebel 1982). Nematodes are capable in activating phenolic compounds (Ohri and Pannu 2010).

From many decades, the association between plants and nematodes leads in the evolution of the plant-parasitic nematode. Nematodes are broadly distributed in vascular plants, causes great loss in yield due to their pathogenic action. More than 4100 species of plant-parasitic nematodes have explored and it has been reported that lead to $80–$118 billion dollars loss every year due to damage in crops. The complex association between the parasitic nematode and plants have concluded in an evolutionary arm race. These "phyto-parasitic nematodes" develop strategies to inhibit the immune system of host and its response required for the development of feeding sites However, plants are producing specific molecules to identify the pathogen and initiates the activation of immune responses.

In plants, invertebrates, and fungi, nematodes are considered as most important pathogens (or agricultural pests) belongs to order of Tylenchida [22]. Among all the nematode groups which are plant-parasites, the sedentary group plays an important role in pathogenesis in plants. They develop a specific feeding mechanism and generate nutrients in order to their complete lifecycle. Sedentary nematodes are having advantages over the migratory ones, in sedentary nematodes transformation of the host cell helps in the development of sustainable feeding structure. There are about 4000 nematodes (plant-parasitic), but among all of these, only some species causes significant losses in crops. A study conducted in US on different crops to find out the major genera of phytoparasitic nematodes which are causing harm to the crops (Heterodera, Hoplolaimus, Meloidogyne, Pratylenchus, Rotylenchulus, and Xiphinema) [23].

16.10 P1

Phenolic compounds are synthesized from two aromatic amino acids L-phenylalanine and L-tyrosine, phenylalanine ammonia-lyase causes breakdown of L-phenylalanine and gets converted into (E)-cinnamic acid. Cinnamic acid-4-hydroxylase further causes its hydroxylation at para position and then converts into para-coumaric acid. Tyrosine ammonia-lyase causes breakdown of L-tyrosine to form para-coumaric acid and also it activated with the help of thioester coupling of acetyl-co-enzyme-A through 4-Coumarate-Coenzyme-A results in the formation of para-Coumarate-CoA. Formation of the secondary metabolites such as hydroxycinnamic acids, flavonoids, tannins, and more occurs through *phenylpropanoid pathway* (PPP) branches in more than one direction (Vogt 2010). However, PPP also helps to form most of the phenolic compounds (e.g., Alkylresorcinols) that are produced by polyketide metabolism (Baerson et al. 2010).

16.11 Hydroxycinnamic Acids

Hydroxycinnamic Acids are the derivatives of (E)-cinnamic acid. HA-para-coumaric acid is one of the major intermediates in the PPP and other Has (e.g., ferulic acid, caffeic acid, or sinapic acid) most abundant in different plants in both forms (pure as well as conjugate form) (Vogt 2010).

Chlorogenic acid is only the phenolic compound that shows defense against the nematodes, and this esteric form of caffeic acid and (-)-quinic acid becomes deposited in both plants (monocot as well as dicot) at the place of PPN infection (Pegard et al. 2005). Chlorogenic acid provides resistance against the nematode in individual species of plants (Hung and Meher et al. 2015). Previous studies reported that the chlorogenic acid accumulation becomes less in popular clones i.e., Populus tremula × Populus alba (in Meloidogyne incognita galls which reveals that repression of biosynthesis of chlorogenic acid may be a pathogenesis strategy (which employed by Meloidogyne incognita) (Baldacci-Cresp et al. 2020). Whereas, chlorogenic acid is having poor nematicidal activity (D'Addabbo et al. 2013). Caffeic acid Quinone is showing nematocidal activity by inducing toxicity to nematodes (Mnviajan et al. 1992), but caffeic acid is formed from chlorogenic acid (by hydrolysis of chlorogenic acid formed quinic and caffeic acid) which can be oxidized to form caffeic acid quinone (Hapiot et al. 1996). Caffeic acid becomes accumulated after the infection of Meloidogyne incognita in a resistant tomato cultivation (but not in three susceptible ones) (Afifah et al. 2019). Moreover, caffeic acid is involved in lignification, and acts as another defense against nematodes (Boerjan et al. 2003; Sato et al. 2019). But it cannot a universal fact that chlorogenic acid involvement in the resistance of nematodes (when the interaction between coffee and Meloidogyne species). Various researchers reported the accumulation of chlorogenic acid for the same degree and resistant cultivar (Machado et al. 2012).

16.12 Stilbenoids and Diarylheptanoids

They are the plant's secondary metabolites small classes that are derived from PPP. In the stilbenoids, the biosynthesis was found to form a phenylpropanoid-CoA to 3-malonyl-coenzyme-A units by performing the continuous process of reaction through stilbene synthase, which converted into 6-C2-C6 stilbene skeleton. In which when the structure formed para-coumaroyl-Co-A then it is trans-resveratrol and similarly when derived from cinnamoyl-CoA known as pinosylvin and when these structures modified and formed many derivatives then it is known as stilbenoids (Jeandet et al. 2010).

On the other hand, "diarylheptanoids" are not well known as stilbenoids but their characters at their initial stages as similar to it. The condensation reaction in between the phenylpropanoid-coenzyme-A and malonyl-coenzyme-A helps to form a diketene intermediate. After sometimes there is a formation of linear diarylheptanoid when condensation occurs between diketene and phenylpropanoid-coenzyme-A and the compound cyclization forms phenylphenalenone backbone. Both linear, as well as diarylheptanoids, move towards the processing which is noted by hydroxylation reported by various researchers (Brand et al. 2006; Munde et al. 2013).

Various researchers also reported in some studies that both (stilbenoids and diarylheptanoids) help to prevent plant diseases and are known to be a key defense compound in plants (Akinwumi et al. 2018). The role of both stilbenoids and diarylheptanoids was found very significant in PPN resistance. Stilbenoids have been applied in pine and grapevines that resist nematodes. Various researchers reported that stilbenoids 3-O-methyl-dihydro pinosylvin are found in *Pinus strobus* bark and wood when infection occurs due to the pinewood nematode (*Bursaphelenchus xylophilus*). It was found nematicidal activity in in-vitro conditions (Hanawa et al. 2001). This 3-O-methyl-dihydro pinosylvin accumulates in *P. strobus* bark and its wood that is resistant towards *B. xylophilus* temporally. It is the nematode penetrates fast at its initial state, but when it reaches approx. I week then its reproduction and movement inhibit. At the same time, the concentration of this 3-O-methyldihydropnosylvin formed at its peak in both (wood and bark) which was essential for the in-vitro study to observe the nematicidal activity. The in-vitro study showed that when 3-O-methyl-dihydro pinosylvin (250 μg/mL) expose for 24 h then it can kill nematodes completely. But in the infected plant (the bark and wood of *P. strobus*) accumulated about 1000–400 μg/g of 3-O-methyl-dihydro pinosylvin. So it can be a result that this compound plays an important role resistance of this plant from *B. xylophilus* (Hanawa et al. 2001).

Wallis (2020) studied the comparative study or rootstock of two *Vitis vinifera*. In this, one was susceptible and another one was resistant to *M. incognita* and it also suggested the important role in the resistance of nematodes for stilbenoids. These two rootstocks were inoculated with *M. incognita* (sampled in 6–12 weeks post-oculation) and in the experiment showed similar results i.e., same total stilbenoids levels. And therefore, in both rootstock, there was no infection caused by nematodes in the content of total stilbenoids by remains un-affected and its profile varied in these two rootstocks (stilbenoids trimer miyabenol-C and hopeaphenol), which are most abundant in rootstock that resistant. They found it to be a compound that acts as phytoacticipins against PPN. However, in an in-vitro study, its anti-nematodal activity does not found.

Radopholus similis is a burrowing nematode, in banana *Phenylphenalenone Phytoalexins* showing a key role in resistance against it. Various researchers reported a study in which roots of bananas were collected after nematode infection (about at 12 weeks). They found that the cyclic diarylheptanoids were more abundant significantly in *R. similis* on its infection site (in a resistant banana) than in susceptible reference cultivar (Hölscher et al. 2014). They also reported that in in-vitro studies, among all the 13 phenylphenalenones showed nematistatic activity found in the extract form of these cultivars (Hölscher et al. 2014). It was further investigated and found that anigorufone showed IC50 on *R. similis* (having 59 and 23 motility after 24 and 72 h respectively) reported by various workers (Hölscher et al. 2014). Hölscher et al. (2014) showed results over anigorufone because in the nematode they form complex forms with lipids. After that, they resulted in large lipid-anigorufone droplets and leads to the death of nematodes. The quantification of anigorufone was attempted and found about 39 mg/g root tissue or infection in *R. similis*. It was reported by different workers that the concentration of anigorufone showed by banana root accumulation biologically (Hölscher et al. 2014).

16.13 Flavonoids

Among all the secondary metabolites, flavonoids are one of the largest secondary metabolites (about 10,000 members) reported by Mathesius (2018). They are used to inhibit the pest and diseases than that of PPN (Treutter 2005, 2006). It can be widely used secondary metabolites concerning PPN resistance.

As stilbenoids bio-synthesis, the biosynthesis of flavonoids is similar. The process differs i.e., Chalcone synthase is used to catalyze in flavonoids synthesis, and in stilbenoids, stilbene synthase is used. Different workers reported that both the Chalcone synthase and stilbene synthase forms high sequence homology. In this Stilbene synthesis evolved from Chalcone synthesis (Tropf et al. 1994). Chalcone skeleton forms when the condensations occur of para-coumaroyl-Coenzyme-A and malonyl-Coenzyme-A units through Chalcone synthase. It further forms the corresponding flavonoids when the isomerization process occurs. Based on flavonoid basic structure they are divided into bioflavonoids, isoflavonoids, and neoflavanoids such as 2-phenylchromen-4-one, 3-phenylchromen-4-one, and 4-phenylcoumarin respectively (McNaught and Wilkinson 1999).

There are so many in-vitro experiments that showed that they also show anti-nematode activity. Among these kaempferol help to inhibit the *R. similis* egg hatching. However, kaempferol, quercetin, and myricetin have properties as a repellent and in some cases nematistatic to *M. incognita* juveniles reported by various workers (Wuyts et al. 2006). Flavonoids were shown very complex on nematode behavior. Their molecular structure and the concentration they can repel or attract towards *M. incognita* juveniles. Various studies were carried out in which they showed the relation of plant nematodes in the Fabaceae family because their members produce isoflavonoids and pterocarpans at the time of infection.

Concerning the nematode resistance, Phytoalexins glyceollin I is one of the most studied pterocarpans. It is produced by soybeans (glycine max). Huang and Barker (1991) studied the accumulation of Glyceollin I in soybean cyst nematodes *H. glycines* (near the head region), post penetration forms within 8 h in resistant soybean cultivar. During post penetration, the level of Glyceollin I is on peak (4–6 days) and after that get decreases. In the case of resistant cultivar root, the Glyceollin I level increased up to 23 μg/g of fresh root. However, it decreases 3 times in susceptible cultivars that showing that there is no "preferential accumulation" near the head of nematodes (Huang and Barker 1991). They also discuss the preferential deposition in the nematode (near the head) that shows elicited response (Huang and Barker 1991). The molecular pattern of this in H. glycinea is still unidentified.

Different workers also reported that in soybean resistant Glyceollin I play an important role in *M. ingonita* (Kaplan et al. 1980a) however its concentration in the root of resistant cultivar forms up to 80 μg/g of roots 7 days post-inoculation. A spatiotemporal correlation was found between both the accumulation of glyceollin I and HR (hypersensitive response) in resistant cultivars. Both forms up to 3 days post-inoculation, so the concentration of glyceollin gets increased in root stele. Hypersensitive response (HR) was observed in the root issue only. Instead of it the mechanism of accumulation of hypersensitive response and glyceollin I is not clear. Glyceollin, I remain un-accumulated when M. incognita is not inoculated (resistant cultivar) with M. javanica (not resistant) reported by various researchers. Hence, the hypothesis shows that the deposition of glyceollin I is very specific to the induced resistant response.

Glyceollin I mechanisms action against nematodes is cleared partially. In in-vitro studies the glyceollin I concentration is very strong towards J2 juveniles (M. incognita) that inhibits their respiration without any harm to it (Kaplan et al. 1980a, 1980b).

There are two isoflavonoids such as daidzein and genistein. They are most abundantly present in soybean roots but do not play a significant role against the H. glycines. However, they both daidzein and genistein accumulated after nematode infection in the same susceptible degree and resistant cultivar reported by various researchers (Kennedy et al. 1999). A repellent effect was observed during the in-vitro studies; however, all these observations show positive results for *R. similis* (Wuyts et al., 2006b).

In common bean (*Phaseolus vulgaris*), the pterocarpans phaseolin accumulation is found in its roots 5 days when Pratylenchus penetrans get penetrated. The concentration of it was 59 μg/g of fresh root. But in in-vitro studies, its concentration showed no effect on the survival and the motility rate of PPN, so phaseolin plays an important role against P. penetrans reported in the study carried out by Abawi and Vanetten (1971).

Furthermore, the nematistatic effect was reported in pterocarpans medicarpin against P. penetrans (having IC50 below 20 μg/mL) reported by various workers (Baldridge et al. 1998). Medicago sativa reported an increase in the several gene expression (found in the biosynthesis of isoflavonoids compared to susceptible cultivar). However, HPLC (high-performance liquid chromatography resulted that the levels of isoflavonoids remaining similar in both (before nematode inoculation and at least two subsequent days). Different authors also reported that there is no correlation between basal medicarpin concentration and nematode resistance in cultivars that were tested (Baldridge et al. 1998). These results showed that they play a minor role or contribution to alfalfa (resistant to nematode).

In white clover, the role of both medicarpin and isoflavonoids formononetin (medicarpin-3-O-glucoside-6-O-malonate and formononetin- 7-O-glucoside-6-O-malonate) in resistance to stem nematode (Ditylenchus dipsaci) was studied by various researchers (Cook et al. 1995). Both (susceptible and resistant) verities of white clove have the same basal level of four metabolites (in roots, leaves, and meristems). However, after the inoculation of D. dipsaci, the resistance variety produces more medicarpin and formononetin (in inoculated meristems). The same reaction goes slow down during the infectious stage (appears in 7–10 days post-inoculation instead of 3 days post-inoculation). The accumulation of systemic flavonoids does not form in both (resistant and susceptible varieties). In meristem, the flavonoids level showed increased in the study. However, no study showed the direct effect of these (medicarpin and formononetin) on *D. dipsaci* (Cook et al. 1995). When infection was caused by *D. dipsaci* then there was no accumulation found in Arial tissues (neither susceptible nor resistant alfalfa) during the work carried out on the same group. However about 2–3 folds were found in root isoflavonoids content showed in resistant cultivar (Edwards et al. 1995) but having no clear evidence regarding this phenomenon.

Pterocarpans, mainly Coumestns found as Phytoalexins in plant nematode interaction. In case when *Phaseolus lunatus* (resistant) and also common beans (susceptible) expose to nematodes (*Pratylenchus scriberini*) then *Phaseolus lunatus* accumulated substantial quantities two coumestrol and psoralidin) (Rich et al. 1977). Coumestrol levels were the same in both beans and *Phaseolus lunatus*. However, the basal levels of psoralidins are higher in *Phaseolus lunatus* about two times. There was no change found in the level of coumestrol in common beans after 2 days infection and 3 times more fold was reported in the roots. Coumestrol and psoralidin firstly accumulated in *Phaseolus lunatus* in HR sites. In in-vitro studies, the concentration of the compounds increased 7–32 times in IC50 towards motility of *P. scriberni* (10–15 μg/mL) reported by various researchers (Rich et al. 1977).

There is not so much study over the flavonoid's role in cereals and nematodes interaction because when varieties of rice were sampled (5 days inoculation) with stem nematodes (*Ditylenchus angustus*), it resulted in a resistant variety of rice. In it, about 13 μg/g of flavonoid phytolexin sakuranetin was found, and not no other varieties were reported by Gill et al. (1996). The effect of sakuranetin on it (*D. angustus)* not has been reported otherwise these results show a correlation.

It was reported by various researchers that in *Avina sativa* (oat) in the concentration of roots and shoot of compound (methanol-soluble), the fold increases up to 2–3 having flavonoids UV-absorbance spectra on the infection that was created by *Heterodera avenae*. Also, foliar treatment occurred with the defense hormone (methyl jasmonate) Soriano et al. (2004). When methyl jasmonate extract was induced in *Avina sativa* (oat) then, it was observed that it was highly nematicidal for *Heterodera avenae*, about 2–3 flavonoid Phytoalexins purified. Among them, one flavonoid phytoalexins show strong nematicidal activity. There are some inducible flavonoids i.e. apigenin-C-hexoside-O-pentoxide, O-methyl-apigenin-C-deoxyhexoside-O-hexoside, and luteolin-C-hexoside-O-pentoside which are non-nematicidal, nematicidal, and not purified respectively.

The plant was treated with methyl jasmonate (3 days prior) for the induction of all these flavonoids that helps to reduce the total no. of nematodes population (on 10 days post-inoculation) for *H. avenae* and *P. neglectus* both. And also used to increase the nematode percentage outside the roots mostly. This situation was also observed in the case of wheat plants (susceptible which was treated with flavonoids rich extract (from induced oats plants) and when taken together, results showed that these are both repellent and nematicidal as well. Only *H. avenae* and *P. neglectus* showed positive results after penetration e.g. pre-treatment of methyl jasmonate. However when *H. avenae* and *P. neglectus* causes infection in untreated susceptible plants then the flavonoids concentration gets increased reported in methyl jasmonate-induced plants reported by various authors (Soriano et al. 2004).

Root extract was analyzed by HPLC-MS from different lines of single seed descent population, in which one can be used by Soriano et al. (2004). They found that there was no correlation between *H. avenae* and flavonoids' basal concentration.

Arabidopsis thaliana transparent testa (tt) mutants, which are impaired in the biosynthesis of flavonoids, have been used to study the role of flavonoids in PPN resistance. One study reported that none of the tested tt mutants differed from their wild type in susceptibility to M. incognita (Wuyts et al. 2006), while another study found that against Heterodera schachtii most tt mutants show either unchanged or slightly increased susceptibility compared to their wild types (Jones et al. 2007). These results suggest that flavonoids play at most a minor role in PPN resistance in A. thaliana.

Various authors studied that the exploitation of flavonoids is due to sedentary nematodes (as a part of their pathogenesis processes (Chin et al. 2018). It was also reported by the researchers that the formation of a feeding site, PPN alter plant auxin homeostasis and different flavonoids reported that inhibit the transport of auxin (Grunewald et al. 2009a, 2009b; Ng et al. 2015). The expression of Chalcone synthase, reported in support of PPN exploitation by flavonoids. The biosynthesis of flavonoids, a report was studied by different authors and resulted that in white clover there was increased auxin response in developing *M. incognita* (in feeding side) (Hutangura et al. 1999). Researchers also re4ported that a flavonoid-deficient

Medicago truncatula line host smaller *M. incognita* than the wild type of it (Wasson et al. 2009).

The results concluded that flavonoids in PPN resistance play an important role depending upon the nematodes and the presence of specific flavonoids. Also glyceollin I in soybean plays an important role for H. glycines and M. incognita.

16.14 Tannins

They are the heterogeneous group of polyphenolic compounds derived in two sup-groups (hydrolyzable and condensed tannins). In the process of hydrolyzable tannins, a polyol core (in which esterification of galloyl group formed). However, Condensed tannins are oligomers. Both hydrolyzable and condensed tannins show variety in the polymerization, monomer composition and in decoration with other phenolic compounds (Barbehenn and Peter Constabel 2011). In in sect herbivory, evolved in resistant to plants (Barbehenn and Peter Constabel 2011). Some studies were carried in which a correlation between tannin accumulation and nematode resistance was found. However, no specific reason presented reported by the authors (Barbehenn and Peter Constabel 2011). They also reported that by inducing protein precipitation tannins help to hinder herbivorous (Salminen and Karonen 2011). But in in-vivo studies, there is no evidence that tannins perform their activities by cytotoxic and anti-nutritive products. It was formed when tannins are oxidized (through Polyphenol Oxidases reported by various workers (Barbehenn and Peter Constabel 2011; Salminen and Karonen 2011). The cultivar was resistant to *R.* similus (which contained more basal tannins concentration) 12 weeks after inoculation than susceptible cultivars (Collingborn et al. 2000). The nematode infection gets increased when the condensed tannins levels increase in all cultivars. The concentration of remains less susceptible than that of the resistant cultivar. Also, the same observation was carried out by various researchers for flavan-3, 4-diols in banana (Collingborn et al. 2000).

The resistant banana cultivar also incorporated propelargonidins alongside the usual procyanidin in its condensed tannins (Collingborn et al. 2000); whether the resistance of the banana cultivar can be attributed to its higher tannin concentration and/or its different tannin composition remains unclear, as the direct effects of banana tannins on R. similis were not evaluated. A putative role for tannins in resistance to the pinewood nematode B. xylophilus has also been proposed. When B. xylophilus was cultured on the phloem sap of eight pine species, its growth rate was negatively correlated to the concentration of condensed tannins in the sap of each species (Pimentel et al. 2017). However, negative correlations were also observed between nematode growth rate and total flavonoid concentration as well as total phenolic compound concentration. This makes it difficult to assess the relative contribution of condensed tannins, flavonoids, and other phenolic metabolites to the inhibitory effect on B. xylophilus (Pimentel et al. 2017).

16.15 Insects

Insects secrete various salivary secretions and cause stress in plants because the salivary secretions of insects contain various enzymes which has an important function in the digestion of food in sucking-piercing insects (Lattanzio et al. 2006). These parasites directly pair to the vascularized system of insect infected plant thus decreasing the water uptake, nutrients and consequently causing enormously decreased biomass and loss in the yield of seed of the insect infected plant (Hegenauer et al. 2017).

16.15.1 Protection from Insects Infection Through Phenolic Compounds

Carya ovate produce a phenolic compound named as Juglone, which is not edible for bark beetle i.e., *Scolytus multistriatus* (Byers 1995).

Lypericum produce a crimson-coloured compound named quinine also called as hypericin, intake of this component leads to severe photosensitivity and irritation on the skin, and sometimes cause blindness, lesions, and ultimately death. This compound is very toxic to insects and mammals as well (Levin 1971).

16.16 Weeds

The loss due to weeds infection was reported above 100 billion US$ annually. Weeds can act as a vector to virus or reservoir to a pathogen; and can considerably affect the incidence of disease (Palanisamy et al. 2020; Swanton et al. 2015). The main weed of wheat crop is the *Phalaris minor*. The cereal crops such as maize, millet, sorghum and rice mainly attacked by witch weeds i.e., *Striga* species. A phytohormone named as Strigolactone exuded from the infected plant cause germination of the parasite seed (Qasem 2006).

16.16.1 Protection from Weeds Infection Through Phenolic Compounds

The higher concentration of catechin decreases the germination and development of *C. maculosa*, thus acting as an inhibitor (Li et al. 2010). The seed extract of bur cucumber and their polyphenolic allelochemical activates the accretion of abscisic acid, salicylic acid and jasmonic acid and hinder the gibberellin pathway, and cause haltering of seed germination (Lee et al. 2015). Polyphenols such as p-hydroxybenzoic, p-coumaric, gallic, syringic, gentisic and vanillic acids are synthetic components of various species of *Eucalyptus* which shows harmful effects on the crops and cause late germination, seedling mortality and decrease in development and productivity of other plants (John and Sarada 2012).

16.17 Conclusion and Future Prospective

Now-a-days changes in the climate and global warming, crops naturally come across increased different abiotic and biotic stress combinations, which generally have an impact on the germination, reproduction, development, and yield of crop plants. Combination of a number of biotic stressors may cause harmful effects to the development of plant. The major biotic stressors are bacteria, viruses, fungi, weeds, nematodes, insects etc. (Lattanzio et al. 2006). All these biotic stresses are very harmful for the development of plant. Among all the biotic stresses fungi cause a serious threat as compared to other biotic stresses. A vast species of fungi is approximately having 8200 fungal species and acts as pathogens (Tarkowski and Vereecke 2014). Likewise, viruses are also uncertain pathogens of plants which destroy agricultural crops more than fungi over the world (Bai et al. 2002). These pathogens are responsible in causing major harm to the plants such as rotting, yellowing or wilting of plants, spots on the leaves, galls formation and destruction of seeds (Anderson et al. 2004). All parts of the plants such as leaves, roots, stem, bark and flowers are greatly affected by these pathogens (Walling 2008; Mann et al. 2012). Weeds are very fast-growing plant parasites due to their fast growing ability. Weeds also have the potential to fastly adapt to an environment and dominate over other plants adaptability (Dass et al. 2017).

References

Abawi GS, Vanetten HD (1971) Phaseollin production induced by Pratylenchus penetrans in Phaseolus vulgaris. In: Abstracts of papers presented at the tenth annual meeting of the Society of Nematologists; August 22–25, 1971. Society of Nematologists, Ottawa, pp 301

Afifah EN, Murti RH, Nuringtyas TR (2019) Metabolomics approach for the analysis of resistance of four tomato genotypes (Solanum lycopersicum L.) to root-knot nematodes (Meloidogyne incognita). Open Life Sci 14:141–149. https://doi.org/10.1515/biol-2019-0016

Akinwumi BC, Bordun KAM, Anderson HD (2018) Biological activities of stilbenoids. Int J Mol Sci 19:792. https://doi.org/10.3390/ijms19030792

Anderson PK, Cunningham AA, Patel NG, Morales FJ, Epstein PR, Daszak P (2004) Emerging infectious diseases of plants: pathogen pollution, climate change and agrotechnology drivers. Trends Ecol Evol 19(10):535–544

Andreasen MF, Christensen LP, Meyer AS, Hansen Å (2000) Content of phenolic acids and ferulic acid dehydrodimers in 17 Rye (Secale c ereale L.) varieties. J Agric Food Chem 48(7): 2837–2842

Asgari Lajayer B, Ghorbanpour M, Nikabadi S (2017) Heavy metals in contaminated environment: destiny of secondary metabolite biosynthesis, oxidative status and phytoextraction in medicinal plants. Ecotoxicol Environ Saf 145:377

Baerson SR, Schröder J, Cook D, Rimando AM, Pan Z, Dayan FE et al (2010) Alkylresorcinol biosynthesis in plants: new insights from an ancient enzyme family? Plant Signal Behav 5: 1286–1289. https://doi.org/10.4161/psb.5.10.13062

Bai FW, Yan J, Qu ZC, Zhang HW, Xu J, Ye MM, Shen DL (2002) Phylogenetic analysis reveals that a dwarfing disease on different cereal crops in China is due to rice black streaked dwarf virus (RBSDV). Virus Genes 25(2):201–206

Bais HP, Weir TL, Perry LG, Gilroy S, Vivanco JM (2006) The role of root exudates in rhizosphere interactions with plants and other organisms. Annu Rev Plant Biol 57:233–266

Baldacci-Cresp F, Behr M, Kohler A, Badalato N, Morreel K, Goeminne G et al (2020) Molecular changes concomitant with vascular system development in mature galls induced by root-knot nematodes in the model tree host Populus tremula × P. alba. Int J Mol Sci 21:406. https://doi.org/10.3390/ijms21020406

Baldridge GD, O'Neill NR, Samac DA (1998) Alfalfa (Medicago sativa L.) resistance to the root-lesion nematode, Pratylenchus penetrans: defense-response gene mRNA and isoflavonoid phytoalexin levels in roots. Plant Mol Biol 38:999–1010. https://doi.org/10.1023/a:1006182908528

Barbehenn RV, Peter Constabel C (2011) Tannins in plant-herbivore interactions. Phytochemistry 72:1551–1565. https://doi.org/10.1016/j.phytochem.2011.01.040

Baskar V, Venkatesh R, Ramalingam S (2018) Flavonoids (antioxidants systems) in higher plants and their response to stresses. In: Antioxidants and antioxidant enzymes in higher plants. Springer, Cham, pp 253–268

Behmann J, Mahlein AK, Rumpf T, Römer C, Plümer L (2015) A review of advanced machine learning methods for the detection of biotic stress in precision crop protection. Precis Agric 16(3):239–260

Bhattacharya A, Sood P, Citovsky V (2010) The roles of plant phenolics in defence and communication during Agrobacterium and Rhizobium infection. Mol Plant Pathol 11(5):705–719

Bi B, Tang J, Han S, Guo J, Miao Y (2017) Sinapic acid or its derivatives interfere with abscisic acid homeostasis during Arabidopsis thaliana seed germination. BMC Plant Biol 17(1):1–12

Boerjan W, Ralph J, Baucher M (2003) Lignin biosynthesis. Annu Rev Plant Biol 54:519–546. https://doi.org/10.1146/annurev.arplant.54.031902.134938

Brand S, Hölscher D, Schierhorn A, Svatoš A, Schröder J, Schneider B (2006) A type III polyketide synthase from Wachendorfia thyrsiflora and its role in diarylheptanoid and phenylphenalenone biosynthesis. Planta 224:413–428. https://doi.org/10.1007/s00425-006-0228-x

Burch-Smith TM, Zambryski PC (2016) Regulation of plasmodesmal transport and modification of plasmodesmata during development and following infection by viruses and viral proteins. In: Plant-virus interactions. Springer, Cham, pp 87–122

Byers JA (1995) Host-tree chemistry affecting colonization in bark beetles. In: Chemical ecology of insects, vol 2. Springer, Boston, pp 154–213

Cheynier V, Comte G, Davies KM, Lattanzio V, Martens S (2013) Plant phenolics: recent advances on their biosynthesis, genetics, and ecophysiology. Plant Physiol Biochem 72:1–20

Chin S, Behm CA, Mathesius U (2018) Functions of flavonoids in plant-nematode interactions. Plants 7:85. https://doi.org/10.3390/plants7040085

Chitwood DJ (2003) Research on plant-parasitic nematode biology conducted by the United States Department of Agriculture–Agricultural Research Service. Pest Manag Sci 59(6–7):748–753

Collingborn FMB, Gowen SR, Mueller-Harvey I (2000) Investigations into the biochemical basis for nematode resistance in roots of three Musa cultivars in response to Radopholus similis infection. J Agric Food Chem 48:5297–5301. https://doi.org/10.1021/jf000492z

Congdon BS, Coutts BA, Renton M, Flematti GR, Jones RAC (2017) Establishing alighting preferences and species transmission differences for Pea seed-borne mosaic virus aphid vectors. Virus Res 241:145–155

Cook R, Tiller SA, Mizen KA, Edwards R (1995) Isoflavonoid metabolism in resistant and susceptible cultivars of white clover infected with the stem nematode Ditylenchus dipsaci. J Plant Physiol 146:348–354. https://doi.org/10.1016/S0176-1617(11)82067-5

Culver JN, Padmanabhan MS (2007) Virus-induced disease: altering host physiology one interaction at a time. Annu Rev Phytopathol 45:221–243

D'Addabbo T, Carbonara T, Argentieri MP, Radicci V, Leonetti P, Villanova L et al (2013) Nematicidal potential of Artemisia annua and its main metabolites. Eur J Plant Pathol 137: 295–304. https://doi.org/10.1007/s10658-013-0240-5

Dar SA, Wani AB, Ganie A, Kandoo AA, Wani MY (2017) Resistance against insect pests by phenolics and their derivative compounds. J Crop Weed 13(2):187–193

Dass A, Shekhawat K, Choudhary AK, Sepat S, Rathore SS, Mahajan G, Chauhan BS (2017) Weed management in rice using crop competition—a review. Crop Prot 95:45–52
Davies K, Spiegel Y (eds) (2011) Biological control of plant-parasitic nematodes: building coherence between microbial ecology and molecular mechanisms, vol. 11. Springer Science & Business Media
Delaye L, García-Guzmán G, Heil M (2013) Endophytes versus biotrophic and necrotrophic pathogens—are fungal lifestyles evolutionarily stable traits? Fungal Divers 60(1):125–135
Dresselhaus T, Hückelhoven R (2018) Biotic and abiotic stress responses in crop plants, 8
Duke SO (2018) Interaction of chemical pesticides and their formulation ingredients with microbes associated with plants and plant pests. J Agric Food Chem 66(29):7553–7561
Easwar Rao D, Divya K, Prathyusha IVSN, Rama Krishna C, Chaitanya KV (2017) Insect-resistant plants. In: Current developments in biotechnology and bioengineering. Elsevier, Amsterdam, pp 47–74
Edwards R, Cook R, Mizen T (1995) Isoflavonoid conjugate accumulation in the roots of lucerne (Medicago sativa) seedlings following infection by the stem nematode Ditylenchus dipsaci. Nematologica 41:51–66. https://doi.org/10.1163/003925995X00044
Escobar C, Barcala M, Cabrera J, Fenoll C (2015) Overview of root-knot nematodes and giant cells. In: Advances in botanical research, vol. 73. Academic, pp 1–32
Fried G, Chauvel B, Reynaud P, Sache I (2017) Decreases in crop production by non-native weeds, pests, and pathogens. In: Impact of biological invasions on ecosystem services. Springer, Cham, pp 83–101
Ghosh P, Bhattacharya A, Char B (2017) Manipulating disease and pest resistance pathways in plants for enhanced crop improvement. Biosci Biotechnol Res Commun 10:631–644
Giebel J (1982) Mechanism of resistance to plant nematodes. Annu Rev Phytopathol 20(1): 257–279
Gill JR, Harbornez JB, Plowright RA, Grayer RJ, Rahman ML (1996) The induction of phenolic compounds in rice after infection by the stem nematode Ditylenchus angustus. Nematologica 42:564–578. https://doi.org/10.1163/004625996X00063
González-Sarrías A, Tomás-Barberán FA, García-Villalba R (2020) Structural diversity of polyphenols and distribution in foods. In: Tomás-Barberán FA, González-Sarrías A, García-Villalba R (eds) Dietary polyphenols: their metabolism and health effects. Wiley, Chichester, pp 1–29
Green ER, Mecsas J (2016) Bacterial secretion systems: an overview. Microbiol Spectr 4(1):4–1
Grunewald W, Cannoot B, Friml J, Gheysen G (2009a) Parasitic nematodes modulate PIN-mediated auxin transport to facilitate infection. PLoS Pathog 5:e1000266. https://doi.org/10.1371/journal.ppat.1000266
Grunewald W, van Noorden G, Van Isterdael G, Beeckman T, Gheysen G, Mathesius U et al (2009b) Manipulation of auxin transport in plant roots during rhizobium symbiosis and nematode parasitism. Plant Cell 21:2553–2562. https://doi.org/10.1105/tpc.109.069617
Guo Q, Major IT, Howe GA (2018) Resolution of growth–defense conflict: mechanistic insights from jasmonate signaling. Curr Opin Plant Biol 44:72–81
Hajam YA, Rai S, Kumar R, Bashir M, Malik JA (2020) Phenolic compounds from medicinal herbs: their role in animal health and diseases—a new approach for sustainable welfare and development. In: Plant phenolics in sustainable agriculture. Springer, Singapore, pp 221–239
Hanawa F, Yamada T, Nakashima T (2001) Phytoalexins from Pinus strobus bark infected with pinewood nematode, Bursaphelenchus xylophilus. Phytochemistry 57:223–228. https://doi.org/10.1016/s0031-9422(00)00514-8
Hapiot P, Neudeck A, Pinson J, Fulcrand H, Neta P, Rolando C (1996) Oxidation of caffeic acid and related hydroxycinnamic acids. J Electroanal Chem 405:169–176. https://doi.org/10.1016/0022-0728(95)04412-4
Hardoim PR, Van Overbeek LS, Berg G, Pirttilä AM, Compant S, Campisano A, Sessitsch A (2015) The hidden world within plants: ecological and evolutionary considerations for defining functioning of microbial endophytes. Microbiol Mol Biol Rev 79(3):293–320

Hegenauer V, Körner M, Albert M (2017) Plants under stress by parasitic plants. Curr Opin Plant Biol 38:34–41
Hirano SS, Upper CD (2000) Bacteria in the leaf ecosystem with emphasis on Pseudomonas syringae—a pathogen, ice nucleus, and epiphyte. Microbiol Mol Biol Rev 64(3):624–653
Holbein J, Grundler FM, Siddique S (2016) Plant basal resistance to nematodes: an update. J Exp Bot 67(7):2049–2061
Hölscher D, Dhakshinamoorthy S, Alexandrov T, Becker M, Bretschneider T, Buerkert A et al (2014) Phenalenone-type phytoalexins mediate resistance of banana plants (Musa spp.) to the burrowing nematode Radopholus similis. Proc Natl Acad Sci U S A 111(1):105–110
Huang JS, Barker KR (1991) Glyceollin I in soybean-cyst nematode interactions: spatial and temporal distribution in roots of resistant and susceptible soybeans. Plant Physiol 96:1302–1307. https://doi.org/10.1104/pp.96.4.1302
Hutangura P, Mathesius U, Jones MGK, Rolfe BG (1999) Auxin induction is a trigger for root gall formation caused by root-knot nematodes in white clover and is associated with the activation of the flavonoid pathway. Aust J Plant Physiol 26:221–231. https://doi.org/10.1071/PP98157
Jamal-u-ddinHajano G, Pathan QA, Mubeen AL (2011) Rice blast-mycoflora, symptomatology and pathogenicity. Sindh Agric Univ Tandojam 5(1):53–63
Jeandet P, Delaunois B, Conreux A, Donnez D, Nuzzo V, Cordelier S et al (2010) Biosynthesis, metabolism, molecular engineering, and biological functions of stilbene phytoalexins in plants. Biofactors 36:331–341. https://doi.org/10.1002/biof.108
John J, Sarada S (2012) Role of phenolics in allelopathic interactions. Allelopath J 29(2):215–230
Jones DL, Hodge A, Kuzyakov Y (2004) Plant and mycorrhizal regulation of rhizodeposition. New Phytol 163(3):459–480
Jones JT, Furlanetto C, Phillips MS (2007) The role of flavonoids produced in response to cyst nematode infection of Arabidopsis thaliana. Nematology 9:671–677. https://doi.org/10.3390/plants9020253
Joshi JR, Khazanov N, Senderowitz H, Burdman S, Lipsky A, Yedidia I (2016) Plant phenolic volatiles inhibit quorum sensing in pectobacteria and reduce their virulence by potential binding to ExpI and ExpR proteins. Sci Rep 6(1):38126
Kaplan DT, Keen NT, Thomason IJ (1980a) Association of glyceollin with the incompatible response of soybean roots to Meloidogyne incognita. Physiol Plant Pathol 16:309–318. https://doi.org/10.1016/S0048-4059(80)80002-6
Kaplan DT, Keen NT, Thomason IJ (1980b) Studies on the mode of action of glyceollin in soybean incompatibility to the root knot nematode, Meloidogyne incognita. Physiol Plant Pathol 16:319–325. https://doi.org/10.1016/S0048-4059(80)80003-8
Kennedy MJ, Niblack TL, Krishnan HB (1999) Infection by Heterodera glycines elevates isoflavonoid production and influences soybean nodulation. J Nematol 31:341–347
Kumar M (2016) Impact of climate change on crop yield and role of model for achieving food security. Environ Monit Assess 188(8):1–14
Kumar A, Verma JP (2018) Does plant—microbe interaction confer stress tolerance in plants: a review? Microbiol Res 207:41–52
Kyndt T, Zemene HY, Haeck A, Singh R, De Vleesschauwer D, Denil S, De Meyer T, Höfte M, Demeestere K, Gheysen G (2017) Below-ground attack by the root knot nematode Meloidogyne graminicola predisposes rice to blast disease. Mol Plant Microbe Interact 30(3):255–266
Lam TBT, Kadoya K, Iiyama K (2001) Bonding of hydroxycinnamic acids to lignin: ferulic and p-coumaric acids are predominantly linked at the benzyl position of lignin, not the β-position, in grass cell walls. Phytochemistry 57(6):987–992
Lattanzio V, Lattanzio VM, Cardinali A (2006) Role of phenolics in the resistance mechanisms of plants against fungal pathogens and insects. Phytochemistry 661(2):23–67
Ledesma-Escobar CA, Priego-Capote F, Luque de Castro MD (2019) Relevance and analysis of citrus flavonoids. In: Polyphenols in plants. Elsevier, London, pp 133–150
Lee SM, Radhakrishnan R, Kang SM, Kim JH, Lee IY, Moon BK, Yoon BW, Lee IJ (2015) Phytotoxic mechanisms of bur cucumber seed extracts on lettuce with special reference to

analysis of chloroplast proteins, phytohormones, and nutritional elements. Ecotoxicol Environ Saf 122:230–237

Levin DA (1971) Plant phenolics: an ecological perspective. Am Nat 105(942):157–181

Li ZH, Wang Q, Ruan X, Pan CD, Jiang DA (2010) Phenolics and plant allelopathy. Molecules 15(12):8933–8952

Lin D, Xiao M, Zhao J, Li Z, Xing B, Li X, Kong M, Li L, Zhang Q, Liu Y, Chen H (2016) An overview of plant phenolic compounds and their importance in human nutrition and management of type 2 diabetes. Molecules 21(10):1374

Lincoln JE, Sanchez JP, Zumstein K, Gilchrist DG (2018) Plant and animal PR1 family members inhibit programmed cell death and suppress bacterial pathogens in plant tissues. Mol Plant Pathol 19(9):2111–2123

Linić I, Šamec D, Grúz J, Bok VV, Strnad M, Salopek-Sondi B (2019) Involvement of phenolic acids in short-term adaptation to salinity stress is species-specific among Brassicaceae. Plants 8: 155

Ludwig-Müller J (2015) Bacteria and fungi controlling plant growth by manipulating auxin: balance between development and defense. J Plant Physiol 172:4–12

Ma XH, Ma Y, Tang JF, He YL, Liu YC, Ma XJ, Shen Y, Cui GH, Lin HX, Rong QX, Guo J (2015) The biosynthetic pathways of tanshinones and phenolic acids in Salvia miltiorrhiza. Molecules 20(9):16235–16254

Maccheroni W, Alegria MC, Greggio CC, Piazza JP, Kamla RF, Zacharias PR, Bar-Joseph M, Kitajima EW, Assumpçao LC, Camarotte G, Cardozo J (2005) Identification and genomic characterization of a new virus (Tymoviridae family) associated with citrus sudden death disease. J Virol 79(5):3028–3037

Machado ART, Campos VPVAC, da Silva WJR, Campos VPVAC, de Mattos Zeri AC, Oliveira DF (2012) Metabolic profiling in the roots of coffee plants exposed to the coffee root-knot nematode, Meloidogyne exigua. Eur J Plant Pathol 134:431–441. https://doi.org/10.1007/s10658-012-0008-3

Malik JA, Bhadauria M, Lone R, Hajam YA (2020) Exploitation of plant phenolics in animal farming. In: Plant phenolics in sustainable agriculture. Springer, Singapore, pp 69–89

Mandal SM, Chakraborty D, Dey S (2010) Phenolic acids act as signaling molecules in plant-microbe symbioses. Plant Signal Behav 5(4):359–368

Mann RS, Ali JG, Hermann SL, Tiwari S, Pelz-Stelinski KS, Alborn HT, Stelinski LL (2012) Induced release of a plant-defense volatile 'deceptively' attracts insect vectors to plants infected with a bacterial pathogen. PLoS Pathog 8(3):e1002610

Marahatta SP (2018) Nematode community associated with recently planted breadfruit, artocarpusaltilis, rhizosphere on Kauai, Hawaii. Nematropica 48(2):179–185

Mathesius U (2018) Flavonoid functions in plants and their interactions with other organisms. Plants 7:30. https://doi.org/10.3390/plants7020030

Mayer AM, Staples RC, Gil-ad NL (2001) Mechanisms of survival of necrotrophic fungal plant pathogens in hosts expressing the hypersensitive response. Phytochemistry 58(1):33–41

McNaught AD, Wilkinson A (1999) Flavonoids (isoflavonoids and neoflavonoids). In: McNaught AD, Wilkinson A (eds) IUPAC compendium of chemical terminology. Blackwell Scientific Publications, Oxford

Meena NL, Verma P, Pande R, Kumar M, Watts A, Gupta OP (2020) Bioavailability and nutritional analysis of flavonoids. In: Plant phenolics in sustainable agriculture. Springer, Singapore, pp 135–156

Meher HC, Gajbhiye VT, Singh G, Chawla G (2015) Altered metabolomic profile of selected metabolites and improved resistance of Cicer arietinum (L.) against Meloidogyne incognita (Kofoid & White) Chitwood following seed soaking with salicylic acid, benzothiadiazole or nicotinic acid. Acta Physiol Plant 37:1–12. https://doi.org/10.1007/s11738-015-1888-6

Melvin P, Bankapalli K, D'Silva P, Shivaprasad PV (2017) Methylglyoxal detoxification by a DJ-1 family protein provides dual abiotic and biotic stress tolerance in transgenic plants. Plant Mol Biol 94(4–5):381–397

Minnis ST, Haydock PPJ, Ibrahim SK, Grove IG, Evans K, Russell MD (2002) Potato cyst nematodes in England and Wales-occurrence and distribution. Ann Appl Biol 140(2):187–195
Mnviajan R, Singh P, Bajaj KL (1992) Nematicidal activity of some phenolic compounds against Meloidogyne incognita. Nematol Mediterr 20:217–219
Mukherjee S, Kutty NN, Bera P, Mitra A (2019) Impact of light and sucrose supplementation on cellular differentiation, metabolic shift and modulation of gene expression in hairy roots of Daucuscarota. Plant Cell Tissue Organ Cult 136(2):383–397
Munde T, Brand S, Hidalgo W, Maddula RK, Svatoš A, Schneider B (2013) Biosynthesis of tetraoxygenated phenylphenalenones in Wachendorfia thyrsiflora. Phytochemistry 91:165–176. https://doi.org/10.1016/j.phytochem.2012.02.020
Ng JLP, Perrine-Walker F, Wasson AP, Mathesius U (2015) The control of auxin transport in parasitic and symbiotic root–microbe interactions. Plants 4:606–643. https://doi.org/10.3390/plants4030606
Ohri P, Pannu SK (2010) Effect of phenolic compounds on nematodes—a review. J Appl Nat Sci 2(2):344–350
Palanisamy CP, Gunasekaran VP, Dominic S, Xuan TD (2020) Phenolic allelochemicals from crops and weed management. In: Plant phenolics in sustainable agriculture. Springer, Singapore, pp 183–199
Parr AJ, Bolwell GP (2000) Phenols in the plant and in man. The potential for possible nutritional enhancement of the diet by modifying the phenols content or profile. J Sci Food Agric 80(7): 985–1012
Patel A, Panchal I, Parmar I, Mishtry B (2017) Synthesis of new flavanoid and chalcone derivatives as antimicrobial agent by green chemistry approach. Int J Pharm Sci Res 8(6):2725–2730
Pegard A, Brizzard G, Fazari A, Soucaze O, Abad P, Djian-Caporalino C (2005) Histological characterization of resistance to different root-knot nematode species related to phenolics accumulation in Capsicum annuum. Phytopathology 95:158–165. https://doi.org/10.1094/PHYTO-95-0158
Peperidou A, Pontiki E, Hadjipavlou-Litina D, Voulgari E, Avgoustakis K (2017) Multifunctional cinnamic acid derivatives. Molecules 22(8):1247
Pimentel CS, Firmino PN, Calvão T, Ayres MP, Miranda I, Pereira H (2017) Pinewood nematode population growth in relation to pine phloem chemical composition. Plant Pathol 66:856–864. https://doi.org/10.1111/ppa.12638
Pina-Pérez MC, Pérez MF (2018) Antimicrobial potential of legume extracts against foodborne pathogens: a review. Trends Food Sci Technol 72:114–124
Pusztahelyi T, Holb IJ, Pócsi I (2017) Plant-fungal interactions: special secondary metabolites of the biotrophic, necrotrophic, and other specific interactions. In: Fungal metabolites, pp 1–58
Qasem JR (2006) Parasitic weeds and allelopathy: from the hypothesis to the proofs. In: Allelopathy. Springer, Dordrecht, pp 565–637
Raman N, Marchillo K, Lee MR, RodríguezLópez ADL, Andes DR, Palecek SP, Lynn DM (2016) Intraluminal release of an antifungal β-peptide enhances the antifungal and anti-biofilm activities of multilayer-coated catheters in a rat model of venous catheter infection. ACS Biomater Sci Eng 2(1):112–121
Rashad Y, Aseel D, Hammad S (2020) Phenolic compounds against fungal and viral plant diseases. In: Plant Phenolics in sustainable agriculture. Springer, Singapore, pp 201–219
Rich JR, Keen NT, Thomason IJ (1977) Association of coumestans with the hypersensitivity of Lima bean roots to Pratylenchus scribneri. Physiol Plant Pathol 10:105–116. https://doi.org/10.1016/0048-4059(77)90014-5
Saddique M, Kamran M, Shahbaz M (2018a) Differential responses of plants to biotic stress and the role of metabolites. In: Plant metabolites and regulation under environmental stress. Elsevier, San Diego, pp 69–87
Saddique M, Kamran M, Shahbaz M (2018b) Differential responses of plants to biotic stress and the role of metabolites. In: Plant metabolites and regulation under environmental stress. Academic, pp 69–87

Salminen JP, Karonen M (2011) Chemical ecology of tannins and other phenolics: we need a change in approach. Funct Ecol 25:325–338. https://doi.org/10.1111/j.1365-2435.2010.01826.x
Saltveit ME (2017) Synthesis and metabolism of phenolic compounds. Fruit Veget Phytochem Chem Human Health 2:115
Santos-Sánchez NF, Salas-Coronado R, Hernández-Carlos B, Villanueva-Cañongo C (2019) Shikimic acid pathway in biosynthesis of phenolic compounds. In: Plant physiological aspects of phenolic compounds, 1
Sato K, Kadota Y, Shirasu K (2019) Plant immune responses to parasitic nematodes. Front Plant Sci 10:1165. https://doi.org/10.3389/fpls.2019.01165
Sharma A, Shahzad B, Rehman A, Bhardwaj R, Landi M, Zheng B (2019) Response of phenylpropanoid pathway and the role of polyphenols in plants under abiotic stress. Molecules 24:2452
Soriano IR, Asenstorfer RE, Schmidt O, Riley IT (2004) Inducible flavone in oats (Avena sativa) is a novel defense against plant-parasitic nematodes. Phytopathology 94:1207–1214. https://doi.org/10.1094/PHYTO.2004.94.11.1207
Sun S, Zhi Y, Zhu Z, Jin J, Duan C, Wu X, Xiaoming W (2017) An emerging disease caused by Pseudomonas syringaepv. phaseolicola threatens mung bean production in China. Plant Dis 101(1):95–102
Swanton CJ, Nkoa R, Blackshaw RE (2015) Experimental methods for crop–weed competition studies. Weed Sci 63(SP1):2–11
Tak Y, Kumar M (2020) Phenolics: a key defence secondary metabolite to counter biotic stress. In: Plant phenolics in sustainable agriculture. Springer, Singapore, pp 309–329
Talhaoui N, Taamalli A, Gómez-Caravaca AM, Fernández-Gutiérrez A, Segura-Carretero A (2015) Phenolic compounds in olive leaves: analytical determination, biotic and abiotic influence, and health benefits. Food Res Int 77:92–108
Tarkowski P, Vereecke D (2014) Threats and opportunities of plant pathogenic bacteria. Biotechnol Adv 32(1):215–229
Thakur H, Jindal SK, Sharma A, Dhaliwal MS (2018) Chilli leaf curl virus disease: a serious threat for chilli cultivation. J Plant Dis Protect 125(3):239–249
Treutter D (2005) Significance of flavonoids in plant resistance and enhancement of their biosynthesis. Plant Biol 7:581–591. https://doi.org/10.1055/s-2005-873009
Treutter D (2006) Significance of flavonoids in plant resistance: a review. Environ Chem Lett 4: 147–157. https://doi.org/10.1007/s10311-006-0068-8
Tripathi P, Dubey NK (2004) Exploitation of natural products as an alternative strategy to control postharvest fungal rotting of fruit and vegetables. Postharvest Biol Technol 32(3):235–245
Tropf S, Lanz T, Rensing SA, Schröder J, Schröder G (1994) Evidence that stilbene synthases have developed from chalcone synthases several times in the course of evolution. J Mol Evol 38:610–618. https://doi.org/10.1007/BF00175881
Ueda H, Kugimiya S, Tabata J, Kitamoto H, Mitsuhara I (2019) Accumulation of salicylic acid in tomato plant under biological stress affects oviposition preference of Bemisiatabaci. J Plant Interact 14(1):73–78
Varbanova M, Porter K, Lu F, Ralph J, Hammerschmidt R, Jones AD, Day B (2011) Molecular and biochemical basis for stress-induced accumulation of free and bound p-coumaraldehyde in cucumber. Plant Physiol 157(3):1056–1066
Vishwanath SJ, Delude C, Domergue F, Rowland O (2015) Suberin: biosynthesis, regulation, and polymer assembly of a protective extracellular barrier. Plant Cell Rep 34(4):573–586
Vogt T (2010) Phenylpropanoid biosynthesis. Mol Plant 3:2–20. https://doi.org/10.1093/mp/ssp106
Wallace KJ, Laughlin DC, Clarkson BD (2017) Exotic weeds and fluctuating microclimate can constrain native plant regeneration in urban forest restoration. Ecol Appl 27(4):1268–1279
Walling LL (2008) Avoiding effective defenses: strategies employed by phloem-feeding insects. Plant Physiol 146(3):859–866

Wallis CM (2020) Grapevine (Vitis spp.) rootstock stilbenoid associations with host resistance to and induction by root knot nematodes, Meloidogyne incognita. BMC Res Notes 13:360. https://doi.org/10.1186/s13104-020-05201-3

Wang L, Sun R, Zhang Q, Luo Q, Zeng S, Li X, Gong X, Li Y, Lu L, Hu M, Liu Z (2019) An update on polyphenol disposition via coupled metabolic pathways. Expert Opin Drug Metab Toxicol 15(2):151–165

Wani SH, Kumar V, Shriram V, Sah SK (2016) Phytohormones and their metabolic engineering for abiotic stress tolerance in crop plants. Crop J 4(3):162–176

War AR, Paulraj MG, Ahmad T, Buhroo AA, Hussain B, Ignacimuthu S, Sharma HC (2012) Mechanisms of plant defense against insect herbivores. Plant Signal Behav 7(10):1306–1320

Wasson AP, Ramsay K, Jones MGK, Mathesius U (2009) Differing requirements for flavonoids during the formation of lateral roots, nodules and root knot nematode galls in Medicago truncatula. New Phytol 183:167–179. https://doi.org/10.1111/j.1469-8137.2009.02850.x

Wuyts N, Swennen R, De Waele D (2006) Effects of plant phenylpropanoid pathway products and selected terpenoids and alkaloids on the behaviour of the plant-parasitic nematodes Radopholus similis, Pratylenchus penetrans and Meloidogyne incognita. Nematology 8:89–101. https://doi.org/10.1163/156854106776179953

Zavaleta ES, Shaw MR, Chiariello NR, Mooney HA, Field CB (2003) Additive effects of simulated climate changes, elevated CO2, and nitrogen deposition on grassland diversity. Proc Natl Acad Sci U S A 100(13):7650–7654

Zernova OV, Lygin AV, Pawlowski ML, Hill CB, Hartman GL, Widholm JM, Lozovaya VV (2014) Regulation of plant immunity through modulation of phytoalexin synthesis. Molecules 19(6):7480–7496

Plant Phenolics in Alleviating Root-Knot Disease in Plants Caused by *Meloidogyne* spp.

17

Semran Parvaiz, Parvaiz Yousuf, Rafiq Lone, and Younis Ahmad Rather

Abstract

Plants are known to produce numerous secondary metabolites, including phenolics. The metabolites included in this group are flavonoids, tannins, isoflavonoids, terpenoids, phenylpropanoids, catechols, cinnamic acid, and coumarins. These phenolic compounds are produced through different metabolic pathways in plants in response to invasion by pathogenic organisms such as bacteria, viruses and fungi, as well as injuries. After the invasion by the pathogens, plants produce such oxidized compounds, which show significant biological activity, thereby aiding in pest resistance. This chapter deals with the use of plant phenolics in alleviating root-knot disease caused by *Meloidogyne* spp. It is revealed that phenolic compounds are promising subjects in dealing with nematode attacks in plants.

Keywords

Plant · Phenolics · Nematode · Root-knot disease · Resistance

S. Parvaiz · P. Yousuf (✉)
Department of Zoology, Central University of Kashmir, Ganderbal, Jammu and Kashmir, India

R. Lone
Department of Botany, Central University of Kashmir, Ganderbal, Jammu and Kashmir, India

Y. A. Rather
Department of Zoology, Government Degree College, Ramban, Jammu and Kashmir, India

R. Lone et al. (eds.), *Plant Phenolics in Biotic Stress Management*,
https://doi.org/10.1007/978-981-99-3334-1_17

17.1 Introduction

Agriculture contributes significantly to Gross Domestic Product or GDP and is considered an important economic activity all over the globe. It is estimated that nearly 65% of working adults earn their livelihood from agricultural activities (The World Bank 2021). Farmers are involved in both subsistence and cash crop farming and export their agricultural produce to other nations or use it locally. Because of the elevated demand for agricultural produce, the amount of land used to grow the crops has increased considerably. However, increasing the percentage of land area for agriculture has not proved sufficient as it has been found that the agricultural yield is not directly proportional to the land area utilized for crop production (M-Farm 2014). Moreover, the technologies used by farmers are either applied inefficiently or are becoming obsolete. Furthermore, another implication is that land used by farmers for crop production is not efficiently used (The World Bank 2021). The growth of the global population is on the rise; this is the reason for the increased demand for food. This means more modern technologies must be employed by farmers to properly use land resources. This is because it may lead to crises all over the world with a pronouncing effect in developing nations. At the same time, crops face a much bigger issue of disease infestation that reduces the crop yield in some vital crops such as millet, maize, potato, wheat, rice, and soybean. Other factors such as the high number and high diversity of pests and pathogens cause huge losses to the crops. More and more pests and pathogens are coevolving with plant species, which has made controlling pests difficult, which will prevent significant crop losses. Several pests that attack crops include insects, birds, and rodents (Fauster Admin 2020).

Crops can be damaged directly or indirectly by insects. Insect activity that injures the crops, such as burrowing holes and feeding on those different tissues, is direct damage (Fauster Admin 2020). The physiological activities of plants are disrupted by the resulting damage to plant tissues, such as water uptake and photosynthesis, leading to decreased yields. Some insects, such as aphids, cause indirect damage to crops by acting as vectors of other parasitic microorganisms, for instance, nematodes. By feeding on different crop tissues, rodents and birds cause direct damage to crops. Rodents' indiscriminate feeding habits challenge farmers as the pest can feed on any crop in the field. Thus, it becomes pivotal for the farmers to identify the pests on time, using effective technologies, thereby improving the yields. Similarly, other diseases can directly or indirectly result from biotic and abiotic factors. Other than affecting the Plant's physiological activity, these diseases lead to great economic losses. This is why a great loss in crop production results. Abiotic factors can result in several abnormal conditions or diseases among plants. For instance, unfavorable growth conditions, such as insufficient light, inadequate nutrients, and mesobiotic factors, result in numerous health conditions (Tjosvold and Koiko 2015).

Entities that exhibit an intermediate state of nonliving and living organisms are mesobiotic factors, and these entities include viruses and viroids. Plant diseases caused by biotic factors are pathogenic and animate. These pathogens include both prokaryotes and eukaryotes. Plant diseases, including wilting of potatoes and soft rot in vegetables, are caused by prokaryotic organisms of bacterial origin, while plant diseases caused by eukaryotes include fungi (rust, smut, and powdery mildew), algae (red rust in papaya and mango), protozoa (hart rot in palm and coconut tree, phloem necrosis in coffee), and nematodes (ear cockle of wheat, root-knot in vegetables, and molya disease in barley and wheat) (Tjosvold and Koiko 2015). Apart from reducing crop yield, plant diseases can also result in natural ecosystems disruption, leading to an imbalance in the ecosystem. One of the most important plant parasites affecting a huge range of plants worldwide is *Meloidogyne*. It includes a genus of species that are polyphagous, sedentary, and endoparasitic root-knot nematodes. Some important members such as *M. incognita*, *M. chitwoodi*, and *M. javanica* (also called southern root-knot nematode) cause huge crop losses, adversely affecting the economic conditions of farmers (University of California n.d.; Bernard et al. 2017). Once the Plant is affected by these parasitic nematodes, both underground and upper ground plant parts are affected. It leads to abnormal plant development and growth, reducing crop yield (Bernard et al. 2017). This is why it is important to control the parasitic infestation caused by *Meloidogyne* spp. to ensure minimal crop losses. This becomes vital to prevent any global food crisis since the economic impact of this spp. is higher. Thus farmers, researchers, and global companies are focusing on developing strategies to control them (Stirling et al. 1999). One of the strategies involves the use of chemical nematicides. However, their use has been questioned from time to time owing to their polluting nature. Their presence in the food chains is detrimental to human health. Therefore, organizations and authorities increasingly discourage their application (Stirling et al. 1999; Poveda et al. 2020). Similarly, farmers also use the concept of crop rotation. However, nematodes form dauer stages that empower them to survive in the soil until they identify a likely host and infect it again. This is why researchers are employing plant phenolics to cope with the root-knot diseases in plants.

One of the main reasons for biotic stress in plants is because of plant–parasitic nematodes (PPNs) (Nicol et al. 2011). Among these nematodes, *Meloidogyne* spp. is responsible for great economic losses. Most of the species in this genus are named root-knot nematodes (RKNs) because of their habit of inducing root knots (Eves-van den Akker et al. 2016). One such important member is *Meloidogyne incognita* (Kofoid and White). A variety of crops with global economic importance are affected by this species (Sikora and Fernandez 2005), including one of the most cultivated vegetables in the world, i.e., tomato crop (FAO 2016; Jones et al. 2013). At the same time, growers do not appreciate PPNs management because of its difficulty and need for crop rotation. In most cases, the nematicides use has proven effective for efficiently controlling the PPNs; however, they may negatively impact nontarget humans and organisms and are not often economically feasible. Therefore, it is important to look for alternative compounds with nematicidal activity as well as with lower toxicity to the environment and humans. Phenolic compounds are promising nematicidal molecules because they influence the relationship between nematodes and plants. For instance, in coffee plants (*Coffea arabica* L.), after the beginning of giant cell formation, the cells surrounding the feeding site of

M. incognita die rapidly, an illustration of hypersensitivity reaction (HR) that prompts coffee resistance to *M. incognita* (Albuquerque et al. 2010). Similarly, Pegard et al. (2005) observed HR occurrence in chili pepper (*Capsicum annuum* L.) attacked RKNs. RKNs resistance has been reported in other cases, such as in tomato (Brueske and Dropkin 1973), rice (Cabasan et al. 2014), and cotton (Mota et al. 2013). When the infection sites were studied, it was observed that phenols were present in the vicinity. It was hypothesized that through the HR reaction, the plant resistance to the pathogens is associated with the phenolic compounds present in the feeding site of RKNs. In addition to the indirect effects, the direct effects of phenolic compounds have also been reported in the infection sites caused by RKNs (Caboni et al. 2016). Gupta et al. (2005) showed *in vitro* nematicidal activity of 3,4-dimethylphenol (a phenolic compound) against *M. incognita*. Similarly, Li et al. (2006) showed nematicidal activity of butyl 2,3-dihydroxy-6-methyl benzoate, 2-hydroxinaphthoic acid (Mahajan et al. 1985), and catechol (Balaji and Kannan 1988). In addition, the immersion of second-stage juveniles (J2) of Meloidogyne Chitwood affected the nematode mobility for 2 h into resorcinol solution at 10,000 μg/mL (Goto et al. 2010). However, the incubation of eggs and *M. incognita* J2 for 72 h into the same solution showed no difference in J2 mortality or in inhibiting the egg hatching (Huang et al. 2013). This means more studies are needed to confirm plant phenolics' role in alleviating plant diseases. In this chapter, we will discuss more plant phenolics and how they can be used to alleviate root-knot disease in plants caused by *Meloidogyne* spp.

17.2 Root-Knot Disease and Its Impact on Plants

Root-knot nematodes that are distributed globally are economically important obligate plant parasites belonging to *Meloidogyne* spp. that have parasitized almost every species of higher plants. 'They are polyphagous sedentary endoparasites, and their success as a parasite depends on inducing the feeding sites in roots of host plants' (Moens et al. 2009; Perry and Moens 2006). Root-knot nematodes, which are the most widespread nematodes, are considered destructive, and over wide areas of cultivated land can cause 25–50% of estimated yield loss. For instance, they specifically affect tomatoes with an estimated yield loss of 24–38%. Besides, RKNs affect a large number of crops all over the world, out of which more than 100 species have been described so far (Taylor and Sasser 1978; Sikora and Fernandez 2005; Hunt and Handoo 2009). Different species of *Meloidogyne* are found in different regions, and the extent of damage caused by them depends upon how favorable the conditions are. The most common species found in tropical regions are *Meloidogyne incognita*, *M. arenaria*, and *M. javanica*. In contrast, *M. hapla*, *M. chitwoodi*, and *M. fallax* are successful in cooler and temperate regions. Still, the damage is more pronounced in tropical climates than temperate climates because of the favorable conditions in tropical climates for nematode multiplication and survival (Karssen and Van Aelst 2001; De Waele and Elsen 2007).

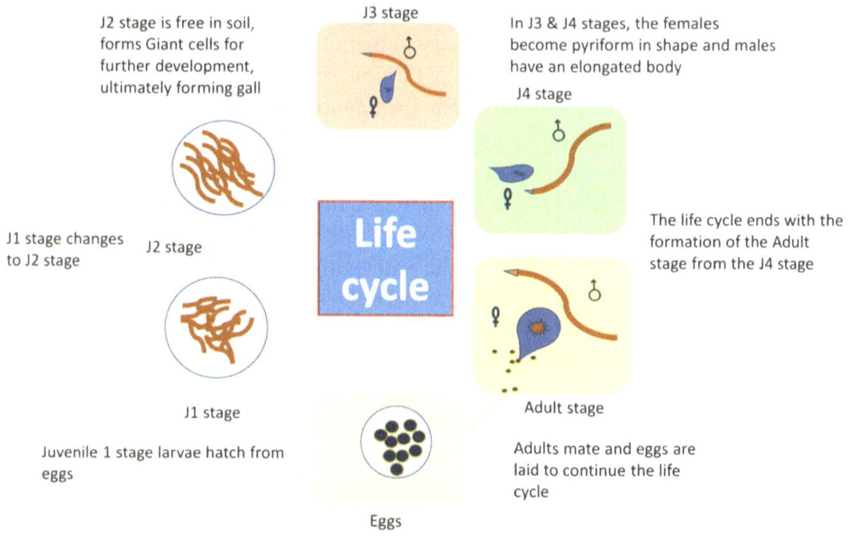

Fig. 17.1 Root-knot nematodes have six-staged life cycles: Egg, J1, J2, J3, J4, and an adult formed after the last molt, which can either be male (appears as a long filiform nematode) or a female (pyriform in shape)

The life cycle of nematodes includes various stages, such as the infective stage juveniles, which migrate into the soil and, by using their style, penetrate root tips of the host plant. Once the juveniles migrate inside the root into the vascular cylinder, they induce the formation of a feeding site intercellularly in case the Plant is susceptible to giant cells formation. Giant cells are multinucleate feeding sites whose formation is induced by the release of pharyngeal secretions, and these giant cells act as suppliers of nutrients to growing nematodes (Karssen 2002; Gheysen and Jones 2006; Abad et al. 2008). Root-knot nematodes have six staged life cycles: Egg, J1, J2, J3, J4, and an adult formed after the last molt, which can either be male (appears as a long filiform nematode) or a female (pyriform in shape). After feeding, J2 juveniles become sedentary and then undergo three molts (J3, J4, and adult) (Fig. 17.1). At times vermiform males, after development, migrate out of the root while pear-shaped females remain sedentary. After feeding, they produce a huge number of eggs in a gelatinous matrix secreted by them. Under unfavorable environmental conditions, this gelatinous matrix protects the eggs from external stress so that embryogenesis initiates from inside. A second stage juvenile hatches after the first molt (Moens et al. 2009; Abad et al. 2008). Moreover, most juveniles develop into males in adverse conditions, while in favorable conditions, most juveniles develop into females. For different root-knot species, the temperatures for survival and reproduction are different. For instance, the optimum temperature range for survival and reproduction of *M. arenaria*, *M. javanica*, and *M. incognita* is 25–30 °C (Taylor and Sasser 1978), which are commonly found in subtropical and tropical countries. Hatching occurs at a temperature lower than 10 °C for cryophiles, that is, *M. naasi*, *M. chitwoodi*, *M. hapla*, along with other temperate Meloidogyne species (Moens et al. 2009). It is necessary to mention that RKNs cannot survive for long without a host plant. Thus, under optimum temperature, the root-knot species mature and feed inside the plant roots. The feeding initiates abnormal enlargement of roots leading to gall formation, and the presence of root galls is the primary evident diagnostic symptom. However, laboratory analysis is necessary for exact species identification. Worldwide, root-knot nematodes limit the production of vegetable crops causing huge losses (Netscher and Sikora 1990), leading to $157 billion annual loss globally (Abad et al. 2008). At the same time, the impact of Meloidogyne species is still underestimated globally, especially in developing nations of Africa. However, it leads to much higher annual losses as no appropriate measures are taken to control the attack of RKNs. In Africa, no comprehensive assessments have been put forward focusing on the economic impact of Meloidogyne species specifically (Coyne et al. 2006). The most recorded hosts for some of the root-knot nematodes like *M. javanica*, *M. incognita*, and *M. arenaria* are vegetable crops, thus affecting the crop yield. This makes vegetable crops more prone to fungal and bacterial attacks (Zhou et al. 2016). Thus, depending on the infection severity, they may cause an annual yield loss of vegetables ranging from about 10–30% (Radwan et al. 2012). Thus, the resistance of plants toward RKNs is unstable and leads to a decreased yield, meaning it becomes difficult to grow vital vegetables like tomatoes in soil infested with RKNs. This is particularly true for the crops grown in semitropical and tropical regions (Williamson et al. 2009; Radwan et al. 2012). Furthermore, feeding of PPNs on plant roots leads to the transmission of plant viruses or creating a passage for the entry of secondary pathogens, making plants prone to secondary infections and pathogens (Rowe and Powelson 2002).

17.3 Plant Phenolics: An Overview

Phenolics are compounds broadly distributed in the plant kingdom possessing one or more aromatic rings with one or more hydroxyl groups. These secondary metabolites have more than 8000 phenolic structures ranging from simple molecules to highly polymerized substances such as phenolic acids to tannins. Plant phenolics are involved in defense against UV radiations or aggression by parasites, predators, and pathogens and contribute to plants' colors. In plant parts, these phenolics are ubiquitous, therefore, an integral part of the human diet. Moreover, phenolics constitute an important part of plant foods, which are partially responsible for their organoleptic properties. For instance, interaction among phenolics contributes to the astringency and bitterness of fruits and their juices. Henceforth, natural phenolics are of great interest for many reasons (astringency, antioxidants, browning reactions, bitterness, color, etc.). At the same time, selecting a proper analytical strategy for extracting and studying phenolics in plant materials relies upon the nature of the sample, the analyte, and the purpose of the study (Robards 2003). The analysis of phenolics is done by using assays that either quantify the whole class of phenolic compounds, a particular group, or those measuring total phenolic content.

Furthermore, choosing an assay with care is vital as it can affect the extraction process. For example, the quantification of phenolic compounds in plant extract is affected by the assay method used, the chemical nature of the analyte, as well as the presence of interfering substances, and the selection of standards (Naczk and Shahidi 2006). In the past, researchers used several methods for determining total phenolic concentration. However, most of them had one or more defects because of the possible interference in the plant materials from other readily oxidized substances and heterogeneity of natural compounds (Singleton et al. 1999). Some methods used include the Folin–Ciocalteu method (F-C), Folin–Denis method (FD), colorimetry with iron salts, ultraviolet absorbance, and permanganate titration. In many cases, the F-C assay has been found preferable over other methods [90]. The F-C assay relies on transferring electrons in an alkaline medium to form blue complexes, possibly PMoW110404. The electrons are transferred from phenolic compounds to phosphotungstic/phosphomolybdic acid complexes, determined spectroscopically at approximately 760 nm (Singleton et al. 1999; Singleton and Rossi 1965). Gallic acid is used as the comparison standard for measuring total phenolic content, and values are generally analyzed as milligram of gallic acid equivalent per kg or liter of extract among samples. Inferring to the nature of F-C chemistry, it is indeed the measure of total phenolics and other oxidation substrates; these oxidation substrates can interfere with measuring total phenolics in an additive, inhibitory, or enhancing manner (Singleton et al. 1999; Singleton and Rossi 1965). The inhibitory effects are caused due to air oxidation after the sample is made alkaline, or it could be due to competing of oxidants with F-C reagent, and this is why ahead of alkali F-C reagent is added [90]. Additive effects occur from aromatic amines, unanticipated phenols, ascorbic acid, or high sugar levels in samples. The effects can be measured before adding alkali to the sample or by a specific method of a known interference which is then subtracted from the F-C value (Singleton et al. 1999). Enhancing effects are caused by common additives of wine that are Sulfites and sulfur dioxide (Singleton et al. 1999). Moreover, the techniques for correcting these factors and the effects of potential interference compounds were discussed by Singleton et al. (1999). However, despite all impediments, the F-C assay is widely used in quantifying phenolic compounds in plant extracts and materials as this assay is simple and reproducible. Moreover, anthocyanins are one of the six subgroups of widespread and large plant polyphenol constituents known as flavonoids that are responsible for red, blue, purple, and orange colors of many vegetables and fruits, for example, berries, apples, onions, and beets. Furthermore, in nature, six common anthocyanidins are found while more than 540 anthocyanin pigments have been identified so far (Andersen et al. 2004), and the easiest assay for quantifying anthocyanins as a group depends on measuring the absorption wavelength ranging from 490 to 550 nm where all anthocyanins show absorption at maximum. However, the absorption band for anthocyanins is far from the band of other phenolic compounds having spectral maxima in the UV region (Fuleki and Francis 1968a, b). Moreover, the browning reactions producing anthocyanin polymerized degradation products are codetermining and leading to overestimation of anthocyanin content. Therefore, an approach is preferable that differentiates anthocyanins from their degradation products. An advantage of structural transformations of anthocyanin chromophore is taken by the pH differential method as a function of pH, and this method is used in measuring the absorption of the sample at pH 4.5 (anthocyanins as colorless hemiketal) as well as at pH 1 (anthocyanins as colored oxonium salts) where anthocyanin degradation pigments do not show any reversible change with pH; thus, they are excluded from absorbance calculation (Fuleki and Francis 1968a, b).

This method is usually based on molecular weight (MW) for calculating monomeric anthocyanin concentration and molar extinction coefficient (ε) of either cyanidin-3-glucoside, the most important common anthocyanin in nature, or the main anthocyanin in the sample. As the differences in the influence of the solvent on ε and the MW of the anthocyanins considerably distort the results; thus, the underlying calculations for quantifying ε and MW should be given (Giusti and Wrolstad 2001). For instance, the quantification for berries containing mainly malvidin glycosides and delphinidin as cyanidin-3-glucoside equivalents gives markedly lower results than "real" values quantified depending up on corresponding standard compounds (Kähkönen et al. 2003). On studying 20 food supplements containing extracts of elderberry, blueberry, chokeberry, and cranberry, the total anthocyanin content (determined as cyanidin-3-glucoside equivalent) obtained with pH differential method were in good concentration as compared with those obtained with the HPLC method (Wrolstad et al. 2002). Moreover, in a collaborative study, seven fruit juices, natural colorants, wine, and beverage samples were analyzed by 11 collaborators representing government, academic and industrial laboratories, demonstrating that the total anthocyanin content can be measured with great agreement among laboratories utilizing the pH differential method. This method is approved as First Action Official Method (Lee et al. 2005). Anthocyanins, liable compounds, are easily condensed and oxidized with other phenolics to make brown polymeric pigments. Somers and Evans developed a method to determine the browning and polymeric color in wines based on the use of a bleaching reagent (sodium sulfite) (Wrolstad et al. 2002). The polymeric anthocyanin degradation products being covalently linked to another phenolic compound are resistant to bleaching by bisulfite, as the 4-position is not available, while the monomeric anthocyanins form a colorless sulfonic acid addition adduct by combining with bisulfite. This technique has been applied to various anthocyanin-rich products and hinges to be widely useful for browning during storage and processing and for monitoring anthocyanin degradation (Giusti and Wrolstad 2001).

17.4 Role of Phenolics in Alleviating Root-Knot Disease

A wide range of biologically active chemicals and secondary metabolites are produced by plants involved in defense against diseases & pests, and the major classes of these secondary metabolites include Terpenoids, Alkaloids, and Phenolic compounds (Fig. 17.2). Phenolic compounds have described a group of structurally diverse secondary metabolites of plants as one of the three groups. Phenolic compounds simply possess an aromatic ring bearing one or more hydroxyl groups

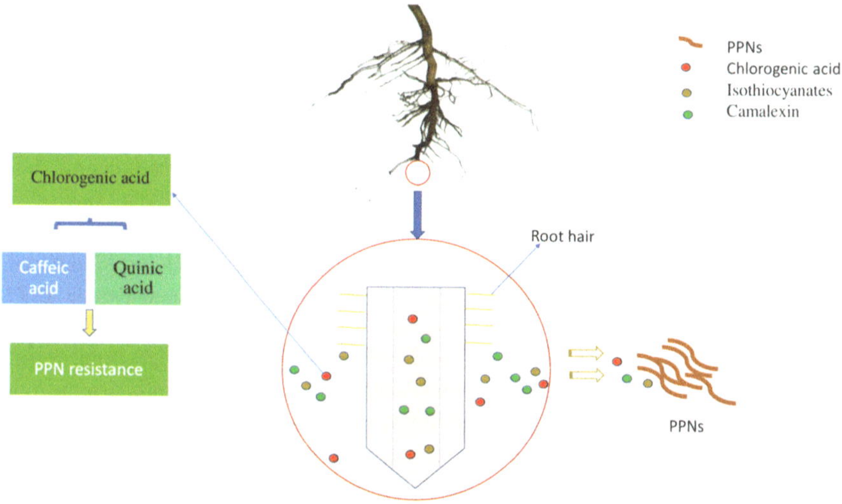

Fig. 17.2 In response to RKNs, several plants produce various phenolic compounds such as Chlorogenic acid, Isothiocyanates, Camalexin, etc.

and polyphenols from the majority of phenolic compounds and are divided into five groups: phenylpropanoids, simple phenols, phenolic acids, tannins, quinone, and flavonoids. In plants, the metabolism and occurrence of phenolic substances in response to the invasion or injury by pathogens, such as bacteria, fungi, and viruses, have been extensively studied (Farkas and Kiraly 1962; Patil et al. 1964). Besides, a great number of phenolic compounds have been found in nematodes, too, with strong nematicidal activity.

17.4.1 Phenols

A number of phenolic compounds have been studied for their nematicidal activity, including dihydroxy, trihydroxy, monohydroxy compounds, aromatic acids such as transcinnamic acid and quinones, and their effect on hatching of the egg of *Meloidogyne incognita* (Mahajan et al. 1985). 2-OH napthoic acid, pyrogallol, transaminic acid, and ethyl gallate were seen to be highly toxic, with a death rate higher than 95%. In contrast, moderate activity ranging from 52 to 66% mortality rate was exhibited by catechin hydrate, dibromoquinone chloroimide, m-hydroxybenzoic acid, 3,4-dihydroxy benzoic acid, and bresorcylic acid. Furthermore, ten phenolic compounds were assayed and oxidized in the same study for nematicidal activity, out of which oxidized ferulic acid, a-resorcylic acid, caffeic acid, and 3,4 dihydroxybenzoic acid showed high mortality effect. Lower activity was indicated by catechin hydrate and chlorogenic acid, while phloroglucinol, p-hydroxycinnamic acid, and vanillic acid induced negligible nematode mortality. With the total suspension of hatching, naringenin was found to be the most effective.

In suppressing the hatching of eggs, transcinnamic acid and 2-OH napthoic acid were also found highly effective. Mahajan et al. (1992) again tested a wide range of phenolic compounds for their nematicidal activity against *M. incognita*. Out of the 55 phenolic compounds tested, few showed high nematicidal activity, including dihydroxy caffeic acid quinone, juglone, coumestrol, gentisic acid, and p-methoxy cinnamic acid vanillic, 7-OH-coumarin, 3-phenyl phenol, protocatechuic acid, syringic, and 2,6-dihydroxy benzoic acid. Host–parasite relationships of *Pratylenchus penetrans* and *M. incognita acrita* have been compared to three closely related tomato cultivars: "Hawaii 7153" moderately resistant to RKNs, "Nemared," resistant to root-knot nematodes, and "B-5," susceptible (Hung and Rohde 1973). The inhibition of a large number of *P. penetrans* and *M. incognita* was reported due to some sort of inhibition, as they never penetrated the tomato resistant variety because of the presence of phenolic compounds. Prior to or after infection in roots, the major phenolic compound is identified as chlorogenic acid. Various substituted phenols, hydrazides, and phenoxyacetic esters had nematicidal activity determined against the J2 stage of seed-gall nematode (*Anguilla tritici*), pigeon pea cyst nematode (*H. cajani*), and root-knot nematode (*M. javanica*). This helps establish the relationship of activity, although having structural variations (Malik et al. 1989). It was reported that phenols with electron-withdrawing substituents are less active than those with electron-donating substituents, especially chloro-substituted phenols. In the root-knot nematode, *M. javanica*, the effects of lawsone, juglone, and naphthoquinones plumbagin from *Lawsonia alba*, *Juglans regia*, and *Plumbago zeylanica*, respectively, have been studied in vitro (Dama 2002). Nematodes were exposed to concentration 200 mg/10 mL water/flask for 24 h, 12 h, and 6 h. The percent mortality was 58.3% in lawsone, 97.9% in juglone, and 100% in plumbagin. For the action of plumbagin on *M. incognita*, the structure–activity relationship was observed. It exhibited a quite structural requirement for the accomplishment of maximal activity. In this manner, lawsone and juglone vary from plumbagin in a way that they don't have a methyl group and were less compelling against *M. javanica* considerably. Similar earlier studies showed that juglone has antiparasitic properties (Dama and Jadhav 1997). Similarly, another compound, plumbagin, inhibited the development of parasitic nematode (*Haemonchus contortus*) (Fetterer and Fleming 1991).

17.4.2 Salicylic Acid

In many plants, a natural phenolic compound salicylic acid is present, and in various plants is associated with induction of resistance. On tomato cv, Pusa Ruby, the foliar application of salicylic acid at 25 and 50 μg/mL impacted the development of *M. incognita* (Ganguly et al. 1999). On the application of 50 μg/mL of salicylic acid, the development of juveniles into the adult male population increased, and the population of females was considerably delayed indicating sex reversal. Reduced root galling (50% or lower) was seen in treated tomato plants over control. On *M. incognita* infesting tomato Pusa Ruby, the impact of exogenous salicylic acid has been worked out (Nandi et al. 2000a). The outcome showed that salicylic acid increased plant development in shoot weight, root length, and shoot length compared with inoculated untreated plants. Nematode population, root galls in the soil and in roots, and protein coat content are essentially reduced in plants treated with salicylic acid compared to the untreated ones. Compared to the postinoculation treatment, lesser intensity and better plant growth were seen in plants pretreated with SA. Comparable outcomes were obtained when the impact of salicylic acid was investigated on *M. incognita* infesting okra and cowpea (Nandi et al. 2000b, 2002, 2003). Relative sensitivity of *Rotylenchus reniformis* and *M. incognita* on okra was demonstrated by Pankaj and Sharma (2003). Salicylic acid, when showered, advanced the plant development fundamentally over control. The number of egg masses and galls created by *M. incognita* were fundamentally lesser. Accordingly, salicylic acid meddled gall information and hindered normal egg mass production and consequently hatching. A study of systemic acquired resistance (SAR) conducted on bare root dip application instigating chemicals in tomato against *M. incognita* was directed by Sirohi and Pankaj (2005). In the study, cv. Pusa Ruby tomato seedling which was 3 weeks old, were treated with 2-hydroxybenzoic acid or salicylic acid, gibberellic acid, and rose Bengal at 25, 100, and 50 μg/mL as a foliar spray, soil drench, and root dip for 10 min with synchronous or delayed nematode inoculation. From the date of sowing, after 60 days, the tomato seedlings were harvested. The outcomes showed that salicylic acid was the most effective molecule for inducing resistance of all three chemicals. However, all the three showed varied effects as an inducer in resistance while the two other chemicals showed similar effects as resistance inducers, but it was not a significant effect. A study conducted by Pankaj et al. (2005) revealed that salicylic acid has a positive role in the resistance mechanism of chickpea against *M. incognita*. Similarly, a study was conducted by Jayakumar et al. (2006) on the evaluation of SA as SAR inducer on tomato cv, Co3 against *M. incognita*. It was also shown that among different dosage levels and techniques used for application of SA, i.e., foliar application and root dip of SA on tomato cv. Co3 at 200, 100, and 50 ppm impacted the growth of *M. incognita*, essentially increased root length, root weight, shoot length, and plant height after 90 days of transplanting. Consequently, it was revealed that the decrease in root-knot nematode population is chiefly because of induced resistance, and such plant development may be prompting the effect of induced exogenous application of Salicylic acid. In relation with a growth hormone Indole3-acetic acid and eight pesticides (Phenomiphos, Phorate, Carbofuran, Metayalaxyl + Mancozeb, Mancozeb, Bitertanol, thiophanate methyl, and Thionazin), the effect of SA on *M. incognita* was demonstrated (Naik and Sharma 2007). Similarly, the effect on larval mortality and egg hatching was observed. Two different concentrations, viz., 0.1% and 0.2%, were taken, and it was observed that as compared to 0.1%, a concentration of 0.2% was more effective. A. thaliana mutants were utilized in signal transduction and SA biosynthesis to examine the role of SA in parasitism inhibition by *H. schachtii*, the beet cyst nematode (Wubben et al. 2008).

17.4.3 Phenylpropanoid

The phenylpropanoid serves as antibiotics, low molecular weight flower pigments, insect repellants, UV protectants, and in plant–microbe interaction act as signal molecules. In plants' defense and wound responses, phenylpropanoids are reported to take part. For better resistance against nematodes, phenylpropanoids have a well-characteristic pathway that constitutes a potential target. In plants, directly or indirectly, enzymes that are functional in the phenylpropanoids pathway are induced in response to pathogen infection, including sedentary endoparasitic nematodes and wounding. In banana roots, the activities of polyphenol oxidase (PPO), peroxidase (PO), and phenylalanine ammonia-lyase (PAL) before were analyzed after 1.3 and 7 days of inoculation with *Radopholus similis* (the burrowing nematode) compared with mechanically wounded roots (Wuyts et al. 2006a). Constitutive activities of PO, PPO, and PAL were higher in susceptible cv. Grande Naine than in resistant cv. Yangambi km 5 (*Musa acuminata* AAA). The response of PAL induction only in *R. Similis*-inoculated roots of the resistant cultivator was different from wound induction. During the experiment, as a response to general stress in plants, the levels of PPO and PO increased in resistant cultivars from those of the susceptible cultivar. Nevertheless, as compared to constitutive ones, the final levels of PPO and PO activity were found higher in *R. similis*-inoculated roots.

17.4.4 Tannins

Tannins that are able to precipitate proteins from solution are basically a group of water-soluble polyphenols. Tannins are found in a huge variety of higher plants, both woody and herbaceous types. In plants, you may find substantial accumulations of vegetable tannins in any part. Tannins protect a few plants against herbivores (Feeny 1976), and to a wide variety of bacteria, fungi, and yeast, they prove toxic (Scalbert 1991). Since the host plant tannins cause severe phototoxicity, studies showing tannins' effects on PPNs are few. For the control of *M. arenaria* (Neal) Chitwood, the effectiveness of four organic amendments, which are common with high tannin and phenolic content, has been investigated (Mian and Rodriguez-Kabana 1982). By applying tannic acid of 0.4% or more, the number of galls/gm of the fresh root was decreased, and for gall index values similar response was observed. The tannic acid application at 0.2% rates or more resulted in the lower shoots, root weights, and shorter shoots. As compared to the values corresponding to plants in untreated soil, the values of the root condition index were lower. The number of seedlings that emerged was not affected by tannic acid. On hatching *Heterodera glycines*, a soybean cyst nematode, the effects of tannic acid were evaluated. The study was also used to determine the prospect of utilizing tannic acid to induce hatching in nematodes. With an increase of tannic acid concentration, hatching percent also increased up to 35–40 mg/L, and then after this concentration with increasing concentration, hatching declined, i.e., high tannic acid concentration inhibits hatching.

In the same way, tannic acids efficacy against *M. arenaria* on tomato as well as its impact on the behavior of *M. incognita*, *M. Arenaria*, *R. similis*, and *H. glycines* was studied (Hewlett et al. 1997). Regardless of methods of application, reduction in galling was observed upon using tannic acid as compared to the untreated control. While doing behavioral studies on water agar, it was found that Meloidogyne J2 got attracted to areas with more tannic acid gradient. Among one of two experiments conducted, R. similis got repelled from the tannic acid gradient, but there was no impact on H. glycines. During the experiment, it was found that the behavioral response of Meloidogyne J2 to tannic acid showed that other phenolic compounds and tannins might be chemical signals that *Meloidogyne* spp. Use to recognize the plant hosts, navigate toward roots, or locate areas for root penetration (Hewlett et al. 1997).

17.4.5 Flavonoids

Flavonoids are secondary metabolites with a low molecular weight that have numerous functions, including UV protection, defense, flower coloring, allelopathy, and auxin transport inhibition. Evidence shows that plants that are nematode resistant have higher constitutive degrees of record for key enzymes associated with biosynthesis of isoflavonoid phytoalexins, which play a part in resistance to fungi and are embroiled in resistance to both migratory and sedentary nematodes. The primary proof was acquired for the association in alfalfa resistance to *P. penatrans* of several know plant defense response genes (Baldridge et al. 1998). In *P. penetrans*-infected resistant alfalfa plants, the legitimacy of phenylpropanoid pathway enzyme transcript suppression was upheld by the concurrent induction of b-1,3-glucanase transcripts in a resistant versus vulnerable host design pattern of numerous another pathogen/plant systems. Additionally, it has been accounted that the roots of soybeans which were resistantly inoculated with *M. incognita* show a hypersensitivity response and collect a product of the isoflavonoid branch of the phenylpropanoid pathway, which is phytoalexin glyceollin. Likewise, glyceollin inhibits respiration, motility, and oxidation of *M. incognita* in vitro and aggregates promptly near the head region of H, glycines (soybean cyst nematode) in resistant but not susceptible soybean root tissue (Kaplan et al. 1980) (Table 17.1).

17.5 Production of Antinematode Compounds in Plants

In response to PPN invasion, plants produce secondary metabolites. For example, a phenolic compound, chlorogenic acid, is produced in various plants, including rice (Plowright et al. 1996), carrots (Knypl et al. 1975), and solanaceous plants (Milne et al. 1965; Hung and Rohde 1973; Pegard et al. 2005), which suggests a common defense response against PPN infection. For *M. incognita*, chlorogenic acid is weakly nematicidal, albeit there is well-correlation in the chlorogenic acid production with PPN resistance levels (Mahajan et al. 1985; D'Addabbo et al. 2013).

Table 17.1 Different plants produce numerous kinds of phenolic compounds, thereby alleviating the response to RKNs

Type of phenolic compound	Plant	Role	References
Phenols	Tomato, lettuce	Affects egg hatching *Meloidogyne incognita*	Mahajan et al. (1985)
Salicylic acid	Tomato, okra, etc.	Lowers egg mass & galls	Sirohi and Pankaj (2005)
Phenylpropanoid	Banana	Resists overall nematode infection	Wuyts et al. (2006a)
Tannins	Both woody and herbaceous types	Lower gall formation	Mian and Rodriguez-Kabana (1982)
Flavonoids	Soybean	Biosynthesis of isoflavonoid phytoalexins	Baldridge et al. (1998), Kaplan et al. (1980)

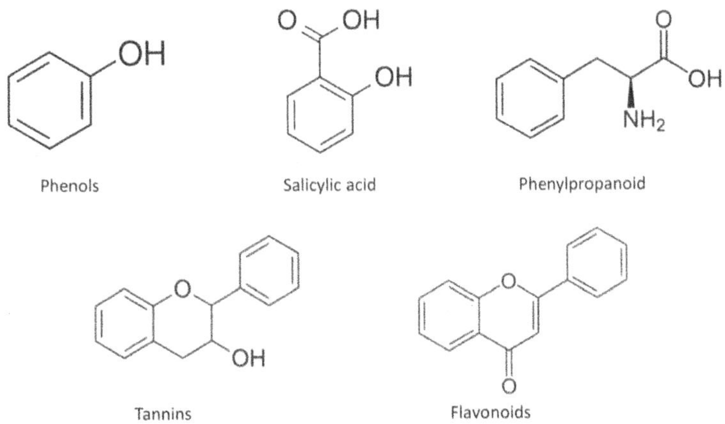

Fig. 17.3 Structure of phenolic compounds

However, it has moderate activity against a false nematode, *Nacobbus aberrans* (López-Martínez et al. 2011). One potential clarification for the absence of correlation between effectiveness and in response target organisms, metabolized products of chlorogenic acid have higher nematicidal activity, but those compounds may be highly toxic or unstable in plants. Chlorogenic acid can be hydrolyzed to caffeic acid and quinic acid, with caffeic acid being additionally oxidized to orthoquinone, which is toxic to PPNs (Mahajan et al. 1985). However, the roles of orthoquinone and caffeic acid against PPNs resistance should be established further (Fig. 17.3). In a resistant banana cultivar (Musa sp.), at the infection sites of *Radopholus similis*, the burrowing nematode phenylphenalenone anigorufone, a phenolic compound, gets accumulated (Dhakshinamoorthy et al. 2014; Hölscher et al. 2014). In the bodies of *R. similis*, anigorufone forms a large lipid–anigorufone complex because of which Anigorufone has high nematicidal activity. The synthesis of Anigorufone is being activated by infection with *Fusarium oxysporum*, a pathogenic fungus. Thus, anigorufone is also named antifungal phytoalexin (Luis et al. 1995). Moreover, in mitochondrial respiratory complex II, anigorufone causes inhibition of succinate dehydrogenase in Leishmania, a human protozoan parasite, and kills it (Luque-Ortega et al. 2004). However, in PPNs, the toxic mechanism of anigorufone and its relationship with the development of large lipid–anigorufone complexes has not yet been determined.

In plants, flavonoids establish a large group of secondary metabolites, and some flavonoids play an important role by functioning as nematicides, repellents, nematic compounds (inhibit their movement instead of killing), or inhibitors of egg hatching in PPN resistance (Chin et al. 2018). Flavonoids with antinematodal activity belong to the classes of isoflavonoids, pterocarpans (e.g., glyceollin, medicarpin), and flavonols (e.g., quercetin, kaempferol, myricetin). Egg hatching of *R. similis* is inhibited by Kaempferol (Wuyts et al. 2006b). Quercetin, myricetin, and kaempferol are nemastic and repellents to *M. incognita* juveniles (Wuyts et al. 2006b). Similarly, in a concentration-dependent manner, the motility of Pratylenchus penetrans is inhibited by medicarpin (Baldridge et al. 1998). Similarly, for infective juveniles of Heterodera zee, a CN, patulitrin, rutin, and paruletin are nematicidal (Faizi et al. 2011). Also, during infection in resistant plants, the synthesis of flavonoids is induced. For instance, glyceollins, a group of specific prenylated pterocarpan phytoalexins which upon infection are expressed, are accumulated by *M. incognita*-resistant soybean cultivars (Kaplan et al. 1980), and moreover, the motility of *M. incognita* is inhibited by glyceollin (Kaplan et al. 1979, 1980). In CN-resistant soybean cultivars, glyceollin accumulation is higher in susceptible ones, and in resistant soybean roots, glyceollin I, one of the glyceollin isomers, accumulates in adjacent tissues to the head of the cyst nematode (Huang and Barker 1991), proposing that glyceollin accumulation is spatiotemporally specific to the site of infection. Several nematode–antagonistic plants like asparagus and marigold produce other nematicidal chemicals apart from phenolic compounds, which in soil help in the reduction of nematode populations. For instance, α-terthienyl is secreted by marigold (Gommers and Bakker 1988; Wang et al. 2007; Faizi et al. 2011), α-terthienyl is a stress-inducing oxidative chemical that penetrates the hypodermis of nematode effectively and exerts nematicidal activity (Nivsarkar et al. 2001; Hamaguchi et al. 2019). In the same way, asparagusic acid is produced by asparagus, which inhibits the hatching of two important CNs, *G. rostochiensis* and *Heterodera glycines* (Takasugi et al. 1975).

In plants of the Brassicaceae family, the broad-spectrum antimicrobial indole glucosinolates and isothiocyanates are considered anti-PPN compounds which effectively inhibit the hatching of RKNs and CNs (Brown et al. 1997; Yu et al. 2005). They are also toxic to the semiendoparasitic nematode *Tylenchulus semipenetrans* and to RKNs (Zasada and Ferris 2003). In Arabidopsis, the synthesis of an indole alkaloid glucosinolatetype phytoalexin, i.e., camalexin is catalyzed by three monooxygenases which are cytochrome P450-dependent, CYP79B3, CYP79B2 (Hull et al. 2000; Mikkelsen et al. 2000, 2004; Bak et al. 2001), and PAD3 (phytoalexin-deficient 3, CYP71B15). In PAD3, instead of wild types, camaxelin-deficient mutants are more susceptible to RKNs (Teixeira et al. 2016), while double mutants cyp79b2/b3 that don't accumulate indolic glucosinolates are more susceptible to CNs (Shah et al. 2017). These outcomes show that on PPNs, few indole glucosinolates, including camalexin, show inhibitory effects but there are no reports so far showing direct toxicity of indolic glucosinolates on PPNs. For limiting or inhibiting PPN infection in addition to metastatic compounds and nematicides, PPN chemotaxis interruption may be an effective plant response too. Ethylene is produced normally during pathogen invasion as well as wounding, reducing the attraction of PPN to the root (Booker and DeLong 2015; Guan et al. 2015; Marhavý et al. 2019). An Arabidopsis mutant overproducing ethylene is less attractive for PPNs, and attractiveness is higher in ethylene-insensitive mutants or in plants treated with inhibitors of ethylene synthesis (Fudali et al. 2013; Hu et al. 2017). These outcomes show that ethylene production is induced by PPN infection, which reduces attractiveness and prevents secondary PPN invasion, and the reduction in attraction may be due to an increase in repellents or a reduction in the secretion of attractants. However, for PPNs, the attractiveness's molecular basis is still largely unknown. Moreover, a number of groups have tried to recognize RKN attractants from seed coat mucilage (Tsai et al. 2019) and root tips (Čepulytė et al. 2018). The recognition of chemorepellents and chemoattractants may put light on how plants in the rhizosphere respond to nematodes both during and before PPN infection.

However, some phenols that did not have any nematicidal effect were also observed by researchers, such as unpublished results of resorcinol on the in vitro mortality of J2 *M. incognita*. Therefore, it seems all the phenolic compounds are not toxic to PPNs. Thus, phenolic compounds may have another role in relation to modifying the physiology of nematodes. RKNs secrete a pool of substances into the plant cell membrane to induce the formation of a nematode feeding site (Williamson and Gleason 2003; Caillaud et al. 2008), and it has been proposed that some phenolic compounds induce such secretions, such as catechol, resorcinol, caffeic acid, and hydroquinone (McClure and Von Mende 1987; Jaubert et al. 2002; Bellafiore et al. 2008). However, there is no clarification about the fact whether J2's exposure to those kinds of phenolic compounds could make the nematodes secrete their substances somewhere else instead of roots, thus preventing the formation of feeding sites in the roots of plants and using a product of this mode of action to control *Meloidogyne* spp. It is unlikely to increase mortality of nematode, which is in agreement with caffeic acid's inactivity against *M. incognita* (Aoudia et al. 2012).

17.6 Conclusion

Nearly 4100 species of PPNs have been identified, among which 100 species are RKNs. Apart from migrational damage and direct feeding, subsequent infestation by secondary pathogens like bacteria and fungi is facilitated by nematode feeding. In terms of resource use efficiency, significant improvements are consequently important. As the proportional production of some commodities steadily shifts, disease management and optimal pest management are necessary for taking crop yields toward an efficiency frontier. Nowadays, globally a large number of crops are affected by RKNs. They are believed to be the most destructive and widespread plant–parasitic nematode, and over wide areas of cultivated land, they can cause an estimated yield loss of 25–50%. The estimated yield loss is about 24–38% on tomatoes.

Henceforth, PPNs are important parasitic pests, although nematode parasitism is rarely fatal. However, the plant–parasitic nematodes cause substantial loss in yield by disrupting water transport, acting as vectors for viruses, diverting nutrients, and increasing susceptibility to secondary infections (Bird and Kaloshian 2003; Nicol et al. 2011). Even though evaluating their effect is difficult, estimates recommend that PPN lessen worldwide yields by 10–25% (Nicol et al. 2011). In the United States, *Heterodera glycines*, the soybean nematode, decreases yield by nearly 10% (Savary et al. 2019). More than 4000 species of PPN have been recognized (Wyss 1997; Decraemer and Hunt 2006; Nicol et al. 2011); some feed on aerial parts but the majority feed on roots (fuller et al. 2008). In spite of their variety, a modest bunch of sedentary PPN genera causes most economic losses, especially the cyst nematodes (*Heterodera* spp. and *Globodera* spp.) and the root-knot nematodes (*Meloidogyne* spp.) (Fuller et al. 2008; Nicol et al. 2011). The effective control of nematodes is challenging and needs an integrated approach that consolidates cultural practices, chemicals, biocontrol, and resistant varieties were accessible (Fuller et al. 2008). To facilitate the improvement of control strategies of novel nematodes, plant nematologists have invested impressive effort and time in concentrating on the plant defense mechanisms against PPN. Production of metabolites with antinematode activity is one of the defense mechanisms.

So, it is clear that plants cooperate with their environment through secondary metabolic pathways. Thus, it is concluded that phenylpropanoid pathways are one of the basic secondary metabolic pathways through which plants produce these phenolic compounds. In turn, they protect the plants against numerous biotic and abiotic factors. Thus, it is clear that phenolic compounds are subsequently involved in plant defense and have a higher potential against nematodes. However, more research is needed to elucidate the mechanism responsible for resistance. More screening of the germplasm and hybrid varieties is demanded to contain RKNs. Using compounds such as plant phenolics will help create better varieties that resist any infection.

Acknowledgments The Authors acknowledge Central University of Kashmir for support

Conflict of Interest The authors declare no financial or non-financial conflict of interest.

Additional Notes None

References

Abad P, Gouzy J, Aury JM, Castagnone-Sereno P, Danchin EG, Deleury E et al (2008) Genome sequence of the metazoan plant-parasitic nematode Meloidogyne incognita. Nat Biotechnol 26(8):909–915

Albuquerque EVS, Carneiro RMDG, Costa PM, Gomes ACMM, Santos M, Pereira AA, Nicole M, Fernandez D, Grossi-de-Sa MF (2010) Resistance to Meloidogyne incognita expresses a hypersensitive-like response in Coffea arabica. Eur J Plant Pathol 127:365–373

Andersen ØM, Fossen T, Torskangerpoll K, Fossen A, Hauge U (2004) Anthocyanin from strawberry (Fragaria ananassa) with the novel aglycone, 5-carboxypyranopelargonidin. Phytochemistry 65(4):405–410

Aoudia H, Ntalli N, Aissani N, Yahiaoui-Zaidi R, Caboni P (2012) Nematotoxic phenolic compounds from Melia azedarach against Meloidogyne incognita. J Agric Food Chem 60(47):11675–11680

Bak S, Tax FE, Feldmann KA, Galbraith DW, Feyereisen R (2001) CYP83B1, a cytochrome P450 at the metabolic branch point in auxin and indole glucosinolate biosynthesis in Arabidopsis. Plant Cell 13(1):101–111. https://doi.org/10.1105/tpc.13.1.101

Balaji A, Kannan S (1988) Impact of different phenolic compounds on hatchability of Meloidogyne incognita. Geobios (Jodhpur, India) 15:143–144

Baldridge GD, O'Neill NR, Samac DA (1998) Alfalfa (Medicago sativa L.) resistance to the root-lesion nematode, Pratylenchus penetrans: defense-response gene mRNA and isoflavonoid phytoalexin levels in roots. Plant Mol Biol 38(6):999–1010

Bellafiore S, Shen Z, Rosso M, Abad P, Shih P, Briggs SP (2008) Direct identification of the Meloidogyne incognita secretome reveals proteins with host cell reprogramming potential. PLoS Pathog 4:1–12

Bernard GC, Egnin M, Bonsi C (2017) The impact of plant-parasitic nematodes on agriculture and methods of control. In: Nematology-concepts, diagnosis and control, 10

Bird DMK, Kaloshian I (2003) Are roots special? Nematodes have their say. Physiol Mol Plant Pathol 62:115–123. https://doi.org/10.1016/S0885-5765(03)00045-6

Booker MA, DeLong A (2015) Producing the ethylene signal: regulation and diversification of ethylene biosynthetic enzymes. Plant Physiol 169(1):42–50. https://doi.org/10.1104/pp.15.00672

Brown PD, Morra MJ, Sparks DL (1997) Control of soil-borne plant pests using glucosinolate-containing plants. Adv Agron 61:167–231. https://doi.org/10.1016/S0065-2113(08)60664-1

Brueske CH, VH D (1973) Free phenols and root necrosis in Nematex tomato infected with the root knot nematode. Phytopatholgy 63:329–334

Cabasan MTN, Kumar A, Bellafiore S, De Waele D (2014) Histopathology of the rice root-knot nematode, Meloidogyne graminicola, on Oryza sativa and O. glaberrima. Nematology 16(1): 73–81

Caboni P, Aissani N, Demurtas M, Ntalli N, Onnis V (2016) Nematicidal activity of acetophenones and chalcones against Meloidogyne incognita and structure–activity considerations. Pest Manag Sci 72(1):125–130

Caillaud MC, Dubreuil G, Quentin M, Perfus-Barbeoch L, Lecomte P, de Almeida Engler J et al (2008) Root-knot nematodes manipulate plant cell functions during a compatible interaction. J Plant Physiol 165(1):104–113

Čepulytė R, Danquah WB, Bruening G, Williamson VM (2018) Potent attractant for root-knot nematodes in exudates from seedling root tips of two host species. Sci Rep 8(1):10847. https://doi.org/10.1038/s41598-018-29165-4

Chin S, Behm CA, Mathesius U (2018) Functions of flavonoids in plant–nematode interactions. Plants (Basel) 7(4):85. https://doi.org/10.3390/plants7040085

Coyne DL, Tchabi A, Baimey H, Labuschagne N, Rotifa I (2006) Distribution and prevalence of nematodes (Scutellonema bradys and Meloidogyne spp.) on marketed yam (Dioscorea spp.) in West Africa. Field Crop Res 96(1):142–150

Dama LB, Jadhav BV (1997) Anthelmintic effect of Juglone on mature and immature Hymenolepis nana in mice. Rivista di Parassitologia 58:303–306

D'Addabbo T, Carbonara T, Argentieri M, Radicci V, Leonetti P, Villanova L et al (2013) Nematicidal potential of Artemisia annua and its main metabolites. Eur J Plant Pathol 137(2): 295–304. https://doi.org/10.1007/s10658-013-0240-5

Dama LB (2002) Effect of naturally occurring napthoquinones on root-knot nematode, Meloidogyne javanica. Indian Phytopathol 55:67–69

De Waele D, Elsen A (2007) Challenges in tropical plant nematology. Annu Rev Phytopathol 45: 457–485

Decraemer W, Hunt DJ (2006) Structure and classification. In: Perry RN, Moens M (eds) Plant nematology. CABI Publishing, Wallingford, pp 3–32

Dhakshinamoorthy S, Mariama K, Elsen A, De Waele D (2014) Phenols and lignin are involved in the defence response of banana (Musa) plants to Radopholus similis infection. Nematology 16: 565–576. https://doi.org/10.1163/15685411-00002788

Eves-van den Akker S, Laetsch DR, Thorpe P, Lilley CJ, Danchin EG, Da Rocha M et al (2016) The genome of the yellow potato cyst nematode, Globodera rostochiensis, reveals insights into the basis of parasitism and virulence. Genome Biol 17(1):1–23

Faizi S, Fayyaz S, Bano S, Iqbal EY, Lubna, Siddiqi H et al (2011) Isolation of nematicidal compounds from Tagetes patula L. yellow flowers: structure–activity relationship studies against cyst nematode Heterodera zeae infective stage larvae. J Agric Food Chem 59(17): 9080–9093. https://doi.org/10.1021/jf201611b

FAO (2016). http://www.fao.org/faostat/en/#data/QC/visualize (Published online)

Farkas GL, Kiraly Z (1962) Role of phenolic compounds in the physiology of plant diseases and disease resistance. Phytopathol J 44:105–150

Fauster Admin (2020) Crop pests and diseases—agriculture form 3 notes [Internet]. https://www.easyelimu.com/highschool-notes/agriculture/form-3/item/2086-crop-pests-and-diseases

Feeny P (1976) Plant apparency and chemical defense. Recent Adv Phytochem 10:1–40

Fetterer RH, Fleming MW (1991) Effect of plumbagin on development of the parasitic nematodes Haemonchus contortus and Ascaris suum. Comp Biochem Physiol 100:539–542

Fudali SL, Wang C, Williamson VM (2013) Ethylene signaling pathway modulates attractiveness of host roots to the root-knot nematode Meloidogyne hapla. Mol Plant Microbe Interact 26(1): 75–86. https://doi.org/10.1094/MPMI-05-12-0107-R

Fuleki T, Francis FJ (1968a) Quantitative methods for anthocyanins. 1. Extraction and determination of total anthocyanin in cranberries. J Food Sci 33(1):72–77

Fuleki T, Francis FJ (1968b) Quantitative methods for anthocyanins. 4. Determination of individual anthocyanins in cranberry and cranberry products. J Food Sci 33(5):471–478

Fuller VL, Lilley CJ, Urwin PE (2008) Nematode resistance. New Phytol 180:27–44. https://doi.org/10.1111/j.1469-8137.2008.02508.x

Ganguly AK, Sirohi A, Pankaj, Singh V (1999) Salicylic acid induced resistance in tomato against Meloidogyne incognita race 1. Indian J Nematol 29:182–184

Gheysen G, Jones JT (2006) Molecular aspects of plant-nematode interactions. In: Plant nematology. CABI, Wallingford UK, pp 234–254

Giusti MM, Wrolstad RE (2001) Characterization and measurement of anthocyanins by UV-visible spectroscopy. Curr Protocol Food Anal Chem 1:F1–F2

Gommers FJ, Bakker J (1988) Mode of action of alpha-terthienyl and related compounds may explain the suppressant effects of tagetes species on populations of free living endoparasitic plant nematodes. Bioact Mol 7:61–69

Goto DB, Fosu-Nyarko J, Sadler J, Flottman-Reid M, Jones MG (2010) In planta observation of live fluorescent plant endoparasitic nematodes during early stages of infection. 日本線虫学会誌 40(1):15–19

Guan R, Su J, Meng X, Li S, Liu Y, Xu J et al (2015) Multilayered regulation of ethylene induction plays a positive role in Arabidopsis resistance against Pseudomonas syringae. Plant Physiol 169(1):299–312. https://doi.org/10.1104/pp.15.00659

Gupta RL, Prasad D, Thukral R (2005) Quantitative structure activity relationship study for the fungicidal and nematicidal activity of phenols. Pestic Res J 17(2):15–20

Hamaguchi T, Sato K, Vicente CSL, Hasegawa K (2019) Nematicidal actions of the marigold exudate α-terthienyl: oxidative stress-inducing compound penetrates nematode hypodermis. Biol Open 8(4):bio038646. https://doi.org/10.1242/bio.038646

Hewlett TE, Hewlett EM, Dickson DW (1997) Response of Meloidogyne spp., Heterodera glycines and Radopholus similis to tannic acid. J Nematol 29:737–741

Hölscher, D., Dhakshinamoorthy, S., Alexandrov, T., Becker, M., Bretschneider, T., Buerkert, A., et al. (2014). Phenalenone-type phytoalexins mediate resistance of banana plants (Musa spp.) to the burrowing nematode Radopholus similis. Proc Natl Acad Sci U S A 111 (1), 105–110. doi: https://doi.org/10.1073/pnas.1314168110

Hu Y, You J, Li C, Williamson VM, Wang C (2017) Ethylene response pathway modulates attractiveness of plant roots to soybean cyst nematode Heterodera glycines. Sci Rep 7:41282. https://doi.org/10.1038/srep41282

Huang JS, Barker KR (1991) Glyceollin I in soybean-cyst nematode interactions: spatial and temporal distribution in roots of resistant and susceptible soybeans. Plant Physiol 96(4): 1302–1307. https://doi.org/10.1104/pp.96.4.1302

Huang Y, Mei M, Shen B, Mao Z, Xie B (2013) RNAi of MiASB caused high mortality of Meloidogyne incognita juveniles and inhibited the nematode disease. Agric Sci 4(9):483

Hull AK, Vij R, Celenza JL (2000) Arabidopsis cytochrome P450s that catalyze the first step of tryptophan-dependent indole-3-acetic acid biosynthesis. Proc Natl Acad Sci U S A 97(5): 2379–2384. https://doi.org/10.1073/pnas.040569997

Hung CL, Rohde RA (1973) Phenol accumulation related to resistance in tomato to infection by root-knot and lesion nematode. J Nematol 5:255–258

Hunt DJ, Handoo ZA (2009) Taxonomy, identification and principal species. In: Root-knot nematodes, 1, pp 55–88

Jaubert S, Laffaire JB, Piotte C, Abad P, Rosso MN, Ledger TN (2002) Direct identification of stylet secreted proteins from root-knot nematodes by a proteomic approach. Mol Biochem Parasitol 121(2):205–211

Jayakumar J, Rajendran G, Ramakrishnan S (2006) Evaluation of salicylic acid as a systemic resistance inducer against Meloidogyne incognita on tomato cv. Co3. Indian J Nematol 36:77–80

Jones JT, Haegeman A, Danchin EG, Gaur HS, Helder J, Jones MG et al (2013) Top 10 plant-parasitic nematodes in molecular plant pathology. Mol Plant Pathol 14(9):946–961

Kähkönen MP, Heinämäki J, Ollilainen V, Heinonen M (2003) Berry anthocyanins: isolation, identification and antioxidant activities. J Sci Food Agric 83(14):1403–1411

Kaplan DT, Thomason IJ, Van Gundy SD (1979) Histological study of the compatible and incompatible interaction of soybeans and Meloidogyne incognita. J Nematol 11(4):338–343

Kaplan DT, Keen NT, Thomason IJ (1980) Studies on the mode of action of glyceollin in soybean incompatibility to the root knot nematode, Meloidogyne incognita. Physiol Plant Pathol 16:319–325

Karssen G (2002) The plant parasitic nematode genus Meloidogyne Göldi, 1892 (Tylenchida) in Europe. Brill

Karssen G, Van Aelst A (2001) Root-knot nematode perineal pattern development: a reconsideration. Nematology 3(2):95–111

Knypl JS, Chylinska KM, Brzeski MW (1975) Increased level of chlorogenic acid and inhibitors of indoyl-3-acetic acid oxidase in roots of carrot infested with the northern root-knot nematode. Physiol Plant Pathol 6(1):51–64. https://doi.org/10.1016/0048-4059(75)90104-6

Lee J, Durst RW, Wrolstad RE, Collaborators: Eisele T, Giusti MM, Hach J, Hofsommer H, Koswig S, Krueger DA, Kupina S, Martin SK, Martinsen BK, Miller TC, Paquette F, Ryabkova A, Skrede G, Trenn U, Wightman JD (2005) Determination of total monomeric anthocyanin pigment content of fruit juices, beverages, natural colorants, and wines by the pH differential method: collaborative study. J AOAC Int 88(5):1269–1278

Li GH, Li L, Duan M, Zhang KQ (2006) The chemical constituents of the fungus Stereum sp. Chem Biodivers 3(2):210–216

López-Martínez N, Colinas-León MT, Peña-Valdivia CB, Salinas-Moreno Y, Fuentes-Montiel P, Biesaga M, Zavaleta-Mejía E (2011) Alterations in peroxidase activity and phenylpropanoid metabolism induced by Nacobbus aberrans Thorne and Allen, 1944 in chilli (Capsicum annuum L.) CM334 resistant to Phytophthora capsici Leo. Plant Soil 338(1):399–409

Luis JG, Fletcher WQ, Echeverri F, Abad T, Kishi MP, Perales A (1995) New phenalenone-type phytoalexins from Musa acuminata (Colla AAA) grand nain. Nat Prod Lett 6(1):23–30. https://doi.org/10.1080/10575639508044083

Luque-Ortega JR, Martínez S, Saugar JM, Izquierdo LR, Abad T, Luis JG et al (2004) Fungus-elicited metabolites from plants as an enriched source for new leishmanicidal agents: antifungal phenyl-phenalenone phytoalexins from the banana plant (Musa acuminata) target mitochondria of Leishmania donovani promastigotes. Antimicrob Agents Chemother 48(5):1534–1540. https://doi.org/10.1128/AAC.48.5.1534-1540.2004

Mahajan R, Singh P, Bajaj KL (1985) Nematicidal activities of some phenolic compounds against Meloidogyne incognita. Revue Nematology 8:161–164

Mahajan R, Kaur DJ, Bajaj KL (1992) Nematicidal activity of phenolic compounds against Meloidogyne incognita. Nematol Mediterr 20:217–219

Malik MS, Pal V, Sangwan NK, Dhindsa KS, Verma KK, Bhatti DS (1989) Nematicidal efficacy of substituted phenols, phenoxyacetic acid esters and hydrazides: a structure-activity relationship study. Nematology 35:366–370

Marhavý P, Kurenda A, Siddique S, Denervaud Tendon V, Zhou F, Holbein J et al (2019) Single-cell damage elicits regional, nematode-restricting ethylene responses in roots. EMBO J 38(10): e100972. https://doi.org/10.15252/embj.2018100972

McClure MA, Von Mende N (1987) Induced salivation in plant-parasitic nematodes. Phytopathology 77(10):1463–1469

M-Farm (2014) Potato production in Kenya [Internet]. https://mfarm.co.ke/blog/post/potatoproduction-in-kenya

Mian IH, Rodriguez-Kabana R (1982) Organic amendments with high tannin and phenolic contents for control of Meloidogyne arenaria in infested soil. Nematropica 12:221–234

Mikkelsen MD, Hansen CH, Wittstock U, Halkier BA (2000) Cytochrome P450 CYP79B2 from Arabidopsis catalyzes the conversion of tryptophan to indole-3-acetaldoxime, a precursor of indole glucosinolates and indole-3-acetic acid. J Biol Chem 275(43):33712–33717. https://doi.org/10.1074/jbc.M001667200

Mikkelsen MD, Naur P, Halkier BA (2004) Arabidopsis mutants in the C-S lyase of glucosinolate biosynthesis establish a critical role for indole-3-acetaldoxime in auxin homeostasis. Plant J 37(5):770–777. https://doi.org/10.1111/j.1365-313X.2004.02002.x

Milne DL, Boshoff DN, Buchan PWW (1965) The nature of resistance of Nicotiana repanda to the root-knot nematode, Meloidogyne javanica. S Afr J Agric Sci 8:557–570

Moens M, Perry RN, Starr JL (2009) Meloidogyne species–a diverse group of novel and important plant parasites. In: Root-knot nematodes, 1, pp 483

Mota FDC, Alves GCS, Giband M, Gomes ACMM, Sousa FR, Mattos VS et al (2013) New sources of resistance to Meloidogyne incognita race 3 in wild cotton accessions and histological characterization of the defence mechanisms. Plant Pathol 62(5):1173–1183

Naczk M, Shahidi F (2006) Phenolics in cereals, fruits and vegetables: occurrence, extraction and analysis. J Pharm Biomed Anal 41(5):1523–1542

Naik VN, Sharma DD (2007) Efficacy of pesticides and growth hormones against root disease complex of mulberry (Morus alba L.). Int J Ind Entomol 15:1–6

Nandi B, Sukul NC, Babu SPS (2000a) Exogenous salicylic acid reduces Meloidogyne incognita infestation of tomato. Allelopath J 7:285–288

Nandi B, Sukul NC, Banerjee N, Babu SPS (2000b) Salicylic acid reduces Meloidogyne infestation of cowpea. Proc Zool Soc 53:93–95

Nandi B, Sukul NC, Banerjee N, Sengupta S, Das P, Babu SPS (2002) Salicylic acid enhances resistance in cowpea against Meloidogyne incognita. Phytopathol Mediterr 41:39–44

Nandi B, Kundu K, Banerjee N, Babu SPS (2003) Salicylic acid induced suppression of Meloidogyne incognita infestation of okra and cowpea. Nematology 5:747–752

Netscher C, Sikora RA (1990) Nematode parasites of vegetables. In: Plant parasitic nematodes in subtropical and tropical agriculture, pp 237–283

Nicol JM, Turner SJ, Coyne DL, den Nijs LJMF, Hockland S, Maafi ZT (2011) Current nematode threats to world agriculture. In: Genomics and molecular genetics of plant-nematode interactions. Springer, Dordrecht, pp 21–43

Nivsarkar M, Cherian B, Padh H (2001) Alpha-terthienyl: a plant-derived new generation insecticide. Curr Sci 81(6):667–672

Pankaj, Sharma HK (2003) Relative sensitivity of Meloidogyne incognita and Rotylenchus reniformis to salicylic acid on okra. Indian J Nematol 33:120–123

Pankaj, Chawla G, Shakil NA, Kishor V, Rohatgi D (2005) Estimation of salicylic acid and its role in resistance mechanism in chickpea against Meloidogyne incognita. Indian J Nematol 35:160–162

Patil SS, Powelson RL, Young RR (1964) Relation of chlorogenic acid and free phenols in potato roots to infection by Verticillium albo-atrum. Phytopathology 54:531–535

Pegard A, Brizzard G, Fazari A, Soucaze O, Abad P, Djian-Caporalino C (2005) Histological characterization of resistance to different root-knot nematode species related to phenolics accumulation in Capsicum annuum. Phytopathology 95(2):158–165

Perry RN, Moens M (eds) (2006) Plant nematology. CABI, Wallingford

Plowright R, Grayer R, Gill J, Rahman M, Harborne J (1996) The induction of phenolic compounds in rice after infection by the stem nematode Ditylenchus angustus. Nematologica 42(5): 564–578. https://doi.org/10.1163/004625996X00063

Poveda J, Abril-Urias P, Escobar C (2020) Biological control of plant-parasitic nematodes by filamentous fungi inducers of resistance: Trichoderma, mycorrhizal and endophytic fungi. Front Microbiol 11:992

Radwan MA, Farrag SAA, Abu-Elamayem MM, Ahmed NS (2012) Biological control of the root-knot nematode, Meloidogyne incognita on tomato using bioproducts of microbial origin. Appl Soil Ecol 56:58–62

Robards K (2003) Strategies for the determination of bioactive phenols in plants, fruit and vegetables. J Chromatogr A 1000(1–2):657–691

Rowe RC, Powelson ML (2002) Potato early dying: management challenges in a changing production environment. Plant Dis 86(11):1184–1193

Savary S, Willocquet L, Pethybridge SJ, Esker P, McRoberts N, Nelson A (2019) The global burden of pathogens and pests on major food crops. Nat Ecol Evol 3:430–439. https://doi.org/10.1038/s41559-018-0793-y

Scalbert A (1991) Antimicrobial properties of tannins. Phytochemistry 30:3875–3883

Shah SJ, Anjam MS, Mendy B, Anwer MA, Habash SS, Lozano-Torres JL et al (2017) Damage-associated responses of the host contribute to defence against cyst nematodes but not root-knot nematodes. J Exp Bot 68(21–22):5949–5960. https://doi.org/10.1093/jxb/erx374

Sikora RA, Fernandez E (2005) Nematode parasites of vegetables. In: Plant parasitic nematodes in subtropical and tropical agriculture, 2nd edn., pp 319–392

Singleton VL, Rossi JA (1965) Colorimetry of total phenolics with phosphomolybdic-phosphotungstic acid reagents. Am J Enol Vitic 16(3):144–158

Singleton VL, Orthofer R, Lamuela-Raventós RM (1999) [14] Analysis of total phenols and other oxidation substrates and antioxidants by means of folin-ciocalteu reagent. Methods Enzymol 299:152–178

Sirohi A, Pankaj (2005) Bare root dip application of SAR inducing chemicals in tomato against root-knot nematode, Meloidogyne incognita. Indian J Nematol 35:195–198

Stirling GR, Nicol J, Reay F (1999) Advisory services for nematode pests. Rural Industries Research & Development Corp

Takasugi M, Yachida Y, Anetai M, Masamune T, Kegasawa K (1975) Identification of asparagusic acid as a nematicide occuring naturally in the roots of asparagus. Chem Lett 1:43–44. https://doi.org/10.1246/cl.1975.43

Taylor AL, Sasser JN (1978) Biology, identification and control of root-knot nematodes. North Carolina State University Graphics, 111

Teixeira MA, Wei L, Kaloshian I (2016) Root-knot nematodes induce pattern-triggered immunity in Arabidopsis thaliana roots. New Phytol 211(1):276–287. https://doi.org/10.1111/nph.13893

The World Bank (2021) Agriculture and food [Internet]. https://www.worldbank.org/en/topic/agriculture/overview#1

Tjosvold S, Koiko S (2015) Distinguishing abiotic and biotic symptoms in plants [Internet]. https://ucanr.edu/blogs/blogcore/postdetail.cfm?postnum=1782

Tsai AY, Higaki T, Nguyen CN, Perfus-Barbeoch L, Favery B, Sawa S (2019) Regulation of root-knot nematode behavior by seed-coat mucilage derived attractants. Mol Plant 12(1):99–112. https://doi.org/10.1016/j.molp.2018.11.008

University of California (n.d.) Nematodes/Potato/Agriculture: pest management guidelines/UC Statewide IPM Program (UC IPM) [Internet]. https://www2.ipm.ucanr.edu/agriculture/potato/nematodes/

Wang KH, Hooks CR, Ploeg A (2007) Protecting crops from nematode pests: using marigold as an alternative to chemical nematicides. In: Plant disease; PD-35. University of Hawaii, Honolulu

Williamson VM, Gleason CA (2003) Plant–nematode interactions. Curr Opin Plant Biol 6(4): 327–333

Williamson VM, Roberts PA, Perry R (2009) 13. Mechanisms and genetics of resistance. In: Root-knot nematodes, pp 301

Wrolstad RE, Durst RW, Giusti MM, Rodriguez-Saona LE (2002) Analysis of anthocyanins in nutraceuticals

Wubben MJE, Jin J, Baum TJ (2008) Cyst nematode parasitism of A. thaliana is inhibited by salicylic acid (S.A.) and elicits uncoupled SA-independent pathogenesis-related gene expression in roots. Mol Plant Microbe Interact 21:424–432

Wuyts N, De Waele D, Swennen R (2006a) Activity of phenylalanine ammonia lyase, peroxidase and polyphenol oxidase in roots of banana (Musa acuminata AAA, CVs Grande Naine and Yangambi Km 5) before and after infection with Radopholus similis. Nematology 8:201–209

Wuyts N, Swennen R, De Waele D (2006b) Effects of plant phenylpropanoid pathway products and selected terpenoids and alkaloids on the behaviour of the plant-parasitic nematodes Radopholus similis, Pratylenchus penetrans and Meloidogyne incognita. Nematology 8:89

Wyss U (1997) Root parasitic nematodes: an overview. In: Fenoll C, Grundler FMW, Ohl SA (eds) Cellular and molecular aspects of plant-nematode interactions. Springer, Dordrecht, pp 5–22

Yu Q, Tsao R, Chiba M, Potter J (2005) Selective nematicidal activity of allyl isothiocyanate. J Food Agric Environ 3(2):218–221

Zasada IA, Ferris H (2003) Sensitivity of Meloidogyne javanica and Tylenchulus semipenetrans to isothiocyanates in laboratory assays. Phytopathology 93(6):747–750. https://doi.org/10.1094/PHYTO.2003.93.6.747

Zhou L, Yuen G, Wang Y, Wei L, Ji G (2016) Evaluation of bacterial biological control agents for control of root-knot nematode disease on tomato. Crop Prot 84:8–13

Plant Phenolics Production: A Strategy for Biotic Stress Management

18

Aqsa Tariq and Ambreen Ahmed

Abstract

Phenolic compounds (polyphenols/phenolics) are the plants' secondary metabolites produced for defense purposes. The ubiquitous nature of phenolics describes their functional diversity. They have noteworthy impact on different physiological and metabolic processes of plant. These compounds act as antimicrobial, antifungal, and antioxidant agents that intercept microbial invasions. Phenolics also play structural role in plants and provide integrity, vigor, and vitality to plant cells. Plant phenolic defense system includes physical alterations (suberization or lignification), metabolic changes (de novo biosynthesis of pathogenesis-related proteins), synthesis, and accretion of phenylpropanoid-derived metabolites (phytoalexins). Phenolics have dual function and act as both attractants and repellents for organisms around plants. They protect plants from the damage caused by pathogens and other microbes. Phenolic compounds have various defensive mechanisms that provide shield to plants for invading pathogens. Moreover, being weak acids, phenolics such as anthocyanins, hydroxycinnamic acid (HCAs), and flavonoids have free radicals scavenging potential that is accumulative in plants under stress and prevents plants from oxidative stress. In addition, phenolic compounds such as flavonoids are an important rhizospheric signaling molecules that attract symbiotic (*Rhizobium*) as well as pathogenic microbes (*Agrobacterium*). Thus, they are important in developing plant–microbial interactions.

Keywords

Phenolics · Flavonoids · Phytoalexins · Antifungal · Antioxidant

---

A. Tariq · A. Ahmed (✉)
Institute of Botany, University of the Punjab, Quaid-e-Azam Campus, Lahore, Pakistan

R. Lone et al. (eds.), *Plant Phenolics in Biotic Stress Management*,
https://doi.org/10.1007/978-981-99-3334-1_18

18.1 Introduction

Phenolics or polyphenols are the plant secondary metabolites with diverse function and distribution. These are mainly derivatives of the shikimate pathway and phenylpropanoid pathway and act as plant defense signaling molecules. They range from simple benzene ring containing one or two hydroxyl (OH) group to highly polymerized compounds such as flavonoids, tannins and anthocyanins (Zhang et al. 2022). Phenolic compounds (PCs) are known for their significant role in providing metabolic plasticity and adaptive advantage to shield plants during various abiotic and biotic stresses. They play remarkable function in various plant metabolic activities and physiological processes (Ahmed et al. 2020; Marchiosi et al. 2020). Phenolics greatly affect plant developmental activities such as seed cell division, plant germination, pigment synthesis etc. Flavonoids are sometimes deposited in nucleus and form complex with DNA to provide protection against DNA damage (Xiao et al. 2019). Phenolic compounds released through plant roots enter in the soil and inhibit pathogenic activities in the root vicinity. Phenolics such as ferulic acid, flavonoids and gallic acid inhibit spore germination of saprophytic fungi (Gautam et al. 2020). Besides this, phenolic compounds also act as signaling molecules that enable plants to communicate with its surrounding environment. These phenolic molecules act as chemoattractants (Shimamura et al. 2022). Soil microbes use plant released phenolics to regulate soil nitrogen mineralization and decomposition of organic matter (Halvorson et al. 2016). In addition, biosynthesis of the phenolics, especially sporopollenins, phenylpropanoids, cuticle, and flavonoids, is evolutionarily significant, which enables plants to survive under unfavorable environments. During evolution, plants have developed wide variations in their organizational patterns leading to phenotypic complexity by experiencing various biotic as well as abiotic stress conditions. Biotic intercessions are primarily due to invasions of pathogenic such as virus, bacteria, fungi, etc. These pathogenic microbes were coevolved with their host plants (Bhar et al. 2022). Due to sessile nature, plants mainly rely on intricate defense signaling mechanisms which involve genetic reprogramming of metabolic activities in response to pathogenic attack. During stress conditions, phenolic compounds get accumulated in subepidermal layer of plant tissues. They are found associated with cell wall, in waxes and on peripheral surfaces (Vishwanath et al. 2014). Phenolic compounds have dual functional properties as attractants and repellents depending on the adjacent environment. Plants either produce allelochemicals and chemoattractants that attract symbiotic microbes or repel pathogenic microbes. Diverse role of plant phenolics drag the attention of scientists to manipulate these phenolic compounds as crop protectors against pathogenic attacks (Pratyusha 2022). The present chapter discusses the function of phenolics under invasion of microbes and various microbicidal and microstatic mechanisms adopted by phenolic compounds during plant defense.

18.2 Phenolics in Plant Environment

Polyphenols are ubiquitous compounds present in entire kingdom plantae. Their existence varies according to plant species and sometimes act as identification markers for plant species. Their function in plant defense mechanisms against abiotic and abiotic stresses is quite evident. Plant phenolic compounds get accumulated within plant vacuoles usually in epidermal, subepidermal, and guard cells of shoots and leaves. Some phenolic compounds covalently link to cell walls or exist on surface of plant organs (Xiao et al. 2019). They have been known for their role in the synthesis of plant pigment, resistance, pollination, and reproduction. Flavonoids such as apigenin, myricetin, kaempferol, quercetin, luteolin, and coloration of flowers and fruits (Brouillard et al. 2010). These compounds provide structural integrity, vigor and vitality. Phytoalexins released from the plant wounds provide protection from insects and pathogenic invaders; however, some pests counteract and nullify the effect of these phytoalexins (Ahmed et al. 2020). Phenolics promote plant growth by supporting cell wall formation. Some hydroxycinnamic acids such as ferulic and p-coumaric acid found in cell wall act as precursor for lignin biosynthesis for lignification. In addition, phenolics like gallic, gentisic, and p-coumaric acids are also responsible for plant movements (nyctinastic) by regulating turgor pressure and water flux (Bhatti et al. 2022). Phenolic acids serve as carbon source for some soil microbes and hence act as chemoattractors. Phenols such as hesperetin, isoflavonoids, naringenin, and chalcones stimulate nod gene expression which stimulate plant–microbe interactions. These molecules regulate the development of legume rhizobial symbiotic relationships (Tariq and Ahmed 2023). In addition, polyphenols also protect nodulating cells from oxidative damage; however, despite all these, some of the phenolic compounds also act as repressors to inhibit nodulation process (Bag et al. 2022). Some phenolics like 4-hydroxybenzaldehyde, protocatechuic acid, and p-coumaric acid enhance production of rhizobacteria auxins that interact with plant root formation process (Chamkhi et al. 2021). In contrast to these beneficial effects, phenolics also act as inhibitors that retard various plant developmental processes, e.g., bud opening, shoot and root elongation, and germination. Phenolics released from plant roots act as allelochemicals and hinder the development of neighboring plants (Hoang Anh et al. 2021).

18.3 Plant Responses to Biotic Stress

The environmental stimulus that disrupts the normal homeostasis of a system is referred as stress. Stress affects plant metabolic pathways by interacting with enzymatic activities. The developmental patterns of plants are influenced by different environmental events. These factors constrain the sustainable agricultural system. Various microbes such as fungi, viruses, bacteria etc. invade and cause infections in plants leading to biotic stresses. These biotic stresses are responsible for 30% loss in the crop productivity, thereby, negatively affect the agriculture efficiency (Kumar and Verma 2018). Among all these pathogens, the fungi are serious threat to the plant species. Fungal parasites can either be necrotrophic or biotrophic. Majority of the diseases in plants (85%) are caused by fungi (Behmann et al. 2014). These organisms cause mottling, necrosis, blights, wilting, tumors, leaf spots, blasts, etc. (Saddique et al. 2018). These pathogens develop competition for nutrients with plant beneficial bacterial and adversely affect plant growth and development influencing crop yield. Viruses also cause localized and systemic desecration to plants causing stunted growth and even chlorosis. The mites and insects act as carriers for viruses and bacteria (Melvin et al. 2017).

Plants develop various stress tolerance mechanisms at biochemical, morphological, molecular, and physiological levels by activating certain signaling molecules such as upregulation of various antioxidants, production of stress-related proteins, and aggregation of compatible solutes. Plant stress responses are regulated by multiple interacting signaling pathways to reprogram cellular activities for the expression of defense-related genes to encounter pathogenic attack (Madani et al. 2019). The initial perception causes the generation of endogenous signals that triggers defense mechanisms to activate and fortify plants to a wide range of microbes. Many defense-related genes encode for the proteins that have antimicrobial activities such as phytoalexins (Lattanzio et al. 2006). Plants have adopted various morphological alterations as a defensive strategy that give strength and rigidity to them. Some defensive barriers such as waxes and cuticles present on plant surface act as front line of defense to avert pathogenic invasions. Plants also produce chemical compounds that act as plant protecting molecules. Plants have developed PAMP-triggered immunity (PTI) in which plant identifies PAMPs (pathogen associated molecular patterns) via PRRs (pattern recognition receptors). Plant identifies specific pathogenic receptors (Avr proteins) and produce pathogenesis-related (PR) proteins which provoke plant defense system (Iqbal et al. 2021). Phytoalexins are synthesized through phenylpropanoid and the terpenoid pathways which is catalyzed by PAL (phenylalanine ammonialyase) and HMGR (hydroxymethyl glutaryl-CoA reductase) enzymes. Thus, ultimate plant response is highly coordinated by various genetic activities empowered by intermingled signaling molecules. Both phytoalexins and PR proteins are involved in plant resistance (Daayf et al. 2012).

18.4 Phenolics Production: A Strategy for Biotic Stress Management in Plants

The stress stimuli lead plants to the formation of stress induced metabolites primarily secondary metabolites. Switching of resource allocations between primary and secondary metabolic pathways allows the smart use of resources under stress conditions (Fig. 18.1). Production of plant secondary metabolites including phenolics uses large number of resources from plant primary metabolism. Stress triggers several biochemical and molecular mechanisms that divert the carbon flux from primary and secondary metabolism leading to the formation of phenolics (Banothu and Uma 2021). Moreover, stress induced accumulation of proline

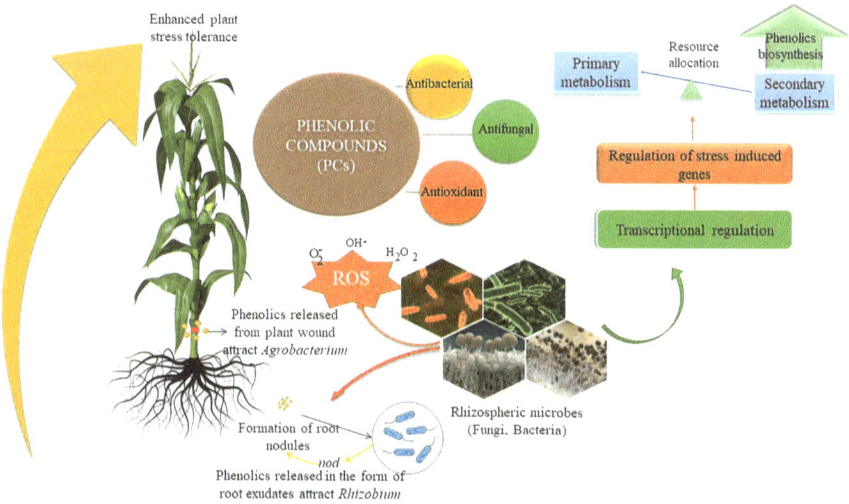

Fig. 18.1 Overview of phenolic-mediated plant response to microbial invasions. Microbial invasion causes disruption of normal metabolism causing the production of reactive oxygen species (ROS) that activates plant defense system by exchanging carbon resources between primary and secondary metabolic pathways resulting in phenolic accumulation within plants. Phenolics released from plant wounds or in the form of root exudates also attract microbes in plant surrounding

transfers energy toward biosynthesis of phenylpropanoid through oxidative pentose phosphate pathway to produce phenolics (Caretto et al. 2015). Stress triggers the expression of various genes involved in phenyl ammonia lyase, chalcone synthase, phosphoenol pyruvate (PEP) carboxylase-specific enzymes, and shikimate dehydrogenase that favors formation of phenolics. Similarly, upregulation of genes involved in production of phenolics like phenylalanine ammonia lyase (PAL) was also observed during pathogenic invasions (Dizengremel et al. 2012). Identification of specific molecular patterns of invaders (PAMPs) trigger the biosynthesis of phenolic compounds in host plants. As a result, microbial infections are counteracted and microbe is cramped before it gets control over the host (Isah 2019).

18.5 Biosynthesis of Phenolic Compounds

The production of plant metabolites is accurately balanced and shares some precursor molecules with other metabolites. Carbon fluxes are consequently switched between the routes. There are three different biosynthetic pathways of phenolics named as phenylpropanoids pathway, polyketide pathway, and mevalonate pathways. Phenylpropanoids pathway termed as shikimate produces the phenyl propanoid derivatives (C_6–C_3). Phenylalanine substrate for the phenyl propanoid pathway can be converted into a variety of phenolics, with aromatic rings ranging from one to six carbons and different substitution patterns. Production of cinnamic acid derivatives begins with the production of chorismic acid from the substrates erythrose-4-phosphate and phosphoenol pyruvate. Stilbenes (C_6–C_2–C_6) and flavonoids (C_6–C_3–C_6) are two categories of metabolites with an aromatic ring associated with phenylpropanoid moiety that is produced by condensation of three C_2 residues and an activated hydroxycinnamic acid product (Tak et al. 2023). In plants, aliphatic and aromatic compounds are biosynthesized through the production of polyketides, the malonate, or polyketide pathway, which makes a significant contribution to this process. Biosynthesis of steroids, acetyl esters, polyketides, amides, terpenoids, aromatic compounds, and fatty acids depletes the metabolic pool of acetyl CoA, which is incessantly replenished by glycolysis and the catabolism of fatty acids and amino acids (Banothu and Uma 2021).

18.6 Phenolics as Plant Defenders

Plant phenolic defense system includes physical alterations (suberization or lignification), metabolic changes (de novo biosynthesis of (PR) pathogenesis-related proteins), synthesis and accumulation of phenylpropanoid-derived secondary metabolites (phytoalexins). Phenolics have binary role, i.e., act as both attractant and repellent for organisms in plant surrounding. They protect plants from the damage caused by pathogens and other microbes. They are synthesized and concentrated in subepidermal layers of plants (Kumar and Goel 2019). Simple phenolic acids, tannins, and resins act as surface protectants' processes, and lower-molecular-weight polyphenols attract symbiotic microbes around root vicinity. Phenolics such as terpenoids, hydroxycinnamates, hydroxynaphthoquinones, hydroxybenzoates, and hydroquinones act as allelochemicals for competitive weeds. Diverse functional nature of phenolics makes them efficient antifungal, antibacterial, antiviral, and antioxidant agents. Their antioxidant nature enables plants to deal with the stress situations. Phenolics such as phytoalexins released from plants act as antibiotics which protect plants from pathogens and make phenolics an antimicrobial agent (Tak and Kumar 2020).

18.6.1 Phenolics as Antifungal Agents

Phenolics are quite effective against various fungal diseases such as *Verticillium dahlia*, *Sphaerotheca fuliginea*, *Fusarium oxysporum*, *Phytophthora infestans*, *Leptosphaeria maculans*, and *Entomosporium mespili* (Daayf et al. 2012). Large numbers of phenolics like phenolic acids, phenols, dihydrochalcones, and flavonols have antifungal ability. These antifungal phenolics are collectively called phytoanticipin. Plants can store some antifungal phenolics in vacuoles as an inactive form that could be hydrolyzed by glycosidases into their active forms in response to pathogenic attack (Lattanzio et al. 2006). These antifungal compounds are distributed differently within plants such as flavone and flavonols are located on plant surfaces whereas some are restricted to cytoplasm or plant epidermis. Catechol and protocatechuic reduces fungal spore germination. Optimum levels of chlorogenic acid are highly efficient against various plant fungus such as *Streptomyces scabies*, *Verticillium alboatrum* and *Phytophthora infestans*. Similarly, benzaldehyde inhibits germination of *Botrytis cinerea* and *Monilia fructicola* (Lattanzio et al. 2006; El-Bilawy et al. 2022). Phenolic acids such as benzoic acid, gallic acid, caffeic acid, ferulic acid and tannic acid are found to produce in plants with disease root rot-wilt caused by *Fusarium oxysporum* and *Rhizoctonia solani* and proved to be efficacious against them (Prasad et al. 2008). Similarly, phenolics such as catechin, flavanols, procyanidin B1, dihydrochalcones, epicatechin, and hydroxycinnamic acids are effective against *Venturia inaequalis* (Slatnar et al. 2016). Effective antifungal phenolics such as apigeninidin, 3-deoxyanthocyanidin phytoalexins, and luteolinidin are found in *Sorghum bicolor* infected with *Colletotrichum sublineolum* (Tugizimana et al. 2019).

Phenolics adopted various antifungal mechanisms depending upon their type and concentration. The antifungal activity of phenolic compounds is commonly due to their ability to disrupt the various genes expression involved in mycotoxicity, cell adhesion, biofilm formation ability and filamentous growth (Zulhendri et al. 2021). Pinocembrin disrupts several cellular processes and thereby inhibits mycelial growth. It reduces cellular energy by lowering cellular ATP in filaments of *Penicillium italicum*. It damages the hyphal structure by causing membrane leakage (Peng et al. 2012). In *Candida* species, pterostilbene causes downregulation of Ras/cAMP pathway and inhibits ergosterol biosynthesis. Resveratrol trigger metacaspase and cytochrome that causes apoptosis (Li et al. 2014; Lee and Lee 2015). Epigallocatechin-3-gallate inhibits dihydrofolate reductase and ergosterol biosynthetic pathway. Gallic acid also inhibits ergosterol biosynthesis by reducing the activity of sterol alpha demethylase and squalene epoxide (Li et al. 2017). Quercetin reduces ergosterol levels and modulate expression of various genes (Bitencourt et al. 2013). Caffeic acid inhibits the activity of enzyme isocitrate lyase. Similarly, proanthocyanidins reduces the biofilm formation ability; thus, they inhibit adherence of fungi attenuating inflammatory response by interfering phosphorylation activity of signal transducing intracellular kinases (Feldman et al. 2012; Cheah et al. 2014). Hydroxycinnamic acids (HCAs) and its derivatives inhibit fungal growth by inhibiting sporulation. Similarly, catechin interacts with fungal proteins and inhibits its release (Gautam et al. 2020).

Thus, these naturally present phenolics showed a new direction to agricultural science. Researchers studied the use of these phenolics for sustainable agricultural purposes with successful applications. Naturally extracted, p-coumaric acid, caffeic acid, chlorogenic acid, gallic acid, catechin and ferulic acid have shown great antifungal potential when applied manually to cherry tomatoes infected with *Alternaria alternata* (Pane et al. 2016). Similarly, various research works are carried out for the treatment of fungal diseases.

18.6.2 Phenolics as Antimicrobial Agents

Large number of bacteria are infecting plants causing damage to most of the economical crops. Phenolics have antimicrobial properties that have significant biological and ecological impact. Various phenolics such as benzoic acids, naphthoquinones, stilbenes, coumarins, flavonoids, and cinnamic acids were examined for their antimicrobial properties (Chibane et al. 2019; Fu et al. 2021). However, microbial susceptibility varies depending on microbial strain, type and concentration of phenolic compounds. Moreover, the mode of action of phenolics also varies. For instance, condensed tannins suppress bacterial growth by binding to its proteins; however, hydrolysable tannins work differently (Jonker and Yu 2017; Zeller 2019). Tannins are capable of inhibiting the growth of wide range of microbes including filamentous fungi. Hydroxycinnamic acids, catechin, gallic acid, quercetins, and sakuranetin are highly active against bacterial pathogens. Sakuranetin is effective against the rice bacterial pathogens such as *Xanthomonas oryzae* and *Burkholderia glumae*. Similarly, phenol-derived alkyl dihydroxybenzoates provides protection against *Xanthomonas citri*. Moreover, hyperaccumulation of polyphenol oxidases were found in plants with pathogenic invasions. Polyphenol oxidases are important in catalyzing oxidation of phenols to quinones (Gautam et al. 2020). All these studies showed that phenolics have strong antimicrobial potential protecting plants from bacterial infections.

The antimicrobial efficiency of phenolics is due to their ability to modify cell permeability that causes formation of cytoplasmic granules that disrupt cytoplasmic membrane disturbing the intracellular functions via generating phenolics and enzyme complex. Phenolics get accumulated on bacterial surface due to the interaction of hydroxyl radicals and cell membrane which causes depolarization of bacterial cell. This forms pH gradient across membrane and reduction in cellular ATP causing bacterial cell death (Zamuz et al. 2021). Phenolic acids generate phenoxyl radical which causes alterations in distribution pattern of phenolics in aqueous and non-aqueous phases that enhance their antibacterial character. Similarly, longer alkyl side chains make caffeic acids more effective against Gram-positive bacterium as compared to Gram-negative (Chibane et al. 2019; Khan et al. 2021). This describes the differential behavior of phenolics compounds despite having similar basic structure.

Phenolics such as eugenol ferulic, thymol, carvacrol, and gallic acids modify the ionic flux causing alterations in intracellular that block energy production and cause cell death. Similarly, gallic and ferulic acids causes intracellular leakage by disruption of membrane integrity. Moreover, coumarins reduces cellular respiration and inhibits bacterial division by interacting with cell division machinery (Adamczak et al. 2019). Flavonoids can modify bacterial cell morphology and inhibit biosynthesis of nucleic acids (Mora-Pale et al. 2015). Similarly, tannins and flavans are capable of binding with enzymes to form complexes; thus, they inactivate metabolic mechanisms and induce damage to cells (Górniak et al. 2019). Quercetin interferes with membrane potential and increases membrane permeability, catechins and apigenin inhibit DNA gyrase activity, and naphthoquinones inhibit efflux pumps and act as oxidative agents (Kalogianni et al. 2020). One of the important reasons of bacterial pathogenicity is their ability to adhere surfaces by forming biofilms. Phenolics at their sublethal concentrations inhibit biofilms by modifying cell wall properties, inhibiting bacterial motility and inhibiting synthesis of extracellular polymeric substances (Vazquez-Armenta et al. 2018). Similarly, phenolics such as chlorogenic, ferulic, gallic and caffeic acids modify bacterial hydrophobicity and inhibit bacterial adhesion. These characters make phenolic compounds a magnificent antimicrobial agent (Zamuz et al. 2021).

18.6.3 Phenolics as Antioxidant Agents

After pathogenic invasion, reactive oxygen species (ROS) are generated that interact and inhibit pathogenic activities. These ROS also damage plant cellular components. Phenolics block the invasion of parasites and reduce plant susceptibility by scavenging the generated free radicals. ROS have activated oxygen that are generated in mitochondria. ROS stimulate cellular oxidative stress by oxidation of cellular compounds (Sharma et al. 2023). Their accumulation can cause damage to various metabolically significant compounds by producing malondialdehyde (MDA). Oxidative stress is a natural phenomenon; however, plants have various mechanisms to deal with this stress. Production of plant secondary metabolites such as carotenoids, tocopherol, glutathione, alkaloids, and flavonoids are one of them (Madani et al. 2019; Contreras-Angulo et al. 2023).

Lower electron donating potential and lesser reactivity makes phenolics strong antioxidant candidates. Thus, phenolics have potential to scavenge radicals but their ability varies. For instance, flavonoids, HCAs and anthocyanins are strong scavengers among other phenolics. Among all anthocyanins, epigallocatechin gallate, anthocyanidins cyanidin, delphinidin, flavanol quercetin, and flavan-3-ols epicatechin gallate are more strong. Electron donating ability depends on phenolic structure; thereby, all phenolics show variable antioxidant potential. For instance, phenolics with ortho-dihydroxy structures are strong antioxidants than phenolics with monohydroxy-7 structures due to their higher electron donating ability (Hajam et al. 2023). The additional hydroxyl group reduces electron donating ability making stable phenoxyl radical. In flavonoids, presence of double bond and hydroxyl groups at variable positions in two different rings enhance their antioxidant ability. Most of the phenolics including gallic, acid, chlorogenic acid and its methyl esters and some higher molecular weight phenolics such as tannins pose antioxidant ability except salicylic acid (Mishra et al. 2023).

Phenolics besides scavenging free radicals directly protect nucleic acids from oxidative damage. Phenolics such as flavonoids and HCAs are efficient chain-breaking antioxidants and metal chelators; thereby, they are effective against lipoprotein oxidation and lipid peroxidation (Grace 2005). In addition, flavonoids oxidize alkyl peroxyl radicals to protect surface lipids from oxidation. Similarly, anthocyanins are also strong antioxidants as well as screening agents and provide shield to plant photosynthetic apparatus. Their antioxidant potentiality depends on their hydrolyzation ability. Among anthocyanins, their glycosylated forms are efficient antioxidants due to reduced electron donating, metal chelating, and delocalizing ability (Liu et al. 2018).

18.7 Phenolics-Mediated Microbe–Plant Interactions: Case Studies of *Agrobacterium* and *Rhizobium* Infections

The quality and quantity of phenolics determine the relationship of microbe with plants. Mostly studied microbe–plant interactions with reference to phenolic acids are of *Agrobacterium* (pathogenic association) and *Rhizobium* (symbiotic association). Plants release various types of phenolic compounds into the surrounding soil that impact soil microflora greatly. *Agrobacterium* and *Rhizobium* have evolved certain mechanisms to counteract and nullify plant defense. In these infections, phenolics influence bacterial chemotaxis, activation of bacterial gene network responsible for nodulation (*nod*) and virulence (*vir*), xenobiotic detoxification, and quorum signaling mechanisms (Palmer et al. 2004). Phenolics contribute significantly in bacterial chemotactic movement in response to plant root exudates which serves as a source of nutrients for both *Agrobacterium* and *Rhizobium*. They serve as signal for the development of plant microbe interactions. These microbes move across concentration gradient developed by phenolics. Phenolics such as acetosyringone umbelliferone, p-hydroxybenzoic acid, vanillyl alcohol, and 3,4-dihydroxybenzoic acid strongly attract *Rhizobium* as compared to naringenin. Although acetosyringone and umbelliferone inhibit *nod* gene inducers in *Rhizobium*, some rhizobia species exhibit strong attraction toward higher concentration of these two phenolics (Samac and Graham 2007). This explains the competitive behavior of these microbes in root vicinity. Various phenolics have direct effect on *vir* (A/G) genes located on Ti plasmid of *Agrobacterium* which in turn greatly impact chromosomal genes related to chemotaxis and flagellar proteins. Even the lower concentrations of phenolics such as acetosyringone and hydroxyac-etosyringone released from plant wounds attract *Agrobacterium* and induce *vir* genes. Contrary, higher concentrations of some phenolics have bactericidal or bacteriostatic effects on *Agrobacterium*. Chemoattractant phenolics regulate the expression of *nod* and *vir* genes. Phenolics such as flavonoids, flavones are potential inducers of *nod* genes (Peck et al. 2006). Despite this, some phenolics negatively influence *nod* gene. For instance, isoflavonoids, genistein, and daidzein induce *nod* genes in *Bradyrhizobium japonicumare,* whereas inhibit *Sinorhizobium meliloti* induced by luteolin. In case of *Agrobacterium*, acetosyringone is widely known to induce *vir* genes (Gelvin 2009). These phenolics are detected by two-component sensor–transducer system encoded by VirA/G. Moreover, various infecting mechanisms of *Rhizobium* and *Agrobacterium* were coordinated and controlled by quorum sensing signals including exopolysaccharide production and adaptation to stress conditions.

Microbes have naturally counteracting ability to nullify plant phenolics defense system. These microbes degrade phenolics for their carbon sources. Bacteria utilize these resistance mechanisms for establishing plant–bacteria interactions. In case of *Agrobacterium*, vir H located on Ti plasmid is involved in triggering the factor for detoxification of harmful phenolic compounds. Phenolics induced vir genes are involved in the inactivation of phenolic compounds. VirH2 has significant xenobiotic detoxification ability and detoxify plant released phenolic compounds. Phenolic-mediated detoxification of glycolylurea or hydantoins by hyuH is another important example (Jiwaji and Dorrington 2009). Moreover, *Agrobacterium* and *Rhizobium* both have hydrophobic flavonoids inducible efflux pumps that actively exclude flavonoids out of the cell and prevent to reach at their toxic levels. For instance, during *Agrobacterium* infections, it actively pumps out coumestrol, formononetin, and medicarpin. In case of *Rhizobium,* it readily degrades plant released genistein, daidzein, and flavonoids quercetin (Burse et al. 2004). These present an attractive opportunity for agricultural researchers to use such biodegradation compounds for rhizosphere manipulations with reference to soil bioremediation.

18.8 Conclusion

Phenolic compounds contribute significantly in plant defense mechanisms as they exhibit strong antimicrobial, antifungal, and antioxidant ability. A number of complex intermingled signaling molecules are involved in the biosynthesis of phenolic compounds that trigger plant defense. Unraveling their interacting mechanisms would provide an insight into their behavior that would be quite helpful for the rhizospheric engineering and their use in sustainable agriculture. Moreover, the plant phenolic patterns are supreme under changing environments; therefore, their exclusive knowledge would provide researchers a new biotechnological tool for agricultural practices to prevent plants against pathogens.

Conflict of Interest Authors declare no conflict of interest.

References

Adamczak A, Ożarowski M, Karpiński TM (2019) Antibacterial activity of some flavonoids and organic acids widely distributed in plants. J Clin Med 9(1):109

Ahmed A, Tariq A, Habib S (2020) Interactive biology of auxins and phenolics in plant environment. In: Plant phenolics in sustainable agriculture. Springer, Singapore, pp 117–133

Bag S, Mondal A, Majumder A, Mondal SK, Banik A (2022) Flavonoid mediated selective cross-talk between plants and beneficial soil microbiome. Phytochem Rev 21:1739–1760

Banothu V, Uma A (2021) Effect of biotic and abiotic stresses on plant metabolic pathways. In: Phenolic compounds-chemistry, synthesis, diversity, non-conventional industrial, pharmaceutical and therapeutic applications. IntechOpen

Behmann J, Mahlein AK, Rumpf T, Römer C, Plümer L (2014) A review of advanced machine learning methods for the detection of biotic stress in precision crop protection. Precis Agric 16(3):239–260

Bhar A, Chakraborty A, Roy A (2022) Plant responses to biotic stress: old memories matter. Plants 11(1):84

Bhatti MZ, Ismail H, Kayani WK (2022) Plant secondary metabolites: therapeutic potential and pharmacological properties. In: Secondary metabolites: trends and reviews, pp 201

Bitencourt TA, Komoto TT, Massaroto BG, Miranda CES, Beleboni RO, Marins M, Fachin AL (2013) Trans-chalcone and quercetin down-regulate fatty acid synthase gene expression and reduce ergosterol content in the human pathogenic dermatophyte *Trichophyton rubrum*. BMC Complement Altern Med 13:229

Brouillard R, Chassaing S, Isorez G, Kueny-Stotz M, Figueiredo P (2010) The visible flavonoids or anthocyanins: from research to applications. In: Recent advances in polyphenol research, vol 2, pp 1–22

Burse A, Weingart H, Ullrich MS (2004) The phytoalexin-inducible multidrug efflux pump AcrAB contributes to virulence in the fire blight pathogen, Erwinia amylovora. Mol Plant Microbe Interact 17:43–54

Caretto S, Linsalata V, Colella G, Mita G, Lattanzio V (2015) Carbon fluxes between primary metabolism and phenolic pathway in plant tissues under stress. Int J Mol Sci 16:26378–26394. https://doi.org/10.3390/ijms161125967

Chamkhi I, Benali T, Aanniz T et al (2021) Plant-microbial interaction: the mechanism and the application of microbial elicitor induced secondary metabolites biosynthesis in medicinal plants. Plant Physiol Biochem 167:269–295

Cheah HL, Lim V, Sandai D (2014) Inhibitors of the glyoxylate cycle enzyme ICL1 in *Candida albicans* for potential use as antifungal agents. PLoS One 9:e95951

Chibane LB, Degraeve P, Ferhout H, Bouajila J, Oulahal N (2019) Plant antimicrobial polyphenols as potential natural food preservatives. J Sci Food Agric 99(4):1457–1474

Contreras-Angulo LA, Emus-Medina A, Gutierrez-Grijalva EP, Heredia JB (2023) Stressed plants: an improved source for bioactive phenolics. In: Plant phenolics in abiotic stress management. Springer, Singapore, pp 195–214

Daayf F, El Hadrami A, El-Bebany AF, Henriquez MA, Yao Z, Derksen H, . . ., Adam LR (2012) Phenolic compounds in plant defense and pathogen counter-defense mechanisms. In: Recent advances in polyphenol research, vol 3, pp 191–208

Dizengremel P, Vaultier MN, Le Thiec D, Cabané M, Bagard M, Gérant D, Gérard J, Dghim AA, Richet N, Afif D, Pireaux JC, Hasenfratz-Sauder MP, Jolivet Y (2012) Phosphoenolpyruvate is at the crossroads of leaf metabolic responses to ozone stress. New Phytol 195:512–517

El-Bilawy EH, Al-Mansori ANA, Alotibi FO, Al-Askar AA, Arishi AA, Teiba II et al (2022) Antiviral and antifungal of Ulva fasciata extract: HPLC analysis of polyphenolic compounds. Sustainability 14(19):12799

Feldman M, Tanabe S, Howell A, Grenier D (2012) Cranberry proanthocyanidins inhibit the adherence properties of *Candida albicans* and cytokine secretion by oral epithelial cells. BMC Complement Altern Med 12:6

Fu Y, Yang L, Zhang J, Hu J, Duan G, Liu X et al (2021) Polydopamine antibacterial materials. Mater Horizons 8(6):1618–1633

Gautam AK, Singh PK, Aravind M (2020) Defensive role of plant phenolics against pathogenic microbes for sustainable agriculture. In: Plant phenolics in sustainable agriculture. Springer, Singapore, pp 579–594

Gelvin SB (2009) *Agrobacterium* in the genomics age. Plant Physiol 150:1665–1676

Górniak I, Bartoszewski R, Króliczewski J (2019) Comprehensive review of antimicrobial activities of plant flavonoids. Phytochem Rev 18(1):241–272

Grace SC (2005) Phenolics as antioxidants. In: Antioxidants and reactive oxygen species in plants, pp 141–168

Hajam YA, Lone R, Kumar R (2023) Role of plant phenolics against reactive oxygen species (ROS) induced oxidative stress and biochemical alterations. In: Plant phenolics in abiotic stress management. Springer, Singapore, pp 125–147

Halvorson JJ, Schmidt MA, Hagerman AE, Gonzalez JM, Liebig MA (2016) Reduction of soluble nitrogen and mobilization of plant nutrients in soils from US northern Great Plains agroecosystems by phenolic compounds. Soil Biol Biochem 94:211–221

Hoang Anh L, Van Quan N, Tuan Nghia L, Dang Xuan T (2021) Phenolic allelochemicals: achievements, limitations, and prospective approaches in weed management. Weed Biol Manag 21(2):37–67

Iqbal Z, Iqbal MS, Hashem A, Abd Allah EF, Ansari MI (2021) Plant defense responses to biotic stress and its interplay with fluctuating dark/light conditions. Front Plant Sci 12:631810

Isah T (2019) Stress and defense responses in plant secondary metabolites production. Biol Res 52: 39

Jiwaji M, Dorrington RA (2009) Regulation of hydantoin-hydrolyzing enzyme expression in Agrobacterium tumefaciens strain RU-AE01. Appl Microbiol Biotechnol 84:1169–1179

Jonker A, Yu P (2017) The occurrence, biosynthesis, and molecular structure of proanthocyanidins and their effects on legume forage protein precipitation, digestion and absorption in the ruminant digestive tract. Int J Mol Sci 18(5):1105

Kalogianni AI, Lazou T, Bossis I, Gelasakis AI (2020) Natural phenolic compounds for the control of oxidation, bacterial spoilage, and foodborne pathogens in meat. Foods 9(6):794

Khan F, Bamunuarachchi NI, Tabassum N, Kim YM (2021) Caffeic acid and its derivatives: antimicrobial drugs toward microbial pathogens. J Agric Food Chem 69(10):2979–3004

Kumar N, Goel N (2019) Phenolic acids: natural versatile molecules with promising therapeutic applications. Biotechnol Rep 24:e00370

Kumar A, Verma JP (2018) Does plant—microbe interaction confer stress tolerance in plants: a review? Microbiol Res 207:41–52

Lattanzio V, Lattanzio VM, Cardinali A (2006) Role of phenolics in the resistance mechanisms of plants against fungal pathogens and insects. Phytochemistry 661(2):23–67

Lee J, Lee DG (2015) Novel antifungal mechanism of resveratrol: apoptosis inducer in *Candida albicans*. Curr Microbiol 70:383–389

Li DD, Zhao LX, Mylonakis E, Hu GH, Zou Y, Huang TK, Yan L, Wang Y, Jiang YY (2014) In vitro and in vivo activities of pterostilbene against *Candida albicans* biofilms. Antimicrob Agents Chemother 58:2344–2355

Li ZJ, Liu M, Dawuti G, Dou Q, Ma Y, Liu HG, Aibai S (2017) Antifungal activity of gallic acid in vitro and in vivo. Phytother Res 31:1039–1045

Liu Y, Tikunov Y, Schouten RE et al (2018) Anthocyanin biosynthesis and degradation mechanisms in solanaceous vegetables: a review. Front Chem 6:52

Madani B, Mirshekari A, Imahori Y (2019) Physiological responses to stress. In: Postharvest physiology and biochemistry of fruits and vegetables. Woodhead Publishing, Sawston, pp 405–423

Marchiosi R, dos Santos WD, Constantin RP, de Lima RB, Soares AR, Finger-Teixeira A et al (2020) Biosynthesis and metabolic actions of simple phenolic acids in plants. Phytochem Rev 19(4):865–906

Melvin P, Bankapalli K, D'Silva P, Shivaprasad PV (2017) Methylglyoxal detoxification by a DJ-1 family protein provides dual abiotic and biotic stress tolerance in transgenic plants. Plant Mol Biol 94(4–5):381–397

Mishra UN, Dey P, Singhal RK, Sahu C, Jena D, Nanda SP, Chauhan J (2023) Plant phenolics: as antioxidants and potent compounds under multiple stresses. In: Plant phenolics in abiotic stress management. Springer, Singapore, pp 215–234

Mora-Pale M, Bhan N, Masuko S, James P, Wood J, McCallum S et al (2015) Antimicrobial mechanism of resveratrol-trans-dihydrodimer produced from peroxidase-catalyzed oxidation of resveratrol. Biotechnol Bioeng 112(12):2417–2428

Palmer MA, Bernhardt E, Chornesky E, Collins S, Dobson A, Duke C, Gold B, Jacobson R, Kingsland S, Kranz R, Mappin M, Martinez ML, Micheli F, Morse J, Pace M, Pascual M, Palumbi S, Reichman OJ, Simons A, Townsend A, Turner M (2004) Ecology for a crowded planet. Science 304:1251–1252

Pane C, Fratianni F, Parisi M, Nazzaro F, Zaccardelli M (2016) Control of alternaria post-harvest infections on cherry tomato fruits by wild pepper phenolic-rich extracts. Crop Prot 84:81–87

Peck MC, Fisher RF, Long SR (2006) Diverse flavonoids stimulate NodD1 binding to nod gene promoters in Sinorhizobium meliloti. J Bacteriol 188:5417–5427
Peng L, Yang S, Cheng YJ, Chen F, Pan S, Fan G (2012) Antifungal activity and action mode of pinocembrin from propolis against Penicillium italicum. Food Sci Biotechnol 21:1533–1539
Prasad D, Singh A, Singh KP, Bist S, Tewari A, Singh UP (2008) The role of phenolic compounds in disease resistance in geranium. Arch Phytopathol Plant Protect 43(7):615–623
Pratyusha S (2022) Phenolic compounds in the plant development and defense: an overview. In: Plant stress physiology—perspectives in agriculture
Saddique M, Kamran M, Shahbaz M (2018) Differential responses of plants to biotic stress and the role of metabolites. In: Plant metabolites and regulation under environmental stress. Elsevier, San Diego, pp 69–87
Samac DA, Graham MA (2007) Recent advances in legume–microbe interactions: recognition, defense response, and symbiosis from a genomic perspective. Plant Physiol 144:582–587
Sharma K, Kumar M, Chandran D (2023) Ameliorative effects of phenolics in oxidative stress management in plants. In: Plant phenolics in abiotic stress management. Springer, Singapore, pp 369–390
Shimamura M, Kumaki T, Hashimoto S, Saeki K, Ayabe SI, Higashitani A et al (2022) Phenolic acids induce nod factor production in Lotus japonicus–Mesorhizobium symbiosis. Microbes Environ 37(1):ME21094
Slatnar A, Mikulič-Petkovšek M, Veberič R, Štampar F (2016) Research on the involvement of phenolics in the defense of horticultural plants. Acta Agric Slov 107(1):183–189
Tak Y, Kumar M (2020) Phenolics: a key defence secondary metabolite to counter biotic stress. In: Plant phenolics in sustainable agriculture. Springer, Singapore, pp 309–329
Tak Y, Kaur M, Gautam C, Kumar R, Tilgam J, Natta S (2023) Phenolic biosynthesis and metabolic pathways to alleviate stresses in plants. In: Plant phenolics in abiotic stress management. Springer, Singapore, pp 63–87
Tariq A, Ahmed A (2023) Genetic basis of phenolics in abiotic stress management. In: Plant phenolics in abiotic stress management. Springer, Singapore, pp 47–62
Tugizimana F, Djami-Tchatchou AT, Steenkamp PA, Piater LA, Dubery IA (2019) Metabolomic analysis of defense-related reprogramming in Sorghum bicolor in response to Colletotrichum sublineolum infection reveals a functional metabolic web of phenylpropanoid and flavanoid pathways. Front Plant Sci 4(9):1840
Vazquez-Armenta FJ, Bernal-Mercado AT, Lizardi-Mendoza J, Silva-Espinoza BA, Cruz-Valenzuela MR, Gonzalez-Aguilar GA et al (2018) Phenolic extracts from grape stems inhibit Listeria monocytogenes motility and adhesion to food contact surfaces. J Adhes Sci Technol 32(8):889–907
Vishwanath SJ, Delude C, Domergue F, Rowland O (2014) Suberin: biosynthesis, regulation, and polymer assembly of a protective extracellular barrier. Plant Cell Rep 34(4):573–586
Xiao L, Carrillo J, Siemann E, Ding J (2019) Herbivore-specific induction of indirect and direct defensive responses in leaves and roots. AoB Plants 11:plz003. https://doi.org/10.1093/aobpla/plz003
Zamuz S, Munekata PE, Dzuvor CK, Zhang W, Sant'Ana AS, Lorenzo JM (2021) The role of phenolic compounds against Listeria monocytogenes in food. A review. Trends Food Sci Technol 110:385–392
Zeller WE (2019) Activity, purification, and analysis of condensed tannins: current state of affairs and future endeavors. Crop Sci 59(3):886–904
Zhang C, Xue J, Yang X, Ke Y, Ou R, Wang Y et al (2022) From plant phenols to novel bio-based polymers. Prog Polym Sci 125:101473
Zulhendri F, Chandrasekaran K, Kowacz M, Ravalia M, Kripal K, Fearnley J, Perera CO (2021) Antiviral, antibacterial, antifungal, and antiparasitic properties of propolis: a review. Foods 10(6):1360

Role of Phenolic Compounds in Disease Resistance to Plants

19

Ashiq Hussain Khanday, Irfan Ashraf Badroo, Nasir Aziz Wagay, and Shah Rafiq

Abstract

Plants produce a diverse range of compounds that have no direct role in their normal growth and development known as secondary metabolites. These metabolites essentially play an important role in plant defense. Many secondary metabolite groups are known, such as terpenes, phenolic compounds, and alkaloids. All these organic compounds are classified on the basis of their biosynthetic origin. In natural systems, plants face plethora potential enemies. All the types of ecosystems contain a wide variety of organisms such as bacteria, viruses, fungi, nematodes, mites, insects, mammals, and other herbivorous animals that can cause damage to plants. However, plants have evolved myriad defense mechanisms for their protection. Secondary metabolites are one of the incredibly known such mechanisms by which plants are able to cope with various kinds of biotic and abiotic stress. Phenolic compounds are one such group of plant secondary metabolites which play an essential key role in defense against environmental stresses and other stresses from pathogen infections, herbivores, and nutrient deficiency. Phenolic compounds act as the primary molecules of defense in plants against pathogens. Phenolic compounds act as the natural defense in

A. H. Khanday
Department of Botany, Sant Gadge Baba Amravati University, Amravati, Maharashtra, India

I. A. Badroo
Department of Zoology, Sant Gadge Baba Amravati University, Amravati, Maharashtra, India

N. A. Wagay (✉)
Department of Botany, Government Degree College Baramulla (Boys), Khawaja Bagh, Jammu and Kashmir, India

S. Rafiq
Plant Tissue Culture Laboratory, Department of Botany, University of Kashmir, Srinagar, Jammu and Kashmir, India

R. Lone et al. (eds.), *Plant Phenolics in Biotic Stress Management*,
https://doi.org/10.1007/978-981-99-3334-1_19

plants against potential fungal attack. Tannins, lignins, and some simple phenolic compound serve as a defense against pathogens. During any kind of stress, phenolic compound concentration in plants increases dramatically. Activity of phenylammonium lyase (PAL) increases during pathogen attack. Many stress conditions such as wounding, viral, fungal, and insect attack increase PAL activity in majority of the plants. Production of lignin and lignin polymers increases the cell wall rigidity and mechanical resistance and also increases the cell wall hydrophobicity against number of diseases caused to carrot, tobacco, Japanese radish root, muskmelon, etc. Phenolic compounds also produce an "allelopathic" effect to decrease the growth of other plants around. Role of phenols was found to be highly effective in apple against *Penicillium expansum*, *Monilinia fructigena*, *Gloeosporium* spp., and *Venturia inequalis*, pear against *Venturia nashicola,* etc. This chapter has focused to cover the effect of the phenols in plant disease resistance, especially fungal infections.

Keywords

Secondary metabolites · Phenolic biosynthesis · Fungal disease resistance

19.1 Introduction

During the course of evolution, the capacities of plant to resist pathogens' attack and colonization have evolved amazingly. Parasites have developed evolved mechanisms to induce the pathogenicity and the host plants have evolved the mechanisms capable of stopping colonization by parasites. Consequently, both the mechanisms led to the evolution of more complex mechanism between the plant–host interactions. Naturally plants are exposed to different environmental conditions and attacked by different infectious pathogens which can pose stress. In order to endure these pathogens and overcome these stress conditions, plants have adapted different signaling pathways as a defense mechanism (Kozlowska and Konieczny 2003; Król and Kepczynska 2008; Dabrowski et al. 2009). Plants synthesize large number of secondary metabolites as an evolved chemical defense against infectious pathogens, because they do not rely on physical mobility for predators escape (Bell 1980a, b). These secondary metabolites effectively act as a defense mechanism against parasitic microbes and herbivores, works as signal compounds to attract pollinators or seed dispersal agents, also protect the plants from ultraviolet radiation and oxidants (Kutchan 2001).

19.2 Phenolic Compounds

Phenolic compounds are the most studied and widely distributed group secondary metabolites throughout the plant kingdom. Presence of phenolic compounds in vegetables and fruits confers them a unique taste, flavor, and health-promoting

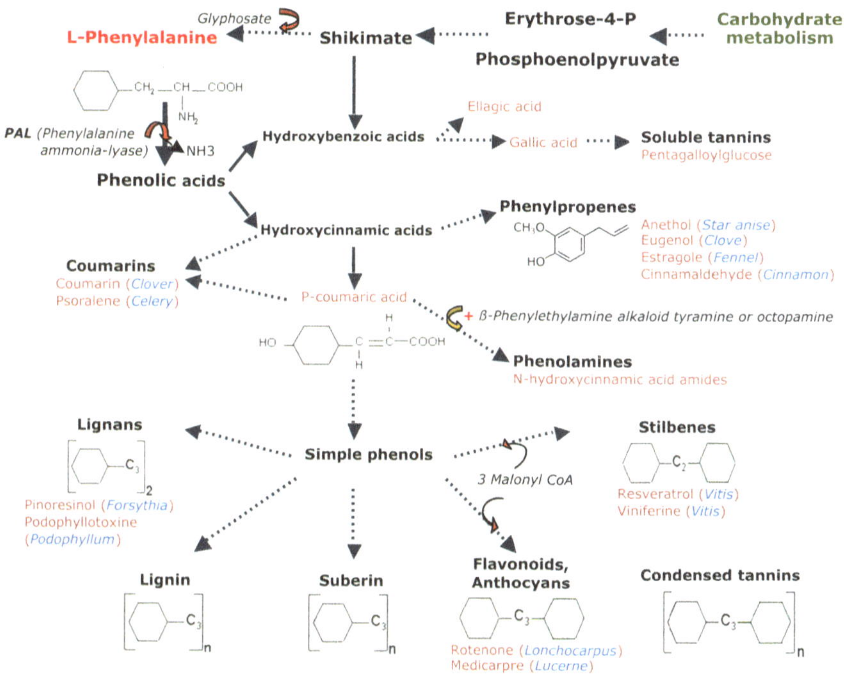

Fig. 19.1 Simplified pathway of phenolic compound synthesis (adapted from Douce 2005)

properties (Tomas-Barberan and Espín 2001). These molecules have in common an aromatic ring bearing at least one phenol hydroxyl substituent and belong to very diverse chemical families. Some phenolic compounds have several hydroxyl group substituents, which can undergo reactions like esterification, methylation, etherification, or glycosylation (Raven et al. 2003; Macheix et al. 2005; Lattanzio et al. 2006).

In plants, phenolic compounds are synthesized from amino acid phenylalanine. Cinnamic acid is the initial molecule for the synthesis of other phenolic compounds (Kozlowska and Konieczny 2003; Knaggs 2003; Bhattacharya et al. 2010). Cinnamic acid and its derivatives also described as phenylpropanoids perform protective functions in plant cells. Tannins, lignin, flavonoids, and some simple phenolic compounds serve as defenses against pathogens. Salicylic acid a nonpolymer derivative of Phenylpropanoid pathway is involved in the induction of plant defense responses (Heil 1999; Kopcewicz and Lewak 2004). In bacteria, algae and fungi the phenolic compounds are very uncommon. Bryophytes usually produce polyphenols including flavonoids but vascular plants produce a full range of polyphenols (Swain 1975).

Plants produce a variety of phenolic compounds synthesized mainly from products of the shikimic acid pathway, playing some important roles in plants (Fig. 19.1). Phenolic compounds synthesized by the plants are directly or indirectly involved in vital functioning of plants. Biochemical interactions and any stress condition in the plants have a direct correlation to the phenol content produced by plants. As per the previous studies, during any biotic or abiotic stress to the plants it has been observed that phenolic content has increased enormously. Different kinds of stresses have a variation of counter production of phenolic compounds in the plants. Plants produce a diversity of phenolic compounds with some specific properties. Phenolic compounds are among the most significant and widely distributed secondary products in the plants and govern disease resistance in many crop plants. Phenolic compounds are involved in major physiological mechanisms of the plant, such as growth, reproduction, pigmentation, rhizogenesis, and resistance to pathogens (Cheynier 2005; Lattanzio et al. 2006), and also play a role in plant genetic resistance against pathogen infection (Treutter 2005). Phenolic compounds and their oxidation products are widely associated with resistance of plants to disease (see reviews by Farkas and Kiraaly 1962).

19.3 Biosynthesis of Phenolic Compounds

The phenolic compound biosynthesis pathway has been well characterized. Phenolic compounds are produced from simple sugars, viz., the well-known shikimate pathway. Carbohydrate transformation of pentose phosphate pathway and glycolysis into erythrose-4-phosphate and phosphoenolpyruvate commences the synthesis of phenylalanine.

Plants when exposed to any pathogen produce many stress compounds (Lavania et al. 2006; Kim et al. 2008). The production of the phenolic compounds usually increases especially during the fungal infections (Gaspar et al. 1982). It has been revealed that these phenolic compounds are capable of forming insoluble complexes with proteins and act as enzyme inhibitors. Many other phenolic associations and their resistance to plant pathogens have been described in numerous crops including sorghum (Dicko et al. 2005), pepper (Baysal et al. 2005), sugarcane (De Armas et al. 2007), banana, chilli (Anand et al. 2009), olives (Markakis et al. 2010), tomato (Baker et al. 2010), and chickpea (Sharma et al. 2011). A significant enhancement in hydrocinnamic and hydroxybenzoic acids in sugarcane infected with smut was reported by De Armas et al. (2007). Similarly, Sahoo et al. (2009) reported the biochemical changes in *Colocasia esculenta* attacked by *Phytophthora colocasiae*. Flavonoids are low-molecular-weight polyphenolic secondary metabolites (Koes et al. 1994). It is the largest groups of simple phenol derived secondary metabolites. Flavonoids have been classified into the following subgroups: chalcones, flavones, flavonols, flavandiols, proanthocyanidins, and their derivatives anthocyanidins and condensed tannins (Falcone-Ferreyra et al. 2012). These compounds are involved in various important functions in plants such as to attract pollinators and seed dispersers, flavonoids are involved in UV-scavenging, fertility, and disease resistance (Elio et al. 2004). Polyphenolic compounds found in plants under stress conditions are significant compounds of vigorous and potent resistance mechanisms against pests and pathogens (Feucht et al. 1994). Phytoalexins are protective compounds which belong to flavonoids and isoflavonoids. Phytoalexins are synthesized *de novo* in plants in response to infection by microorganisms (bacteria, fungi, and viruses) and nematodes (Kozlowska and Konieczny 2003).

Enzymes such as phenylalanine ammonialyase (PAL), peroxidase, and polyphenol oxidase activity increase during the establishment of a pathogen in host tissues. These enzymes has been observed in plants treated with different biotic and abiotic inducers of resistance (Huang and Backhouse 2005; Raghavendra et al. 2007). Many stress conditions such as wounding, viral, fungal, and insect attacks increase PAL activity in various plants (Camm and Towers 1973; Jones 1984; Morelló et al. 2005). These enzymes produce fungitoxic quinines and make the medium adverse for the further development of pathogens (Lattanzio et al. 2006). PAL plays a key role in phenolic compound metabolism (Dixon et al. 2002). The inhibition of PAL led to rigorous scab symptoms due to reduced flavonol accumulation (Mayr et al. 1996). Peroxidase involved in phenol metabolism also catalyzes the condensation of phenol into lignin (Passardi et al. 2004). Polyphenol oxidase is responsible for oxidation of constitutive plant phenols into quinines. Quinines having fungicidal and bactericidal are also responsible for detoxification of pathogen phytotoxins (Macheix et al. 1990; Yoruk and Marshall 2003).

Lignins are the phenolic polymers involved in solute conductance, mechanical support and plant defense mechanism. During the pathogenic infection and other stresses, the deposition of lignins and lignin polymers has been observed. These molecules enhance cell wall rigidity and mechanical resistance and also increase the cell hydrophobicity. Therefore, lignin functions as a physical barrier against pathogenic incursion. In addition, lignin deposits reduce the transmission of enzymes and toxins that the pathogen releases in order to facilitate host tissue colonization. Deprivation of nutrients necessary for proliferation of pathogen is mainly done by lignin (Macheix et al. 2005; Lattanzio et al. 2006). Among various members of the *Poaceae* (Higuchi et al. 1967; Vance et al. 1980; Ride 1983), carrot (Garrod et al. 1982; Heale and Sharman 1977), tobacco and Japanese radish root (Hermann et al. 1987), muskmelon (Grand and Rossignol 1982), and intense red-colored patches around fungal penetration sites were observed by using the Wiesner test. This clarifies the deposition of lignin in response to the infection. Role of lignin in plant resistance was established in the experiment by treating wheat (*Triticum* L.) with lignin synthesis inhibitor, which resulted in the reduction of resistance to a plant pathogen *Puccinia graminis* (Boudet 2000).

Plants encounter any pathogenic attack leads to the resistance to the future attacks. Usually plants after survival from any infection develop resistance for the subsequent infections and survive with defense against a wide range of infections. This phenomenon, called systemic acquired resistance (SAR) (Fig. 19.2).

19.4 Phenolic Compounds in Plant Defense

Preformed or *de novo* synthesis of phenolic compounds is associated with plant defense. Preformed phenols in plants like flavonols, dihydrochalcones, and phenolic acids are called phytoanticipins. Phytoanticipins are distinguished from phytoalexins

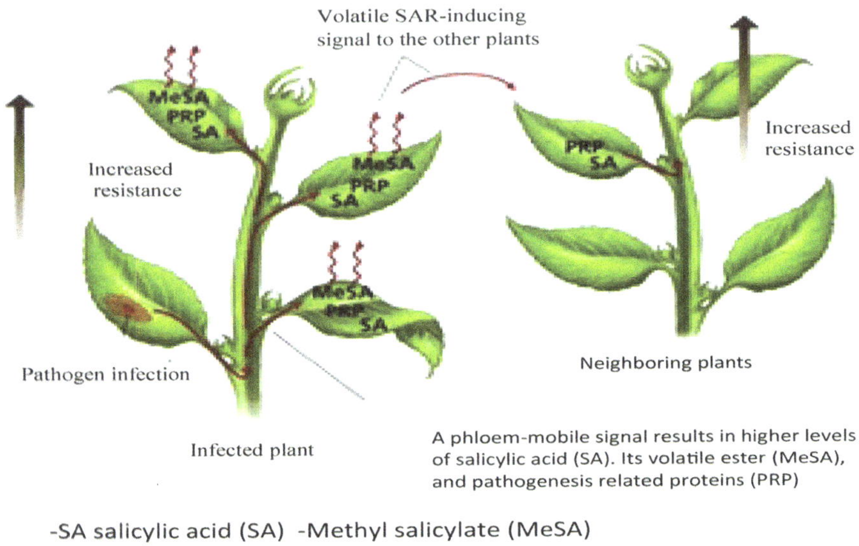

Fig. 19.2 Resistance to future pathogen attack through development of systemic acquired resistance (SAR) initiated from initial pathogen infection (Taiz and Zeiger 2010)

because of their responsive synthesis after pathogenic attack (Lattanzio et al. 2006). A broad range of microorganisms can be inhibited by phytoalexins in response to plant infection (Macheix et al. 2005; Lattanzio et al. 2006).

Researchers from time have evidenced increase in the amount of phenolic content in plants during the encounter of any kind of biotic or abiotic stress. Many such reports have been discussed in this chapter to clear the role of phenols to disease resistance in plants:

Ndoumou et al. (1996) have observed the increased volume of phenolic compounds in Coca pods when infected with *Phytopthora megakary* (Table 19.1).

It can be well assessed from the readings in the table that the phenolic concentration is very high in the infected plants. All the three infected clones have highest phenolic content when compared with the same intact and wounded clones when observed at different intervals. Also the concentration has increased the after the number of days passed after the treatment.

Schovankova and Opatova (2011) have studied the defensive reaction of *Malus domestica* Borkh., against *Penicillium expansum*, *Monilinia fructigena*, and *Gloeosporium* spp. Enhanced phenolic concentration and phenylalanine-ammonia lyase activity was observed in the place of fungal infection mainly in the apple peel, which is the first barrier against fungal attack. In the fruit flesh and peel, the response to the infection was different. Phenylalanine-ammonia lyase activity and total phenol content presented better correlation (r = 0.76–0.98). Significant amounts of Chlorogenic acid and phloridzin has been also examined to divulge their important role in defensive mechanism.

Table 19.1 Soluble phenolic compound (mg g day wt−") in *Phytophthora megakarya*-infected coca pod cortex. Each value is mean^3 SE (n = 5)

Treatment Time (days) and clone	Intact	Wounded	Infected
SNK10			
0	9.68 ± 0.41	9.25 ± 0.31	9.20 ± 0.61
2	9.89 ± 0.24	28.64 ± 0.23	33.36 ± 0.45
4	11.52 ± 0.62	26.32 ± 0.44	29.46 ± 0.96
CS95			
0	260.12 ± 0.82	26.32 ± 0.32	26.25 ± 0.62
2	240.32 ± 0.43	38.14 ± 0.44	45.14 ± 0.86
4	230.14 ± 0.63	36.16 ± 0.63	46.32 ± 0.84
SNK 413			
0	29.24 ± 0.63	29.42 ± 0.23	29.21 ± 0.62
2	30.64 ± 0.84	39.42 ± 0.22	50.65 ± 0.45
4	30.48 ± 0.44	37.26 ± 0.35	48.12 ± 0.84

Kumar et al. suggested that the increasing amounts of total phenols and orthodihydroxy-phenols cause the hyperphenolicity despite increased activities of peroxidase and polyphenol oxidase in infected resistant host tissues. In diseased tissues, the total amino acids and free proline contents increased by manifolds as compared to healthy ones. Similarly, resistant cultivars higher amount of total and OD-phenols have been reported than in susceptible ones (Lily and Ramadasan 1979; Sharma et al. 1983).

Yanbin Hua et al. (2014) investigated the total phenols in the leaves of pear cultivars in relation to pear scab resistance. The contents of total phenol in the immune cultivars of pear were higher than those in other disease-resistant cultivars. Luo and Leng (1990) also showed that the invasion of the fungus *V. nashicola* boosted the production of free phenols in the leaves of immune pear cultivars. However, significantly higher phenolic content has been found in the apple fruit of scab-resistant cultivars than in susceptible cultivars (Treutter and Feucht 1990a, b).

Li et al. (1969) have studied the effect on growth of *Poria weirii* isolates *in vitro* by phenolic and other compounds. Phenolic compounds used with varied molar concentrations of Coumarin, 4-hydroxycoumarin, or 7-hydroxycoumarin compounds showed the inhibition of fungus at more or less concentrations. The higher concentrations of compounds like phenylacetic acid, o-catechol, benzoic acid, salicylic acid, ferulic acid, and o-coumaric acid were inhibitory. Compounds like hydroquinone, p-hydroxybenzoic acid, gallic acid, phenylacetic acid, chlorogenic acid, etc. either stimulated growth of the fungus inhibited or had no effect. Inhibitory compounds present in host tissue could account for resistance to *P. weirii* by plants such *as Alnusrubra* Bong. These results generally diverge to those of found by Christie (1965), for *Phytophthora cactorian* and *Phytophthora parasitica* in which strong inhibition of both *Phytophthora* species by p-hydroxy-

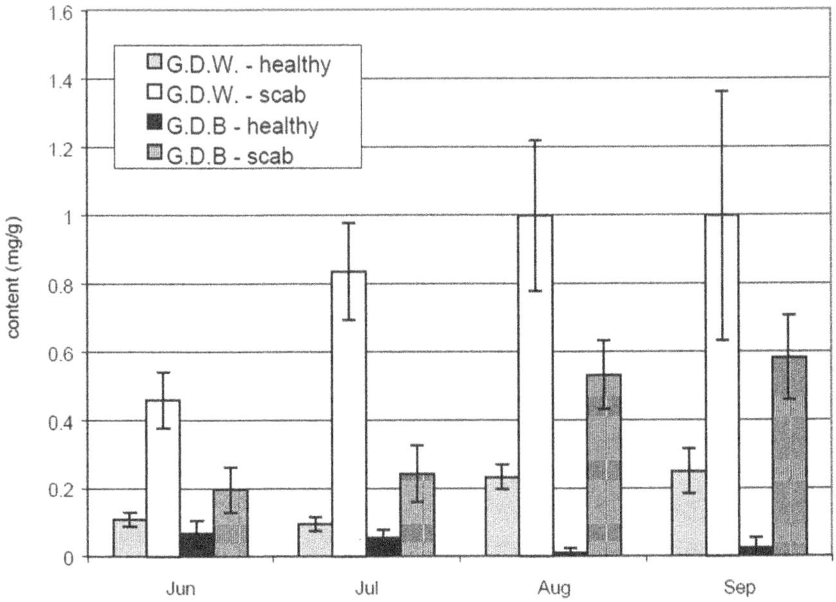

Fig. 19.3 Chlorogenic acid (mg/g) content in healthy and scab-infected apple leaves of "Golden Delicious Weinsberg" (G.D.W.) and "Golden Delicious Clone B" (G.D.B) (Petkovšek et al. 2003)

benzoic acid, p-coumaric acid, and o-hydroxyphenylacetic acid, all of which were non effective or stimulatory to *P. weirii.*

Larger quantities of chlorogenic acid were observed in Goldrush leaves than scab-resistant and scab-susceptible cultivars (Petkovšek et al. 2003). During the growing period, the chlorogenic acid content changed and accumulated with the degree of scab infection of the leaves of scab susceptible cultivars (Fig. 19.3).

Increased phenolic content in apple leaves infected with *Venturia inequalis* has resulted in an enhanced accumulation of phenolic compounds as 2- to 6-fold of chlorogenic acid, a 1.4- to 6.2-fold of flavonols, and a 1.4- to 2.4-fold of Folin–Ciocalteu values in comparison with healthy leaves (Figs. 19.4 and 19.5) (Petkovšek et al. 2008). Similarly, the content of caffeic, *p*-coumaric, and ferulic acids increased by 1.5–2.8 times in the infected leaves with scab infection as compared with healthy controls. Role of Phenolics, flavanols in the resistance of apples to *Venturia inaequalis* has also been revealed by Treutter and Feucht (1990a, b). However, the hydroxycinnamic acids hamper the growth and sporulation of *Venturia inequalis* (Williams and Kuc 1969). In some cases, the resistance of apples to *Venturia inaequalis* resulted in the accumulation of derivatives of chlorogenic and coumaric acids (Bennet and Wallsgrove 1994), and resistant varieties contained more *p*-coumaric acid in comparison with susceptible varieties (Picinelli et al. 1995). Also the accumulation of flavan-3-ols was observed in apple tree tissues infected by *Venturia inequalis* (Treutter and Feucht 1990a), and in the same year, they also reported a six-fold increase of extractable flavan-3-ols in the margin zones of pear

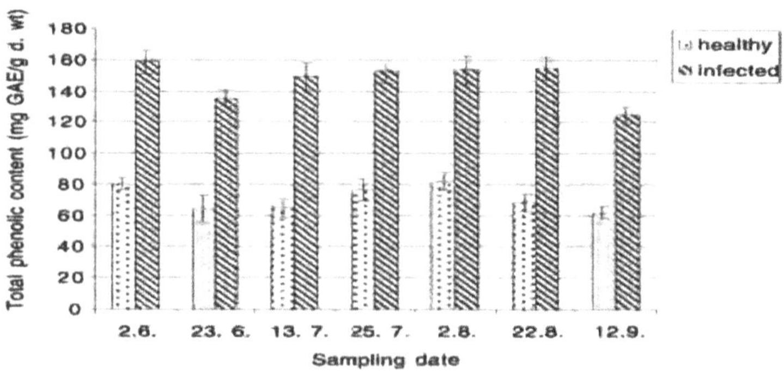

Fig. 19.4 Total phenolic compounds in healthy and infected leaves of cv. Golden delicious determined at various times in Ljubljana

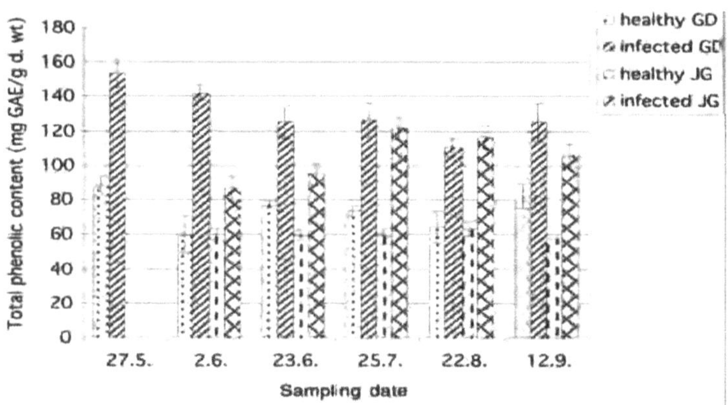

Fig. 19.5 Total phenolic compounds in healthy and infected leaves of cv. Golden delicious (GD) and cv. Jonagold (JG) determined at various times in Maribor

leaves infected by *Gymnosporangium sabinae* in comparison with healthy tissues. Reyes and Cisneros-Zevallos (2003), while treating *Solanum tuberosum* with several abiotic stresses, has observed that mechanical damage induced the accumulation of phenolic compounds, total antioxidant capacity, and increase of PAL activity.

From the above graph, it is quite obvious that the concentration of phenols has increased due to the infection. There may be the difference in the amount of concentration when comparing two different varieties, but that cannot change the fact that phenol concentration has been increased due to the infection. Hyper accumulation of phenolic compounds in the plants during the encounter of any pathogenic infections may suppress the development by directly isolating infection or respond to the pathogen by bringing up the cell death (Heath 1980; Mansfield 1990). Phenolic compounds like benzoic acids and the phenylpropanoids are usually produced initially to respond the infections (Kurosaki et al. 1986; Niemann et al. 1991). Esterification of ferulic acid to the host cell wall is a common host response

(Matern and Kneusel 1988), and it has been suggested that crosslinking of such phenylpropanoid esters leads to the formation of lignin like polymers (Fry 1986).

Researchers proposed that plants present their defense mechanism in two stage responses: First at the infection site plants show a rapid accumulation of phenolic compounds to halt the pathogen growth and allow the plant to prepare for the second stage defense. Secondary response would involve the activation of *de novo* synthesis of phytoalexins or other stress-related substances. In the argument, it was suggested that in primary defense, rapid *de novo* transcription and translation of genes are not possible, which would be a possible characteristic of the second level of defense (Bell 1980a, b; Deverall 1982; Mansfield 1982; Matern and Kneusel 1988).

Prior to the active defense, phenolic compounds accumulate in parsley leaves in response to inoculation with the fungal species *Phytophthora megasperma* f. sp. *glycinea* and *Alternaria carthami* Chowdhury. Treatment with *phytophthora megasperma* elicitor and the cell suspensions of parsley result in the large accumulation of linear furanocoumarins, psoralen, xanthotoxin, benzodipyrandione, graveolen, and small amounts of the furanocoumarin bergapten. Treatment with *Alternaria carthami* elicitor largely accumulates bergapten and the furanocoumarin isopimpinellin, as well as small amounts of graveolen (Tietjen et al. 1983).

Rapid changes in the composition of *Phaseolus vulgari* cell wall when treated with an elicitor prepared from pathogenic fungus *Colletotrichum lindemthianum* resulted in the rapid increase of phenolic material bond to cellulose and hemicellulosic fractions of cell wall and increase in wall-associated hydroxyproline. It is also followed by rapid decrease of intercellular levels of free hydroxycinnamic acid and transient increases in the extractable activities of L-phenylalanine ammonia-lyase and cinnamic acid 4-hydroxylase. Therefore, biosynthesis of all these biomolecules during early events is closely linked to the initial interaction between plant cell and fungal elicitor (Bolwell et al. 1985). These metabolic changes related to cell wall components followed rapid kinetics similar to those involved in the formation of the phytoalexin kievitone in the elicited cultures (Robbins et al. 1985).

Fungi usually attack the plant for their carbon and energy source. Plants produce some phenolic compounds which are antifungal in nature. Antifungal phenolics within plants are usually tissue specific. The distribution of preformed antifungal phenolics within plants is often tissue specific and there is a tendency for many lipophilic compounds such as flavone and flavonols methyl ethers to be located at the plant surface or in the cytoplasmic parts within the epidermal cells, signifying that they may certainly act as deterrents to pathogens. However, preformed antifungal phenolics are generally sequestered in conjugated form with glycosidic attachments in vacuoles or organelles in healthy plants (Nicholson and Hammerschmidt 1992; Wink 1997; Grüner et al. 2003; Beckman 2000).

Phenolics are antifungal and antibacterial compounds in plants (Laks et al. 1988; Sivaprakasan and Vidhyasekaran 1993; Schlösser 1994). Fungal mycelium of Petri disease causes the blockage of xylem vessels of vine plants. *Phaeoacremonium aleophilum* has a greater capacity for degrading the xylem wall because of its high specific activity of laccase and lignin peroxidase and hence initiates the formation of tyloses inside the xylem vessels in vines with Petri disease. Certain phenols inhabit the degrading activity of an oxidative nature exhibited by these enzymes. For example, *p*-coumaric acid and the manganese peroxidase reaction medium reduced activity of that enzyme. Catechin, caffeic acid, tannin, and other vine phenolic compounds also inhibited manganese peroxidase. All these investigations suggest the inhibitory nature of these phenolic compounds that degrade the cell wall. An active participation of phenolic compounds *p*-coumaric acid, catechin, caffeic acid and tannins against petri disease of grape vine was demonstrated. Reduction of Petri-disease symptoms and growth enhancement was observed when SO_4 vines affected by Petri disease were treated with Brotomax (a product that stimulates synthesis of phenolic compounds) (Del Río et al. 2004). Many other reports also suggest that Brotomax enhances disease tolerance or resistance of plants such as citrus, olive and cotton by increasing phenolic compound concentration in plants (Fuster et al. 1995; Ortuño et al. 1997; Botía et al. 2001; Del Río et al. 2001a, b).

Role of some specific phenolic compounds against the fungi discussed by different scientists from time to time are well elaborated (Table 19.2).

19.5 Metabolic Changes After Pathogenic Attack

A number of metabolic changes occur in plants upon the contact with the pathogenic microorganisms. Such changes include reduction in total protein synthesis (Roby et al. 1985), the release of ethylene (Ecker and Davis 1987), the reinforcement of the cell walls, including formation of papillae (Bonhoff et al. 1987), the production of pathogenesis-related proteins of lytic enzymes, and hydroxyproline-rich glycoproteins (Hahlbrock et al. 1987), and the accumulation of low-molecular-weight defense chemicals, the phytoalexins (Bailey and Mansfield 1982).

19.6 Phenolic Compound and Pathogenic Resistance in Forest Trees

A variety of secondary metabolites which provide resistance against natural enemies are present in Pines and spruces, like resin acids, terpenes and phenolics (Honkanen et al. 1999; Sallas et al. 2001). Johanna and Juan in a review article have discussed role of phenolic compounds of some trees in resistance against pathogens, pathogen-induced responses in phenolic compounds in selected timber and landscape trees, and effect of phenolic compounds on *in vitro* growth of forest-tree pathogens in selected studies. The general picture arising from the literature studied concludes: phenols have a positive role in the disease resistance to confine the infection, may provide indirect protection against the damage caused by pathogens. The relationship between tree phenolics and resistance to pathogens is ambiguous (Tables 19.3, 19.4, and 19.5).

Table 19.2 Fungi and their potential antifungal phenolic compounds

Compound	Fungus	References
Benzaldehyde, Ethylbenzoate	*Botrytis cinerea*	Wilson and Wisniewski (1989)
	Monilinia fructicola	
3,4-Dihydroxybenzaldehyde	*Gloeosporium musarum*	Friend (1979)
Catechol, Protocatechuic acid	*Colletotrichum circinans*	Walker and Stahmann (1955)
2,5-Dimethoxybenzoic acid	*Botrytis cinerea*	Lattanzio et al. (1996a, b)
	Rhizopus stolonifer	
Salicylic acid	*Eutypa lata*	Amborabé et al. (2002)
Vanillic acid, 4-Hydroxybenzoic acid	*Phytophthora infestans*	Harborne (1980)
Chlorogenic acid	*Verticillium albo-atrum*	Lee and LeTourneau (1958)
	Phytophthora infestans	Valle (1957)
	Phlyctaena vagabunda	Lattanzio et al. (2001)
Chlorogenic acid, Rutin	*Fusarium oxysponun*	Carrasco et al. (1978)
p-Coumaric acid, Cyanidin	*Gloeosporium perennas*	Hulme and Edney (1960)
3-and 7-Hydroxyflavone	*Penicillium glabrum*	Martini et al. (1997)
	Cladosporium herbarum	
Dihydroquercetin	*Fusarium spp.*	Skadhauge et al. (1997)
Naringenin, Kaempferol	*Pyricularia oryzae*	Padmavati et al. (1997)
Naringin, Tangeretin	*Penicillium digitatum*	Arcas et al. (2000)
Phloridzin, Phloretin	*Venturia inaequalis*	Overeem (1976)
Flavone, Flavanone	*Aspergillus sp.*	Weidenbörner et al. (1990)
Cirsiliol, Cirsunaritin, Hispidulin	*Cladosporium sphaerospermum*	Alcerito et al. (2002)
7,4'-Dihydroxynavan, 5,8-Dihydroxy-6,7-dimethoxyflavan	*Bobytis cinerea*	Saini and Ghosal (1984)
	Fusarium oxysporum	Weidenbörner et al. (1990)
	Helmintho sporiumoryzae	
Oleuropein	*Phytophtora spp.*	Del Río et al. (2003)
Nobiletin	*Phoma tracheiphila*	Friend (1979)
Genistein, Biochanin	*Monilinia fracticola*	Johnson et al. (1976)
	Cercospora bieticola	
Hordatine A and B	*Helminthosporium sativum*	Overeem (1976)

Table 19.3 Role of phenolic compounds in resistance against pathogens of selected timber and landscape trees

Tree	Pathogen	Phenolic compounds	Type of phenolics	Plant part or tissue	Analysis method	Role in resistance
Pinus sylvestris	*Coniophora puteana*	PS, PSME	S	Heartwood	HPLC	+
	Ophiostoma brunneo-ciliatum	Taxifolin	F	Phloem	HPLC, TLC	+
	Leptographium wingfieldii	Taxifolin glucoside	F		HPLC	–
		p-Coumaric acid esters	HCA			–
		Acetophenone glycoside	AcPh			–
		Taxifolin glycoside	F			+
Picea abies	*Ceratocystis polonica*	Isorhapotin	S	Needles	HPLC	–
	Sirococeus conigenus	Astringin	S		HPLC	n
		Picein	AcPh			n
		Taxifolin glycoside	F			n
		(+)-Catechin	F			n
		Picein	AcPh			+
		(+)-Catechin	F			+
Picea sitchensis	*Heterobasidion annosum, Phaeolus schweinitzii*	Isorhapotin	S	Bark	HPLC	n
		Astringin	S			n
Populus tremuloides	*Phellinus tremulae*	Benzoic acid	PhA	Stem (galls)	GC, MS, NMR, TLC, FT-AR, FC	+
		p-Hydroxybenzoic acid	PhA			n
		Trans-cinnamic acid	HCA			n
		p-Hydroxycinnamic acid	HCA			n
		Naringenin	F			n
		7′-Methyl-3-hydronaringin	F			n
		Taxifolin	F			n

(continued)

Table 19.3 (continued)

Tree	Pathogen	Phenolic compounds	Type of phenolics	Plant part or tissue	Analysis method	Role in resistance
Salix myrsinifolia	*Melampsora sp.*	(+)-Catechin	F	Leaf	HPLC	(+)
		Luteolin-7-glucoside	F			n
		Chlorogenic acid	HCA			(+)
		Salicin	SaD			(+)
		Salicortin	SaD			n

aPS pinosylvin, *PSME* pinosylvin monomethylether, *bPhA* phenolic acids (C6–C1), *AcPh* acetophenone (C6–C2), *HCA* hydroxycinnamic acids (C6–C3), *S* stilbenes (C6–C2–C6), *F* flavonoids (C6–C3–C6), *SaD* salicin derivatives, *PP* polyphenolics, *cHPLC* high-performance liquid chromatography, *TLC* two-dimensional thin-layer chromatography, *GC* mass chromatography, *MS* mass spectrometry, *NMR* nuclear magnetic resonance, *FT-IR* Fourier transform infrared spectrometry, *FC* flash chromatography, *d+* high constitutive concentrations related to enhanced resistance, – low constitutive concentrations related to enhanced resistance, *(+)* high constitutive concentrations related in some cases to enhanced resistance, *n* no role in resistance

Table 19.4 Examples of pathogen-induced responses in phenolic compounds in selected timber and landscape trees

Host	Pathogen	Induced phenolics[a]	Type of phenolics[b]	Response[c]	Analyzed plant part or tissue[d]	Analysis method
Pinus sylvestris	*Heterobasidion annosum* *Heterobasion parviporum* *Ophiostoma brunneo-ciliatum* *Leptographium wingfieldii*	PS, PSME Lignin Catechin PS, PSME PS, PSME Pinocembrin p-Coumaric acid esters Acetophenone glycoside Taxifolin Taxifolin glucoside PS, PSME Pinocembrin Taxifolin	S PP F S S F HCA HCA F F S F F	+ + – + + + – – +/– +/– + + +	L (roots) S (needles) L (roots) L (stem and root sapwood) L (stem phloem)	HPLC Spectrophotometry HPLC, histochemistry HPLC, TLC HPLC
Pinus nigra	*Sphaeropsis sapinea* *(=Diploidia pinea)*	PS, PSME Taxifolin-like compound Ferulic acid Taxifolin Naringenin-like compound Lignin Catechin-like compound Ferulic acid glucoside	S F PhA F F PP F PhA	+ + – + + + – –	L (stem phloem), S (stem phloem) L (stem phloem)	HPLC

(continued)

Table 19.4 (continued)

Host	Pathogen	Induced phenolics[a]	Type of phenolics[b]	Response[c]	Analyzed plant part or tissue[d]	Analysis method
Picea abies	*Ceratocystis polonica*	Isorhapontigenin	S	+	L (stem phloem)	GC, MS
	Gremmeniella abietina	Stilbene glycosides	S	–	L (branch phloem)	HPLC
	Sirococcus conigenus	(+)-Catechin	F	+	S (branch phloem)	HPLC,
	Pythium dimorphum	Phenolic acids	PhA	+	L (needles)	spectrophotometry
		p-Coumaric acid	HCA	+	L (root cortex)	HPLC
		Lignin	PP	+		Histochemistry
		p-Hydroxybenzoic	PhA	+		
		acid	F	+		
		(+)-Catechin	AcPh	+		
		Picein	S	+		
		Isorhapontin	S	+		
		Astringin	F	+		
		Kaempferol-3-	PP	+		
		glucoside				
		Lignin				
Populus tremuloides	*Entoleuca mammata*	Lignin-like phenolics	PP	+	L (stem phloem)	

*a**PS* pinosylvin, *PSME* pinosylvin monomethylether, *b**PhA* phenolic acids (C6–C1), *AcPh* acetophenones (C6–C2), *HCA* hydroxycinnamic acids (C6–C3), *S* stilbenes (C6–C2–C6), *F* flavonoids (C6–C3–C6), *SaD* salicin derivatives, *PP* polyphenolics, *c*+ increased concentration, – decreased concentration, +/– inconsistent variation, *dL* local response, *S* systemic response, *eHPLC* high-performance liquid chromatography, *TLC* thin-layer chromatography, *GC* gas chromatography, *MS* mass spectrometry

Table 19.5 Effect of phenolic compounds on *in vitro* growth of forest-tree pathogens in selected studies

Pathogen	Main host(s)	Phenolic compounds	Type of phenolics	Media	Growth inhibition
Heterobasidion annosum	*Pinus, Picea*	PS, PSME	S	L	+
		Vanillin	F	MA	–
		Catechin	S	L	–
		Isorhapontin	F	–	–
		"Soluble phenolics"	S	–	–
		(+)-Catechin	HCA	MA	–
		Trans-resveratrol	F	NM/	–
		p-coumaric acid	PhA	L	+
		(–)-Epicatechin	AcPh		+
		p-Hydroxybenzoic	S		SD
		acid	PhAc		SD
		4-	F		SD
		Hydroxyacetophenone	HCA		SD
		Piceatannol	HCA		SD
		Protocatechuic acid	HCA		+
		(4)-Catechin			+
		Ferulic acid +			+
		p-Coumaric acid			
		Caffeic acid			
Phaeolus schweinitzii		Isorhapontin	S	MA	+
Fomotsis pinicola		Isorhapontin	S	MA	+
Fusarium avenaceum		Ferulic acid	HCA	NM	+
		p-Coumaric acid	HCA	L	+
		Caffeic acid	HCA		+
Trametes versicolor		Isorhapontin	S	MA	–
Armillaria ostoyae		Isorhapontin	S	MA	–
Ceratocystis polonica	*Picea abies*	Soluble phenolics	F	L	+
		(+)-Catechin	S		–
		Trans-resveratrol			–
Ophiostoma penicillatum	*Picea abies*	Soluble phenolics	F	L	+
		(+)-Catechin	S		–
		Trans-resveratrol			–
Ophiostoma piceaperdum		Resveratrol	S	MA	+
		(+)-Catechin	F		–
Ophiostoma bicolor		Resveratrol	S	MA	+
		(+)-Catechin	F		–
Sphaeropsis sapinea (=Diploidia pinea)		PS	S	WA	+
		Tannic acid	PP		–
Gremmeniella abietina		PS	S	MA	–
		Resveratrol	S		–

(continued)

Table 19.5 (continued)

Pathogen	Main host(s)	Phenolic compounds	Type of phenolics	Media	Growth inhibition
Phellinus tremulae	*Populus*	Benzoic acid	PhA	CA	+
		p-Hydroxybenzoic acid	PhA		–
		Trans-cinnamic acid	HCA		–
		p-Hydroxycinnamic acid	HCA		–

aPS pinosylvin, *PSME* pinosylvin monomethylether, *bPhA* phenolic acids (C6–C1), *AcPh* acetophenones (C6–C2), *HCA* hydroxycinnamic acids (C6–C3), *S* stilbenes (C6–C2–C6), *F* flavonoids (C6–C3–C6), *SaD* salicin derivatives, *PP* polyphenolics, *cL* liquid medium, *MA* malt agar, *NM* Norkran's minimal medium, *WA* water agar, *CA* carrot agar, *TLC* thin layer chromatography bioassay, *d+* compound inhibited fungal growth, – compound did not affected or enhanced fungal growth, *SD* strain-dependent reaction

19.6.1 Antifungal Role of Phenolic Compounds in Grapevine Wood Decay

Wood afflictions occur worldwide are very destructive grapevine diseases (Bertsch et al. 2009a, b). Many fungal species of Botryosphaeriaceae such as *Neofusicoccum parvum* and *Diplodia seriata* are also associated with a wide range of grapevine decline symptoms including Esca syndrome (Niekerk et al. 2006). The knowledge about onset of fungal infection in grapevine is not clear as it can happen via many channels (Gramaje and Armengol 2011). Owing to this lack of knowledge and treatment to control grapevine wood diseases, it is thought that preformed defenses induced by induced phenolic compounds could form a chemical barrier that limits fungal pathogen growth and restrain the development of these pathogens in the wood before disease symptoms appearance (Shigo and Marx 1977).

Phenolic compound levels increase especially those of the hydroxystilbenes, trans-resveratrol, and ε-viniferin (Amalfitano et al. 2000) and stilbenoids such as ampelopsins A, B, H, leachianols F, G, hopeaphenol, isohopeaphenol, and pallidol in infected wood (Amalfitano et al. 2000, 2005). Phenolic compound accumulation was also studied in plants with *P. aleophilum* and *P. chlamydospora* (Del Rio et al. 2001; Troccoli et al. 2001). Major grapevine trunk disease agents (*P. chlamydospora*, *P. aleophilum*, *F. mediterranea*, *E. lata*, *D. seriata* strain Bo F99-1, and *N. parvum* strain Bp0014) were screened for growth on phenolic compounds and acids which demonstrated the difference in sensitivity of phenolic compounds toward fungi. However, majority of the phenolic compounds could display an antifungal activity, and some enhance the growth of some pathogens, while two molecules triggered both an inhibition and stimulation (Lambert et al. 2012).

The effect of phenolic compound concentrations is important to carry out its functional role. Treatment of fungal pathogens using different concentration of phenolic compounds reveals different results. Resveratrol restricted the growth of *Diplodia seriata* about 1.6-fold, *Neofusicoccum parvum* 1.1-fold, and *Eutypa lata*

1.7-fold (Lambert et al. 2012). Similarly, resveratrol inhibited *E. lata* growth about 1.7-fold (Coutos-Thévenot et al. 2001). Whereas, Bruno and Sparapano (2006) found resveratrol to inhabit the growth of *Phaeoacremonium aleophilum* at concentration of 220 μM.

19.6.2 Physiological Changes in Wheat During Development of Loose Smut

Ustilago tritici infections cause the production of phenolic acids in wheat plant. The diseased root showed higher concentration of phenol increment over healthy roots. Stem portion had about 56% rise in phenolic compounds in infected plants in comparison to the healthy ones. The enhancement in peroxides was also observed in the infected plant. Anionic isoenzymes of peroxidase elevated level in basal zone of wheat roots infected with stinking smut was also reported (Khairullin et al. 2000). After inoculation with *Fusarium culmorum,* wheat cultivars showed a sequential increase in free phenolics in glumes, lemmas, and palea (Siranidou et al. 2002).

19.7 Conclusion

It is quite explicable from the earlier studies that phenolic compounds are very important in natural plant defense mechanism. Many of these physiologically active compounds are now used as insecticides, fungicides, or pharmaceuticals. Use of hybrid-resistant varieties with increased levels of phenolic compounds can potentially reduce the need for various pesticides. The mechanism of action against disease, determination of gene activity, characterization of some developmental stages, and other physiological events of phenolic compounds have been revealed. However, further analysis is needed to reveal the underlying mechanism of defensive nature of each individual phenolic compounds. Indeed, each phenolic compound has a specific role in a particular type of disease; even some phenolic compounds enhance the pathogenic infection.

References

Alcerito T, Barbo FE, Negri G, Santos DYAC, Meda CI, Young MCM, Chávez D, Blatt CTT (2002) Biochem Syst Ecol 30:677

Amalfitano C, Evidente A, Surico G, Tegli S, Bertelli E, Mugnai L (2000) Phenols and stilbene polyphenols in the wood of esca-diseased grapevines [Vitis vinifera L.]. Phytopathologia Mediterr 39(1):178–183

Amalfitano C, Peduto F, Mugnai L, Evidente A, Surico G (2005) Chemical characterisation of stilbenic polyphenols from esca-diseased wood and their role in defense mechanisms. In: 4th international workshop on grapevine trunk diseases. Phytopathol Mediterr 44:99

Amborabé B-E, Fleurat-Lessard P, Chollet J-F, Roblin G (2002) Plant Physiol Biochem 40:1051

Anand T, Bhaskaran R, Raguchander T, Samiyappan R, Prakasam V, Gopalakrishnan C (2009) Defence responses of chilli fruits to Colletotrichum capsici and Alternaria alternata. Biol Plant 53(3):553

Arcas MC, Botia JM, Ortuño AM, Del Rio JA (2000) Eur J Plant Pathol 106:617

Bailey JA, Mansfield JW (1982) Phytoalexins. Includes phytoalexins from leguminosae, from Solanaceae, from other families, role in disease resistance. Papers not separately indexed. Review article General article, Path_anat (PMBD 185701442). Glasgow & London, Blackie, 334 pp

Baker CJ, Owens RA, Whitaker BD, Mock NM, Roberts DP, Deahl KL, Aver'yanov AA (2010) Effect of viroid infection on the dynamics of phenolic metabolites in the apoplast of tomato leaves. Physiol Mol Plant Pathol 74(3–4):214–220

Baysal O, Turgut C, Mao G (2005) Acibenzolar-S-methyl induced resistance to Phytophthora capsici in pepper leaves. Biol Plant 49(4):599–604

Beckman CH (2000) Phenolic-storing cells: keys to programmed cell death and periderm formation in wilt disease resistance and in general defence responses in plants? Physiol Mol Plant Pathol 57(3):101–110

Bell AA (1980a) The time sequence of defense. In: Horsfall JG, Cowling EB (eds) Plant disease, vol 5. Academic, New York, pp 53–73

Bell EA (1980b) Secondary plant products

Bennet RN, Wallsgrove RM (1994) Secondary metabolites in plant defence mechanisms. New Phytol 127:617–633

Bertsch C, Beuve M, Dolja VV, Wirth M, Pelsy F, Herrbach E, Lemaire O (2009a) Retention of the virus-derived sequences in the nuclear genome of grapevine as a potential pathway to virus resistance. Biol Direct 4(1):21

Bertsch C, Larignon P, Farine S, Clément C, Fontaine F (2009b) The spread of grapevine trunk disease. Science 324(5928):721–721

Bhattacharya A, Sood P, Citovsky V (2010) The roles of plant phenolics in defence and communication during Agrobacterium and Rhizobium infection. Mol Plant Pathol 11(5):705–719

Bolwell GP, Robbins MP, Dixon RA (1985) Metabolic changes in elicitor-treated bean cells: Enzymic responses associated with rapid changes in cell wall components. Eur J Biochem 148(3):571–578

Bonhoff A, Rieth B, Golecki J, Grisebach H (1987) Race cultivar-specific differences in callose deposition in soybean roots following infection with Phytophthora megasperma f. sp. glycinea. Planta 172(1):101–105

Botía JM, Ortuño A, Benavente-García O, Báidez A, Frías J, Marcos D, Del Río JA (2001) Modulation of the biosynthesis of some phenolic compounds in *Olea europaea* L. fruits: their influence on olive oil quality. J Agric Food Chem 49:355–358

Boudet AM (2000) Lignins and lignification: selected issues. Plant Physiol Biochem 38(1–2):81–96

Bruno G, Sparapano L (2006) Effects of three esca-associated fungi on Vitis vinifera L.: III. Enzymes produced by the pathogens and their role in fungus-to-plant or in fungus-to-fungus interactions. Physiol Mol Plant Pathol 69(4–6):182–194

Camm EL, Towers GN (1973) Phenylalanine ammonia lyase. Phytochemistry 12(5):961–973

Carrasco A, Boudet AM, Marigo G (1978) Physiol Plant Pathol 12:225

Cheynier V (2005) Polyphenols in foods are more complex than often thought. Am J Clin Nutr 81(1):223S–229S

Christie T (1965) The effects of some phenolic compounds upon the growth of two species of Phytophthora. N Z J Agric Res 8(3):630–635

Coutos-Thévenot P, Poinssot B, Bonomelli A, Yean H, Breda C, Buffard D, Esnault R, Hain R, Boulay M (2001) In vitro tolerance to Botrytis cinerea of grapevine 41B rootstock in transgenic plants expressing the stilbene synthase Vst 1 gene under the control of a pathogen-inducible PR 10 promoter. J Exp Bot 52(358):901–910

Dabrowski S, Glowacki S, Macioszek VK, Kononowicz AK (2009) Reaktywne formy tlenu w odpowiedzi obronnej roslin na grzyby nekrotroficzne. Postępy Biologii Komórki Suplement 25:163–176

De Armas R, Santiago R, Legaz ME, Vicente C (2007) Levels of phenolic compounds and enzyme activity can be used to screen for resistance of sugarcane to smut (Ustilago scitaminea). Australas Plant Pathol 36(1):32–38

Del Rio JA, Gonzales A, Fuster MD, Botia JM, Gomez P, Ortuno A, Frias V (2001) Tylose formation and changes in phenolic compounds of grapevine roots infected with Phaeomoniella chlamydospora and Phaeoacremonium species [Vitis vinifera L.]. Phytopathol Mediterr 40(3): 394–399

Del Río JA, González A, Fuster MD, Botía JM, Gómez P, Frías V, Ortuño A (2001a) Tyloses formation and changes in phenolic compounds of grape roots infected with *Phaeomoniella chlamydospora* and *Phaeoacremonium* species. Phytopathol Mediterr 40:S394–S399

Del Río JA, González A, Gómez P, Botía JM, Fuster MD, Ortuño A, Frías V, Gutiérrez JC, López M (2001b) Modulation of cotton phenolic compounds by Brotomax. Effect on resistance against *Verticillium* on yield. In: 8th international verticillium symposium, Córdoba, Spain, pp 54

Del Río JA, Báidez AG, Botía JM, Ortuño A (2003) Food Chem 83:75

Del Río JA, Gómez P, Báidez A, Fuster MD, Ortuño A, Frías V (2004) Phenolic compounds have a role in the defence mechanism protecting grapevine against the fungi involved in petri disease. Phytopathol Mediterr 43(1):87–94

Deverall BJ (1982) Introduction. See ref 4, pp 1–20

Dicko MH, Gruppen H, Traoré AS, van Berkel WJ, Voragen AG (2005) Evaluation of the effect of germination on phenolic compounds and antioxidant activities in sorghum varieties. J Agric Food Chem 53(7):2581–2588

Dixon RA, Achnine L, Kota P, Liu CJ, Reddy MS, Wang L (2002) The phenylpropanoid pathway and plant defence—a genomics perspective. Mol Plant Pathol 3(5):371–390

Douce R (2005) Les plantes supérieures: divines et/ou diaboliques. Conférences et débats sur les grands défis du 21e siècle. Institut de France, Académie des Sciences, Paris

Ecker JR, Davis RW (1987) Plant defense genes are regulated by ethylene. Proc Natl Acad Sci U S A 84(15):5202–5206

Elio GWM, Schijlen CH, de Vos R, van Tunen AJ, Bovy AG (2004) Modification of flavonoid biosynthesis in crop plants. Phytochemistry 65:2631–2648

Falcone-Ferreyra ML, Casas MI, Questa J, Herrera L, Wang J, Jackson D, Grotewold E, Casati P (2012) Evolution and expression of tandem duplicated maize flavonol synthase genes. Front Plant Sci 3:101

Farkas GL, Kiraaly Z (1962) Role of phenolic compounds in the physiology of plant diseases and disease resistance. J Phytopathol 44(2):105–150

Feucht W, Treutter D, Christ E (1994) Accumulation of flavanols in yellowing beech leaves from forest decline sites. Tree Physiol 14(4):403–412

Friend J (1979) Recent advances in phytochemistry. In: Swain T, Harborne JB, Van Sumere CF (eds) Biochemistry of plant phenolics. Plenum Press, New York, p 557

Fry SC (1986) Cross-linking of matrix polymers in the growing cell walls of angiosperms. Annu Rev Plant Physiol 37:165–186

Fuster MD, García D, Ortuño A, Botía JM, Sabater F, Porras I, García-Lidón A, Del Río JA (1995) Selection of citrics highly productive in secondary metabolites of industrial interest. Modulation of synthesis and/or accumulation processes. In: García VC, Castañer M, Gil MI, Ferreres F, Tomás Barberán FA (eds) Current trends in fruit and vegetables phytochemistry. CSIC, Madrid, pp 81–85

Garrod B, Lewis BG, Brittain MJ, Davies WP (1982) Studies on the contribution of lignin and suberin to the impedance of wounded carrot root tissue to fungal invasion. New Phytol 90(1): 99–108

Gaspar T, Penel C, Thorpe T, Greppin H (1982) Peroxidases 1970-1980. A survey of their biochemical and physiological roles in higher plants

Gramaje D, Armengol J (2011) Fungal trunk pathogens in the grapevine propagation process: potential inoculum sources, detection, identification, and management strategies. Plant Dis 95(9):1040–1055

Grand C, Rossignol M (1982) Changes in the lignification process induced by localized infection of muskmelons with Colletotrichum lagenarium. Plant Sci Lett 28(1):103–110

Grüner R, Strompen G, Pfitzner AJ, Pfitzner UM (2003) Salicylic acid and the hypersensitive response initiate distinct signal transduction pathways in tobacco that converge on the as-1-like element of the PR-1a promoter. Eur J Biochem 270(24):4876–4886

Hahlbrock K, Cretin C, Cuypers B, Fritzemeier KH, Hauffe KD, Jahnen W, Kombrink E, Rohwer F, Scheel D, Schmelzer E (1987) Tissue specificity and dynamics of disease resistance responses in plants. In: NATO advanced science institutes series: series a: life sciences (USA)

Harborne JB (1980) Secondary plant products. In: Bell EA, Charlwood BW (eds) Encyclopedia of plant physiology, new series, vol 8. Springer, Berlin, pp 329

Heale JB, Sharman S (1977) Induced resistance to Botrytis cinerea in root slices and tissue cultures of carrot (Daucus carota L.). Physiol Plant Pathol 10(1):51–61

Heath MC (1980) Reactions of nonsuscepts to fungal pathogens. Annu Rev Phytopathol 18(1): 211–236

Heil M (1999) Systemic acquired resistance: available information and open ecological questions. J Ecol 87(2):341–346

Hermann C, Legrand M, Geoffroy P, Fritig B (1987) Enzymatic synthesis of lignin: purification to homogeneity of the three O-methyltransferases of tobacco and production of specific antibodies. Arch Biochem Biophys 253(2):367–376

Higuchi T, Ito Y, Kawamura I (1967) p-Hydroxyphenylpropane component of grass lignin and role of tyrosine-ammonia lyase in its formation. Phytochemistry 6(6):875–881

Honkanen T, Haukioja E, Kitunen V (1999) Responses of *Pinus sylvestris* branches to stimulated herbivory are modified by tree sink/source dynamics and by external resources. Funct Ecol 13: 126–140. https://doi.org/10.1046/j.1365-2435.1999.00296.x

Hua Y, Guo H, Zhou XG, Li X, Yang S, Song Y, Ma N, Chai C, Qiao X, Li L (2014) Correlations between soluble sugar and phenol contents in leaves and pear scab resistance. J Plant Pathol Microbiol 5(1):1–4

Huang LD, Backhouse D (2005) Induction of defence responses in roots and mesocotyls of sorghum seedlings by inoculation with Fusarium thapsinum and F. proliferatum, wounding and light. J Phytopathol 153(9):522–529

Hulme AC, Edney KL (1960) Phenolics in plant and disease. In: Pridham JB (ed) Pergamon Press, London, pp 87

Johnson G, Maag DD, Johnson DK, Thomas RD (1976) Physiol Plant Pathol 8:225

Jones DH (1984) Phenylalanine ammonia-lyase: regulation of its induction, and its role in plant development. Phytochemistry 23(7):1349–1359

Khairullin RM, Yusupova ZR, Troshina NB (2000) Protective responses of wheat treated with fungal pathogens: 2. Activation of anionic peroxidase isoforms in wheat seedlings induced by inoculation with Tilletia caries spores. Russian J Plant Physiol 47(1):103–108

Kim YH, Kim CY, Song WK, Park DS, Kwon SY, Lee HS, Bang JW, Kwak SS (2008) Overexpression of sweetpotato swpa4 peroxidase results in increased hydrogen peroxide production and enhances stress tolerance in tobacco. Planta 227(4):867–881

Knaggs AR (2003) The biosynthesis of shikimate metabolites. Nat Prod Rep 20(1):119–136

Koes RE, Quattrocchio F, Mol JN (1994) The flavonoid biosynthetic pathway in plants: function and evolution. BioEssays 16(2):123–132

Kopcewicz J, Lewak S (2004) Fizjologia roślin. PWN, Warszawa

Kozlowska M, Konieczny G (2003) Biologia odpornosci roslin na patogeny i szkodniki. Wydawnictwo Akademii Rolniczej imienia A. Cieszkowskiego, Poznan

Król P, Kepczynska E (2008) Rola jasmonianow w indukowanej systemicznej odpornosci roslin przeciwko patogenom. Biotechnologia 1:122–135

Kurosaki F, Tashiro N, Nishi A (1986) Induction of chitinase and phenylalanine ammonia-lyase in cultured carrot cells treated with fungal mycelial walls. Plant Cell Physiol 27(8):1587–1591

Kutchan TM (2001) Ecological arsenal and developmental dispatcher. The paradigm of secondary metabolism. Plant Physiol 125(1):58–60

Laks PE, Mckaig PA, Hemingway RW (1988) Flavonoid biocides: wood preservatives based on condensed tannins. Holzforschung 42:299–306

Lambert C, Bisson J, Waffo-Téguo P, Papastamoulis Y, Richard T, Corio-Costet MF, Mérillon JM, Cluzet S (2012) Phenolics and their antifungal role in grapevine wood decay: focus on the Botryosphaeriaceae family. J Agric Food Chem 60(48):11859–11868

Lattanzio V, Di Venere D, Linsalata V, Lima G, Ippolito A, Salerno M (1996a) Postharvest Biol Technol 9(3):325

Lattanzio V, Cardinali A, Linsalata V, Perrino P, Ng NQ (1996b) Genet Resour Crop Evol 43(6): 493

Lattanzio V, Di Venere D, Linsalata V, Bertolini P, Ippolito A, Salerno M (2001) J Agric Food Chem 49(12):5817

Lattanzio V, Lattanzio VM, Cardinali A (2006) Role of phenolics in the resistance mechanisms of plants against fungal pathogens and insects. Phytochemistry 661(2):23–67

Lavania M, Chauhan PS, Chauhan SVS, Singh HB, Nautiyal CS (2006) Induction of plant defense enzymes and phenolics by treatment with plant growth–promoting rhizobacteria Serratia marcescens NBRI1213. Curr Microbiol 52(5):363–368

Lee S, LeTourneau DJ (1958) Phytopathology 48:268

Li CY, Lu KC, Nelson EE, Bollen WB, Trappe JM (1969) Effect of phenolic and other compounds on growth of Poria weirii in vitro. Microbios 3:305–311

Lily VG, Ramadasan A (1979) Changes in phenolic content of coconut leaf in relation to the development of leaf rot. Indian Phytopathol 32:112–113

Luo WH, Leng HQ (1990) Studies on resistance of pear varieties to pear scab (*Venturia Nashicola*). J Sichuan Agric Univ 8:141–146

Macheix J, Fleuriet A, Billot J (1990) Phenolic compounds in fruit processing. In: Macheix J, Fleuriet A, Billot J (eds) Fruit phenolics. CRC Press, Boca Raton, pp 239–312

Macheix JJ, Fleuriet A, Jay-Allemand C (2005) Les composés phénoliques des végétaux: un exemple de métabolites secondaires d'importance économique. PPUR presses polytechniques

Mansfield JW (1982) The role of phytoalexins in disease resistance. See ref 4, pp 253–288

Mansfield J (1990) Recognition and response in plant fungus interactions. In: Fraser RSS (ed) Recognition and response in plant virus interactions. Springer, Berlin, pp 31–52

Markakis EA, Tjamos SE, Antoniou PP, Roussos PA, Paplomatas EJ, Tjamos EC (2010) Phenolic responses of resistant and susceptible olive cultivars induced by defoliating and nondefoliating Verticillium dahliae pathotypes. Plant Dis 94(9):1156–1162

Martini H, Weidenborner M, Adams S, Kunz B (1997) Mycol Res 101(8):920

Matern U, Kneusel RE (1988) Phenolic compounds in plant disease resistance. Phytoparasitica 16: 153–170

Mayr U, Treutter D, Feucht W (1996) Phenolic compounds of apple fruit. Erwerbsobstbau 38:8–12

Morelló JR, Vuorela S, Romero MP, Motilva MJ, Heinonen M (2005) Antioxidant activity of olive pulp and olive oil phenolic compounds of the Arbequina cultivar. J Agric Food Chem 53(6): 2002–2008

Ndoumou OD, Ndzomo TG, Djocgoue PF (1996) Changes in carbohydrate, amino acid and phenol contents in cocoa pods from three clones after infection with Phytophthora megakarya Bra. and Grif. Ann Bot 77(2):153–158

Nicholson RL, Hammerschmidt R (1992) Phenolic compounds and their role in disease resistance. Annu Rev Phytopathol 30(1):369–389

Niekerk JM, Fourie PH, Hallenn F, Crous P (2006) Botryosphaeria spp. as grapevine trunk disease pathogens. Phytopathol Mediterr 45(4):43–54

Niemann GJ, van der Kerk A, Niessen WM, Versluis K (1991) Free and cell wall-bound phenolics and other constituents from healthy and fungus-infected carnation (Dianthus caryophyllus L.) stems. Physiol Mol Plant Pathol 38(6):417–432

Ortuño A, Botía JM, Fuster MD, Porras I, García-Lidón A, Del Río JA (1997) Effect of Scoparone (6,7-Dimethoxycoumarin) biosynthesis on the resistance of tangelo Nova, *Citrus paradisi* and *Citrus aurantium* fruits against *Phytophthora parasitica*. J Agric Food Chem 45:2740–2743

Overeem JC (1976) Biochemical aspects of plant parasite relationships. In: Friend J, Threlfall DR (eds) Academic, London, pp 195

Padmavati M, Sakyhivel N, Thara KV, Reddy AR (1997) Phytochemistry 46(3):499

Passardi F, Penel C, Dunand C (2004) Performing the paradoxical: how plant peroxidases modify the cell wall. Trends Plant Sci 9(11):534–540

Petkovšek MM, Usenik V, Štampar F (2003) The role of chlorogenic acid in the resistance of apples to apple scab (Venturia inaequalis (Cooke) G. Wind.Aderh.). Zbornik Biotechniške Fakultete Univerze v Ljubljani Kmetijstvo 81(2):233–242

Petkovšek MM, Stampar F, Veberic R (2008) Increased phenolic content in apple leaves infected with the apple scab pathogen. J Plant Pathol 90:49–55

Picinelli A, Dapena E, Mangas JJ (1995) Polyphenolic pattern in apple tree leaves in relation to scab resistance. A preliminary study. J Agric Food Chem 4:2273–2278

Raghavendra VB, Lokesh S, Govindappa M, VasanthKumar T (2007) Dravya—as an organic agent for the management of seed-borne fungi of sorghum and its role in the induction of defense enzymes. Pestic Biochem Physiol 89(3):190–197

Raven PH, Evert RF, Eichhorn SE, Bouharmont J (2003) Biologie végétale. Éditions De Boeck Université, Bruxelles

Reyes LF, Cisneros-Zevallos L (2003) Wounding stress increases the phenolic content and antioxidant capacity of purple-flesh potatoes (Solanum tuberosum L.). J Agric Food Chem 51(18): 5296–5300

Ride JP (1983) Cell wall and other structural barriers in defense. Biochem Plant Pathol, pp 215–236

Robbins MP, Bolwell GP, Dixon RA (1985) Metabolic changes in elicitor-treated bean cells: selectivity of enzyme induction in relation to phytoalexin accumulation. Eur J Biochem 148(3):563–569

Roby D, Toppan A, Esquerré-Tugayé MT (1985) Cell surfaces in plant-microorganism interactions: v. elicitors of fungal and of plant origin trigger the synthesis of ethylene and of cell wall hydroxyproline-rich glycoprotein in plants. Plant Physiol 77(3):700–704

Sahoo MR, Kole PH, Dasgupta M, Mukherjee A (2009) Changes in phenolics, polyphenol oxidase and its isoenzyme pattern in relation to resistance in taro against Phytophthora colocasiae. J Phytopathol 157:145–153

Saini KS, Ghosal S (1984) Phytochemistry 23(11):2415

Sallas L, Kainulainen P, Utriainen J, Holopainen T, Holopainen JK (2001) The influence of elevated O3 and CO2 concentrations on secondary metabolites of Scots pine (*Pinus sylvestris* L.) seedlings. Glob Change Biol 7:303–311. https://doi.org/10.1046/j.1365-2486.2001.00408.x

Schlösser E (1994) Preformed phenols as resistance factors. Acta Hortic 381:615–630

Schovankova J, Opatova H (2011) Changes in phenols composition and activity of phenylalanine-ammonia lyase in apples after fungal infections. Hortic Sci 38(1):1–10

Sharma SG, Narayan R, Lal S, Chaturvedi C (1983) Role of phenolic compounds in resistance of maize to leaf blight caused by Drechslera state of Cochliobolus heterostrophus. Indian Phytopathol

Sharma A, Joshi N, Sharma P, Sharma S, Maheshwari A (2011) HPLC determination of phenolic acids in infected plants of chickpea. Asian J Chem 23(9):3979

Shigo AL, Marx HG (1977) Compartmentalization of decay in trees. Agric Inf Bull 405. US Department of Agriculture, Forest Service, Washington. 73 p, 405, pp 1–73

Siranidou E, Kang Z, Buchenauer H (2002) Studies on symptom development, phenolic compounds and morphological defence responses in wheat cultivars differing in resistance to Fusarium head blight. J Phytopathol 150(4–5):200–208

Sivaprakasan K, Vidhyasekaran P (1993) Phenylalanine ammonia lyase gene for crop disease management. In: Vidhyasekaran P (ed) Genetic engineering, molecular biology and tissue culture for crop pest and disease management. Daya Publishing House, Delhi, pp 113–122

Skadhauge B, Thomsen K, von Wettstein D (1997) Hereditas 126:147

Swain T (1975) The flavonoids. In: Harborne JB, Mabry TJ, Mabry H (eds). Chapman & Hall, London, pp 1096

Taiz L, Zeiger E (2010) Plant physiology, 5th edn. Sinauer Assosiates, Sunderland

Tietjen KG, Hunkler D, Matern U (1983) Differential response of cultured parsley cells to elicitors from two non-pathogenic strains of fungi: 1. Identification of induced products as Coumarin derivatives. Eur J Biochem 131(2):401–407

Tomas-Barberan FA, Espín JC (2001) Phenolic compounds and related enzymes as determinants of quality in fruits and vegetables. J Sci Food Agric 81(9):853–876

Treutter D (2005) Significance of flavonoids in plant resistance and enhancement of their biosynthesis. Plant Biol 7(6):581–591

Treutter D, Feucht W (1990a) The pattern of flavan-3-ols in relation to scab resistance of apple cultivars. J Hortic Sci 65:511–517

Treutter D, Feucht W (1990b) Accumulation of flavan-3-ols in fungus-infected leaves of Rosaceae. Zeitschrift fürPflanzenkrankheiten und Pflanzenschutz 97:634–641

Troccoli L, Mugnai L, Surico G, Calamassi R, Mori B (2001) Phaeomoniella chlamydospora-grapevine interaction. Histochemical reactions to fungal infection [Vitis vinifera L.]. Phytopathol Mediterr 40:S400–S406

Valle E (1957) On anti-fungal factors in potato leaves. Acta Chem Scand 11(2):395–397

Vance CP, Kirk TK, Sherwood RT (1980) Lignification as a mechanism of disease resistance. Annu Rev Phytopathol 18(1):259–288

Walker JC, Stahmann MA (1955) Annu Rev Plant Physiol 6:351

Weidenbörner M, Hindorf H, Jha HC, Tsotsonos P (1990) Phytochemistry 29:1103

Williams EB, Kuc J (1969) Resistance in malus to *Venturia inaequalis*. Annu Rev Phytopathol 7: 223–246

Wilson CL, Wisniewski ME (1989) Annu Rev Phytopathol 27:425

Wink M (1997) Adv Bot Res 25

Yoruk R, Marshall MR (2003) Physicochemical properties and function of plant polyphenol oxidase: a review. J Food Biochem 27(5):361–422

Plant Phenolics Compounds and Stress Management: A Review

20

Azharuddin B. Daphedar, Salim Khan, Siddappa Kakkalamel, and Tarikere C. Taranath

Abstract

Plant polyphenols are naturally occurring bioactive compounds that are present in a variety of plant species. There are two major pathways, viz., shikimic acid or phenylpropanoid and the malonic acid pathway through which phenolic compounds are synthesized. Phenylalanine is a crucial compound that originates from the phenylpropanoid metabolites of plants. Secondary metabolites apparently act as plant resistance against (microorganisms including pathogens) the gauntlet of biotic and abiotic stresses. Several varieties of large known phenolic compounds have been synthesized by higher plants. The compounds of plant polyphenols generally have a limited response to certain chemicals such as tannins and phytoalexins, signal compounds like salicylic acid and flavonoids, ultraviolet screens like flavonoids, and structural polymers like lignin, and fascinates like carotenoids and flavonoids can act as antioxidants. The therapeutic potential of these compounds lessens the risk of life-threatening diseases like cardiovascular, diabetes mellitus, hypertension, obesity, and neurodegenerative epidemics. Hence, this chapter covers different sources of polyphenols, and their

A. B. Daphedar (✉)
Department of Botany, Anjuman Arts, Science and Commerce College, Vijayapura, Karnataka, India

S. Khan
Department of Botany and Microbiology, King Saud University, Riyadh, Saudi Arabia

S. Kakkalamel
Department of Botany, Davangere University, Davangere, Karnataka, India

T. C. Taranath
Environmental Biology Laboratory, P. G. Department of Studies in Botany, Karnataka University, Dharwad, Karnataka, India

R. Lone et al. (eds.), *Plant Phenolics in Biotic Stress Management*,
https://doi.org/10.1007/978-981-99-3334-1_20

chemical nature and its advantageous pharmacological properties have been reported.

Keywords

Phenolic compounds · Types · Plant defense system · Stress management · Biological activity

20.1 Introduction

The heterogeneous group of secondary metabolites known as phenolics consists of a wide array of molecules found in the plant kingdom (Haminiuk et al. 2012; Gugleva et al. 2021). According to the WHO (World Health Organization), around 21,000 higher plant species have been evaluated for medicinal uses throughout the world (Samantaray and Ngangkham 2017). It is estimated that more than 100,000 to 200,000 secondary metabolites exist in the world (Pereira et al. 2009; Tressera-Rimbau et al. 2017). Currently, more than 8000 different structures of plant phenolics have been identified (Tungmunnithum et al. 2018). Plant phenolic compounds, which predominantly feature hydroxy or hydroxyl functional groups, are largely utilized to mitigate the effects of high light, low temperatures, UV-B radiations, heavy metals, pathogen contagion, herbivore damage, and nutrient deficiency (Lattanzio 2013; Hamid et al. 2023).

Plant metabolites comprise a variety of small, distinct chemical compounds that are produced by different component of the plant in diverse forms. They were divided into two major groups including primary and secondary metabolites (Chikezie et al. 2015). The substances like fatty acids, proteins, carbohydrates and nucleic acids are called primary metabolites are essential for sustaining plant growth and development (Wu and Chappell 2008). The secondary metabolites are much more varied structurally and chemically than the primary metabolites. The specialized cells are not directly involved in the formation of compounds owing to its photosynthetic and respiratory metabolism. Plants require essential substances, referred to collectively as nutrients, in order to survive in the environment (Vuolo et al. 2019). Due to their greater participation, secondary metabolites are directly involved in morphological and physiological processes; hence, it is also known as phenolics reported by Działo et al. (2016). Most important human dietary and nutritional requirements are derived from polyphenols which are present in fruits, vegetables, and beverages. It has been observed that dietary polyphenols ameliorate brain functions via molecular signaling pathways (Gonzalez-Gallego et al. 2010).

Microorganisms produce distinct potent polyphenolic compounds. It has been observed that cryptogams (for instance bryophytes), are consistent producers of flavonoids and a wide range of polyphenols is found in vascular plants (Naikoo et al. 2019). Plant phenolics derived from various natural sources demonstrate a variety of beneficial properties, including antioxidant, antiallergic, anti-inflammatory, antihypertensive, anticarcinogenic, antiarthritic, cardioprotective, and antimicrobial activities (Dai and Mumper 2010; Bhuyan and Basu 2017). Recently, polyphenols have gained much scientific attention owing to their persuasive antioxidant and various beneficial effects on human health (Ganseen and Xu 2017). Polyphenols fabricated from plants are useful in curing diseases like Alzheimer's, atherosclerosis, cancer, cardiovascular (CVD), and diabetes mellitus (Del Rio et al. 2013; Lamuela-Raventós et al. 2014; Rahman Md. et al. 2022). Most pronounced phenolic compounds found in plants, for instance green tea, gingerol, resveratrol, curcumin, genistein, rosmarinic acid, silymarin, and apigenin, are mainly used in the treatment of both radiation therapy and chemotherapy (Wang et al. 2012). However, phenolic compounds and their beneficial activities are still unknown. Hence, this review aims to summarize phenolic compounds, strong potential biological activity and their beneficial effects to the human and environment.

20.2 Types of Plant Phenolics and Structure

Despite the significant variety of phenolic compounds, they can be divided into flavonoids and nonflavonoids (Fig. 20.1). Flavonoid compounds structurally comprise two aromatic rings associated with two bridge and three carbon atoms (C_6–C_3–

Fig. 20.1 Major classes of phenolic compounds

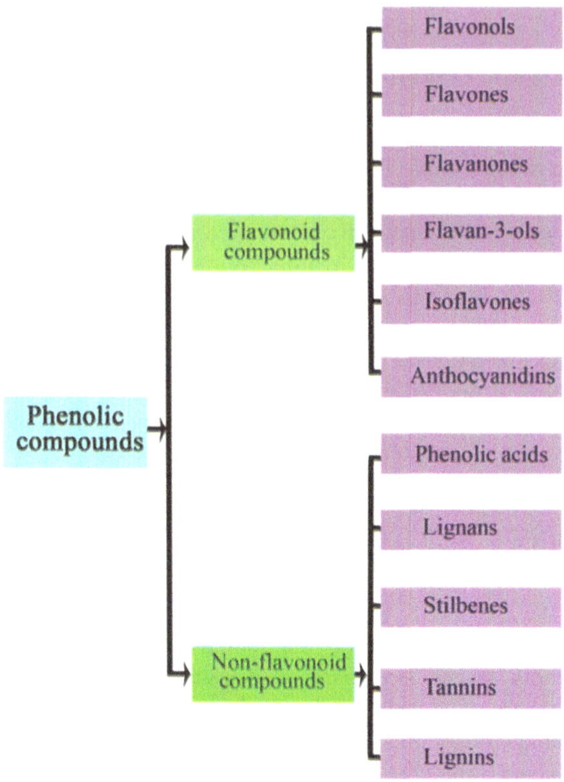

C_6) (Działo et al. 2016). Further, flavonoids are classified as six foremost subclasses, including flavonols, flavones, flavanones, flavan-3-ols, isoflavones, and anthocyanidins, whereas nonflavonoids into phenolic acids, lignans, stilbenes, tannins, and lignins.

20.2.1 Flavonols

Flavonols (polyphenols) are the largest class of flavonoids with ketone group, extensively distributed in the surface of leaves in higher plants. Moreover, about 450 diverse types of aglycones have been well documented so far in the higher plants. Quercetin and kaempferol are most studied flavonols and occur abundantly in a variety of vegetables, legumes, and fruits (Manach et al. 2004). The most abundant sources of flavonols are apple, blueberries, onion, broccoli, leeks, and sweet pepper (D'Archivio et al. 2007). According literature, plant-derived flavonols play vital role in the healing of infectious diseases like leukemia, breast, pancreatic, cervical, prostate, and urothelial cancers (Giovinazzo et al. 2020). The regular intake of flavonols reduces the risk of cancers by 10–60% (Batra and Sharma 2013). This protective activity results in inhibition of cellular proliferation and stimulates anti-oxidant defenses (Table 20.1).

20.2.2 Flavones

Flavones belonging to the class flavonoid; structurally contains C–2 and C–3 position double bonds, which is very akin to flavonols (Gutiérrez-Grijalva et al. 2020). Flavones are soluble in water and ethanol. The concentrations of flavones occur most abundantly in matured rice grain, mostly accumulated in rice embryo, and show a very stable profile during seed germination of rice (Galland and Rajjou 2015). Typically, these are present in plants as glycosides. Flavones exhibit several intervention enzymatic activities including catalase, superoxide dismutase, glutathione peroxidase, and glutathione reductase (Nielsen et al. 1999). Lakshmi et al. (2018) reported that *Alpinia galanga* (L.) extract possesses 5,7-dihydroxy flavones were used in the curing of ascetic lymphoma and human lung cancer which can lead to cell necrosis or cell death has been reported.

20.2.3 Flavan-3-ols

The flavonoids like Flavan-3-ols are consumed routinely in the US diet. The polyphenolic polymers are found in the different parts of plants like rhizomes, roots, leaves, stem bark, seed testa, and fruits (Hammerbacher et al. 2014). The richest of sources of flavon-3-ols are green tea, chocolate (cocoa), and apple ubiquitously present in plant foods. Nutritional and pharmacological properties of flavan-3-ols pose large variety of biological activities which includes antioxidant,

Table 20.1 Flavonoid and nonflavonoid plant phenolics and their biological activity

Classes of phenolic compounds	Major phytochemicals	Biological activity	References
Flavonoid compounds			
Flavonols	Quercetin, isorhamnetin, kaempferol, myricetin	Antioxidant and anticancer	Lea (2015)
Flavones	Baicalein, chrysin, eupalin, apigenin, luteolin	Antibacterial, antiallergic, antioxidant, antitumor	Verma and Pratap (2010)
Flavanones	Eriodictyol, hesperetin, naringenin	Antiviral, antimicrobial, and anti-inflammatory activities	Tomás-Navarro et al. (2014)
Flavan-3-ols	Epigallocatechin, (+)-gallocatechin, proanthocyanidins, theaflavins, thearubigins	Antifungal, antiviral, anti-inflammatory, anticancer, and antiangiogenic	Saito (2017)
Isoflavones	Daidzein, genistein, glycitein, biochanin A, formononetin	Estrogenic	Vitale et al. (2013)
Anthocyanidins	Cyaniding, delphinidin, malvidin	Antioxidant and anti-inflammatory	Blando et al. (2018)
Nonflavonoid compounds			
Phenolic acids	p-hydroxybenzoic acid, hydroxycinnamic acid	Antioxidant, anti-inflammation, antiulcer, anticancer, antidiabetic and hepatoprotective	Saibabu et al. (2015)
Lignans	Secoisolariciresinol, matairesinol, lariciresinol, pinoresinol	Antioxidant, antileishmanial, anticancer, anti-HIV-1 and anti-inflammatory	Zhang et al. (2014), Teponno et al. (2016)
Stilbenes	Resveratrol	Anticancer, anti-inflammation, Cardioprotection, neuroprotection, dipigmentation, and antidiabetic properties	Sirerol et al. (2016), Akinwumi et al. (2018)
Tannins	Geraniin, corilagin, Amarulone, Furosin, corilagin, melatonin, phyllanthus D (ellagitannin)	Antioxidant and antimicrobial	Maisetta et al. (2019)
Lignins	Hydroxyphenylpropane	Antioxidant, antimicrobial, antiradical, and antimutagenic	Espinoza-Acosta et al. (2016), Alzagameem et al. (2019)

antidiabetic, antiatherogenic, and anticarcinogenic effects (Song and Chun 2008) depicted in Table 20.1. The dietary intervention studies suggest that flavan-3-ols could directly affect metabolic and cell signaling pathways which leads to inhibition of cell proliferation, reticence of invasion, metastasis, and angiogenesis apoptosis (Silva and Costa 2014).

20.2.4 Isoflavones

Isoflavones are commonly known as plant bioactive nonsteroidal polyphenolic metabolites, found typically in vegetables, fruits, and other nuts (USDA 2008), which is considered as a part of human diet across many countries. Soy (*Glycine max* L.) food is the richest source of isoflavones, structurally they are very close with estradiol-17β and possess estrogenic receptor (ER) modulators (Kim 2021). Isoflavones can cause either estrogenic or antiestrogenic effects based on the hormone status of a person owing to binding of estrogen receptors (ER) in the body. In addition, isoflavones provide pre- or postmenopausal relief, in conjunction with preventing some types of cancer and protecting the heart diseases. Isoflavones exhibits disease curing splendid beneficial properties including antibacterial, antiviral, anticancer, antiosteoporotic, anti-inflammatory, and estrogenic as well (Table 20.1). Current evidence suggests that soy/isoflavone consumption up to three serve per day can regulate enzyme systems related to malignant activity and also induce apoptosis in cancer cells, without distributing event of the cell cycle.

20.2.5 Anthocyanidins

Anthocyanidins known for their brilliant colors (like blue, red or purple) occur naturally in food plants mainly vegetables, flowers, and fruits. Anthocyanidins have bright attractive shade, water solubility and are easily soluble into aqueous food system so they are the best potential replacement for synthetic colors. Flavylium cation (2-phenylbenzopyrilium) is basic chemical structure of anthocyanidins, which links to one or more sugars containing hydroxyl (−OH) and/or methoxyl ($-OCH_3$) groups presented in Fig. 20.2. These sugar-free molecules are counterparts of anthocyanidins. The carotenoids and polyphenols are two significant secondary plant metabolites obtained from various food sources which influence industrial processes as well as agricultural practices. Anthocynins are rich in antioxidants and other phenolics are also considered as major food supplementary, used in our day today life. On the other hand, their beneficial biological activities exhibit such as insulin secretion ability, anti-inflammatory, antimicrobial, anticarcinogenic, and neuroprotective activities (Abdin et al. 2020).

Based on their carbon skeleton, nonflavonoid phenolics can be categorized into following subclasses comprising phenolic acids, lignans, stilbenes, tannins, and lignins.

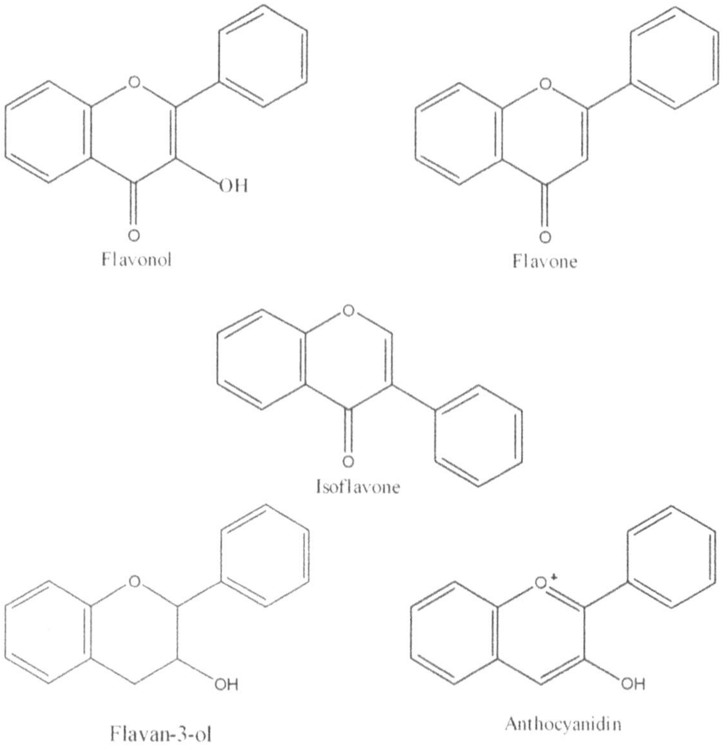

Fig. 20.2 Basic chemical structure of major classes of flavonoids

20.2.6 Phenolic Acids

Phenolic acids are polyphenolic nonflavonoid phytochemical compound occur naturally in plant imitative foods for instance, berries, toadstools, kiwis, karonda, apples, pears, chicory, potatoes, and coffee (Heleno et al. 2015). Phenolic acids are classified based on their biosynthetic routes into two classes which include hydroxybenzoic and hydroxycinnamic acids (Fig. 20.3). These are hydroxy derivatives of aromatic carboxylic acids: benzoic acids (with 7 carbon atoms (C_6–C_1)) and cinnamic acids (with 9 carbon atoms (C_6–C_3)) reported by Khadem and Marles (2010). Moreover, *p*-hydroxybenzoic, protocatechuic, vanillic, and syringic acids are the examples of hydroxybenzoic acids (Kumar and Goel 2019), whereas sinapic, ferulic, and caffeic acids are the examples of hydroxycinnamic acids (Bento-Silva et al. 2019). Phenolic acids derived from two aromatic amino acids such as L-phenylalanine and/or L-tyrosine which are responsible for biosynthesis via shikimate/phenylpropanoid pathway that occurs ubiquitously in plants. Although some studies has shown high potent antioxidant anticancer, anti-inflammatory, antimicrobial, antidiabetic, anticholesterolemic, antimutagenic, and antihypertensive properties (Shin et al. 2018; Al-Jitan et al. 2018).

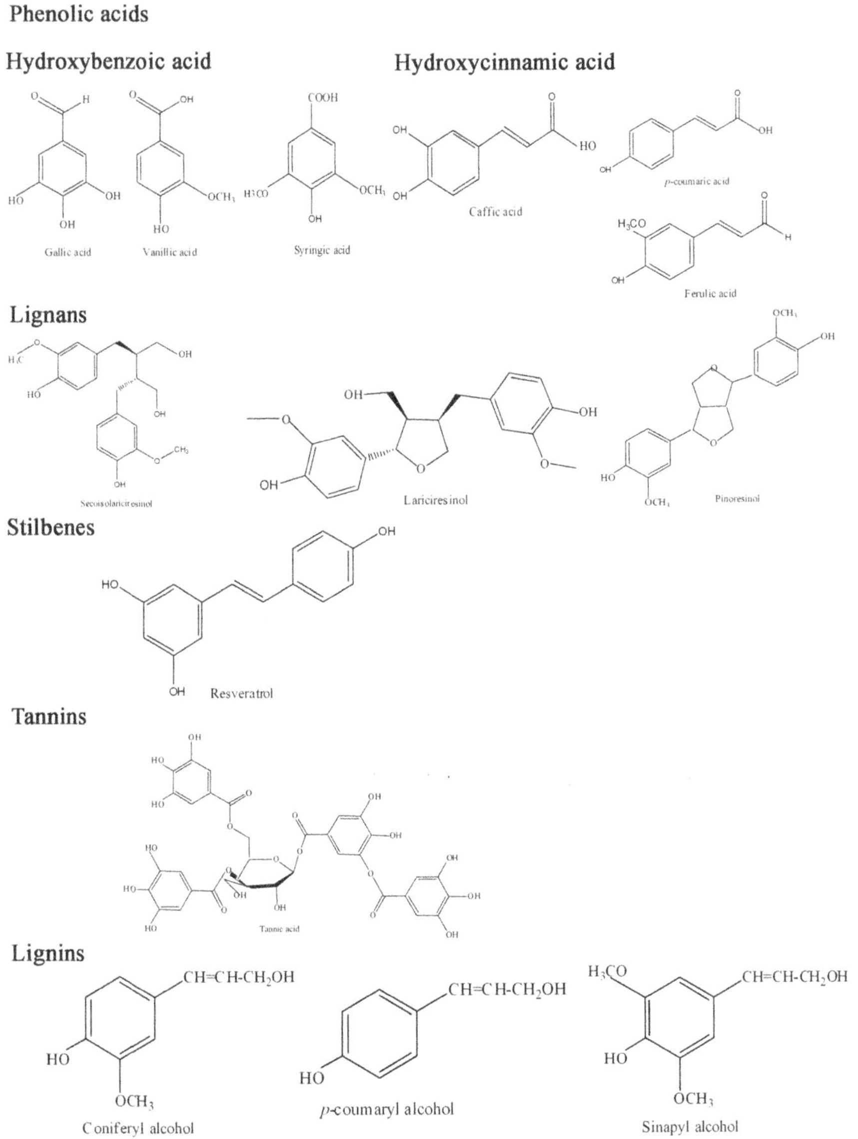

Fig. 20.3 Basic molecular structures of nonflavonoid phenolic compounds

20.2.7 Lignans

Lignans are polyphenols widely distributed in diverse plant-mediated groceries such as fruits, vegetables, seaweed, flax seeds, legumes, and whole grains (Meagher and Beecher 2000). Basically their chemical structure comprises two phenylpropane C_6–

C_3 units corresponding to carbon–carbon link between the central atoms of the relevant side chains (position 8 or β) (Fig. 20.3), which mainly derived from biosynthetic pathway of shikimic acid (Durazzo et al. 2019). Lignan acts as defense molecules against various pathogenic microbes due of their rich antioxidant properties. Other than antioxidant properties, lignin also exhibits anti-inflammatory, antituberculosis, and antitumor properties (Ionkova 2011). Additionally, epidemiological and physiological studies have shown positive impact in the averting of lifestyle disorders for instance diabetes type II diseases, cardiovascular and cancers diseases (Table 20.1) (Rodríguez-Garcia et al. 2019).

20.2.8 Stilbenes

Stilbenes are well-known secondary metabolites of nonflavonoid groups that are derived from the phenylpropanoids pathway. The structure of stilbenes (1,2-diphenylethylene) backbone containing C_6–C_2–C_6 with two aromatic rings linked by a methylene bridge (Maru et al. 2014). Further, stilbenes are classified into monomeric and oligomeric stilbenes. Phytoalexins produced by plants that are typically biofabricated only in response to disease infection (viral pathogens) and stressors. The stilbenes are isolated from different plant species including grapes, berry, pine, peanut, soya, sorghum and other plant sources (Kristina et al. 2017). However, due to their bioactive properties, plant stilbenes are used in the inhibition of various life-threatening diseases and also extend the existence of variety of organisms (Baur and Sinclair 2006; Chong et al. 2009). Resveratrol (3,4′,5-trihydroxystil-bene) is one of the most widely studied stilbene which has been used for various therapeutic purposes such as antioxidative, anticarcinogenic, phytoestrogenic, and cardioprotective activities (Kaushik et al. 2015; Pathak et al. 2018).

20.2.9 Tannins

Tannins are also called tannic acid, derived from polyphenols typically soluble in water. The major sources of tannins predominantly found in plant components like seeds, stem bark, wood, foliage, fruit skins (rind), rhizomes and even the galls of plants (Singh and Kumar 2019). Tannins containing metabolites can be divided into two major groups such as hydrolysable and nonhydrolyzable (or condensed tannins). The hydrolysable tannins are responsible for gallic acid and hexahydroxydiphenyl moieties, both are the subunits of ellagitannins. In other words, proanthocyanidins are also called condensed tannins which comprises dimmers, oligomers of flavan-3-ol (catechin monomers) and/or flavan-3,4-dio (catechin polymers), usually linked by C–C bonds (Smeriglio et al. 2017). Maximum content of tannins were found in unripe fruits and level decline when fruits mature and ripen. There astringent taste affects directly on nutritional quality, palatability and also reduces digestibility (Lattanzio 2013). The immune modulatory, cardio-protective, anticancer, antioxidant, and antithrombotic properties of tannins enhances several health benefits of human health and other living organisms (Sieniawska 2015; Smeriglio et al. 2017).

20.2.10 Lignins

Lignin [(C_6–C_3) n] is belongs to the class polymeric phenols. It is one of the major organic polymers found in the structure of plants. Structurally they made up of numerous complicated materials like monolignol (coumaryl, coniferyl, and sinapyl alcohol) and lignin (dimmers of monolignols) units (Fig. 20.3). Generally, they found in several integral parts of the plants (such as root, stem, leaves), and also all vascular plants for instance herbaceous species. Coniferyl alcohol and *p*-coumaryl alcohol lignin were derived from gymnosperms, whereas sinapyl alcohols and coniferyl from angiosperms respectively. The content of lignin roughly present more or less in equal proportions of both gymnosperm and angiosperm plant species reported by Vanholme et al. (2010). The most significant functions of lignin in the plant cell membrane is to envelop structural support, minimize the evaporation of water, allow transport (of organic and inorganic solvents) and also protect plant tissues from invasion by pathogenic microorganisms (Lu et al. 2017).

20.3 Role of Phenolics and Plant Defense System

Studies on plant phenolic compounds, such as antibiotics, and their diverse mechanisms of action have been extensively investigated. The phenolics are involved in protecting plants from pathogens which attacks them and cause diseases. When the disease becomes a pandemic, it spreads rapidly in a given population in various communities. The biological or biocontrol methods have been effectively used against a number of pest diseases and chemical pesticides. In addition, plant phenolic compounds perform dual function against the repulsion and attraction of various organisms in the plant's milieu. For instance, phenolics help in treating various organisms' such as herbivores, phytophagous insects, nematodes, and microbial pathogens against pesticides and natural animal toxicants (Bhattacharya et al. 2010). Usually, phenolics are accumulating in various parts of the plant mainly central vacuole of guard cells and subepidermal tissues of leaves and shoots (Cle et al. 2008). Certain bioactive compounds occur on the external surface of plant organs as well as cell membrane, which are covalently linked to phenolic compounds (Hutzler et al. 1998).Waxes are water repellents, which reduces the menace of infections, while cell wall reinforcement, reduction of lignin, and suberin content, which act as a cell wall barrier against fungal pathogens (Daayf et al. 2012).

On the contrary, different stages of plant tissues were phenols penetrating mainly pigmentation, development, reproduction, growth, resistance to pathogens, and many other functions. The hyper accumulation of phenolics is more in noninfected plant tissues, as preformed antimicrobial compounds, which impede the growth of

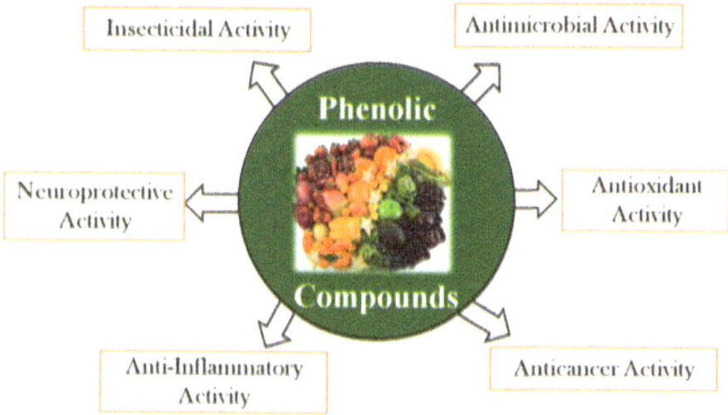

Fig. 20.4 Biological activities of plant phenolics

fungi and helps to frontier pathogen attachment, invasion and infection. The absence of compounds could facilitate the establishment of pathogens, which are considered to be a major defense mechanism against resistance. Resistance mechanism may lead to recognition results into complex signaling reactions that trigger a wide range of biochemical defense responses. Some genes express specifically during transcriptome analysis which identifies compatible and incompatible interactions in higher plants. Sheng and Citovsky (1996) reported that (*virA/G)* genes of Ti plasmid affects directly or indirectly on phenolic compounds of different *Agarobacterium* species. For example, fungus *Magnaporthe oryzae* as biotrophy-associated secreted (BAS1–4) proteins exhibits distinct compatible and incompatible patterns during interaction of transcriptome analysis in rice blast diseases (Mosquera et al. 2009). Certain phenolic compounds are toxic to the majority of microorganisms; however, certain pathogens can counteract or nullify these defenses and even use them to their own advantage (Fig. 20.4). Plant-derived phenolic compounds present compelling opportunities for pharmaceutical and biomedical applications like antibacterial, anti-inflammation, antioxidants, and cardioprotective, and promote immune system and skin protection against UV rays (Tungmunnithum et al. 2018; Emus-Medina et al. 2023).

Inflammation is an intricate biological route that often occurs with ache redness, swelling, heat, and loss of functions. This is an increase in protein denaturation, vascular permeability, and alteration of membrane (Ruiz-Ruiz et al. 2017). Once inflammatory cells become injured, they release prostaglandins, leukotrienes, histamine, and kinins which affect the immune system. According to the report of Abdulkhaleq et al. (2018), inflammatory cells release specialized compounds such as vasoactive amines and peptides, which act as chemical mediators that draw natural defense cells of the body to the area. This phenomenon is also known as chemotaxis. Much attention has been given to plant-mediated natural phenolic and flavonoid compounds for their potential to prevent illnesses associated with oxidative stress (Alvarez-Suarez et al. 2013). The primary function of these compounds is to protect various macromolecules (like amino acids, fats and nucleic acids) against oxidative DNA damage and oxidative stress, caused by free radicals generated from biochemical reactions (Abbas et al. 2014). For instance, phenolics and flavonoids isolated from honey *Melipona beecheii* exhibit greater antioxidant and anti-inflammatory activities by reducing lipid peroxidation, inhibiting free radicals, inhibiting protease, inhibiting protein denaturation, and performing membrane stabilization assay (Ruiz-Ruiz et al. 2017). Biomolecules like proteins, carbohydrates, lipids and DNA damaged by strong oxidizing species under stress conditions resulting in generation of reactive oxygen species (ROS) which is commonly produced by proliferation of oxygen containing radicals like OH· and nonradicals like H_2O_2 (Sharma et al. 2012; Hajam et al. 2023). Several plant species are rich content of antioxidants, which are present in the form of both enzymatic and nonenzymatic reactions. Superoxide dismutase (SOD) and catalase peroxidase are the examples of enzymatic antioxidants whereas glutathione-S-transferase and ascorbate peroxidase are the nonenzymatic reactions (Thakur and Sohal 2013). These oxidative enzymes play significant role in protecting cells against oxidative damage at the sites of increased ROS production.

Polyphenols, found abundantly in plant products like fruits, vegetables, nuts, and whole cereals, are the major sources of fiber, trace metals, vitamins, and essential oils that play an extremely vital role in the inhibition of degenerative disorders like cancer and many more. Plants have been very helpful in synthesizing vast range of secondary metabolites which are effective against cancer cells. Additionally, secondary metabolites which have been derived from more than 60% of the plant extract were used for anticancer properties like antiproliferation and the apoptotic cell death activity. There is plenty of evidence to suggest that the edible herb of *Piper sarmentosum* used as spices, which have the potential to reduce NSCLC (nonsmall-cell lung carcinoma) cells as well as cell viability reported by Yong et al. (2013). According to Epifano et al. (2013), anticancer properties like inhibition of the growth of both wild-type and chemo-resistant colon NSCLC cells and prevention of the recurrence of cancer stem cells has been done by Auraptene (7-geranyloxycoumarin, AUR) compound which is isolated from Rutaceae family. For example, *Curcuma longa* and other species of *Curcuma* have exhibit anticancer therapeutic properties which have been mainly derived from secondary metabolites. The growth suppression of HCT-116 tumor xenografts and permanence of SIRT1 (an NAD+-dependent histone/protein deacetylase) is analyzed by cancer cells resulting inhibition of oncogenicity of human colon NSCLC cells has been reported (Lestari and Indrayanto 2014; Lee et al. 2018).

20.4 Interaction of Phenolic Compounds to Various Food Components in the Cell

Phenolics and flavonoids are by-products of hydroxycinnamic acid occur commonly in edible plants as simple phenols these are called bio-mediated bioactive compounds. Generally, healthy unprocessed foods are rich content of nutrition, vitamins and proteins when compared to processed foods. During food/beverage processing, disruption of tissues and cellular breakdown causes the release of intercellular and extracellular metabolites that were initially inadequate. The phenolic compounds contain enzymatic browning reactions catalyzed by polyphenoloxidase that take place in majority of the fruits, vegetables, and their other products resulting change in color, texture, and loss of nutrients during processing. For instance, polyphenol oxidase content of strawberry juices used in browning of grapes whereas, catechin is required for browning of yams (Moon et al. 2020; Salazar-Orbea et al. 2021; Daniels et al. 2021). The browning of fruits/ beverages certain plant derived antioxidants have been used such as cysteine hydrochloride, N-acetyl cysteine (NAC), hexylresorcinol, glutathione, ascorbic acid, and erythorbic acid. Whereas, in food and beverage industries, sulfides were used as an antibrowning agents. Ozturk and Hakeem (2019) have reported in his studies that the sulfating agents pose risk to the human health and FDA prohibits sulfites from being used in certain food industries. For instance, plant-fabricated polyphenols are used in cocoa fermentation process.

It has been observed that the phenolic compounds in Black Tea (BT) contribute to the presence of pigments due to theaflavins (TF) which produce a reddish orange pigment and thearubigins (TB), which bring forth brown pigments. The oxidation of process of BT causes certain polyphenolic compound like polyphenol oxidase dependent oxidative polymerization during crushed tea leaves (Jiang et al. 2018). Analysis of numerous food products' taste through the use of phenolic compounds is considered to be a powerful precursor to determine aroma, flavor, and oxygen stability. Similarly, Neto and Vinson (2010) reported that the associated bitterness in the cranberries fruits is due to the presence of hydroxycinnamic acids, phenolics, and other derivatives. Further, phenolic substance also contributes one-third of the flavor of vanilla, which is most abundantly extracted from the vanilla pods. The vanillins are simple phenolics like p-hydroxybenzaldehyde, guaiacol, and anise alcohol (Sinha et al. 2008). Of all the flavors, the top flavor of ice cream is vanilla which is used to prepare confectioneries items like Chocó's, soft cakes, beverages, liqueurs, and pharmaceutical products and also in nutraceuticals (Menon and Nayeem 2013). Synthetic fruit extracts such as vanillin and ethyl vanillin are used for chocolate flavorings, while eugenol, methyleugenol, and elimicin come from ripened banana extracts; meanwhile, ethyl salicylic, methyl cinnamic, and ethyl benzoic acids are derived from strawberry extracts. Hence, the phenolic compounds play vital role for giving flavor in various spices and herbs because of some chemical constituents such as eugenol, carvacrol, anethole, estragole, and thymol. Despite Eugenol being one of the finest flavoring compounds found in essential oils, it is extracted from the bark or leaf oil of cinnamon (Khalil et al. 2017). The polyethylene glycols (PEG) and phenolic compounds can cause the energy associated in phenol with PEG resulting in incompatibility during drug preparations, which controls the active concentration of chemical and physical stability as well as drug bioavailability properties (Fuchs and Rupperecht 1983). Recently, Shahidi et al. (2019) suggest fruit by-products have higher antioxidant activity, which is attributed to their rich sources of carotenoids, ascorbic acid, and polyphenolic compounds. Additionally, apple extracts suppress the antiproliferative activity of human breast cancer cells reported by Yang and Liu (Yang and Liu 2009). The inhibition of LDL in vitro oxidation process is only possible by synergistically of soya and alfalfa interactions with phytoestrogen and *acerola* cherry extracts (Hwang et al. 2001). Parker et al.

(2010) demonstrated that ORAC is enhanced due to the interaction of certain compounds with the mixtures of ascorbic acid, p-coumaric acid, abscisic acid, rutin, and mixture of monomers analyzed by using electron paramagnetic resonance (EPR). Besides, synergistic interaction of phenolics found to be more effective in food/beverage storages, against microbial growth as well as in other oxidative conditions has been reported by Freeman et al. (2010).

Phenolics are natural antioxidants that have been derived from diverse plant by-products. In reality, natural phenolics can assist in the curing of life-threatening illnesses which include chronic, heart diseases, brain dysfunction, mutagenesis, and carcinogenesis (Ho 1992; Zhu 2015; Vazquez-Olivo et al. 2023). On the other hand, antioxidation, antiproliferation, and anti-inflammation activities have been corresponding by phenolic compounds (Velderrain-Rodríguez et al. 2014). The phenolic compounds have anticancer properties that comprise lignans, acid phenolics, quinines, coumarins, curcuminoids, flavonoids, and others. Phenolic compounds possess a vast array of bioactivities and are employed to induce cell death by disrupting cell cycle progression. The carcinogen regulates the expression of ontogenesis resulting migration, proliferation, and blocking signaling pathways (Huang et al. 2010).

20.5 Plant Phenolics and Signaling

The production of metabolites is augmented when the phytochemical content of plants is increased. Adaptive ability enables plant life to cope with challenging and changing conditions, allowing it to survive in stressful environments. They may involve in construction of all types of complex chemical compounds and responsible for the affinity and stability of the interaction by signaling functions (Isah 2019). The biosynthesis of secondary metabolites from primary metabolites and their accumulation in plant tissues can be stimulated under in-vitro conditions using elicitors. Cell cultures are treated with biotic and abiotic factors via signaling pathways in order to induce this process (Narayani and Srivastava 2018). Furthermore, light-derived metabolic reprogramming has a major role in modulating phytochemicals through signaling pathways (Bain et al. 2016). Compounds that are produced by different acid pathways, and methyl erythritol phosphate pathway and the synthesis of lignin, suberin, tannins, flavonoids, and coumarins are biosynthesized by phenolic compounds (Saltveit 2009). Phenolics protect against stress responses and damage caused by Ultraviolet-B (UV-B) irradiation (Radhiga et al. 2016), lignin enhances mechanical strength (Liu et al. 2018), suberin and cutin produced by epidermal cells facilitate the minimization of water and solute loss, as well as enable the control of gaseous exchange when stomata are either closed or open (Nawrath 2002).

Abscisic acid (ABA) commonly known as "stress hormone" which implicated in phenolic compounds accretion, particularly in the regulation of anthocyanin biosynthesis in leaves of *Aristotelia chilensis* plant (mol) an endemic berry in Chile under drought stress (González-Villagra et al. 2018). The ABA regulates opening and closing of stomata under stress tolerance through signal transduction pathway.

Additionally, phenolic compounds are divided into phenolic acids and their derivatives made by plants. Phenolic acid in concentration with plant microbes via shikimic acid through the phenylpropanoid pathway plays multifunctional roles in rhizospheres. The agglomeration of acid phenolics in foliage of transgenic *Nicotina tobaccum* plants causes morphological changes when the transcription factor is over expressed (Tamagnone et al. 1998). Another study reveals increased production of lipid peroxidase and malondialdehyde is the result of growth of chlorogenic acid (CGA) in tobacco (*Nicotina tabacuma*) which also leads to rapid death of mature leaves of tobacco plant (Tamagnone et al. 1998). Recently, Ma et al. (2017) reported that CGA may be useful for slowing down the development of diseases in apple fruits after infection by *Botrytis cinerea.*

Interaction/symbiosis of plant–microbes play dynamic role in plants which are derived from bioactive compounds that act as signaling molecules. For example, arbuscular mycorrhizal (AM) fungi initiate legume rhizobia within the root nodule which will act as defensive agents in the plant cells (Mandal et al. 2010). *Rhizobium* spp. involves regulation of nodule initiation, differentiation, structure, and functioning by symbiosis which requires exchange of signal by diverse molecules between the infecting microbes and plants; hence, symbiosis entails signaling and detection by both the partners (Wang et al. 2018). Inhibition of nod gene expression causes chemoattraction of rhizobia, inhibition of pathogens, enhancement of mycorrhizal spore germination, and regulation of phytoalexin resistance in rhizobia; all these mechanisms are done by flavonoids which depend on the structure to function properly in the rhizosphere (Cooper 2004). For the induction of the desired pathways, rhizosphere undergoes regulation of the flavonoid secretion and alteration of expression to rhizosphere signals, hence resulting in manipulation of flavonoids (Hassan and Mathesius 2012). Synthesis of nod factor also called a signal, which is lipo-chito-oligosaccharidic is necessary for initiating most legumes in symbiotic development. But due to decreasing nitrogen conditions legume root nodules secrete a mixture of flavonoid compound which helps to activate the bacterial nodulation (nod) and gene expression (Oldroyd et al. 2011). Phenolic substances are often regulating an invading pathogen, impose stress and also provide developmental signals detected by fungi. Accumulation of benzoic acid, cinnamic acid, thymol, salicylic acid, and dihydroxybenzaldehydes leads to inhibition of growth of several *Candida sps* and *C. neoformans* (Shalaby and Horwitz 2014). These are present in some phenolic compounds as well as in flavonoids which act as antifungal in nature. Previous study by Vicente and Boscaiu (2018) reported that the flavonoids serve as signaling molecules in interaction between plants and microbes.

During abiotic stress, phenolic compounds play a critical role in protecting plant cells from oxidative damage caused by reactive oxygen species (ROS) through their scavenging activity. Additionally, the generation of ROS triggers the expression of a number of genes that control development processes such as cell expansion, systematic signaling, and pathogen defense; these, in turn, lead to apoptosis or cell necrosis. Narayanana et al. (1999) mentioned in their studies that ellagic acid arrests events of cell cycle, hinders cell proliferation, and stimulates cervical carcinoma cell necrosis in a dose- and duration-dependent manner. Numerous diseases like osteoporosis, heart-related disease, and neurodegenerative diseases offer protection and have been benefitted as dietary antioxidants (Pandey and Rizvi 2009). Therefore, phenolic compounds have a substantial role in protecting and inhibiting the growth of A549 and H1299 human lung cancer cell lines via *Curcumin* in a time- and concentration-dependent manner, leading to either apoptosis or necrosis reported by Pillai et al. (2004).

20.6 Conclusions

In recent times, plant phenolics have gained attention owing to their low toxicity, and its natural origin has been elucidated. Current scientific data suggest that the regular consumption of plants in high amounts including fruits, vegetable, and beverages could potentially reduce the risk of life threatened diseases. Plants are subjected to a wide range of environmental stresses such as wounding, pathogen attack, mineral deficiencies, and temperature stress which reduces and limits the productivity of plants. Moreover, plant secondary metabolites or polyphenols have been extensively reported for its antimicrobial, anticancer, insecticidal, and neuroprotective activities.

In this chapter, we discussed the main classes of plant phenolic compounds, their plant secondary metabolites produced via shikimic acid pathways. Additionally, phenolics derived from plant extract are known to serve as detoxification agents for phytoalexins and interfere with signal regulatory pathway as well as defense-related metabolites have been reported. Overall, from the observation of this theme, it can be concluded that the plant-fabricated natural metabolites play a major role in alleviating several aliments in the human health system.

Acknowledgments The author Dr. Azharuddin Daphedar is thankful to the HOD of Botany and Principal, Anjuman Degree College, Vijayapura, Karnataka, India, for their support.

References

Abbas ZK, Saggu S, Sakeran MI, Zidan N, Rehman H, Ansari AA (2014) Phytochemical, antioxidant and mineral composition of hydroalcoholic extract of chicory (*Cichorium intybus* L.) leaves. Saudi J Biol Sci 22:322–326

Abdin M, Hamed YS, Akhtar HMS, Chen D, Chen G, Wan P, Zeng X (2020) Antioxidant and anti-inflammatory activities of target anthocyanins di-glucosides isolated from *Syzygium cumini* pulp by high speed counter-current chromatography. J Food Biochem 44:1050. https://doi.org/10.1111/jfbc.13209

Abdulkhaleq LA, Assi MA, Abdullah R, Zamri-Saad M, Taufiq-Yap YH, Hezmee MNM (2018) The crucial roles of inflammatory mediators in inflammation: a review. Vet World 11(5):627–635

Akinwumi BC, Bordun KM, Anderson HD (2018) Biological activities of stilbenoids. Int J Mol Sci 19(3):792

Al-Jitan S, Alkhoori SA, Yousef LF (2018) Phenolic acids from plants: extraction and application to human health. Stud Nat Prod Chem 58:389–417

Alvarez-Suarez JM, Giampieri F, Battino M (2013) Honey as a source of dietary antioxidants: structures, bioavailability and evidence of protective effects against human chronic diseases. Curr Med Chem 30(5):621–638

Alzagameem A, Klein SE, Bergs M, Do XT, Korte I, Dohlen S, Huwe C, Kreyenschmidt J, Kamm B, Larkins M, Schulze M (2019) Antimicrobial activity of lignin and lignin-derived cellulose and chitosan composites against selected pathogenic and spoilage microorganisms. Polymers 11:670

Bain P, Milfont T, Kashima Y, Bilewicz M, Doron G, Garðarsdóttir RB, Gouveia VV, Guan Y, Johansson LO, Pasquali C, Corral-Verdugo C, Aragones JI, Utsugi A, Demarque C, Otto S, Park J, Soland M, Steg L, González R, Lebedeva N, Madsen OJ, Wagner C, Akotia CS, Kurz T, Saiz JL, Schultz PW, Einarsdóttir G, Saviolidis NM (2016) Co-benefits of addressing climate change can motivate action around the world. Nat Clim Chang 6:154–157

Batra P, Sharma AK (2013) Anti-cancer potential of flavonoids: recent trends and future perspectives. 3 Biotech 3(6):439–459

Baur JA, Sinclair DA (2006) Therapeutic potential of resveratrol: the *in vivo* evidence. Nat Rev Drug Discov 5:493–506

Bento-Silva A, Koistinen VM, Mena P, Bronze MR, Hanhineva K, Sahlstrom S, Kitryte V, Moco S, Aura AM (2019) Factors affecting intake, metabolism and health benefits of phenolic acids: do we understand individual variability? Eur J Nutr 59:1275. https://doi.org/10.1007/s00394-019-01987-6

Bhattacharya A, Sood P, Citovsky V (2010) The roles of plant phenolics in defense and communication during *Agrobacterium* and *Rhizobium* infection. Mol Plant Pathol 11(5):705–719

Bhuyan DJ, Basu A (2017) Phenolic compounds potential health benefits and toxicity. Utilization of bioactive compounds from agricultural and food production waste. 33. isbn:9781315151540

Blando F, Calabriso N, Berland H, Maiorano G, Gerardi C, Carluccio MA, Andersen OM (2018) Radical scavenging and anti-inflammatory activities of representative anthocyanin groupings from pigment-rich fruits and vegetables. Int J Mol Sci 19:169

Chikezie PC, Ibegbulem CO, Mbagwu FN (2015) Bioactive principles from medicinal plants. Res J Phytochem 9(3):88–115. https://doi.org/10.3923/rjphyto.2015.88.115

Chong J, Poutaraud A, Hugueney P (2009) Metabolism and roles of stilbenes in plants. Plant Sci 177(3):143–155

Cle C, Hill LM, Niggeweg R, Martin CR, Guisez Y, Prinsen E, Jansen MAK (2008) Modulation of chlorogenic acid biosynthesis in *Solanum lycopersicum*; consequences for phenolic accumulation and UV-tolerance. Phytochemistry 69:2149–2156

Cooper J (2004) Multiple responses of *rhizobia* to flavonoids during legume root infection. Adv Bot Res 41:1–62

D'Archivio M, Filesi C, Di Benedetto R, Gargiulo R, Giovannini C, Masella R (2007) Polyphenols, dietary sources and bioavailability. Ann Ist Super Sanita 43(4):348–361

Daayf F, El Hadrami A, El-Bebany AF, Henriquez MA, Yao Z, Derksen H, El-Hadrami I, Adam LR (2012) Phenolic compounds in plant defense and pathogen counter-defense mechanisms. In: Recent advances in polyphenol research, pp 191–208

Dai J, Mumper RJ (2010) Plant phenolics: extraction, analysis and their antioxidant and anticancer properties. Molecules 15(10):7313–7352

Daniels AJ, Poblete-Echeverría C, Nieuwoudt HH, Botha N, Opara UL (2021) Classification of browning on intact table grape bunches using near-infrared spectroscopy coupled with partial least squares-discriminant analysis and artificial neural networks. Front Plant Sci 12:768046

Del Rio D, Rodriguez-Mateos A, Spencer JPE, Tognolini M, Borges G, Crozier A (2013) Dietary (poly)phenolics in human health: structures, bioavailability, and evidence of protective effects against chronic diseases. Antioxid Redox Signal 18(14):1818–1892

Durazzo A, Lucarini M, Souto EB, Cicala C, Caiazzo E, Izzo AA, Novellino E, Santini A (2019) Polyphenols: a concise overview on the chemistry, occurrence, and human health. Phytother Res 33:2221–2243

Działo M, Mierziak J, Korzun U, Preisner M, Szopa J, Kulma A (2016) The potential of plant phenolics in prevention and therapy of skin disorders. Int J Mol Sci 17:160

Emus-Medina A, Contreras-Angulo LA, Ambriz-Perez DL, Vazquez-Olivo G, Heredia JB (2023) UV light stress induces phenolic compounds in plants. In: Plant phenolics in abiotic stress management. https://doi.org/10.1007/978-981-19-6426-8_19

Epifano F, Genovese S, Miller R, Majumdar AP (2013) Auraptene and its effects on the re-emergence of colon cancer stem cells. Phytother Res 27:784–786

Espinoza-Acosta JL, Torres-Chávez PI, Ramírez-Wong B, López-Saiz CM, Montaño-Leyva B (2016) Antioxidant, antimicrobial, and antimutagenic properties of technical lignins and their applications. BioRes 11(2):5452–5481

Freeman BL, Eggett DL, Parker TL (2010) Synergistic and antagonistic interactions of phenolic compounds found in navel oranges. J Food Sci 75(6):570–576

Fuchs G, Rupperecht H (1983) Interaction of phenolic compounds with polyethylene glycols. Colloids Surf 6:175–187

Galland M, Rajjou L (2015) Regulation of mRNA translation controls seed germination and is critical for seedling vigor. Front Plant Sci 6:284

Ganseen K, Xu B (2017) Polyphenol-rich dry common beans (Phaseolus vulgaris L.) and their health benefits. Int J Mol Sci 18(11):2331

Giovinazzo JA, Thomson RP, Khalizova N, Zager PJ, Malani N, Rodriguez-Boulan E, Raper J, Schreiner R (2020) Apolipoprotein L-1 renal risk variants form active channels at the plasma membrane driving cytotoxicity. Elife 9:e51185. https://doi.org/10.7554/eLife.51185

Gonzalez-Gallego J, Garcia-Mediavilla MV, Sanchez-Campos S, Tunon MJ (2010) Fruit polyphenols, immunity and inflammation. Br J Nutr 104(3):15–27

González-Villagra J, Cohen JD, Reyes-Díaz M (2018) Abscisic acid (ABA) is involved in phenolic compounds biosynthesis, mainly anthocyanins, in leaves of *Aristotelia chilensis* plants (Mol.) subjected to drought stress. Physiol Plant 165:855–866. https://doi.org/10.1111/ppl.12789

Gugleva V, Ivanova N, Sotirova Y, Andonova V (2021) Dermal drug delivery of phytochemicals with phenolic structure via lipid-based nanotechnologies. Pharmaceuticals 14:837. https://doi.org/10.3390/ph14090837

Gutiérrez-Grijalva EP, Santos-Zea L, Ambriz-Pérez DL, López-Martínez LX, Heredia JB (2020) Flavones and flavonols: bioactivities and responses under light stress in herbs. In: Plant phenolics in sustainable agriculture. https://doi.org/10.1007/978-981-15-4890-1_4

Hajam YA, Lone R, Kumar R (2023) Role of plant phenolics against reactive oxygen species (ROS) induced oxidative stress and biochemical alterations. In: Plant phenolics in abiotic stress management. https://doi.org/10.1007/978-981-19-6426-8_7

Hamid S, Yatoo AM, Meer MY, Ali S, Mohamed HI (2023) Historical perspective of plant phenolics. In: Plant phenolics in abiotic stress management. https://doi.org/10.1007/978-981-19-6426-8_1

Haminiuk CWI, Maciel GM, Plata-Oviedo MSV, Peralta RM (2012) Phenolic compounds in fruits—an overview. Food Sci Technol 47(10):2023

Hammerbacher A, Paetz C, Wright LP, Fischer TC, Bohlmann J, Andrew J, Davis AJ, Fenning TM, Gershenzon J, Schmidt A (2014) Flavan-3-ols in Norway spruce: biosynthesis, accumulation, and function in response to attack by the bark beetle-associated fungus *Ceratocystis polonica*. Front Plant Sci 164(4):2107

Hassan S, Mathesius U (2012) The role of flavonoids in root–rhizosphere signalling: opportunities and challenges for improving plant–microbe interactions. J Exp Bot 63(9):3429–3444

Heleno SA, Martins A, Queiroz MJ, Ferreira IC (2015) Bioactivity of phenolic acids: metabolites versus parent compounds: a review. Food Chem 173:501–513

Ho CT (1992) Phenolic compounds in food. ACS symposium series; American Chemical Society, Washington 1, pp 2–7. issn:0097-6156

Huang WY, Yi-Zhong C, Yanbo Z (2010) Natural phenolic compounds from medicinal herbs and dietary plants: potential use for cancer prevention. Nutr Cancer 62(1):1–20

Hutzler P, Fischbach R, Heller W, Jungblut TP, Reuber S, Schmitz R, Veit M, Weissenbo G (1998) Tissue localization of phenolic compounds in plants by confocal laser scanning microscopy. J Exp Bot 49(323):953–965

Hwang J, Hodis HN, Sevanian A (2001) Soy and alfalfa phytoestrogen extracts become potent low-density lipoprotein antioxidants in the presence of acerola cherry extract. J Agric Food Chem 49:308–314

Ionkova I (2011) Anticancer lignans-from discovery to biotechnology. Mini Rev Med Chem 10: 843–856

Isah T (2019) Stress and defense responses in plant secondary metabolites production. Biol Res 52: 39

Jiang Y, Hua J, Wang B, Yuan H, Ma H (2018) Effects of variety, season, and region on Theaflavins content of fermented Chinese congou black tea. J Food Qual 2018:5427302

Kaushik P, Andujar I, Vilanova S, Plazas M, Gramazio P, Herraiz FJ, Brar NS, Prohens J (2015) Breeding vegetables with increased content in bioactive phenolic acids. Molecules 20:18464–18481

Khadem S, Marles RJ (2010) Monocyclic phenolic acids; hydroxy- and polyhydroxybenzoic acids: occurrence and recent bioactivity studies. Molecules 15:7985–8005

Khalil AA, Rahman U, Khan MR, Sahar A, Mehmood T, Khan M (2017) Essential oil eugenol: sources, extraction techniques and nutraceutical perspectives. RSC Adv 7:32669–32681

Kim IS (2021) Current perspectives on the beneficial effects of soybean isoflavones and their metabolites for humans. Antioxidants (Basel) 10:1064. https://doi.org/10.3390/antiox10071064

Kristina SA, Wiedyaningsih C, Widyakusuma NN, Aditama H (2017) Extemporaneous compounding practice by pharmacists: a systematic review. Int J Pharm Pharm Sci 9(2):42

Kumar N, Goel N (2019) Phenolic acids: natural versatile molecules with promising therapeutic applications. Biotechnol Rep 24:e00370

Lakshmi BKM, Muni Kumar D, Hemalatha KPJ (2018) Purification and characterization of alkaline protease with novel properties from Bacillus cereus strain S8. J Genet Eng 16:12. https://doi.org/10.1016/j.jgeb.2018.05.009

Lamuela-Raventós RM, Vallverdú-Queralt A, Jáuregui O, Martínez-Huélamo M, Quifer-Rada P (2014) Improved characterization of polyphenols using liquid chromatography. In: Polyphenols in plants, vol 14, pp 261–292

Lattanzio V (2013) Phenolic compounds: introduction. In: Natural products, pp 1543–1580

Lea MA (2015) Flavonol regulation in tumor cells. Cell Biochem 116:1190–1194

Lee YH, Song NY, Suh J, Kim DH, Kim W, Ann J, Lee J, Baek JH, Na HK, Surh YJ (2018) Curcumin suppresses oncogenicity of human colon NSCLC cells by covalently modifying the cysteine 67 residue ofsirt1. Cancer Lett 431:219–229

Lestari ML, Indrayanto G (2014) Curcumin. Profiles of drug substances, excipients and related methodology 39, pp 113–204

Liu Q, Luo L, Zheng L (2018) Lignins: biosynthesis and biological functions in plants. Int J Mol Sci 19(2):335

Lu Y, Lu YC, Hu HQ, Xie FJ, Wei XY, Fan X (2017) Structural characterization of lignin and its degradation products with spectroscopic methods. J Spectroscopy 15:8951658

Ma L, He J, Liu H, Zhou H (2017) The phenylpropanoid pathway affects apple fruit resistance to *Botrytis cinerea*. Phytopathology 166(3):1–10

Maisetta G, Batoni G, Caboni P, Esin S, Rinaldi AC, Zucca P (2019) Tannin profile, antioxidant properties, and antimicrobial activity of extracts from two Mediterranean species of parasitic plant Cytinus. BMC Complement Altern Med 19(1):82

Manach C, Scalbert A, Morand C, Rémésy C, Jiménez L (2004) Polyphenols: food sources and bioavailability. Am J Clin Nutr 79(5):727–747

Mandal SM, Chakraborty D, Dey S (2010) Phenolic acids act as signaling molecules in plant-microbe symbioses. Plant Signal Behav 5:359–368

Maru GB, Kumar G, Ghantasala S, Tajpara P (2014) Polyphenol-mediated in vivo cellular responses during carcinogenesis. In: Polyphenols in human health and disease, vol 2, pp 1141–1179

Meagher LP, Beecher GR (2000) Assessment of data on the lignan content of foods. J Food Compos Anal 13(6):935–947
Menon S, Nayeem N (2013) Vanilla Planifolia: a review of a plant commonly used as flavouring agent. Int J Pharm Sci Rev Res 20(2):225–228
Moon KM, Kwon EB, Lee B, Kim CY (2020) Recent trends in controlling the enzymatic browning of fruit and vegetable products. Molecules 25:2754
Mosquera G, Giraldo MC, Khang CH, Coughlan S, Valen B (2009) Interaction transcriptome analysis identifies *Magnaportheoryzae* BAS1-4 as biotrophy-associated secreted proteins in Rice blast disease. Plant Cell 21:1273–1290
Naikoo MI, Dar MI, Raghib F, Jaleel H, Ahmad B, Raina A, Khan FA, Naushin F (2019) Role and regulation of plants phenolics in abiotic stress tolerance: an overview. In: Plant signaling molecules, pp 157–168
Narayanana BA, Geoffroy O, Willingham MC, Re GG, Nixon DW (1999) P53/p21(WAF1/CIP1) expression and its possible role in G1 arrest and apoptosis in ellagic acid treated cancer cells. Cancer Lett 36:215–221
Narayani M, Srivastava S (2018) Elicitation: a stimulation of stress in in vitro plant cell/tissue cultures for enhancement of secondary metabolite production. Phytochem Rev 16(6): 1227–1252
Nawrath C (2002) The biopolymers Cutin and Suberin. Arabidopsis Book 1:e0021. https://doi.org/10.1199/tab.0021
Neto CC, Vinson JA (2010) Cranberry herbal medicine: biomolecular and clinical aspects. 2nd edn. Boca Raton. https://www.ncbi.nlm.nih.gov/books/NBK92762/
Nielsen DC, Anderson RL, Bowman RA, Aiken RM, Vigil MF, Benjamin JG (1999) Winter wheat and Proso millet yield reduction due to sunflower in rotation. J Prod Agric 12(2):193–197
Oldroyd GE, Murray JD, Poole PS, Downie JA (2011) The rules of engagement in the legume-rhizobial symbiosis. Annu Rev Genet 45:119–144
Ozturk M, Hakeem KR (2019) Plant and human health, vol 2. Phytochemistry and molecular. https://doi.org/10.1007/978-3-319-93997-1
Pandey KB, Rizvi SI (2009) Plant polyphenols as dietary antioxidants in human health and disease. Oxid Med Cell Longev 2(5):270–278
Parker TL, Miller SA, Myers LE, Miguez FE, Engeseth NJ (2010) Evaluation of synergistic antioxidant potential of complex mixtures using oxygen radical absorbance capacity (ORAC) and electron paramagnetic resonance (EPR). J Agric Food Chem 58:209–217
Pathak S, Kesavan P, Banerjee A, Banerjee A, Bissi L, Marotta F (2018) Metabolism of dietary polyphenols by human gut microbiota and their health benefits. In: Mechanisms of action in human health and disease, 2nd edn., pp 347–359
Pereira DM, Valentão P, Pereira JA, Andrade PB (2009) Phenolics: from chemistry to biology. Molecules 14:2202–2211. https://doi.org/10.3390/molecules14062202
Pillai GR, Srivastava AS, Hassanein TI, Chauhan DP, Carrier E (2004) Induction of apoptosis in human lung cancer cells by curcumin. Cancer Lett 208:163–117
Radhiga T, Agilan B, Muzaffer U, Karthikeyan R, Kanimozhi G, Paul VI, Prasad NR (2016) Phytochemicals as modulators of ultraviolet-b radiation induced cellular and molecular events: a review. J Radiat Cancer Res 7:2–12
Rahman Md. M, Rahaman Md. S, Islam Md. R, FirozaRahman FMM, Taha A, Almikhlafi MA, Alghamdi SQ, Alruwaili AS, Hossain MS, Ahmed M, Das R, Emran TB, Uddin MS (2022) Role of phenolic compounds in human disease: current knowledge and future prospects. Molecules 27(1):233
Rodríguez-Garcia C, Sanchez-Quesada C, Toledo E, Delgado-Rodríguez M, Gaforio JJ (2019) Naturally lignan-rich foods: a dietary tool for health promotion. Molecules 24(5):917
Ruiz-Ruiz JC, Matus-Basto AJ, Acereto-Escoffié P, Segura-Campos MB (2017) Antioxidant and anti-inflammatory activities of phenolic compounds isolated from *Melipona beecheii* honey. Food Agric Immunol 28(6):1424–1437

Saibabu V, Fatima Z, Khan LA, Hameed S (2015) Therapeutic potential of dietary phenolic acids. Adv Pharm Sci 10:1
Saito A (2017) Challenges and complexity of functionality evaluation of flavan-3-ol derivatives. Biosci Biotechnol Biochem 81(6):1055–1060
Salazar-Orbea GL, García-Villalba R, Tomás-Barberán FA, Sánchez-Siles LM (2021) High–pressure processing vs. thermal treatment: effect on the stability of polyphenols in strawberry and apple products. Foods 10(12):2919
Saltveit ME (2009) Synthesis and metabolism of phenolic compounds. In: Fruit and vegetable phytochemicals, pp 89–100
Samantaray S, Ngangkham U (2017) Evaluation of genetic diversity in *Chlorophytum borivilianum* (Santp. and Fernan.) using molecular markers: an endangered medicinal plant. Active ingredients from aromatic and medicinal plants. https://doi.org/10.5772/66536
Shahidi FV, Ramraj C, Sod-Erdene O, Hildebrand V, Siddigi A (2019) The impact of social assistance programs on population health: a systematic review of research in high-income countries. BMC Public Health 19:2
Shalaby S, Horwitz BA (2014) Plant phenolic compounds and oxidative stress: integrated signals in fungal–plant interactions. Curr Genet 61(3):347–357
Sharma P, Jha AB, Dubey RS, Mohammad P (2012) Reactive oxygen species, oxidative damage, and antioxidative defense mechanism in plants under stressful conditions. J Bot 2012:217037
Sheng J, Citovsky V (1996) Agrobacterium-plant cell DNA transport: have virulence proteins, will travel. Plant Cell 8(10):1699–1710
Shin SA, Moon SY, Kim WY, Paek SM, Park HH, Lee CS (2018) Structure-based classification and anti-cancer effects of plant metabolites. Int J Mol Sci 19:2651
Sieniawska E (2015) Activities of tannins—from in vitro studies to clinical trials. Nat Prod Commun 10:1877–1884
Silva LR, Costa R (2014) Health benefits of nongallated and gallated flavan-3-ols: a prospectus. In: Kinsey AL (ed) Recent advances in gallate research. Nova Science Publishers, Hauppauge, pp 1–191
Singh AP, Kumar S (2019) Applications of tannins in industry. In: Structural properties, biological properties and current knowledge, pp 1–19
Sinha AK, Sharma UK, Sharma N (2008) A comprehensive review on vanilla flavor: extraction, isolation and quantification of vanillin and others constituents. Int J Food Sci Nutr 59(4): 299–326
Sirerol JA, Rodríguez ML, Mena S, Asensi MA, Estrela JM, Ortega AL (2016) Role of natural stilbenes in the prevention of cancer. Oxid Medi Cell Longev 2016:3128951. https://doi.org/10.1155/2016/3128951
Smeriglio A, Barreca D, Bellocco E, Trombetta D (2017) Proanthocyanidins and hydrolysable tannins: occurrence, dietary intake and pharmacological effects. Br J Pharmacol 174(11): 1244–1262
Song WO, Chun OK (2008) Tea is the major source of flavan-3-ol and flavonol in the U.S. diet. J Nutr 138:1543–1547
Tamagnone L, Merida A, Stacey N, Plaskitt K, Parr A, Chang C, Lynn D, Maxwell DJ, Roberts K, Martin C (1998) Inhibition of phenolic acid metabolism results in precocious cell death and altered cell morphology in leaves of transgenic tobacco plants. Plant Cell 10:1801–1816
Teponno RB, Kusari S, Spiteller M (2016) Recent advances in research on lignans and neolignans. Nat Prod Rep 33:1044
Thakur M, Sohal BS (2013) Role of elicitors in inducing resistance in plants against pathogen infection: a review. ISRN Biochem 2013:762412
Tomás-Navarro M, Vallejo F, Tomás-Barberán FA (2014) Bioavailability and metabolism of citrus fruit beverage flavanones in humans. In: Polyphenols in human health and disease, pp 537–551
Tressera-Rimbau A, Arranz S, Eder M, Vallverdú-Queralt A (2017) Dietary polyphenols in the prevention of stroke. Oxid Med Cell Longev 2017:7467962. https://doi.org/10.1155/2017/7467962

Tungmunnithum D, Thongboonyou A, Pholboon A, Yangsabai A (2018) Flavonoids and other phenolic compounds from medicinal plants for pharmaceutical and medical aspects: an overview. Medicines (Basel) 5(3):93

United States Department of Agriculture (2008) A Database for the Isoflavone Content of Selected Foods, Release 2.0. September

Vanholme R, Demedts B, Morreel K, Ralph J, Boerjan W (2010) Lignin biosynthesis and structure. Plant Physiol 153(3):895–905

Vazquez-Olivo G, Cota-Pérez JL, García-Carrasco M, Zamudio-Sosa VE, Heredia JB (2023) Antioxidant phenolics from vegetable by-products. In: Plant phenolics in abiotic stress management. https://doi.org/10.1007/978-981-19-6426-8_5

Velderrain-Rodríguez GR, Palafox-Carlos H, Wall-Medrano A, Ayala-Zavala JF, Chen CYO, Robles-Sánchez M, Astiazaran-García H, Alvarez-Parrilla E, González-Aguilar GA (2014) Phenolic compounds: their journey after intake. Food Funct 5:189–197

Verma AK, Pratap R (2010) The biological potential of flavones. Nat Prod Rep 27:1571–1593

Vicente O, Boscaiu M (2018) Flavonoids: antioxidant compounds for plant defence and for a healthy human diet. Not Bot Horti Agrobot 46(1):14–21

Vitale DC, Piazza C, Melilli B, Drago F, Salomone S (2013) Isoflavones: estrogenic activity, biological effect and bioavailability. Eur J Drug Metab Pharmacokinet 38(1):15–25. https://doi.org/10.1007/s13318-012-0112-y

Vuolo MM, Lima VS, Maróstica Junior MR (2019) Phenolic compounds: structure, classification, and antioxidant power. In: Phenolic compounds, pp 33–50

Wang H, Khor TO, Shu L, Su ZY, Fuentes F, Lee JH, Kong AN (2012) Plants vs. cancer: a review on natural phytochemicals in preventing and treating cancers and their druggability. Anticancer Agents Med Chem 12(10):1281–1305

Wang Q, Liu J, Zhu H (2018) Genetic and molecular mechanisms underlying symbiotic specificity in legume-rhizobium interactions. Front Plant Sci 9:313

Wu S, Chappell J (2008) Metabolic engineering of natural products in plants; tools of the trade and challenges for the future. Curr Opin Biotechnol 19(2):145–152

Yang J, Liu RH (2009) Synergistic effect of apple extracts and quercetin 3-β-D-glucoside combination on antiproliferative activity in MCF-7 human breast cancer cells in vitro. J Agric Food Chem 57:8581–8586

Yong Y, Matthew S, Wittwer J, Pan L, Shen Q, Kinghorn AD, Swanson SM, De Blanco EJ (2013) Dichamanetin inhibits cancer cell growth by affecting Ros-related signaling components through mitochondrial-mediated apoptosis. Anticancer Res 33:5349–5355

Zhang J, Chen J, Liang Z, Zhao C (2014) New Lignans and their biological activities. Chem Biodivers 11(1):1–54

Zhu F (2015) Interactions between starch and phenolic compound. Trends Food Sci Technol 43:129–143

21 Phenolic Compounds and Nanotechnology: Application During Biotic Stress Management in Agricultural Sector and Occupational Health Impacts

Deepsi Rathore, Nibedita Naha, and Shraddha Singh

Abstract

Current agricultural practices are becoming unsustainable due to extreme phase of climatic change and growing population. Phenolic compounds are phytochemicals, having aromatic benzene ring with one hydroxyl group that secretes from plants in response to biotic stress; the rate of secretion is solely dictated by the types of stress conditions. Besides the sanative role of phenolics on plant growth in the agriculture sector, an array of biological properties also impacts the health of the farmers in a promoting manner. Due to its natural origin and low toxicity, phenolic compounds are considered as the promising tools in eliminating the causes and effects of various diseases such as skin aging, delay of wound healing, burning, and life-threatening disease like cancer. Nowadays, nanobiotechnology is a novel tool to manipulate and enhance the overall production in the agriculture sector in order to maintain the needs and the protection from different kinds of biotic stressors. Smart site-responsive delivery of nanomaterials and active compounds in a sustained release fashion is an upcoming agritech rebellion that ensures a more efficient and sustainable agricultural system, including agricultural production, which is ultimately beneficial to not only the health of consumers but also the farmer and other people engaged in this particular occupation.

Keywords

Phenolic compounds · Secondary metabolites · Biotic stress · Nanomaterials · Agricultural sector · Occupational health

D. Rathore · N. Naha (✉) · S. Singh
Biochemistry Department, Biological Sciences Division, ICMR-National Institute of Occupational Health (NIOH), Ahmedabad, Gujarat, India
e-mail: naha.n@gov.in

R. Lone et al. (eds.), *Plant Phenolics in Biotic Stress Management*,
https://doi.org/10.1007/978-981-99-3334-1_21

Abbreviations

$[Ca^{2+}]_c$	Cytoplasmic calcium ion concentrations
8-OHdG	8-hydroxy-2′-deoxyguanosine
Ag/Cs	Alginate/chitosan
AgNPs	Silver nanoparticles
Al_2O_3NPs	Aluminum oxide nanoparticles
AlNPs	Aluminum nanoparticles
ATP	Adenosine triphosphate
AuNPs	Gold nanoparticles
BtMV	Beet mosaic virus
Ca^{2+}	Calcium
CaMV	Cauliflower mosaic virus
CeNPs	Cerium nanoparticles
CeONPs	Cerium oxide nanoparticles
CMC	Carboxymethyl cellulose
CNPs	Carbon nanoparticles
CNS	Central nervous system
CNTs	Carbon nanotubes
Co/Cr	Cobalt/chrome
CoA	Coenzyme-A
CPDs	Cyclobutane pyrimidine dimers
CsNPs	Cesium nanoparticles
CuNPs	Copper nanoparticles
CuONPs	Copper oxide nanoparticles
CVD	Cardiovascular diseases
EHMT2	Euchromatic Histone Lysine Methyltransferase 2
EPPO	European and Mediterranean Plant Protection Organization
ETI	Effective triggered immunity
FAO	Food and Agriculture Organization
Fe_2O_3NPs	Iron oxide nanoparticles
Fe-Cr	Ferrochromium
HSV-1	Herpes simplex virus type-1
ILOSTAT	International Labour Organization Statistics
JA	Jasmonic acids
KLF2	Kruppel-Like Factor 2
MAPK	Mitogen-activated protein kinase
MECP2	Methyl CpG binding protein 2
MgONPs	Magnesium oxide nanoparticles
NSD1	Nuclear receptor binding SET Domain Protein 1
OGAs	Oligogalacuronoids
PAL	Phenylalanine ammonia-lyase
PAMPs	Pathogen-associated molecular patterns
PCs	Phenolic compounds

PEBBLE	Photonic explorer for bioanalysis with biologically localized embedding
PEG	Polyethylene glycol
PFE	Pressurized fluid extraction
PPP	Pentose phosphate pathway
PR	Pathogenesis-related
PRPs	Pathogen resistance proteins
PRRs	Pattern recognition receptors
PTI	PAMP-triggered immunity
RNA	Ribonucleic acid
ROS	Reactive oxygen species
r-RNA	Ribosomal-RNA
SA	Salicylic acid
SAR	Systemic acquired resistance
SFE	Supercritical fluid extraction
SiNPs	Silicon nanoparticles
SiO_2	Silicon dioxide
SiO_2NPs	Silicon oxide nanoparticles
UNESCO	United Nations Educational, Scientific and Cultural Organization
TiO_2NPs	Titanium oxide nanoparticles
TMV	Tobacco mosaic virus
TuMV	Turnip mosaic virus
ZnNPs	Zinc nanoparticles
ZnONPs	Zinc oxide nanoparticles

21.1 Introduction

As per Dagoudo et al. (2022), the worldwide population is projected to rise nearly 9.6 billion by 2050, which challenge the growth of agricultural production up to 70–100% to meet up the worldwide food demand and food security. However, nowadays, natural resources are either remaining same, or decreasing day-by-day due to irrational use of human that causes enormous stress on environment beyond sustainability (Food and Agriculture Organization (FAO) 2022). Global sustainable development describes "a development that meets current demands without jeopardizing the ability of future generations to fulfill their own needs" (United Nations Educational, Scientific and Cultural Organization (UNESCO) 2022). As a consequence, agriculture's long-term viability has emerged as "a set of strategies especially, management, which try to improve, or maintain the quality of food without compromising the environment, quality of life, or interfering with long-term development of food production (crops) considering the demand of the consumers in a more sustainable way" (Thornton 2010).

The agricultural production is directly depending on several parameters, including biotic and environmental factors that affect growth, productivity and nutrient efficiency (Pandey et al. 2017). Therefore, it is necessary to provide safeguard from these factors through several morphological, biochemical, and molecular mechanisms by cross-talk of the respective signaling pathways. Further, as the global estimated loss of agricultural productivity is recorded by 16% due to several pathogens (Koirala et al. 2021), farmers and auxiliary workers in the agricultural sector are also exposed to variety of parasites (insects, nematodes, birds, rodents), weeds, as well as microbes (viruses, bacteria, fungi) during their routine work, which ultimately imposes pre- and post-harvest loss to crops (Oerke 2005). Moreover, the global agricultural occupation involves 26.692% workforce (International Labour Organization Statistics (ILOSTAT) 2022), which also added more burden to this sector. Thus, the total loss in agricultural yield is subjected to intensity and severity of pathogen infection, pathogen population, and innate capacity of crops to fight against the biotic stresses (Ficke et al. 2018). The yield loss data globally estimates for five major crops, i.e., wheat (21.5%), rice (30.0%), maize (22.5%), potato (17.2%), and soybean (21.4%) (Savary et al. 2019). It has also been documented that the highest rate of loss comes from the regions with rapidly growing populations like India, Uganda, Syria, and Niger, where food-deficit and re-emergence of pathogens are the common problems (Rohr et al. 2019).

Nanotechnology has been acknowledged as a "key enabling technology" by European Commission due to its wide-ranging application in different areas of science and technology. However, nanobiotechnology has just begun as a next great frontier for sustainable agriculture. The use of chemicals in management of biotic stress nowadays has been restricted because of their adverse impact on the environment and human health. As a consequence of various threats to nonsubjected entities (organisms, or other types of crops), scientists have sought distinctive and environmental-amiable strategies for the management of biotic stress with the aid of nanotechnology development. Hence, in this context, the present book chapter highlights the impact of phenolic compounds and nanomaterials in the management of biotic stress in a more sustainable and efficient way to promote occupational health in agriculture and ensure food security for the future generation.

21.2 Biotic Stress

Biotic stress is known as decay, contributes 30% worldwide loss of agriculture productivity, cause by various microorganisms such as viruses, bacteria, fungi, and other pests, parasites and nematodes since ancient times (Pandey et al. 2017). Considering all the biotic stressors, fungi, bacteria, nematodes and viruses are primarily responsible agricultural pathogens, among which 85% of diseases are carried out by fungi, or fungal-like organisms (Nazarov et al. 2020). Another biotic stressor that affects development, vigor, and productivity primarily in economically significant plants is weeds, which either directly destroy plants, or significantly raise competitive pressure for area and nutrient uptake (Dahiya et al. 2020). Further, the intensity of biotic stress depends on whether condition, cropping system, farmer cultivation conditions, crop variants, as well as resistance levels. When plants undergo biotic stress, various physiological changes are visible such as wilting, spotting, molds, pustules, rot, deformation, mummification, hypertrophy, or hyperplasia, discoloration, or destruction of tissues, and so on (Nazarov et al. 2020).

The classification of living biotic stress producers, are divided into seven major categories (Boote 1983), which are considered as one of the major environmental factors, described below:

- Those reduce standing crop (*fungi, damping-off pathogens*).
- Those have negative impact on photosynthetic activity (*bacteria, fungi*).
- Those promote leaf aging (*viruses, numerous pathogens*).
- Those affect and reduce light exposure (*weed, insect herbivores*).
- Those pick soluble assimilates out of plant cells (*nematodes, sucking pests*).
- Those eat vascular tissue (*herbivores, bacteria, necrotrophic pathogens*).
- Those reduce necessary turgor pressure (*bacteria, fungi*).

Plants have to compete with these adverse factors, as well as pathogens for nutrients and other essential commodities for own survival, which impose serious threat to health, nutrient efficiency and occupational hazards in the agricultural sector. Thus, it has become the need of the hour to prevent enormous use of pests and pathogens, both in pre- and post-harvest periods, and ensure safe health to agricultural practitioners through an alternative manner like smart use of phenolic compounds either alone, or in conjugation with nanobiotechnology.

21.2.1 Infection Strategies of Biotic Stressors

Pests and pathogens induce biotic stress, which is classified based on the strategies of infection that are harmful to the host crops (Laluk and Mengiste 2010; Sergeant and Renaut 2010):

21.2.1.1 Necrotrophic

Necrotrophic kill their host organism by secretion of certain toxins (*Rotting bacteria*).

21.2.1.2 Biotrophic

Biotrophic though allow their host organisms to live but constantly feed upon them for food and nutrition; thereby depriving the host from essential nutrients, and inducing cankers in host plants, leaf spots and vascular wilts (*viruses, nematodes*).

21.2.1.3 Hemibiotrophic

They are biotrophic in early stages of infection and later kill their host organism; thereby acting as necrotrophic pathogens (*Phytophthora*).

21.2.2 Classification of Biotic Stressors

21.2.2.1 Viruses and Viroids

In tropical and subtropical regions, viral infections can cause up to 98% of crop loss (Czosnek and Laterrot 1997). Viruses are dynamic entities and manifest themselves in variety of ways to impose maximum damage on crops. Virus infection causes numerous metabolic and biochemical disturbance in crop (Culver and Padmanabhan 2007). The contamination spreads through lesions, or damaged parts of the crop, because they provide a conductive path for its inherit material to cross the plant cell wall, manifesting as a series of symptoms that eventually disrupt host physiology (Iqbal et al. 2021):

- Growth suppression: Reduction in normal growth of crop, or its shoot parts (*Tobacco mosaic virus, TMV*).
- Discoloration: Mosaic pattern on leaves, chlorotic rings, chlorosis in leaves, and variegation (*Rice dwarf virus*).
- Deformation: Wrinkling of leaves, corrugation in leaves, thin and thread-like leaves (*Wheat streak mosaic virus*).
- Necrosis: Death of crops (*Tobacco necrosis virus*).
- Impaired reproduction: Sterility in flowers, parthenocarpy, and shedding of premature flowers and ovaries (*Cucumber mosaic virus*).

Furthermore, a research of 18 different vegetable varieties from 14 different farms in Kenya found that cabbage, collard greens, cauliflower, kale and Swiss chard were all infected with an alliance of cauliflower mosaic virus (CaMV), as well as turnip mosaic virus (TuMV), while beet mosaic virus (BtMV) infected 80% of Swiss chard (Spence et al. 2007). Besides that stunting and reductions in leaf width (kale) and head circumference (cabbage), as well as chlorotic leaf mottle and drastic leaf distortion, are also identified (Spence et al. 2007).

Viroids are pathogenic organisms with circular ribonucleic acid (RNA) but without any capsid that cause various diseases in crops, manifested by reduction of growth of entire plants, or any specific part of plants, discoloration and deformation of plant organs (Nazarov et al. 2020) though the complete molecular mechanism is yet to be explored. Viroids can alter the phosphorylation of gene products through binding to certain cellular kinases (Kovalskaya and Hammond 2014). They can also affect the expression of certain genes responsible for growth, stress, development and protection of plants, and can successfully induce proteins that are associated with enhanced pathogenesis during infection (Ali et al. 2018). Viroids also cause posttranscriptional suppression of gene expression by RNA interference, impairment of RNA splicing and inducing demethylation of ribosomal-RNA (r-RNA) genes (Méndez et al. 2015).

21.2.2.2 Bacteria

Bacterial pathogens are widely studied, which hindrance the nutrient and water transport to different parts of plants, resulting in massive destruction of crops (Cabral

2010). Also, it has been alarmed in recent times that due to increase in global annual temperature from 20 °C to 45 °C, the environment goes favorable for several bacteria (*Ralstonia sp.*, *Arthrobacter sp.*), pests (*cabbage worms, aphids*) and pathogens (*nematodes, fungi*) such as all complex cells; thereby enabling them to cause much more damage than observed before (Flury et al. 2016). Overall, two broad-spectrum diseases of bacterial origin are:

- **Systemic bacterial blight:** Affects vascular systems and penetrates conductive bundles and adjacent tissues; thereby disrupting normal water consumption of plants, cause by *Xanthomonas oryzae* (Purahong et al. 2018).
- **Local bacterial blight:** Induces damage in parenchymal tissue and plant organs separately, symbolizing wilting, chlorosis, necrosis, rot, overgrowth and scabs upon infected by *Pseudomonas syringae* (Purahong et al. 2018).

Bacterial infection can also cause mosaic-like symptoms resembling viral infections, reflected by gigantic malformation like as galls, or tumors (Rubio et al. 2020), as well as other early signs such as patches on leaf and fruit, leaf blight and tissue deadening on leaf, stems and plants root rot, leading to substantial loss of productivity and quality. In India's Tarai region, bacterial soft rot caused by *Erwinia chrysanthemi* has been associated to massive losses in banana production, affecting the socioeconomic situation of local farmers, and auxiliary workers contribute in this occupation (Akkopru et al. 2018).

21.2.2.3 Fungi

Almost 1.5 million fungi species are classified as biotic stressors and are categorized into two parts: biotrophs and necrotrophs (Lattanzio et al. 2006). Because fungi are not photosynthetic, they must evolve new tactics to acquire food from living, or nonliving life forms. Biotrophic spores obtain nutrients from living host tissues via haustoria that form inside infected cells, whereas necrotrophic fungal pathogens destroy their host tissue through the production of toxins and also obtain nutrients from dead cells (Laluk and Mengiste 2010).

However, there are few fungi like microorganism, which are biotrophic in their early stage of infection and necrotrophic in later stage of pathogen such as *Phytopthora*, a plant pathogenic fungus that causes the death of one million people in Ireland due to potato blight disease (Yoshida et al. 2013). Overall, fungal diseases are characterized by wilting, spotting, molds on affected parts, over growth, deformations of leaves, rotting, pustules (i.e., accumulations of fungal spores) and mummification (i.e., shrinkage, darkening and compaction of infected tissue) as revealed by Nazarov et al. (2020).

21.2.2.4 Nematodes

Nematodes are costly burden on crop production, accounting for approximately 14% of global annual losses totaling nearly $100 billion US dollars (Chitwood 2003). Nematodes, as ubiquitous in nature and associated with nearly every important agriculture crop, develop a constraint on global food security (Laluk and Mengiste

2010). The wide variety of interaction of nematodes is either ecto-, or endo-parasitic, depending on plant tissues they feed by penetrating through stylet (a needle-like protrusible oral structure) (Sato et al. 2019) and secretion of specific digestive enzymes for dissoluting tissues and exploring both inter- and intra-cellular movement of endoparasites, resulting in death of the target crops (Borji et al. 2011). Biotrophic and sedentary species include root-knot (*Meloidogyne sp.*) and citrus (*Tylenchulus semipenetrans*). Root-knot nematode causes most damage below the ground by feeding crop roots and reducing flow of nutrient and water, resulting in galls/root-knots, increase branching of roots, wounds of root tips and impairment of root growth. Root-knot nematodes also collaborate with some other pathogens to cause a variety of diseases including *Fusarium wilt* and *Thielaviopsis basicola* (Adam et al. 2014). However, the use of a range of market products (nematicides) is not a long-term management solution since they are not cost effective, and carcinogenic and hazardous to farmers, animals, and the environment. Therefore, for the welfare of living being and environment, nematicides are mostly banned, or highly restricted worldwide.

In contrast, bio-fertilizers (formulations comprising living microbes, either of single strain, or multiple strains, help in promoting plant growth organically: *Rhizobium, Azotobacter*), organic amendments (organic compositions/waste-derived from living biomass in order to improve soil health: *neem extract, castor oil plant, velvet bean*) and some microbial-based products (*Abamectin, BioNem, Deny, DiTera, Nemout, Royal 350, Xianchongbike*) have also gain interest in nematode management, though exhibit some common constrain in field performance such as variable responses on different types of soils, inefficiency in mass production, ease of use, shelf-life, etc. (Pulavarty et al. 2021). Therefore, the present science of agronomy with technology gives stress on the successful and sustainable management of nematode infestation.

Table 21.1 briefly exhibits some of the biotic stressors in the agricultural sector as described above.

21.2.3 Mechanism of Action of Crops in Biotic Stress Management

Plants have a variety of physical and chemical barriers that prevent nearly all undesirable plant biotic stressor interactions from forming (Sergeant and Renaut 2010). The accumulation of secondary metabolites like, phytoalexins and phytoanticipins is a toxic chemical defense that can inhibit the pathogen's enzymatic function. While all plants strengthen physical barriers such as cuticle and extracellular cell walls, to prevent unpleasant biotic interactions, the form and magnitude may vary depending on species and environment (Sergeant and Renaut 2010). To safeguard themselves against pathogenic and herbivore attacks, plants also use various kinds of constitutive and induced defense responses (Walling 2000). Biotic stress in plants also reduces plant's ability to absorb and assimilate nutrients essential for their optimum growth, long-time deprivation of which ultimately causes death of the plants. Hence, counteracting these sorts of stresses by certain internal strategies,

Table 21.1 Role of biotic stressors on plant host system and specific causative agents in the agricultural sector

Biotic Stressors	Crops	Diseases	Causative Agents	References
Bacteria	Tomato	Bacterial spot disease	*X. euvesicatoria* *X. gardneri* *X. perforans*	European and Mediterranean Plant Protection Organization (EPPO) Bulletin (2013)
	Pepper	Bacterial spot disease	*X. euvesicatoria* *X. gardneri*	
Fungi	Wheat	Septoria leaf blotch	*Mycosphaerella graminicola*	Zaffarano et al. (2008)
	Rice	Rice blast	*Magnaporthe oryzae*	
	Potato	Potato blight disease	*Phytopthora*	Yoshida et al. (2013)
Virus	Cabbage, Cauliflower, Kale, Swiss chard	Cauliflower mosaic disease	*Cauliflower mosaic virus* *Turnip mosaic virus*	Spence et al. (2007)
	Swiss chard	Beet mosaic disease	*Beet mosaic virus*	
	Brinjal,			
	Tomato, Potato,			
	Chili, Cucumber,	Root-knot	*Meloidogyne sp.*	
Nematodes	Pumpkin, Cabbage,			Bellé et al. (2021)
	Okra			
	Cotton,			
	Tomato,	Reniform	*Rotylenchulus sp.*	
	Cowpea			

crops exhibit several resistant genes that are later transcribed in large quantity on exposure to different biotic stressors (Parmar et al. 2017), as described below:

21.2.3.1 Pathogen-Associated Molecular Patterns (PAMPs)

The first basal response of a plant's defense system to pathogen/biotic stressors indicated using pathogen-associated molecular patterns, or PAMPs (Nazarov et al. 2020). Pattern recognition receptors (PRRs) on cell surfaces recognize pathogen-associated specific molecular patterns, enabling this first level of pathogen recognition. PAMP-triggered immunity, or PTI, is a type of plant immune function (Monaghan and Zipfel 2012).

21.2.3.2 Pathogen Resistance Proteins (PRPs)

This is second type of plant's response to invade pathogens/biotic stressors. This sort of pathogen recognition emphasizes on particular plant resistance proteins, or PRPs to pathogen that could able to identify a specific receptor from a pathogen, also called pattern recognition receptors, or PRRs (Dangl and McDowell 2006; Gouveia et al. 2017; Abdul Malik et al. 2020). Whenever there is an interaction between a plant and a pathogen, a specific type of receptor-mediated resistance is activated that enhances factors to combat the pathogen, or pathogenic factors, responsible for increase pathogenicity of a particular pathogen on specific host plant. This is highly effective plant's defense mechanism against pests involving effective triggered immunity (ETI). ETI stimulates hypersensitive immune response in crops and programmed cell death of the infected cells, or cells surrounding these infected cells (Mur et al. 2007; Spoel and Dong 2012).

21.2.3.3 Calcium (Ca^{2+}) Signaling Perturbations

Perturbations in cytoplasmic levels of calcium ion concentrations ($[Ca^{2+}]_c$) can be a signal for exposure of crops to various kinds of biotic stressors as $[Ca^{2+}]_c$ dominate plant's immune signaling pathways (Seybold et al. 2014; Aldon et al. 2018). Rapid shifts and volatile perturbations in $[Ca^{2+}]_c$ are important for generating an adequate response in plants through a specific gene reprograming (Reddy et al. 2011). Moreover, plants have unique Ca^{2+} signatures, which also provide differentiating immune responses.

21.2.3.4 Reactive Oxygen Species (ROS) and Mitogen-Activated Protein Kinase (MAPK)

Generation of ROS and activation of MAPK signaling are equally important response mechanism to pathogen/biotic stressors encounter with crop production in the agricultural sector (Muthamilarasan and Prasad 2013). MAPK are signaling molecules playing a crucial role in plant defense against pathogens, which are closely related to mammalian signal-regulated kinases. Once a pathogen attacks, MAPK are activated due to sensation of plants regarding a threat of pathogen, or PAMPs, and initiated defense responses mechanism such as closing of stomata, generation of ROS, upregulation of stress hormones, cell wall strengthening, and many more (Meng and Zhang 2013). Generation of ROS has also equivalent protective roles against biotic stressors by promoting cell wall strengthening and activation of cytosolic Ca^{2+} channels; thereby increasing $[Ca^{2+}]_c$ (Asai and Yoshioka 2008).

21.2.3.5 Pathogenesis-Related (PR) Genes

Considerable role of PR genes in plant defense responses as a part of biotic stressors has also been noted (Ali et al. 2018). PR genes are important pillars of plant immune responses, which are translated into proteins that activate once a plant encounters pathological condition (Tang et al. 2017). There are 17 families of PR proteins categorized on the basis of their biochemical and molecular characteristics (Van Loon et al. 2006). Hamamouch et al. (2010) investigated the role of five PR genes in plant biotic stress management mechanisms in *A. thaliana*. In *A. Thaliana*, the PR-1, PR-2 and PR-5 genes are involved in the salicylic acid (SA)-dependent SAR reaction, whereas the PR-3 and PR-4 genes are engaged in the jasmonic acid (JA)-dependent SAR reaction (Thomma et al. 1998; Hamamouch et al. 2010).

21.2.3.6 Systemic Acquired Resistance (SAR) Genes and Other Factors

Pest attack on crops also activates some local defense responses involving JA, oligogalactruonoids (OGAs) and hydrogen peroxide signaling pathways (Fürstenberg-Hägg et al. 2013). Plants also produce volatile compounds such as acrolein, methacrolein, nonanal, benzenoids, etc. that discourage attacking insects and repel them by inducing defense responses through lipoxygenase and terpernoid signaling pathways (Pichersky and Gershenzon 2002). Plants' production of specific defensive polypeptides, which are classified as protein antagonists, lectins, chitinases, alpha-amylase antagonists and polyphenol oxidases, is an important downstream defense mechanism (Lee and Jun 2019; Fürstenberg-Hägg et al. 2013). Another pivotal downstream signaling pathway of defense mechanism against pest includes ETI- and PTI-induced signaling pathways, involvement of SA, JA and ethylene. SA pathway is reported to stimulate resistance against biotrophic and hemibiotrophic pathogens by activating SAR that further enhances PR genes expression and provides long-term defense against wide-spectrum pathogens/biotic stressors. JA and ethylene are involved in combating necrotrophic pathogens such as *Botrytis cinerea, Erwinia carotovora,* and other chewing insects by activating JA-ethylene signaling pathway (Gimenez et al. 2018).

Moreover, plants have the ability to increase their level of basal resistance against future pathogen attack by definite and accurate stimulation of appropriate pathways, known as induced resistance. There is a huge diversity among such signaling pathways and the levels of their effectiveness. The classic form includes SAR, which occurs in systemic plant parts, are infected by necrosis-inducing pathogens (Ryals et al. 1996). There are also certain resistant genes that help plants to thrive in negative environments and maximize their chances of survival against biotic stressors. Overall, two basic categories of plant defense responses are recorded, i.e., performed and inducible. The inducible defense responses are activated against living pathogens, herbivores and weeds through genetic expression for synthesis of several secondary metabolites and proteins associated with defense response against biotic stress producers (Mithöfer and Boland 2012; Bhattacharya et al. 2010). In contrast, performed defense responses are noticeable against alfalfa and TMV after injecting arabidopsis and tobacco plant with SA, resulting in higher resistance against several biotic stressors (Vlot et al. 2009).

Figure 21.1 shows the overall mechanism of actions of crops against different biotic stressors in the agricultural sector as discussed above.

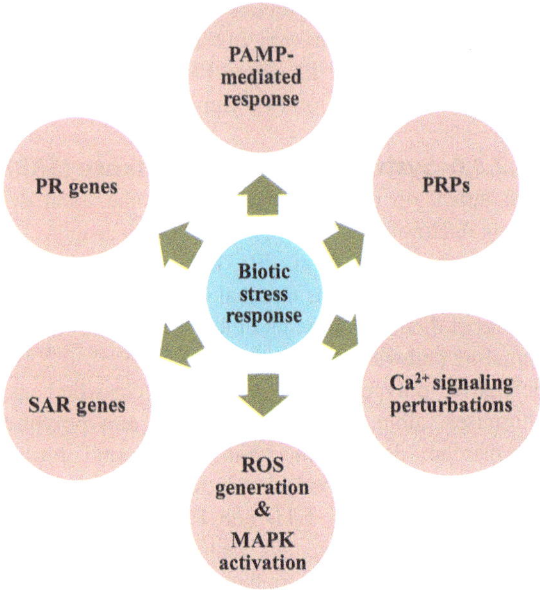

Fig. 21.1 Various mechanisms of actions during biotic stress management in the agricultural sector

21.3 Role of Biotic Stressors in Occupational Health Hazards

Biotic stressors not only affect crop systems but also human health especially, farmers as they are directly exposed to a variety of elements during biotic stress management. Farmers, as well as the auxiliary workers have a high health risk factor due to a number of biological, physical and chemical agents during day-to-day practice in the agricultural sector (Donham et al. 2021). Biological agents such as micro- and macro-organisms are the foremost reason of several occupational diseases among the farmers that can be pathogens (*Bacillus anthracis, Orientia tsutsugamushi*), allergens (*Tyrophagus putrescentiae, Lepidoglyphus destructor*), biological toxins (*carbamates, triazines*), carcinogens (*malathion, dimethoate*) and biological vectors (*mosquitoes*) (Rahman et al. 2020; Hage-Hamsten et al. 1985; Nicolopoulou-Stamati et al. 2016; Gasnier et al. 2009; Alwis et al. 1999).

From epidemiological studies, it is clear that pneumoconioses and parasitic diseases are the most common occupational diseases in the agricultural sector (Moleocznik 2004; Żukiewicz-Sobczak et al. 2013; Wormser et al. 2006). According to Zielińska-Jankiewicz et al. (2005), biological risk factors have four classes depending upon the severity of disease, or degree of danger:

- **First group of risk factors:** Do not pose a serious hazard (*e.g. Allergies caused by mites, or bacteria*).

- **Second group of risk factors:** Transmissible pathogens; efficient prevention/ treatment measures are available for them (*e.g. Mumps and Influenza*).
- **Third group of risk factors:** Treatable but if untreated, or spread, then can cause serious illness (*e.g. Moulds and Yeast causing respiratory disorders*).
- **Fourth group of risk factors:** Factors result in serious hazards with high mortality rate, for which there are no viable preventative, or treatment approaches (*e.g. Viral infections like HIV, or tuberculosis*).

Further, animal and vegetable products, dusts, human excreta, and trashes are the common sources of biological agents in workplace, which are transmitted through dusts, droplets, aerosols, or mucous membrane, or by ingestion among the farmers and other workers in the agricultural sector (Lacey and Dutkiewicz 1994). The small airborne particles cause the most common allergic illness such as asthma, irritation of mucous membrane, infectious diseases and some serious issues, leading to development of cancers (Lacey and Dutkiewicz 1994; Żukiewicz-Sobczak et al. 2013). Farmers are also experienced to heavy noise and vibrations due to mechanical load associated with their day-to-day work (i.e., physical risk factors), and also include organic and inorganic toxins (i.e., chemical risk factors). Organic toxins contain molds and bacteria, which are extremely harmful for health, for example, *Aspergillus* and *Penicillium* sensitivity are found among 5–80% of the allergic patients (Mousavi et al. 2016). Apart from allergic diseases, farmers also suffer from the biological vectors. Lyme disease is a well-known hazard, cause by *Borrelia burgdorferi*, which cannot be diagnosed, or treated in early stages (Shapiro 2014).

Chemical risk factors are basically due to use of pesticides, herbicides, or insecticides in the field. They contain one, or more active ingredients, or adjuvants with potential hazards to farmers' health (Watterson and Thomas 1992). Agricultural and forestry workers had a rate of occupational illnesses of 418.5 out of 100,000 workers (Wilczyńska et al. 2013). Epidemiological studies also show that due to use of these combined causative factors, the health of current and older agricultural population is worse than that of the different occupational groups of other sectors (Żukiewicz-Sobczak et al. 2013; Moleocznik 2004; Szeszenia-Dąbrowska et al. 2016). The fact is that farmers are currently underserved by a system for detecting, diagnosing, and recognizing hazardous occupational diseases; thus, the implementation of preventive and educational programs, awareness campaigns, etc. through different Govt./non-Govt. agencies for farmers appears to be critical, which cannot be delayed anymore.

21.4 Phenolic Compounds

Phenolic compounds (polyphenols), or secondary metabolites originate from the primary metabolites in a sequential manner (Guo et al. 2018) when crops encounter pathogens/pests, which are potentially harmful for growth, development and normal functioning. To fight against these biotic stressors, plants develop smart machinery, which can synthesize chemicals/compounds that provide strong defense mechanism

(Boudet 2007) though the purpose of synthesis of these compounds is not entirely focused on protective mechanisms against biotic stress producers but also have various other important functions, like attracting pollinators and seed dispersing agents, acting as signal molecules, role in allelopathy, etc. (Iqbal et al. 2021). Phenolic compounds accumulate in subepidermal layers of tissues attached to cell walls, or periphery of organs, the favorable parts that are exposed to biotic stressors (Clé et al. 2008; Schmitz-Hoerner and Weissenböck 2003; Vishwanath et al. 2014), leading to increase in metabolism of host crop in response to infection; thereby inhibiting the pathogen from causing further damage to the host (Mayer et al. 2001) through a hypersensitive reaction that triggers cell death in infected and nearby areas, retarding pathogen growth, as well as development (Lincoln et al. 2018).

The concentration of phenolic compounds depends on various factors such as stage of development, phase of growth, or seasons (Lynn and Chang 1990; Ozyigit et al. 2007). External factors like trauma, wounds, draught, etc. are also affecting the synthesis and accumulation of phenolic compounds at particular sites in crops (Kefeli et al. 2003). Besides, cell wall thickening, production of certain hormones, reproduction, fruit flavoring, protection and pigmentation, osmotic regulation, anti-microbial activity and protection from ultraviolet rays (Dixon 2001), contribution to growth and developmental processes in plants by lignin biosynthesis, etc. (Ndakidemi and Dakora 2003) are the overall crucial role of phenolic compounds in the agricultural sector. Lignin also acts as a chemical and physical barrier for pathogens. Apart from lignin, suberin and pollen sporopollenin are the examples of phenolic compounds in plant kingdom (Cheynier et al. 2013).

Further, low-molecular-weight phenyl propanol/derivatives have positive role in pigmentation and scents; thereby promote symbiotic microbes, pollinators and other animals to help in dispersal of fruits. Phenolics also act as repelling agents, inhibitors and natural toxicants for herbivores, and as bio-pesticides for invading pathogens/organisms (Dakora and Phillips 1996; Lattanzio et al. 2006). Common examples are some phenolic acids, tannins, resins on plant surface that deter digestive abilities of birds by interacting with their gut microflora (Bhattacharya et al. 2010). Polyphenols like catechins affect bacterial species by interfering plasma membrane permeability and generating ROS (Wang et al. 2018). Phytoalexins also secrete after injury, or by damage organ and repel several potentially harmful microorganisms (Dar et al. 2017). There are other applications of plant phenolic compounds against biotic stressors like structural integrity and scaffolding (Mandal et al. 2010), bioremediation, promoting plant growth and as allelochemicals (Seybold et al. 2014), increase uptake and absorb ions (*magnesium, potassium, zinc, calcium, iron, manganese*) and soil porosity to improve their mobilization (Aldon et al. 2018).

21.4.1 Classification of Phenolic Compounds

A benzene ring with one, or more hydroxyl groups is the fundamental structure of phenolic compounds. Phenolics are ubiquitous secondary metabolites, synthesize from primary structure. Phenolic compounds are typically classified based on the carbon atom numbers present in the intramolecular hydrogen (Caleja et al. 2017). Vascular plant contains esters, amides, glycosylated flavonoids, glycosides of hydroxycinnamic acid, etc. in their leaves, which are examples of defensive secondary metabolites. Phenolic compounds are mainly classified into three groups based on their abundance in human diet, mentioned below:

21.4.1.1 Phenolic Acids

Phenolic acids are nonflavonoids having complex structure with high molar mass, color durability and antioxidant capacity that are proportionally related to the hydroxyl group numbers present in the structure (Vuong 2017). They are subclassified into two groups: hydroxybenzoic and hydroxycinnamic acids. Hydroxybenzoic acids are the simplest form of glycosylated phenolic acids found in nature, present in mostly nonedible part of plants (Caleja et al. 2017). Hydroxycinnamic acids are complex, present in edible part such as apples and peaches. The most common pigment of both the subgroups is gallic acid and caffeic acid, respectively (Vuolo et al. 2019). Lignins are composed by two cinnamic acid residues and are present in vascular plants like soybean. They are phytoestrogens having low concentration in fruits and vegetables, for example, secoisolariciresinol diglucoside (Falcone Ferreyra et al. 2012). Stilbenes are also nonflavonoids in grapes and wines, biosynthesized by the same route as flavonoids, and are well known for their antibacterial and anticarcinogenic properties. Resveratrol is the commonly found pigment of this group (Carocho and Ferreira 2013; Koushki et al. 2018).

21.4.1.2 Flavonoids

The most prevalent forms of phenolic compound are flavonoids, being responsible for carotenoids and chlorophylls in plants for blue, purple, yellow, orange and red color. Structurally, they have C_6–C_3–C_6 skeleton along with aromatic ring attached by a C_3 link, and are subcategorized according to the number and location of hydroxyl groups along with their alkylation and glycosylation (Khoddami et al. 2013). The change in the structural property is directly proportional to the antioxidant activity of flavonoids (Caleja et al. 2017). Subgroups of flavonoids are anthocyanins, isoflavonoids, flavonols, flavones, flavanones and chalcones, and out of them, anthocyanins are highly water-soluble pigments responsible for providing a wide range of colors (red to purple) in vegetables. The most common pigments are myricetin and genistein (Vuolo et al. 2019). Isoflavones are phytoestrogenic compounds with estrogenic properties, found in leguminous plants. Daidzein is the principal member of isoflavonoid group (Carocho and Ferreira 2013). Flavonols are present in myriads of flavonoids like in fruit and vegetables. Hyperoside is the principal compound in this subclass. Flavonols are the rich source of high radical scavenging activity in grapes, red wine and chocolates (Vuong 2017). Flavones consist of a basic flavonoid structure and are present in fruits and vegetables. Apigenin is the principal pigment of this subgroup (Falcone Ferreyra et al. 2012). Flavanones are the nanosubclass among other flavonoids subgroups, and are limitedly present in citrus fruits responsible for sour and sweet taste such as naringin and naringenin (Carocho and Ferreira 2013). Chalcones are an intermediate product, which is formed during biosynthesis of the flavonoids, and are present in apples such as phloretin (Rudrapal et al. 2021).

21.4.1.3 Tannins

Tannins are the astringent phenolic compounds that bind and precipitate proteins, which are subclassified into hydrolyzable and condensed tannin groups. The hydrolysable tannins are found in fruits and berries, having ellagitannins and cyaniding-3-glucoside as a principal pigment (Calvo-Castro et al. 2013). On the other hand, condensed, or complex tannins are found in woody part of the plants such as catechins, proanthocyanidins, and anthocyanidins (Perde-Schrepler et al. 2013).

21.4.2 Mechanism of Action of Phenolic Compounds in Agriculture

Biosynthesis of the stress-inducible phenolic compounds occurs by the precursors of phenylpropanoids, which are synthesized during primary metabolic pathways such as shikimic acid pathway and malonic pathway (Sarkar and Shetty 2014). Carbohydrate metabolites phosphoenolpyruvate and erythrose 4-phosphate produce during pentose phosphate pathway and glycolysis, respectively, are indirectly utilized for synthesis of various stress-inducible phenolic compounds such as lignins and flavonoids as mentioned above (Dixon and Paiva 1995; Wink 2015).

The phenylpropanoid biosynthetic pathway marks the transition from primary to secondary metabolism and is essential for structural support, vascular integrity, and microorganism resistance in plants (Cosme et al. 2020). It produces monolignols that are essential for plants structural and vascular integrity, and pathogen resistance, and are used in lignin biosynthesis (Sarkar and Shetty 2014). The enzyme phenylalanine ammonia-lyase (PAL) and its subtype play a significant role in the modification of phenylalanine as the final product of p-coumaroyl Coenzyme-A (CoA) (Boerjan et al. 2003) that is crucial in biosynthesis of flavonoids, phenolic acids and lignins via enzyme-mediated pathway (Fraser and Chapple 2011), as its overexpression can promotes the formation of protective phenolic compounds with antibacterial properties (Shetty 2004). Coumarins and hydroxycoumarins are produced in plants from trans-p-coumaric acid and transcinnamic acid, respectively; but the entire mechanism of their production is yet to be known. However, Wang et al. (2018) suggested that the biosynthetic route for coumarin could be via hydroxylation to produce coumaric acid, followed by glycosylation to generate transcoumaric acid-2-O-glucoside.

Phenolic compounds provide structural integrity and scaffolding with respect to protection against biotic stressors, as mentioned above, by esterification of phenols to cell walls (Mandal et al. 2010) such as benzoic acid and phenyl propanoids are some low-molecular-weight phenols, which are formed as an initial response to an infection. Initiation of hypersensitivity reaction, as well as rapid necrotic death of surrounding cells at the place of disease, prevents the infection from spreading systemically. Ion flux changes, lipid hyperoxidation, protein phosphorylation, and nitric oxide (ROS) and phytoalexin generation characterize these changes in the agricultural field (Van Loon et al. 2006). Plants have a nonspecific response at an individual level for developing resistance against these stresses, known as SAR, which gets activated without prior treatment to biotic stressors and reinforces plants' defense system against future encounter with the same kind of pathogens.

Another proven mechanism of phenolic compounds is allelopathy, a mechanism encompassing the direct, or indirect, positive, or negative consequences of one plant on another plant via chemical release (Li et al. 2014). Metal chelating processes absorb ortho-substitution in phenolics like salicylic and o-coumaric acids, along with dihydrosubstitution of phenolics like protocatechuic and caffeic acids (Capasso et al. 1992). Thus, available phenolic compounds may rack up in soils swamped with waters derived from vegetable waste, influencing the formation and availability of soil nutrients, as well as the rates of nutrient cycling: all of which affect agricultural health growth and development (Li et al. 2014). Protocatechuic acid and catechol moiety from onions are two of the most well-known examples of phenolics' protective function, which are known to help in the regression of *Colletotrichum circinaus* infection (Capasso et al. 1992). Further, studies have shown various extraction strategies of plant phenolics such as pressurized fluid extraction (PFE) and supercritical fluid extraction (SFE) (Ruiz-Ruiz et al. 2020), solid–liquid extraction (Alara et al. 2021), enzyme-assisted extraction (Xue and Farid 2015), pulsed electric field extraction (Chan et al. 2011), etc., and their applications in order to prevent various diseases (Puri et al. 2012) (Table 21.2; Fig. 21.2); however, the extraction strategies due to use of chemicals may impact a negative role in the health of the workers in the agricultural field, which requires further in-depth study.

21.5 Phenolic Compounds and Health Impacts

Phenolic compounds act as a ligand to restore the body's function and wellness like neurotransmitters, peptides, steroids, hormones, enzymes, or vitamins, as well as a rich source of novel therapeutics because of low cost, easy accessibility and safer application (Cory et al. 2018: Rafiq et al. 2016). Besides phenolic compounds also possess other direct and indirect biological effects on human such as antimicrobial, anti-inflammatory, anticancer, cardioprotective activities, and antioxidant by attenuating ROS production (Lin et al. 2016). Phenolic compounds like kaempferols and fisetin show ROS attenuation with concomitant stimulation of glutathione, resulting in protection against oxidative stress-induced cellular death in heart and liver (Cosme et al. 2020).

The pioneer work of Wattenberg (1992) demonstrates the interference of phenolic compounds with the initiation, promotion, progression and metastasis of cancer cells; thus, reducing the risk of cancer. A report by Kris-Etherton et al. (2002) suggests that high intake of fruits and vegetables have been linked to a lower risk of human cancers of breast, lung, colon and prostate. Flavonoids are highly bioactive compounds that confer chemoprotective function. Drinking tea and its polyphenolic constituents protect mice against skin cancer, caused by ultraviolet radiation, or

Table 21.2 Phenolic compounds and subclasses possess protective and therapeutic effects on human health

Phenolics	Types	Examples	Human health Impacts	References
Phenolic acids	Hydroxy-benzoic acid	Gallic acid	Block Herpes simplex virus type 1 (HSV1) & Parainfluenza type 3	Özçelik et al. (2011)
	Hydroxy-cinnamic acid	Caffeic acid	Psoriasis treatment	Pluemsamran et al. (2012); Dudonné et al. (2011)
		Coumarin	Wound healing	Albaayit et al. (2014)
		Licoarylcoumarin, Glycycoumarin, Gancaonin	Antibacterial	Eerdunbayaer et al. (2014)
Stilbenes	1,1-Diphenylethylene	Resveratrol	Anticarcinogenic & antibacterial	Del Valle et al. (2016)
Tannins	Hydrolysable	Ellagitannins, Cyanidin-3-glucoside	Reduction of DNA damage including cyclobutane pyrimidine dimers (CPDs) & 8-hydroxy-2′-deoxyguanosine (8-OHdG) formation	Calvo-Castro et al. (2013)
	Condensed	Catechins, Proanthocyanidins, Anthocyanidins	Reduce apoptosis, DNA damage & oxidative stress	Perde-Schrepler et al. (2013)
		Epigallocatechin gallate	Block *Hepatitis B virus* in primary hepatocytes	Huang et al. (2014)
Flavonoids	Anthocyanins	Myricetin, Genistein	Anticarcinogenic, chlorogenic & anti-acne	Kazi et al. (2003); Tonho et al. (2010)
	Flavonols	Hyperoside	Radical scavenging	Liu et al. (2016)
	Flavones	Apigenin	Decrease α-synuclein accumulation & protection against neural apoptosis	Anusha et al. (2017)
	Flavanones	Naringin, Naringenin	Antibacterial	Ahmed et al. (2017); Cosme et al. (2020)
		Flavanone, Naringenin	Potency on *Dengue virus* replication	Sanhueza et al. (2017)

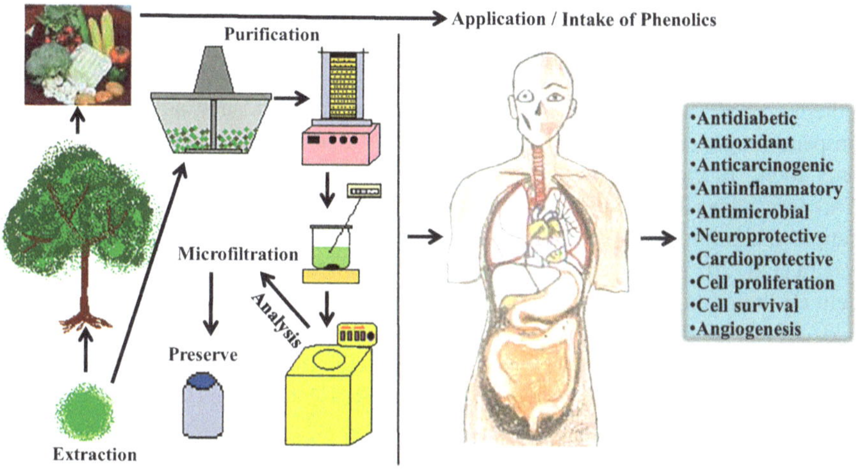

Fig. 21.2 An overview of biological effects of phenolic compounds' use in the agricultural sector

exposure to chemical carcinogens like benzene, or nickel (Mukhtar and Ahmad 2000). In human, consumption of phenolic compounds in long-term results in the protection against certain cardiovascular diseases (CVD), gastrointestinal issues, type II diabetes and lung damage (Orgogozo et al. 1997; Xiao and Hogger 2015). Polyphenols as antioxidant further treat CVD and platelet aggregation (Lin et al. 2016). Cohort studies found that eating whole grain foods either slow, or prevent the development of type II diabetes (Belobrajdic and Bird 2013). Anti-inflammation and adhesion, which include peroxidation, and other cellular and membrane damage, are important because they have prominent physiological role such as lowering of blood pressure (Ruiz-Ruiz et al. 2020). Flavonoids and phenolic acids also help to boost overall health in human by lowering the risk of metabolic syndrome and consequences associated with type II diabetes (Lin et al. 2016), and regulate immune response, promote cellular growth and confer protection against oxidative stress (Sroka and Cisowski 2003; Saeidnia and Abdollahi 2013).

Free radicals and oxidative stress have been thought to be an important factor in ageing, or age-related degenerative diseases as antioxidant load in the body decline with age and natural dietary compounds content. For example, lycopene provides protection against hydrogen peroxides by regulating p53 expression, which influences the survival of endothelial cells in human. Kaempferol also confers protection against glucose-induced oxidative damage and cell death in human pancreatic cells (Pan et al. 2012). Moreover, phenolic compounds are reported protective agents against various neurodegenerative diseases such as dementia and Parkinson's. A survey based study in Bordeaux demonstrates that daily consumption of approximately 500 mL wine containing resveratrol decreases the incidence of Alzheimer's by 80% as compared to the nondrinkers (Orgogozo et al. 1997). Catechin and anthocyanins display their involvement in the regulation of obesity and inflammation (Hursel et al. 2011; Pojer et al. 2013). Phenolics help to prevent the negative effects of fungal toxins (*mycotoxins*), while also aiding in detoxification (Kris-Etherton et al. 2002). Many volatile phenolic compounds such as hydroxyphenyl propene (*eugenol*), phenolic terpene (*carvacrol*), or curcuminoids (*curcumin*) are only found in the rhizomes of *Curcuma longa* (*turmeric*) (Beekrum et al. 2003). Curcumin plays an important role in preventing and treating diseases ranging from cancer to autoimmune, neurological, cardiovascular and diabetes (Amiot et al. 2016; Vieira et al. 2019). However, the health benefits of phenolic compounds are dependent on their conjugation with other phenolic groups, individual molecular weight, and degree of glycosylation, acylation and solubility (Kumar and Goel 2019). Overall, there are substantial evidences that specific phenolic compounds are beneficial for the anticipation and control of certain disease in human (Cosme et al. 2020).

21.6 Nanotechnology

The main concern of incorporating nanotechnology in the agricultural field is to augment the growth and productivity of crops without any harm to specific occupational groups (farmers and auxiliary workers), as well as environment, and to provide protection against various biotic stressors as discussed above. In this respect, nanoparticles are a wide class of materials having distinct physicochemical properties (crystallinity, catalytic properties, porosity, aggregating properties), sizes (1–100 nm) and bioactive nature. Due to their unique characteristics, nanoparticles such as gold nanoparticles (AuNPs), cerium oxide nanoparticles (CeONPs), zinc oxide nanoparticles (ZnONPs), etc. can cross cellular barriers and exhibit their impact on living organisms; thus, they have various applications during biotic stress management in the agricultural sector. For example, nanomaterial-based sulfhydryl groups in fungal extracellular matrix nanofertilizers increase the absorption of essential elements like nitrogen and potassium to seeds; thereby confirm the influence of nanotechnology in present days' agricultural application (Sekhon 2014).

21.6.1 Types and Mechanism of Nanomaterials

The following segment briefly discusses some major nanoparticles and underlying mechanism based on their role in the agriculture sector:

21.6.1.1 Silver Nanoparticles (AgNPs)

The involvement of AgNPs in enhancing plant growth and productivity, and photosynthetic activity has previously been reported (Sadak 2019; Sharma et al. 2012). AgNPs due to high microbial activity, can be function as an antimicrobial agent throughout crop protection (Saber et al. 2017). Duhan et al. (2017) reported a potential antifungal activity of AgNPs by transmembrane disorientation, electron transport chain and adenosine triphosphate (ATP) metabolism through inactivation of sulfhydryl groups in fungal extracellular matrix. Biosynthesized AgNPs are effective against most of the gram negative and gram positive bacteria, which proves their antibacterial activity. The mechanism of AgNPs over bacteria is to neutralizing the electric charges of their outer cell membrane, resulting in loss of permeability, leading to cell death (Prasad et al. 2017). Besides, AgNPs also have great influence on growth and development of crops especially, for seed germination, root growth and elongation, as well as senescence inhibition due to their high surface area, which provide increase sequestering activity for nutrient ions from soil (Jhanzab et al. 2019). Overall, the potency of AgNPs relay on particle size, shape, surface coating, duration of exposure, concentration and particular species of the crop itself (Jhanzab et al. 2019). Triangular AgNPs show higher response than spherical and rod-shaped particles (Zhang et al. 2016). Thus, the application of AgNPs in agriculture field is a feasible, effective and safe alternate mode as they amplify the competence of uptake and translocate more nutrients, as well as combat various environmental stresses while consequently improving crop yield.

21.6.1.2 Zinc Nanoparticles (ZnNPs)

Zinc is one of the most essential micronutrients for living systems. Zinc deficiency has a negative impact on crop yield in highly alkaline environment with calcium carbonate (Duhan et al. 2017). In combination to calcium carbonate as a nanofertilizer, zinc improves its dissipation and biodistribution in soil and enhances its diffusion from fertilizer to plant tissue (Sharma et al. 2012). ZnNPs posses' very small size (<100 nm) but having high surface-to-volume ratio, due to which, ZnNPs interact with different bacterial strains and fungi, resulting in bactericidal activity against these pathogens (Xie et al. 2012). ZnNPs derived from *Parthenium hysterophorus* leaf extract, have antifungal property against *Aspergillus favus* and *Aspergillus niger* (Rajiv et al. 2013). Furthermore, ZnNPs synthesized from *Moringa oleifera* leaf extract, of 16–20 nm size shown antimicrobial activity against various bacterial strains such as *Escherichia coli*, *Staphylococcus aureus* and *Bacillus subtilis*, as well as fungal stains such as *Candida albicans*, *Aspergillus flavus*, *Fusarium moniliforme*, *Rhizopus stolonifera*, and so on (Bhalodia and Shukla 2011). Overall, plant-based ZnNPs synthesis is eco-friendly and more cheap than the chemically synthesized nanoparticles and devoid of any side-effects (Kalpana and Devi Rajeswari 2018). Therefore, the use of ZnNPs in agronomy is encouraged due to wide-spectrum favorable outputs including augmentation of crop growth.

21.6.1.3 Carbon Nanotubes (CNTs)

CNTs have been a hollow cylindrical carbon tubes with a two-dimensional graphene sheet wrapped around them (Zaytseva and Neumann 2016). They are classified into two types based on the total count of concentric graphene layer sheets, namely single- and multi-walled nanotubes. CNTs soak up contaminants in water such as toxic organic solvents, dichlorobenzene, fertilizers and pesticides; thus, causing reduction of harmful effects on plants (Prasad et al. 2017). Silver-coated single-walled CNTs shown strong antifungal activity (Zaytseva and Neumann 2016). Further, CNTs have been proven to enhance the growth and development of crops with respect to increase germination rate, biomass accumulation, and nutrient and water uptake (Tripathi et al. 2011; Khodakovskaya et al. 2012; Mukherjee et al. 2016).

21.6.1.4 Silicon Nanoparticle (SiNPs)

Silica can be found easily in nature and in all forms of life due to their abundance. SiNPs promotes cell expansion and development in sponges and corals (Alsaeedi et al. 2017). SiNPs also regulate plant defense mechanism against biotic stressors like insects and various diseases by reducing their sensitivity for toxic organisms, and strengthens the activity of antioxidants enzymes (Mateos-Naranjo et al. 2013; Tripathi et al. 2015). Therefore, silica has universally used to form SiNPs to enhance crop productivity and ensure food quality (Lu and Radisic 2021; Siddiqui and Al-Whaibi 2014). The combined use of silica with AgNPs stabilizes with polymers and reveals profound antifungal effects on crops (Lee and Jun 2019). SiNPs treatment also causes high rate of insect mortality (Debnath et al. 2011). The exogenous use of SiNPs in the agricultural sector is expected to increase crop production by increasing plant nutrition, uptake of water, seed germination, root growth and elongation, senescence inhibition, as well as protection of crops against insects (Mateos-Naranjo et al. 2013).

21.6.1.5 Titanium Oxide Nanoparticle (TiO_2NPs)

TiO_2 are biocompatible and photochemically active nanomaterials. On exposure to near-ultraviolet rays of 310 and 400 nm, TiO_2NPs absorb photon of energy equal to, or higher than its own band gap, which is 3.0 eV and generate superoxide radicals (Li et al. 2014). This light-induced biocidal effect of TiO_2 is used against harmful microorganisms such as bacteria (*Escherichia coli*) and mold (*Aspergillus niger*) (Yu et al. 2013; Wu et al. 2010). Photochemically active TiO_2 is first reported as bactericidal by Matsunaga et al. (1985). Paret et al. (2013) reported that TiO_2NPs have high photocatalytic activity and antibacterial potential against *X. perforans*, causing spot disease in tomato plants. Employing photochemically active TiO_2NPs in the agricultural sector is not only effective on *X. perforans* but also shows other advantages such as eco-friendly, self-cleaning capability and self-regeneration capability (Cui et al. 2010).

Several nanomaterials with broad spectrum activities that are alternative for chemical herbicides/pesticides, or fertilizers for management of biotic stressors and related diseases in the agricultural sector, are summarized in Table 21.3.

21.6.2 Nanotechnology in Biotic Stress Management

Nanoparticles in different formats efficiently interact with biological system either chemically, or mechanically due to their small sizes and large surface areas. Use of engineered nanomaterials for sustainable agriculture can be concluded as a great breakthrough in agro-technology as they promote plant growth by increase nutrient

Table 21.3 Nanomaterials in biotic stress management including positive impact on hosts

Nanomaterials	Species Exploited	Positive Impact on Hosts	References
ZnONPs	*Aspergillus flavus, Aspergillus fumigates*	Potential antifungal activity by affecting cellular function of fungal hyphae	Jones et al. (2008)
	Aspergillus niger, Penicillium expansum		Rajiv et al. (2013)
	Aspergillus flavus, Trichoderma harzianum, Rhizopus stolonifer		Gunalan et al. (2012)
	Fusarium graminearum		Dimkpa et al. (2013)
	Pseudomonas aeruginosa		Jayaseelan et al. (2013)
	Xanthomonas hortorum	Antibacterial activity by improving antioxidant enzyme activity of infected plant biomass	Norman and Chen (2011)
	Alternaria alternate, Fusarium oxysporum, Rhizopus stolonifer	Inhibit fungal spore germination	Wani and Shah (2012)
	Ralstonia solanacearum	Antibacterial activity by improving plant biomass	Khan et al. (2021)
	Aspergillus niger	Antibacterial activity through increase host's membrane permeability & antifungal activity by inhibiting spore growth	Patra et al. (2012)
ZnO + AuNPs	*Escherichia coli*	Leakage of intracellular components by generating holes in host species	Attar and Altikatoglu (2018)
AuNPs	*Staphylococcus aureus*	Antibacterial activity by increase in ROS-induced oxidative cell destruction	Yuan et al. (2017)
	Escherichia coli	Leakage of intracellular components by generating holes in host species	Dangl and McDowell (2006)
	Candida albicans	Antifungal activity by inhibiting cell growth	Aljabali et al. (2018)
	Aspergillus flavus, Aspergillus niger, Candida albicans	Antifungal activity by inhibiting uncoiling & transcription of DNA	Jayaseelan et al. (2013)
	Alternaria alternate, Cladosporium cucumerinum,	Antifungal activity by suppression of fungi growth & structures	Kim et al. (2012)

(continued)

Table 21.3 (continued)

Nanomaterials	Species Exploited	Positive Impact on Hosts	References
	Botrytis cinerea, Stemphylium lycopersici		
AgNPs	*Bacillus megaterium, Pseudomonas syringe, Fusarium graminearum*	Potent inhibitor through modulation of microbial signal transduction	Gautam et al. (2019)
	Bacillus subtilis	Inhibitory effect by disrupting cell membrane	Wang et al. (2017)
	Pseudomonas aeruginosa	Antibacterial activity by binding to cell membrane	Mohanta et al. (2017)
	Alternaria alternate, Penicillium digitatum	Antifungal activity by destruction of fungi's membrane integrity	Salaheldin (2016)
Ag + Cesium nanoparticles (CsNPs)	*Pseudomonas syringaepv*	Destabilize enzyme for oxygen utilization for antibacterial activity	Shahryari et al. (2020)
Magnesium oxide nanoparticles (MgONPs)	*Ralstonia solanacearum, Phomopsis vexans*	Inactivate bactericidal activity by surface ROS generation & antifungal activity by inhibiting pore germination	Sharma et al. (2012)
SiNPs	*Alternaria alternate*	Antifungal activity by facile breakdown of cell wall	Derbalah et al. (2018)
Silicon oxide nanoparticles (SiO_2NPs)	*Fusarium oxysporum*	Antifungal activity by inhibiting spore germination	Akpinar et al. (2021)
TiO_2NPs	*Pseudomonas syringaepv, Lachrymans sp., Pseudomonas cubensis, Xanthomonas hortorum*	Reduce pathogen infection by direct oxidation of intracellular co-enzyme & cellular degradation	Cui et al. (2010) Norman and Chen (2011)

and water uptake as well as provide protection against various pathogens/biotic stressors.

21.6.2.1 Nanopesticides

Nanopesticides formulation is known to increase the solubility of poorly soluble ingredients, slow/targeted release of nanopesticides and protection against premature degradation of free pesticides (Kah et al. 2012). Pesticides, which include weed killers, pesticides, nematicides, hervicides, bactericides and biocides, are used to control pests and pathogens; hence, overall management of biotic stressors in the agricultural sector (Kumar et al. 2021). Engineered metal nanoparticles such as AgNPs, ZnONPs, copper nanoparticles (CuNPs), cerium nanoparticles (CeNPs) and aluminum nanoparticles (AlNPs) serve as nanopesticides, and exhibit antibacterial and pest control functions. According to Alghuthaymi et al. (2015), AgNPs not just obstruct *Raffaelea* sp. hyphae nutrient absorption, but also inhibit them from growth and attenuate fungal diseases during agricultural production. Besides, these nanoencapsulations in which nanometer size active pesticides are entrapped, are extensively used in agriculture to enhance insecticidal assessment by controlling active ingredient release; thereby improve efficacy and provide defense against numerous biotic stressors (Nuruzzaman et al. 2016).

21.6.2.2 Nanoherbicides

Herbicide resistance develops as a result of repeated use, lowering the production efficiency of essential crops. Nanomaterial-based herbicides formulation in an eco-friendly manner improves the overall efficacy of the herbicides (Chaud et al. 2021). Nanoparticle-encapsulated herbicide molecules aim to specific receptors at the roots of specific weed and translocate to inhibit glycolysis of the targeted root system. This action results in the starvation and death of the particular weed (Choudhary et al. 2020). The extensive use of nanoherbicides are reported in the rain-fed areas, where regular herbicides profligate due to insufficient soil moisture (Sekhon 2014). Polymeric nanoparticles concocted as nanospheres (altered-starch nanospheres), or nanocapsules (liposomal nanocapsules) are used to incorporate atrazine herbicides like poly-(epsilon)-caprolactone, which are distinguished by size, polydispersity index, pH and entrapment efficiency (Jiménez et al. 2004). Use of poly-(epsilon)-caprolactone is also advantageous due to excellent physico-chemical characteristics, as well as enhance bioavailability and biocompatibility, which increases the activity of herbicides molecule (atrazine) due to its slow release in environment while reducing its mobility for the benefit of biodiversity (Duhan et al. 2017) against target plant like *Brassica sp.* (Pereira et al. 2014). Another popular combination for nanoparticle encapsulation is Alginate/Chitosan (Ag/Cs) with parquet herbicides. The engineered nanoparticles encapsulated parquet herbicide shows a significant positive difference by averting the negative effects as induced by the herbicides itself (Chaud et al. 2021). However, in spite of research, till date, reports highlight the role of comparatively few polymeric nanoparticles-encapsulated herbicides in the agricultural sector, which provide a safer use due to low adverse environmental impacts on biodiversity as compared to the free herbicide molecules (Mittal et al. 2020) that further ensures in-depth study in this field. Carboxymethyl cellulose (CMC) received as forthcoming nanoherbicides in fertilizers in the form of a hydrogel to overcome the issues of splattering traditional fertilizers in the agricultural sector as dietary deficit possess major environmental risks if traditional modes do not work well (Mali et al. 2020).

21.6.2.3 Nanofertilizers

Nanofertilizers have been found to influence the structure of biotic stressors (microbial community) and have variable effects on targeted microbial groups. The application of nanofertilizers on the growth of *Acidobacteria* has been found to be suppressed by influencing their nitrogen and carbon cycles (Khan et al. 2021). Nanoparticles such as aluminum oxide nanoparticles (Al_2O_3NPs), SiO_2NPs, and AgNPs with a particular concentration of 50 mg/kg show significant decrease in the relative abundance of *Acidobacteria, Verrucoicrobia* and *Planctomycetes* population (Grün and Emmerling 2018). Li et al. (2014) reported that *Verrucomicrobia* and *Chloroflexi* bacteria lack an outer membrane, which facilitates iron (III)-silver nanomaterials to regulate microbial growth and population. Wang et al. (2018) reports AgNPs as nanofertilizer, control *Acidobacteria, Nitrospira, Cyanobacteria* and *Actinobacteria* populations. A report by Mishra et al. (2020) suggests that the exposure duration of nanofertilizer also plays an important role over the microbial community. Nanofertilizers also mitigate the fungal infection caused by specific biotic stressors such as *Bipolaris sorokiniana*, *Magnaporthe grisea, Alternaria alternate* and *Botrytis cinerea*. The application of nanofertilizer to soil has a negative impact on the uniqueness of ectomycorrhizal fungi on the root system of *Pinus muricata*, also known as "Bishop pine" (Sweet and Singleton 2015). Similarly, iron oxide nanoparticles (Fe_2O_3NPs) as nanofertilizers significantly reduce the mycorrhizal biomass and glomalin content in clover plants (Kalwani et al. 2022). Without a doubt, nanofertilizers have the ability to change current agricultural practices because of their distinct physicochemical properties, which can significantly conquer many of the drawbacks of conventional fertilizers such as atrazine, zinc fertilizers, and so on.

21.6.2.4 Nanoinsecticides

Polymer-based nanoencapsulated insecticides have several advantages over natural and chemical insecticides, including increase efficiency through sustainable release, reduce environmental contamination due to lower application rates, high photocatalysis susceptibility, and safer and easier handling for farmers and other agricultural workers (Huang et al. 2018). For the synthesis of engineered nanoinsecticides, various polysaccharides such as chitosan, starch, alginates, lignin, polyesters and polyamide have been considered (Torres and Pisani 2019). Nanofibers, nanospheres, nanogels, nanomicelles, nanocapsules and nanoemulsions are examples of polymer- and nonpolymer-based nanoinsecticide formulations. Among all of these categories, nanocapsules are most commonly used to control the release of insecticides. Many organic insecticides such as margosa oil, turmeric, lemon grass, henna and clove are also used as nanoformulations to ensure environmental eco-friendly, or green technology (Chaudhary 2017). A study by Deka et al. (2021) reports the use of polyethylene glycol (PEG) as copolymers for formulation of nanoinsecticides. PEG-based copolymers have higher biodegradability and are easier to process (Shakil et al. 2010). In comparison to chemical insecticides such as thiamethoxam (Sarkar and Shetty 2014), imidacloprid (Adak et al. 2012) and carbofuran (Shakil et al. 2010; Pankaj et al. 2012), PEG polymer nanoformulations-based insecticides demonstrate significantly slower rate of nanoparticle release and higher rate of efficacy. As nanoinsecticides are prepared using thermo-, light-, humidity-, soil-, pH- and enzyme-sensitive high polymer materials to deliver the pesticides, they possess slow release and protection performance. These formulations improve droplet adhesion on crop surfaces, resulting in improved dispersion and bioactivity of target pesticide molecules (Deka et al. 2021); thereby nanoinsecticides outperform conventional insecticides in biotic stress management.

21.6.2.5 Nanocomposites

Nanocomposites have been considered as composite materials with at least two characteristics: matrix and continuous phase. The classification of nanocomposite depends on the matrix material, which can be metal matrix composite such as ferrochromium (Fe-Cr)/Al_2O_3, cobalt/chrome (Co/Cr), iron/magnesium oxide (Fe/MgO), or ceramic matrix composite (Al_2O_3/SiO_2, Al_2O_3/CNTs, Al_2O_3/TiO_2), or polymer matrix composite (polymer/CNTs, polyester/TiO_2) (Camargo et al. 2009; Gupta 2018). Nanocomposite is important in plant expansion, progression and pest management. Metal nanocomposite such as AgNPs and CuNPs act as antimicrobial agent by interacting with bacterial surface and modifying their cell wall properties, leading to destruction of bacterial cell with leaky intracellular stuffs (Rai and Ingle 2012; Sekhon 2014). Metal nanocomposites interact with bacterial cellular organelles to perturb their metabolic processes and ultimately cause cellular death (Navarro et al. 2008). *Escherichia coli* and *Micrococcus luteus* colonies suppress after application of copper nanocomposites in soda lime glass powder, which represents its antibacterial activity (Esteban-Tejeda et al. 2009). Silver nanocomposites, on the other hand, have antimicrobial property against rhizospheric microbial populations (Arora et al. 2016). Polymer/CNTs have also been used as nanocomposites due to their multifunctional structures (Peigney et al. 2000). For example, chitosan/CNT nanocomposites regulate the broad-spectrum of insecticide azinphosmethyl and safeguard citrus fruits like *Citrus limon* and *Prunus persica* from pests and biotic stressors (Bibi et al. 2016).

21.6.2.6 Nanobiosensors

Nanobiosensors are analytical units containing sensitizing element derived from biological source like bacterial, or viral DNA, or proteins, and are attached with physiochemical transducers (for example, electrochemical, piezoelectric, or optical transducers). Nanobiosensors are capable to detect a range of microorganisms and many other analytes like insecticides and herbicides (You et al. 2009), and are perfect for field and laboratory settings due to ease-to-carry property. Nanobiosensors support the idea of sustainable agriculture in improving the crop productivity (Khanna 2008). Varied range of nanobiosensors is available that can be a promising tool for sustainable agriculture such as ion channel switch nanobiosensor, viral nanobiosensor, mechanical nanobiosensor, optical nanobiosensor, electronic nanobiosensor, photonic explorer for bioanalysis with biologically localized embedding (PEBBLE) biosensor, nanoshell biosensor and nanowire (Naresh and Lee 2021). Use of multiarrayed nanobiosensor for detection of plant pathogen is also established (Maurer et al. 2012). Nanobiosensors are also functioning in detection of antibacterial activity in certain plant species like *Malus domestica,* or *Aradiopsis* (Litescu et al. 2011). Besides, nanobiosensors are capable of detecting different biotoxins, carcinogens, agrochemical residues, as well as allergens (Ma et al. 2008; Rai et al. 2012; Zhang et al. 2007), resulting as a future for early diagnosis of diseases.

Overall, in the agricultural sector, nanotechnology has directly targeted the sustainable production by increasing crop growth and development, efficiency of water and nutrient uptake, and protection against biotic stressors like insects, pests, fungi, bacteria, etc. as illustrated in Fig. 21.3.

21.7 Impact of Nanotechnology on Occupational Health During Biotic Stress Management

Since the rapid increase of world population, day-by-day the demand for higher agricultural yields is rising; thus, it has become the need of the hour to evolve the effective strategies in order to optimize and maximize the agricultural production. For this purpose, nanotechnology is a promising technology especially, during biotic stress management (Gogos et al. 2012; Prasad et al. 2017) with an aim to maximize the agricultural output/crop yields, and prevent unnecessary crop damage, as well as lower input cost through different targeted approaches (Fraceto et al. 2016; Bose 2022; Neme et al. 2021; Servin et al. 2015). However, the use of nanomaterials in agricultural activities shows certain detrimental effects on targeted occupational groups (farmers, auxiliary workers) in spite of their promising phase in the agricultural sector such as TiO_2NPs penetrating dermis, leading to oxidative damage of skin (Naseer et al. 2018). TiO_2NPs as food additives cause unbalanced redox homeostasis-induced genetic mutations in human gastric epithelial cells (Botelho et al. 2014). It has also been found that nondegradable, or gradually degradable nanoparticles accumulate in different organs of human while use in the agricultural sector like tracheobronchial and alveolar regions of lungs, causing pulmonary and systemic inflammation, and onset of lung tumors (De Matteis 2017).

Further, phagocytes become overload with nanoparticles especially, AgNPs and SiNPs, and therefore, activate stress reactions in body, resulting in inflammation and weakening the defense mechanism against pathogen (Rajput et al. 2020). Nanoparticles when exposed to tissues and body fluids, uptake macromolecules (proteins, lipids, sugars) on their surface and form "nanoparticle-macromolecule complex" through adsorption, which affect the regulatory mechanisms of several enzymes and proteins (Saptarshi et al. 2013). Nanoparticles due to their unique physicochemical properties (shape, size, surface charge, solubility, hydrophilicity, etc.), interact with serum/plasma proteins like albumin and fibrinogen (Saptarshi et al. 2013). DNA methylation, histone modification and posttranslational changes of specific gene expression such as cardiac genes like euchromatic histone lysine methyltransferase 2 (EHMT2), or nuclear receptor binding SET domain Protein 1 (NSD1), endothelial genes like methyl CpG binding protein 2 (MECP2), or

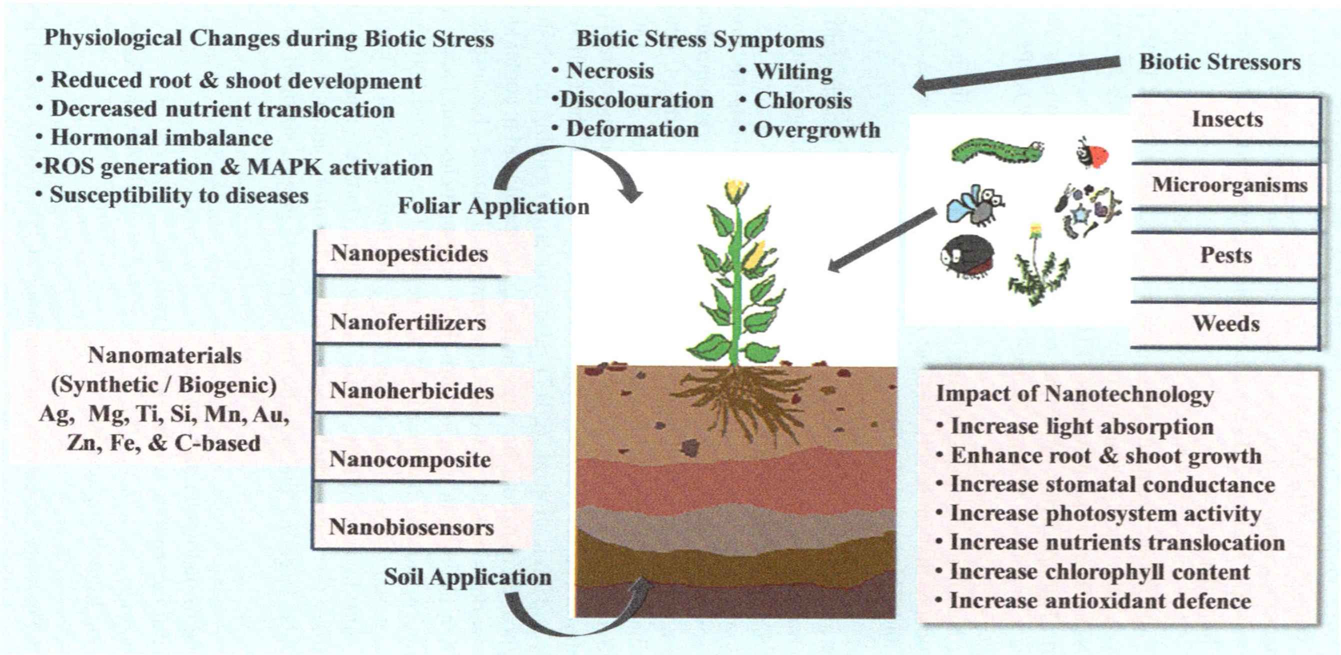

Fig. 21.3 A model of biotic stress' effects on crop health management and sustainable growth using nanobiotechnology in the agricultural sector

Kruppel-like factor 2 (KLF2), etc. are the proven health hazards caused by the enormous use of nanoparticles in the agricultural sector (Naseer et al. 2018; Yang et al. 2021; Gogos et al. 2012; Kah et al. 2012).

Overall, nanomaterials as discussed here, entre the human body through the following routes (De Matteis 2017), which facilitates generation of "free radicals", leading to cellular damage.

- **Dermal:** Nanoparticles remain on skin for an extended period of time, penetrate into the body; the most common route for pesticide poisoning among regular users (AgNPs) (Barua and Mitragotri 2014).
- **Inhalation:** Nanoparticles enter the body through inhalation, or respiratory exposure, cause serious damage to nose, throat and lungs (TiO_2NPs) (Sonwani et al. 2021).
- **Ingestion:** Nanoparticles that enter the body through mouth is the most frequent way of exposure that cause serious illness, severe injury, or even death (AgNPs) (Damalas and Koutroubas 2016).

Thomas et al. (2012) revealed that nanoparticles, once in bloodstream, become competent to traverse the blood-brain barrier, leading to neurological disorders such as Parkinson's, Alzheimer's and Huntington's.

The potential uses of nanoparticles are listed here, responsible for crop disease protection, prevention against biotic stressors, nanofertilizers, nanosensors and nanopesticides, encapsulation of enzymes (horseradish peroxidase), detection of food contamination and food adulteration, etc. (Iavicoli et al. 2017; Liu and Lal 2015) that provide hope to the world community in order to combat current problems in the agricultural sector. However, Kah (2015), Scott and Chen (2013), Thomas et al. (2012), and Iavicoli et al. (2017) raise concern regarding hassle-free wide-spectrum application of nanotechnology in the agricultural field in context to farmers and auxiliary workers' health, as well as environmental hazards, because agricultural workers are first to come in contact with these nanoparticles and facing high risk. Most of these engineered nanoparticles are reported xenobiotic and agricultural workers are exposed for a substantial period during their job tasks (Vimercati et al. 2020; Leso and Iavicoli 2018). Thus, it is important to ensure a complete safety guidelines for nanoparticles prior to their handling and use in large-scale especially, when targeting the specific occupational groups (Iavicoli et al. 2017) involved in:

- Engineering of nanomaterials.
- Nanomaterial incorporation into agricultural goods.
- Mixing, loading, and transporting nanoformulated pesticides and fertilizers.
- Licensed and trained applicators.

Long-term exposure to carbon nanoparticles (CNPs) gives rise to pulmonary inflammation, epithelial damage and vascular impairments such as reduce heart rate and alteration in heart rate variability (Zheng et al. 2018). TiO_2NPs enter the body through dermis and interacts with lymph nodes, leading to oxidative damage to skin by generation of free hydroxyl radicals (Naseer et al. 2018). AgNPs and TiO_2NPs induce cytotoxic and genotoxic effects to human, as well as experimental model (mice) by generation of ROS, leading to cell proliferation, DNA damage, mitochondrial membrane damage, apoptosis and cell membrane perforation, which is manifested by inflammation, carcinogenesis, genotoxicity and fibrosis (Gopinath et al. 2008; Mohamed and Hussien 2016; Wan et al. 2012; Abdal Dayem et al. 2017). *In vivo* studies of SiO_2NPs demonstrate lung inflammation, granuloma and focal emphysema in the workers of chemical industries, food packaging industries, cosmetics, mining and agricultural sector (Leung et al. 2012; Lee et al. 2020). AuNPs have wide applications in food packaging, which cause chromatin damage in human lung fibroblasts, manifested by changes in gene expression and epigenetic changes that may be a potential risk of lung cancers in the farmers and other agricultural field workers (Attarilar et al. 2020).

AgNPs in toxicological studies demonstrate ROS-induced damage of cell membrane, DNA and chromosomes while entering through intestinal mucus barrier (Attarilar et al. 2020). The cytotoxicity of AgNPs depends on its concentration, duration of exposure, extent of particle aggregation and surface oxidation while forming silver oxides (Akter et al. 2018). Toxicological effects of these nanoparticles on body with higher concentrations give rise to issues such as abnormal cell damage, shrinkage, apoptosis and skin cancers (Duncan 2011). Short-term toxicological studies also demonstrate that nanomaterials have a proinflammatory influence on human health by affecting the respiratory, cardiovascular, nervous and endocrine systems (Iavicoli et al. 2013; Thomas et al. 2012; Lu and Radisic 2021), whereas long-term exposure causes adverse effects on vital organs like liver, kidney, lungs, etc. (Naseer et al. 2018). Further, Momin et al. (2013) demonstrate the potency of nanoparticles-induce oxidative stress in brain inducing apoptosis and DNA damage while passing through lungs to blood, cell nuclei and central nervous system (CNS), leading to risk of inflammations in neurons as manifested by DNA impairment, and Parkinson's and Alzheimer's disease.

So, in nutshell, from an occupational health perspective, nanotechnology in the agricultural sector demands an urgent action as workers including farmers and auxiliary workers are the most vulnerable groups, who might be in contact with these xenobiotic prior to others (consumers) and that too for longer periods of time due to their unique job profile. Occupational safety is thus required to be ensured sincerely in this field before large-scale production of nanomaterials and application of nanotechnology in biotic stress management (Maynard and Kuempel 2005; Seaton et al. 2010). Additionally, researches suggest that the toxicological impact of nanomaterial exposure should be prioritized first. It is also important to assess all possible occupational health implications of nanomaterials in the agricultural sector and simultaneously suitable approaches for risk assessment, management and prevention should be investigated. Hence, large-scale development of nanomaterials with utmost care and responsibility are the critical needs of the hour with an ultimate goal for management of biotic stressors in agriculture.

21.8 Conclusion and Future Perspective of Research

In conclusion, it can be said that in spite of advancement of science and technology, till today due to limited resources and vast climatic changes, we are facing numerous challenges to feed growing population of the world, 9.9% of which comes from the developing and underdeveloped countries (Action Against Hunger 2022). In addition, the present agriculture practices and extensive use of chemicals contaminate soil, underground water and food supplies with toxic pollutants that possess severe impact on environment and human health either directly, or indirectly, which cannot be left unnoticed in the need of increasing agricultural demands day-by-day. Globally, around 192.8 m chemical fertilizers are used during 2016–2021. Hence, nanotechnology and/or combination with phenolic compounds may act as suitable and cost-effective carriers, allowing human population to eventually see the dream of achieving sustainable and environmentally friendly agricultural yield in one hand, and combating with biotic stressors in the other hand. Worldwide, many developed and developing countries; thus, predict the potential role of nanotechnology as an alternative in the agricultural sector that can reduce crop diseases and increase the yield; but at the same time, exposure to nanomaterials in the agricultural field causes several diseases among the farmers and auxiliary workers (Sekhon 2014). Moreover, extensive use of nanotechnology in this field plays critical role in environmental hazards that indicate implication of nanotechnology in agriculture should have three major criteria, i.e., eco-friendly access, hazards prevention and minimum, or no risk associated with it. Therefore, future research should consider this aspect along with generation of public awareness is needed through the Govt./non-Govt. not-for-profit organizations to develop the technology on a large-scale basis in one hand for achieving sustainable development in true sense, and in other hand promoting and maintaining the green and clean environment.

Author Contribution **NN**, corresponding author; conceptualized and edited the book chapter. **DR** and **SS** performed literature survey and drew illustration **DR, SS,** and **NN** wrote the book chapter and approved the final version of the book chapter. **DR** and **NN** performed proof correction. **SS** contributed during her one and half-year tenure at ICMR-NIOH.

Acknowledgment Thanks to ICMR-NIOH, for basic infrastructure support; Ms. Vishakha Shrimali, PhD student of Biochemistry Department, for bibliography editing; and Dr. A Banerjee, BITS-Pilani (Goa campus), for plagiarism checking.

Fund Disclaimer The authors declare no grant received from any funding agency in a public, commercial, or nonprofit sector.

Conflict of Interest The authors declare no actual, or potential competing financial interest; and confirm that the institute had no influence over the study design, content of the article, or selection of the publishing house.

Declaration The authors declare that the manuscript, or portions thereof, is not under consideration by another book/journal, and has not been previously published. The research performed is in accordance with the statement of ethical standards of the publishing house. Further, the editors of the book have the right of reviewing and editing the submission; hence, authors transfer, assign, or otherwise convey all copyright ownership to the publisher.

References

Abdal Dayem A, Hossain M, Lee S, Kim K, Saha S, Yang G, Choi H, Cho S (2017) The role of reactive oxygen species (ROS) in the biological activities of metallic nanoparticles. Int J Mol Sci 18(1):120. https://doi.org/10.3390/ijms18010120

Abdul Malik N, Kumar I, Nadarajah K (2020) Elicitor and receptor molecules: orchestrators of plant defense and immunity. Int J Mol Sci 21(3):963. https://doi.org/10.3390/ijms21030963

Action Against Hunger (2022) Action Against Hunger—Ending world hunger & malnutrition. [online]. https://www.actionagainsthunger.org/. Accessed 10 May 2022

Adak T, Kumar J, Shakil N, Walia S (2012) Development of controlled release formulations of imidacloprid employing novel nano-ranged amphiphilic polymers. J Environ Sci Health B 47(3):217–225. https://doi.org/10.1080/03601234.2012.634365

Adam M, Heuer H, Hallmann J (2014) Bacterial antagonists of fungal pathogens also control root-knot Nematodes by induced systemic resistance of tomato plants. PLoS One 9(2):e90402. https://doi.org/10.1371/journal.pone.0090402

Ahmed O, Hassan M, Abdel-Twab S, Abdel Azeem M (2017) Navel orange peel hydroethanolic extract, naringin and naringenin have anti-diabetic potentials in type 2 diabetic rats. Biomed Pharmacother 94:197–205. https://doi.org/10.1016/j.biopha.2017.07.094

Akkopru A, Çakar K, Husseini A (2018) Effects of endophyte bacteria on disease and plant growth in plants under biotic stress. Yüzüncü Yıl Univ J Agric Sci 28(2):200–208. https://doi.org/10.29133/yyutbd.418070

Akpinar I, Unal M, Sar T (2021) Potential antifungal effects of silver nanoparticles (AgNPs) of different sizes against phytopathogenic Fusarium oxysporum f. sp. radicis- lycopersici (FORL) strains. SN Appl Sci 3(4):506. https://doi.org/10.1007/s42452-021-04524-5

Akter M, Sikder M, Rahman M, Ullah A, Hossain K, Banik S, Hosokawa T, Saito T, Kurasaki M (2018) A systematic review on silver nanoparticles-induced cytotoxicity: physicochemical properties and perspectives. J Adv Res 9:1–16. https://doi.org/10.1016/j.jare.2017.10.008

Alara O, Abdurahman N, Ukaegbu C (2021) Extraction of phenolic compounds: a review. Curr Res Food Sci 4:200–214. https://doi.org/10.1016/j.crfs.2021.03.011

Albaayit SFA, Abba Y, Abdullah R, Abdullah N (2014) Evaluation of Antioxidant Activity and Acute Toxicity of Clausena excavata Leaves Extract. Evid Based Complementary Altern Med 2014:975450. https://doi.org/10.1155/2014/975450

Aldon D, Mbengue M, Mazars C, Galaud J (2018) Calcium signalling in plant biotic interactions. Int J Mol Sci 19(3):665. https://doi.org/10.3390/ijms19030665

Alghuthaymi M, Almoammar H, Rai M, Said-Galiev E, Abd-Elsalam K (2015) Myconanoparticles: synthesis and their role in phytopathogens management. Biotechnol Biotechnol Equip 29(2): 221–236. https://doi.org/10.1080/13102818.2015.1008194

Ali S, Ganai B, Kamili A, Bhat A, Mir Z, Bhat J, Tyagi A, Islam S, Mushtaq M, Yadav P, Rawat S, Grover A (2018) Pathogenesis-related proteins and peptides as promising tools for engineering plants with multiple stress tolerance. Microbiol Res 212–213:29–37. https://doi.org/10.1016/j.micres.2018.04.008

Aljabali A, Akkam Y, Al Zoubi M, Al-Batayneh K, Al-Trad B, Abo Alrob O, Alkilany A, Benamara M, Evans D (2018) Synthesis of gold nanoparticles using leaf extract of Ziziphus zizyphus and their antimicrobial activity. Nanomaterials 8(3):174. https://doi.org/10.3390/nano8030174

Alsaeedi A, El-Ramady H, Alshaal T, El-Garawani M, Elhawat N, Almohsen M (2017) Engineered silica nanoparticles alleviate the detrimental effects of Na+ stress on germination and growth of common bean (Phaseolus vulgaris). Environ Sci Pollut Res 24(27):21917–21928. https://doi.org/10.1007/s11356-017-9847

Alwis K, Mandryk J, Hocking A (1999) Exposure to biohazards in wood dust: bacteria, fungi, endotoxins, and (1 3)-beta-D-glucans. Appl Occup Environ Hyg 14(9):598–608. https://doi.org/10.1080/104732299302404

Amiot M, Riva C, Vinet A (2016) Effects of dietary polyphenols on metabolic syndrome features in humans: a systematic review. Obes Rev 17(7):573–586. https://doi.org/10.1155/2017/5812401

Anusha C, Sumathi T, Joseph L (2017) Protective role of apigenin on rotenone induced rat model of Parkinson's disease: suppression of neuroinflammation and oxidative stress mediated apoptosis. Chem Biol Interact 269:67–79. https://doi.org/10.1016/j.cbi.2017.03.016

Arora P, Mehta C, Srivastava R, Arora S, Sharma A (2016) Impact assessment of silver nanoparticles on plant growth and soil bacterial diversity. 3 Biotech 6(2):254. https://doi.org/10.1007/s13205-016-0567-7

Asai S, Yoshioka H (2008) The role of radical burst via MAPK signaling in plant immunity. Plant Signal Behav 3(11):920–922. https://doi.org/10.4161/psb.6601

Attar A, Altikatoglu M (2018) Biomimetic synthesis, characterization and antibacterial efficacy of ZnO and Au nanoparticles using echinacea flower extract precursor. Mater Res Express 5(5): 055403. https://doi.org/10.1088/2053-1591/aac05f

Attarilar S, Yang J, Ebrahimi M, Wang Q, Liu J, Tang Y, Yang J (2020) The toxicity phenomenon and the related occurrence in metal and metal oxide nanoparticles: a brief review from the biomedical perspective. Front Bioeng Biotechnol 8:822. https://doi.org/10.3389/fbioe.2020.00822

Barua S, Mitragotri S (2014) Challenges associated with penetration of nanoparticles across cell and tissue barriers: a review of current status and future prospects. Nano Today 9(2):223–243. https://doi.org/10.1016/j.nantod.2014.04.008

Beekrum S, Govinden R, Padayachee T, Odhav B (2003) Naturally occurring phenols: a detoxification strategy for fumonisin B1. Food Addit Contam 20(5):490–493. https://doi.org/10.1080/0265203031000098678

Bellé C, dos Santos P, Kaspary T (2021) First report of rice root-knot nematode, *Meloidogyne graminicola*, infecting *Juncus microcephalus* in Brazil. J Nematol 53(1):1–4. https://doi.org/10.21307/jofnem-2021-031

Belobrajdic D, Bird A (2013) The potential role of phytochemicals in wholegrain cereals for the prevention of type-2 diabetes. Nutr J 12(1):62. https://doi.org/10.1186/1475-2891-12-62

Bhalodia N, Shukla V (2011) Antibacterial and antifungal activities from leaf extracts of Cassia fistula l.: an ethnomedicinal plant. J Adv Pharma Technol Res 2(2):104. https://doi.org/10.4103/2231-4040.82956

Bhattacharya A, Sood P, Citovsky V (2010) The roles of plant phenolics in defence and communication during Agrobacterium and Rhizobium infection. Mol Plant Pathol 11(5):705–719. https://doi.org/10.1111/j.1364-3703.2010.00625.x

Bibi S, Nawaz M, Yasin T, Riaz M (2016) Chitosan/CNTs nanocomposite as green carrier material for pesticides controlled release. J Polym Res 23(8):1–7. https://doi.org/10.1007/s10965-016-1055-9

Boerjan W, Ralph J, Baucher M (2003) Lignin biosynthesis. Annu Rev Plant Biol 54(1):519–546. https://doi.org/10.1146/annurev.arplant.54.031902.134938

Bose P (2022) How agricultural nanotechnology will influence the future of farming sustainability. [online] AZoNano.com. https://www.azonano.com/article.aspx?ArticleID=5647. Accessed 22 Mar 2022

Boote K (1983) Coupling pests to crop growth stimulators to predict yield reductions. Phytopathology 73(11):1581. https://doi.org/10.1094/PHYTO-73-1581

Borji H, Razmi G, Ahmadi A, Karami H, Yaghfoori S, Abedi V (2011) A survey on endoparasites and ectoparasites of stray cats from Mashhad (Iran) and association with risk factors. J Parasit Dis 35(2):202–206. https://doi.org/10.1007/s12639-011-0057-0

Botelho M, Costa C, Silva S, Costa S, Dhawan A, Oliveira P, Teixeira J (2014) Effects of titanium dioxide nanoparticles in human gastric epithelial cells in vitro. Biomed Pharmacother 68(1): 59–64. https://doi.org/10.1016/j.biopha.2013.08.006

Boudet A (2007) Evolution and current status of research in phenolic compounds. Phytochemistry 68(22–24):2722–2735. https://doi.org/10.1016/j.phytochem.2007.06.012

Cabral J (2010) Water microbiology. Bacterial pathogens and water. Int J Environ Res Public Health 7(10):3657–3703. https://doi.org/10.3390/ijerph7103657

Caleja C, Ribeiro A, Barreiro M, Ferreira I (2017) Phenolic compounds as nutraceuticals or functional food ingredients. Curr Pharm Des 23(19):2787–2806. https://doi.org/10.2174/1381612822666161227153906

Calvo-Castro L, Syed D, Chamcheu J, Vilela F, Pérez A, Vaillant F, Rojas M, Mukhtar H (2013) Protective effect of tropical highland blackberry juice (*Rubus adenotrichos/Schltdl.*) against UVB-mediated damage in human epidermal keratinocytes and in a reconstituted skin equivalent model. Photochem Photobiol 89(5):1199–1207. https://doi.org/10.1111/php.12104

Camargo P, Satyanarayana K, Wypych F (2009) Nanocomposites: synthesis, structure, properties and new application opportunities. Mater Res 12(1):1–39. https://doi.org/10.1590/s1516-14392009000100002

Capasso R, Cristinzio G, Evidente A, Scognamiglio F (1992) Isolation, spectroscopy and selective phytotoxic effects of polyphenols from vegetable waste waters. Phytochemistry 31(12): 4125–4128. https://doi.org/10.1016/0031-9422(92)80426-F

Carocho M, Ferreira I (2013) The role of phenolic compounds in the fight against cancer—a review. Anticancer Agents Med Chem 13(8):1236–1258. https://doi.org/10.2174/18715206113139990301

Chan C, Yusoff R, Ngoh G, Kung F (2011) Microwave-assisted extractions of active ingredients from plants. J Chromatography A 1218(37):6213–6225. https://doi.org/10.1016/j.chroma.2011.07.040

Chaud M, Souto E, Zielinska A, Severino P, Batain F, Oliveira-Junior J, Alves T (2021) Nanopesticides in agriculture: benefits and challenge in agricultural productivity, toxicological risks to human health and environment. Toxics 9(6):131. https://doi.org/10.3390/toxics9060131

Chaudhary S (2017) Progress on Azadirachta indica based biopesticides in replacing synthetic toxic pesticides. Front Plant Sci 8:610. https://doi.org/10.3389/fpls.2017.00610

Cheynier V, Comte G, Davies K, Lattanzio V, Martens S (2013) Plant phenolics: recent advances on their biosynthesis, genetics, and ecophysiology. Plant Physiol Biochem 72:1–20. https://doi.org/10.1016/j.plaphy.2013.05.009

Chitwood D (2003) Research on plant-parasitic nematode biology conducted by the United States Department of Agriculture-Agricultural Research Service. Pest Manag Sci 59(6–7):748–753. https://doi.org/10.1002/ps.684

Choudhary S, Kumar A, Kumar R (2020) Novel nanotechnological tools for weed management—a review. Chem Sci Rev Lett 9(36):886–894. https://doi.org/10.37273/chesci.CS205105136

Clé C, Hill L, Niggeweg R, Martin CR, Guisez Y, Prinsen E, Jansen M (2008) Modulation of chlorogenic acid biosynthesis in Solanum lycopersicum; consequences for phenolic accumulation and UV-tolerance. Phytochemistry 69(11):2149–2156. https://doi.org/10.1016/j.phytochem.2008.04.024

Cory H, Passarelli S, Szeto J, Tamez M, Mattei J (2018) The role of polyphenols in human health and food systems: a mini-review. Front Nutr 5:87. https://doi.org/10.3389/fnut.2018.00087

Cosme P, Rodríguez A, Espino J, Garrido M (2020) Plant phenolics: bioavailability as a key determinant of their potential health-promoting applications. Antioxidants 9(12):1263. https://doi.org/10.3390/antiox9121263

Cui H, Jiang J, Gu W, Sun C, Wu D, Yang T, Yang G (2010) Photocatalytic inactivation efficiency of anatase nano-TiO2 sol on the H9N2 avian influenza virus. Photochem Photobiol *86*(5): 1135–1139. https://doi.org/10.1111/j.1751-1097.2010.00763.x

Culver J, Padmanabhan M (2007) Virus-induced disease: altering host physiology one interaction at a time. Annu Rev Phytopathol 45(1):221–243. https://doi.org/10.1146/annurev.phyto.45.062806.094422

Czosnek H, Laterrot H (1997) A worldwide survey of tomato yellow leaf curl viruses. Arch Virol 142(7):1391–1406. https://doi.org/10.1007/s007050050168

Dagoudo M, Qiang J, Solevo M (2022) Status in science and technology developments in Benin's aquaculture industry: a review. Aquacult Int 9(47):49. https://doi.org/10.1007/s10499-022-00853-3

Dahiya A, Chahar K, Sindhu S (2020) The rhizosphere microbiome and biological control of weeds: a review. Spanish J Agric Res 17(4):e10R01. https://doi.org/10.5424/sjar/2019174-15073

Dakora F, Phillips D (1996) Diverse functions of isoflavonoids in legumes transcend anti- microbial definitions of phytoalexins. Physiol Mol Plant Pathol 49(1):1–20. https://doi.org/10.1006/pmpp.1996.0035

Damalas C, Koutroubas S (2016) Farmers' exposure to pesticides: toxicity types and ways of prevention. Toxics 4(1):1. https://doi.org/10.3390/toxics4010001

Dangl J, McDowell J (2006) Two modes of pathogen recognition by plants. Proc Natl Acad Sci U S A 103(23):8575–8576. https://doi.org/10.1073/pnas.0603183103

Dar S, Rather B, Wani A, Ganie M (2017) Resistance against insect pests by plant phenolics and their derivative compounds. Chem Sci Rev Lett *6*(23):1941–1949. issn:2278-6783

De Matteis V (2017) Exposure to inorganic nanoparticles: routes of entry, immune response, biodistribution and in vitro/in vivo toxicity evaluation. Toxics 5(4):29. https://doi.org/10.3390/toxics5040029

Debnath N, Das S, Seth D, Chandra R, Bhattacharya S, Goswami A (2011) Entomotoxic effect of silica nanoparticles against Sitophilus oryzae (L.). J Pest Sci 84(1):99–105. https://doi.org/10.1007/s10340-010-0332-3

Deka B, Babu A, Baruah C, Barthakur M (2021) Nanopesticides: a systematic review of their prospects with special reference to tea pest management. Front Nutr 8:686131. https://doi.org/10.3389/fnut.2021.686131

Del Valle P, García-Armesto M, de Arriaga D, González-Donquiles C, Rodríguez- Fernández P, Rúa J (2016) Antimicrobial activity of kaempferol and resveratrol in binary combinations with parabens or propyl gallate against Enterococcus faecalis. Food Control 61:213–220. https://doi.org/10.1016/j.foodcont.2015.10.001

Derbalah A, Shenashen M, Hamza A, Mohamed A, El Safty S (2018) Antifungal activity of fabricated mesoporous silica nanoparticles against early blight of tomato. Egypt J Basic Appl Sci 5(2):145–150. https://doi.org/10.1016/j.ejbas.2018.05.002

Dimkpa C, McLean J, Britt D, Anderson A (2013) Antifungal activity of ZnO nanoparticles and their interactive effect with a biocontrol bacterium on growth antagonism of the plant pathogen Fusarium graminearum. Biometals 26(6):913–924. https://doi.org/10.1007/s10534-013-9667-6

Dixon R (2001) Natural products and plant disease resistance. Nature 411(6839):843–847. https://doi.org/10.1038/35081178

Dixon R, Paiva N (1995) Stress-induced phenylpropanoid metabolism. Plant Cell 7(7):1085. https://doi.org/10.1105/tpc.7.7.1085

Donham K, Rautiainen R, Schuman S, Lay J (2021) Agricultural health and safety: recent advances. CRC Press, Boca Raton. https://doi.org/10.1201/9781003248958

Dudonné S, Poupard P, Coutière P, Woillez M, Richard T, Mérillon J, Vitrac X (2011) Phenolic composition and antioxidant properties of poplar bud (*Populus nigra*) extract: individual antioxidant contribution of phenolics and transcriptional effect on skin aging. J Agric Food Chem 59(9):4527–4536. https://doi.org/10.1021/jf104791t

Duhan JS, Kumar R, Kumar N, Kaur P, Nehra K, Duhan S (2017) Nanotechnology: the new perspective in precision agriculture. Biotechnol Rep 15:11–23. https://doi.org/10.1016/j.btre.2017.03.002

Duncan T (2011) Applications of nanotechnology in food packaging and food safety: barrier materials, antimicrobials and sensors. J Colloid Interface Sci 363(1):1–24. https://doi.org/10.1016/j.jcis.2011.07.017

Eerdunbayaer, Orabi MAA, Aoyama H, Kuroda T, Hatano T (2014) Structures of two new flavonoids and effects of licorice phenolics on vancomycin-resistant Enterococcus species. Molecules 19(4):3883–3897. https://doi.org/10.3390/molecules19043883

Esteban-Tejeda L, Malpartida F, Esteban-Cubillo A, Pecharromán C, Moya J (2009) The antibacterial and antifungal activity of a soda-lime glass containing silver nanoparticles. Nanotechnology 20(8):085103. https://doi.org/10.1088/0957-4484/20/8/085103

European and Mediterranean Plant Protection Organization (EPPO) Bulletin (2013) PM 7/119 (1) Nematode extraction. 43(3):471–495. https://doi.org/10.1111/epp.12077.

Falcone Ferreyra M, Rius S, Casati P (2012) Flavonoids: biosynthesis, biological functions, and biotechnological applications. Front Plant Sci 3:222. https://doi.org/10.3389/fpls.2012.00222

Ficke A, Cowger C, Bergstrom G, Brodal G (2018) Understanding yield loss and pathogen biology to improve disease management: *Septoria Nodorum* blotch—a case study in wheat. Plant Dis 102(4):696–707. https://doi.org/10.1094/pdis-09-17-1375-fe

Flury P, Aellen N, Ruffner B, Péchy-Tarr M, Fataar S, Metla Z, Dominguez-Ferreras A, Bloemberg G, Frey J, Goesmann A, Raaijmakers J, Duffy B, Höfte M, Blom J, Smits T, Keel C, Maurhofer M (2016) Insect pathogenicity in plant-beneficial pseudomonads: phylogenetic distribution and comparative genomics. ISME J 10(10):2527–2542. https://doi.org/10.1038/ismej.2016.5

Food and Agriculture Organization (FAO) (2022) The State of Food and Agriculture 2022. Leveraging automation in agriculture for transforming agrifood systems. FAO Accessed 20 March 2022. https://doi.org/10.4060/cb9479en

Fraceto L, Grillo R, de Medeiros G, Scognamiglio V, Rea G, Bartolucci C (2016) Nanotechnology in agriculture: which innovation potential does it have? Front Environ Sci 4:20. https://doi.org/10.3389/fenvs.2016.00020

Fraser C, Chapple C (2011) The phenylpropanoid pathway in Arabidopsis. Arabidopsis Book 9: e0152. https://doi.org/10.1199/tab.0152

Fürstenberg-Hägg J, Zagrobelny M, Bak S (2013) Plant defense against insect herbivores. Int J Mol Sci 14(5):10242–10297. https://doi.org/10.3390/ijms140510242

Gasnier C, Dumont C, Benachour N, Clair E, Chagnon M, Séralini G (2009) Glyphosate-based herbicides are toxic and endocrine disruptors in human cell lines. Toxicology 262(3):184–191. https://doi.org/10.1016/j.tox.2009.06.006

Gautam N, Salaria N, Thakur K, Kukreja S, Yadav N, Yadav R, Goutam U (2019) Green silver nanoparticles for phytopathogen control. Proc Natl Acad Sci India Sect B Biol Sci 90(2): 439–446. https://doi.org/10.1007/s40011-019-01115-8

Gimenez E, Salinas M, Manzano-Agugliaro F (2018) Worldwide research on plant defense against biotic stresses as improvement for sustainable agriculture. Sustainability 10(2):391. https://doi.org/10.3390/su10020391

Gogos A, Knauer K, Bucheli T (2012) Nanomaterials in plant protection and fertilization: current state, foreseen applications, and research priorities. J Agric Food Chem 60(39):9781–9792. https://doi.org/10.1021/jf302154y

Gopinath P, Gogoi S, Chattopadhyay A, Ghosh S (2008) Implications of silver nanoparticle induced cell apoptosis for in vitro gene therapy. Nanotechnology 19(7):075104. https://doi.org/10.1088/0957-4484/19/7/075104

Gouveia B, Calil I, Machado J, Santos A, Fontes E (2017) Immune receptors and co-receptors in antiviral innate immunity in plants. Front Microbiol 7:2193. https://doi.org/10.3389/fmicb.2016.02139

Grün A, Emmerling C (2018) Long-term effects of environmentally relevant concentrations of silver nanoparticles on major soil bacterial phyla of a loamy soil. Environ Sci Europe 30(1): 1–13. https://doi.org/10.1186/s12302-018-0160-2

Gunalan S, Sivaraj R, Rajendran V (2012) Green synthesized ZnO nanoparticles against bacterial and fungal pathogens. Progress Nat Sci Mater Int 22(6):693–700. https://doi.org/10.1016/j.pnsc.2012.11.015

Guo Q, Major I, Howe G (2018) Resolution of growth–defense conflict: mechanistic insights from jasmonate signaling. Curr Opin Plant Biol 44:72–81. https://doi.org/10.1016/j.pbi.2018.02.009

Gupta H (2018) Role of nanocomposites in agriculture. In: Nano hybrids and composites. Trans Tech Publications Ltd. 20:81–89. https://doi.org/10.4028/www.scientific.net/nhc.20.81.

Hamamouch N, Li C, Seo P, Park C, Davis E (2010) Expression of Arabidopsis pathogenesis-related genes during nematode infection. Mol Plant Pathol 12(4):355–364. https://doi.org/10.1111/j.1364-3703.2010.00675.x

Huang H, Tao M, Hung T, Chen J, Lin Z, Huang C (2014) (−)-Epigallocatechin-3- gallate inhibits entry of hepatitis B virus into hepatocytes. Antiviral Res 111:100–111. https://doi.org/10.1016/j.antiviral.2014.09.009

Huang B, Chen F, Shen Y, Qian K, Wang Y, Sun C, Zhao X, Cui B, Gao F, Zeng Z, Cui H (2018) Advances in targeted pesticides with environmentally responsive controlled release by nanotechnology. Nanomaterials 8(2):102. https://doi.org/10.3390/nano8020102

Hursel R, Viechtbauer W, Dulloo A, Tremblay A, Tappy L, Rumpler W, Westerterp-Plantenga M (2011) The effects of catechin rich teas and caffeine on energy expenditure and fat oxidation: a meta-analysis. Obes Rev 12(7):e573–e581. https://doi.org/10.1111/j.1467-789x.2011.00862.x

Iavicoli I, Fontana L, Leso V, Bergamaschi A (2013) The effects of nanomaterials as endocrine disruptors. Int J Mol Sci 14(8):16732–16801. https://doi.org/10.3390/ijms140816732

Iavicoli I, Leso V, Beezhold D, Shvedova A (2017) Nanotechnology in agriculture: opportunities, toxicological implications, and occupational risks. Toxicol Appl Pharmacol 329:96–111. https://doi.org/10.1016/j.taap.2017.05.025

International Labour Organization Statistics (ILOSTAT) (2022) Country profiles—ILOSTAT. [online]. https://ilostat.ilo.org/data/country-profiles/. Accessed 27 Mar 2022.

Iqbal Z, Iqbal M, Hashem A, Abd Allah E, Ansari M (2021) Plant defense responses to biotic stress and its interplay with fluctuating dark/light conditions. Front Plant Sci 12:631810. https://doi.org/10.3389/fpls.2021.631810

Jayaseelan C, Ramkumar R, Rahuman A, Perumal P (2013) Green synthesis of gold nanoparticles using seed aqueous extract of Abelmoschus esculentus and its antifungal activity. Indus Crops Prod 45:423–429. https://doi.org/10.1016/j.indcrop.2012.12.019

Jhanzab H, Razzaq A, Bibi Y, Yasmeen F, Yamaguchi H, Hitachi K, Tsuchida K, Komatsu S (2019) Proteomic analysis of the effect of inorganic and organic chemicals on silver nanoparticles in wheat. Int J Mol Sci 20(4):825. https://doi.org/10.3390/ijms20040825

Jiménez M, Pelletier J, Bobin M, Martini M, Fessi H (2004) Poly-epsilon-caprolactone nanocapsules containing octyl methoxycinnamate: preparation and characterization. Pharma Dev Technol 9(3):329–339. https://doi.org/10.1081/PDT-200031456

Jones N, Ray B, Ranjit K, Manna A (2008) Antibacterial activity of ZnO nanoparticle suspensions on a broad spectrum of microorganisms. FEMS Microbiol Lett 279(1):71–76. https://doi.org/10.1111/j.1574-6968.2007.01012.x

Kah M (2015) Nanopesticides and nanofertilizers: emerging contaminants or opportunities for risk mitigation? Front Chem 3:64. https://doi.org/10.3389/fchem.2015.00064

Kah M, Beulke S, Tiede K, Hofmann T (2012) Nanopesticides: state of knowledge, environmental fate, and exposure modeling. Crit Rev Environ Sci Technol 43(16):1823–1867. https://doi.org/10.1080/10643389.2012.671750

Kalpana V, Devi Rajeswari V (2018) A review on green synthesis, biomedical applications, and toxicity studies of ZnO NPs. Bioinorg Chem Appl 2018:1–12. https://doi.org/10.1155/2018/3569758

Kalwani M, Chakdar H, Srivastava A, Pabbi S, Shukla P (2022) Effects of nanofertilizers on soil and plant-associated microbial communities: emerging trends and perspectives. Chemosphere 287:132107. https://doi.org/10.1016/j.chemosphere.2021.132107

Kazi A, Daniel K, Smith D, Kumar N, Dou Q (2003) Inhibition of the proteasome activity, a novel mechanism associated with the tumor cell apoptosis-inducing ability of genistein. Biochem Pharmacol 66(6):965–976. https://doi.org/10.1016/s0006-2952(03)00414-3

Kefeli V, Kalevitch M, Borsari B (2003) Phenolic cycle in plants and environment. J Cell Mol Biol *2*(1):13–18. https://doi.org/10.2174/97816080516871l101010075

Khan R, Tang Y, Naz I, Alam S, Wang W, Ahmad M, Najeeb S, Rao C, Li Y, Xie B, Li Y (2021) Management of *Ralstonia solanacearum* in tomato using ZnO nanoparticles synthesized through *Matricaria chamomilla*. Plant Dis 105(10):3224–3230. https://doi.org/10.1094/PDIS-08-20-1763-RE

Khanna V (2008) New-generation nano-engineered biosensors, enabling nanotechnologies and nanomaterials. Sensor Rev 28(1):39–45. https://doi.org/10.1108/02602280810850017

Khodakovskaya M, De Silva K, Biris A, Dervishi E, Villagarcia H (2012) Carbon nanotubes induce growth enhancement of tobacco cells. ACS Nano 6(3):2128–2135. https://doi.org/10.1021/nn204643g

Khoddami A, Wilkes M, Roberts T (2013) Techniques for analysis of plant phenolic compounds. Molecules *18*(2):2328–2375. https://doi.org/10.3390/molecules18022328

Kim S, Jung J, Lamsal K, Kim Y, Min J, Lee Y (2012) Antifungal effects of silver nanoparticles (AgNPs) against various plant pathogenic fungi. Mycobiology 40(1):53–58. https://doi.org/10.5941/myco.2012.40.1.053

Koirala M, Tiwari S, Kunwar N, Poudel B (2021) Field evaluation of the resistance in potato varieties against Oriental army ant (Dorylus orientalis Westwood) (Hymenoptera: Formicide). J Entomol Res 45(suppl):891–896. https://doi.org/10.5958/0974-4576.2021.00139.0

Koushki M, Amiri-Dashatan N, Ahmadi N, Abbaszadeh H, Rezaei-Tavirani M (2018) Resveratrol: a miraculous natural compound for diseases treatment. Food Sci Nutr 6(8):2473–2490. https://doi.org/10.1002/fsn3.855

Kovalskaya N, Hammond R (2014) Molecular biology of viroid–host interactions and disease control strategies. Plant Sci *228*:48–60. https://doi.org/10.1016/j.plantsci.2014.05.006

Kris-Etherton P, Hecker K, Bonanome A, Coval S, Binkoski A, Hilpert K, Griel A, Etherton T (2002) Bioactive compounds in foods: their role in the prevention of cardiovascular disease and cancer. Am J Med 113(9):71–88. https://doi.org/10.1016/s0002-9343(01)00995-0

Kumar N, Goel N (2019) Phenolic acids: natural versatile molecules with promising therapeutic applications. Biotechnol Rep 24:e00370. https://doi.org/10.1016/j.btre.2019.e00370

Kumar J, Ramlal A, Mallick D, Mishra V (2021) An overview of some biopesticides and their importance in plant protection for commercial acceptance. Plants 10(6):1185. https://doi.org/10.3390/plants10061185

Lacey J, Dutkiewicz J (1994) Bioaerosols and occupational lung disease. J Aerosol Sci 25(8): 1371–1404. https://doi.org/10.1016/0021-8502(94)90215-1

Laluk K, Mengiste T (2010) Necrotroph attacks on plants: wanton destruction or covert extortion? Arabidopsis Book 8:e0136. https://doi.org/10.1199/tab.0136

Lattanzio V, Lattanzio V, Cardinali A (2006) Role of phenolics in the resistance mechanisms of plants against fungal pathogens and insects. Phytochemistry 661(2):23–67. https://www.researchgate.net/publication/303270594

Lee S, Jun B (2019) Silver nanoparticles: synthesis and application for nanomedicine. Int J Mol Sci 20(4):865. https://doi.org/10.3390/ijms20040865

Lee K, Su C, Fang K, Wu C, Wu C, Chen Y (2020) Ultrafine silicon dioxide nanoparticles cause lung epithelial cells apoptosis via oxidative stress-activated PI3K/Akt- mediated mitochondria- and endoplasmic reticulum stress-dependent signaling pathways. Sci Rep 10(1):9928. https://doi.org/10.1038/s41598-020-66644-z

Leso V, Iavicoli I (2018) Palladium nanoparticles: toxicological effects and potential implications for occupational risk assessment. Int J Mol Sci 19(2):503. https://doi.org/10.3390/ijms19020503

Leung C, Yu I, Chen W (2012) Silicosis. Lancet 379(9830):2008–2018. https://doi.org/10.1016/S0140-6736(12)60235-9

Li M, Yin J, Wamer W, Lo Y (2014) Mechanistic characterization of titanium dioxide nanoparticle-induced toxicity using electron spin resonance. J Food Drug Anal 22(1):76–85. https://doi.org/10.1016/j.jfda.2014.01.006

Lin D, Xiao M, Zhao J, Li Z, Xing B, Li X, Kong M, Li L, Zhang Q, Liu Y, Chen H (2016) An overview of plant phenolic compounds and their importance in human nutrition and management of type 2 diabetes. Molecules 21(10):1374. https://doi.org/10.3390/molecules21101374

Lincoln J, Sanchez J, Zumstein K, Gilchrist D (2018) Plant and animal PR1 family members inhibit programmed cell death and suppress bacterial pathogens in plant tissues. Mol Plant Pathol 19(9):2111–2123. https://doi.org/10.1111/mpp.12685

Litescu S, Eremia S, Diaconu M, Tache A, Radu G (2011) Biosensors applications on assessment of reactive oxygen species and antioxidants. Environ Biosensors 1:35–40. https://doi.org/10.5772/16250

Liu R, Lal R (2015) Potentials of engineered nanoparticles as fertilizers for increasing agronomic productions. Sci Total Environ 514:131–139. https://doi.org/10.1016/j.scitotenv.2015.01.104

Liu Z, Luo Z, Jia C, Wang D, Li D (2016) Synergistic effects of *Potentilla fruticosa* L. leaves combined with green tea polyphenols in a variety of oxidation systems. J Food Sci 81(5): C1091–C1101. https://doi.org/10.1111/1750-3841.13292

Lu R, Radisic M (2021) Organ-on-a-chip platforms for evaluation of environmental nanoparticle toxicity. Bioactive Mater 6(9):2801–2819. https://doi.org/10.1016/j.bioactmat.2021.01.021

Lynn D, Chang M (1990) Phenolic signals in cohabitation: implications for plant development. Annu Rev Plant Biol 41(1):497–526. https://doi.org/10.1146/annurev.pp.41.060190.002433

Ma Y, Jiao K, Yang T, Sun D (2008) Sensitive PAT gene sequence detection by nano-SiO2/p-aminothiophenol self-assembled films DNA electrochemical biosensor based on impedance measurement. Sensors Actuat B Chem 131(2):565–571. https://doi.org/10.1016/j.snb.2007.12.046

Mali CS, Raj S, Trivedi R (2020) Nanotechnology a novel approach to enhance crop productivity. Biochem Biophys Rep 24:100821. https://doi.org/10.1016/j.bbrep.2020.100821

Mandal S, Chakraborty D, Dey S (2010) Phenolic acids act as signaling molecules in plant- microbe symbioses. Plant Signal Behav 5(4):359–368. https://doi.org/10.4161/psb.5.4.10871

Mateos-Naranjo E, Andrades-Moreno L, Davy A (2013) Silicon alleviates deleterious effects of high salinity on the halophytic grass *Spartina densiflora*. Plant Physiol Biochem 63:115–121. https://doi.org/10.1016/j.plaphy.2012.11.015

Matsunaga T, Tomoda R, Nakajima T, Wake H (1985) Photoelectrochemical sterilization of microbial cells by semiconductor powders. FEMS Microbiol Lett 29(1–2):211–214. https://doi.org/10.1111/j.1574-6968.1985.tb00864.x

Maurer E, Comfort K, Hussain SM, Schlager J, Mukhopadhyay S (2012) Novel platform development using an assembly of carbon nanotube, nanogold and immobilized RNA capture element towards rapid, selective sensing of bacteria. Sensors 12(6):8135–8144. https://doi.org/10.3390/s120608135

Mayer A, Staples R, Gil-ad N (2001) Mechanisms of survival of necrotrophic fungal plant pathogens in hosts expressing the hypersensitive response. Phytochemistry *58*(1):33–41. https://doi.org/10.1016/s0031-9422(01)00187-x

Maynard A, Kuempel E (2005) Airborne nanostructured particles and occupational health. J Nanoparticle Res 7(6):587–614. https://doi.org/10.1007/s11051-005-6770-9

Méndez C, Ahlenstiel C, Kelleher A (2015) Post-transcriptional gene silencing, transcriptional gene silencing and human immunodeficiency virus. World J Virol 4(3):219. https://doi.org/10.5501/wjv.v4.i3.219

Meng X, Zhang S (2013) MAPK cascades in plant disease resistance signaling. Annu Rev Phytopathol 51(1):245–266. https://doi.org/10.1146/annurev-phyto-082712-102314

Mishra S, Yang X, Singh H (2020) Evidence for positive response of soil bacterial community structure and functions to biosynthesized silver nanoparticles: an approach to conquer nanotoxicity? J Environ Manag 253:109584. https://doi.org/10.1016/j.jenvman.2019.109584

Mithöfer A, Boland W (2012) Plant defense against herbivores: chemical aspects. Annu Rev Plant Biol 63(1):431–450. https://doi.org/10.1146/annurev-arplant-042110-103854

Mittal D, Kaur G, Singh P, Yadav K, Ali S (2020) Nanoparticle-based sustainable agriculture and food science: recent advances and future outlook. Front Nanotechnol 2:10. https://doi.org/10.3389/fnano.2020.579954

Mohamed H, Hussien N (2016) Genotoxicity studies of titanium dioxide nanoparticles (TiO2NPs) in the brain of mice. Scientifica 2016:1–7. https://doi.org/10.1155/2016/6710840

Mohanta Y, Panda S, Bastia A, Mohanta T (2017) Biosynthesis of silver nanoparticles from *Protium serratum* and investigation of their potential impacts on food safety and control. Front Microbiol 8:626. https://doi.org/10.3389/fmicb.2017.00626

Moleocznik A (2004) Time of farmers' exposure to biological factors in agricultural working environment. Ann Agric Environ Med 11(1):85–89. PMID: 15236503

Momin J, Jayakumar C, Prajapati J (2013) Potential of nanotechnology in functional foods. Emirates J Food Agric *25*(1):10–19. https://doi.org/10.9755/ejfa.v25i1.9368

Monaghan J, Zipfel C (2012) Plant pattern recognition receptor complexes at the plasma membrane. Curr Opin Plant Biol 15(4):349–357. https://doi.org/10.1016/j.pbi.2012.05.006

Mousavi B, Hedayati M, Hedayati N, Ilkit M, Syedmousavi S (2016) Aspergillus species in indoor environments and their possible occupational and public health hazards. Curr Med Mycol 2(1): 36–42. https://doi.org/10.18869/acadpub.cmm.2.1.36

Mukherjee A, Majumdar S, Servin A, Pagano L, Dhankher O, White J (2016) Carbon nanomaterials in agriculture: a critical review. Front Plant Sci 7:172. https://doi.org/10.3389/fpls.2016.00172

Mukhtar H, Ahmad N (2000) Tea polyphenols: prevention of cancer and optimizing health. Am J Clin Nutr 71(6):1698S–1702S. https://doi.org/10.1093/ajcn/71.6.1698S

Mur L, Kenton P, Lloyd A, Ougham H, Prats E (2007) The hypersensitive response; the centenary is upon us but how much do we know? J Exp Bot 59(3):501–520. https://doi.org/10.1093/jxb/erm239

Muthamilarasan M, Prasad M (2013) Plant innate immunity: an updated insight into defense mechanism. J Biosci 38(2):433–449. https://doi.org/10.1007/s12038-013-9302-2

Naresh V, Lee N (2021) A review on biosensors and recent development of nanostructured materials-enabled biosensors. Sensors 21(4):1109. https://doi.org/10.3390/s21041109

Naseer B, Srivastava G, Qadri O, Faridi S, Islam R, Younis K (2018) Importance and health hazards of nanoparticles used in the food industry. Nanotechnol Rev 7(6):623–641. https://doi.org/10.1515/ntrev-2018-0076

Navarro E, Baun A, Behra R, Hartmann NB, Filser J, Miao AJ, Quigg A, Santschi PH, Sigg L (2008) Environmental behavior and ecotoxicity of engineered nanoparticles to algae, plants, and fungi. Ecotoxicology 17(5):372–386. https://doi.org/10.1007/s10646-008-0214-0

Nazarov P, Baleev D, Ivanova M, Sokolova L, Karakozova M (2020) Infectious plant diseases: etiology, current status, problems and prospects in plant protection. Acta Nat 12(3):46–59. https://doi.org/10.32607/actanaturae.11026

Ndakidemi P, Dakora F (2003) Legume seed flavonoids and nitrogenous metabolites as signals and protectants in early seedling development. Funct Plant Biol *30*(7):729–745. https://doi.org/10.1071/fp03042

Neme K, Nafady A, Uddin S, Tola Y (2021) Application of nanotechnology in agriculture, postharvest loss reduction and food processing: food security implication and challenges. Heliyon 7(12):e08539. https://doi.org/10.1016/j.heliyon.2021.e08539

Nicolopoulou-Stamati P, Maipas S, Kotampasi C, Stamatis P, Hens L (2016) Chemical pesticides and human health: the urgent need for a new concept in agriculture. Front Public Health 4:148. https://doi.org/10.3389/fpubh.2016.00148

Norman D, Chen J (2011) Effect of foliar application of titanium dioxide on bacterial blight of geranium and xanthomonas leaf spot of poinsettia. Hortscience 46(3):426–428. https://doi.org/10.21273/hortsci.46.3.426

Nuruzzaman M, Rahman M, Liu Y, Naidu R (2016) Nanoencapsulation, nano-guard for pesticides: a new window for safe application. J Agric Food Chem 64(7):1447–1483. https://doi.org/10.1021/acs.jafc.5b05214

Oerke E (2005) Crop losses to pests. J Agric Sci 144(1):31–43. https://doi.org/10.1017/s0021859605005708

Orgogozo J, Dartigues J, Lafont S, Letenneur L, Commenges D, Salamon R, Renaud S, Breteler MB (1997) Wine consumption and dementia in the elderly: a prospective community study in the Bordeaux area. Rev Neurol 153(3):185–192. PMID: 9296132

Özçelik B, Kartal M, Orhan I (2011) Cytotoxicity, antiviral and antimicrobial activities of alkaloids, flavonoids, and phenolic acids. Pharm Biol 49(4):396–402. https://doi.org/10.3109/13880209.2010.519390

Ozyigit I, Kahraman M, Ercan O (2007) Relation between explant age, total phenols and regeneration response in tissue cultured cotton (Gossypium hirsutum L.). Afr J Biotechnol 6(1):3–8. https://doi.org/10.5897/AJB07.393

Pan M, Lai C, Tsai M, Wu J, Ho C (2012) Molecular mechanisms for anti-aging by natural dietary compounds. Mol Nutr Food Res 56(1):88–115. https://doi.org/10.1002/mnfr.201100509

Pandey P, Irulappan V, Bagavathiannan M, Senthil-Kumar M (2017) Impact of combined abiotic and biotic stresses on plant growth and avenues for crop improvement by exploiting physio-morphological traits. Front Plant Sci 8:537. https://doi.org/10.3389/fpls.2017.00537

Pankaj, Shakil N, Kumar J, Singh M, Singh K (2012) Bioefficacy evaluation of controlled release formulations based on amphiphilic nano-polymer of carbofuran against *Meloidogyne incognita* infecting tomato. J Environ Sci Health B 47(6):520–528. https://doi.org/10.1080/03601234.2012.665667

Paret M, Vallad G, Averett D, Jones J, Olson S (2013) Photocatalysis: effect of light-activated nanoscale formulations of TiO on *Xanthomonas perforans* and control of bacterial spot of tomato. Phytopathology 103(3):228–236. https://doi.org/10.1094/PHYTO-08-12-0183-R

Parmar N, Singh K, Sharma D, Singh L, Kumar P, Nanjundan J, Khan Y, Chauhan D, Thakur A (2017) Genetic engineering strategies for biotic and abiotic stress tolerance and quality enhancement in horticultural crops: a comprehensive review. 3 Biotech 7(4):239. https://doi.org/10.1007/s13205-017-0870-y

Patra P, Mitra S, Debnath N, Goswami A (2012) Biochemical-, biophysical-, and microarray-based antifungal evaluation of the buffer-mediated synthesized nano zinc oxide: an in vivo and in vitro toxicity study. Langmuir 28(49):16966–16978. https://doi.org/10.1021/la304120k

Peigney A, Laurent C, Flahaut E, Rousset A (2000) Carbon nanotubes in novel ceramic matrix nanocomposites. Ceramics Int 26(6):677–683. https://doi.org/10.1016/s0272-8842(00)00004-3

Perde-Schrepler M, Chereches G, Brie I, Tatomir C, Postescu I, Soran L, Filip A (2013) Grape seed extract as photochemopreventive agent against UVB-induced skin cancer. J Photochem Photobiol B Biol 118:16–21. https://doi.org/10.1016/j.jphotobiol.2012.10.008

Pereira A, Grillo R, Mello N, Rosa A, Fraceto L (2014) Application of poly(epsilon-caprolactone) nanoparticles containing atrazine herbicide as an alternative technique to control weeds and reduce damage to the environment. J Hazard Mater 268:207–215. https://doi.org/10.1016/j.jhazmat.2014.01.025

Pichersky E, Gershenzon J (2002) The formation and function of plant volatiles: perfumes for pollinator attraction and defense. Curr Opin Plant Biol 5(3):237–243. https://doi.org/10.1016/s1369-5266(02)00251-0

Pluemsamran T, Onkoksoong T, Panich U (2012) Caffeic acid and ferulic acid inhibit UVA-induced matrix metalloproteinase-1 through regulation of antioxidant defense system in keratinocyte HaCaT cells. Photochem Photobiol 88(4):961–968. https://doi.org/10.1111/j.1751-1097.2012.01118.x

Pojer E, Mattivi F, Johnson D, Stockley C (2013) The case for anthocyanin consumption to promote human health: a review. Comp Rev Food Sci Food Saf 12(5):483–508. https://doi.org/10.1111/1541-4337.12024

Prasad R, Bhattacharyya A, Nguyen Q (2017) Nanotechnology in sustainable agriculture: recent developments, challenges, and perspectives. Front Microbiol 8:1014. https://doi.org/10.3389/fmicb.2017.01014

Pulavarty A, Egan A, Karpinska A, Horgan K, Kakouli-Duarte T (2021) Plant parasitic nematodes: a review on their behaviour, host interaction, management approaches and their occurrence in two sites in the republic of Ireland. Plants 10(11):2352. https://doi.org/10.3390/plants10112352

Purahong W, Orrù L, Donati I, Perpetuini G, Cellini A, Lamontanara A, Michelotti V, Tacconi G, Spinelli F (2018) Plant microbiome and its link to plant health: host species, organs and Pseudomonas syringae pv. actinidiae infection shaping bacterial phyllosphere communities of kiwifruit plants. Front Plant Sci 9:1563. https://doi.org/10.3389/fpls.2018.01563

Puri M, Sharma D, Barrow C (2012) Enzyme-assisted extraction of bioactives from plants. Trends Biotechnol 30(1):37–44. https://doi.org/10.1016/j.tibtech.2011.06.014

Rafiq S, Kaul R, Sofi S, Bashir N, Nazir F, Ahmad Nayik G (2016) Citrus peel as a source of functional ingredient: a review. J Saudi Soc Agric Sci 17(4):351–358. https://doi.org/10.1016/j.jssas.2016.07.006

Rahman M, Sobur M, Islam M, Ievy S, Hossain M, El Zowalaty M, Rahman A, Ashour H (2020) Zoonotic diseases: etiology, impact, and control. Microorganisms 8(9):1405. https://doi.org/10.3390/microorganisms8091405

Rai M, Ingle A (2012) Role of nanotechnology in agriculture with special reference to management of insect pests. Appl Microbiol Biotechnol 94(2):287–293. https://doi.org/10.1007/s00253-012-3969-4

Rai V, Acharya S, Dey N (2012) Implications of nanobiosensors in agriculture. J Biomater Nanobiotechnol 3(2):315–324. https://doi.org/10.4236/jbnb.2012.322039

Rajiv P, Rajeshwari S, Venckatesh R (2013) Bio-fabrication of zinc oxide nanoparticles using leaf extract of Parthenium hysterophorus L. and its size-dependent antifungal activity against plant fungal pathogens. Spectrochim Acta A Mol Biomol Spectrosc 112:384–387. https://doi.org/10.1016/j.saa.2013.04.072

Rajput V, Minkina T, Mazarji M, Shende S, Sushkova S, Mandzhieva S, Burachevskaya M, Chaplygin V, Singh A, Jatav H (2020) Accumulation of nanoparticles in the soil-plant systems and their effects on human health. Ann Agric Sci 65(2):137–143. https://doi.org/10.1016/j.aoas.2020.08.001

Reddy A, Ali G, Celesnik H, Day I (2011) Coping with stresses: roles of calcium- and calcium/calmodulin-regulated gene expression. Plant Cell 23(6):2010–2032. https://doi.org/10.1105/tpc.111.084988

Rohr J, Barrett C, Civitello D, Craft M, Delius B, DeLeo G, Hudson P, Jouanard N, Nguyen K, Ostfeld R, Remais J, Riveau G, Sokolow S, Tilman D (2019) Emerging human infectious diseases and the links to global food production. Nat Sustain 2(6):445–456. https://doi.org/10.1038/s41893-019-0293-3

Rubio L, Galipienso L, Ferriol I (2020) Detection of plant viruses and disease management: relevance of genetic diversity and evolution. Front Plant Sci 11:1092. https://doi.org/10.3389/fpls.2020.01092

Rudrapal M, Khan J, Dukhyil A, Alarousy R, Attah E, Sharma T, Khairnar S, Bendale A (2021) Chalcone scaffolds, bioprecursors of flavonoids: chemistry, bioactivities, and pharmacokinetics. Molecules 26(23):7177. https://doi.org/10.3390/molecules26237177

Ruiz-Ruiz J, del Carmen Esapadas Aldana G, Cruz A, Segura-Campos M (2020) Antioxidant activity of polyphenols extracted from hop used in craft beer. Biotechnol Progress Beverage Consump 19:283–310. https://doi.org/10.1016/B978-0-12-816678-9.00009-6

Ryals J, Neuenschwander U, Willits M, Molina A, Steiner H, Hunt M (1996) Systemic acquired resistance. Plant Cell 8(10):1809–1819. https://doi.org/10.1105/tpc.8.10.1809

Saber H, Alwaleed E, Ebnalwaled K, Sayed A, Salem W (2017) Efficacy of silver nanoparticles mediated by Jania rubens and Sargassum dentifolium macroalgae; Characterization and biomedical applications. Egypt J Basic Appl Sci 4(4):249–255. https://doi.org/10.1016/j.ejbas.2017.10.006

Sadak M (2019) Impact of silver nanoparticles on plant growth, some biochemical aspects, and yield of fenugreek plant (Trigonella foenum-graecum). Bull Natl Res Centre 43(1):48. https://doi.org/10.1186/s42269-019-0077-y

Saeidnia S, Abdollahi M (2013) Antioxidants: friends or foe in prevention or treatment of cancer: the debate of the century. Toxicol Appl Pharmacol 271(1):49–63. https://doi.org/10.1016/j.taap.2013.05.004

Salaheldin T (2016) Silver nanoparticles as a potent fungicide for citrus phytopathogenic fungi. J Nanomed Res 3(5):00065. https://doi.org/10.15406/jnmr.2016.03.00065

Sanhueza L, Melo R, Montero R, Maisey K, Mendoza L, Wilkens M (2017) Synergistic interactions between phenolic compounds identified in grape pomace extract with antibiotics of different classes against Staphylococcus aureus and Escherichia coli. PLoS One 12(2):e0172273. https://doi.org/10.1371/journal.pone.0172273

Saptarshi S, Duschl A, Lopata A (2013) Interaction of nanoparticles with proteins: relation to bio-reactivity of the nanoparticle. J Nanobiotechnol 11(1):26. https://doi.org/10.1186/1477-3155-11-26

Sarkar D, Shetty K (2014) Metabolic stimulation of plant phenolics for food preservation and health. Annu Rev Food Sci Technol 5(1):395–413. https://doi.org/10.1146/annurev-food-030713-092418

Sato K, Kadota Y, Shirasu K (2019) Plant immune responses to parasitic nematodes. Front Plant Sci 10:1165. https://doi.org/10.3389/fpls.2019.01165

Savary S, Willocquet L, Pethybridge S, Esker P, McRoberts N, Nelson A (2019) The global burden of pathogens and pests on major food crops. Nat Ecol Evol 3(3):430–439. https://doi.org/10.1038/s41559-018-0793-y

Schmitz-Hoerner R, Weissenböck G (2003) Contribution of phenolic compounds to the UV-B screening capacity of developing barley primary leaves in relation to DNA damage and repair under elevated UV-B levels. Phytochemistry 64(1):243–255. https://doi.org/10.1016/s0031-9422(03)00203-6

Scott N, Chen H (2013) Nanoscale science and engineering for agriculture and food systems. Indus Biotechnol 9(1):17–18. https://doi.org/10.1089/ind.2013.1555

Seaton A, Tran L, Aitken R, Donaldson K (2010) Nanoparticles, human health hazard and regulation. J R Soc Interface 7(suppl_1):119–129. https://doi.org/10.1098/rsif.2009.0252.focus

Sekhon B (2014) Nanotechnology in agri-food production: an overview. Nanotechnol Sci Appl 7: 31–53. https://doi.org/10.2147/NSA.S39406

Sergeant K, Renaut J (2010) Plant biotic stress and proteomics. Curr Proteomics 7(4):275–297. https://doi.org/10.2174/157016410793611765

Servin A, Elmer W, Mukherjee A, De la Torre-Roche R, Hamdi H, White J, Bindraban P, Dimkpa C (2015) A review of the use of engineered nanomaterials to suppress plant disease and enhance crop yield. J Nanoparticle Res 17(2):92. https://doi.org/10.1007/s11051-015-2907-7

Seybold H, Trempel F, Ranf S, Scheel D, Romeis T, Lee J (2014) Ca2+ signalling in plant immune response: from pattern recognition receptors to Ca2+ decoding mechanisms. New Phytol 204(4):782–790. https://doi.org/10.1111/nph.13031

Shahryari F, Rabiei Z, Sadighian S (2020) Antibacterial activity of synthesized silver nanoparticles by sumac aqueous extract and silver-chitosan nanocomposite against Pseudomonas syringae pv. syringae. J Plant Pathol 102(2):469–475. https://doi.org/10.1007/s42161-019-00478-1

Shakil N, Singh M, Pandey A, Kumar J, Pankaj, Parmar V, Singh M, Pandey R, Watterson A (2010) Development of poly (ethylene glycol) based amphiphilic copolymers for controlled release delivery of carbofuran. J Macromol Sci A 47(3):241–247. https://doi.org/10.1080/10601320903527038

Shapiro E (2014) *Borrelia burgdorferi* (Lyme disease). Pediatric Rev 35(12):500–509. https://doi.org/10.1542/pir.35-12-500

Sharma A, Patni B, Shankhdhar D, Shankhdhar S (2012) Zinc—an indispensable micronutrient. Physiol Mol Biol Plants 19(1):11–20. https://doi.org/10.1007/s12298-012-0139-1

Shetty K (2004) Role of proline-linked pentose phosphate pathway in biosynthesis of plant phenolics for functional food and environmental applications: a review. Process Biochem 39(7):789–804. https://doi.org/10.1016/s0032-9592(03)00088-8

Siddiqui M, Al-Whaibi M (2014) Role of nano-SiO2 in germination of tomato (*Lycopersicum esculentum* seeds Mill.). Saudi J Biol Sci 21(1):13–17. https://doi.org/10.1016/j.sjbs.2013.04.005

Sonwani S, Madaan S, Arora J, Suryanarayan S, Rangra D, Mongia N, Vats T, Saxena P (2021) Inhalation exposure to atmospheric nanoparticles and its associated impacts on human health: a review. Front Sustain Cities 3:690444. https://doi.org/10.3389/frsc.2021.690444

Spence N, Phiri N, Hughes S, Mwaniki A, Simons S, Oduor G, Chacha D, Kuria A, Ndirangu S, Kibata G, Marris G (2007) Economic impact of Turnip mosaic virus, Cauliflower mosaic virus and Beet mosaic virus in three Kenyan vegetables. Plant Pathol 56(2):317–323. https://doi.org/10.1111/j.1365-3059.2006.01498.x

Spoel S, Dong X (2012) How do plants achieve immunity? Defence without specialized immune cells. Nat Rev Immunol 12(2):89–100. https://doi.org/10.1038/nri3141

Sroka Z, Cisowski W (2003) Hydrogen peroxide scavenging, antioxidant and anti-radical activity of some phenolic acids. Food Chem Toxicol 41(6):753–758. https://doi.org/10.1016/s0278-6915(02)00329-0

Sweet M, Singleton I (2015) Soil contamination with silver nanoparticles reduces Bishop pine growth and ectomycorrhizal diversity on pine roots. J Nanoparticle Res 17(11):448. https://doi.org/10.1007/s11051-015-3246-4

Szeszenia-Dąbrowska N, Świątkowska B, Wilczyńska U (2016) Occupational diseases among farmers in Poland. Medycyna Pracy 67(2):163–171. https://doi.org/10.13075/mp.5893.00303

Tang D, Wang G, Zhou J (2017) Receptor kinases in plant-pathogen interactions: more than pattern recognition. Plant Cell 29(4):618–637. https://doi.org/10.1105/tpc.16.00891

Thomas S, Al-Mutairi E, De S (2012) Impact of nanomaterials on health and environment. Arab J Sci Eng 38(3):457–477. https://doi.org/10.1007/s13369-012-0324-0

Thomma B, Eggermont K, Penninckx I, Mauch-Mani B, Vogelsang R, Cammue B, Broekaert W (1998) Separate jasmonate-dependent and salicylate-dependent defense- response pathways in *Arabidopsis* are essential for resistance to distinct microbial pathogens. Proc Natl Acad Sci U S A 95(25):15107–15111. https://doi.org/10.1073/pnas.95.25.15107

Thornton P (2010) Livestock production: recent trends, future prospects. Philos Trans R Soc B Biol Sci 365(1554):2853–2867. https://doi.org/10.1098/rstb.2010.0134

Tonho H, Horii C, Fuse T, Okonogi A, Yomoda S (2010) Evaluation of estrogen receptor β binding of pruni cortex and its constituents. Yakugaku Zasshi 130(7):989–997. https://doi.org/10.1248/yakushi.130.989

Torres FG, Troncoso OP, Pisani A, Gatto F, Bardi G (2019) Natural polysaccharide nanomaterials: an overview of their immunological properties. Int J Mol Sci 20(20):5092. https://doi.org/10.3390/ijms20205092

Tripathi S, Sonkar S, Sarkar S (2011) Growth stimulation of gram (Cicer arietinum) plant by water soluble carbon nanotubes. Nanoscale 3(3):1176. https://doi.org/10.1039/c0nr00722f

Tripathi D, Singh V, Prasad S, Chauhan D, Dubey N (2015) Silicon nanoparticles (SiNp) alleviate chromium (VI) phytotoxicity in Pisum sativum (L.) seedlings. Plant Physiol Biochem 96:189–198. https://doi.org/10.1016/j.plaphy.2015.07.026

United Nations Educational, Scientific and Cultural Organization (UNESCO) (2022) UNESCO 2021 World conference on education for sustainable development. [online]. https://en.unesco.org/events/esdfor2030. Accessed 20 Mar 2022

Hage-Hamsten MV, Johansson S, Hoglund S, Tull P, Wiren A, Zetterstrom O (1985) Storage mite allergy is common in a farming population. Clin Allergy 15(6):555–564. https://doi.org/10.1111/j.1365-2222.1985.tb02309.x

Van Loon L, Rep M, Pieterse C (2006) Significance of inducible defense-related proteins in infected plants. Annu Rev Phytopathol 44(1):135–162. https://doi.org/10.1146/annurev.phyto.44.070505.143425

Vieira R, Souto BS, Sánchez-López E, Machado L, Severino P, Jose S, Santini A, Fortuna A, García L, Silva A, Souto E (2019) Sugar-lowering drugs for type 2 diabetes mellitus and metabolic syndrome—review of classical and new compounds: Part-I. Pharmaceuticals 12(4):152. https://doi.org/10.3390/ph12040152

Vimercati L, Cavone D, Caputi A, De Maria L, Tria M, Prato E, Ferri G (2020) Nanoparticles: an experimental study of zinc nanoparticles toxicity on marine crustaceans. General overview on the health implications in humans. Front Public Health 8:192. https://doi.org/10.3389/fpubh.2020.00192

Vishwanath S, Delude C, Domergue F, Rowland O (2014) Suberin: biosynthesis, regulation, and polymer assembly of a protective extracellular barrier. Plant Cell Rep 34(4):573–586. https://doi.org/10.1007/s00299-014-1727-z

Vlot A, Dempsey D, Klessig D (2009) Salicylic acid, a multifaceted hormone to combat disease. Annu Rev Phytopathol 47(1):177–206. https://doi.org/10.1146/annurev.phyto.050908.135202

Vuolo M, Lima V, Maróstica Junior M (2019) Phenolic compounds. Bioactive Comp:33–50. https://doi.org/10.1016/B978-0-12-814774-0.00002-5

Vuong Q (2017) Utilisation of bioactive compounds from agricultural and food production waste. CRC Press, Boca Raton. isbn:9780367781958

Walling L (2000) The myriad plant responses to herbivores. J Plant Growth Regul 19(2):195–216. https://doi.org/10.1007/s003440000026

Wan R, Mo Y, Feng L, Chien S, Tollerud D, Zhang Q (2012) DNA damage caused by metal nanoparticles: involvement of oxidative stress and activation of ATM. Chem Res Toxicol 25(7):1402–1411. https://doi.org/10.1021/tx200513t

Wang L, Hu C, Shao L (2017) The antimicrobial activity of nanoparticles: present situation and prospects for the future. Int J Nanomedicine 12:1227–1249. https://doi.org/10.2147/IJN.S121956

Wang T, Li Q, Bi K (2018) Bioactive flavonoids in medicinal plants: Structure, activity and biological fate. Asian J Pharm Sci 13(1):12–23. https://doi.org/10.1016/j.ajps.2017.08.004

Wani AH, Shah MA (2012) A unique and profound effect of MgO and ZnO nanoparticles on some plant pathogenic fungi. J Appl Pharm Sci 2(3):40–44. issn:2231-3354

Wattenberg L (1992) Inhibition of carcinogenesis by minor dietary constituents. Cancer Res 52(7 Suppl):2085s–2091s. PMID: 1544145

Watterson A, Thomas H (1992) Acute pesticide poisoning in the UK and information and training needs of general practitioners: recording a conundrum. Public Health 106(6):473–480. https://doi.org/10.1016/s0033-3506(05)80462-7

Wilczyńska U, Sobala W, Szeszenia-Dąbrowska N (2013) Occupational diseases in Poland, 2012. Medycyna Pracy 61(4):369–379. https://doi.org/10.13075/mp.5893.2013.0027

Wink M (2015) Modes of action of herbal medicines and plant secondary metabolites. Medicines 2(3):251–286. https://doi.org/10.3390/medicines2030251

Wormser G, Dattwyler R, Shapiro E, Halperin J, Steere A, Klempner M, Krause P, Bakken J, Strle F, Stanek G, Bockenstedt L, Fish D, Dumler J, Nadelman R (2006) The clinical assessment, treatment, and prevention of Lyme disease, human granulocytic anaplasmosis, and babesiosis: clinical practice guidelines by the Infectious Diseases Society of America. Clin Infect Dis 43(9):1089–1134. https://doi.org/10.1086/508667

Wu P, Xie R, Imlay K, Shang J (2010) Visible-light-induced bactericidal activity of titanium dioxide codoped with nitrogen and silver. Environ Sci Technol 44(18):6992–6997. https://doi.org/10.1021/es101343c

Xiao J, Hogger P (2015) Dietary polyphenols and type 2 diabetes: current insights and future perspectives. Curr Med Chem 22(1):23–38. https://doi.org/10.2174/0929867321666140706130807

Xie Y, Wang Y, Zhang T, Ren G, Yang Z (2012) Effects of nanoparticle zinc oxide on spatial cognition and synaptic plasticity in mice with depressive-like behaviors. J Biomed Sci 19(1):14. https://doi.org/10.1186/1423-0127-19-14

Xue D, Farid M (2015) Pulsed electric field extraction of valuable compounds from white button mushroom (Agaricus bisporus). Innov Food Sci Emerg Technol 29:178–186. https://doi.org/10.1016/j.ifset.2015.03.012

Yang Y, Luan Y, Yuan R, Luan Y (2021) Histone methylation related therapeutic challenge in cardiovascular diseases. Front Cardiovasc Med 8:710053. https://doi.org/10.3389/fcvm.2021.710053

Yoshida K, Schuenemann V, Cano L, Pais M, Mishra B, Sharma R, Lanz C, Martin F, Kamoun S, Krause J, Thines M, Weigel D, Burbano H (2013) The rise and fall of the *Phytophthora infestans* lineage that triggered the Irish potato famine. eLife 2:e00731. https://doi.org/10.7554/elife.00731

You C, Bhagawati M, Brecht A, Piehler J (2009) Affinity capturing for targeting proteins into micro and nanostructures. Anal Bioanal Chem 393(6–7):1563–1570. https://doi.org/10.1007/s00216-008-2595-6

Yu K, Huang Y, Yang S (2013) The antifungal efficacy of nano-metals supported TiO2 and ozone on the resistant *Aspergillus niger* spore. J Hazard Mater 261:155–162. https://doi.org/10.1016/j.jhazmat.2013.07.029

Yuan C, Huo C, Gui B, Cao W (2017) Green synthesis of gold nanoparticles using *Citrus maxima* peel extract and their catalytic/antibacterial activities. IET Nanobiotechnol 11(5):523–530. https://doi.org/10.1049/iet-nbt.2016.0183

Zaffarano P, McDonald B, Linde C (2008) Rapid speciation following recent host shifts in the plant pathogenic fungus rhynchosporium. Evolution 62(6):1418–1436. https://doi.org/10.1111/j.1558-5646.2008.00390.x

Zaytseva O, Neumann G (2016) Carbon nanomaterials: production, impact on plant development, agricultural and environmental applications. Chem Biol Technol Agric 3(1):17. https://doi.org/10.1186/s40538-016-0070-8

Zhang W, Tang H, Geng P, Wang Q, Jin L, Wu Z (2007) Amperometric method for rapid detection of Escherichia coli by flow injection analysis using a bismuth nano-film modified glassy carbon electrode. Electrochem Commun 9(4):833–838. https://doi.org/10.1016/j.elecom.2006.11.019

Zhang X, Liu Z, Shen W, Gurunathan S (2016) Silver nanoparticles: synthesis, characterization, properties, applications, and therapeutic approaches. Int J Mol Sci 17(9):1534. https://doi.org/10.3390/ijms17091534

Zheng W, McKinney W, Kashon M, Pan D, Castranova V, Kan H (2018) The effects of inhaled multi-walled carbon nanotubes on blood pressure and cardiac function. Nanoscale Res Lett 13:189. https://doi.org/10.1186/s11671-018-2603-5

Zielińska-Jankiewicz K, Kozajda A, Szadkowska-Stańczyk I (2005) Occupational exposure to biological agents and the protection of exposed workers according to the new legal regulations. Medycyna Pracy 56(4):319–323. PMID: 16457370

Żukiewicz-Sobczak W, Chmielewska-Badora J, Wróblewska P, Zwoliński J (2013) Farmers' occupational diseases of allergenic and zoonotic origin. Adv Dermatol Allergol 5:311–315. https://doi.org/10.5114/pdia.2013.38361

The manufacturer's authorised representative in the EU is Springer Nature Customer Service Centre GmbH, Europaplatz 3, 69115 Heidelberg, Germany. If you have any concerns regarding our products, please contact ProductSafety@springernature.com

Printed and bound by CPI Group (UK) Ltd, Croydon, CR0 4YY
30/09/2026
02225314-0019